T0211243

ATOMS IN INTENSE LASER FIELDS

The development of lasers capable of producing high-intensity pulses has opened a new area in the study of light–matter interactions. The corresponding laser fields are strong enough to compete with the Coulomb forces in controlling the dynamics of atomic systems, and give rise to multiphoton processes. This book presents a unified account of this rapidly developing field of physics.

The first part describes the fundamental phenomena occurring in intense laser–atom interactions and gives the basic theoretical framework to analyze them. The second part contains a detailed discussion of Floquet theory, the numerical integration of the wave equations and approximation methods for the low- and high-frequency regimes. In the third part, the main multiphoton processes are discussed: multiphoton ionization, high harmonic and attosecond pulse generation and laser-assisted electron–atom collisions. Aimed at graduate students in atomic, molecular and optical physics, the book will also interest researchers working on laser interactions with matter.

C. J. JOACHAIN is Emeritus Professor of Theoretical Physics at the University of Brussels. His research activities concern quantum collision theory and the interaction of intense laser fields with matter. He has received many scientific awards and distinctions, in particular the Alexander von Humboldt Prize.

N. J. KYLSTRA is currently employed in the financial services industry, and is a research fellow at the University of Brussels. His research includes the interaction of atoms with high-intensity laser fields.

R. M. POTVLIEGE is currently Reader in Physics at the University of Durham. His research interests cover the theoretical aspects of multiphoton processes in intense laser fields.

ATOMS IN INTENSE
LASER FIELDS

C. J. JOACHAIN
University of Brussels

N. J. KYLSTRA
University of Brussels

R. M. POTVLIEGE
University of Durham

CAMBRIDGE
UNIVERSITY PRESS

CAMBRIDGE
UNIVERSITY PRESS

University Printing House, Cambridge CB2 8BS, United Kingdom

Published in the United States of America by Cambridge University Press, New York

Cambridge University Press is part of the University of Cambridge.

It furthers the University's mission by disseminating knowledge in the pursuit of
education, learning and research at the highest international levels of excellence.

www.cambridge.org
Information on this title: www.cambridge.org/9781107424777

© C. Joachain, N. Kylstra and R. Potvliege 2012

First published 2012
First paperback edition 2014

A catalogue record for this publication is available from the British Library

Library of Congress cataloguing in Publication data
Joachain, C. J. (Charles Jean)
Atoms in intense laser fields / C.J. Joachain, N. J. Kylstra, and R. M. Potvliege.
p.   cm.
Includes bibliographical references and index.
ISBN 978-0-521-79301-8 (hardback)
1. Electron-atom collisions.   2. Floquet theory.   3. Laser pulses, Ultrashort.   4. Multiphoton processes.
5. Multiphoton ionization.   I. Kylstra, N. J.   II. Potvliege, R. M.   III. Title.
QC793.5.E628J63 2011
539.7'2–dc23   2011024273

ISBN 978-0-521-79301-8 Hardback
ISBN 978-1-107-42477-7 Paperback

Felix qui potuit rerum cognoscere causas.
(Virgil, Georgics, II, 490)

Contents

Preface

The availability of intense laser fields over a wide frequency range, in the form of short pulses of coherent radiation, has opened a new domain in the study of light–matter interactions. The peak intensities of these laser pulses are so high that the corresponding laser fields can compete with the Coulomb forces in controlling the dynamics of atomic systems. Atoms interacting with such intense laser fields are therefore exposed to extreme conditions, and new phenomena occur which are known as multiphoton processes. These phenomena generate in turn new behaviors of bulk matter in strong laser fields, with wide-ranging applications.

The purpose of this book is to give a self-contained and unified presentation of high- intensity laser–atom physics. It is primarily aimed at physicists studying the interaction of laser light with matter at the microscopic level, although it is hoped that any scientist interested in laser–matter interactions will find it useful.

The book is divided into three parts. The first one contains two chapters, in which the basic concepts are presented. In Chapter 1, we give a general overview of the new phenomena discovered by studying atomic multiphoton processes in intense laser fields. In Chapter 2, the theory of laser–atom interactions is expounded, using a semi-classical approach in which the laser field is treated classically, while the atom is described quantum mechanically. The wave equations required to study the dynamics of atoms interacting with laser fields are discussed, starting with the non-relativistic time-dependent Schrödinger equation in the dipole approximation, then moving to the description of non-dipole effects and finally to relativistic wave equations.

The second part, containing five chapters, is devoted to a detailed discussion of the most important theoretical methods used to solve the wave equations given in Chapter 2. We begin, in Chapter 3, by considering perturbation theory, which can only be employed for laser fields having moderate intensities and for non-resonant multiphoton processes. In the next four chapters we discuss non-perturbative methods, which must be used when atoms interact with strong laser fields. In Chapter 4, we review the Floquet theory, in particular the Sturmian-Floquet and R-matrix–Floquet methods. Chapter 5 is devoted to the numerical solution of the wave equations. Approximation methods appropriate to investigate the interaction of atoms with low-frequency and high-frequency laser fields are considered in Chapters 6 and 7, respectively. It is remarkable that in these two distinct frequency regimes, simple theoretical considerations provide considerable insight into the physics of intense laser–atom interactions.

In the third part of the book, which contains the final three chapters, the methods discussed in the second part are applied to the analysis of the three most important atomic multiphoton processes in intense laser fields: multiphoton ionization, harmonic generation and laser-assisted electron–atom collisions. Thus, in Chapter 8 we discuss successively multiphoton single and double ionization of atoms. In Chapter 9, after analyzing the emission of harmonics by atoms, we review the generation and characterization of attosecond pulses, and their use in the new field of attophysics. Finally, in Chapter 10, we begin our theoretical analysis of laser-assisted electron–atom collisions by considering the simple case for which the target atom is modeled by a potential. We then turn our attention to collisions with real atoms having an internal structure.

We wish to thank our colleagues and students for numerous helpful discussions and suggestions. One of us (C.J.J.) would like to acknowledge the hospitality of the Max-Planck Institut für Quantenoptik in Garching, where he was the guest of Professor H. Walther and more recently of Professor F. Krausz. We would also like to thank Professor H. Joachain-Bukowinski and Professor N. Vaeck for their help in preparing the diagrams.

Part I

Basic concepts

1

High-intensity laser–atom physics

In recent years, intense laser fields have become available, over a wide frequency range, in the form of short pulses. Such laser fields are strong enough to compete with the Coulomb forces in controlling the dynamics of atomic systems. As a result, atoms in intense laser fields exhibit new properties that have been discovered via the study of *multiphoton processes*. After some introductory remarks in Section 1.1, we discuss in Section 1.2 how intense laser fields can be obtained by using the "chirped pulse amplification" method. In the remaining sections of this chapter, we give a survey of the new phenomena discovered by studying three important multiphoton processes in atoms: multiphoton ionization, harmonic generation and laser-assisted electron–atom collisions.

1.1 Introduction

If radiation fields of sufficient intensity interact with atoms, processes of higher order than the single-photon absorption or emission play a significant role. These higher-order processes, called multiphoton processes, correspond to the net absorption or emission of more than one photon in an atomic transition. It is interesting to note that, in the first paper he published in *Annalen der Physik* in the year 1905, his "Annus mirabilis," Einstein [1] not only introduced the concept of "energy quantum of light" – named "photon" by Lewis [2] in 1926 – but also mentioned the possibility of multiphoton processes occurring when the intensity of the radiation is high enough, namely "if the number of energy quanta per unit volume simultaneously being transformed is so large that an energy quantum of emitted light can obtain its energy from several incident energy quanta." Multiphoton processes were also considered in the pioneering work of Göppert-Mayer [3].

There are several types of multiphoton processes. For instance, an atom can undergo a transition from a bound state to another bound state of higher energy via the absorption of n photons ($n \geq 2$), a process known as *multiphoton excitation*.

Also, an atom in an excited state can emit n photons in a transition to a state of lower energy, a process called *multiphoton de-excitation*, either by spontaneous emission (which does not require the presence of an external radiation field) or by stimulated emission. Another example is the *multiphoton ionization* (MPI) of an atom, a process in which the atom absorbs n photons, and one or several of its electrons are ejected. An atom interacting with a strong laser field can also emit radiation at higher-order multiples, or harmonics, of the frequency of the laser; this process is known as *harmonic generation*. Finally, radiative collisions involving the exchange (absorption or emission) of n photons can occur in *laser-assisted atomic collisions* such as electron–atom or atom–atom collisions in the presence of a laser field.

Except for spontaneous emission, which will not be considered here, the observation of multiphoton transitions requires relatively large laser intensities. Typically, intensities of the order of $10^8 \, \mathrm{W \, cm^{-2}}$ are required to observe multiphoton transitions in laser-assisted electron–atom collisions, while intensities of $10^{10} \, \mathrm{W \, cm^{-2}}$ are the minimum necessary for the observation of multiphoton ionization in atoms. In fact, such intensities are now considered to be rather modest. Indeed, as we shall see in the following section, laser fields have become available in the form of short pulses having intensities of the order of, or exceeding, the atomic unit of intensity

$$I_a = \frac{1}{2}\epsilon_0 c \mathcal{E}_a^2 \simeq 3.5 \times 10^{16} \, \mathrm{W \, cm^{-2}}, \tag{1.1}$$

where c is the velocity of light in vacuo, ϵ_0 is the permittivity of free space and $\mathcal{E}_a$ is the atomic unit of electric field strength, namely

$$\mathcal{E}_a = \frac{e}{(4\pi\epsilon_0)a_0^2} \simeq 5.1 \times 10^9 \, \mathrm{V \, cm^{-1}}, \tag{1.2}$$

where e is the absolute value of the electron charge and a_0 is the first Bohr radius of atomic hydrogen. Atomic units (a.u.) are discussed in the Appendix. We note that $\mathcal{E}_a$ is the strength of the Coulomb field experienced by an electron in the first Bohr orbit of the hydrogen atom. Laser fields having intensities of the order of, or larger than, I_a are strong enough to compete with the Coulomb forces in governing the dynamics of atoms. Thus, while multiphoton processes involving laser fields with intensities $I \ll I_a$ can be studied by using perturbation theory, the effects of laser fields with intensities of the order of, or exceeding, I_a must be analyzed by using non-perturbative approaches.

In Chapter 2, we shall discuss the theory of laser–atom interactions based on a semi-classical approach which provides the framework for studying atomic multiphoton processes in intense laser fields. In particular, we shall introduce the *dipole approximation*, in which the laser field is described by a spatially homogeneous electric-field component, while its magnetic-field component vanishes. The dipole approximation is fully adequate to investigate atomic multiphoton processes over a wide range of laser frequencies and intensities. However, as the intensity increases beyond critical values that depend on the frequency, *non-dipole* effects due to the magnetic-field component of the laser field, and eventually *relativistic* effects, must be taken into account [4–7].

The theoretical methods required to solve the quantum-mechanical wave equations introduced in Chapter 2 will be developed in the second part of this book (Chapters 3–7). We shall discuss powerful *ab initio* methods such as the Sturmian-Floquet method [8, 9], the R-matrix–Floquet method [10, 11] and the numerical solution of the time-dependent Schrödinger equation [12, 13]. In this second part, we shall also examine *methods of approximation* which can be used to analyze multiphoton processes at low or at high laser frequencies, respectively. All of these methods will be applied in the third part (Chapters 8–10) to analyze atomic multiphoton processes.

The subject of atoms in intense laser fields has been covered in the volumes edited by Gavrila [14] and by Brabec [15], in the review articles by Burnett, Reed and Knight [16], Joachain [17], Kulander and Lewenstein [18], Protopapas, Keitel and Knight [19], Joachain, Dörr and Kylstra [20], Milosevic and Ehloztky [21], and also in the books by Faisal [22], Mittleman [23], Delone and Krainov [24] and Grossmann [25].

1.2 High-intensity lasers

To obtain high-intensity laser fields, one must concentrate large amounts of energy into short periods of time, and then focus the laser light onto small areas. In an intense laser system, the oscillator produces a train of pulses of short duration. The amplifier then increases the energy of the pulses, which are subsequently focused. A very successful method of amplification, called "chirped pulse amplification" (CPA), was devised in 1985 by Strickland and Mourou [26]. This method, which is illustrated in Fig. 1.1, consists in the following three steps. Firstly, the short laser pulse to be amplified (produced by the oscillator) is stretched in time into its frequency components by a dispersive system such as a pair of diffraction gratings, so that a *chirped* pulse is generated. This stretching in time of the pulse greatly reduces its peak intensity, so that in the second step the frequency components of the chirped pulse can be sent in succession through a laser amplifier without distortions and damage. In the third step, the amplified chirped pulse is compressed in time by another pair of diffraction gratings, which recombine the dispersed frequencies, thus producing a short pulse with a very large peak intensity. Finally, the resulting amplified short pulse is tightly focused onto a small area. After focusing, intensities of the order of the atomic unit of intensity I_a can be readily obtained. An important advantage of the CPA method is that it can yield very intense, short pulses by using

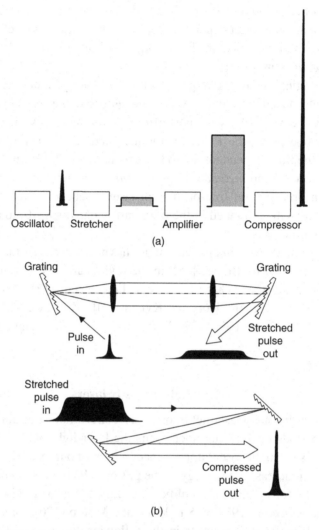

Oscillator   Stretcher   Amplifier   Compressor

(a)

(b)

Figure 1.1. Chirped pulse amplification (CPA) method. (a) An oscillator produces a short pulse, which is then chirped (stretched in time into its frequency components). In this way, the peak intensity of the pulse is lowered, so that amplification can take place without damage or distortions. The amplified chirped pulse is then compressed in time, resulting in a short pulse with a very high intensity. (b) The matched stretcher and compressor of the CPA method. The stretcher (top) consists of a telescope of magnification unity placed between two antiparallel gratings. In this configuration, the low-frequency components of the pulse have a shorter optical path than the high-frequency ones. Conversely, the compressor (bottom) consists of a pair of parallel gratings, so that the optical path for the high-frequency components of the pulse is shorter than for the low-frequency ones. (From G. A. Mourou, C. P. J. Barty and M. D. Perry, *Phys. Today*, **Jan.**, 22 (1998).)

a "table-top" laser system. A review of the CPA method has been given by Mourou, Tajima and Bulanov [27].

The first CPA high-intensity lasers to be constructed used Nd:glass as the amplifying medium. In the system developed during the 1990s at Imperial College, London, a pulse from a Nd oscillator, of wavelength $\lambda = 1064$ nm, duration 1 ps $(10^{-12}$ s) and energy 1 nJ was stretched by diffraction gratings to about 25 ps. It was then amplified by using Nd:glass as the amplifying medium to an energy of about 1 J. This amplified chirped pulse was subsequently compressed to a duration close to its initial picosecond value by diffraction gratings, so that output powers of around 1 TW $(10^{12}$ W) could be obtained. By focusing over an area having a diameter of 10 μm, intensities of the order of 10^{18} W cm^{-2} were reached. Lasers of this kind have a repetition rate of about one shot per minute. More recently, CPA laser systems employing Ti:sapphire for the oscillator and the amplifying medium have been used extensively, because they can generate very short pulses with high repetition rates. If only moderate intensities ($\sim 10^{14}$ W cm^{-2}) are required, such laser systems can produce pulses having a duration of about 30 fs, with a repetition rate of 300 kHz. CPA Ti:sapphire lasers can also yield very intense pulses. For example, at the ATLAS laser facility in the Max-Planck Institut für Quantenoptik in Garching, pulses of 100 fs in duration and 1 nJ in energy have been stretched, amplified and compressed, giving output pulses of wavelength $\lambda = 790$ nm, duration 150 fs and energy 220 mJ at a repetition rate of 10 Hz. After focusing on a spot 6 μm in diameter, the intensity available from this laser reached 4×10^{18} W cm^{-2}. More recently, intensities up to 10^{22} W cm^{-2} have been achieved using the Hercules Ti:sapphire laser at the University of Michigan [28].

The CPA concept was originally developed for the amplification of short laser pulses with laser amplifiers based on laser gain media. However, it was subsequently realized that it can also be used with optical parametric amplifiers (OPA), in which case it is known as the *optical parametric chirped pulse amplification* (OPCPA) method [29, 30]. Optical parametric amplification [31, 32] is a second-order phenomenon of non-linear optics, arising from the fact that crystal materials lacking inversion symmetry can display a $\chi^{(2)}$ non-linearity, where $\chi^{(n)}$ denotes the nth-order susceptibility [33]. Apart from other effects (frequency doubling, generation of sum and difference frequencies), this gives rise to parametric amplification, in which a weak signal beam of angular frequency ω_1 and an intense pump beam of angular frequency $\omega_3 > \omega_1$ generate two intense beams with angular frequencies ω_1 and $\omega_2 = \omega_3 - \omega_1$. Indeed, as the signal beam and the pump beam propagate together through the crystal, photons of the pump beam, having energy $\hbar\omega_3$, are converted into lower-energy signal photons of energy $\hbar\omega_1$ and an equal number of "idler" photons of energy $\hbar\omega_2 = \hbar(\omega_3 - \omega_1)$, where $\hbar = h/(2\pi)$ and h is Planck's constant. A schematic diagram of an optical parametric amplifier is shown in Fig. 1.2.

Basic concepts

Figure 1.2. Optical parametric amplifier.

Recent technological advances in ultra-fast optics have allowed the generation of intense laser pulses comprising only a few optical cycles (that is, laser periods, where the laser period is defined as $T = 2\pi/\omega$) of the laser field [34, 35]. In particular, the development of Ti:sapphire laser systems using the CPA method has made possible the generation of such high-intensity "few-cycle pulses" in the near infra-red region of the electromagnetic spectrum, with a central wavelength around 800 nm, corresponding to a photon energy of 1.55 eV and an optical cycle of 2.7 fs.

A successful way of obtaining intense few-cycle laser pulses relies on external bandwidth broadening of amplified pulses in gas-filled capillaries [36, 37], using the hollow-fiber technique [38] and chirped-mirror technology [39]. For example, using a Ti:sapphire laser system and a hollow-fiber chirped-mirror compressor, Sartania *et al.* [37] demonstrated the generation of 0.1 TW, 5 fs laser pulses at a repetition rate of 1 kHz. However, the method of gas-filled capillaries involves important energy losses and is difficult to scale to very high energies and peak powers. Several OPCPA systems delivering very intense, few-cycle 800 nm laser pulses have also been reported [40–43].

In addition to very high peak intensities and high repetition rates, the laser systems delivering few-cycle pulses must also provide reliable control over the *carrier-envelope phase* (CEP) φ, namely the phase of the carrier wave with respect to the maximum of the laser pulse envelope, since the CEP sensitively determines the variation of the electric field [44].

As an example, we show in Fig. 1.3 the wave form of the electric field of a linearly polarized laser pulse whose carrier wavelength is $\lambda = 800$ nm and whose intensity profile is proportional to $F(t)\cos(\omega t + \varphi)$, where $F(t)$ is a sech envelope function. In Fig. 1.3(a), the CEP is $\varphi = 0$, corresponding to a "cosine-like" pulse, while in Fig. 1.3(b) the CEP is $\varphi = -\pi/2$, corresponding to a "sine-like" pulse.

Intense few-cycle laser pulses with stabilized CEP have been obtained by using CPA Ti:sapphire laser systems [45–56]. A few-cycle OPCPA system producing infra-red laser pulses at a wavelength of 2.1 μm with a stable CEP has also been demonstrated [57]. One of the major goals is to perform a *single-shot* determination

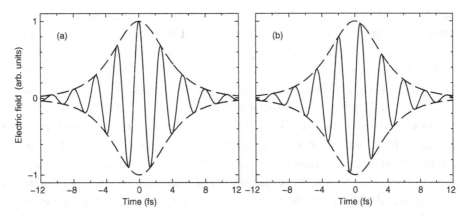

Figure 1.3. Wave form of the electric field (solid curves) of a linearly polarized laser pulse, taken to be proportional to $F(t)\cos(\omega t + \varphi)$, where $F(t)$ is a sech envelope function such that $F^2(t)$ is a sech^2 function of 5 fs full width at half maximum (dashed curves). The carrier wavelength is $\lambda = 800$ nm and the carrier-envelope phase is (a) $\varphi = 0$ for the "cosine-like" pulse and (b) $\varphi = -\pi/2$ for the "sine-like" pulse.

of the CEP of the laser system, while using only a relatively small fraction of the available laser pulse energy.

Intense few-cycle laser pulses with a stable CEP play an important role in high-intensity laser–matter interactions. Indeed, with such pulses, complete control of the electric field wave form of the laser pulse is obtained, since the pulse shape, the carrier wavelength and the carrier-envelope phase can all be determined. As a result, these pulses provide a new way to study the electron dynamics in intense laser–atom processes. They can exert a controlled force on electrons that may vary on atomic scales, not only in strength, but also in time.

Most of the work in the area of high-intensity laser–matter interactions has been restricted to infra-red, visible and ultra-violet radiation [58,59]. With the advent of *free electron lasers*, another source has become available to perform experiments over a wide range of wavelengths extending from the millimeter to the X-ray domains [60].

In a free-electron laser (FEL), an electron beam moving at a relativistic velocity passes through a periodic, transverse magnetic field produced by arranging magnets with alternating poles along the beam path. This array of magnets is called an undulator or wiggler because it forces the electrons to acquire a wiggle motion in the plane orthogonal to the magnetic field. This transverse acceleration produces spontaneous longitudinal emission of electromagnetic radiation of the synchrotron radiation type. Laser action is due to the fact that the electron motion is in phase with the electromagnetic field of the radiation already emitted, so that the fields add coherently and further emission is stimulated. Free-electron lasers have many attractive properties such as wide tunability and high laser power. However, they are large and expensive, since they involve using electron beam accelerators. The first FEL was demonstrated at a wavelength of 3.4 μm using the Stanford Linear Accelerator [61]. Since then, several FELs have been operated at wavelengths ranging from the millimeter to the soft X-ray region.

In the short-wavelength (VUV and X-ray) region, the lack of appropriate mirrors prevents the operation of an FEL oscillator. As a result, there must be suitable amplification during a single pass of the electron beam through the undulator. It is worth noting that even if the initial electromagnetic field is zero, laser action can still occur in the FEL through the process of "self-amplified spontaneous emission" (SASE), whereby shot noise in the electron beam causes a noisy signal to be initially radiated. This noise then acts as a seed for the FEL, so that the amplification process develops and intense coherent radiation is produced in a narrow band around the resonance wavelength. The first observation of the SASE process was reported at the Free Electron Laser in Hamburg (FLASH) at a wavelength of 109 nm [62]. Also at the FLASH facility, short VUV laser pulses of wavelengths in the range 95–105 nm, durations of 30–100 fs and peak powers at the gigawatt level have been generated [63]. More recently, lasing was observed by the FLASH team at a wavelength of 6.5 nm, in the soft X-ray domain. The European X-ray Free Electron Laser (XFEL) in Hamburg and the Linac Coherent Light Source (LCLS) at Stanford, both under development, will operate at wavelengths down to around 0.1 nm, well into the X-ray region.

1.3 Multiphoton ionization and above-threshold ionization

In this section, we give a survey of the basic features of the *multiphoton ionization* (MPI) process, starting with the multiphoton *single* ionization reaction

$$n\hbar\omega + A^q \rightarrow A^{q+1} + e^-, \tag{1.3}$$

where q is the charge of the target atomic system A, expressed in atomic units, $\hbar\omega$ is the photon energy and n is a positive integer.

This process was first observed in 1963 by Damon and Tomlinson [64], who used a ruby laser to ionize helium, argon and a neutral air mixture. In subsequent investigations, Voronov and Delone [65] used a ruby laser to induce seven-photon ionization of xenon, and Hall, Robinson and Branscomb [66] recorded two-photon electron detachment from the negative ion I^-. In later years, important results were obtained by several experimental groups, in particular at Saclay, where the dependence of the ionization rates on the laser intensity were studied. For the intensities $I \ll I_a$ available at that time, it was observed that the total n-photon ionization rate was proportional to I^n. As we shall see in Chapter 3, this result is in agreement with the prediction of the lowest (non-vanishing) order of perturbation theory (LOPT). At that time, the phenomenon of resonantly enhanced multiphoton ionization (REMPI) was also studied.

A crucial breakthrough was made when experiments detecting the energy-resolved photoelectrons were performed. In this way, Agostini *et al.* [67] discovered in 1979 that at sufficiently high intensities ($I > 10^{11}\,\mathrm{W\,cm^{-2}}$), the ejected electron can absorb photons in excess of the minimum number required for ionization to occur. This phenomenon is called "above-threshold ionization" (ATI). The photoelectron spectra were seen to consist of several peaks, separated by the photon energy $\hbar\omega$, and appearing at energies E_s satisfying the generalized Einstein equation

$$E_s = (n_0 + s)\hbar\omega - I_\mathrm{P}, \qquad (1.4)$$

where n_0 is the minimum number of photons needed to exceed the field-free ionization potential I_P of the atom and $s = 0, 1, \ldots$ is the number of excess photons (or "above-threshold" photons) absorbed by the atom.

A typical example of ATI photoelectron energy spectra, measured in 1988 by Petite, Agostini and Muller [68], is shown in Fig. 1.4. Pulses of 130 ps duration obtained from a Nd:YAG laser of wavelength $\lambda = 1064\,\mathrm{nm}$ were focused into a xenon vapor, and the electron energy spectrum was recorded using a time of flight spectrometer, with a 25 meV resolution. At relatively weak intensities, the

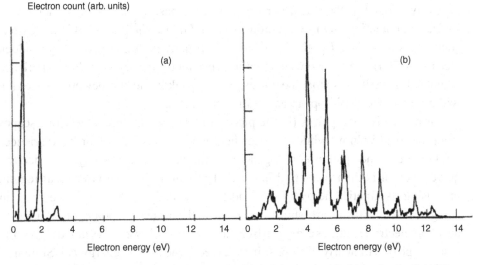

Figure 1.4. Electron energy spectra showing "above-threshold ionization" (ATI) of xenon at a laser wavelength $\lambda = 1064\,\mathrm{nm}$. (a) $I = 2.2 \times 10^{12}\,\mathrm{W\,cm^{-2}}$, (b) $I = 1.1 \times 10^{13}\,\mathrm{W\,cm^{-2}}$. (From G. Petite, P. Agostini and H. G. Muller, *J. Phys. B* **21**, 4097 (1988).)

intensity dependence of the peaks follows the LOPT prediction according to which the ionization rate for an $(n_0 + s)$-photon process is proportional to I^{n_0+s} (see Fig. 1.4(a)). As the intensity increases, peaks at higher energies appear (see Fig. 1.4(b)), whose intensity dependence does not follow the I^{n_0+s} prediction of LOPT.

Another remarkable feature of the ATI spectrum in Fig. 1.4(b) is that as the intensity increases, the low-energy peaks are reduced in magnitude. The reason for this *peak suppression* is that the energies of the atomic states are *Stark-shifted* in the presence of a laser field. For low laser frequencies (for example, a Nd:YAG laser with photon energy $\hbar\omega = 1.17$ eV), the AC Stark shifts of the lowest bound states are small in magnitude. On the other hand, the induced Stark shifts of the Rydberg and continuum states are essentially given by the electron *ponderomotive energy* U_p, which is the cycle-averaged kinetic energy of a quivering electron in a laser field. For a monochromatic laser field, it is given by

$$U_p = \frac{e^2 \mathcal{E}_0^2}{4m\omega^2},   \tag{1.5}$$

where m is the mass of the electron and $\mathcal{E}_0$ is the electric field strength. It is worth noting that the ponderomotive energy U_p is proportional to I/ω^2, and may become quite large. For example, in the case of a Nd:YAG laser of wavelength $\lambda = 1064$ nm, the ponderomotive energy U_p given by Equation (1.5) becomes equal to the laser photon energy $\hbar\omega = 1.17$ eV at an intensity $I \simeq 10^{13}$ W cm^{-2}. Since the energies of the Rydberg and continuum states are shifted upwards relative to the lower bound states by about U_p, there is a corresponding increase in the intensity-dependent ionization potential $I_P(I)$ of the atom, so that $I_P(I) \simeq I_P + U_p$. If this increase is such that $n\hbar\omega < I_P + U_p$, then ionization by n photons is energetically forbidden (see Fig. 1.5). However, atoms interacting with smoothly varying pulses experience a range of intensities, so that the corresponding peak in the photoelectron spectrum will not completely disappear, as seen in Fig. 1.4(b).

For relatively long pulses (in the picosecond range), the photoelectron escapes from the focal volume while the laser field is still present, so that it experiences a force due to the spatial inhomogeneity of the laser field intensity. The electron quiver motion is then converted into radial motion out of the laser focal region, increasing its kinetic energy by U_p, and hence exactly canceling the decrease in energy caused by the (Stark-shifted) increase in the ionization potential. As a result, the photoelectron energies are given by Equation (1.4). However, as noted above, the first ATI peak will nearly disappear if U_p exceeds the photon energy $\hbar\omega$. Similarly, the first two peaks will be weakened if U_p exceeds $2\hbar\omega$, etc. It should be noted that in this long-pulse limit, the photoelectrons have a kinetic energy at least equal to U_p once they have left the laser beam.

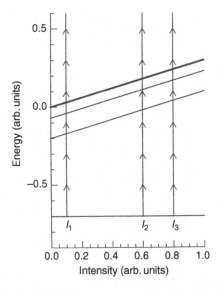

Figure 1.5. Mechanism responsible for the suppression of low-energy peaks in ATI spectra. For low laser frequencies, the intensity-dependent ionization potential of the atom, $I_P(I)$, is such that $I_P(I) \simeq I_P + U_p$, and hence increases linearly with the intensity I. Ionization by four photons, which is possible at the intensity I_1, for which $4\hbar\omega \geq I_P + U_p$, is prohibited at the higher intensities I_2 and I_3, where five photons are needed to ionize the atom. Also illustrated is the mechanism responsible for the resonantly induced structures appearing in ATI spectra for short laser pulses. At the intensity I_2, a Rydberg state has shifted into four-photon resonance with the ground state.

For short (sub-picosecond) laser pulses, the laser field turns off before the photoelectron can escape from the focal volume. In this case, the quiver energy is returned to the laser field and the ATI spectrum becomes more complicated. The observed photoelectron energies are given by the values

$$\tilde{E}_s = (n_0 + s)\hbar\omega - (I_P + U_p) \tag{1.6}$$

relative to the *shifted* ionization potential $I_P + U_p$. Photoelectrons originating from different regions of the focal volume are thus emitted with different ponderomotive shifts. As a result, the ATI peaks exhibit a substructure which, as seen from Fig. 1.5, arises from the fact that the intensity-dependent Stark shifts bring different states of the atom into multiphoton resonance during the laser pulse. An example of such substructure due to the REMPI phenomenon, observed by Freeman *et al.* [69], is shown in Fig. 1.6. This substructure is not seen in long-pulse experiments because in that case, as explained above, the photoelectrons regain their ponderomotive energy deficit from the laser field as they escape from the focal volume.

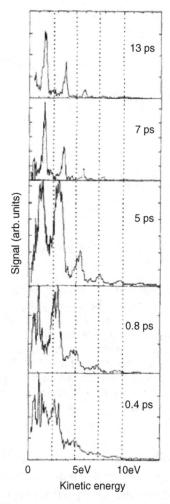

Figure 1.6. Kinetic energy of photoelectrons emitted from xenon as a function of the laser pulse width. The pulse energy is held roughly constant for all runs, so that the intensity increases from about $1.2 \times 10^{13}\,\mathrm{W\,cm^{-2}}$ for 13 ps to about $3.9 \times 10^{14}\,\mathrm{W\,cm^{-2}}$ for 0.4 ps. For the shortest-pulse widths, the individual ATI peaks break up into a narrow fine structure. (From R. R. Freeman *et al.*, *Phys. Rev. Lett.* **59**, 1092 (1987).)

If the frequency is low enough and the laser field is sufficiently strong, ionization can be interpreted by using a quasi-static model in which the bound electrons experience an effective potential formed by adding to the atomic potential the contribution due to the instantaneous laser electric field (see Fig. 1.7(a)). The "instantaneous" ionization rate may then be approximated by the static-limit tunneling rate which can be calculated for hydrogenic systems using the formula given by

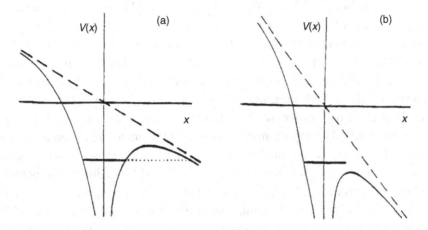

Figure 1.7. One-dimensional model showing (a) tunneling ionization and (b) over-the-barrier ionization. The dashed lines correspond to the contribution to the potential energy due to the instantaneous laser electric field. The solid lines correspond to the full effective potential energy. The position of a bound energy level (in the absence of the laser field) is indicated.

Ammosov, Delone and Krainov [70]. In the case of a hydrogenic system in its ground state, the tunneling rate is given in atomic units by [71]

$$W_{\text{ion}} = \frac{4(2I_P)^{5/2}}{\mathcal{E}} \exp\left(-\frac{2(2I_P)^{3/2}}{3\mathcal{E}}\right), \tag{1.7}$$

where $\mathcal{E}$ is the static electric field strength. Because of the exponential factor, tunneling occurs predominantly at the peaks of the electric field during the half-cycle when it lowers the potential barrier. As a result, the photoelectron wave packets are emitted in periodic bursts in time. In the energy domain, for sufficiently long pulses, this periodicity gives rise to the ATI spectrum.

A more general quasi-static theory was developed by Keldysh [72] to describe multiphoton ionization in the low-frequency limit, and was pursued by Faisal [73] and Reiss [74]. In the approach of Keldysh, the *strong field approximation* (SFA) is made, whereby it is assumed that an electron, after having being ionized, interacts only with the laser field and not with its parent core. As will be discussed in Chapter 6, an important quantity in this theory is the *Keldysh parameter* γ_K, defined as the ratio of the laser and tunneling frequencies, which is given by

$$\gamma_K = \left(\frac{I_P}{2U_p}\right)^{1/2}. \tag{1.8}$$

For $\gamma_K \lesssim 1$, tunneling dynamics dominates, while $\gamma_K \gtrsim 1$ is referred to as the multiphoton ionization regime. It is interesting to note that early evidence of quasi-static tunneling was found in 1974 by Bayfield and Koch [75] in the microwave ionization of highly excited Rydberg atoms [76].

At low frequencies, as the laser intensity is increased, the barrier in the effective potential becomes narrower and lower, and the sharp ATI peaks of the photoelectron spectra gradually blur into a continuous distribution. Eventually, above a critical intensity I_c (also called "appearance" intensity), the electron can classically "flow over the top" of the barrier (see Fig. 1.7(b)). This is known as "over-the-barrier" ionization (OBI). The approximations based on the tunneling formulae then break down and the atom ionizes quickly. The critical intensity, I_c, at which the maximum of the effective potential is lowered to a value equal to the ionization potential of the bound electron, is $1.4 \times 10^{14} \, \mathrm{W \, cm^{-2}}$ for atomic hydrogen in the ground state and $1.5 \times 10^{15} \, \mathrm{W \, cm^{-2}}$ for helium, also in the ground state. Augst *et al.* [77] and Mevel *et al.* [78] have studied the ionization of noble gases in this intensity regime. As an illustration, the energy spectrum of photoelectrons ejected from helium at the critical intensity I_c and the wavelength $\lambda = 617 \, \mathrm{nm}$, as obtained by Mevel *et al.* [78] is shown in Fig. 1.8.

In order to develop a successful model of strong field phenomena at low frequencies, it is necessary to go beyond tunneling or over-the-barrier ionization, and take into account the possibility that the ionized electron will return to the vicinity of its parent ion or atom core. The semi-classical "recollision model" developed by Corkum [79] and by Kulander, Schafer and Krause [80] is based on the idea that ionization by strong laser fields at low frequencies proceeds via several steps. In the first ("bound–free") step, the active electron is detached from its parent core by tunneling or over-the-barrier ionization. In the second ("free–free") step, the unbound electron interacts mainly with the laser field, so that its dynamics are

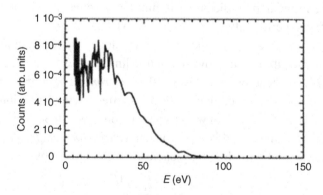

Figure 1.8. Electron energy spectrum from helium at a laser wavelength $\lambda = 617 \, \mathrm{nm}$ and the critical intensity $I_c = 1.5 \times 10^{15} \, \mathrm{W \, cm^{-2}}$. The duration of the laser pulse is 100 fs. There is no structure above 30 eV. (From E. Mevel *et al.*, *Phys. Rev. Lett.* **70**, 406 (1993).)

essentially those of a free electron in the field, and can be treated to a good approximation by using classical mechanics. An earlier version of this approach has been used by Kuchiev [81] and van Linden van den Heuvell and Muller [82], and is known as the "simple man's" model of strong field phenomena. As the electric field component of the laser field changes sign, the electron can be accelerated back toward its parent core. If the electron does not return to the core, single ionization will occur. If it does return to the core, then a third step takes place in which a collision of the electron with the core leads to single or multiple ionization while radiative recombination leads to the process of harmonic generation, which will be described in Section 1.4. As we shall see in Chapter 8, the semi-classical "recollision model" has been very useful for explaining a number of novel features observed in multiphoton ionization experiments performed with low-frequency lasers. For example, experiments using kilohertz-repetition rate, high-intensity lasers have allowed precise measurements of photoelectron total yields and energy and angle-differential spectra over many orders of magnitude in yield to be carried out. In particular, the experiments of Paulus *et al.* [83] have revealed the existence of a "plateau" in the ATI photoelectron energy spectra (see Fig. 1.9) which, as we shall see in Chapter 8, is due to recollisions of electrons with their parent cores.

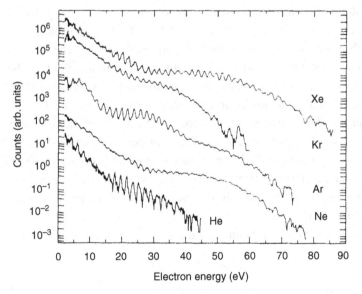

Figure 1.9. Photoelectron counts as a function of photoelectron energy, for various noble gases, at a laser wavelength $\lambda = 630\,\text{nm}$ and an intensity $I \simeq 2 \times 10^{14}\,\text{W cm}^{-2}$ ($3 \times 10^{14}\,\text{W cm}^{-2}$ for He). (From G. G. Paulus *et al.*, *Phys. Rev. Lett.* **72**, 2851 (1994).)

More recently, the investigation of ATI has benefited from the rapid progress made in generating few-cycle laser pulses with peak intensities exceeding the atomic unit I_a. Already at intensities one or more orders of magnitude lower than I_a, atoms can ionize on the time scale of a few optical periods. Hence, the only way to expose atoms to high-intensity laser fields is to irradiate them with the shortest possible laser pulses. Grasbon *et al.* [84] have measured ATI photoelectron spectra for noble gas atoms ionized with intense few-cycle laser pulses. They also found an extended plateau-like structure in the photoelectron energy spectra that can be explained by using the semi-classical recollision model. With respect to the atomic response, the use of sub 10 fs laser pulses makes a significant difference in comparison with 30 fs pulses because all Rydberg states have much longer orbit times than the pulse duration. As a result the REMPI phenomenon, which plays a major role for longer pulses, is much less important for few-cycle pulses.

Using laser pulses having a duration of approximately 6 fs, Paulus *et al.* [85] were able to demonstrate the influence of the carrier-envelope phase (CEP) of an ultra-short pulse on the emission of ATI photoelectrons. They ionized krypton atoms and recorded the photoelectrons with two opposite detectors perpendicular to the laser beam, as illustrated in Fig. 1.10(a). They detected an anticorrelation in the number of electrons emitted to the left or to the right (see Fig. 1.10(b)), which is the signature of the carrier-envelope phase. Indeed, if this phase has a value such that the maximum of the electric field points to the left, more photoelectrons will be counted at the right detector than the left detector. Longer pulses do not exhibit this anticorrelation since the influence of the CEP averages out.

In subsequent work, Paulus *et al.* [47] and Verhoef *et al.* [53] made use of intense few-cycle laser pulses with a stabilized CEP to demonstrate that the direction of emission of ATI photoelectrons can be controlled by varying the CEP of the laser field, thus providing a tool for an accurate determination of the CEP. More recently, Kling *et al.* [56] reported sub-femtosecond control of the electron emission in ATI of the noble gases Ar, Kr and Xe in intense, few-cycle laser fields. Using a velocity-map imaging (VMI) technique, where electrons are projected onto a two-dimensional position-sensitive detector, they were able to measure full-momentum distributions of ATI, and also to determine the CEP from the angular distribution of the emitted electrons.

Let us now consider the multiphoton *double* ionization process

$$n\hbar\omega + A^q \rightarrow A^{q+2} + 2e^-. \tag{1.9}$$

The double ionization of helium from its ground state,

$$n\hbar\omega + \mathrm{He}(1^1\mathrm{S}) \rightarrow \mathrm{He}^{2+} + 2e^-, \tag{1.10}$$

(a)

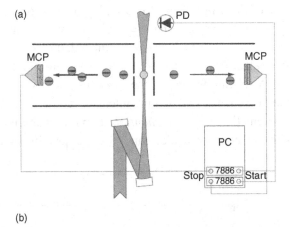

(b)

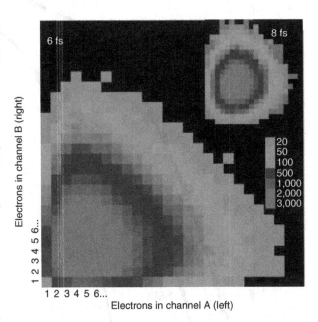

Figure 1.10. (a) Laser pulses of about 6 fs duration are focused onto a gas jet of krypton atoms. The number of electrons emitted to the left and right directions depends on the carrier-envelope phase (CEP) of the laser pulse. They are recorded by two opposite multichannel plate (MCP) detectors perpendicular to the laser beam. The start signal is generated by a fast photodiode (PD) and the electron's time of flight is recorded by a computer (PC). (b) The signature of the carrier-envelope phase is an anticorrelation in the number of electrons emitted to the left or to the right. A graphical representation of the anticorrelation can be obtained by characterizing each laser pulse according to the number of electrons registered in the left and the right detectors. The anticorrelation is manifested in the comparatively high number of laser pulses that produce a strongly asymmetric number of photoelectrons. This leads to contours perpendicular to the diagonal. Longer pulses do not exhibit this anticorrelation, as seen in the upper right corner for a laser pulse of 8 fs duration. (From G. G. Paulus *et al.*, *Nature* **414**, 182 (2001).)

is the most basic of these processes. In 1967, Byron and Joachain [86] proved that in the case of *one*-photon ionization ($n = 1$), the double ionization process (1.10) is very sensitive to *electron correlation effects*. Indeed, this process could not occur in the absence of the electron–electron interaction. The multiphoton case is more complex to analyze. A striking feature of the experimental results obtained by Walker *et al.* [87] is the existence of two distinct intensity regimes (see Fig. 1.11). The first one is at high intensities, where *sequential* double ionization (SDI) dominates in accordance with a "single active electron" (SAE) approximation. The other regime

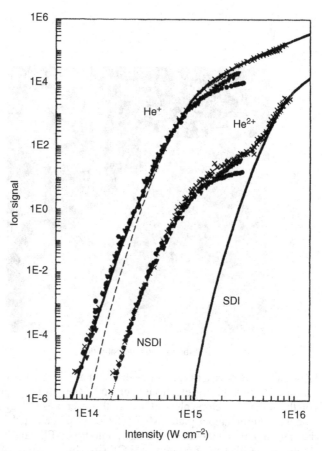

Figure 1.11. Measured He^+ and He^{2+} yields as a function of intensity for linearly polarized laser pulses of wavelength $\lambda = 780$ nm and 160 fs duration. The solid lines show SAE calculations and the dashed line depicts the tunneling theories. The solid curve on the right corresponds to the calculated He^{2+} sequential double ionization (SDI) yield. The symbol NSDI refers to the non-sequential double ionization yield. (From B. Walker *et al.*, *Phys. Rev. Lett.* **73**, 1227 (1994).)

is at lower intensities, where mainly simultaneous double ionization, or in other words *non-sequential* double ionization (NSDI), takes place and the SAE approximation predicts double ionization yields that are several orders of magnitude lower than the experimental data. An important test for theories of multiphoton processes going beyond the SAE approximation (that is, including electron correlation effects) is to calculate accurately double (and, more generally, multiple) ionization yields for multielectron systems in intense laser fields. Advances in experimental techniques based on cold target recoil ion momentum spectroscopy (COLTRIMS) [88] and intense few-cycle laser pulses have allowed multiphoton double ionization to be investigated in more detail [89, 90]. These experimental developments, as well as the theoretical approaches incorporating electron correlation effects, will be discussed in Chapter 8.

1.4 Harmonic generation and attosecond pulses

Matter interacting with a sufficiently intense laser field can emit radiation at higher-order multiples, or *harmonics*, of the angular frequency ω of the "pump" laser. Optical harmonic generation was first observed in a quartz crystal in 1961 by Franken *et al.* [91], who used a ruby laser producing approximately 3 J of 694.3 nm light in a 1 ms pulse, corresponding to an intensity of about $10^6\,\mathrm{W\,cm^{-2}}$, to generate the second harmonic. They pointed out that "the possibility of exploiting this extraordinary intensity for the production of optical harmonics from suitable non-linear materials is most appealing." Since then, harmonic generation has become a phenomenon used in a variety of applications to extend the range of laser light sources to shorter wavelengths. In the case of atoms interacting with laser pulses comprising many optical cycles and having intensities such that non-dipole and relativistic effects can be neglected, the harmonic angular frequencies Ω are only emitted at odd multiples of the laser angular frequency because of the inversion symmetry of the atom in the field. Hence,

$$\Omega = q\omega, \quad q = 3, 5, \ldots \tag{1.11}$$

When the driving laser pulse comprises only a few optical cycles, the photons are emitted with a continuous distribution of frequencies, not at discrete harmonic frequencies. An effective "harmonic order" is then defined to be the ratio Ω/ω of the angular frequency of the emitted photon to that of the pump laser. High-order harmonic generation (HHG) has attracted considerable interest, since it provides a source of very bright, short-pulse, high-frequency coherent radiation. It has been the subject of several review articles [92–94].

The observation of the third harmonic in noble gases was made in 1967 by New and Ward [95]. Harmonic generation experiments were then performed with long-pulse infra-red lasers [96–98] or ultra-violet pump fields [99–101], but were limited by various effects such as the ionization of the medium or the absorption of the generated radiation in optically thick media.

It was only in the late 1980s that the availability of intense, short-pulse laser fields made it possible to observe high-order harmonics. Experiments performed at the University of Illinois by McPherson *et al.* [102] and Rosman *et al.* [103] showed the generation of up to the 17th harmonic of a 248 nm KrF laser in a neon vapor. At Saclay, Ferray *et al.* [104] observed the harmonic $q = 33$ in argon, and Li *et al.* [105] observed the harmonics $q = 29$ in krypton and $q = 21$ in xenon using a 1064 nm Nd:YAG laser delivering pulses of about 30 ps duration and intensity 3×10^{13} W cm^{-2}. These results were extended in a number of experiments performed with shorter pulses, and at higher intensities [106–112]. For example, Sarukura *et al.* [106] observed the 25th harmonic in He and Ne using a 500 fs, 248 nm KrF laser, a result interpreted as harmonic emission by the ions. Miyazaki and Sakai [107] reported the observation of the 41st harmonic in He with a 800 fs, 616 nm dye laser. In 1993, Macklin, Kmetec and Gordon [108] observed the generation of up to the 109th harmonic in Ne, using a 125 fs, 806 nm Ti:sapphire laser. Also in 1993, L'Huillier and Balcou [109] detected HHG in noble gases, using 1 ps pulses from a Nd:glass laser at intensities around 10^{15} W cm^{-2}. The number of photons recorded per laser pulse in xenon, argon, neon and helium at an intensity $I \simeq 1.5 \times 10^{15}$ W cm^{-2} is shown in Fig. 1.12 as a function of the harmonic order. Of special interest is the existence of a plateau of nearly constant conversion efficiency, which is particularly long for helium and neon. We shall see below that non-perturbative theories are required to explain the occurrence of such plateaus. L'Huillier and Balcou detected up to the 135th harmonic in neon at an energy of 160 eV, being then limited by the resolution of their monochromator. The harmonic emission was observed to be directional and of short pulse duration (shorter than the pump pulse). The instantaneous power generated at, for example, 20 eV (the 17th harmonic in Xe) reached about 30 kW, with a conversion efficiency of 10^{-6}. The instantaneous brightness was 10^{22} photons/(Å s), a number which is several orders of magnitude higher than that obtained with conventional light sources in this domain of the electromagnetic spectrum (but of course over a restricted time period).

The use of intense ultra-short laser pulses with peak intensities higher than 10^{15} W cm^{-2} [34] offers new perspectives for the generation of coherent, tunable, high-frequency (UV or X-ray) pulses by HHG in gases. Atoms or ions exposed to such pulses experience only a few optical cycles, and hence can withstand much stronger laser fields before ionizing than would be possible with longer pulses.

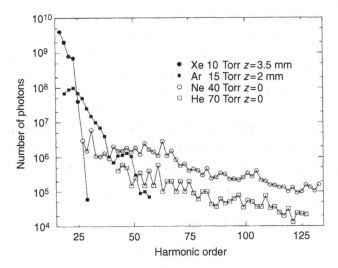

Figure 1.12. Harmonic emission spectra of various noble gases for a "pump" laser of wavelength $\lambda = 1053$ nm and intensity $I \simeq 1.5 \times 10^{15}$ W cm^{-2}. (From A. L'Huillier and P. Balcou, *Phys. Rev. Lett.* **70**, 774 (1993).)

This in turn permits the generation of photons of much higher energies. The resulting high-frequency laser pulses have durations in the sub-femtosecond range. The highest harmonic frequencies and harmonic orders have been observed under these conditions [113–119]. For example, Chang *et al.* [116] used a Ti:sapphire laser of wavelength $\lambda = 800$ nm delivering pulses of 26 fs duration and peak intensities of about 6×10^{15} W cm^{-2} to generate coherent soft X-ray harmonics reaching into the water window spectral region around a wavelength of 2.7 nm (corresponding to an energy of 460 eV) in helium, and 5.2 nm (239 eV) in neon. In helium, they observed harmonic peaks up to order $q = 221$ and unresolved harmonic emission up to order $q = 297$. Schnürer *et al.* [117] reported the generation of coherent X-rays with wavelengths down to 2.5 nm (corresponding to an energy of 500 eV) in a helium gas irradiated by sub 10 fs pulses of peak intensity 4×10^{15} W cm^{-2} generated by a Ti:sapphire laser of wavelength $\lambda = 770$ nm at a 1 kHz repetition rate. In later experiments, Seres *et al.* [118] observed the production of harmonic photons with energies extending to 1.3 keV for helium atoms irradiated by 5 fs pulses, and Chen *et al.* [119] have generated phase-matched high harmonics spanning the water window spectral region.

The theoretical treatment of harmonic generation by an intense laser pulse focused onto a gaseous medium has two main aspects which will be discussed in more detail in Chapter 9. First, the *microscopic single-atom* response to the laser field must be analyzed. Because different atoms in the laser focus experience different peak intensities and phases, the single-atom response must be calculated over a range of laser field intensity profiles. In order to obtain the *macroscopic* harmonic spectrum generated by the coherent photon emission by all of the atoms in the laser focus, the single-atom responses must be combined by solving Maxwell's equations. In this way propagation and interference effects are accounted for. Unless *phase matching* occurs, these effects can lead to the suppression of the harmonic emission signal.

Let us now consider briefly the theoretical aspects of the problem. The emission of harmonics by the atom is due to the electron oscillations caused by the intense laser field. Let us introduce the *laser-induced atomic dipole moment*

$$\mathbf{d}(t) = \langle \Psi(t)| - e\mathbf{R}|\Psi(t)\rangle, \tag{1.12}$$

which is the expectation value of the electric dipole operator

$$\mathbf{D} = -e\mathbf{R}, \tag{1.13}$$

where

$$\mathbf{R} = \sum_{i=1}^{N} \mathbf{r}_i \tag{1.14}$$

is the sum of the coordinates $\mathbf{r}_i$ of the N atomic electrons. In Equation (1.12), $|\Psi(t)\rangle$ denotes the state vector of the atom in the presence of the laser field. The propagation equations that must be solved to obtain the spectrum of harmonics generated by the medium have source terms which are proportional to the Fourier components, $\mathbf{d}(\Omega)$, of $\mathbf{d}(t)$, namely

$$\mathbf{d}(\Omega) = (2\pi)^{-1/2} \int_{-\infty}^{\infty} \exp(-i\Omega t)\,\mathbf{d}(t)\,dt. \tag{1.15}$$

Due to phase matching effects, the strength of the harmonics emitted by the medium may vary with Ω in a different way than $|\mathbf{d}(\Omega)|^2$. For the case of a single atom, the emitted power spectrum is proportional to the quantity $|\mathbf{a}(\Omega)|^2$, where

$$\mathbf{a}(\Omega) = (2\pi)^{-1/2} \int_{-\infty}^{\infty} \exp(-i\Omega t)\,\mathbf{a}(t)\,dt \tag{1.16}$$

is the Fourier transform of the acceleration of the laser-induced atomic dipole moment,

$$\mathbf{a}(t) = \frac{d^2}{dt^2}\mathbf{d}(t) \equiv \ddot{\mathbf{d}}(t). \tag{1.17}$$

For weak laser fields, the harmonic emission rates can be calculated by the perturbation theory which will be developed in Chapter 3. It is then found that in general the harmonic intensity decreases from one order to the next. In contrast, we note from Fig. 1.12 that at high laser intensities, and for linearly polarized pump fields, the harmonic intensity distribution exhibits a rapid decrease over the first few harmonics, followed by a plateau of approximately constant intensity, and then a cut-off corresponding to an abrupt decrease of harmonic intensity. The existence of the plateau can only be explained by using non-perturbative approaches. In particular, by solving numerically the time-dependent Schrödinger equation (TDSE), Krause, Schafer and Kulander [120] found that the cut-off angular frequency ω_c of the harmonic spectrum is given approximately by

$$\omega_c \simeq (I_P + 3U_p)/\hbar. \tag{1.18}$$

Using the semi-classical recollision model [79, 80], it will be shown in Chapter 6 that the maximum kinetic energy of a classical electron recolliding with the atomic core is $3.17U_p$, so that the highest energy that can be radiated is $I_P + 3.17U_p$, in good agreement with the TDSE calculations and with experiment. A quantum-mechanical theory of HHG, based on a low-frequency strong-field approximation (SFA), has been developed by Lewenstein *et al.* [121, 122] and will be discussed in Chapter 6. It embodies the semi-classical recollision model, and also accounts for quantum effects such as tunneling ionization, wave-packet spreading and interferences.

It also follows from the SFA that the Fourier component of the induced atomic dipole moment corresponding to the qth harmonic has a phase, denoted by ϕ_q, that in first approximation is proportional to the product $U_p\tau$, where τ is the time spent by the ionized electron in the continuum before it returns and recombines radiatively with its parent ion. Using Equation (1.5), we see that the phase ϕ_q is a linear function of the intensity, with a slope depending on the time τ. The main contributions to the emission of each harmonic in the plateau region come from two electron trajectories. Electrons following these two trajectories have the same kinetic energy when they return to the atomic core. The two trajectories are referred to, respectively, as the "short" trajectory and "long" trajectory. The corresponding times spent by the electron in the continuum, τ_1 and τ_2, are shorter than one laser field period $T = 2\pi/\omega$, so that

$$0 < \tau_1 < \tau_2 < T. \tag{1.19}$$

The variation of the dipole phase ϕ_q with intensity, in the strong-field regime, plays an important role in the macroscopic aspect of harmonic emission [93].

An important new development is the possibility of using high-order harmonics to generate laser pulses having durations in the attosecond (10^{-18} s) range. This subject has been reviewed by Agostini and DiMauro [123], Scrinzi *et al.* [124], Niikura and Corkum [125] and Krausz and Ivanov [126]. Fourier synthesis was proposed by Hänsch [127], by Farkas and Toth [128] and by Harris, Macklin and

Hänsch [129] as a method to produce pulses having durations of a few attoseconds. The basic idea is to generate a "comb" of equidistant frequencies with controlled relative phases. The principle is analogous to that of the mode-locked laser [130]: if N spectral modes within the gain bandwidth are phase-locked, then the temporal profile is a sequence of pulses separated by the cavity round-trip time, each with a duration proportional to N^{-1}. Several methods using non-linear processes to obtain a wide sequence of equidistant frequencies have been proposed. Hänsch [127] suggested using sum and frequency mixing to generate six frequencies, while Farkas and Toth [128] recognized that high harmonic generation in atoms could readily produce a comb of odd-order harmonics of nearly equal amplitudes over a large frequency range within the plateau region. A similar idea was proposed by Harris, Macklin and Hänsch [129]. If these harmonics were emitted in phase, the corresponding temporal profile (which is the Fourier transform of the periodic spectrum of the harmonics) would consist of a train of ultra-short pulses separated by half the laser period, the duration of each pulse being proportional to the inverse of the number of harmonics. However, calculations of the single-atom response, performed by Antoine, L'Huillier and Lewenstein [131], showed that the harmonics in the plateau region were in general not in phase, due to the interference of various energetically allowed electronic trajectories leading to the harmonic emission. In fact, in the plateau region there are at least two trajectories per half laser field period (in particular the short one and the long one) that lead to the emission of a given harmonic. As a result, the temporal profile exhibits two dominant peaks per half-period of the laser field that can be attributed to these trajectories. This is illustrated in Fig. 1.13, where the single-atom response for the temporal superposition of harmonics 41 to 61 generated in neon calculated by Antoine, L'Huillier and Lewenstein [131] is shown. However, Antoine, L'Huillier and Lewenstein also showed that the propagation in the atomic medium could select one of these trajectories. Indeed, under certain conditions, phase matching strongly depends on the phase ϕ_q, which exhibits a different intensity dependence for the different trajectories, as mentioned above. For example, as illustrated in Fig. 1.13, Antoine, L'Huillier and Lewenstein found that the macroscopic temporal profile for laser focusing before the gas jet – which selects the short electron trajectories – yields a single peak, of 300 as duration, per half-period of the laser field. This prediction has been confirmed qualitatively by a measurement of the relative phases of a group of harmonics generated in argon [132]. The phases are consistent with the emission of a train of 250 as pulses.

The generation of a *train* of attosecond pulses by Fourier synthesis of harmonics does not require particularly short pump laser pulses. However, for many applications, it is desirable to obtain *isolated* attosecond pulses. A first method, proposed by Corkum, Burnett and Ivanov [133], Ivanov *et al.* [134] and Platonenko and

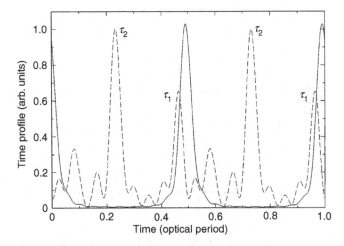

Figure 1.13. Theoretical harmonic intensity time profile obtained by summing the ten harmonics between the 41st to the 61st (solid line) emitted by a gas jet located before the laser focus. The peak intensity of the driving laser pulse is 6.6×10^{14} W cm^{-2}. The dashed line denotes the corresponding single-atom response. The labels τ_1 and τ_2 refer to the short and the long electron trajectories, respectively. The figure shows that the intensity peaks corresponding to the long electron trajectories are suppressed during propagation through the atomic medium. (From P. Antoine, A. L'Huillier and M. Lewenstein, *Phys. Rev. Lett.* **77**, 1234 (1996).)

Strelkov [135], uses the high sensitivity of the harmonic efficiency to the laser field polarization [136], and is therefore called *polarization gating*. It follows from the three-step recollision model that harmonics are essentially produced when the polarization of the laser field is linear. Indeed, the probability of an ionized electron returning to its parent ion is significantly reduced when the laser field is no longer linearly polarized. By creating a laser pulse whose polarization is linear only during a short time (close to a laser period) the harmonic emission may be limited to this interval, so that single attosecond pulses are produced. There are different techniques for implementing a polarization gate. For example, using two chirped laser pulses delayed in time, Altucci *et al.* [137] have reduced photon emission to a few femtoseconds. Using two delayed counter-circularly polarized laser pulses, a temporal gating has also been demonstrated [138–140].

A second method to obtain isolated attosecond pulses from HHG is to use an ultra-short few-cycle pump pulse. Theoretical calculations have predicted the possibility of generating a single XUV burst [34, 141]. Using few-cycle (<7 fs) linearly polarized laser pump pulses with stabilized carrier-envelope phase, isolated attosecond pulses have been produced by selecting the high-energy (cut-off) harmonics (~90 eV) generated in neon [48, 142, 143].

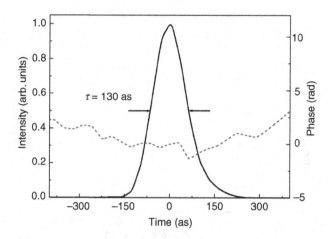

Figure 1.14. Reconstruction of the temporal intensity profile (solid curve) and phase (dotted curve) of the attosecond pulses generated using 5 fs laser pulses ($\lambda = 750$ nm) with stabilized carrier-envelope phase interacting with Ar atoms. (From G. Sansone *et al.*, *Science* **314**, 443 (2006).)

Another approach for generating isolated attosecond pulses uses a combination of the two aforementioned methods. It is based on the use of phase-stabilized few-cycle driving pulses in combination with the polarization gating technique. Sansone *et al.* [144] have generated in this way single-cycle isolated 130 as pulses around 36 eV by using 5 fs driving pulses, as shown in Fig. 1.14.

Let us now consider photon emission by *positive ions* interacting with very intense, ultra-short (few-cycle) pulses. Positive ions can survive higher laser intensities because of their higher binding energies, and can therefore emit more energetic photons [112, 145–147]. However, the HHG conversion efficiency begins to decrease as the laser intensity increases. This is a consequence of non-dipole effects, due essentially to the magnetic field component of the laser field [148–152]. As an illustration, we show in Fig. 1.15 the magnitude squared of the Fourier transform of the dipole acceleration of a Be^{3+} ion as a function of the emitted photon energy (in units of $\hbar\omega$) for an 800 nm four-cycle laser pulse with a peak intensity of 3.6×10^{17} W cm^{-2}. The influence of the magnetic field on photon emission polarized along the polarization direction of the laser field (taken to be the x-axis) can be seen by comparing the dipole and non-dipole results. This reduction is easy to understand within the semi-classical three-step recollision model: the magnetic field component of the laser pulse induces a displacement of the electron along the laser field propagation direction which causes returning electrons to "miss" the core. Another non-dipole effect is the emission of photons polarized along the laser field propagation direction (taken to be the z-axis), which is forbidden in the dipole

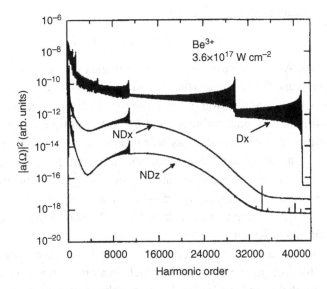

Figure 1.15. Magnitude squared of the Fourier transform of the SFA dipole acceleration (in arbitrary units) of Be^{3+} as a function of the harmonic order (the emitted photon energy in units of laser photon energy). Dipole results (Dx) are shown, as well as non-dipole results for photons polarized along the laser field polarization direction $\hat{x}$ (NDx) and polarized along the laser field propagation direction $\hat{z}$ (NDz). The incident laser pulse has a total duration of four optical cycles, a carrier wavelength of 800 nm and a peak intensity of $3.6 \times 10^{17}\,W\,cm^{-2}$. (From C. J. Joachain, N. J. Kylstra and R. M. Potvliege, *J. Mod. Opt.* **50**, 313 (2003).)

approximation. As seen from Fig. 1.15, this emission is typically two orders of magnitude lower than emission along the laser field polarization direction.

1.5 Laser-assisted electron–atom collisions

An electron scattered by an atom in the presence of a laser field can absorb or emit radiation. Since these radiative collisions involve continuum states of the electron–atom system, they are often called "free–free transitions" (FFT). In weak laser fields, only one-photon processes have a large enough probability to be observed. However, as the field strength is increased, multiphoton processes become important. Examples of laser-assisted electron–atom collisions are "elastic" collisions:

$$e^- + A(i) + n\hbar\omega \rightarrow e^- + A(i); \tag{1.20}$$

inelastic collisions:

$$e^- + A(i) + n\hbar\omega \rightarrow e^- + A(f); \tag{1.21}$$

and single ionization (e, 2e) collisions:

$$e^- + A(i) + n\hbar\omega \rightarrow A^+(f) + 2e^-, \qquad (1.22)$$

where $A(i)$ and $A(f)$ denote an atom or ion A in the initial state i and the final state f, respectively, and $A^+(f)$ means the A^+ ion in the final state f. Positive values of n correspond to photon absorption (inverse bremsstrahlung), negative ones to photon emission (stimulated bremsstrahlung) and $n = 0$ to a collision process in the laser field without net absorption or emission of photons. In contrast to *laser-induced* processes, such as multiphoton ionization and harmonic generation, *laser-assisted* processes can take place in the absence of the laser field, but are modified by its presence. Reviews of laser-assisted electron–atom collisions have been given by Gavrila [153], Francken and Joachain [154], Mason [155], Joachain [17] and Ehlotzky, Jaron and Kaminski [156].

Information on laser-assisted electron–atom collision processes is obtained by performing three-beam experiments, in which an atomic beam is crossed in coincidence by a laser beam and an electron beam, and the scattered electrons are detected. It is worth pointing out that the laser intensity should not be too high ($I < I_a$), since otherwise the atom would be ionized. Several experiments of this kind have been carried out, in which the exchange of photons between the electron–atom system and the laser field has been observed in laser-assisted elastic [157–159], inelastic [160, 161] and ionization processes [162]. As an illustration, we show in Fig. 1.16 the results of Weingartshofer *et al.* [158] for laser-assisted elastic electron–argon scattering. Even at the modest intensity of $10^8\,\text{W cm}^{-2}$, as many as 11 photon emission and absorption transitions were observed. As seen from Fig. 1.16, the relative intensities of two successive peaks are of the same order of magnitude, which indicates that perturbation theory cannot be used to analyze these results.

A detailed treatment of laser-assisted electron–atom collisions will be given in Chapter 10. We remark that this problem is in general very complex, not only from the experimental point of view, but also on the theoretical side. Indeed, in addition to the difficulties associated with the treatment of field-free electron–atom collisions, the presence of the laser field introduces new parameters (laser frequency, intensity, polarization,...) which influence the collision. It is therefore of interest to consider a much simpler problem in which the target atom is modeled by a center of force, namely a static potential, and hence does not exhibit any internal structure. Within the framework of this model, general expressions for the required collision cross sections can be obtained [163, 164], and the first Born approximation result of Bunkin and Fedorov [165] as well as the low-frequency (or soft-photon) approximation of Kroll and Watson [166] derived.

As in the field-free case [167, 168], the theoretical treatment of electron collisions with real atoms in the presence of a laser field depends on the energy of the projectile

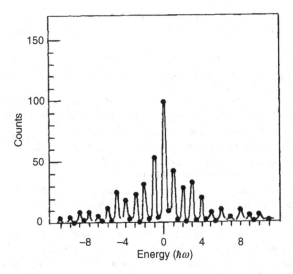

Figure 1.16. Energy spectrum of electrons scattered by argon atoms in the presence of a laser field of photon energy $\hbar\omega = 0.117\,\text{eV}$ and intensity $I = 10^8\,\text{W cm}^{-2}$ from a CO_2 laser. The circles correspond to the experimental data; the full line is drawn to guide the eye. The abscissa gives the final electron energy in units of the photon energy, with the origin fixed at the initial electron energy of 15.8 eV. The scattering angle $\theta = 155°$. (From A. Weingartshofer *et al.*, *J. Phys. B* **16**, 1805 (1983).)

electron. For fast incident electrons, having an energy of at least 100 eV, the semi-perturbative theory of Byron and Joachain [169] can be applied. In this approach, the interaction of the laser field with the unbound electron(s) is treated "exactly." On the other hand, the projectile electron–target atom interaction is treated perturbatively by using the Born series [167]. Finally, the interaction of the laser field with the target atom, responsible for *target-dressing effects,* can be treated perturbatively for laser intensities $I < I_a$ and for non-resonant processes.

For slow incident electrons, a fully non-perturbative treatment is required. The *R*-matrix–Floquet theory [10, 11] provides such a treatment. We point out that the semi-perturbative theory mentioned above and the *R*-matrix–Floquet theory are complementary. The former breaks down for slow incident electrons, where the Born series cannot be used to treat the electron–atom interaction. The latter is difficult to apply for fast incident electrons, where many partial waves are required to calculate the cross sections.

References

[1] A. Einstein, *Annalen der Physik* **17**, 132 (1905).
[2] G. N. Lewis, *Nature* **118**, 132 (1926).
[3] M. Göppert-Mayer, *Ann. Phys.* **9**, 273 (1931).
[4] C. J. Joachain and N. J. Kylstra, *Laser Phys.* **11**, 212 (2001).

[5] C. J. Joachain, N. J. Kylstra and R. M. Potvliege, *J. Mod. Opt.* **50**, 313 (2003).

[6] Y. I. Salamin, S. X. Hu, K. Z. Hatsagortsyan and C. H. Keitel, *Phys. Rep.* **427**, 41 (2006).

[7] A. Maquet, R. Taïeb and V. Véniard, Relativistic laser-atom physics. In T. Brabec, ed., *Strong Field Laser Physics*, Springer Series in Optical Sciences 134 (New York: Springer, 2009), p. 477.

[8] A. Maquet, S. I. Chu and W. P. Reinhardt, *Phys. Rev. A* **27**, 2946 (1983).

[9] R. M. Potvliege and R. Shakeshaft, *Phys. Rev. A* **38**, 1098 (1988).

[10] P. G. Burke, P. Francken and C. J. Joachain, *Europhys. Lett.* **13**, 617 (1990).

[11] P. G. Burke, P. Francken and C. J. Joachain, *J. Phys. B* **24**, 761 (1991).

[12] K. C. Kulander, *Phys. Rev. A* **36**, 2726 (1987).

[13] H. G. Muller, *Phys. Rev. A* **60**, 1341 (1999).

[14] M. Gavrila, ed., *Atoms in intense laser fields, Adv. At. Mol. Phys. Suppl.* **1** (1992).

[15] T. Brabec, ed., *Strong Field Laser Physics*, Springer Series in Optical Science 134 (New York: Springer, 2009).

[16] K. Burnett, V. C. Reed and P. L. Knight, *J. Phys. B* **26**, 561 (1993).

[17] C. J. Joachain, Theory of laser-atom interactions. In R. M. More, ed., *Laser Interactions with Atoms, Solids and Plasmas* (New York: Plenum Press, 1994), p. 39.

[18] K. C. Kulander and M. Lewenstein, *Springer Handbook of Atomic, Molecular and Optical Physics*, chapter Multiphoton and Strong-Field Processes (New York: Springer Science, 2006), p. 1077.

[19] M. Protopapas, C. H. Keitel and P. L. Knight, *Rep. Progr. Phys.* **60**, 389 (1997).

[20] C. J. Joachain, M. Dörr and N. J. Kylstra, *Adv. At. Mol. Opt. Phys.* **42**, 225 (2000).

[21] D. B. Milosevic and F. Ehlotzky, *Adv. At. Mol. Opt. Phys.* **49**, 373 (2003).

[22] F. H. M. Faisal, *Theory of Multiphoton Processes* (New York: Plenum Press, 1987).

[23] M. H. Mittleman, *Introduction to the Theory of Laser–Atom Interactions*, 2nd edn (New York: Plenum Press, 1993).

[24] N. B. Delone and V. P. Krainov, *Multiphoton Processes in Atoms* (Berlin: Springer-Verlag, 1999).

[25] F. Grossmann, *Theoretical Femtosecond Physics* (Berlin: Springer-Verlag, 2008).

[26] D. Strickland and G. Mourou, *Opt. Commun.* **56**, 219 (1985).

[27] G. A. Mourou, T. Tajima, and S. V. Bulanov, *Rev. Mod. Phys.* **78**, 309 (2006).

[28] S. Bahk, P. Rousseau, T. A. Planchon *et al.*, *Optics Lett.* **29**, 2837 (2004).

[29] A. Dubietis, G. Jonusauskas and A. Piskarkas, *Opt. Commun.* **88**, 437 (1992).

[30] I. N. Ross, P. Matousek, M. Towrie, A. J. Langley and J. L. Collier, *Opt. Commun.* **144**, 125 (1997).

[31] R. A. Baumgartner and R. L. Byer, *IEEE J. Quantum Electron.* **QE-15(6)**, 432 (1979).

[32] I. N. Ross, Optical parametric amplification techniques. In T. Brabec, ed., Strong Field Laser Physics, *Springer Series in Optical Sciences* 134 (New York: Springer, 2009), p. 35.

[33] P. Meystre and M. Sargent III, *Elements of Quantum Optics* (Berlin: Springer-Verlag, 1990).

[34] T. Brabec and F. Krausz, *Rev. Mod. Phys.* **72**, 545 (2000).

[35] U. Keller, *Nature* **424**, 831 (2003).

[36] M. Nisoli, S. Stagira, S. De Silvestri *et al. Appl. Phys. B* **65**, 189 (1997).

[37] S. Sartania, Z. Cheng, M. Lenzner *et al. Optics Lett.* **22**, 1562 (1997).

[38] M. Nisoli, S. De Silvestri and O. Svelto, *Appl. Phys. Lett.* **68**, 2793 (1996).

[39] R. Szipöcs, K. Ferencz, C. Spielmann and F. Krausz, *Opt. Lett.* **19**, 201 (1994).

[40] C. P. Hauri, P. Schlup, G. Arisholm, J. Biegert and U. Keller, *Opt. Lett.* **29**, 1369 (2004).

[41] R. T. Zinkstok, S. Witte, W. Hogervorst and K. S. E. Eikema, *Opt. Lett.* **30**, 78 (2005).

[42] N. Ishii, L. Turi, V. S. Yakovlev *et al.*, *Opt. Lett.* **30**, 567 (2005).

[43] S. Witte, R. T. Zinkstok, W. Hogervorst and K. S. E. Eikema, *Opt. Express* **13**, 4903 (2005).

[44] S. T. Cundiff, F. Krausz and T. Fuji, Carrier-envelope phase of ultrashort pulses. In T. Brabec, ed., *Strong Field Laser Physics*, Springer Series in Optical Sciences 134 (New York: Springer, 2009), p. 61.

[45] D. J. Jones, S. A. Diddams, J. K. Ranka *et al.*, *Science* **288**, 635 (2000).

[46] A. Apolonski, A. Poppe, G. Tempea *et al.*, *Phys. Rev. Lett.* **85**, 740 (2000).

[47] G. G. Paulus, F. Lindner, H. Walther *et al.*, *Phys. Rev. Lett.* **91**, 253004 (2003).

[48] A. Baltuška, T. Udem, M. Uiberacker *et al.*, *Nature* **421**, 611 (2003).

[49] E. Goulielmakis, M. Uiberacker, R. Kienberger *et al.*, *Science* **305**, 1267 (2004).

[50] A. Apolonski, P. Dombi, G. G. Paulus *et al.*, *Phys. Rev. Lett.* **92**, 073902 (2004).

[51] T. Fuji, J. Rauschenberger, C. Gohle *et al.*, *New J. Phys.* **7**, 116 (2005).

[52] E. Gagnon, I. Thomann, A. Paul *et al.*, *Opt. Lett.* **31**, 1866 (2006).

[53] A. Verhoef, A. Fernández, M. Lezius, K. O'Keeffe, M. Uiberacker and F. Krausz, *Opt. Lett.* **31**, 3520 (2006).

[54] M. Kress, T. Loffler, M. D. Thomson *et al.*, *Nat. Phys.* **2**, 327 (2006).

[55] C. Haworth, L. Chipperfield, J. Robinson, P. Knight, J. Marangos and J. Tisch, *Nat. Phys.* **3**, 52 (2007).

[56] M. F. Kling, J. Rauschenberger, A. Verhoef *et al.*, *New J. Phys* **10**, 025024 (2008).

[57] T. Fuji, N. Ishii, C. Y. Teisset *et al.*, *Opt. Lett.* **31**, 1103 (2006).

[58] D. Batani, C. J. Joachain, S. Martellucci and A. N. Chester, eds, *Atoms, Solids and Plasmas in Super-Intense Laser Fields*, (New York: Kluwer Academic – Plenum Publishers, 2001).

[59] D. Batani, C. J. Joachain and S. Martellucci, eds, *Atoms and Plasmas in Super-Intense Laser Fields*, vol. 88 (Bologna: Italian Physical Society, 2004).

[60] J. Feldhaus and B. Sonntag, Free-electron lasers – high-intensity X-ray sources. In T. Brabec, ed., *Strong Field Laser Physics*, Springer Series in Optical Sciences 134 (New York: Springer, 2009), p. 91.

[61] D. A. G. Deacon, L. R. Elias, J. M. J. Madley, G. J. Ramian, H. A. Schwettman and T. I. Smith, *Phys. Rev. Lett.* **38**, 892 (1977).

[62] J. Andruszkow, B. Aune, V. Ayvazyan *et al.*, *Phys. Rev. Lett.* **85**, 3825 (2000).

[63] V. Ayvazyan, N. Baboi, I. Bohnet *et al.*, *Phys. Rev. Lett.* **88**, 104802 (2002).

[64] E. K. Damon and R. G. Tomlinson, *Appl. Opt.* **2**, 546 (1963).

[65] G. S. Voronov and N. B. Delone, *Sov. Phys. JETP Lett.* **1**, 66 (1965).

[66] J. L. Hall, E. J. Robinson and L. M. Branscomb, *Phys. Rev. Lett.* **14**, 1013 (1965).

[67] P. Agostini, F. Fabre, G. Mainfray, G. Petite and N. K. Rahman, *Phys. Rev. Lett.* **42**, 1127 (1979).

[68] G. Petite, P. Agostini and H. G. Muller, *J. Phys. B* **21**, 4097 (1988).

[69] R. R. Freeman, P. H. Bucksbaum, H. Milchberg, S. Darack, D. Schumacher and M. E. Geusic, *Phys. Rev. Lett.* **59**, 1092 (1987).

[70] M. V. Ammosov, N. B. Delone and V. P. Krainov, *Sov. Phys. JETP* **64**, 1191 (1986).

[71] L. D. Landau and E. M. Lifshitz, *Quantum Mechanics* (Reading, Mass.: Addison Wesley, 1958).

[72] L. V. Keldysh, *Sov. Phys. JETP* **20**, 1307 (1965).

[73] F. H. M. Faisal, *J. Phys. B* **6**, L89 (1973).

[74] H. R. Reiss, *Phys. Rev. A* **22**, 1786 (1980).

[75] J. E. Bayfield and P. M. Koch, *Phys. Rev. Lett.* **33**, 258 (1974).

[76] T. F. Gallagher, *Rydberg Atoms* (Cambridge: Cambridge University Press, 1984).

[77] S. Augst, D. Strickland, D. Meyerhofer, S. L. Chin and J. H. Eberly, *Phys. Rev. Lett.* **63**, 2212 (1989).

[78] E. Mevel, P. Breger, R. Trainham *et al.*, *Phys. Rev. Lett.* **70**, 406 (1993).

[79] P. B. Corkum, *Phys. Rev. Lett.* **71**, 1994 (1993).

[80] K. C. Kulander, K. J. Schafer and J. L. Krause. In B. Piraux, A. L'Huillier, and K. Rzążewski, eds, *Super-Intense Laser-Atom Physics* (New York: Plenum Press, 1993), p. 95.

[81] M. Y. Kuchiev, *JETP Lett.* **45**, 404 (1987).

[82] H. B. van Linden van den Heuvell and H. G. Muller. In S. J. Smith and P. L. Knight, eds., *Multiphoton Processes*, (Cambridge: Cambridge University Press, 1988).

[83] G. G. Paulus, W. Nicklich, H. Xu, P. Lambropoulos and H. Walther, *Phys. Rev. Lett.* **72**, 2851 (1994).

[84] F. Grasbon, G. G. Paulus, H. Walther *et al.*, *Phys. Rev. Lett.* **91**, 173003 (2003).

[85] G. G. Paulus, F. Grasbon, H. Walther *et al.*, *Nature* **414**, 182 (2001).

[86] F. W. Byron and C. J. Joachain, *Phys. Rev.* **164**, 1 (1967).

[87] B. Walker, B. Sheehy, L. F. DiMauro, P. Agostini, K. J. Schafer and K. Kulander, *Phys. Rev. Lett.* **73**, 1227 (1994).

[88] J. Ullrich, R. Moshammer, A. Dorn, R. Dörner, L. P. Schmidt and H. Schmidt-Böcking, *Rep. Prog. Phys.* **66**, 1463 (2003).

[89] A. Rudenko, K. Zrost, B. Feuerstein *et al.*, *Phys. Rev. Lett.* **93**, 253001 (2004).

[90] X. Liu, H. Rottke, E. Eremina *et al.*, *Phys. Rev. Lett.* **93**, 263001 (2004).

[91] P. A. Franken, A. E. Hill, C. W. Peters and G. Weinreich, *Phys. Rev. Lett.* **7**, 118 (1961).

[92] A. L'Huillier, L. A. Lompré, G. Mainfray and C. Manus, *Adv. At. Mol. Opt. Phys. Suppl.* **1**, 139 (1992).

[93] P. Salières, A. L'Huillier, P. Antoine and M. Lewenstein, *Adv. At. Mol. Opt. Phys.* **41**, 83 (1999).

[94] P. Salières, In D. Batani, C. J. Joachain and S. Martellucci, eds. *Atoms and Plasmas in Super-Intense Laser Fields*, vol. 88 (Bologna: Italian Physical Society, 2004), p. 303.

[95] G. H. C. New and J. F. Ward, *Phys. Rev. Lett.* **19**, 556 (1967).

[96] D. I. Metchkov, V. M. Mitev, L. I. Pavlov and K. V. Stamenov, *Opt. Commun.* **21**, 391 (1977).

[97] M. G. Groseva, D. I. Metchkov, V. M. Mitev, L. I. Pavlov and K. V. Stamenov, *Opt. Commun.* **23**, 77 (1977).

[98] J. Wildenauer, *J. Appl. Phys.* **62**, 41 (1987).

[99] J. Reintjes, C. Y. She and R. C. Eckard, *IEEE J. Quantum Electron.* **QE-14**, 581 (1978).

[100] J. Reintjes, L. L. Tankersley and R. Christensen, *Opt. Commun.* **39**, 334 (1981).

[101] J. Bokor, P. H. Bucksbaum and R. R. Freeman, *Opt. Lett.* **8**, 217 (1983).

[102] A. McPherson, G. Gibson, H. Jara *et al.*, *J. Opt. Soc. Am. B* **4**, 595 (1987).

[103] R. Rosman, G. Gibson, K. Boyer *et al.*, *J. Opt. Soc. Am. B* **5**, 1237 (1988).

[104] M. Ferray, A. L'Huillier, X. F. Li, L. A. Lompré, G. Mainfray and C. Manus, *J. Phys. B* **21**, L31 (1988).

[105] X. F. Li, A. L'Huillier, M. Ferray, L. A. Lompré and G. Mainfray, *Phys. Rev. A* **39**, 5751 (1989).

[106] N. Sarukura, K. Hata, T. Adachi, R. Nodomi, M. Watanabe and S. Watanabe, *Phys. Rev. A* **43**, 1669 (1991).

[107] K. Miyazaki and H. Sakai, *J. Phys. B* **25**, L83 (1992).

[108] J. J. Macklin, J. D. Kmetec and C. L. Gordon, *Phys. Rev. Lett.* **70**, 766 (1993).

[109] A. L'Huillier and P. Balcou, *Phys. Rev. Lett.* **70**, 774 (1993).

[110] K. Miyazaki and H. Takada, *Phys. Rev. A* **52**, 3007 (1995).

[111] Y. Nagata, K. Midorikawa, M. Obara and K. Toyoda, *Opt. Lett.* **21**, 15 (1996).

[112] S. G. Preston, A. Sampera, M. Zepf *et al.*, *Phys. Rev. A* **53**, R31 (1996).

[113] J. Zhou, J. Peatross, M. M. Murnane, H. C. Kapteyn and I. P. Christov, *Phys. Rev. Lett.* **76**, 752 (1996).

[114] I. P. Christov, J. P. Zhou, J. Peatross, A. Rundquist, M. M. Murnane and H. C. Kapteyn, *Phys. Rev. Lett.* **77**, 1743 (1996).

[115] C. Spielmann, N. H. Burnett, S. Sartania *et al.*, *Science* **278**, 661 (1997).

[116] Z. Chang, A. Rundquist, H. Wang, M. M. Murnane and H. C. Kapteyn, *Phys. Rev. Lett.* **79**, 2967 (1997).

[117] M. Schnürer, C. Spielmann, P. Wobrauschek *et al.*, *Phys. Rev. Lett.* **80**, 3236 (1998).

[118] J. Seres, E. Seres, A. Verhoef *et al.*, *Nature* **433**, 596 (2005).

[119] M. C. Chen, P. Arpin, T. Popmintchev *et al.*, *Phys. Rev. Lett.* **105**, 173901 (2010).

[120] J. L. Krause, K. J. Schafer and K. C. Kulander, *Phys. Rev. Lett.* **68**, 3535 (1992).

[121] A. L'Huillier, M. Lewenstein, P. Salières *et al.*, *Phys. Rev. A* **48**, R3433 (1993).

[122] M. Lewenstein, P. Balcou, M. Y. Ivanov, A. L'Huillier and P. B. Corkum, *Phys. Rev. A* **49**, 2117 (1994).

[123] P. Agostini and L. F. DiMauro, *Rep. Prog. Phys.* **67**, 813 (2004).

[124] A. Scrinzi, M. Ivanov, R. Kienberger and D. Villeneuve, *J. Phys. B* **39**, R1 (2006).

[125] H. Niikura and P. B. Corkum, *Adv. At. Mol. Opt. Phys.* **54**, 511 (2007).

[126] F. Krausz and M. Ivanov, *Rev. Mod. Phys.* **81**, 163 (2009).

[127] T. W. Hänsch, *Opt. Commun.* **80**, 70 (1990).

[128] G. Farkas and C. Toth, *Phys. Lett. A* **168**, 447 (1992).

[129] S. E. Harris, J. J. Macklin and T. W. Hänsch, *Opt. Commun.* **100**, 487 (1993).

[130] O. Svelto, *Principles of Lasers*, 4th edn (New York: Plenum Press, 1998).

[131] P. Antoine, A. L'Huillier and M. Lewenstein, *Phys. Rev. Lett.* **77**, 1234 (1996).

[132] P. M. Paul, E. S. Toma, P. Breger *et al.*, *Science* **292**, 1689 (2001).

[133] P. B. Corkum, N. H. Burnett and M. Y. Ivanov, *Opt. Lett.* **19**, 1870 (1994).

[134] M. Ivanov, P. B. Corkum, T. Zuo and A. Bandrauk, *Phys. Rev. Lett* **74**, 2933 (1995).

[135] V. T. Platonenko and V. V. Strelkov, *J. Opt. Soc. Am. B* **16**, 435 (1999).

[136] K. S. Budil, P. Salières, A. L'Huillier, T. Ditmire and M. D. Perry, *Phys. Rev. A* **48**, R3437 (1993).

[137] C. Altucci, C. Delfin, L. Roos *et al.*, *Phys. Rev. A* **58**, 3934 (1998).

[138] M. Kovacev, Y. Mairesse, E. Priori *et al.*, *Eur. Phys. J. D* **26**, 79 (2003).

[139] O. Tcherbakoff, E. Mével, D. Descamps, J. Plumridge and E. Constant, *Phys. Rev. A* **68**, 043804 (2003).

[140] H. Mashiko, S. Gilbertson, C. Li *et al.*, *Phys. Rev. Lett.* **100**, 103906 (2008).

[141] I. P. Christov, M. M. Murnane and H. C. Kapteyn, *Phys. Rev. Lett.* **78**, 1251 (1997).

[142] M. Hentschel, R. Kienberger, C. Spielmann *et al.*, *Nature* **414**, 509 (2001).

[143] R. Kienberger, E. Gouliemakis, M. Uiberacker *et al.*, *Nature* **427**, 817 (2004).

[144] G. Sansone, E. Benedetti, F. Calegari *et al.*, *Science* **314**, 443 (2006).

[145] M. Casu, C. Szymanowski, S. Hu and C. H. Keitel, *J. Phys. B* **33**, L411 (2000).

[146] R. M. Potvliege, N. J. Kylstra and C. J. Joachain, *J. Phys. B* **33**, L743 (2000).

[147] D. M. Gaudiosi, B. Reagan, T. Popmintchev *et al.*, *Phys. Rev. Lett.* **96**, 203001 (2006).

[148] M. W. Walser, C. H. Keitel, A. Scrinzi and T. Brabec, *Phys. Rev. Lett.* **85**, 5082 (2000).

[149] N. J. Kylstra, R. M. Potvliege and C. J. Joachain, *J. Phys. B* **34**, L55 (2001).

[150] D. B. Milosevic, S. X. Hu and W. Becker, *Phys. Rev. A* **63**, R011403 (2001).

[151] N. J. Kylstra, R. M. Potvliege and C. J. Joachain, *Laser Phys.* **12**, 409 (2002).

[152] C. C. Chirilă, N. J. Kylstra, R. M. Potvliege and C. J. Joachain, *Phys. Rev. A* **66**, 063411 (2002).

[153] M. Gavrila, Free-free transitions of electron-atom systems in intense radiation fields. In F. A. Gianturco, ed., *Collision Theory for Atoms and Molecules* (New York: Plenum Press, 1989), p. 139.

[154] P. Francken and C. J. Joachain, *J. Opt. Soc. Am. B* **7**, 554 (1990).

[155] N. J. Mason, *Rep. Progr. Phys.* **56**, 1275 (1993).

[156] F. Ehlotzky, A. Jaron and J. Z. Kaminski, *Phys. Rep.* **297**, 63 (1998).

[157] A. Weingartshofer, J. K. Holmes, G. Caudle, E. M. Clarke and H. Kruger, *Phys. Rev. Lett.* **39**, 269 (1977).

[158] A. Weingartshofer, J. K. Holmes, J. Sabbagh and S. L. Chin, *J. Phys. B* **16**, 1805 (1983).

[159] B. Wallbank and J. K. Holmes, *J. Phys. B* **27**, 1221 (1994).

[160] N. J. Mason and W. R. Newell, *J. Phys. B* **20**, L323 (1987).

[161] S. Luan, R. Hippler and H. O. Lutz, *J. Phys. B* **24**, 3241 (1991).

[162] C. Höhr, A. Dorn, B. Najjari, D. Fischer, C. Schröter and J. Ullrich, *Phys. Rev. Lett.* **94**, 153201 (2005).

[163] N. J. Kylstra and C. J. Joachain, *Phys. Rev. A* **58**, R26 (1998).

[164] N. J. Kylstra and C. J. Joachain, *Phys. Rev. A* **60**, 2255 (1999).

[165] F. V. Bunkin and M. V. Fedorov, *Sov. Phys. JETP* **22**, 844 (1966).

[166] N. M. Kroll and K. M. Watson, *Phys. Rev. A* **8**, 804 (1973).

[167] C. J. Joachain, *Quantum Collision Theory*, 3rd edn (Amsterdam: North Holland, 1983).

[168] B. H. Bransden and C. J. Joachain, *Physics of Atoms and Molecules*, 2nd edn (Harlow, UK: Prentice Hall-Pearson, 2003).

[169] F. W. Byron and C. J. Joachain, *J. Phys. B* **17**, L295 (1984).

2

Theory of laser–atom interactions

In this chapter, we shall discuss the theory of laser–atom interactions, using a *semi-classical* method in which the laser field is treated classically, while the atom is studied by using quantum mechanics. This semi-classical approach constitutes an excellent approximation for intense laser fields, since in that case the number of photons per laser mode is very large [1, 2]. In addition, spontaneous emission can be neglected. We begin therefore by giving in Section 2.1 a classical description of the laser field in terms of electric- and magnetic-field vectors satisfying Maxwell's equations. We start by considering plane wave solutions of these equations. Then general solutions describing laser pulses are introduced. The dynamics of a classical electron in the laser field, and in particular the *ponderomotive energy* and *force*, are discussed in Section 2.2. Neglecting first relativistic effects, we write down in Section 2.3 the *time-dependent Schrödinger equation* (TDSE), which is the starting point of the theoretical study of atoms in intense laser fields, and introduce the *dipole approximation*. In the subsequent two sections, we study the behavior of the TDSE under *gauge transformations* and the *Kramers frame transformation*. In view of the central role that the time evolution operator plays in the development of the theory of laser–atom interactions, some general properties of this operator are reviewed in Section 2.6. In Section 2.7, the TDSE is solved explicitly for the simple case of a "free" electron in a laser field to obtain the *non-relativistic Gordon–Volkov wave functions*. In Section 2.8, we discuss the *non-relativistic, non-dipole* regime of laser–atom interactions. Finally, in Section 2.9, *relativistic effects* in laser–atom interactions are taken into account, and the appropriate relativistic wave equations are discussed.

2.1 Classical description of a laser field

The classical electromagnetic field generated by a laser is described in vacuo by electric and magnetic fields, $\mathcal{E}(\mathbf{r}, t)$ and $\mathcal{B}(\mathbf{r}, t)$, which satisfy Maxwell's equations without sources:

$$\nabla \cdot \mathcal{E} = 0,$$

$$\nabla \cdot \mathcal{B} = 0,$$

$$\nabla \times \mathcal{E} = -\frac{\partial \mathcal{B}}{\partial t}, \tag{2.1}$$

$$\nabla \times \mathcal{B} = \frac{1}{c^2} \frac{\partial \mathcal{E}}{\partial t},$$

where c is the velocity of light in vacuo. The electric and magnetic fields can be generated from scalar and vector potentials, $\phi(\mathbf{r}, t)$ and $\mathbf{A}(\mathbf{r}, t)$, respectively, by the following relations:

$$\mathcal{E} = -\nabla \phi - \frac{\partial \mathbf{A}}{\partial t} \tag{2.2}$$

and

$$\mathcal{B} = \nabla \times \mathbf{A}. \tag{2.3}$$

In addition, from Equations (2.1)–(2.3) it follows that the vector potential $\mathbf{A}$ satisfies the homogeneous wave equation (as do ϕ, $\mathcal{E}$ and $\mathcal{B}$)

$$\nabla^2 \mathbf{A} - \frac{1}{c^2} \frac{\partial^2 \mathbf{A}}{\partial t^2} = 0. \tag{2.4}$$

The potentials ϕ and $\mathbf{A}$ are not uniquely defined by these equations, since the fields $\mathcal{E}$ and $\mathcal{B}$ are invariant under the (classical) *gauge transformation*

$$\begin{aligned} \mathbf{A} \rightarrow \mathbf{A}' &= \mathbf{A} + \nabla f, \\ \phi \rightarrow \phi' &= \phi - \partial f / \partial t, \end{aligned} \tag{2.5}$$

where f is an arbitrary real, differentiable function of $\mathbf{r}$ and t. The freedom implied by the *gauge invariance* (2.5) means that one can choose a set of potentials $(\phi, \mathbf{A})$ which satisfy the *Lorentz condition*

$$\nabla \cdot \mathbf{A} + \frac{1}{c^2} \frac{\partial \phi}{\partial t} = 0. \tag{2.6}$$

The potentials satisfying this condition are said to belong to the *Lorentz gauge*.

Another useful gauge for the potentials is the *Coulomb* (or *radiation*) *gauge*, which is defined by the condition

$$\nabla \cdot \mathbf{A} = 0. \tag{2.7}$$

The Coulomb gauge is often used when no sources are present. Then $\phi = 0$, and the fields are given by

$$\mathcal{E} = -\frac{\partial \mathbf{A}}{\partial t} \tag{2.8}$$

and

$$\mathcal{B} = \nabla \times \mathbf{A}. \tag{2.9}$$

2.1.1 Plane wave solutions of Maxwell's equations

A monochromatic plane wave solution of Equation (2.4) corresponding to the angular frequency ω, i.e. to the frequency $\nu = \omega/(2\pi)$ and wavelength $\lambda = c/\nu$, is given by

$$\mathbf{A}(\mathbf{r}, t) = \hat{\boldsymbol{\epsilon}} A_0 \sin(\mathbf{k}_L \cdot \mathbf{r} - \omega t - \varphi), \tag{2.10}$$

where $\mathbf{k}_L$ is the *propagation vector* of the laser field, φ is a real constant phase and

$$\omega = k_L c. \tag{2.11}$$

The Coulomb gauge condition (2.7) is satisfied if

$$\mathbf{k}_L \cdot \hat{\boldsymbol{\epsilon}} = 0 \tag{2.12}$$

so that $\hat{\boldsymbol{\epsilon}}$ is perpendicular to $\mathbf{k}_L$ and the wave is said to be *transverse*.

The corresponding electric field is given by

$$\mathcal{E}(\mathbf{r}, t) = \hat{\boldsymbol{\epsilon}} \mathcal{E}_0 \cos(\mathbf{k}_L \cdot \mathbf{r} - \omega t - \varphi), \tag{2.13}$$

with the electric-field strength given by $\mathcal{E}_0 = \omega A_0$. The quantity A_0, which we take to be positive, is the amplitude of the vector potential. Both the vector potential $\mathbf{A}$ and the electric field $\mathcal{E}$ are in the direction of the real unit vector $\hat{\boldsymbol{\epsilon}}$, which is called the *polarization vector*. Using Equations (2.9) and (2.11), the magnetic field arising from the vector potential (2.10) is given by

$$\mathcal{B}(\mathbf{r}, t) = \frac{\mathcal{E}_0}{c} (\hat{\mathbf{k}}_L \times \hat{\boldsymbol{\epsilon}}) \cos(\mathbf{k}_L \cdot \mathbf{r} - \omega t - \varphi). \tag{2.14}$$

From Equations (2.12)–(2.14) it follows that the vectors $\mathcal{E}$, $\mathcal{B}$ and $\mathbf{k}_L$ are mutually orthogonal. Moreover, we see that

$$\frac{|\mathcal{B}|}{|\mathcal{E}|} = \frac{1}{c}. \tag{2.15}$$

An electromagnetic plane wave described by Equations (2.13) and (2.14), for which the electric-field vector points in a *fixed* (time-independent) direction $\hat{\boldsymbol{\epsilon}}$, is said to be *linearly polarized*. A *general state of polarization* for a plane wave propagating in the direction $\hat{\mathbf{k}}_L$ can be described by combining two independent linearly polarized plane waves with real unit polarization vectors $\hat{\boldsymbol{\epsilon}}_a$ and $\hat{\boldsymbol{\epsilon}}_b$ perpendicular to $\mathbf{k}_L$, where the phases of the two component waves are, in general, different. The corresponding vector potential and electric field are given by

$$A(\mathbf{r},t) = \hat{\boldsymbol{\epsilon}}_a A_{0,a} \sin(\mathbf{k}_L \cdot \mathbf{r} - \omega t - \varphi_a) + \hat{\boldsymbol{\epsilon}}_b A_{0,b} \sin(\mathbf{k}_L \cdot \mathbf{r} - \omega t - \varphi_b) \quad (2.16)$$

and

$$\mathcal{E}(\mathbf{r},t) = \hat{\boldsymbol{\epsilon}}_a \mathcal{E}_{0,a} \cos(\mathbf{k}_L \cdot \mathbf{r} - \omega t - \varphi_a) + \hat{\boldsymbol{\epsilon}}_b \mathcal{E}_{0,b} \cos(\mathbf{k}_L \cdot \mathbf{r} - \omega t - \varphi_b), \quad (2.17)$$

where $A_{0,a}$ and $A_{0,b}$ are positive quantities, $\mathcal{E}_{0,a} = \omega A_{0,a}$ and $\mathcal{E}_{0,b} = \omega A_{0,a}$. It is always possible to find a phase φ, a real number ξ such that $-1 \leq \xi \leq 1$, and two unit vectors $\hat{\boldsymbol{\epsilon}}_1$ and $\hat{\boldsymbol{\epsilon}}_2$ forming with $\hat{\mathbf{k}}_L$ a right-handed orthogonal coordinate system, so that the vector potential (2.16) can also be written in the form

$$A(\mathbf{r},t) = \frac{A_0}{(1+\xi^2)^{1/2}} \left[\hat{\boldsymbol{\epsilon}}_1 \sin(\mathbf{k}_L \cdot \mathbf{r} - \omega t - \varphi) - \xi \hat{\boldsymbol{\epsilon}}_2 \cos(\mathbf{k}_L \cdot \mathbf{r} - \omega t - \varphi) \right].$$

$$(2.18)$$

The electric field corresponding to the vector potential (2.18) can be expressed as

$$\mathcal{E}(\mathbf{r},t) = \frac{\mathcal{E}_0}{(1+\xi^2)^{1/2}} \left[\hat{\boldsymbol{\epsilon}}_1 \cos(\mathbf{k}_L \cdot \mathbf{r} - \omega t - \varphi) + \xi \hat{\boldsymbol{\epsilon}}_2 \sin(\mathbf{k}_L \cdot \mathbf{r} - \omega t - \varphi) \right],$$

$$(2.19)$$

with $\mathcal{E}_0 = \omega A_0$.

The constant ξ is the *ellipticity parameter* of the radiation field, which, upon varying in the range $-1 \leq \xi \leq 1$, describes all possible cases of polarization. The value $\xi = 0$ corresponds to linear polarization. For $\xi = \pm 1$, the monochromatic plane wave is said to be *circularly polarized*. At a fixed point in space, the electric-field vector $\mathcal{E}$ is constant in magnitude and, as a function of time, traces out a circle at an angular frequency ω in the plane of the vectors $\hat{\boldsymbol{\epsilon}}_1$ and $\hat{\boldsymbol{\epsilon}}_2$ perpendicular to the propagation vector $\mathbf{k}_L$. As a function of $\mathbf{r}$, for a fixed time, the vector $\mathcal{E}$ traces a helix. If $\xi = -1$, the rotation of $\mathcal{E}$ as a function of time at a fixed point is counter-clockwise for an observer facing into the oncoming wave (looking into the direction $-\hat{\mathbf{k}}_L$), while at a fixed time the helix is left-handed. This wave is said to be *left-circularly polarized*.

Such a wave is also said to have *positive helicity* because it has a positive projection of angular momentum on the propagation direction $\hat{\mathbf{k}}_L$. If $\xi = 1$, the rotation of the vector $\mathcal{E}$ as a function of time at a fixed point in space is clockwise for an observer facing into the incoming wave and the helix is right-handed. This wave is said to be *right-circularly polarized*; it is said to have *negative helicity*, because it has a negative projection of angular momentum on the propagation direction $\hat{\mathbf{k}}_L$. When $0 < |\xi| < 1$, the vector $\mathcal{E}$, at a fixed point of space, traces out an ellipse as a function of time in the plane of the vectors $\hat{\boldsymbol{\epsilon}}_1$ and $\hat{\boldsymbol{\epsilon}}_2$. Such a wave is said to be *elliptically polarized*.

In order to explore in more detail the concept of polarization, we write the electric-field vector (2.19) corresponding to a monochromatic plane wave of arbitrary polarization in the following form:

$$\mathcal{E}(\mathbf{r},t) = \mathcal{E}_0 \mathrm{Re}\left\{\hat{\boldsymbol{\epsilon}}_c \exp\left[i(\mathbf{k}_L \cdot \mathbf{r} - \omega t - \varphi)\right]\right\}. \tag{2.20}$$

In the above equation, $\hat{\boldsymbol{\epsilon}}_c$ is a *complex* unit polarization vector such that

$$\hat{\boldsymbol{\epsilon}}_c^* \cdot \hat{\boldsymbol{\epsilon}}_c = 1. \tag{2.21}$$

This vector can be written as a linear combination of the two basis vectors $\hat{\boldsymbol{\epsilon}}_1$ and $\hat{\boldsymbol{\epsilon}}_2$, namely

$$\hat{\boldsymbol{\epsilon}}_c = c_1 \hat{\boldsymbol{\epsilon}}_1 + c_2 \hat{\boldsymbol{\epsilon}}_2, \tag{2.22}$$

where the complex coefficients c_1 and c_2 satisfy the equation

$$|c_1|^2 + |c_2|^2 = 1. \tag{2.23}$$

We note that Equation (2.20) reduces to Equation (2.19) if we take

$$c_1 = \frac{1}{(1+\xi^2)^{1/2}}, \qquad c_2 = \frac{-i\xi}{(1+\xi^2)^{1/2}}. \tag{2.24}$$

Let us write the two complex coefficients c_1 and c_2 in the form

$$c_1 = |c_1|\exp(i\alpha), \qquad c_2 = |c_2|\exp(i\beta). \tag{2.25}$$

If the phases α and β are equal (modulo π), so that $\alpha = \beta + m\pi$ (with $m = 0, \pm1, \pm2, \dots$), the electric-field vector (2.20) can be written in the form

$$\begin{aligned}\mathcal{E}(\mathbf{r},t) &= \mathcal{E}_0 \mathrm{Re}\left\{(|c_1|\hat{\boldsymbol{\epsilon}}_1 \pm |c_2|\hat{\boldsymbol{\epsilon}}_2)\exp\left[i(\mathbf{k}_L \cdot \mathbf{r} - \omega t - \varphi + \alpha)\right]\right\} \\ &= \mathcal{E}_0(|c_1|\hat{\boldsymbol{\epsilon}}_1 \pm |c_2|\hat{\boldsymbol{\epsilon}}_2)\cos(\mathbf{k}_L \cdot \mathbf{r} - \omega t - \varphi + \alpha),\end{aligned} \tag{2.26}$$

and we see that the direction of the vector $\mathcal{E}$ is independent of time, so that the monochromatic plane wave (2.26) is linearly polarized. If the amplitudes of the coefficients c_i are equal, so that $|c_1| = |c_2| = 1/\sqrt{2}$, but the phases α and β differ by $\pi/2$ (modulo 2π), i.e. $\beta = \alpha \pm \pi/2 + 2m\pi$ (with $m = 0, \pm1, \pm2, \dots$), then the electric field can be written as

$$\begin{aligned}\mathcal{E}(\mathbf{r},t) &= \frac{\mathcal{E}_0}{\sqrt{2}}\mathrm{Re}\left\{(\hat{\boldsymbol{\epsilon}}_1 \pm i\hat{\boldsymbol{\epsilon}}_2)\exp\left[i(\mathbf{k}_L \cdot \mathbf{r} - \omega t - \varphi + \alpha)\right]\right\} \\ &= \frac{\mathcal{E}_0}{\sqrt{2}}\left\{\hat{\boldsymbol{\epsilon}}_1 \cos(\mathbf{k}_L \cdot \mathbf{r} - \omega t - \varphi + \alpha) \mp \hat{\boldsymbol{\epsilon}}_2 \sin(\mathbf{k}_L \cdot \mathbf{r} - \omega t - \varphi + \alpha)\right\},\end{aligned}$$

$$\tag{2.27}$$

and the monochromatic plane wave (2.27) is left-circularly polarized if the upper sign is selected and right-circularly polarized if the lower sign is selected. If the complex coefficients c_1 and c_2 in Equation (2.22) do not correspond to either linear or circular polarization, the wave is elliptically polarized.

In the above discussion, we have described a general state of polarization for a monochromatic plane wave by using two linearly polarized waves to form a set of basis fields, with real, orthogonal unit polarization vectors $\hat{\epsilon}_1$ and $\hat{\epsilon}_2$ in a plane perpendicular to $\hat{\mathbf{k}}_L$. The complex polarization vector $\hat{\epsilon}_c$ was then expanded in terms of these basis vectors. The two circularly polarized waves (2.27) constitute an equally acceptable set of basic fields for the description of an arbitrary state of polarization. We introduce the two complex orthogonal unit basis vectors

$$\hat{\epsilon}_l = \frac{1}{\sqrt{2}}\left(\hat{\epsilon}_1 + i\hat{\epsilon}_2\right), \qquad \hat{\epsilon}_r = \frac{1}{\sqrt{2}}\left(\hat{\epsilon}_1 - i\hat{\epsilon}_2\right), \tag{2.28}$$

which correspond to left- and right-circular polarization, respectively. These vectors are such that

$$\hat{\epsilon}_l^* \cdot \hat{\epsilon}_l = \hat{\epsilon}_r^* \cdot \hat{\epsilon}_r = 1 \tag{2.29}$$

and

$$\hat{\epsilon}_l^* \cdot \hat{\epsilon}_r = \hat{\epsilon}_r^* \cdot \hat{\epsilon}_l = \hat{\epsilon}_l^* \cdot \hat{\mathbf{k}}_L = \hat{\epsilon}_r^* \cdot \hat{\mathbf{k}}_L = 0. \tag{2.30}$$

A general state of polarization can then be specified by a complex unit polarization vector $\hat{\epsilon}_c$ such that

$$\hat{\epsilon}_c = c_l \hat{\epsilon}_l + c_r \hat{\epsilon}_r, \tag{2.31}$$

where c_l and c_r are complex coefficients satisfying the equation

$$|c_l|^2 + |c_r|^2 = 1. \tag{2.32}$$

Throughout this book, we will assume that the polarization of the laser field is known, so that partially polarized fields will not be considered.

The energy density of the monochromatic electromagnetic field is given by

$$\frac{1}{2}\left(\epsilon_0 |\mathcal{E}|^2 + \mu_0^{-1}|\mathcal{B}|^2\right) = \frac{\epsilon_0 \mathcal{E}_0^2}{1 + \xi^2}\left[\cos^2(\mathbf{k}_L \cdot \mathbf{r} - \omega t - \varphi) + \xi^2 \sin^2(\mathbf{k}_L \cdot \mathbf{r} - \omega t - \varphi)\right], \tag{2.33}$$

where ϵ_0 and μ_0 are the permittivity and permeability of free space, respectively, and $\epsilon_0\mu_0 = c^{-2}$. Averaging the energy density over a period $T = 2\pi/\omega$, and using the fact that

$$\frac{1}{T}\int_0^T \sin^2(\mathbf{k}_L \cdot \mathbf{r} - \omega t - \varphi)\mathrm{d}t = \frac{1}{T}\int_0^T \cos^2(\mathbf{k}_L \cdot \mathbf{r} - \omega t - \varphi)\mathrm{d}t = \frac{1}{2}, \tag{2.34}$$

the cycle-averaged energy density at the given angular frequency ω is found to be

$$\rho(\omega) = \frac{1}{2}\epsilon_0 \mathcal{E}_0^2 = \frac{1}{2}\epsilon_0 \omega^2 A_0^2. \tag{2.35}$$

It is interesting to relate this result to the photon density, keeping in mind that each photon at a frequency ν carries a quantum of energy $h\nu = \hbar\omega$. If $\mathcal{N}(\omega)$ denotes the number of photons of angular frequency ω within a volume V, the energy density is given by

$$\rho(\omega) = \frac{\hbar\omega\mathcal{N}(\omega)}{V}. \tag{2.36}$$

From Equations (2.35) and (2.36), we find that

$$\mathcal{E}_0^2 = \frac{2\rho(\omega)}{\epsilon_0} = \frac{2\hbar\omega\mathcal{N}(\omega)}{\epsilon_0 V}. \tag{2.37}$$

The Poynting vector

$$\boldsymbol{S} = \frac{1}{\mu_0}(\boldsymbol{\mathcal{E}} \times \boldsymbol{B}) \tag{2.38}$$

is in the direction of the propagation vector $\mathbf{k_L}$. Its magnitude is the rate of energy flow through a unit area normal to the direction of propagation. Averaged over a period, T, this quantity gives the intensity $I(\omega)$ associated with the monochromatic plane wave (2.19) of angular frequency ω. That is,

$$I(\omega) = \frac{1}{2}\epsilon_0 c \mathcal{E}_0^2 = \frac{1}{2}\epsilon_0 c \omega^2 A_0^2. \tag{2.39}$$

It should be noted that the definition of the electric-field strength introduced in Equation (2.19) implies that the intensity (2.39) does not depend on the ellipticity parameter ξ. From Equations (2.36), (2.37) and (2.39) we also have

$$I(\omega) = \rho(\omega)c = \frac{\hbar\omega\mathcal{N}(\omega)c}{V}. \tag{2.40}$$

Finally, the photon flux of the field is given by

$$\Phi(\omega) = \frac{I(\omega)}{\hbar\omega}. \tag{2.41}$$

As an example, let us consider a linearly polarized laser field generated by a Nd:YAG laser with photon energy $\hbar\omega = 1.17\,\text{eV}$. Even for a modest intensity $I = 10^{12}\,\text{W cm}^{-2}$, the number of photons in a coherence volume $V = \lambda^3$ (with $\lambda = 2\pi c/\omega = 1064\,\text{nm}$), as obtained from Equation (2.40), is

$$\mathcal{N} = \frac{IV}{\hbar\omega c} \simeq 2 \times 10^8, \tag{2.42}$$

which is very large. A classical description of the laser field is therefore justified. We point out that a more rigorous justification of this statement can be given by using the fact that, to a good approximation, the radiation generated by a laser is in a coherent state, which is the quantum electrodynamic state approximating most closely the classical state of the field [3–5]. For large values of the average number of photons in the coherent state, quantum corrections only cause small fluctuations about the classical field.

2.1.2 Laser pulses

Thus far, we have considered only monochromatic plane wave solutions of Equation (2.4). As we will see later, the electromagnetic field can often be taken to be monochromatic when describing the interaction of an atom with a "long" laser pulse, lasting tens of optical cycles or more. However, this is not the case for shorter laser pulses, in particular for ultra-short pulses lasting a few optical cycles or even less. In what follows, we shall give a brief theoretical description of laser pulses.

A general laser pulse can be formed by superimposing monochromatic plane waves with appropriate amplitudes, frequencies and phases. It is useful to distinguish between pulses that are *spatially homogeneous* in the plane perpendicular to the laser propagation vector $\mathbf{k}_L$ and *focused* laser pulses. We begin by discussing the former. We shall consider the simple case where each plane wave component has the same direction of propagation $\hat{\mathbf{k}}_L$ and is linearly polarized in the direction $\hat{\boldsymbol{\epsilon}}$. The vector potential in the Coulomb gauge can then be written as

$$\mathbf{A}(\mathbf{r},t) = \hat{\boldsymbol{\epsilon}} \int_0^\infty A_0(\omega') \sin[\mathbf{k}_L' \cdot \mathbf{r} - \omega't - \varphi(\omega')] d\omega', \qquad (2.43)$$

where $\varphi(\omega')$ is the phase associated with the angular frequency ω' and $\mathbf{k}_L' = (\omega'/c)\hat{\mathbf{k}}_L$. Since the laser pulse is localized in space and time, the amplitude $A_0(\omega')$ is peaked about the pulse *carrier angular frequency* ω, while its "width" is inversely proportional to the pulse duration. If the laser field is monochromatic, then $A_0(\omega')$ reduces to $A_0\delta(\omega' - \omega)$, where $\delta(x)$ denotes the Dirac delta function. Both $\varphi(\omega')$ and $A_0(\omega')$ are in general smooth functions. However, for ultra-short pulses $A_0(\omega')$ may vary in a complicated way and extend over a wide range of frequencies.

Using the plane wave expansion (2.43) for the vector potential, the electric-field component of the laser pulse is found to be

$$\boldsymbol{\mathcal{E}}(\mathbf{r},t) = \hat{\boldsymbol{\epsilon}} \int_0^\infty \mathcal{E}_0(\omega') \cos[\mathbf{k}_L' \cdot \mathbf{r} - \omega't - \varphi(\omega')] d\omega', \qquad (2.44)$$

where the quantity $\mathcal{E}_0(\omega') = \omega' A(\omega')$ has the dimensions of an electric-field amplitude per unit angular frequency.

The total energy flux passing through a plane of unit area perpendicular to the propagation vector $\mathbf{k}_L$ must be finite. Using Equation (2.38), together with Equations (2.44) and (2.15), we have

$$\int_{-\infty}^{\infty} \hat{\mathbf{k}}_L \cdot \mathbf{S}(t) \, dt = \pi \epsilon_0 c \int_0^{\infty} \mathcal{E}_0^2(\omega') d\omega' < \infty \tag{2.45}$$

so that $\mathcal{E}_0^2(\omega')$ must decrease faster than $(\omega')^{-1}$ as ω' goes to infinity. In addition, due to the finite size of the laser oscillator, the laser pulse cannot contain angular frequencies that are smaller than some minimum cut-off angular frequency $\omega_{\min} > 0$. Therefore

$$\mathcal{E}_0(\omega') = 0, \quad \omega' < \omega_{\min}. \tag{2.46}$$

The existence of this cut-off allows one to deduce two properties of a laser pulse that is described by the vector potential (2.43). Let us consider the following integral over the electric-field component of the laser pulse:

$$\lim_{t \to \infty} \int_{-t}^{t} \boldsymbol{\mathcal{E}}(\mathbf{r}, t') dt' = \lim_{t \to \infty} [\mathbf{A}(\mathbf{r}, -t) - \mathbf{A}(\mathbf{r}, t)]$$

$$= \lim_{t \to \infty} 2\hat{\boldsymbol{\epsilon}} \int_0^{\infty} \frac{\mathcal{E}_0(\omega')}{\omega'} \sin(\omega' t) \cos[\mathbf{k}_L' \cdot \mathbf{r} - \varphi(\omega')] d\omega'. \tag{2.47}$$

Treating t as a parameter, we first calculate this limit by making use of the following representation of the Dirac delta function:

$$\delta(x) = \lim_{\epsilon \to 0^+} \frac{\sin(x/\epsilon)}{\pi x}. \tag{2.48}$$

Then, using Equation (2.46), it follows that

$$\lim_{t \to \infty} \int_{-t}^{t} \boldsymbol{\mathcal{E}}(\mathbf{r}, t') dt' = 0. \tag{2.49}$$

A similar calculation shows that

$$\lim_{t \to \infty} \int_{-t}^{t} \mathbf{A}(\mathbf{r}, t') dt' = 0. \tag{2.50}$$

We will see in Section 2.2 that the two results in Equations (2.49) and (2.50) have important implications for the dynamics of free electrons in laser fields.

In studying the interaction of laser pulses with atoms, the spatial profile of the laser pulse can be assumed to remain constant over atomic dimensions. In this case, it is convenient to use the fact that any function that depends on position and time through the quantity

$$\eta = \omega t - \mathbf{k}_L \cdot \mathbf{r} \qquad (2.51)$$

is a solution of Equation (2.4). A linearly polarized laser pulse can then be described by a vector potential of the form

$$\mathbf{A}(\eta) = \mathbf{A}(\mathbf{r}, t) = -\hat{\epsilon} \frac{\mathcal{E}_0}{\omega} \int_{-\infty}^{\eta} F(\eta') \cos(\eta' + \varphi) d\eta', \qquad (2.52)$$

where F is a non-negative function that describes the temporal profile (or shape or envelope) of the pulse. It should be noted that, in contrast to Equation (2.43), the vector potential (2.52) is restricted to the description of non-dispersive laser pulses.

From Equation (2.8), the electric field of the laser pulse is found to be

$$\mathcal{E}(\eta) = \mathcal{E}(\mathbf{r}, t)$$

$$= \hat{\epsilon} \, \mathcal{E}_0 F(\eta) \cos(\eta + \varphi). \qquad (2.53)$$

In this way, the pulse is simply described as a single carrier wave, of angular frequency ω, modulated by the envelope function $F(\eta)$. It is convenient to take $F(\eta)$ to vary between 0 and 1, so that $\mathcal{E}_0$ is the peak electric-field strength of the pulse. As noted in Chapter 1, the carrier-envelope phase (CEP) is the phase of the carrier wave with respect to the maximum of the laser pulse envelope. Thus, if this maximum occurs at $\eta = 0$, which we will assume in the following, the carrier-envelope phase is φ. For long pulses, and as far as the single-atom response is concerned, this phase has no importance. On the contrary, its value does matter for ultra-short pulses. In fact, as we shall see in Chapters 8 and 9, for few-cycle pulses both multiphoton ionization and harmonic generation depend sensitively on the carrier-envelope phase.

At the maximum value of $F(\eta)$, which is unity, we can take the quantity

$$I = \frac{1}{2} \epsilon_0 c \mathcal{E}_0^2 \qquad (2.54)$$

as defining the "peak intensity" of a laser pulse of the form (2.53). It is often useful to introduce an "instantaneous intensity"

$$I(\eta) = I F^2(\eta), \qquad (2.55)$$

which varies in space and time like the square of the envelope function $F(\eta)$. Since the amplitude of oscillation of the electric field changes throughout the pulse, this quantity is in general not equal to the average over one optical period of the magnitude of the Poynting vector. However, in practice the difference is usually negligible if the pulse encompasses more than a few optical cycles. The cycle-averaged intensity is not a relevant parameter for shorter pulses.

A laser pulse of arbitrary polarization with ellipticity parameter ξ is readily described in a similar manner. The electric field is now given by

$$\mathcal{E}(\eta) = \frac{\mathcal{E}_0}{(1+\xi^2)^{1/2}} F(\eta) \left[\hat{\boldsymbol{\epsilon}}_1 \cos(\eta+\varphi) - \xi \hat{\boldsymbol{\epsilon}}_2 \sin(\eta+\varphi) \right], \tag{2.56}$$

where F is the pulse shape function, and the corresponding vector potential is

$$\mathbf{A}(\eta) = -\frac{\mathcal{E}_0}{\omega(1+\xi^2)^{1/2}} \int_{-\infty}^{\eta} F(\eta') \left[\hat{\boldsymbol{\epsilon}}_1 \cos(\eta'+\varphi) - \xi \hat{\boldsymbol{\epsilon}}_2 \sin(\eta'+\varphi) \right] d\eta'. \tag{2.57}$$

As an example, let us consider a linearly polarized laser pulse having an electric-field envelope described by the Gaussian function

$$F(\eta) = \exp\left\{ -\eta^2 / [2(\omega\tau)^2] \right\}, \tag{2.58}$$

where ω is the carrier angular frequency of the pulse. The parameter τ characterizes the pulse duration. The distribution of angular frequencies in the pulse about ω is obtained by performing a Fourier transform, with the following result:

$$\mathcal{E}(\mathbf{r}, \omega') = (\tau \mathcal{E}_0/2) \hat{\boldsymbol{\epsilon}} \exp\left[-\tau^2(\omega'-\omega)^2/2 \right] \exp\left[i(\mathbf{k}'_L \cdot \mathbf{r} - \varphi) \right]$$

$$+ (\tau \mathcal{E}_0/2) \hat{\boldsymbol{\epsilon}} \exp\left[-\tau^2(\omega'+\omega)^2/2 \right] \exp\left[i(-\mathbf{k}'_L \cdot \mathbf{r} + \varphi) \right]. \tag{2.59}$$

The function $\mathcal{E}(\mathbf{r}, \omega')$ is peaked about $\omega' = \omega$ and differs appreciably from zero only in a narrow region whose width is inversely proportional to τ. The second term on the right-hand side of Equation (2.59) is thus negligible. We remark that the presence of this term, which corresponds to plane waves traveling in the direction opposite to $\hat{\mathbf{k}}_L$, is due to the fact that we have assumed that the potential vector can be written in the form given by Equation (2.52). We also point out that phases $\pm(\hat{\mathbf{k}}_L \cdot \mathbf{r})\omega'/c \mp \varphi$ of the angular frequency components vary linearly with ω'. Pulses having this property are called "Fourier-transform-limited" (or "bandwidth-limited"); they have no chirp (that is, the angular frequency does not depend on time) and their frequency distribution is the narrowest allowed given the laser pulse duration.

For a pulse generated by a Ti:sapphire laser, whose intensity profile has a width of 50 fs at half maximum, the full width at half maximum $\Delta\omega_{1/2}$ of $|\mathcal{E}(\mathbf{r}, \omega')|$ is $0.033\,\omega$, corresponding to a width in energy $\hbar\Delta\omega_{1/2}$ of 0.05 eV. This energy spread is comparable to the energy difference between the $n = 8$ and $n = 9$ excited states of atomic hydrogen. Due to their large frequency bandwidth, ultra-short pulses cannot populate highly excited states individually. However, they can excite groups of neighboring states coherently and hence create long-lived bound wave packets [6].

Table 2.1. Full width at half maximum in electric field and full width at half maximum in intensity of the sech pulse defined by Equation (2.60), of the Gaussian pulse defined by Equation (2.58), and of the finite-duration pulse defined by Equation (2.61); F is the envelope function of the electric field.

Envelope profile of $\mathcal{E}(\eta)$	FWHM of F	FWHM of F^2
sech	2.63τ	1.76τ
Gaussian	2.35τ	1.67τ
$\cos^2$	1.57τ	1.14τ

The large bandwidth of ultra-short laser pulses also makes it possible to change their temporal shape almost arbitrarily. This can be done, for example, by dispersing the pulse using a grating, making it pass through a mask which selectively attenuates or dephases groups of spectral components, and then resynthesizing it with a second grating. Pulse shaping can be used, for instance, to enhance particular multiphoton transitions or particular molecular reactions [7].

Without reshaping, the intensity profile of the ultra-short pulses produced directly by a CPA system can often be better approximated by a sech^2 function than a Gaussian function. The corresponding envelope function is given by

$$F(\eta) = \text{sech}[\eta/(\omega\tau)]. \tag{2.60}$$

We show, in Table 2.1, how the parameter τ is related to the full width at half maximum of such pulses. This parameter must be chosen such that Equations (2.49) and (2.50) are satisfied. Also given in this table are the corresponding relations for pulses with a Gaussian temporal profile and for laser pulses with the following temporal profile, which is convenient in numerical calculations:

$$F(\eta) = \begin{cases} \cos^2[\eta/(\omega\tau)], & |\eta| \leq \pi\omega\tau/2, \\ 0, & |\eta| > \pi\omega\tau/2. \end{cases} \tag{2.61}$$

It should be noted that the spectrum and the temporal profile of a laser pulse may change upon propagation through a medium, sometimes considerably. An example of the distortion in the profile of a few-cycle pulse due to the ionization of a helium gas target is given in Fig. 2.1. The results were obtained using the method described by Geissler *et al.* [8]. Such changes may have to be taken into account when modeling experiments. It is also worth noting that the central, high-intensity peak of ultra-short pulses is preceded and followed by a "pedestal," during which the

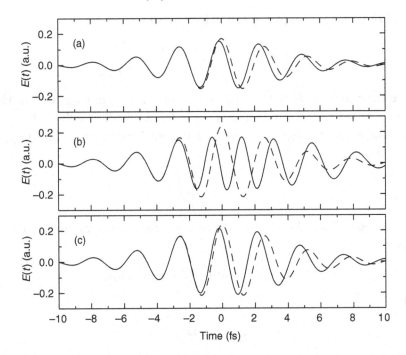

Figure 2.1. Electric-field component along the polarization direction of the 5 fs laser pulse of Fig. 1.3(a) before (dashed curves) and after (solid curves) propagation through a 1 mm thick He gas target. The peak intensity of the pulse and the pressure of the gas are: (a) 1×10^{15} W cm^{-2} and 500 Torr; (b) 2×10^{15} W cm^{-2} and 500 Torr; (c) 2×10^{15} W cm^{-2} and 250 Torr.

laser field is somewhat weaker but non-zero. Unless the intensity in the pulse's initial pedestal remains low enough, significant ionization may happen in the target before it is exposed to the peak intensity of the pulse.

When analyzing or modeling experiments, it is not only important to describe the temporal profile of the pulse correctly, but also the intensity distribution of the pulse in the interaction region. In order to obtain high intensities, pulses are typically focused to near the diffraction limit. This implies that the peak intensity experienced by an atom will depend strongly on both its distance from and its position along the beam axis. This spatial dependence is usually modeled by assuming that in the interaction region the pulse has the form of a Gaussian beam. With this dependence taken into account, the electric field of a laser pulse freely propagating in vacuo can be represented approximately by the following expression:

$$\mathcal{E}(\mathbf{r}, t) = \mathcal{E}_0 \mathrm{Re} \left\{ \hat{\boldsymbol{\epsilon}}_\mathrm{c} F(\eta) G(\mathbf{r}) \exp[-\mathrm{i}(\eta + \varphi)] \right\}, \tag{2.62}$$

which generalizes Equation (2.20). The function $G(\mathbf{r})$ defines the spatial profile of the pulse. In the usual case of a transverse electromagnetic TEM$_{00}$ Gaussian beam [9], $G(\mathbf{r})$ is a complex function having the form

$$G(\mathbf{r}) = \left(\frac{1}{1 + iz/z_R}\right) \exp\left(-\frac{\rho^2/\rho_0^2}{1 + iz/z_R}\right), \qquad (2.63)$$

where the z-axis is taken to be along the direction of propagation. In Equation (2.63), ρ is the distance to that axis, z_R is the Rayleigh length of the beam (i.e. the distance from the focus at $z = 0$ such that the beam intensity has dropped by a factor 2) and

$$\rho_0 = \left(\frac{2z_R}{k_L}\right)^{1/2} \qquad (2.64)$$

is the "waist radius." We note that the complex function $G(\mathbf{r})$ can be written in the form

$$G(\mathbf{r}) = |G(\mathbf{r})| \exp[-i(\varphi_G + \varphi_c)], \qquad (2.65)$$

with

$$\varphi_G = \arctan(z/z_R) \qquad (2.66)$$

and

$$\varphi_c = -\frac{z}{z_R} \frac{\rho^2/\rho_0^2}{1 + (z/z_R)^2}. \qquad (2.67)$$

The phase φ_G, which is known as the *Gouy phase* (or phase shift) [10], varies from $-\pi/2$ to $\pi/2$ across the focus. The small phase φ_c arises from the curvature of the wave fronts; it vanishes on the beam axis ($\rho = 0$), at the beam waist ($z = 0$) and for $z \to \pm\infty$. With the spatial variation of the laser field taken into account, the carrier-envelope phase is therefore $\varphi + \varphi_G + \varphi_c$. Outside the $z = 0$ plane, this phase varies from point to point. The influence of the Gouy phase on focused few-cycle laser pulses has been measured by Lindner *et al.* [11], who found that the CEP undergoes a smooth variation over a few Rayleigh lengths.

It should be noted that the exact solutions of Maxwell's equations differ from the approximate expression (2.62) in two respects. Firstly, the electric-field $\mathcal{E}(\mathbf{r}, t)$ cannot have the same orientation everywhere if it varies transversally, since this would imply that $\nabla \cdot \mathcal{E} \neq 0$. In particular, the electric-field vector of a focused pulse has, almost everywhere, a non-zero longitudinal component. In general, this component is much smaller than the radial component, and for few-cycle Ti:sapphire laser pulses it can be neglected for intensities up to at least 10^{16} W cm^{-2}. However, it can become significant in the focal region at higher intensities. Secondly, one cannot exactly factor the dependence on η and the dependence on $\mathbf{r}$ of an electromagnetic pulse into the two functions $F(\eta)$ and $G(\mathbf{r})$, as was assumed in

Equation (2.62). Indeed, diffraction affects the low-frequency spectral components of a pulse more than the high-frequency ones, leading to different spectral profiles (and therefore different temporal profiles) at different points in space. The resulting coupling between the temporal and spatial variations of the field leads to various effects, such as a delay in the arrival time of the laser pulse and a frequency shift, both of which are position-dependent, as well as distortions in the shape of the pulse as it propagates. However, in practice, these effects can be neglected if the pulse duration is longer than a few optical cycles. Indeed, they are of relative importance only far from the focus, where the laser field is usually negligibly weak.

2.1.3 Fluctuations of the laser field

One can usually distinguish between two kinds of fluctuations of the laser field, characterized by different time (or space) scales. First, there are slow fluctuations associated with the laser pulse shape and beam focusing. The net effect of these inhomogeneities on multiphoton processes can often be taken into account by simply averaging, for example ionization rates or cross sections, with respect to the temporal or spatial dependence of the field intensity. Laser fields can also exhibit fast fluctuations of the laser parameters (amplitude, phase, frequency) that take place on time scales that are comparable to atomic times. As far as multiphoton processes in strong laser pulses are concerned, the influence of these fast fluctuations are usually small. Theoretical approaches based on stochastic Markov processes have been developed to take them into account [12]. Several Markov models of fluctuating laser fields have been proposed. They have been reviewed by Daniele *et al.* [13], Bivona *et al.* [14] and Francken and Joachain [15].

A first model is that of a *chaotic field*, which undergoes both *Gaussian amplitude* and *phase* fluctuations [16, 17], and corresponds to a pulsed multimode laser field with a large number of uncorrelated modes. A special case of this model is that of a *Gaussian-amplitude* field in which the laser field is assumed to undergo only amplitude fluctuations, and is used to describe laser fields that have strong amplitude fluctuations. In the *phase diffusion* model the laser field undergoes only phase fluctuations[18, 19], and corresponds to an intensity-stabilized single-mode laser field.

An alternative to the above-mentioned models has been proposed by Francken and Joachain [20, 21] who considered discrete Markov chains [12, 22]. The advantages of this approach are that it (i) permits a unified description of amplitude, phase and frequency fluctuations of the field, (ii) reduces most of the calculations to linear algebra and (iii) offers great flexibility in modeling the field fluctuations. The simplest example of a discrete Markov chain is the *random telegraph signal*. In this case it is assumed that the random process can take only two values and switches randomly from one to another. Two important generalizations of this model are the *Kubo–Anderson process* [23], which can be considered as an N-state random-telegraph signal, and the pre-Gaussian process [24], which is the superposition of N independent random-telegraph signals and reduces to Gaussian statistics in the limit $N \to \infty$, as a consequence of the central-limit theorem. From the mathematical point of view, these processes are fully characterized by the magnitude of the fluctuations, their correlation time, an initial probability distribution function $g(\alpha)$, where α labels the possible states of the Markov chain, and a conditional probability density function $P(\alpha, t | \alpha_0, t_0)$. The latter satisfies a Chapman–Kolmogorov equation of the form [25]

$$\frac{\partial}{\partial t} P(\alpha, t | \alpha_0, t_0) = \sum_{\beta} W_{\alpha}^{\beta} P(\beta, t | \alpha_0, t_0), \qquad (2.68)$$

where the coefficients W_{α}^{β} are time-independent.

Finally, we note that Trombetta, Ferrante and Zoller [26] have provided a non-Markovian treatment of the phase-diffusion model.

2.2 Classical non-relativistic electron dynamics in a laser field

The Lorentz equation governing the classical dynamics of an electron of charge $q = -e$ in a laser field is given by [27]

$$\frac{d}{dt} \mathbf{p}_{cl} = -e \left[\boldsymbol{\mathcal{E}}(\mathbf{r}, t) + \mathbf{v} \times \boldsymbol{\mathcal{B}}(\mathbf{r}, t) \right]. \qquad (2.69)$$

The electron momentum $\mathbf{p}_{cl}$ is related to its velocity $\mathbf{v}$ by $\mathbf{p}_{cl} = \gamma m \mathbf{v}$, where m is the mass of the electron and the Lorentz factor γ is given by

$$\gamma = (1 - v^2/c^2)^{-1/2}. \qquad (2.70)$$

The back-reaction of the motion of the electron on the radiation field is ignored, as this effect is very small [28]. We work in the Coulomb gauge and the scalar potential ϕ is taken to be zero, so that the electric and magnetic components of the laser field are obtained from the vector potential $\mathbf{A}(\mathbf{r}, t)$ according to Equations (2.8) and (2.9).

We first consider the microscopic motion of a classical electron in the field. As we shall see in subsequent chapters, this motion plays an important role in understanding a number of strong-field phenomena. For a large range of applications, the intensity of the laser field is such that a non-relativistic description of the electron motion is justified. In order to obtain the non-relativistic limit of the Lorentz equation, we first note that, according to Equation (2.15), the second term on the right-hand side of Equation (2.69) is a factor v/c smaller than the first term. In the non-relativistic regime, we have $v/c \ll 1$, allowing this second term to be neglected. In addition, we can set $\gamma = 1$, so that $\mathbf{p}_{cl} = m\mathbf{v}$. Finally, we restrict ourselves to the case in which the initial velocity of the electron is such that its displacement from its initial position $\mathbf{r}_0$ along the propagation direction remains much smaller than the carrier wavelength. As a result, in the non-relativistic, long-wavelength limit, the Lorentz equation (2.69) reduces to

$$\frac{\mathrm{d}}{\mathrm{d}t}\mathbf{v} = -\frac{e}{m}\mathcal{E}(\mathbf{r}_0, t). \tag{2.71}$$

2.2.1 Quiver and drift motions

In the non-relativistic, long-wavelength approximation, the acceleration imparted to the electron by the laser field is simply along the direction of $\mathcal{E}(\mathbf{r}_0, t)$. From Equation (2.71), the velocity of an electron having an initial velocity $\mathbf{v}_0$ at time t_0 is given by

$$\mathbf{v}(t) = -\frac{e}{m}\int_{t_0}^{t}\mathcal{E}(\mathbf{r}_0, t')\mathrm{d}t' + \mathbf{v}_0$$

$$= \frac{e}{m}\mathbf{A}(\mathbf{r}_0, t) - \frac{e}{m}\mathbf{A}(\mathbf{r}_0, t_0) + \mathbf{v}_0. \tag{2.72}$$

The quantity $m\mathbf{v}(t)$ is called the kinetic momentum of the electron. Introducing its canonical momentum as

$$\mathbf{p}(t) = m\mathbf{v}(t) - e\mathbf{A}(\mathbf{r}_0, t) \tag{2.73}$$

allows Equation (2.72) to be written as

$$\mathbf{p}(t) = \mathbf{p}(t_0), \tag{2.74}$$

which expresses the conservation of the electron canonical momentum in the long-wavelength approximation. The general classical motion of an electron in the laser field is therefore a superposition of a *quiver* motion and a *drift* motion. The former is characterized by the quiver velocity, $e\mathbf{A}(\mathbf{r}_0, t)/m$, and the latter by the drift velocity,

$$\mathbf{v}_d = \mathbf{p}(t_0)/m = \mathbf{v}_0 - \frac{e}{m}\mathbf{A}(\mathbf{r}_0, t_0). \tag{2.75}$$

The position of an electron, located at $\mathbf{r}_0$ at time t_0, is given by

$$\mathbf{r}(t) = \frac{e}{m}\int_{t_0}^{t}\mathbf{A}(\mathbf{r}_0, t')\mathrm{d}t' + \mathbf{v}_d(t - t_0) + \mathbf{r}_0$$

$$= \boldsymbol{\alpha}(t, t_0) + \mathbf{v}_d(t - t_0) + \mathbf{r}_0, \tag{2.76}$$

where

$$\boldsymbol{\alpha}(t, t_0) = \frac{e}{m} \int_{t_0}^{t} \mathbf{A}(\mathbf{r}_0, t') dt' \tag{2.77}$$

is the displacement vector of the electron due to its quiver motion in the laser field.

For the special case of a linearly polarized monochromatic laser field whose electric field is given by Equation (2.13), an electron at the position $\mathbf{r}_0$ at time $t = t_0$ and with no drift velocity undergoes simple harmonic motion. The corresponding displacement is given by

$$\boldsymbol{\alpha}(t, t_0) = \hat{\boldsymbol{\epsilon}} \alpha_0 [\cos(\mathbf{k}_L \cdot \mathbf{r}_0 - \omega t - \varphi) - \cos(\mathbf{k}_L \cdot \mathbf{r}_0 - \omega t_0 - \varphi)], \tag{2.78}$$

where

$$\alpha_0 = \frac{e \mathcal{E}_0}{m \omega^2} = \left(\frac{2}{\epsilon_0 c} \right)^{1/2} \frac{e}{m} \frac{I^{1/2}}{\omega^2} \tag{2.79}$$

is called the *quiver*, or *excursion*, amplitude of the electron in the laser field. The electron trajectory is in this case a straight-line segment along the polarization vector $\hat{\boldsymbol{\epsilon}}$, extending from $-\alpha_0$ to α_0.

For a monochromatic laser field of arbitrary polarization, we have

$$\boldsymbol{\alpha}(t, t_0) = \frac{\alpha_0}{(1 + \xi^2)^{1/2}} \{ \hat{\boldsymbol{\epsilon}}_1 [\cos(\mathbf{k}_L \cdot \mathbf{r}_0 - \omega t - \varphi) - \cos(\mathbf{k}_L \cdot \mathbf{r}_0 - \omega t_0 - \varphi)]$$

$$+ \xi \hat{\boldsymbol{\epsilon}}_2 [\sin(\mathbf{k}_L \cdot \mathbf{r}_0 - \omega t - \varphi) - \sin(\mathbf{k}_L \cdot \mathbf{r}_0 - \omega t_0 - \varphi)] \}. \tag{2.80}$$

The trajectory followed by the extremity of the vector $\boldsymbol{\alpha}(t, t_0)$ is in general an ellipse. It reduces to a straight-line segment for linear polarization ($\xi = 0$) and to a circle with radius $\alpha_0/\sqrt{2}$ for circular polarization ($\xi = \pm 1$).

Referring to Equations (2.49) and (2.50), and setting $t_0 = -t$ in Equations (2.72) and (2.77), we have

$$\lim_{t \to \infty} \mathbf{v}(t) = \mathbf{v}_0 \tag{2.81}$$

and

$$\lim_{t \to \infty} \boldsymbol{\alpha}(t, -t) = 0. \tag{2.82}$$

Thus, a free electron that interacts with a laser pulse described in the long-wavelength approximation by a spatially homogeneous electric field cannot acquire a net displacement or a net velocity from the field. In Section 2.2.2, we will see that these properties are no longer true when the spatial dependence of the electric field is taken into account. Moreover, the foregoing analysis must be modified if the electron is moving in a polarized medium.

2.2.2 Ponderomotive energy and force

The *ponderomotive energy* is defined as the cycle-averaged quiver energy of a "free" electron in the laser field, namely the cycle-averaged kinetic energy of a "free" electron that has zero drift velocity in the laser field. Assuming that on the short time scale of the laser period the field can be taken to be monochromatic, the ponderomotive energy is given by

$$U_p = \frac{e^2 \mathcal{E}_0^2}{4m\omega^2} = \frac{e^2}{2\epsilon_0 cm} \frac{I}{\omega^2}. \tag{2.83}$$

When analyzing the *microscopic* dynamics of a non-relativistic electron in a laser pulse, a spatially homogeneous laser field can be assumed. However, in order to describe correctly the *macroscopic* dynamics it is often important to take into account the spatial variations of the intensity of the laser pulse. In particular, in order to interpret experimental ATI spectra it is necessary to understand the dynamics of ionized electrons as they leave the laser focus.

Let us consider an ionized electron in the laser pulse, and neglect the interaction of the electron with its parent ion. On time scales much larger than one laser period the electron motion is influenced by spatial inhomogeneities in the electric field due primarily to the focusing of the pulse in the laser–atom interaction region and the finite duration of the pulse. This gives rise to the *ponderomotive force* that acts to expel the electron from the laser focus, as will now be demonstrated.

Let us suppose that at some time t_0 during the laser pulse the electron, which we describe classically, is located at $\mathbf{r}_0$. In addition, we assume that the laser-pulse duration is long enough so that the envelope of the pulse changes little per optical cycle. We seek to obtain an approximate equation governing the motion of the electron in the vicinity of $\mathbf{r}_0$ which takes into account the slowly varying spatial dependence of the electric field of the laser pulse. To this end, we expand the electric field around $\mathbf{r}_0$. Retaining only the first-order spatial dependence of the electric field gives

$$\mathcal{E}(\mathbf{r}, t) = \mathcal{E}(\mathbf{r}_0, t) + (\delta\mathbf{r} \cdot \nabla)\mathcal{E}(\mathbf{r}, t)\big|_{\mathbf{r}=\mathbf{r}_0}, \tag{2.84}$$

where

$$\delta\mathbf{r} = \mathbf{r} - \mathbf{r}_0. \tag{2.85}$$

We remark that since $k_L = \omega/c$ the second term on the right-hand side in Equation (2.84) is of order $1/c$ with respect to the first term. Using Equation (2.15), we see that to the same level of approximation we can set

$$\mathcal{B}(\mathbf{r}, t) = \mathcal{B}(\mathbf{r}_0, t). \tag{2.86}$$

Next, referring to Equation (2.76), and assuming for simplicity that the drift velocity $v_d = 0$, we express the position $\mathbf{r}(t)$ of the electron in terms of a microscopic contribution $\boldsymbol{\alpha}(t, t_0)$ due to the quiver motion in the laser field and an additional macroscopic displacement $\Delta \mathbf{r}(t)$ arising from the spatial variation of the laser field intensity in the laser focus. That is,

$$\mathbf{r}(t) = \mathbf{r}_0 + \boldsymbol{\alpha}(t, t_0) + \Delta \mathbf{r}(t). \tag{2.87}$$

The velocity is then given by

$$\mathbf{v}(t) = \frac{d}{dt}\boldsymbol{\alpha}(t, t_0) + \Delta \mathbf{v}(t), \tag{2.88}$$

where the displacement velocity is $\Delta \mathbf{v}(t) = d[\Delta \mathbf{r}(t)]/dt$.

If we now use the expansion (2.84), with $\delta \mathbf{r} = \boldsymbol{\alpha}(t, t_0) + \Delta \mathbf{r}$, and Equation (2.88) in the Lorentz equation (2.69), we obtain for the displacement velocity, to order $1/c$, the following equation:

$$\frac{d}{dt}\Delta \mathbf{v}(t) = -\frac{e}{m}\left([\boldsymbol{\alpha}(t, t_0) \cdot \nabla]\mathcal{E}(\mathbf{r}, t)\big|_{\mathbf{r}=\mathbf{r}_0} + \left[\frac{d}{dt}\boldsymbol{\alpha}(t, t_0)\right] \times \mathcal{B}(\mathbf{r}_0, t)\right)$$
$$- \frac{e}{m}\left([\Delta \mathbf{r}(t) \cdot \nabla]\mathcal{E}(\mathbf{r}, t)\big|_{\mathbf{r}=\mathbf{r}_0} + \Delta \mathbf{v}(t) \times \mathcal{B}(\mathbf{r}_0, t)\right). \tag{2.89}$$

The last two terms on the right-hand side in this equation can be ignored since they give a negligible contribution on average over the duration of one optical cycle, provided the electron displacement over this time interval is small compared to the length scale over which the electric field varies. Furthermore, we have

$$\left[\frac{d}{dt}\boldsymbol{\alpha}(t, t_0)\right] \times \mathcal{B}(\mathbf{r}_0, t) = -\boldsymbol{\alpha}(t, t_0) \times \frac{d}{dt}\mathcal{B}(\mathbf{r}_0, t) + \frac{d}{dt}[\boldsymbol{\alpha}(t, t_0) \times \mathcal{B}(\mathbf{r}_0, t)]$$

$$= \boldsymbol{\alpha}(t, t_0) \times \nabla \times \mathcal{E}(\mathbf{r}, t)\big|_{\mathbf{r}=\mathbf{r}_0} + \frac{d}{dt}[\boldsymbol{\alpha}(t, t_0) \times \mathcal{B}(\mathbf{r}_0, t)].$$
$$\tag{2.90}$$

The condition that the envelope of the pulse remains nearly constant on the time scale of a laser period allows us to express Equation (2.77) in the form

$$\boldsymbol{\alpha}(t, t_0) = \frac{e}{m\omega^2}\mathcal{E}(\mathbf{r}_0, t) - \frac{e}{m\omega^2}\mathcal{E}(\mathbf{r}_0, t_0). \tag{2.91}$$

Next, using the result from vector calculus

$$\nabla(V^2) = 2(\mathbf{V} \cdot \nabla)\mathbf{V} + 2\mathbf{V} \times \nabla \times \mathbf{V}, \tag{2.92}$$

where $\mathbf{V}$ is a vector and $V = |\mathbf{V}|$, we find that

$$\frac{d}{dt}\Delta\mathbf{v}(t) = -\frac{e^2}{2m^2\omega^2}\left(\nabla\mathcal{E}^2(\mathbf{r}, t)\big|_{\mathbf{r}=\mathbf{r}_0}\right)$$

$$-\frac{e^2}{m^2\omega^2}\left([\mathcal{E}(\mathbf{r}_0, t_0)\cdot\nabla]\mathcal{E}(\mathbf{r}, t)\big|_{\mathbf{r}=\mathbf{r}_0} + \mathcal{E}(\mathbf{r}_0, t_0)\times\nabla\times\mathcal{E}(\mathbf{r}, t)\big|_{\mathbf{r}=\mathbf{r}_0}\right)$$

$$-\frac{e}{m}\frac{d}{dt}[\boldsymbol{\alpha}(t, t_0)\times\mathcal{B}(\mathbf{r}_0, t)]. \tag{2.93}$$

The first term on the right-hand side of this equation contains a contribution that varies slowly over an optical cycle as well as a contribution that oscillates with angular frequency 2ω. The remaining terms oscillate with angular frequency ω. On average, the dominant contribution to the displacement velocity is given by the slowly varying term, allowing us to write to good approximation

$$m\frac{d}{dt}\Delta\mathbf{v}(t) = -\frac{e^2}{2m\omega^2}\nabla\overline{\mathcal{E}^2(\mathbf{r}, t)}\big|_{\mathbf{r}=\mathbf{r}_0}$$

$$= -\nabla U_p(\mathbf{r})\big|_{\mathbf{r}=\mathbf{r}_0}, \tag{2.94}$$

where the bar denotes a time-average over an optical cycle. In the vicinity of $\mathbf{r}_0$, the electron thus moves, on average, as if its potential energy were equal to its ponderomotive energy at $\mathbf{r}_0$. For this reason, $U_p(\mathbf{r})$ is called the *ponderomotive potential*.

For a monochromatic laser field, the cycle-average of the magnitude squared of the electric field can be related to the field intensity at $\mathbf{r}_0$ using Equation (2.39), so that

$$U_p(\mathbf{r}_0) = \frac{e^2}{2\epsilon_0 cm\omega^2}I(\mathbf{r}_0). \tag{2.95}$$

The ponderomotive potential therefore acts to expel the electron from regions of high intensity, namely the laser focus. This result has important consequences for interpreting experimental data, as we saw in Section 1.3.

2.3 Time-dependent Schrödinger equation

2.3.1 Interaction of an electron with an electromagnetic field

The non-relativistic Hamiltonian of an electron in an electromagnetic field described, in an arbitrary gauge, by the scalar potential $\phi(\mathbf{r}, t)$ and the vector potential $\mathbf{A}(\mathbf{r}, t)$ is

$$H(t) = \frac{1}{2m}(\mathbf{p} + e\mathbf{A})^2 - e\phi$$

$$= \frac{\mathbf{p}^2}{2m} + \frac{e}{2m}(\mathbf{A} \cdot \mathbf{p} + \mathbf{p} \cdot \mathbf{A}) + \frac{e^2}{2m}\mathbf{A}^2 - e\phi, \qquad (2.96)$$

where small spin-dependent terms have been neglected. In the above equations, $\mathbf{p}$ is the canonical momentum operator of the electron. Working in the Schrödinger picture and using the position representation, in which $\mathbf{p} = -i\hbar\nabla$, the corresponding time-dependent Schrödinger equation (TDSE) is as follows:

$$i\hbar\frac{\partial}{\partial t}\Psi(\mathbf{r}, t) = \left[-\frac{\hbar^2}{2m}\nabla^2 - i\hbar\frac{e}{2m}(\mathbf{A} \cdot \nabla + \nabla \cdot \mathbf{A}) + \frac{e^2}{2m}\mathbf{A}^2 - e\phi \right]\Psi(\mathbf{r}, t). \quad (2.97)$$

Adopting the Coulomb gauge defined by Equation (2.7), and remembering that in empty space we have $\phi = 0$, the TDSE (2.97) reduces to

$$i\hbar\frac{\partial}{\partial t}\Psi(\mathbf{r}, t) = \left[-\frac{\hbar^2}{2m}\nabla^2 - i\hbar\frac{e}{m}\mathbf{A} \cdot \nabla + \frac{e^2}{2m}\mathbf{A}^2 \right]\Psi(\mathbf{r}, t), \qquad (2.98)$$

where we have used the fact that, in the Coulomb gauge,

$$\nabla \cdot (\mathbf{A}\Psi) = \mathbf{A} \cdot (\nabla\Psi) + (\nabla \cdot \mathbf{A})\Psi$$

$$= \mathbf{A} \cdot (\nabla\Psi). \qquad (2.99)$$

2.3.2 Interaction of an atomic system with an electromagnetic field

We now consider the interaction of an electromagnetic field with an atomic system (atom or ion) composed of a nucleus of atomic number Z and N electrons. We neglect relativistic effects which, as we shall see in Section 2.9, occur when atoms interact with ultra-strong laser fields. We also neglect small effects due to the interaction between the nucleus and the electromagnetic field. In the same spirit we neglect reduced mass effects and take the nucleus to be the origin of the coordinates.

Let us first consider the case $N = 1$, corresponding to hydrogenic atoms and ions. We must then include in the Hamiltonian the electrostatic Coulomb potential $-Ze^2/(4\pi\epsilon_0 r)$ between the electron and the nucleus. It is convenient to regard this electrostatic interaction as an additional potential energy term, while the radiation field is described in the Coulomb gauge in terms of a vector potential $\mathbf{A}(\mathbf{r}, t)$ alone, as discussed above. The TDSE then reads

$$i\hbar\frac{\partial}{\partial t}\Psi(\mathbf{r}, t) = \left[-\frac{\hbar^2}{2m}\nabla^2 - \frac{Ze^2}{(4\pi\epsilon_0)r} - i\hbar\frac{e}{m}\mathbf{A} \cdot \nabla + \frac{e^2}{2m}\mathbf{A}^2 \right]\Psi(\mathbf{r}, t). \quad (2.100)$$

We can rewrite this equation in the form

$$i\hbar\frac{\partial}{\partial t}\Psi(\mathbf{r},t) = [H_0 + H_{\text{int}}(t)]\Psi(\mathbf{r},t), \tag{2.101}$$

where

$$H_0 = -\frac{\hbar^2}{2m}\nabla^2 - \frac{Ze^2}{(4\pi\epsilon_0)r} \tag{2.102}$$

is the time-independent hydrogenic Hamiltonian describing the hydrogenic atom (ion) in the absence of the radiation field, and

$$H_{\text{int}}(t) = -i\hbar\frac{e}{m}\mathbf{A}\cdot\nabla + \frac{e^2}{2m}\mathbf{A}^2$$

$$= \frac{e}{m}\mathbf{A}\cdot\mathbf{p} + \frac{e^2}{2m}\mathbf{A}^2 \tag{2.103}$$

is the Hamiltonian describing the interaction of the hydrogenic atom with the radiation field.

We now generalize these results to the case of an N-electron atom (ion). The TDSE in the Schrödinger picture and in the position representation is then given by

$$i\hbar\frac{\partial}{\partial t}\Psi(X,t) = H(t)\Psi(X,t), \tag{2.104}$$

where $X \equiv (q_1, q_2, \ldots, q_N)$ denotes the ensemble of the coordinates of the N electrons and $q_i \equiv (\mathbf{r}_i, \sigma_i)$ are the space and spin coordinates of the ith electron. The semi-classical Hamiltonian $H(t)$ describing the atomic system in the presence of the radiation field is readily obtained by an extension of the arguments given above for the case $N = 1$. Using the Coulomb gauge, we have

$$H(t) = H_0 + H_{\text{int}}(t), \tag{2.105}$$

where

$$H_0 = \frac{1}{2m}\sum_{i=1}^{N}\mathbf{p}_i^2 + V = -\frac{\hbar^2}{2m}\sum_{i=1}^{N}\nabla_{\mathbf{r}_i}^2 + V \tag{2.106}$$

is the time-independent Hamiltonian of the N-electron atom (ion) in the absence of the electromagnetic field. In the above equation $\mathbf{p}_i = -i\hbar\nabla_{\mathbf{r}_i}$ is the momentum operator of the electron i and V denotes the sum of all the interactions within the atomic system in the absence of the radiation field. For example, if all but the Coulomb interactions are neglected, we have

$$V = -\sum_{i=1}^{N}\frac{Ze^2}{(4\pi\epsilon_0)r_i} + \sum_{i<j=1}^{N}\frac{e^2}{(4\pi\epsilon_0)r_{ij}}, \tag{2.107}$$

where $r_{ij} = |\mathbf{r}_i - \mathbf{r}_j|$. The interaction Hamiltonian $H_{\text{int}}(t)$ appearing in Equation (2.105) is given by generalizing Equation (2.103) to an N-electron atom. That is,

$$H_{\text{int}}(t) = \frac{e}{m} \sum_{i=1}^{N} \mathbf{A}(\mathbf{r}_i, t) \cdot \mathbf{p}_i + \frac{e^2}{2m} \sum_{i=1}^{N} \mathbf{A}^2(\mathbf{r}_i, t). \tag{2.108}$$

The explicit form of the TDSE (2.104) is therefore as follows:

$$i\hbar \frac{\partial}{\partial t} \Psi(X, t) = \left[\frac{1}{2m} \sum_{i=1}^{N} \mathbf{p}_i^2 + V + \frac{e}{m} \sum_{i=1}^{N} \mathbf{A}(\mathbf{r}_i, t) \cdot \mathbf{p}_i + \frac{e^2}{2m} \sum_{i=1}^{N} \mathbf{A}^2(\mathbf{r}_i, t) \right]$$

$$\times \Psi(X, t). \tag{2.109}$$

2.3.3 Dipole approximation

Let us now assume that the wavelength λ of the laser field is large compared with the size of the atomic system under consideration and that the laser field intensity is not too high. When these two conditions are fulfilled, the *dipole approximation* can be made, which consists in *neglecting the spatial variation* of the radiation field across the atom. In this approximation, for an atom whose nucleus is located at the position $\mathbf{r}_0$, the vector potential $\mathbf{A}(\mathbf{r}_0, t) = \mathbf{A}(t)$ is *spatially homogeneous* (i.e. it depends only on the time variable) so that the Coulomb gauge condition (2.7) is automatically satisfied. Moreover, using Equation (2.8), we see that the electric field is given by

$$\mathcal{E}(t) = -\frac{d}{dt} \mathbf{A}(t) \tag{2.110}$$

and hence depends only on time, while the magnetic field $\mathcal{B} = \nabla \times \mathbf{A}$ vanishes. The vector potential $\mathbf{A}(t)$ appearing in Equation (2.110) is not a solution of the wave equation (2.4), and is used only in the context of an approximation to the true vector potential appearing in the interaction Hamiltonian $H_{\text{int}}(t)$. For very strong fields *non-dipole* effects (due to the magnetic-field component of the laser field) must be taken into account and eventually, for ultra-strong fields, relativistic effects must be included. The equations needed to study the non-dipole and relativistic regimes of laser–atom interactions will be introduced in Sections 2.8 and 2.9, respectively.

Allowing for arbitrary polarization, the electric-field component of a laser pulse in the dipole approximation is obtained by setting $\mathbf{r} = \mathbf{r}_0$ in Equation (2.56). Without loss of generality, we set $\mathbf{r}_0 = 0$. Including into the phase φ any phase shift, such as the Gouy phase arising from the propagation of the beam, we have

$$\mathcal{E}(t) = \frac{\mathcal{E}_0}{(1+\xi^2)^{1/2}} F(t) \left[\hat{\boldsymbol{\epsilon}}_1 \cos(\omega t + \varphi) - \xi \hat{\boldsymbol{\epsilon}}_2 \sin(\omega t + \varphi) \right]. \tag{2.111}$$

For simplicity, we denote by $F(t)$ the envelope function $F(\omega t - \mathbf{k}_L \cdot \mathbf{r}_0)$ taken at $\mathbf{r}_0 = 0$.

The corresponding vector potential in the dipole approximation, obtained from Equation (2.57), is as follows:

$$\mathbf{A}(t) = -\frac{\mathcal{E}_0}{(1+\xi^2)^{1/2}} \int_{-\infty}^{t} F(t') \left[\hat{\boldsymbol{\epsilon}}_1 \cos(\omega t' + \varphi) - \xi \hat{\boldsymbol{\epsilon}}_2 \sin(\omega t' + \varphi) \right] dt'.$$

(2.112)

While the phase φ in the above formulae can be considered to be constant as far as the calculation of the response of the atom to the laser field is concerned, it varies from point to point across the laser focus and has only a local meaning.

The displacement vector of a classical electron from its oscillation center due to its quiver motion in the laser field is now given by

$$\boldsymbol{\alpha}(t) = \frac{e}{m} \int_{-\infty}^{t} \mathbf{A}(t') dt'.$$

(2.113)

For a monochromatic laser field of arbitrary polarization, we have $F(t) = 1$ so that in the dipole approximation the electric field and vector potential, respectively, are

$$\boldsymbol{\mathcal{E}}(t) = \frac{\mathcal{E}_0}{(1+\xi^2)^{1/2}} \left[\hat{\boldsymbol{\epsilon}}_1 \cos(\omega t + \varphi) - \xi \hat{\boldsymbol{\epsilon}}_2 \sin(\omega t + \varphi) \right]$$

(2.114)

and

$$\mathbf{A}(t) = -\frac{A_0}{(1+\xi^2)^{1/2}} \left[\hat{\boldsymbol{\epsilon}}_1 \sin(\omega t + \varphi) + \xi \hat{\boldsymbol{\epsilon}}_2 \cos(\omega t + \varphi) \right]$$

(2.115)

with $A_0 = \mathcal{E}_0/\omega$. The corresponding displacement vector is given, from Equation (2.113), by

$$\boldsymbol{\alpha}(t) = \frac{\alpha_0}{(1+\xi^2)^{1/2}} \left[\hat{\boldsymbol{\epsilon}}_1 \cos(\omega t + \varphi) - \xi \hat{\boldsymbol{\epsilon}}_2 \sin(\omega t + \varphi) \right],$$

(2.116)

where α_0 is the electron quiver amplitude given by Equation (2.79). We have assumed that the laser field is "switched on" adiabatically at $t_0 = -\infty$.

We also note that, for the case of a linearly polarized laser pulse, the electric field can be written in the dipole approximation as

$$\boldsymbol{\mathcal{E}}(t) = \hat{\boldsymbol{\epsilon}} \mathcal{E}_0 F(t) \cos(\omega t + \varphi),$$

(2.117)

while, for a linearly polarized monochromatic laser field,

$$\boldsymbol{\mathcal{E}}(t) = \hat{\boldsymbol{\epsilon}} \mathcal{E}_0 \cos(\omega t + \varphi),$$

(2.118)

where $\hat{\boldsymbol{\epsilon}}$ is the polarization vector. For a chirped pulse, either ω or φ is time-dependent.

Let us return to the interaction Hamiltonian (2.108). We see that in the dipole approximation it takes the simple form

$$H_{\text{int}}(t) = \frac{e}{m}\mathbf{A}(t)\cdot\mathbf{P} + \frac{e^2 N}{2m}\mathbf{A}^2(t), \qquad (2.119)$$

where

$$\mathbf{P} = \sum_{i=1}^{N}\mathbf{p}_i \qquad (2.120)$$

is the total momentum operator. The TDSE for an N-electron atom (ion) in an electromagnetic field can therefore be written in the dipole approximation as

$$i\hbar\frac{\partial}{\partial t}\Psi(X,t) = \left[H_0 + \frac{e}{m}\mathbf{A}(t)\cdot\mathbf{P} + \frac{e^2 N}{2m}\mathbf{A}^2(t)\right]\Psi(X,t), \qquad (2.121)$$

where the field-free Hamiltonian H_0 is given by Equation (2.106).

2.4 Gauge transformations

As we noted in Section 2.1, the invariance of the fields $\mathcal{E}$ and $\mathcal{B}$ under the classical gauge transformation (2.5) allows further conditions on the potentials ϕ and $\mathbf{A}$ to be imposed. For example, one can choose the Coulomb gauge such that $\nabla\cdot\mathbf{A} = 0$, with $\phi = 0$ in empty space. In this section, we shall examine gauge transformations in the context of the semi-classical theory of atomic systems interacting with electromagnetic fields.

2.4.1 Gauge invariance of the Schrödinger equation

Let us return to the TDSE (2.97), which we wrote down for an electron in an electromagnetic field described by the scalar potential $\phi(\mathbf{r},t)$ and the vector potential $\mathbf{A}(\mathbf{r},t)$. An important property of this equation is that its form is *unchanged* under the quantum mechanical *gauge transformation*

$$\mathbf{A} \to \mathbf{A}' = \mathbf{A} + \nabla f,$$

$$\phi \to \phi' = \phi - \frac{\partial f}{\partial t}, \qquad (2.122)$$

$$\Psi \to \Psi' = \exp(-ief/\hbar)\Psi,$$

where f is an arbitrary real, differentiable function of $\mathbf{r}$ and t. That is, the wave function $\Psi'(\mathbf{r}, t)$ satisfies the equation

$$i\hbar\frac{\partial}{\partial t}\Psi'(\mathbf{r}, t) = \left[-\frac{\hbar^2}{2m}\nabla^2 - i\hbar\frac{e}{2m}(\mathbf{A}'\cdot\nabla + \nabla\cdot\mathbf{A}') + \frac{e^2}{2m}\mathbf{A}'^2 - e\phi'\right]\Psi'(\mathbf{r}, t). \tag{2.123}$$

This property is readily generalized to the case of atomic systems interacting with electromagnetic fields. Thus, the TDSE for an N-electron atom interacting with a radiation field described by the scalar potential $\phi(\mathbf{r}, t)$ and the vector potential $\mathbf{A}(\mathbf{r}, t)$ is *invariant* under the gauge transformation (2.122), where f is now a real, differentiable function of the spatial coordinates $\mathbf{r}_i$ of the N electrons and of the time t. Since, as seen from Equation (2.122), a gauge transformation is a particular case of a unitary transformation, measurable quantities (such an expectation values or transition probabilities) calculated in different gauges should in principle be the same. In practice, of course, differences arise because of the approximations used in the calculations. The differences will be small if accurate wave functions are used.

2.4.2 The velocity and length gauges

The property of gauge invariance allows us to simplify the TDSE by an appropriate choice of gauge. We have already seen above that it is often convenient to work in the Coulomb gauge, such that $\nabla\cdot\mathbf{A} = 0$, with $\phi = 0$. We shall now show that, *in the dipole approximation*, the interaction term in the TDSE (2.121) can be simplified by performing gauge transformations.

We first note that the term in $\mathbf{A}^2$ in Equation (2.121) can be eliminated by extracting from the wave function a time-dependent phase factor according to

$$\Psi^V(X, t) = \exp\left[\frac{i}{\hbar}\frac{e^2 N}{2m}\int_{-\infty}^{t}\mathbf{A}^2(t')dt'\right]\Psi(X, t), \tag{2.124}$$

which amounts to choosing the function f of the gauge transformation (2.122) to be

$$f = -\frac{eN}{2m}\int_{-\infty}^{t}\mathbf{A}^2(t')dt'. \tag{2.125}$$

We see from Equations (2.122) and (2.125) that, with this choice, we have

$$\mathbf{A}' = \mathbf{A},$$

$$\phi' = \frac{eN}{2m}\mathbf{A}^2. \tag{2.126}$$

The resulting TDSE for the new wave function $\Psi'(X,t) \equiv \Psi^V(X,t)$ is

$$i\hbar \frac{\partial}{\partial t}\Psi^V(X,t) = \left[H_0 + \frac{e}{m}\mathbf{A}(t)\cdot\mathbf{P}\right]\Psi^V(X,t), \tag{2.127}$$

which is said to be in the *velocity gauge*, since the interaction Hamiltonian

$$H_{\text{int}}^V(t) = \frac{e}{m}\mathbf{A}(t)\cdot\mathbf{P} \tag{2.128}$$

couples the vector potential $\mathbf{A}(t)$ to the operator $\mathbf{P}/m$.

Another form of the TDSE in the dipole approximation can be obtained by returning to Equation (2.121) and performing a gauge transformation specified by taking

$$f = -\mathbf{A}(t)\cdot\mathbf{R}, \tag{2.129}$$

where

$$\mathbf{R} = \sum_{i=1}^{N}\mathbf{r}_i \tag{2.130}$$

is the sum of the coordinates of the N electrons. We then see from Equations (2.122) and (2.129) that

$$\mathbf{A}' = 0,$$
$$\phi' = (\partial\mathbf{A}/\partial t)\cdot\mathbf{R} \tag{2.131}$$
$$= -\mathcal{E}(t)\cdot\mathbf{R},$$

and the new wave function is given by

$$\Psi^L(X,t) = \exp\left[\frac{ie}{\hbar}\mathbf{A}(t)\cdot\mathbf{R}\right]\Psi(X,t). \tag{2.132}$$

It satisfies the TDSE

$$i\hbar\frac{\partial}{\partial t}\Psi^L(X,t) = [H_0 + e\mathcal{E}(t)\cdot\mathbf{R}]\Psi^L(X,t), \tag{2.133}$$

which is said to be in the *length gauge* because the interaction Hamiltonian

$$H_{\text{int}}^L(t) = e\mathcal{E}(t)\cdot\mathbf{R} \tag{2.134}$$

couples the electric field $\mathcal{E}(t)$ to the operator $\mathbf{R}$. Introducing the electric dipole moment operator of the atom

$$\mathbf{D} = \sum_{i=1}^{N}(-e)\mathbf{r}_i = -e\mathbf{R}, \tag{2.135}$$

we see that the interaction Hamiltonian (2.134) can also be written in the form

$$H_{\text{int}}^{\text{L}}(t) = -\mathcal{E}(t) \cdot \mathbf{D}. \tag{2.136}$$

We also remark that, for a monochromatic field treated in the dipole approximation, the interaction Hamiltonian can be written in the velocity or length gauges in the form

$$H_{\text{int}}(t) = H_+ \exp(-i\omega t) + H_- \exp(i\omega t), \tag{2.137}$$

where H_+ and H_- are time-independent operators and $H_- = H_+^\dagger$. As we shall see in Chapter 3, the first term on the right-hand side of Equation (2.137) can be interpreted as being responsible for the *absorption* of a photon, while the second term corresponds to the *emission* of a photon. For instance, for a linearly polarized, spatially homogeneous and monochromatic laser field of the form (2.118), we have

$$H_+ = H_-^\dagger = \frac{e\mathcal{E}_0}{2} \hat{\boldsymbol{\epsilon}} \cdot \mathbf{R} \exp(-i\varphi) \tag{2.138}$$

in the length gauge and

$$H_+ = H_-^\dagger = \frac{e\mathcal{E}_0}{2m\omega i} \hat{\boldsymbol{\epsilon}} \cdot \mathbf{P} \exp(-i\varphi) \tag{2.139}$$

in the velocity gauge.

We will refer to Equation (2.121) as the original (untransformed) TDSE.

2.5 The Kramers transformation

As we shall see later, in the high-intensity and high-frequency regime it is useful to study the interaction of an atomic system with a laser field in an accelerated frame (A), called the *Kramers–Henneberger* (K–H) frame. Starting from the TDSE (2.127) in the dipole approximation and in the velocity gauge, we perform the unitary transformation

$$\Psi^{\text{A}}(X, t) = \exp\left[\frac{i}{\hbar}\boldsymbol{\alpha}(t) \cdot \mathbf{P}\right] \Psi^{\text{V}}(X, t), \tag{2.140}$$

where $\boldsymbol{\alpha}(t)$ is the displacement vector of a classical electron, given by Equation (2.113).

The K–H transformation (2.140) therefore corresponds to a spatial translation, characterized by the vector $\boldsymbol{\alpha}(t)$, to a new frame (the K–H frame) oscillating with respect to the laboratory frame in the same way as a "classical" electron in the electric field $\mathcal{E}(t)$. In this accelerated K–H frame, the new TDSE for the wave function $\Psi^A(X, t)$ is

$$i\hbar \frac{\partial}{\partial t} \Psi^A(X, t) = \left[\frac{1}{2m} \sum_{i=1}^N \mathbf{p}_i^2 + V[\mathbf{r}_1 + \boldsymbol{\alpha}(t), \dots, \mathbf{r}_N + \boldsymbol{\alpha}(t)] \right] \Psi^A(X, t), \quad (2.141)$$

so that the interaction with the laser field is now incorporated via $\boldsymbol{\alpha}(t)$ into the potential V, which becomes time-dependent.

The space-translated form of the TDSE was discovered by Pauli and Fierz [29] and used by Kramers [30] to study the renormalization of quantum electrodynamics (QED). It was rediscovered by Henneberger [31] within the framework of the semi-classical theory of laser–atom interactions.

Due to the space translation characterized by the vector $\boldsymbol{\alpha}(t)$, we can interpret the new TDSE (2.141) as describing the dynamics of the electrons in a moving frame of reference (the K–H frame) that follows the classical quiver motion $\boldsymbol{\alpha}(t)$ of the electrons. We note that the center of force, the atomic nucleus, which is fixed in the laboratory frame, has a quiver motion $-\boldsymbol{\alpha}(t)$ in the K–H frame, as indicated by the shift of origin of the potential in Equation (2.141).

2.6 The evolution operator

In this section, we give a brief review of the *evolution operator*. As we will see in subsequent chapters, several methods of solution of the TDSE rely on obtaining approximate expressions for the evolution operator.

When discussing the evolution operator, it is convenient to make use of Dirac's bra and ket notation. The wave function $\Psi(X, t_0)$ at time t_0 is related to the state vector $|\Psi(t_0)\rangle$ of the system at that time according to

$$\Psi(X, t_0) \equiv \langle X | \Psi(t_0) \rangle. \quad (2.142)$$

Let us now introduce the evolution operator in the Schrödinger picture, $U(t, t_0)$, such that

$$|\Psi(t)\rangle = U(t, t_0) |\Psi(t_0)\rangle, \quad (2.143)$$

so that when acting on the state vector $|\Psi(t_0)\rangle$ it yields a new state vector $|\Psi(t)\rangle$. The evolution operator satisfies the initial condition

$$U(t_0, t_0) = I, \quad (2.144)$$

where I is the unit operator. Applying twice the definition (2.143), we also obtain

$$U(t, t_0) = U(t, t_1) U(t_1, t_0), \quad (2.145)$$

and from Equations (2.144) and (2.145) we also have

$$U^{-1}(t, t_0) = U(t_0, t), \tag{2.146}$$

so that the evolution operator exhibits the group properties. In addition, conservation of probability requires that the norm of the state vector be conserved, so that

$$U^{\dagger}(t, t_0)U(t, t_0) = U(t, t_0)U^{\dagger}(t, t_0) = I \tag{2.147}$$

and hence $U(t, t_0)$ is a unitary operator. It follows from Equations (2.146) and (2.147) that

$$U^{\dagger}(t, t_0) = U^{-1}(t, t_0) = U(t_0, t). \tag{2.148}$$

The state vector $|\Psi(t)\rangle$ is a solution of the TDSE equation

$$i\hbar \frac{\partial}{\partial t}|\Psi(t)\rangle = H(t)|\Psi(t)\rangle, \tag{2.149}$$

and it follows from Equations (2.143) and (2.149) that the evolution operator $U(t, t_0)$ satisfies the equation

$$i\hbar \frac{\partial}{\partial t}U(t, t_0) = H(t)U(t, t_0). \tag{2.150}$$

The differential equation (2.150) together with the initial condition (2.144) can be replaced by an integral equation of the Volterra type, namely

$$U(t, t_0) = I - \frac{i}{\hbar}\int_{t_0}^{t} H(t_1)U(t_1, t_0)dt_1. \tag{2.151}$$

Making use of Equation (2.148), we see that $U(t, t_0)$ also satisfies the equation

$$U(t, t_0) = I - \frac{i}{\hbar}\int_{t_0}^{t} U(t, t_1)H(t_1)dt_1. \tag{2.152}$$

The evolution operator is therefore entirely determined by the Hamiltonian $H(t)$ of the system. It is clear from the above equations that solving the TDSE is equivalent to obtaining the evolution operator $U(t, t_0)$.

It is also convenient to introduce *propagators*, which will allow us to construct solutions of the TDSE satisfying time-retarded (causal) or time-advanced (anti-causal) boundary conditions. These operators, $K^{(\pm)}(t, t_0)$, are defined by the equations

$$\left[i\hbar \frac{\partial}{\partial t} - H(t)\right]K^{(\pm)}(t, t_0) = \delta(t - t_0) \tag{2.153}$$

with the initial condition

$$K^{(+)}(t, t_0) = 0 \quad t < t_0 \tag{2.154}$$

for the time-retarded (causal) propagator $K^{(+)}(t, t_0)$ and

$$K^{(-)}(t, t_0) = 0 \quad t > t_0 \tag{2.155}$$

for the time-advanced (anti-causal) propagator $K^{(-)}(t, t_0)$. Comparing Equations (2.153) and (2.150), we see that the time-retarded propagator is given by

$$K^{(+)}(t, t_0) = (i\hbar)^{-1} \Theta(t - t_0) U(t, t_0), \tag{2.156}$$

where $\Theta(x)$ is the Heaviside step function [32] such that

$$\Theta(x) = \begin{cases} 1, & x > 0 \\ \frac{1}{2}, & x = 0 \\ 0, & x < 0 \end{cases} \tag{2.157}$$

and its "derivative" is defined as

$$\lim_{\epsilon \to 0^+} \frac{\Theta(x + \epsilon) - \Theta(x)}{\epsilon} = \delta(x). \tag{2.158}$$

Thus, from Equations (2.143) and (2.156),

$$|\Psi(t)\rangle = i\hbar K^{(+)}(t, t_0) |\Psi(t_0)\rangle \tag{2.159}$$

for $t > t_0$, while $|\Psi(t)\rangle$ vanishes for $t < t_0$. Hence, in order to ensure that the state $|\Psi(t)\rangle$ can only be determined from a state $|\Psi(t_0)\rangle$ with $t > t_0$, so that the evolution of the state is causal, the time-retarded propagator $K^{(+)}$ must be employed. Similarly, the time-advanced propagator can be expressed as

$$K^{(-)}(t, t_0) = -(i\hbar)^{-1} \Theta(t_0 - t) U(t, t_0). \tag{2.160}$$

A detailed account of the relationship between ingoing and outgoing wave boundary conditions and causality has been given by Joachain [33].

2.7 Gordon–Volkov wave functions

Let us consider the simple case of a "free" electron in the presence of a laser field described in the dipole approximation by the vector potential $\mathbf{A}(t)$. The electron motion is then governed by the TDSE

$$i\hbar \frac{\partial}{\partial t} \chi(\mathbf{r}, t) = H_F(t) \chi(\mathbf{r}, t)$$

$$= \frac{1}{2m} [\mathbf{p} + e\mathbf{A}(t)]^2 \chi(\mathbf{r}, t), \tag{2.161}$$

where $H_F(t)$ denotes the Hamiltonian of the free electron in the laser field. It is convenient to perform the gauge transformation

$$\chi^V(\mathbf{r}, t) = \exp\left[\frac{i}{\hbar} \frac{e^2}{2m} \int_{-\infty}^{t} \mathbf{A}^2(t') dt'\right] \chi(\mathbf{r}, t), \tag{2.162}$$

which gives, for $\chi^V(\mathbf{r}, t)$, the TDSE in the velocity gauge

$$i\hbar\frac{\partial}{\partial t}\chi^V(\mathbf{r}, t) = H_F^V(t)\chi^V(\mathbf{r}, t)$$

$$= \left[\frac{\mathbf{p}^2}{2m} + \frac{e}{m}\mathbf{A}(t)\cdot\mathbf{p}\right]\chi^V(\mathbf{r}, t). \tag{2.163}$$

The Hamiltonian $H_F^V(t)$ commutes with the operator $\mathbf{p} = -i\hbar\nabla$, and since $\exp(i\mathbf{k}\cdot\mathbf{r})$ is an eigenfunction of $\mathbf{p}$ corresponding to the eigenvalue $\hbar\mathbf{k}$, we can look for solutions of Equation (2.161) having the form

$$\chi_\mathbf{k}^V(\mathbf{r}, t) = (2\pi)^{-3/2}\exp(i\mathbf{k}\cdot\mathbf{r})f_\mathbf{k}(t). \tag{2.164}$$

Substituting Equation (2.164) into Equation (2.163), we obtain for the function $f_\mathbf{k}(t)$ the first-order differential equation

$$i\hbar\frac{\mathrm{d}}{\mathrm{d}t}f_\mathbf{k}(t) = \left[\frac{\hbar^2k^2}{2m} + \frac{e\hbar}{m}\mathbf{k}\cdot\mathbf{A}(t)\right]f_\mathbf{k}(t), \tag{2.165}$$

which is readily solved to give

$$f_\mathbf{k}(t) = C\exp[-iE_kt/\hbar - i\mathbf{k}\cdot\boldsymbol{\alpha}(t)]. \tag{2.166}$$

In this equation, $\boldsymbol{\alpha}(t)$ is given by Equation (2.113), C is a constant and $E_k = \hbar^2k^2/(2m)$ is the electron kinetic energy. Substituting Equation (2.166) into Equation (2.164), we find that the solution of Equation (2.163) is the non-relativistic Gordon–Volkov wave function [34, 35]

$$\chi_\mathbf{k}^V(\mathbf{r}, t) = (2\pi)^{-3/2}\exp\{i\mathbf{k}\cdot[\mathbf{r} - \boldsymbol{\alpha}(t)] - iE_kt/\hbar\}, \tag{2.167}$$

where we have made the choice $C = 1$ so that the wave function (2.167) is normalized according to

$$\langle\chi_{\mathbf{k}'}^V|\chi_\mathbf{k}^V\rangle = \delta(\mathbf{k} - \mathbf{k}'). \tag{2.168}$$

The Gordon–Volkov wave function (2.167) is a "dressed" free-particle wave function which describes the superposition of the quiver motion of the electron in the laser field and a drift of constant momentum $\hbar\mathbf{k}$ and energy E_k. It is the non-relativistic counterpart of the relativistic expressions derived by Gordon [34] and Volkov [35] for spinless and spin-1/2 charged particles, respectively.

Unless otherwise stated, the normalization condition (2.168) will be used. An alternative choice is the "energy normalization," which is defined by the condition

$$\langle\chi_{\mathbf{k}'}^V|\chi_\mathbf{k}^V\rangle = \delta(E_k - E_{k'})\delta(\hat{\mathbf{k}} - \hat{\mathbf{k}}'). \tag{2.169}$$

Here $\hat{\mathbf{k}}$ and $\hat{\mathbf{k}}'$ are unit vectors in the directions of $\mathbf{k}$ and $\mathbf{k}'$ and $E_{k'} = \hbar^2k'^2/(2m)$. This normalization corresponds to the choice $C = (mk/\hbar^2)^{1/2}$ in Equation (2.166).

When normalized in this way, the Gordon–Volkov wave function (2.164) takes on the form

$$\chi_{\mathbf{k}}^{V}(\mathbf{r}, t) = (2\pi)^{-3/2} \left(mk/\hbar^2\right)^{1/2} \exp\{i\mathbf{k} \cdot [\mathbf{r} - \boldsymbol{\alpha}(t)] - iE_k t/\hbar\}. \qquad (2.170)$$

Using Equations (2.162) and (2.167), the Gordon–Volkov solution $\chi_{\mathbf{k}}(\mathbf{r}, t)$ of the original (untransformed) TDSE (2.161) is given by

$$\chi_{\mathbf{k}}(\mathbf{r}, t) = (2\pi)^{-3/2} \exp\left\{i\mathbf{k} \cdot [\mathbf{r} - \boldsymbol{\alpha}(t)] - iE_k t/\hbar - \frac{i}{\hbar}\frac{e^2}{2m} \int_{-\infty}^{t} \mathbf{A}^2(t')dt'\right\},$$
$$(2.171)$$

while the Gordon–Volkov wave function in the length gauge is, from Equation (2.132),

$$\chi_{\mathbf{k}}^{L}(\mathbf{r}, t) = (2\pi)^{-3/2}$$

$$\times \exp\left\{\frac{ie}{\hbar}\mathbf{A}(t) \cdot \mathbf{r} + i\mathbf{k} \cdot [\mathbf{r} - \boldsymbol{\alpha}(t)] - iE_k t/\hbar - \frac{i}{\hbar}\frac{e^2}{2m} \int_{-\infty}^{t} \mathbf{A}^2(t')dt'\right\}$$

$$= (2\pi)^{-3/2} \exp\left\{\frac{i}{\hbar}[\hbar\mathbf{k} + e\mathbf{A}(t)] \cdot \mathbf{r} - \frac{i}{2m\hbar} \int_{-\infty}^{t} [\hbar\mathbf{k} + e\mathbf{A}(t')]^2 dt'\right\}$$
$$(2.172)$$

so that $\chi_{\mathbf{k}}^{L}(\mathbf{r}, t)$ is a particular solution of the TDSE

$$i\hbar\frac{\partial}{\partial t}\chi^{L}(\mathbf{r}, t) = H_{F}^{L}(t)\chi^{L}(\mathbf{r}, t)$$

$$= \left[\frac{\mathbf{p}^2}{2m} + e\boldsymbol{\mathcal{E}}(t) \cdot \mathbf{r}\right]\chi^{L}(\mathbf{r}, t). \qquad (2.173)$$

Finally, we note that in the K–H frame the Gordon–Volkov wave function is simply the plane wave

$$\chi_{\mathbf{k}}^{A}(\mathbf{r}, t) = (2\pi)^{-3/2} \exp[i(\mathbf{k} \cdot \mathbf{r} - E_k t/\hbar)]. \qquad (2.174)$$

The Gordon–Volkov evolution operator that transforms a Gordon–Volkov state at time t_0 to the corresponding state at time t can be readily obtained. Using Dirac's bra and ket notation, we first recall that the evolution operator for a free electron in the absence of a laser field can be written in terms of plane wave states $|\mathbf{k}\rangle$, normalized so that $\langle\mathbf{k}'|\mathbf{k}\rangle = \delta(\mathbf{k} - \mathbf{k}')$, as follows:

$$U_{\text{free}}(t, t_0) = \int d\mathbf{k}\exp[-iE_k(t - t_0)/\hbar]|\mathbf{k}\rangle\langle\mathbf{k}|. \qquad (2.175)$$

Let us denote by $U_F^V(t, t_0)$ the Gordon–Volkov evolution operator in the velocity gauge. It is a solution of the equation

$$i\hbar \frac{\partial}{\partial t} U_F^V(t, t_0) = H_F^V(t) U_F^V(t, t_0) \tag{2.176}$$

with $U_F^V(t_0, t_0) = I$. We write the operator $U_F^V(t, t_0)$ in terms of the field-free plane wave states $|\mathbf{k}\rangle$ and an unknown function $f_\mathbf{k}(t, t_0)$ as

$$U_F^V(t, t_0) = \int d\mathbf{k} \, f_\mathbf{k}(t, t_0) |\mathbf{k}\rangle\langle\mathbf{k}|. \tag{2.177}$$

Substituting this expansion into Equation (2.176), the function $f_\mathbf{k}(t, t_0)$ is seen to be a solution of the first-order differential equation (2.165) satisfied by $f_\mathbf{k}(t)$ with the initial condition $f_\mathbf{k}(t_0, t_0) = 1$. Therefore,

$$f_\mathbf{k}(t, t_0) = \exp\{-iE_k(t - t_0)/\hbar - i\mathbf{k} \cdot [\boldsymbol{\alpha}(t) - \boldsymbol{\alpha}(t_0)]\}. \tag{2.178}$$

The Gordon–Volkov evolution operator in the velocity gauge $U_F^V(t, t_0)$ can be obtained from the field-free evolution operator (2.175) by replacing the plane wave states $|\mathbf{k}\rangle \exp(-iE_k t/\hbar)$ and $\exp(iE_k t_0/\hbar)\langle\mathbf{k}|$ with Gordon–Volkov states in the velocity gauge, namely

$$U_F^V(t, t_0) = \int d\mathbf{k} |\chi_\mathbf{k}^V(t)\rangle\langle\chi_\mathbf{k}^V(t_0)|. \tag{2.179}$$

Similarly, we obtain the Gordon–Volkov evolution operator in the length gauge as

$$U_F^L(t, t_0) = \int d\mathbf{k} |\chi_\mathbf{k}^L(t)\rangle\langle\chi_\mathbf{k}^L(t_0)|. \tag{2.180}$$

We can also define the Gordon–Volkov evolution operator corresponding to the original Hamiltonian $H_F(t)$ in the untransformed TDSE (2.161). This operator is given by

$$U_F(t, t_0) = \int d\mathbf{k} |\chi_\mathbf{k}(t)\rangle\langle\chi_\mathbf{k}(t_0)|. \tag{2.181}$$

The Gordon–Volkov wave functions have a number of applications in multiphoton physics. For example, they will be used in Chapter 6 within the low-frequency strong-field approximation to calculate approximate ionization rates. Moreover, they play a key role in describing laser-assisted electron–atom collisions, as will be shown in Chapter 10.

2.8 Non-dipole effects

The dipole approximation introduced in Section 2.3 can be used to study atomic multiphoton processes for a wide range of angular frequencies and intensities of the laser field. However, as is well known, in the low-intensity regime, at high angular frequencies, this approximation is no longer adequate, so that *non-dipole* effects give rise to magnetic dipole and electric quadrupole transitions [36]. This is the case, for example, of X-ray radiation emitted or absorbed in "bound–bound" transitions involving the inner electrons of atomic systems with large Z, or in the photoionization of atoms by high-frequency photons.

When the intensity becomes sufficiently high, non-dipole effects due to the magnetic-field component of the laser field, and eventually *relativistic* effects, must be taken into account [28, 37–39]. As will be shown below, the onset of non-dipole as well as relativistic effects depends both on the intensity and on the angular frequency of the laser field.

In this section, we shall consider non-dipole effects that arise for high laser intensities, while relativistic effects will be treated in Section 2.9. For simplicity, only one-electron atoms will be discussed.

2.8.1 Classical electron dynamics beyond the dipole approximation

Let us consider a classical electron interacting with a laser pulse that is linearly polarized along $\hat{\mathbf{x}}$ and propagating along $\hat{\mathbf{z}}$ with wave vector $\mathbf{k}_L = k_L\hat{\mathbf{z}}$. We describe the laser pulse by the vector potential

$$\mathbf{A}(\eta) = \hat{\mathbf{x}}A(\eta), \tag{2.182}$$

where, from Equation (2.51), $\eta = \omega t - \mathbf{k}_L \cdot \mathbf{r} = \omega(t - z/c)$. The corresponding electric and magnetic fields are given by

$$\boldsymbol{\mathcal{E}}(\eta) = -\frac{\partial}{\partial t}\mathbf{A}(\eta) = \hat{\mathbf{x}}\mathcal{E}(\eta), \tag{2.183}$$

and

$$\boldsymbol{\mathcal{B}}(\eta) = \nabla \times \mathbf{A}(\eta) = \hat{\mathbf{y}}\mathcal{B}(\eta), \tag{2.184}$$

and we recall that $|\boldsymbol{\mathcal{B}}|/|\boldsymbol{\mathcal{E}}| = 1/c$. The Lorentz equation (2.69) for an electron in this laser pulse is

$$\frac{d}{dt}\mathbf{p}_{cl} = -e[\boldsymbol{\mathcal{E}}(\eta) + \mathbf{v} \times \boldsymbol{\mathcal{B}}(\eta)]. \tag{2.185}$$

In the non-relativistic, long-wavelength limit, this equation reduces to Equation (2.71), which in the present case can be written as

$$\frac{d}{dt}v_x = -\frac{e}{m}\mathcal{E}(\omega t), \tag{2.186}$$

where $\mathcal{E}(\omega t)$ is a spatially homogeneous electric field.

Including terms to order $1/c$ in Equation (2.185), the electron dynamics is found to be two-dimensional in the x-z plane, with the motion along the polarization axis $\hat{\mathbf{x}}$ given by Equation (2.186), while the motion along the propagation direction $\hat{\mathbf{k}}_L$ is governed by the equation

$$\frac{d}{dt} v_z = -\frac{e}{m} v_x \mathcal{B}(\omega t). \tag{2.187}$$

We note that, at this level of approximation, the magnetic field $\mathcal{B}$ is also spatially homogeneous. The motion along $\hat{\mathbf{x}}$ is unaffected by the magnetic field. For an electron initially at rest at the origin, it follows from Equation (2.186) that the displacement along the polarization direction $\hat{\mathbf{x}}$ is

$$x(t) = \alpha(t) = \frac{e}{m} \int_{-\infty}^{t} A(\omega t') dt', \tag{2.188}$$

where $\alpha(t)$ is the displacement of the classical electron from its oscillation center due to its quiver motion in the electric field $\mathcal{E}(\omega t)$.

Using Equation (2.187), we see that the electron displacement along the propagation direction $\hat{\mathbf{z}}$, originating from the magnetic component of the laser pulse, is given by

$$z(t) = \frac{1}{2c} \int_{-\infty}^{t} v_x^2(t') dt' \tag{2.189}$$

and increases monotonically during the pulse.

The motion of the electron along the propagation direction $\hat{\mathbf{z}}$ consists of a positive drift velocity that increases from zero to a maximum value at the pulse peak intensity, and then decreases again to zero at the end of the pulse. Superimposed on this drift motion, the electron oscillates at an angular frequency that is equal to twice the carrier angular frequency of the pulse. Therefore, for a linearly polarized laser pulse, the electron's motion is two-dimensional. In a reference frame in which the electron is on average at rest, the electron undergoes a "figure-of-eight" motion in the x-z plane, with the major axis of the eight along the polarization direction $\hat{\mathbf{x}}$. For elliptically, and in particularly circularly, polarized light the electron motion is three-dimensional.

As an example, let us consider a linearly polarized step pulse of electric-field strength $\mathcal{E}_0$ arriving at the origin at $t = 0$. We then have, for $t > 0$,

$$x(t) = \alpha(t) = \alpha_0 \sin(\omega t), \tag{2.190}$$

where the electron quiver amplitude α_0 is given by Equation (2.79) and

$$z(t) = \frac{e^2 \mathcal{E}_0^2}{2m^2 c \omega^2} \left[\frac{t}{2} + \frac{\sin(2\omega t)}{4\omega} \right]. \tag{2.191}$$

The drift per cycle in the propagation direction is

$$\chi_0 = \frac{\pi e^2 \mathcal{E}_0^2}{2m^2 c \omega^3}. \tag{2.192}$$

For an angular frequency $\omega = 0.057$ a.u. (corresponding to a Ti:sapphire laser of wavelength $\lambda = 800$ nm), we have $\chi_0 \simeq 1$ a.u. when the intensity $I \simeq 10^{15}$ W cm^{-2}, while for $\omega = 1$ a.u. one has $\chi_0 \simeq 1$ a.u. when the intensity $I \simeq 5 \times 10^{18}$ W cm^{-2}. Classically, the magnetic-field component of a laser pulse should therefore be expected to play a role if the peak intensity exceeds these estimates. However, for some processes, such as harmonic generation by ions driven by a laser pulse with a low carrier angular frequency, we shall see in Section 9.2 that the spreading of the electron wave packet over an optical cycle is considerable, which means that higher intensities are required for non-dipole effects to be significant. For example, in the case of a Ti:sapphire laser pulse, intensities of the order of 10^{17} W cm^{-2} are needed for non-dipole effects to play a role, instead of the estimate of 10^{15} W cm^{-2} quoted above.

2.8.2 Quantum treatment

The non-relativistic, non-dipole TDSE for a one-electron atom (ion) in a laser field is given by

$$i\hbar \frac{\partial}{\partial t} \Psi(\mathbf{r}, t) = \left\{ \frac{1}{2m} [-i\hbar \nabla + e\mathbf{A}(\mathbf{r}, t)]^2 + V(r) \right\} \Psi(\mathbf{r}, t), \tag{2.193}$$

where $V(r)$ is the potential binding the electron.

Letting the laser field propagation direction be along $\hat{\mathbf{z}}$ and neglecting dispersion, so that $\mathbf{A}(\mathbf{r}, t) = \mathbf{A}(\eta)$, we have, to lowest order in $1/c$,

$$\mathbf{A}(\eta) \simeq \mathbf{A}(t) + (\mathbf{r} \cdot \nabla)\mathbf{A}(\eta)\big|_{\mathbf{r}=0}$$

$$= \mathbf{A}(t) + \frac{z}{c} \mathcal{E}(t). \tag{2.194}$$

Transforming to the "length gauge" by writing

$$\Psi^L(\mathbf{r}, t) = \exp\left[\frac{ie}{\hbar} \mathbf{A}(t) \cdot \mathbf{r} \right] \Psi(\mathbf{r}, t), \tag{2.195}$$

we obtain for $\Psi^L(\mathbf{r}, t)$ the non-dipole TDSE

$$i\hbar \frac{\partial}{\partial t} \Psi^L(\mathbf{r}, t) = \left\{ -\frac{\hbar^2}{2m} \nabla^2 + e\left[\mathbf{r} - i\frac{\hbar z}{mc} \nabla \right] \cdot \mathcal{E}(t) + V(r) \right\} \Psi^L(\mathbf{r}, t). \tag{2.196}$$

This equation can be written in the form

$$i\hbar \frac{\partial}{\partial t} \Psi^L(\mathbf{r}, t) = H(t)\Psi^L(\mathbf{r}, t), \tag{2.197}$$

where $H(t) = H_0 + H_{\text{int}}^{\text{ND}}(t)$, $H_0 = -\hbar^2 \nabla^2/2m + V(r)$ is the field-free Hamiltonian and

$$H_{\text{int}}^{\text{ND}}(t) = e\left[\mathbf{r} - i\frac{\hbar z}{mc}\nabla\right] \cdot \boldsymbol{\mathcal{E}}(t) \tag{2.198}$$

is the non-dipole interaction Hamiltonian.

An exact solution of Equation (2.197) can be obtained when $V(r) = 0$. The corresponding non-relativistic, non-dipole Gordon–Volkov wave function is given by [40]

$$\chi_{\mathbf{k}}^{\text{L, ND}}(\mathbf{r}, t) = (2\pi)^{-3/2} \exp\left(\frac{i}{\hbar}\boldsymbol{\pi}(\mathbf{k}, t) \cdot \mathbf{r} - \frac{i}{2m\hbar}\int_{-\infty}^{t}[\boldsymbol{\pi}(\mathbf{k}, t')]^2 dt'\right), \tag{2.199}$$

where

$$\boldsymbol{\pi}(\mathbf{k}, t) = \hbar\mathbf{k} + e\mathbf{A}(t) + \frac{e}{mc}\left[\hbar\mathbf{k} \cdot \mathbf{A}(t) + \frac{e}{2}\mathbf{A}^2(t)\right]\hat{\mathbf{z}}. \tag{2.200}$$

The wave function (2.199) describes the dynamics of a free electron interacting with a laser pulse in the intermediate intensity regime between the non-relativistic region (where non-dipole effects can be neglected) and the fully relativistic regime. This wave function reduces to the dipole Gordon–Volkov wave function in the length gauge (2.172) when $1/c \to 0$.

2.9 Relativistic effects

Relativistic effects must be taken into account when describing target atoms or ions with high atomic number Z, where spin-orbit effects are important, as well as relativistic corrections to the atomic electrons' kinetic and potential energy terms [41].

Relativistic effects also occur when atoms or ions interact with ultra-strong laser fields. In this case, they are expected to become important when the electron's ponderomotive energy $U_p = e^2\mathcal{E}_0^2/(4m\omega^2)$ is of the order of its rest mass energy mc^2, so that the quantity

$$q = \frac{U_p}{mc^2} = \frac{e^2\mathcal{E}_0^2}{4m^2\omega^2c^2} \tag{2.201}$$

is of the order of unity. Using atomic units,

$$q = 1.33 \times 10^{-5}\frac{\mathcal{E}_0^2}{\omega^2}. \tag{2.202}$$

Thus, if $\omega = 0.057$ a.u. (corresponding to a Ti:sapphire laser), we see that $q = 1$ when $\mathcal{E}_0 = 15.6$ a.u., that is, when the intensity $I = 8.6 \times 10^{18}\,\text{W cm}^{-2}$. For a laser of higher angular frequency $\omega = 1$ a.u., we have $q = 1$ when $\mathcal{E}_0 = 274$ a.u., corresponding to the very large intensity $I = 2.6 \times 10^{21}\,\text{W cm}^{-2}$.

2.9.1 Classical relativistic electron dynamics

The relativistic relationship between the energy E and the momentum $\mathbf{p}$ of a free particle of mass m is given by

$$E = \left(m^2 c^4 + c^2 \mathbf{p}^2\right)^{1/2}.$$

(2.203)

In terms of the velocity $\mathbf{v}$ of the particle, we have

$$\mathbf{p} = \gamma m \mathbf{v}$$

(2.204)

and

$$E = \gamma m c^2,$$

(2.205)

where γ is the Lorentz factor defined by Equation (2.70), and we note that

$$v = \frac{\mathrm{d}E}{\mathrm{d}p} = \frac{c^2 p}{E}.$$

(2.206)

Relativistic phenomena are described within the framework of space-time, with an "event," or world point, being specified by the set of four space-time coordinates

$$x^\mu \equiv (x^0, x^1, x^2, x^3), \qquad \mu = 0, 1, 2, 3,$$

(2.207)

where $x^0 = ct$, $x^1 = x$, $x^2 = y$ and $x^3 = z$. We shall also write the four-vector x^μ as

$$x^\mu \equiv (x^0, \mathbf{x})$$

(2.208)

where $\mathbf{x} \equiv (x^1, x^2, x^3)$ is an ordinary three-component vector.

The metric tensor $g_{\mu\nu}$ is such that

$$g_{00} = 1, \quad g_{11} = g_{22} = g_{33} = -1, \quad g_{\mu\nu} = 0, \quad \mu \neq \nu.$$

(2.209)

If a^μ is a contravariant vector (transforming like x^μ), the corresponding covariant vector (which transforms like $\partial/\partial x^\mu$) is given by

$$a_\mu = g_{\mu\nu} a^\nu,$$

(2.210)

where a summation over repeated indices is implied. We have

$$g_{\mu\nu} = g^{\mu\nu}.$$

(2.211)

We also define the covariant gradient operator as

$$\partial_\mu \equiv \frac{\partial}{\partial x^\mu} \equiv \left(\frac{\partial}{\partial x^0}, \frac{\partial}{\partial x^1}, \frac{\partial}{\partial x^2}, \frac{\partial}{\partial x^3} \right) \equiv \left(\frac{\partial}{\partial ct}, \nabla \right)$$

(2.212)

and the contravariant gradient operator as

$$\partial^{\mu} \equiv g^{\mu\nu}\partial_{\nu} \equiv \left(\frac{\partial}{\partial ct}, -\nabla\right). \tag{2.213}$$

In the presence of a laser field, the trajectory of an electron is obtained by solving the Lorentz equation (2.69). In order to cast this equation in covariant form, we first observe that we can write

$$m\frac{d\gamma}{dt} = -\frac{e}{c^2}\mathbf{v}\cdot\mathcal{E}. \tag{2.214}$$

Introducing the proper time τ of the electron according to the definition

$$d\tau = \left(dx^{\mu}dx_{\mu}\right)^{1/2} = dt/\gamma \tag{2.215}$$

and defining the four-velocity

$$w^{\mu} \equiv \frac{dx^{\mu}}{d\tau} = \gamma(c, \mathbf{v}), \quad w^{\mu}w_{\mu} = 1, \tag{2.216}$$

we see that Equations (2.69) and (2.214) can be expressed in covariant form as

$$\frac{dw^{\mu}}{d\tau} = -\frac{e}{m}F^{\mu\nu}w_{\nu}, \tag{2.217}$$

where $F^{\mu\nu}$ is the antisymmetric tensor

$$F^{\mu\nu} = \partial^{\mu}A^{\nu} - \partial^{\nu}A^{\mu}, \tag{2.218}$$

with

$$A^{\mu} \equiv (\phi, \mathbf{A}). \tag{2.219}$$

For an electron in an electromagnetic field described by a plane wave, the equations (2.217) can be formally solved [28,42]. The relativistic electron dynamics is similar to the non-dipole, non-relativistic dynamics discussed above. In particular, for a linearly polarized laser field, the trajectory followed by an electron being initially at rest is a "figure-of-eight" in the reference frame in which the electron is on average at rest, while in the laboratory frame the electron exhibits a drift motion along the laser propagation direction. An important relativistic effect is the *relativistic mass shift*, or "dressing of the electron mass," which limits the electron quiver motion along the laser polarization direction and its drift along the laser propagation direction due to the fact that the velocity of light c is finite. Averaging over one laser cycle, one finds that the corresponding "dressed mass" is given by

$$m^* = m\,(1+2q)^{1/2}, \tag{2.220}$$

where the quantity q is given by Equation (2.201). From this equation, it follows that the dressed mass increases with the intensity of the laser field and decreases with its angular frequency.

2.9.2 The relativistic Schrödinger and Klein–Gordon equations

We shall now discuss relativistic wave equations for particles moving in given external fields. In general, when relativistic effects are important, particle creation and annihilation must also be taken into account so that a quantum field theory description of these phenomena is needed. Nevertheless, the relativistic wave equations are very useful in studying the interaction of atoms with very strong laser fields.

We begin by considering the *relativistic Schrödinger equation* and the *Klein–Gordon equation*, which are relevant for studying the relativistic dynamics of spinless particles. Because such particles have no internal degrees of freedom, their wave function Ψ in configuration space can only depend on the variables $\mathbf{r}$ and t. In the present context, the relativistic Schrödinger and Klein–Gordon equations can be used to study the relativistic dynamics of electrons if spin effects are neglected.

Using the correspondence rule

$$E \to i\hbar \frac{\partial}{\partial t}, \quad \mathbf{p} \to -i\hbar\nabla, \tag{2.221}$$

and letting both sides of Equation (2.203) operate on the wave function $\Psi(\mathbf{r}, t)$, we obtain

$$i\hbar \frac{\partial}{\partial t} \Psi(\mathbf{r}, t) = \left(m^2 c^4 - \hbar^2 c^2 \nabla^2 \right)^{1/2} \Psi(\mathbf{r}, t), \tag{2.222}$$

which we shall call the *relativistic Schrödinger equation for a free particle*. This equation has serious drawbacks. Firstly, we must interpret the square-root operator on the right-hand side. By expanding it in a power series, we obtain a differential operator of infinite order, which is difficult to handle. Secondly, the resulting wave equation has the unattractive feature that the space and time coordinates appear in an unsymmetrical way, so that relativistic invariance is not clearly exhibited.

To avoid these difficulties, we remove the square root by starting from the classical relationship

$$E^2 = m^2 c^4 + c^2 \mathbf{p}^2 \tag{2.223}$$

obtained by squaring the energy–momentum relation (2.203). Post-multiplying both sides of the above equation by the wave function $\Psi(\mathbf{r}, t)$ and making the substitutions (2.221), we obtain the *Klein–Gordon equation* for a free particle, i.e.

$$-\hbar^2 \frac{\partial^2}{\partial t^2} \Psi(\mathbf{r}, t) = \left(m^2 c^4 - c^2 \hbar^2 \nabla^2 \right) \Psi(\mathbf{r}, t). \tag{2.224}$$

It is worth noting that this is a second-order differential equation with respect to time, in contrast to the non-relativistic and relativistic Schrödinger equations, which are of first-order in $\partial/\partial t$. Equation (2.224) has plane wave solutions of the form

$$\Psi_{\mathbf{k}}(\mathbf{r}, t) = (2\pi)^{-3/2} \exp(i\mathbf{k} \cdot \mathbf{r} - iE_{\pm}t/\hbar), \tag{2.225}$$

the normalization constant being chosen so that

$$\langle \Psi_{\mathbf{k}'} | \Psi_{\mathbf{k}} \rangle = \delta(\mathbf{k} - \mathbf{k}'). \tag{2.226}$$

The plane waves (2.225) are also eigenfunctions of the operators $i\hbar\partial/\partial t$ and $-i\hbar\nabla$ with eigenvalues $E_\pm$ and $\hbar\mathbf{k}$, respectively, and

$$E_\pm = \pm(m^2c^4 + \hbar^2 k^2 c^2)^{1/2}. \tag{2.227}$$

Let us now consider a spinless particle having an electric charge q, which is moving in an electromagnetic field described by a vector potential $\mathbf{A}(\mathbf{r}, t)$ and a scalar potential $\phi(\mathbf{r}, t)$. In particular, we want to study the relativistic dynamics of electrons, for which $q = -e$, neglecting spin effects. Starting from the relativistic relationship (2.203) for a free particle, we make the replacements

$$E \to E + e\phi, \quad \mathbf{p} \to \mathbf{p} + e\mathbf{A}, \tag{2.228}$$

which yields the classical relationship

$$E + e\phi = \left[m^2 c^4 + c^2 (\mathbf{p} + e\mathbf{A})^2 \right]^{1/2}. \tag{2.229}$$

Using the correspondence rule (2.221), we then obtain the relativistic Schrödinger equation for a spinless electron in an electromagnetic field, namely

$$\left(i\hbar\frac{\partial}{\partial t} + e\phi \right) \Psi(\mathbf{r}, t) = \left[m^2 c^4 + c^2 (-i\hbar\nabla + e\mathbf{A})^2 \right]^{1/2} \Psi(\mathbf{r}, t)$$

$$= W\Psi(\mathbf{r}, t). \tag{2.230}$$

By squaring both sides of the classical relation (2.229), and proceeding in a similar manner, the Klein–Gordon equation for a spinless electron in an electromagnetic field is obtained. That is,

$$\left(i\hbar\frac{\partial}{\partial t} + e\phi \right)^2 \Psi(\mathbf{r}, t) = \left[m^2 c^4 + c^2 (-i\hbar\nabla + e\mathbf{A})^2 \right] \Psi(\mathbf{r}, t)$$

$$= W^2\Psi(\mathbf{r}, t). \tag{2.231}$$

While in principle it is important to distinguish between Equations (2.230) and (2.231), in the applications that will be considered here this distinction is not important. To see why this is the case, let us consider a solution Ψ of the relativistic Schrödinger equation (2.230). Pre-multiplying both sides of this equation by $(W + e\phi)$, we find that

$$(W + e\phi)i\hbar\frac{\partial}{\partial t}\Psi(\mathbf{r}, t) = \left(W^2 + [e\phi, W] - e^2\phi^2 \right)\Psi(\mathbf{r}, t) \tag{2.232}$$

or

$$\left(i\hbar\frac{\partial}{\partial t}+e\phi\right)^2\Psi(\mathbf{r},t)=\left[W^2+[e\phi,W]+i\hbar\left(\frac{\partial W}{\partial t}\right)\right]\Psi(\mathbf{r},t). \qquad (2.233)$$

Let us turn to the particular case for which the electron is moving in a static potential $-e\phi$. In the laser frequency and intensity regime where a single-particle relativistic description is still valid, that is when the probability of pair production in the electromagnetic field is small, the term containing the commutator $[e\phi, W]$ can be neglected. If the last term on the right is also small, the wave function Ψ is then seen to be an approximate solution of the Klein–Gordon equation.

We now investigate the non-relativistic limit of the Klein–Gordon equation (2.231). To this end, we introduce the new wave function $\tilde{\Psi}(\mathbf{r},t)$, which is related to $\Psi(\mathbf{r},t)$ by

$$\tilde{\Psi}(\mathbf{r},t)=\exp(imc^2t/\hbar)\Psi(\mathbf{r},t). \qquad (2.234)$$

This wave function is a solution of the equation

$$-\hbar^2\frac{\partial^2\tilde{\Psi}}{\partial t^2}+2i\hbar(mc^2+e\phi)\frac{\partial\tilde{\Psi}}{\partial t}+\left[e\phi(2mc^2+e\phi)+i\hbar e\frac{\partial\phi}{\partial t}\right]\tilde{\Psi}$$
$$=c^2\left[-\hbar^2\nabla^2-2i\hbar e\mathbf{A}\cdot\nabla-i\hbar e(\nabla\cdot\mathbf{A})+e^2\mathbf{A}^2\right]\tilde{\Psi}. \qquad (2.235)$$

In the limit in which $|e\phi|\ll mc^2$, $|[\hbar/(2mc^2)]\partial\phi/\partial t|\ll|\phi|$ and $|[\hbar/(2mc^2)]\partial^2\tilde{\Psi}/\partial t^2|\ll|\partial\tilde{\Psi}/\partial t|$, this equation reduces to

$$i\hbar\frac{\partial}{\partial t}\tilde{\Psi}(\mathbf{r},t)=\left[-\frac{\hbar^2}{2m}\nabla^2-i\hbar\frac{e}{2m}(\mathbf{A}\cdot\nabla+\nabla\cdot\mathbf{A})+\frac{e^2}{2m}\mathbf{A}^2-e\phi\right]\tilde{\Psi}(\mathbf{r},t),$$
$$(2.236)$$

which is the non-relativistic Schrödinger equation (2.97) for a spinless electron in an electromagnetic field.

Let us now consider the Klein–Gordon equation (2.231) for the particular case of a "free" spinless electron interacting with a laser field described by the vector potential $\mathbf{A}(\eta)$ in the Coulomb gauge. This equation is readily solved to obtain the Klein–Gordon–Volkov wave functions. We start by postulating a positive-energy solution of the form

$$\chi_{\mathbf{k}}^{\text{KGV}}(\mathbf{r},t)=\Psi_{\mathbf{k}}(\mathbf{r},t)f(\eta)$$
$$=(2\pi)^{-3/2}\exp(i\mathbf{k}\cdot\mathbf{r}-iE_+t/\hbar)f(\eta), \qquad (2.237)$$

where the wave function $\Psi_{\mathbf{k}}(\mathbf{r},t)$ is given in Equation (2.225). Inserting the ansatz (2.237) into the Klein–Gordon equation (2.231), the function $f(\eta)$ is seen to satisfy the ordinary differential equation

$$i\frac{d}{d\eta} f(\eta) = \frac{c^2}{2(\hbar\omega E_+ - c^2\hbar^2 \mathbf{k}_L \cdot \mathbf{k})} \left[2e\hbar\mathbf{k} \cdot \mathbf{A}(\eta) - e^2\mathbf{A}^2(\eta) \right] f(\eta). \quad (2.238)$$

This function must satisfy the initial condition $f(\eta \to -\infty) = 1$. Hence the required solution of Equation (2.238) is

$$f(\eta) = \exp\left(-i\frac{c^2}{2(\hbar\omega E_+ - c^2\hbar^2 \mathbf{k}_L \cdot \mathbf{k})} \int_{-\infty}^{\eta} d\eta' \left[2e\hbar\mathbf{k} \cdot \mathbf{A}(\eta') - e^2\mathbf{A}^2(\eta') \right] \right).$$
$$(2.239)$$

The wave function (2.237), with $f(\eta)$ given above, is both a solution of the Klein–Gordon equation (2.231) as well as the relativistic Schrödinger equation (2.230). In the limit $c \to \infty$, the non-relativistic Gordon–Volkov wave function (2.171) is recovered.

One of the difficulties encountered in dealing with the Klein–Gordon equation is the presence of *negative energy states*. To illustrate this point, let us return to the free-particle Klein–Gordon equation (2.224). We have seen that this equation has plane wave solutions of the form (2.225) with energies $E_\pm$ given by Equation (2.227). The positive and negative square roots in this equation correspond to an ambiguity in the sign of the energy which is also present in the classical relation (2.223). Thus, in order to obtain the "simple" Klein–Gordon equation (2.224), we have introduced extraneous *negative energy solutions*. As a result, the energy spectrum is no longer bounded from below, and it appears that arbitrary large amounts of energy could be extracted from the system if an external perturbation allows it to make a transition between the positive and negative energy states. For a free particle initially at rest, this would happen if the external perturbation allowed it to jump over the energy gap $\Delta E = 2mc^2$ between the positive and negative energy continua.

This difficulty was only overcome in 1934, when Pauli and Weisskopf [43] reinterpreted the Klein–Gordon equation as a field equation and quantized it using the formalism of quantum field theory. The Klein–Gordon equation then becomes a relativistic wave equation for spinless particles within the framework of a *many-particle* theory in which the negative energy states are interpreted in terms of antiparticles. It is worth noting that the number of particles is not conserved because of the possibility of *creation* and *destruction* of particle–antiparticle pairs.

In the Dirac equation, which we shall now consider, the presence of negative-energy solutions will again imply that the Dirac theory must ultimately be interpreted as a many-particle theory.

2.9.3 The Dirac equation

To describe a particle of spin 1/2, we require a wave function having two components that allow for the two spin states, the z-component of the spin angular momentum taking on the values $m_s\hbar$, where $m_s = \pm 1/2$. In the non-relativistic limit, such a two-component wave function is called a Pauli spinor wave function. However, in the relativistic case, with every spin-1/2 particle there is an associated antiparticle having the same mass and spin, but opposite charge, so that we need a four-component wave function, called a Dirac spinor wave function.

The *Dirac equation* for a free electron reads [41]

$$i\hbar\frac{\partial}{\partial t}\Psi(\mathbf{r},t) = H_D^{(0)}\,\Psi(\mathbf{r},t), \tag{2.240}$$

where the Dirac Hamiltonian $H_D^{(0)}$ is given by

$$H_D^{(0)} = c\boldsymbol{\alpha}\cdot\mathbf{p} + \beta mc^2. \tag{2.241}$$

Here $\boldsymbol{\alpha} \equiv (\alpha_x, \alpha_y, \alpha_z)$ and β are 4×4 matrices given in the Dirac representation by

$$\boldsymbol{\alpha} = \begin{pmatrix} 0 & \boldsymbol{\sigma} \\ \boldsymbol{\sigma} & 0 \end{pmatrix}, \qquad \beta = \begin{pmatrix} I & 0 \\ 0 & -I \end{pmatrix}, \tag{2.242}$$

where I is the unit 2×2 matrix and $\boldsymbol{\sigma} \equiv (\sigma_x, \sigma_y, \sigma_z)$ are the three Pauli matrices

$$\sigma_x = \begin{pmatrix} 0 & 1 \\ 1 & 0 \end{pmatrix}, \qquad \sigma_y = \begin{pmatrix} 0 & -i \\ i & 0 \end{pmatrix}, \qquad \sigma_z = \begin{pmatrix} 1 & 0 \\ 0 & -1 \end{pmatrix}. \tag{2.243}$$

The Dirac equation for an electron in an electromagnetic field described by the scalar potential $\phi(\mathbf{r}, t)$ and the vector potential $\mathbf{A}(\mathbf{r}, t)$ is

$$i\hbar\frac{\partial}{\partial t}\Psi(\mathbf{r},t) = H_D\,\Psi(\mathbf{r},t), \tag{2.244}$$

where the Dirac Hamiltonian is now

$$H_D = c\boldsymbol{\alpha}\cdot(\mathbf{p} + e\mathbf{A}) - e\phi + \beta mc^2. \tag{2.245}$$

As pointed out above, the Dirac equation, like the Klein–Gordon equation, admits negative-energy solutions. In 1930, Dirac proposed a way out of this difficulty for electrons by formulating his *hole theory*, which led to the prediction of the existence of the *positron*, the antiparticle of the electron. In fact, the hole theory involves an infinite number of particles and leaves many questions unanswered. It has been superseded by quantum field theory, in which the Dirac equation, like the Klein–Gordon equation, is interpreted as a field equation.

The Dirac equation can be solved analytically for the case of an electron in an electromagnetic field [34, 35], as we will now show. In order to simplify the calculations, we first express the Dirac equation (2.244) in covariant form. To this end, we introduce the matrices

$$\gamma^\mu \equiv (\gamma^0, \boldsymbol{\gamma}) \equiv (\beta, \beta\boldsymbol{\alpha}) \tag{2.246}$$

and pre-multiply both sides of Equation (2.244) by $c^{-1}\gamma^0$, with the result

$$\left[c^{-1}\gamma^0 \left(i\hbar \frac{\partial}{\partial ct} + e\phi \right) - \boldsymbol{\gamma} \cdot (\mathbf{p} + e\mathbf{A}) - mc \right] \Psi(\mathbf{r}, t) = 0. \tag{2.247}$$

Defining the four-momentum as

$$p_\mu \equiv \left(i\hbar \frac{\partial}{\partial ct}, -\mathbf{p} \right)$$

$$\equiv i\hbar \partial_\mu, \tag{2.248}$$

using the definition (2.219), and remembering that $A_\mu \equiv (\phi, -\mathbf{A})$, allows us to define the generalized four-momentum as

$$P_\mu = p_\mu + eA_\mu$$

$$= i\hbar \partial_\mu + eA_\mu. \tag{2.249}$$

The covariant form of the Dirac equation (2.244) is then given by

$$\left(\gamma^\mu P_\mu - mc \right) \Psi(\mathbf{r}, t) = 0. \tag{2.250}$$

Let us now consider the second-order Dirac equation

$$\left(\gamma^\mu P_\mu + mc \right) \left(\gamma^\mu P_\mu - mc \right) \Psi(\mathbf{r}, t) = 0. \tag{2.251}$$

Provided that the solution of the second-order Dirac equation satisfies the initial condition that it is a solution of the (first-order) Dirac equation, then the solution of the second-order Dirac equation will be a solution of the Dirac equation for all subsequent times [44].

For the particular case of an electron interacting with both a time-independent potential and a laser field, we have

$$A_\mu = \left(\phi(\mathbf{r}), -\mathbf{A}(\eta) \right), \tag{2.252}$$

so that Equation (2.251) becomes

$$\left(P^\mu P_\mu - m^2 c^2 + \gamma^0 \boldsymbol{\gamma} \cdot [\mathbf{p}, e\phi] - ie(\gamma^\mu (k_{\mathrm{L}})_\mu) \boldsymbol{\gamma} \cdot (d\mathbf{A}/d\eta) \right) \Psi(\mathbf{r}, t) = 0, \tag{2.253}$$

where

$$(k_{\mathrm{L}})_\mu = (\omega/c, -\mathbf{k}_{\mathrm{L}}). \tag{2.254}$$

A solution of this equation with $\phi = 0$, referred to as a Dirac–Volkov (DV) wave function [35], can be obtained by making the ansatz

$$\chi_{\mathbf{k}}^{\mathrm{DV}}(\mathbf{r}, t) = g(\eta) u_k \chi_{\mathbf{k}}^{\mathrm{KGV}}(\mathbf{r}, t). \tag{2.255}$$

Here u_k is a four-component spinor such that $u_k \exp(-ik^\mu x_\mu)$ satisfies the first-order field-free Dirac equation. That is,

$$\left(\gamma^\mu p_\mu - mc\right) u_k \exp(-ik^\mu x_\mu) = 0, \tag{2.256}$$

with

$$k^\mu = (E_+/(c\hbar), \mathbf{k}). \tag{2.257}$$

In Equation (2.255), $g(\eta)$ is a matrix function. In the limit $\eta \to -\infty$, we have $\mathbf{A}(\eta) \to 0$ and $g(\eta) \to I$. Inserting Equation (2.255) into Equation (2.253) yields for $g(\eta)$ a system of first-order differential equations, whose solution is given by

$$g(\eta) = I - \frac{e}{2\hbar k^\mu (k_{\mathrm{L}})_\mu} \gamma^\mu (k_{\mathrm{L}})_\mu \gamma^\mu A_\mu(\eta). \tag{2.258}$$

For the particular case in which $g(\eta) \simeq I$, each component of the Dirac spinor wave function is a solution of the Klein–Gordon equation (2.231) with $\phi = 0$. It follows from Equation (2.253) that this property also holds for $\phi \neq 0$, provided that the commutator $[\mathbf{p}, e\phi]$ can be neglected. As noted above, this condition is fulfilled when the probability of pair creation in the laser field is negligible.

2.9.3.1 Non-relativistic limit of the Dirac equation; the Pauli equation

The non-relativistic limit of the Dirac equation (2.244) can be analyzed by defining first the new Dirac spinor

$$\tilde{\Psi}(\mathbf{r}, t) = \Psi(\mathbf{r}, t) \exp(imc^2 t/\hbar), \tag{2.259}$$

which satisfies the equation

$$i\hbar \frac{\partial}{\partial t} \tilde{\Psi}(\mathbf{r}, t) = \left[c\boldsymbol{\alpha} \cdot (-i\hbar\nabla + e\mathbf{A}) - e\phi + (\beta - I)mc^2\right] \tilde{\Psi}(\mathbf{r}, t), \tag{2.260}$$

where I is the 4×4 unit matrix.

Let us write the four-component Dirac spinor $\tilde{\Psi}$ in terms of two-component spinors as

$$\tilde{\Psi} = \left(\begin{array}{c} \Psi_A \\ \Psi_B \end{array} \right), \tag{2.261}$$

where

$$\Psi_A = \left(\begin{array}{c} \tilde{\Psi}_1 \\ \tilde{\Psi}_2 \end{array} \right), \qquad \Psi_B = \left(\begin{array}{c} \tilde{\Psi}_3 \\ \tilde{\Psi}_4 \end{array} \right). \tag{2.262}$$

Using the Dirac representation (2.242) of the matrices $\boldsymbol{\alpha}$ and β, the two-component spinors Ψ_A and Ψ_B are found to satisfy the two coupled equations

$$i\hbar\frac{\partial}{\partial t}\Psi_A = [c\boldsymbol{\sigma}\cdot(-i\hbar\nabla + e\mathbf{A})]\Psi_B - e\phi\Psi_A \tag{2.263}$$

and

$$i\hbar\frac{\partial}{\partial t}\Psi_B = [c\boldsymbol{\sigma}\cdot(-i\hbar\nabla + e\mathbf{A})]\Psi_A - (e\phi + 2mc^2)\Psi_B. \tag{2.264}$$

This pair of equations is still exact. In the non-relativistic limit we can solve Equation (2.264) approximately to obtain

$$\Psi_B \simeq \frac{1}{2mc}\boldsymbol{\sigma}\cdot(-i\hbar\nabla + e\mathbf{A})\Psi_A, \tag{2.265}$$

and we see that Ψ_B is smaller than Ψ_A by a factor of order $p/(mc)$, i.e. v/c, where v is the magnitude of the electron velocity. The two-component spinors Ψ_A and Ψ_B are known in this case as the *large* and *small components*, respectively, of the four-component spinor $\tilde{\Psi}$. Substituting Equation (2.265) into Equation (2.263), we find that

$$i\hbar\frac{\partial}{\partial t}\Psi_A = \frac{1}{2m}[\boldsymbol{\sigma}\cdot(-i\hbar\nabla + e\mathbf{A})]^2\Psi_A - e\phi\Psi_A. \tag{2.266}$$

Let us now use the fact that the Pauli spin matrices satisfy the identity [41]

$$(\boldsymbol{\sigma}\cdot\mathbf{A})(\boldsymbol{\sigma}\cdot\mathbf{B}) = \mathbf{A}\cdot\mathbf{B} + i\boldsymbol{\sigma}\cdot(\mathbf{A}\times\mathbf{B}), \tag{2.267}$$

where $\mathbf{A}$ and $\mathbf{B}$ are two vectors, or two vector operators whose components commute with those of $\boldsymbol{\sigma}$. We can then write the first term on the right-hand side of Equation (2.266) as

$$\frac{1}{2m}[\boldsymbol{\sigma}\cdot(-i\hbar\nabla + e\mathbf{A})]^2\Psi_A = \frac{1}{2m}(-i\hbar\nabla + e\mathbf{A})^2\Psi_A + \frac{e\hbar}{2m}\boldsymbol{\sigma}\cdot\boldsymbol{\mathcal{B}}\Psi_A, \tag{2.268}$$

since $\nabla\times\mathbf{A} = \boldsymbol{\mathcal{B}}$, where $\boldsymbol{\mathcal{B}}$ is the magnetic field. Equation (2.266) now becomes

$$i\hbar\frac{\partial}{\partial t}\Psi_A(\mathbf{r}, t) = \left[\frac{1}{2m}(-i\hbar\nabla + e\mathbf{A})^2 + \frac{e\hbar}{2m}(\boldsymbol{\sigma}\cdot\boldsymbol{\mathcal{B}}) - e\phi\right]\Psi_A(\mathbf{r}, t). \tag{2.269}$$

This equation is known as the *Pauli equation*. It differs from the corresponding non-relativistic form (2.236) of the Klein–Gordon equation for spinless particles in predicting an interaction between the magnetic field $\boldsymbol{\mathcal{B}}$ and the Pauli spin operator $\mathbf{S} = (\hbar/2)\boldsymbol{\sigma}$ of the electron. We note that the term $e\hbar\boldsymbol{\sigma}\cdot\boldsymbol{\mathcal{B}}/(2m)$ corresponds to an interaction $-\boldsymbol{\mathcal{M}}_s\cdot\boldsymbol{\mathcal{B}}$ between the magnetic field $\boldsymbol{\mathcal{B}}$ and an intrinsic *magnetic moment* $\boldsymbol{\mathcal{M}}_s$ of the electron, due to its spin, with

$$\boldsymbol{\mathcal{M}}_s = -\frac{e\hbar}{2m}\boldsymbol{\sigma} = -\mu_B\boldsymbol{\sigma} = -\frac{e}{m}\mathbf{S}, \tag{2.270}$$

where

$$\mu_B = \frac{e\hbar}{2m} \tag{2.271}$$

is the Bohr magneton. We may also write Equation (2.270) as

$$\mathcal{M}_s = -g_s \mu_B \mathbf{S}/\hbar = -g_s \frac{e}{2m} \mathbf{S} \tag{2.272}$$

where the spin gyromagnetic ratio g_s of the electron has the value $g_s = 2$, in very good agreement with experiment. Small corrections to the Dirac value $g_s = 2$, due to the anomalous magnetic moment of the electron, have been calculated very accurately by using quantum electrodynamics.

2.9.3.2 The Foldy–Wouthuysen transformation

We have shown that to lowest order in v/c the Dirac theory is equivalent to the two-component Pauli theory. To investigate higher-order corrections in v/c, it is convenient to use a systematic procedure developed by Foldy and Wouthuysen [45]. Their approach consists in performing a suitably chosen unitary transformation on the wave function and operators of the Dirac theory. The aim is to achieve in the new representation (called the Foldy–Wouthuysen representation) a decoupling of the Dirac equation into two two-component equations in such a way that the first one reduces to the Pauli equation in the non-relativistic limit, while the second one describes the negative-energy states.

As a first illustration of the Foldy–Wouthuysen transformation, let us consider the Dirac equation for a free electron, the corresponding Dirac Hamiltonian being given by Equation (2.241). We look for a unitary operator U_F which will remove from the Dirac equation (2.240) all operators such as $\boldsymbol{\alpha}$ which couple the large to the small components. Operators of this kind are called "odd" in this context, while operators like $\mathbf{p}$, $\mathbf{r}$ or $\boldsymbol{\sigma}$ which do not couple large and small components are called "even." The required unitary operator is given in this case by [46]

$$
\begin{aligned}
U_F &= \left(\frac{2E_+}{mc^2 + E_+} \right)^{1/2} \frac{1}{2} \left[1 + \beta \frac{H_D^{(0)}}{E_+} \right] \\
&= \left(\frac{mc^2 + E_+}{2E_+} \right)^{1/2} + \beta \frac{c\boldsymbol{\alpha} \cdot \mathbf{p}}{\left[2E_+ (mc^2 + E_+) \right]^{1/2}},
\end{aligned} \tag{2.273}
$$

where we recall that $E_\pm = \pm(m^2 c^4 + \hbar^2 k^2 c^2)^{1/2}$. The transformed Hamiltonian in the Foldy–Wouthuysen representation is then

$$
\begin{aligned}
H_F^{(0)} &= U_F H_D^{(0)} U_F^\dagger \\
&= \beta (m^2 c^4 + c^2 \mathbf{p}^2)^{1/2}.
\end{aligned} \tag{2.274}
$$

This Hamiltonian is nearly the same as the one which appears in the relativistic Schrödinger equation for a free particle (2.222), except that the presence of the factor β means that now the negative energies are also accepted. Since $H_F^{(0)}$ is an even operator, the large components Φ_A and the small components Φ_B in the Foldy–Wouthuysen representation are decoupled to all orders in v/c in the equations of motion. That is,

$$i\hbar \frac{\partial \Phi_A}{\partial t} = E_+ \Phi_A \tag{2.275}$$

and

$$i\hbar \frac{\partial \Phi_B}{\partial t} = E_- \Phi_B. \tag{2.276}$$

Moreover, if we limit ourselves to the positive-energy solutions, the Dirac theory is completely equivalent to the two-component theory embodied in Equation (2.275).

Let us now consider the case of an electron interacting with a central potential $V(r)$ and a laser field described by the vector potential $\mathbf{A}(\mathbf{r}, t)$. The corresponding Dirac Hamiltonian H_D is given by Equation (2.245) with $-e\phi = V(r)$, and in the general case it is not possible to construct an operator U_F such that the transformed Hamiltonian H_F in the Foldy–Wouthuysen representation is exactly even. However, a non-relativistic expansion of the transformed Hamiltonian in a power series in $(mc^2)^{-1}$ can be obtained. The result can be written in the form [47]

$$H_F = \sum_{i=1}^{5} H_F^{(i)}, \tag{2.277}$$

where

$$H_F^{(1)} = \beta \left[mc^2 + \frac{1}{2m} (\mathbf{p} + e\mathbf{A})^2 - \frac{\mathbf{p}^4}{8m^3 c^2} \right],$$

$$H_F^{(2)} = V,$$

$$H_F^{(3)} = \beta \frac{e\hbar}{2m} (\boldsymbol{\sigma} \cdot \mathcal{B}), \tag{2.278}$$

$$H_F^{(4)} = \frac{ie\hbar}{8m^2 c^2} \boldsymbol{\sigma} \cdot (\nabla \times \mathcal{E}) + \frac{\hbar}{4m^2 c^2} \boldsymbol{\sigma} \cdot [(\nabla V + e\mathcal{E}) \times \mathbf{p}],$$

$$H_F^{(5)} = \frac{\hbar^2}{8m^2 c^2} \left(\nabla^2 V \right).$$

The five individual terms $H_F^{(i)}$ have the following physical interpretation. The terms in the square bracket on the right-hand side of Equation (2.278) for the first term

$H_{\rm F}^{(1)}$ correspond to the expansion

$$\left[m^2c^4 + c^2(\mathbf{p}+e\mathbf{A})^2\right]^{1/2} = mc^2 + \frac{1}{2m}(\mathbf{p}+e\mathbf{A})^2 - \frac{\mathbf{p}^4}{8m^3c^2} + \cdots \quad (2.279)$$

through order v^2/c^2; they exhibit the relativistic mass increase. The second term $H_{\rm F}^{(2)}$ is the electrostatic energy, while the third term $H_{\rm F}^{(3)}$ yields the magnetic dipole energy. The two terms that make up the Hermitian operator $H_{\rm F}^{(4)}$ give the spin-orbit energy. We note that for a spherically symmetric potential, $\nabla V(r) = \hat{\mathbf{r}}(dV/dr)$, so that

$$\frac{\hbar}{4m^2c^2}\boldsymbol{\sigma}\cdot(\nabla V \times \mathbf{p}) = \frac{\hbar}{4m^2c^2}\frac{1}{r}\frac{dV}{dr}\mathbf{L}\cdot\boldsymbol{\sigma} = \frac{1}{2m^2c^2}\frac{1}{r}\frac{dV}{dr}\mathbf{L}\cdot\mathbf{S}, \quad (2.280)$$

which is the familiar spin-orbit interaction. The last term $H_{\rm F}^{(5)}$ arises from the so-called "zitterbewegung," since the electron coordinate fluctuates over a distance $\delta r \simeq \hbar/(mc)$, and is known as the Darwin term. We note that for the Coulomb potential $V(r) = -Ze^2/(4\pi\epsilon_0 r)$ the Darwin term becomes

$$\frac{\hbar^2}{8m^2c^2}\nabla^2 V(r) = \frac{\pi\hbar^2}{2m^2c^2}\left(\frac{Ze^2}{4\pi\epsilon_0}\right)\delta(\mathbf{r}). \quad (2.281)$$

In view of the fact that obtaining numerical solutions of the time-dependent Dirac equation for atoms or ions in very intense laser fields is a very difficult task, the use of the Foldy–Wouthuysen transformation represents an interesting alternative approach [28, 48].

2.9.3.3 Pair production; the Schwinger limit

To conclude this chapter, we briefly discuss the possibility of producing electron–positron pairs from the vacuum using ultra-intense laser fields. Already in 1931, Sauter [49] predicted that (e^+, e^-) pairs can be created in a *constant* electric field if the electric field strength $\mathcal{E}_0$ exceeds the critical value 1.3×10^{16} V cm^{-1}. Expressed in atomic units this condition reads $\mathcal{E}_0 > c^3$, where $c \simeq 137$ a.u. Schwinger [50, 51] derived this result using quantum electrodynamics (QED) and also pointed out that the pair-creation process cannot take place in a plane-wave field. Pair production can, however, occur in the presence of at least two plane-wave fields with different propagation directions, thereby ensuring conservation of the four-momentum in the process.

Brezin and Itzykson [52] considered pair production in an alternating electric field which can be produced by two counter-propagating laser fields. They obtained an estimate of the (e^+, e^-) production probability in terms of the solution of the one-electron Dirac equation. The pairs can be considered as bound in the vacuum with binding energy $V_0 \simeq 2mc^2$. Using a version of the Wentzel–Kramers–Brillouin

(WKB) approximation employed in the context of tunneling ionization [53, 54], Brezin and Itzykson found that for low-frequency laser fields the (e^+, e^-) pair production probability is proportional to the quantity

$$\exp[-\pi m^2 c^3/(e\hbar \mathcal{E}_0)]. \tag{2.282}$$

This result confirms that, in order to observe the effect, laser fields of intensity $I \simeq 10^{29}\,\mathrm{W\,cm^{-2}}$, known as the *Schwinger limit*, are required.

It is also worth pointing out that (e^+, e^-) production can occur if an ultra-strong laser field interacts with high-Z nuclei. This phenomenon is referred to as the non-linear Bethe–Heitler process [55, 56].

References

[1] M. H. Mittleman, *Introduction to the Theory of Laser–Atom Interactions*, 2nd edn. (New York: Plenum Press, 1993).

[2] C. J. Joachain, Theory of laser-atom interactions. In R. M. More, ed., *Laser Interactions with Atoms, Solids and Plasmas* (New York: Plenum Press, 1994), p. 39.

[3] R. J. Glauber, *Phys. Rev.* **130**, 2529 (1963).

[4] R. J. Glauber, *Phys. Rev.* **131**, 2766 (1963).

[5] L. Mandel and E. Wolf, *Optical Coherence and Quantum Optics* (Cambridge: Cambridge University Press, 1995).

[6] J. Parker and C. R. Stroud, *Phys. Rev. Lett.* **56**, 716 (1986).

[7] P. Král, I. Thanopoulos and M. Shapiro, *Rev. Mod. Phys.* **79**, 53 (2007).

[8] M. Geissler, G. Tempea, A. Scrinzi, M. Schnürer, F. Krausz and T. Brabec, *Phys. Rev. Lett.* **83**, 2930 (1999).

[9] A. E. Siegmann, *Lasers*, (Mill Valley, Calif.: University Science, 1986).

[10] L. G. Gouy, *Comptes Rendus Acad. Sci. Paris* **110**, 1251 (1890).

[11] F. Lindner, G. G. Paulus, H. Walther *et al.*, *Phys. Rev. Lett.* **92**, 113001 (2004).

[12] A. T. Bharucha-Reid, *Elements of the Theory of Markov Processes and their Applications*, 3rd edn (New York: McGraw Hill, 1960).

[13] R. Daniele, G. Ferrante, F. Morales and F. Trombetta, *Fundamentals of Laser Interactions*, Lecture Notes in Physics, 229 (Berlin: Springer-Verlag, 1985).

[14] S. Bivona, R. Burlon, R. Zangara and G. Ferrante, *J. Phys. B* **18**, 3149 (1985).

[15] P. Francken and C. J. Joachain, *J. Opt. Soc. Am. B* **7**, 554 (1990).

[16] P. Zoller, *J. Phys. B* **13**, L249 (1980).

[17] R. Daniele, F. H. M. Faisal and G. Ferrante, *J. Phys. B* **16**, 3831 (1983).

[18] P. Zoller, *J. Phys. B* **11**, 805 (1978).

[19] F. Trombetta, G. Ferrante, K. Wodkiewicz and P. Zoller, *J. Phys. B* **18**, 2915 (1985).

[20] P. Francken and C. J. Joachain, *Europhys. Lett.* **3**, 11 (1987).

[21] P. Francken and C. J. Joachain, *Europhys. Lett.* **9**, 517 (1989).

[22] B. W. Shore, *J. Opt. Soc. Am. B* **1**, 176 (1984).

[23] A. Brissaud and U. Frisch, *J. Math. Phys.* **15**, 524 (1974).

[24] K. Wodkiewicz, B. W. Shore and J. H. Eberly, *J. Opt. Soc. Am. B* **1**, 398 (1984).

[25] N. G. Van Kampen, *Stochastic Processes in Physics and Chemistry* (Amsterdam: North Holland, 1981).

[26] F. Trombetta, G. Ferrante and P. Zoller, *Opt. Commun.* **60**, 213 (1986).

[27] J. D. Jackson, *Classical Electrodynamics*, 3rd edn (New York: Wiley, 1998).

[28] Y. I. Salamin, S. X. Hu, K. Z. Hatsagortsyan and C. H. Keitel, *Phys. Rep.* **427**, 41 (2006).

[29] W. Pauli and M. Fierz, *Nuovo Cimento* **15**, 1167 (1938).
[30] H. A. Kramers, *Collected Scientific Papers* (Amsterdam: North Holland, 1956), p. 272.
[31] W. C. Henneberger, *Phys. Rev. Lett.* **21**, 838 (1968).
[32] M. Abramowitz and I. A. Stegun, *Handbook of Mathematical Functions* (New York: Dover, 1970).
[33] C. J. Joachain, *Quantum Collision Theory*, 3rd edn (Amsterdam: North Holland, 1983).
[34] W. Gordon, *Z. Phys.* **40**, 117 (1926).
[35] D. M. Volkov, *Z. Phys.* **94**, 250 (1935).
[36] B. H. Bransden and C. J. Joachain, *Physics of Atoms and Molecules*, 2nd edn (Harlow, UK: Prentice Hall-Pearson, 2003).
[37] C. J. Joachain and N. J. Kylstra, *Laser Phys.* **11**, 212 (2001).
[38] C. J. Joachain, N. J. Kylstra and R. M. Potvliege, *J. Mod. Opt.* **50**, 313 (2003).
[39] A. Maquet, R. Taïeb and V. Véniard, Relativistic laser–atom physics. In T. Brabec, ed. *Strong Field Laser Physics*, Springer Series in Optical Sciences 134, (New York: Springer, 2009), p. 477.
[40] N. J. Kylstra, R. M. Potvliege and C. J. Joachain, *J. Phys. B* **34**, L55 (2001).
[41] B. H. Bransden and C. J. Joachain, *Quantum Mechanics* 2nd edn (Harlow, UK: Prentice Hall-Pearson, 2000).
[42] F. V. Hartemann and A. K. Kerman, *Phys. Rev. Lett.* **76**, 624 (1996).
[43] W. Pauli and V. Weisskopf, *Helv. Phys. Acta* **7**, 709 (1934).
[44] L. S. Brown and T. W. B. Kibble, *Phys. Rev.* **133A**, 705 (1964).
[45] L. L. Foldy and S. A. Wouthuysen, *Phys. Rev.* **78**, 29 (1950).
[46] A. Messiah, *Quantum Mechanics* (Amsterdam: North Holland, 1968).
[47] J. D. Bjorken and S. D. Drell, *Relativistic Quantum Mechanics* (New York: McGraw Hill, 1964).
[48] S. X. Hu and C. H. Keitel, *Phys. Rev. Lett.* **83**, 4709 (1999).
[49] F. Sauter, *Z. Phys.* **69**, 742 (1931).
[50] J. Schwinger, *Phys. Rev.* **82**, 664 (1951).
[51] J. Schwinger, *Phys. Rev.* **93**, 615 (1954).
[52] E. Brezin and C. Itzykson, *Phys. Rev. D* **2**, 1191 (1970).
[53] A. M. Perelomov, V. S. Popov and M. V. Terent'ev, *Sov. Phys. JETP* **23**, 924 (1966).
[54] A. M. Perelomov, V. S. Popov and M. V. Terent'ev, *Sov. Phys. JETP* **24**, 207 (1967).
[55] H. A. Bethe and W. Heitler, *Proc. R. Soc. London A* **146**, 83 (1934).
[56] C. Müller, A. B. Voitkiv and N. Grün, *Phys. Rev. A* **67**, 063407 (2003).

Part II

Theoretical methods

Part II

Representational methods

3

Perturbation theory

For a laser pulse which is not too intense, so that its peak electric-field strength is much smaller than the electric fields due to the Coulomb interactions experienced by the atomic electrons, the interaction of the atom with the laser field can be treated as a perturbation. In Section 3.1, we discuss the time-dependent perturbation theory for the evolution operator, and obtain expressions for the non-vanishing lowest-order perturbation theory (LOPT) contribution to multiphoton processes involving the absorption or emission of n photons, and for the dynamic (AC) Stark shift. In Section 3.2, we analyze an alternative approach to time-dependent perturbation theory, which is Dirac's method of variation of the constants. This method is used in particular to obtain expressions for the atomic wave functions "dressed" by the laser field. It is also used to study a two-level system submitted to a harmonic perturbation. Finally, in Section 3.3, we describe semi-perturbative methods, in which some of the interactions are treated in a non-perturbative way, and the remaining ones are treated by using perturbation theory. Reviews of the perturbative or semi-perturbative approaches for multiphoton processes discussed in this chapter have been given by Lambropoulos [1], Faisal [2], Crance [3], Lambropoulos and Tang [4] and Joachain, Dörr and Kylstra [5].

3.1 Time-dependent perturbation theory for the evolution operator

3.1.1 Basic equations

We have seen in Chapter 2 that if relativistic effects are neglected, an atom in a laser field can be described in the Schrödinger picture by the TDSE (2.149), namely

$$i\hbar \frac{\partial}{\partial t} |\Psi(t)\rangle = H(t)|\Psi(t)\rangle, \tag{3.1}$$

where $|\Psi(t)\rangle$ is the state vector of the system, which we assume to be normalized to unity. We write the Hamiltonian of the atom in the laser field as follows:

$$H(t) = H_0 + H_{\text{int}}(t), \tag{3.2}$$

where H_0 is the time-independent, non-relativistic field-free atomic Hamiltonian, and the time-dependent Hamiltonian $H_{\text{int}}(t)$ describes the interaction of the atom with the laser field. In this chapter, we shall call H_0 the "unperturbed" Hamiltonian and suppose that $H_{\text{int}}(t)$ is a small enough time-dependent perturbation, so that time-dependent perturbation theory can be used.

We assume that the eigenvalues E_k of the unperturbed Hamiltonian H_0 are known, together with the corresponding energy eigenstates $|\psi_k\rangle$, which we suppose to be orthonormal and to form a complete set. Thus, since

$$H_0|\psi_k\rangle = E_k|\psi_k\rangle, \tag{3.3}$$

the unperturbed TDSE

$$i\hbar\frac{\partial}{\partial t}|\Psi_0(t)\rangle = H_0|\Psi_0(t)\rangle \tag{3.4}$$

has particular solutions which are stationary states of the form

$$|\phi_k(t)\rangle = |\psi_k\rangle \exp(-iE_kt/\hbar). \tag{3.5}$$

The general solution of the unperturbed TDSE (3.4) is therefore given by the linear superposition of stationary states:

$$\begin{aligned}|\Psi_0(t)\rangle &= \sum_k c_k^{(0)}|\phi_k(t)\rangle \\ &= \sum_k c_k^{(0)}|\psi_k\rangle \exp(-iE_kt/\hbar),\end{aligned} \tag{3.6}$$

where the coefficients $c_k^{(0)}$ are constants and the summation symbol implies a sum over the complete set (discrete plus continuous) of energy eigenstates $|\psi_k\rangle$ of H_0. Assuming that $|\Psi_0(t)\rangle$ is normalized to unity, we can interpret $|c_k^{(0)}|^2$ as the probability of finding the system in the unperturbed energy eigenstate $|\psi_k\rangle$ in the absence of perturbation.

Let us return to the TDSE (3.1). In Section 2.6, we introduced the evolution operator $U(t, t_0)$ in the Schrödinger picture by Equation (2.143), namely

$$|\Psi(t)\rangle = U(t, t_0)|\Psi(t_0)\rangle. \tag{3.7}$$

This evolution operator satisfies Equations (2.144)–(2.148). In particular, it is unitary and satisfies the TDSE (2.150). That is,

$$i\hbar\frac{\partial}{\partial t}U(t, t_0) = H(t)U(t, t_0), \tag{3.8}$$

with the initial condition $U(t_0, t_0) = I$. We also saw in Section 2.6 that the evolution operator $U(t, t_0)$ satisfies integral equations of the Volterra type, namely

$$U(t, t_0) = I - \frac{i}{\hbar} \int_{t_0}^{t} H(t_1) U(t_1, t_0) dt_1 \tag{3.9}$$

and

$$U(t, t_0) = I - \frac{i}{\hbar} \int_{t_0}^{t} U(t, t_1) H(t_1) dt_1, \tag{3.10}$$

and we recall that the evolution operator $U(t, t_0)$ is entirely determined when the Hamiltonian $H(t)$ is known.

Let us now introduce the unperturbed (field-free) evolution operator $U_0(t, t_0)$ corresponding to the Hamiltonian H_0. This evolution operator satisfies the TDSE

$$i\hbar \frac{\partial}{\partial t} U_0(t, t_0) = H_0 U_0(t, t_0) \tag{3.11}$$

with the initial condition $U_0(t_0, t_0) = I$. It is clear that $U_0(t, t_0)$ also satisfies the integral equations (3.9) and (3.10), with U replaced by U_0 and H by H_0. Since the unperturbed Hamiltonian H_0 is time-independent, the solution of the TDSE (3.11) satisfying the initial condition $U_0(t_0, t_0) = I$ is given by

$$U_0(t, t_0) = \exp[-iH_0(t - t_0)/\hbar], \tag{3.12}$$

as is readily verified by using the fact that

$$\exp[-iH_0(t - t_0)/\hbar] = \sum_{n=0}^{\infty} \frac{1}{n!} \left(-\frac{i}{\hbar} \right)^n H_0^n (t - t_0)^n. \tag{3.13}$$

3.1.2 The evolution operator in the interaction picture

We shall now obtain an expression for the evolution operator which takes advantage of the splitting (3.2) of the full Hamiltonian of the system into an unperturbed Hamiltonian H_0 and a perturbation $H_{int}(t)$, and also uses the fact that the unperturbed evolution operator $U_0(t, t_0)$ is known, being given by Equation (3.12). To this end, we perform on the Schrödinger state vector $|\Psi(t)\rangle$ the unitary transformation

$$|\Psi_I(t)\rangle = U_0^\dagger(t, t_0)|\Psi(t)\rangle$$
$$= \exp[iH_0(t - t_0)/\hbar]|\Psi(t)\rangle, \tag{3.14}$$

which defines the new state vector $|\Psi_I(t)\rangle$ in the interaction picture. We note that $|\Psi_I(t_0)\rangle = |\Psi(t_0)\rangle$ so that the Schrödinger and interaction pictures coincide at the time $t = t_0$. Using Equations (3.1) and (3.14), we obtain for $|\Psi_I(t)\rangle$ the Tomonaga–Schwinger equation [6,7]

$$i\hbar \frac{\partial}{\partial t}|\Psi_I(t)\rangle = H_{int}^I(t)|\Psi_I(t)\rangle, \tag{3.15}$$

where

$$H_{int}^I(t) = U_0^\dagger(t, t_0) H_{int}(t) U_0(t, t_0)$$
$$= \exp[iH_0(t - t_0)/\hbar] H_{int}(t) \exp[-iH_0(t - t_0)/\hbar]. \tag{3.16}$$

We see that in the interaction picture the time-dependence of the state vector $|\Psi_I(t)\rangle$ is entirely due to the interaction. Since the passage from the Schrödinger picture state vector $|\Psi(t)\rangle$ to the interaction picture state vector $|\Psi_I(t)\rangle$ is accomplished by means of a unitary transformation, all physical quantities such as eigenvalues and expectation values of observables as well as transition probabilities are identical when calculated in either picture [8].

We now define an evolution operator in the interaction picture $U_I(t, t_0)$ by the equation

$$|\Psi_I(t)\rangle = U_I(t, t_0)|\Psi_I(t_0)\rangle. \tag{3.17}$$

This operator possesses all the properties of an evolution operator (see Equations (2.144)–(2.148)). In particular, it is a unitary operator. From Equations (3.15) and (3.17), we also see that $U_I(t, t_0)$ satisfies the Tomonaga–Schwinger equation

$$i\hbar \frac{\partial}{\partial t} U_I(t, t_0) = H_{int}^I(t) U_I(t, t_0). \tag{3.18}$$

This equation, together with the initial condition $U_I(t_0, t_0) = I$, can be replaced by integral equations of the Volterra type, namely

$$U_I(t, t_0) = I - \frac{i}{\hbar} \int_{t_0}^t H_{int}^I(t_1) U_I(t_1, t_0) dt_1 \tag{3.19}$$

and

$$U_I(t, t_0) = I - \frac{i}{\hbar} \int_{t_0}^t U_I(t, t_1) H_{int}^I(t_1) dt_1. \tag{3.20}$$

The relation between the Schrödinger picture evolution operator $U(t, t_0)$ and the interaction picture evolution operator $U_I(t, t_0)$ is readily established. We have, from Equations (3.7) and (3.14),

$$|\Psi_I(t)\rangle = U_0^\dagger(t, t_0)|\Psi(t)\rangle$$
$$= U_0^\dagger(t, t_0) U(t, t_0)|\Psi(t_0)\rangle$$
$$= U_0^\dagger(t, t_0) U(t, t_0)|\Psi_I(t_0)\rangle, \tag{3.21}$$

where we have used the fact that $|\Psi_I(t_0)\rangle = |\Psi(t_0)\rangle$. Upon comparison of Equations (3.17) and (3.21), we deduce that

$$U_I(t, t_0) = U_0^\dagger(t, t_0)U(t, t_0). \tag{3.22}$$

Multiplying both sides of this equation on the left by $U_0(t, t_0)$, and using the fact that the operator $U_0(t, t_0)$ is unitary, we find that

$$U(t, t_0) = U_0(t, t_0)U_I(t, t_0). \tag{3.23}$$

This important relation shows that if we know the operator $U_0(t, t_0)$ corresponding to the unperturbed Hamiltonian H_0, then the evolution operator $U(t, t_0)$ in the Schrödinger picture will be obtained if we can determine the evolution operator $U_I(t, t_0)$ in the interaction picture.

Using Equations (3.16) and (3.23), together with the integral equations (3.19) and (3.20) satisfied by the operator $U_I(t, t_0)$, we readily obtain for the operator $U(t, t_0)$ the two integral equations

$$U(t, t_0) = U_0(t, t_0) - \frac{i}{\hbar} \int_{t_0}^{t} U_0(t, t_1) H_{int}(t_1) U(t_1, t_0) dt_1 \tag{3.24}$$

and

$$U(t, t_0) = U_0(t, t_0) - \frac{i}{\hbar} \int_{t_0}^{t} U(t, t_1) H_{int}(t_1) U_0(t_1, t_0) dt_1. \tag{3.25}$$

Let us now return to the integral equations (3.19) and (3.20) satisfied by the evolution operator $U_I(t, t_0)$. The solution of these equations can at least formally be obtained by iteration. For example, if we consider the integral equation (3.19), we have, to zeroth order in H_{int}^I,

$$U_I^{(0)}(t, t_0) = I. \tag{3.26}$$

To first order in H_{int}^I, we then have

$$U_I^{(1)}(t, t_0) = I - \frac{i}{\hbar} \int_{t_0}^{t} H_{int}^I(t_1) dt_1. \tag{3.27}$$

Similarly, we find that, to second order in H_{int}^I,

$$U_I^{(2)}(t, t_0) = I - \frac{i}{\hbar} \int_{t_0}^{t} H_{int}^I(t_1) dt_1 + \left(-\frac{i}{\hbar}\right)^2 \int_{t_0}^{t} dt_1 \int_{t_0}^{t_1} dt_2 H_{int}^I(t_1) H_{int}^I(t_2). \tag{3.28}$$

Successive iterations give for $U_I(t, t_0)$ the formal series

$$U_I(t, t_0) = \sum_{n=0}^{\infty} \bar{U}_I^{(n)}(t, t_0), \tag{3.29}$$

where

$$\bar{U}_{\mathrm{I}}^{(0)}(t, t_0) = U_{\mathrm{I}}^{(0)}(t, t_0) = I, \tag{3.30}$$

while for $n \geq 1$

$$\bar{U}_{\mathrm{I}}^{(n)}(t, t_0) = \left(-\frac{\mathrm{i}}{\hbar}\right)^n \int_{t_0}^t \mathrm{d}t_1 \int_{t_0}^{t_1} \mathrm{d}t_2 \dots \int_{t_0}^{t_{n-1}} \mathrm{d}t_n \, H_{\mathrm{int}}^{\mathrm{I}}(t_1) H_{\mathrm{int}}^{\mathrm{I}}(t_2) \dots H_{\mathrm{int}}^{\mathrm{I}}(t_n).$$

$$\tag{3.31}$$

If we choose $t_0 < t$, it is clear that the time ordering

$$t \geq t_1 \geq t_2 \dots \geq t_{n-1} \geq t_n \geq t_0 \tag{3.32}$$

is imposed. This ordering is important since in general the operators $H_{\mathrm{int}}^{\mathrm{I}}(t_i)$ and $H_{\mathrm{int}}^{\mathrm{I}}(t_j)$ do not commute when $t_i \neq t_j$.

The expansion (3.29) is a power series in the perturbation $H_{\mathrm{int}}^{\mathrm{I}}(t)$, which converges more rapidly as the magnitude of the perturbation decreases.

3.1.3 Perturbation theory in the Schrödinger picture

We can obtain a perturbative series for the evolution operator in the Schrödinger picture, $U(t, t_0)$, by using Equations (3.16), (3.23) and (3.29)–(3.32). Alternatively, we can solve the integral equations (3.24) and (3.25) by iteration, using $U_0(t, t_0)$ as our zeroth-order approximation. The result is

$$U(t, t_0) = U_0(t, t_0) + \sum_{n=1}^{\infty} \bar{U}^{(n)}(t, t_0), \tag{3.33}$$

where $U_0(t, t_0)$ is given by Equation (3.12) and

$$\bar{U}^{(n)}(t, t_0) = \left(-\frac{\mathrm{i}}{\hbar}\right)^n \int_{t_0}^t \mathrm{d}t_1 \int_{t_0}^{t_1} \mathrm{d}t_2 \dots \int_{t_0}^{t_{n-1}} \mathrm{d}t_n$$

$$\times U_0(t, t_1) H_{\mathrm{int}}(t_1) U_0(t_1, t_2) H_{\mathrm{int}}(t_2) \dots U_0(t_{n-1}, t_n) H_{\mathrm{int}}(t_n) U_0(t_n, t_0), \quad n \geq 1,$$

$$\tag{3.34}$$

with the time ordering (3.32). The integrand on the right-hand side of Equation (3.34) can be interpreted in a simple way. When the operator $\bar{U}^{(n)}(t, t_0)$ acts on the initial state $|\Psi(t_0)\rangle$ of the system, this state evolves in time according to the unperturbed, field-free evolution operator $U_0(t_n, t_0)$. The atom then interacts with the laser field at time t_n. The new state subsequently evolves in time according to the field-free evolution operator and interacts with the laser field at time t_{n-1}, and so on, so that the atom interacts n times with the laser field. In the dipole approximation, and either in the length gauge or the velocity gauge, each laser–atom interaction can be interpreted as the absorption or emission of a photon by the atom. The operator $\bar{U}^{(n)}(t, t_0)$ is then obtained by summing over all the possible times $t_n, t_{n-1}, \ldots, t_1$ satisfying the time ordering (3.32). The expansion (3.33), which is a power series in the perturbation $H_{\text{int}}(t)$, converges faster for smaller perturbations or, in other words, when the unperturbed field-free evolution operator $U_0(t, t_0)$ is closest to $U(t, t_0)$.

Returning to Equation (3.7) and using the series expansion (3.33) for $U(t, t_0)$, we see that the state vector at time t is given by

$$|\Psi(t)\rangle = \left[U_0(t, t_0) + \sum_{n=1}^{\infty} \bar{U}^{(n)}(t, t_0) \right] |\Psi(t_0)\rangle. \tag{3.35}$$

Let us assume that there is no perturbation at times $t \leq t_0$. As a result, the state vector $|\Psi(t_0)\rangle$ is a solution of the unperturbed TDSE (3.4) and we have, using Equation (3.6),

$$|\Psi(t_0)\rangle \equiv |\Psi_0(t_0)\rangle = \sum_k c_k^{(0)} |\phi_k(t_0)\rangle$$

$$= \sum_k c_k^{(0)} |\psi_k\rangle \exp(-iE_k t_0/\hbar). \tag{3.36}$$

In what follows, we shall suppose that the system is initially (for $t \leq t_0$) in a particular unperturbed (field-free) atomic energy eigenstate $|\psi_i\rangle$ of H_0 corresponding to the energy E_i. Then

$$c_k^{(0)} = \delta_{ki} \qquad \text{or} \qquad c_k^{(0)} = \delta(k - i) \tag{3.37}$$

according to whether the state $|\psi_i\rangle$ is discrete or continuous, and we can write the initial state vector as

$$|\Psi_i(t_0)\rangle = |\phi_i(t_0)\rangle$$

$$= |\psi_i\rangle \exp(-iE_i t_0/\hbar). \tag{3.38}$$

If the perturbation H_{int} is applied at the time t_0, we see from Equations (3.35)–(3.38) that the state vector $|\Psi_i(t)\rangle$ obtained at time t from the initial state vector $|\Psi_i(t_0)\rangle$ is given by

$$|\Psi_i(t)\rangle = \left[U_0(t, t_0) + \sum_{n=1}^{\infty} \bar{U}^{(n)}(t, t_0) \right] |\phi_i(t_0)\rangle$$

$$= \sum_{n=0}^{\infty} |\bar{\Psi}_i^{(n)}(t)\rangle. \tag{3.39}$$

where, for $n = 0$,

$$|\bar{\Psi}_i^{(0)}(t)\rangle = U_0(t, t_0)|\phi_i(t_0)\rangle$$
$$= |\phi_i(t)\rangle$$
$$= |\psi_i\rangle \exp(-iE_i t/\hbar) \tag{3.40}$$

and, for $n \geq 1$,

$$|\bar{\Psi}_i^{(n)}(t)\rangle = \left(-\frac{i}{\hbar}\right)^n \int_{t_0}^{t} dt_1 \int_{t_0}^{t_1} dt_2 \ldots \int_{t_0}^{t_{n-1}} dt_n$$
$$\times U_0(t, t_1) H_{\text{int}}(t_1) U_0(t_1, t_2) H_{\text{int}}(t_2) \ldots U_0(t_{n-1}, t_n) H_{\text{int}}(t_n)|\phi_i(t_n)\rangle. \tag{3.41}$$

These states can be obtained recursively using the relation

$$|\bar{\Psi}_i^{(n)}(t)\rangle = -\frac{i}{\hbar} \int_{t_0}^{t} U_0(t, t') H_{\text{int}}(t')|\bar{\Psi}_i^{(n-1)}(t')\rangle dt'. \tag{3.42}$$

In practice, the series expansion (3.39) for the state vector $|\Psi_i(t)\rangle$ must be truncated after a finite number of terms. If the terms from $n = 0$ to $n = N$ are retained in the expansion, we shall write

$$|\Psi_i^{(N)}(t)\rangle = \sum_{n=0}^{N} |\bar{\Psi}_i^{(n)}(t)\rangle, \tag{3.43}$$

and we will refer to this approximate solution of the TDSE (3.1) as the *Nth-order dressed state* of the atom in the presence of the laser field, calculated through order N in the perturbation $H_{\text{int}}(t)$.

3.1.4 Perturbation series for propagators

In Section 2.6, we introduced time-retarded (causal) and time-advanced (anti-causal) propagators, which we denoted by $K^{(+)}(t, t_0)$ and $K^{(-)}(t, t_0)$, respectively. We recall that these operators satisfy Equation (2.153), namely

$$\left[i\hbar \frac{\partial}{\partial t} - H(t)\right] K^{(\pm)}(t, t_0) = \delta(t - t_0) \tag{3.44}$$

with the initial conditions

$$K^{(+)}(t, t_0) = 0, \quad t < t_0, \tag{3.45}$$

and

$$K^{(-)}(t, t_0) = 0, \quad t > t_0. \tag{3.46}$$

Let us now define the unperturbed (field-free) propagators $K_0^{(\pm)}(t, t_0)$ by the equation

$$\left[i\hbar \frac{\partial}{\partial t} - H_0 \right] K_0^{(\pm)}(t, t_0) = \delta(t - t_0) \tag{3.47}$$

with the initial conditions

$$K_0^{(+)}(t, t_0) = 0, \quad t < t_0, \tag{3.48}$$

and

$$K_0^{(-)}(t, t_0) = 0, \quad t > t_0. \tag{3.49}$$

From Equations (3.2) and (3.44) it follows that

$$\left[i\hbar \frac{\partial}{\partial t} - H_0 \right] K^{(\pm)}(t, t_0) = \delta(t - t_0) + H_{\text{int}}(t) K^{(\pm)}(t, t_0). \tag{3.50}$$

Using the initial conditions (3.45) and (3.48), and assuming that $H_{\text{int}}(t) = 0$ for $t \le t_0$, we obtain for the time-retarded propagator $K^{(+)}(t, t_0)$ the integral equation

$$K^{(+)}(t, t_0) = K_0^{(+)}(t, t_0) + \int_{t_0}^{t} K_0^{(+)}(t, t_1) H_{\text{int}}(t_1) K^{(+)}(t_1, t_0) dt_1. \tag{3.51}$$

In the same way, we find that the time-advanced propagator $K^{(-)}(t, t_0)$ satisfies the integral equation

$$K^{(-)}(t, t_0) = K_0^{(-)}(t, t_0) + \int_{t}^{t_0} K_0^{(-)}(t, t_1) H_{\text{int}}(t_1) K^{(-)}(t_1, t_0) dt_1. \tag{3.52}$$

We see that, when $t > t_0$, Equation (3.51) satisfied by the time-retarded propagator $K^{(+)}(t, t_0)$ is very similar to Equation (3.24) satisfied by $U(t, t_0)$, the evolution operator in the Schrödinger picture. A solution of Equation (3.51) giving $K^{(+)}(t, t_0)$ in the form of a perturbative series can therefore be constructed in the same way as for $U(t, t_0)$. Similarly, a solution of Equation (3.52) giving a perturbative series for the time-advanced propagator $K^{(-)}(t, t_0)$ can be obtained.

3.1.5 Transition probabilities and transition rates

Let us assume again that initially (for $t \le t_0$) the system is in a particular unperturbed (field-free) atomic eigenstate $|\psi_i\rangle$ of H_0 corresponding to the energy E_i. According to Equation (3.38), the initial state vector of the system is then given by $|\Psi_i(t_0)\rangle = |\phi_i(t_0)\rangle = |\psi_i\rangle \exp(-iE_i t_0/\hbar)$. The state vector obtained at time t from this initial state is $|\Psi_i(t)\rangle$. The *probability amplitude* for a transition from the initial state $|\phi_i(t_0)\rangle$ to the final state $|\phi_f(t)\rangle = |\psi_f\rangle \exp(-iE_f t/\hbar)$, where $|\psi_f\rangle$

is an unperturbed (field-free) eigenstate of H_0 corresponding to the energy E_f, is given by projecting the state vector $|\Psi_i(t)\rangle$ on the final state $|\phi_f(t)\rangle$, namely

$$a_{fi}(t) = \langle \phi_f(t) | \Psi_i(t) \rangle$$
$$= \langle \phi_f(t) | U(t, t_0) | \phi_i(t_0) \rangle. \tag{3.53}$$

The *transition probability* from the initial state $|\phi_i(t_0)\rangle$ to the final state $|\phi_f(t)\rangle$ is then given by

$$P_{fi}(t) = |a_{fi}(t)|^2 \tag{3.54}$$

and the *transition probability per unit time*, or *transition rate*, is given by

$$W_{fi}(t) = \frac{\mathrm{d}}{\mathrm{d}t} P_{fi}(t). \tag{3.55}$$

Using Equations (3.39)–(3.41), the probability amplitude (3.53) can be expressed in the form of the perturbative series

$$a_{fi}(t) = \sum_{n=1}^{\infty} \bar{a}_{fi}^{(n)}(t), \tag{3.56}$$

where

$$\bar{a}_{fi}^{(n)}(t) = \left(-\frac{\mathrm{i}}{\hbar} \right)^n \int_{t_0}^{t} \mathrm{d}t_1 \int_{t_0}^{t_1} \mathrm{d}t_2 \ldots \int_{t_0}^{t_{n-1}} \mathrm{d}t_n$$
$$\times \langle \phi_f(t_1) | H_{\mathrm{int}}(t_1) U_0(t_1, t_2) \ldots U_0(t_{n-1}, t_n) H_{\mathrm{int}}(t_n) | \phi_i(t_n) \rangle. \tag{3.57}$$

Let us express the unperturbed evolution operator in terms of field-free atomic eigenstates of H_0 as

$$U_0(t, t_0) = \exp[-\mathrm{i}H_0(t - t_0)/\hbar]$$
$$= \sum_k \exp[-\mathrm{i}E_k(t - t_0)/\hbar] |\psi_k\rangle\langle\psi_k|, \tag{3.58}$$

where the summation runs over all the discrete- and continuous-energy eigenstates $|\psi_k\rangle$ of H_0 and we have used the closure relation

$$\sum_k |\psi_k\rangle\langle\psi_k| = I. \tag{3.59}$$

Using Equations (3.5) and (3.58), we can rewrite Equation (3.57) for $\bar{a}_{fi}^{(n)}$ in the form

$$\bar{a}_{fi}^{(n)}(t) = \left(-\frac{\mathrm{i}}{\hbar} \right)^n \sum_{k_1} \sum_{k_2} \cdots \sum_{k_{n-1}} \int_{t_0}^{t} \mathrm{d}t_1 \int_{t_0}^{t_1} \mathrm{d}t_2 \ldots \int_{t_0}^{t_{n-1}} \mathrm{d}t_n$$
$$\times \exp\left(\mathrm{i}E_f t_1/\hbar\right) \langle \psi_f | H_{\mathrm{int}}(t_1) | \psi_{k_1} \rangle \exp\left[-\mathrm{i}E_{k_1}(t_1 - t_2)/\hbar\right] \langle \psi_{k_1} | H_{\mathrm{int}}(t_2) | \psi_{k_2} \rangle$$
$$\times \cdots \times \exp\left[-\mathrm{i}E_{k_{n-1}}(t_{n-1} - t_n)/\hbar\right] \langle \psi_{k_{n-1}} | H_{\mathrm{int}}(t_n) | \psi_i \rangle \exp\left(-\mathrm{i}E_i t_n/\hbar\right). \tag{3.60}$$

In particular, to first order in the perturbation $H_{int}(t)$, the contribution to the probability amplitude is given by

$$\bar{a}_{fi}^{(1)}(t) = -\frac{i}{\hbar} \int_{t_0}^t \langle \psi_f | H_{int}(t_1) | \psi_i \rangle \exp(i\omega_{fi}t_1) dt_1, \qquad (3.61)$$

where

$$\omega_{fi} = \frac{E_f - E_i}{\hbar} \qquad (3.62)$$

is the *Bohr angular frequency* corresponding to the transition $i \to f$.

The second-order contribution to the probability amplitude is

$$\bar{a}_{fi}^{(2)}(t) = \left(-\frac{i}{\hbar}\right)^2 \sum_{k_1} \int_{t_0}^t dt_1 \int_{t_0}^{t_1} dt_2 \exp\left(iE_f t_1/\hbar\right)$$

$$\times \langle \psi_f | H_{int}(t_1) | \psi_{k_1} \rangle \exp\left[-iE_{k_1}(t_1 - t_2)/\hbar\right] \langle \psi_{k_1} | H_{int}(t_2) | \psi_i \rangle \exp\left(-iE_i t_2/\hbar\right). \qquad (3.63)$$

By keeping the first N terms of the expansion (3.56), we obtain the probability amplitude to order N in the perturbation $H_{int}(t)$. Denoting this quantity by $a_{fi}^{(N)}(t)$, we have therefore

$$a_{fi}^{(N)}(t) = \sum_{n=1}^{N} \bar{a}_{fi}^{(n)}(t). \qquad (3.64)$$

The corresponding transition probability to order N, which we shall call $P_{fi}^{(N)}(t)$, is obtained by substituting Equation (3.64) into Equation (3.54). Thus

$$P_{fi}^{(N)}(t) = |a_{fi}^{(N)}(t)|^2. \qquad (3.65)$$

In particular, the first-order transition probability is given by

$$P_{fi}^{(1)}(t) = \hbar^{-2} \left| \int_{t_0}^t \langle \psi_f | H_{int}(t_1) | \psi_i \rangle \exp(i\omega_{fi}t_1) dt_1 \right|^2, \qquad (3.66)$$

and we note that

$$P_{fi}^{(1)}(t) = P_{if}^{(1)}(t). \qquad (3.67)$$

This relation does not hold in general for approximations of order $N \geq 2$.

We also remark that we have obtained expressions for the probability amplitude $a_{fi}(t)$ and the transition probability $P_{fi}(t)$ by working in the Schrödinger picture, using Equations (3.33) and (3.34) for the Schrödinger evolution operator $U(t, t_0)$. Identical results are obtained by working in the interaction picture, using Equations

(3.29)–(3.31) for the evolution operator $U_I(t, t_0)$. Indeed, the Schrödinger and interaction pictures are connected by a unitary transformation, which leaves invariant scalar products, and therefore probability amplitudes and transition probabilities.

Experimental studies of non-resonant multiphoton ionization of atoms performed at moderate laser intensities, where perturbation theory is applicable, often use laser pulses whose durations are relatively long (of the order of 1 ps or more). This implies that the most important laser pulse parameters are the peak electric-field strength and the angular frequency of the laser field. The details of the pulse shape can usually be neglected, so that we shall assume the laser field to be monochromatic. For a linearly polarized laser field in the dipole approximation, the electric field is given by Equation (2.118), namely

$$\boldsymbol{\mathcal{E}}(t) = \hat{\boldsymbol{\epsilon}}\, \mathcal{E}_0 \cos(\omega t + \varphi), \tag{3.68}$$

and the corresponding interaction Hamiltonian is given in the length gauge by

$$H_{\text{int}}^{\text{L}}(t) = -\frac{\mathcal{E}_0}{2}\{\exp[\mathrm{i}(\omega t + \varphi)] + \exp[-\mathrm{i}(\omega t + \varphi)]\}\hat{\boldsymbol{\epsilon}} \cdot \mathbf{D}, \tag{3.69}$$

where $\mathbf{D} = -e\mathbf{R}$ is the electric dipole moment operator of the atom, defined by Equation (2.135), and $\mathbf{R}$ is given by Equation (2.130). We shall suppose that the laser field is turned on suddenly at time t_0 so that $H_{\text{int}}^{\text{L}}(t) = 0$ for $t \le t_0$. Introducing the dipole coupling matrix element

$$M_{ji} = \langle \psi_j | \hat{\boldsymbol{\epsilon}} \cdot \mathbf{D} | \psi_i \rangle, \tag{3.70}$$

we find that, to first order in the perturbation,

$$\bar{a}_{fi}^{(1)}(t) = \frac{\mathrm{i}\mathcal{E}_0}{2\hbar} M_{fi} \int_{t_0}^{t} \mathrm{d}t_1$$

$$\times \{\exp[\mathrm{i}(\omega_{fi} + \omega)t_1]\exp(\mathrm{i}\varphi) + \exp[\mathrm{i}(\omega_{fi} - \omega)t_1]\exp(-\mathrm{i}\varphi)\}, \tag{3.71}$$

while, for $n \ge 2$,

$$\bar{a}_{fi}^{(n)}(t) = \left(\frac{\mathrm{i}\mathcal{E}_0}{2\hbar}\right)^n \sum_{k_1}\sum_{k_2}\cdots\sum_{k_{n-1}} M_{fk_1} M_{k_1 k_2}\ldots M_{k_{n-1}i}$$

$$\times \int_{t_0}^{t} \mathrm{d}t_1 \exp\big[\mathrm{i}(E_f - E_{k_1})t_1/\hbar\big]\{\exp[\mathrm{i}(\omega t_1 + \varphi)] + \exp[-\mathrm{i}(\omega t_1 + \varphi)]\}$$

$$\times \int_{t_0}^{t_1} \mathrm{d}t_2 \exp\big[\mathrm{i}(E_{k_1} - E_{k_2})t_1/\hbar\big]\{\exp[\mathrm{i}(\omega t_2 + \varphi)] + \exp[-\mathrm{i}(\omega t_2 + \varphi)]\}$$

$$\times \cdots \times \int_{t_0}^{t_{n-1}} \mathrm{d}t_n \exp\big[\mathrm{i}(E_{k_{n-1}} - E_i)t_1/\hbar\big]$$

$$\times \{\exp[\mathrm{i}(\omega t_n + \varphi)] + \exp[-\mathrm{i}(\omega t_n + \varphi)]\}. \tag{3.72}$$

The integrals over the time variables $t_1, t_2, \ldots, t_n$ can be readily performed.

It is important to note that the quantity $\bar{a}_{fi}^{(n)}(t)$ is proportional to $\mathcal{E}_0^n$, and hence to $I^{n/2}$, where I is the intensity of the laser field. Since both $\mathcal{E}_0$ and I are constants, it follows from Equations (3.54) and (3.55) that the transition probability and the transition rate for an n-photon absorption or stimulated emission process given by lowest-order perturbation theory (LOPT) are proportional to I^n.

As an illustration, we shall evaluate the LOPT contribution to the n-photon transition rates of an atom which is initially in the state $|\psi_i\rangle$.

3.1.6 One-photon absorption and stimulated emission

We begin by considering the case $n = 1$. Assuming that the interaction is turned on suddenly at time $t_0 = 0$, we find by using Equation (3.71) that the first-order probability amplitude is given by

$$\bar{a}_{fi}^{(1)}(t) = \frac{\mathcal{E}_0}{2\hbar} M_{fi} \left\{ \frac{\exp\left[i\left(\omega_{fi} + \omega\right)t\right] - 1}{\omega_{fi} + \omega} \exp(i\varphi) \right.$$
$$\left. + \frac{\exp\left[i\left(\omega_{fi} - \omega\right)t\right] - 1}{\omega_{fi} - \omega} \exp(-i\varphi) \right\} \tag{3.73}$$

and the corresponding first-order transition probability is $P_{fi}^{(1)}(t) = |\bar{a}_{fi}^{(1)}(t)|^2$. It is clear from Equation (3.73) that the probability of finding the system in the state f will only be appreciable if the denominator of one or the other of the two terms on the right-hand side is close to zero. Moreover, assuming that $E_f \neq E_i$, both denominators cannot simultaneously be close to zero. Thus, a good approximation is to neglect the interference between the two terms in calculating the transition probability $P_{fi}^{(1)}(t)$, a point to which we shall return below. Thus, if the energy E_f lies in a small band about the value $E_i + \hbar\omega$, only the second term in Equation (3.73) will have an appreciable magnitude. The corresponding first-order transition probability is then given by

$$P_{fi}^{(1)}(t) = \left(\frac{\mathcal{E}_0}{2\hbar}\right)^2 |M_{fi}|^2 \left| \frac{\exp\left[i\left(\omega_{fi} - \omega\right)t\right] - 1}{\omega_{fi} - \omega} \exp(-i\varphi) \right|^2$$
$$= \frac{\mathcal{E}_0^2}{2\hbar^2} |M_{fi}|^2 F\left(t, \omega_{fi} - \omega\right), \tag{3.74}$$

where we have introduced the function

$$F(t, \bar{\omega}) = \frac{1 - \cos(\bar{\omega}t)}{\bar{\omega}^2} = 2\frac{\sin^2(\bar{\omega}t/2)}{\bar{\omega}^2} \tag{3.75}$$

and we note that $F(t, \bar{\omega} = 0) = t^2/2$. This function is plotted in Fig. 3.1 as a function of $\bar{\omega}$, for fixed t. We see that it exhibits a sharp peak about the value $\bar{\omega} = 0$.

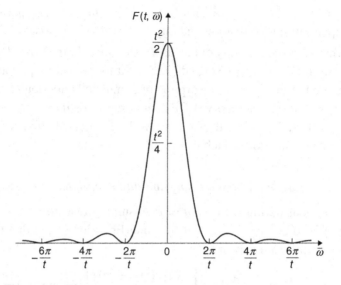

Figure 3.1. The function $F(t, \bar{\omega})$ of Equation (3.75), for fixed t.

The height of this peak is proportional to t^2, while its width at half-maximum is approximately equal to $2\pi/t$. Setting $x = \bar{\omega}t/2$, we note that

$$\int_{-\infty}^{\infty} F(t, \bar{\omega})\mathrm{d}\bar{\omega} = t \int_{-\infty}^{\infty} \frac{\sin^2 x}{x^2}\,\mathrm{d}x = \pi t. \tag{3.76}$$

It follows from the properties of the function $F(t, \bar{\omega})$ that the transition probability (3.74) will only be significant if E_f is located in an interval of width $2\pi\hbar/t$ about the value $E_i + \hbar\omega$. As a result, the first-order transition probability (3.74) will be appreciable if the system has *absorbed* a photon whose energy is given (to within $2\pi\hbar/t$) by $\hbar\omega = E_f - E_i$ (with $E_f > E_i$). When this condition is exactly satisfied, *resonance* is said to occur, and we see from Equations (3.74) and (3.75) that the first-order transition probability is given in this case by

$$P_{fi}^{(1)}(t) = \frac{\mathcal{E}_0^2}{2\hbar^2}\left|M_{fi}\right|^2 F(t, \bar{\omega} = 0)$$

$$= \frac{\mathcal{E}_0^2}{4\hbar^2}\left|M_{fi}\right|^2 t^2 \tag{3.77}$$

and hence increases quadratically with time t. Clearly, the result (3.77) is valid only for small enough times and small enough perturbations since in order for our perturbative treatment to be valid we must have

$$P_{fi}^{(1)}(t) \ll 1. \tag{3.78}$$

It is also interesting to examine the result (3.74) in the limit $t \to \infty$. Using Equation (3.75) and the Dirac delta function representation

$$\delta(x) = \lim_{\epsilon \to 0^+} \frac{\epsilon}{\pi} \frac{1 - \cos(x/\epsilon)}{x^2}, \qquad (3.79)$$

we have

$$F(t, \bar{\omega}) \xrightarrow[t \to \infty]{} \pi t \, \delta(\bar{\omega}). \qquad (3.80)$$

Hence, we find that in this limit the first-order transition probability (3.74) is given by

$$P_{fi}^{(1)}(t) = \frac{2\pi}{\hbar} \frac{\mathcal{E}_0^2}{4} \left| M_{fi} \right|^2 t \, \delta(E_f - E_i - \hbar\omega). \qquad (3.81)$$

The corresponding first-order transition probability per unit time, or transition rate, is therefore given by

$$\begin{aligned} W_{fi}^{(1)} &= \frac{\mathrm{d}}{\mathrm{d}t} P_{fi}^{(1)}(t) \\ &= \frac{2\pi}{\hbar} \frac{\mathcal{E}_0^2}{4} \left| M_{fi} \right|^2 \delta(E_f - E_i - \hbar\omega) \end{aligned} \qquad (3.82)$$

and is seen to be *time-independent*.

We note in Equations (3.81) and (3.82) the appearance of a Dirac delta function imposing energy conservation between an initial state i and a final state f having both well defined (infinitely "sharp") energies E_i and E_f, respectively. The presence of the Dirac delta function is due to the fact that we have assumed the interaction time t to be infinite, which is of course an idealization. It leads to a transition probability (3.81) or a transition rate (3.82) which are either infinite (for $E_f = E_i + \hbar\omega$) or zero (for $E_f \neq E_i + \hbar\omega$) and are therefore *unphysical*. In fact, the probability of detecting a final state f with an infinitely sharp energy $E_f > E_i$ is of measure zero. We must therefore instead consider transitions to a *group* of final states k whose energy E_k lies within an interval $(E_f - \eta, E_f + \eta)$ about the value $E_f = E_i + \hbar\omega$. Let $\rho_k(E_k)$ be the density of levels on the energy scale, so that $\rho_k(E_k)\mathrm{d}E_k$ is the number of final states k in an interval $\mathrm{d}E_k$ containing the energy E_k. The first-order transition probability $P_{fi}^{(1)}(t)$ from the initial state i to the group of states k having energies in the interval $(E_f - \eta, E_f + \eta)$ centered about $E_f = E_i + \hbar\omega$ is obtained by multiplying $P_{ki}^{(1)}(t)$ by $\rho_k(E_k)\mathrm{d}E_k$ and integrating with respect to E_k. Thus, for transitions in which the system absorbs a photon of energy $\hbar\omega = E_f - E_i$, we find by using Equation (3.74) that

$$P_{fi}^{(1)}(t) = \frac{\mathcal{E}_0^2}{2\hbar^2} \int_{E_f - \eta}^{E_f + \eta} \left| M_{ki} \right|^2 F(t, \omega_{ki} - \omega) \rho_k(E_k) \mathrm{d}E_k, \qquad (3.83)$$

with $\omega_{ki} = (E_k - E_i)/\hbar$. Assuming that η is small enough so that M_{ki} and $\rho_k(E_k)$ are nearly constant within the integration range, we have

$$P_{fi}^{(1)}(t) = \frac{\mathcal{E}_0^2}{2\hbar^2}|M_{fi}|^2 \rho_f(E_f) \int_{E_f-\eta}^{E_f+\eta} F(t, \omega_{ki} - \omega)\mathrm{d}E_k. \qquad (3.84)$$

We also suppose that t is large enough so that $\eta \gg 2\pi\hbar/t$. Moreover, it is clear from the form of the function $F(t, \bar{\omega})$, as seen from Fig. 3.1, that the integral on the right of Equation (3.84) is small except for transitions which conserve energy (within $\delta E = 2\pi\hbar/t$). We can then write

$$\int_{E_f-\eta}^{E_f+\eta} F(t, \omega_{ki} - \omega)\mathrm{d}E_k = \hbar \int_{-\infty}^{\infty} F(t, \omega_{ki} - \omega)\mathrm{d}\omega_{ki} = \pi\hbar t, \qquad (3.85)$$

where we have used the result (3.76). Hence, Equation (3.84) reduces to

$$P_{fi}^{(1)}(t) = \frac{2\pi}{\hbar} \frac{\mathcal{E}_0^2}{4}|M_{fi}|^2 \rho_f(E_f)t \qquad (3.86)$$

with $E_f \simeq E_i + \hbar\omega$. We see that the transition probability $P_{fi}^{(1)}(t)$ increases linearly with the time t for energy-conserving transitions to a group of states. It is worth stressing that the result (3.86) is soundly based only if the condition (3.78) required by our perturbative approach is satisfied.

Using Equations (3.55) and (3.86), we find that for transitions in which the system absorbs a photon of energy $\hbar\omega \simeq E_f - E_i$, the first-order transition rate is *time-independent* and is given by

$$W_{fi}^{(1)} = \frac{2\pi}{\hbar} \frac{\mathcal{E}_0^2}{4}|M_{fi}|^2 \rho_f(E_f), \qquad (3.87)$$

with $E_f \simeq E_i + \hbar\omega$. This result is a particular case of the "Golden Rule" of perturbation theory [9, 10].

Looking back at Equation (3.73), we see that if the energy E_f lies in a small band about the value $E_i - \hbar\omega$, only the first term on the right-hand side of this equation will be significant. The corresponding first-order transition probability is then given by

$$P_{fi}^{(1)}(t) = \frac{\mathcal{E}_0^2}{2\hbar^2}|M_{fi}|^2 F(t, \omega_{fi} + \omega) \qquad (3.88)$$

and will only be appreciable if the system has *emitted* a photon whose energy is given (to within $2\pi\hbar/t$) by $\hbar\omega = E_i - E_f$ (with $E_i > E_f$). We remark that for the laser pulses with relatively moderate intensities and relatively long durations (comprising a large number of optical cycles) considered in this chapter, the time t is large enough ($t \gg 2\pi/\omega$) so that the two bands of width $2\pi\hbar/t$ about the values

$E_i + \hbar\omega$ and $E_i - \hbar\omega$ do not overlap. Our neglect of the interference between the two terms on the right of Equations (3.73) is therefore justified.

Let us now return to Equation (3.88). We first note that if the condition $\hbar\omega = E_i - E_f$ (with $E_i > E_f$) is exactly satisfied, *resonance* occurs, in which case the transition probability (3.88) increases quadratically with time, as in Equation (3.77). Secondly, in the limit $t \to \infty$, we find by using Equations (3.75) and (3.80) with $\bar{\omega} = \omega_{fi} + \omega$ that

$$P_{fi}^{(1)}(t) = \frac{2\pi}{\hbar}\frac{\mathcal{E}_0^2}{4}|M_{fi}|^2 \, t \, \delta(E_f - E_i + \hbar\omega), \tag{3.89}$$

and the corresponding first-order transition rate for stimulated emission of a photon is given by the time-independent expression

$$W_{fi}^{(1)} = \frac{2\pi}{\hbar}\frac{\mathcal{E}_0^2}{4}|M_{fi}|^2 \delta(E_f - E_i + \hbar\omega). \tag{3.90}$$

As in the case of one-photon absorption discussed above, the transition probability (3.89) and the transition rate (3.90) for stimulated emission of one photon are unphysical because of the presence of the Dirac delta function $\delta(E_f - E_i + \hbar\omega)$ imposing energy conservation between an initial state i and a final state f having infinitely "sharp" energies E_i and E_f, respectively. As explained for the case of one-photon absorption, one must in fact consider transitions to a *group* of final states. Under conditions similar to those discussed above for one-photon absorption, we find that if the system *emits* a photon of energy $\hbar\omega \simeq E_i - E_f$, the transition rate is given by

$$W_{fi}^{(1)} = \frac{2\pi}{\hbar}\frac{\mathcal{E}_0^2}{4}|M_{fi}|^2 \rho_f(E_f), \tag{3.91}$$

with $E_f \simeq E_i - \hbar\omega$.

So far, we have considered the case for which the interaction is turned on suddenly at time $t_0 = 0$. We shall now examine the case for which a laser pulse of *infinite duration* is applied *adiabatically*. The standard way of dealing with this case relies on modifying the interaction Hamiltonian according to

$$H_{\text{int}}(t) \to \exp(-\epsilon|t|) H_{\text{int}} \tag{3.92}$$

and taking the limit $\epsilon \to 0^+$ at the end of the calculation. Using Equation (3.71), we have

$$\bar{a}_{fi}^{(1)}(t) = \frac{i\mathcal{E}_0}{2\hbar} M_{fi} \lim_{\epsilon \to 0^+} \int_{-\infty}^{t} dt_1 \exp(-\epsilon|t_1|)$$

$$\times \{\exp[i(\omega_{fi} + \omega)t_1]\exp(i\varphi) + \exp[i(\omega_{fi} - \omega)t_1]\exp(-i\varphi)\}, \tag{3.93}$$

and therefore

$$\bar{a}_{fi}^{(1)}(t) = \frac{\mathcal{E}_0}{2\hbar} M_{fi} \left\{ \frac{\exp[i(\omega_{fi} + \omega)t]}{\omega_{fi} + \omega} \exp(i\varphi) + \frac{\exp[i(\omega_{fi} - \omega)t]}{\omega_{fi} - \omega} \exp(-i\varphi) \right\}$$

$$(3.94)$$

is the first-order transition amplitude for finite values of t.

To obtain the corresponding amplitude in the limit $t \to \infty$, we return to Equation (3.93) and use the representation of the Dirac delta function

$$\delta(x) = \lim_{\epsilon \to 0^+} (2\pi)^{-1} \int_{-\infty}^{+\infty} \exp(-\epsilon|u|) \exp(iux) du.$$

$$(3.95)$$

We then find that $\bar{a}_{fi}^{(1)}$ is given by the time-independent expression

$$\bar{a}_{fi}^{(1)} = i\pi \mathcal{E}_0 M_{fi} [\delta(E_f - E_i + \hbar\omega) \exp(i\varphi) + \delta(E_f - E_i - \hbar\omega) \exp(-i\varphi)].$$

$$(3.96)$$

The two energy-conserving Dirac delta functions appearing on the right of Equation (3.96) correspond, respectively, to the stimulated emission of a photon of energy $\hbar\omega = E_i - E_f$ (with $E_i > E_f$) and to the absorption of a photon of energy $\hbar\omega = E_f - E_i$ (with $E_f > E_i$). The presence of these Dirac delta functions is a consequence of taking the limit $t \to \infty$ in Equation (3.93). It leads to probability amplitudes $\bar{a}_{fi}^{(1)}$ for stimulated emission or absorption of one photon which are unphysical since they are infinite when the energy conservation relation is satisfied, and zero otherwise. Using Equations (3.54) and (3.96), we see that the first-order transition probabilities for the absorption or the stimulated emission of one photon are given by

$$P_{fi}^{(1)} = \pi^2 \mathcal{E}_0^2 |M_{fi}|^2 \delta^2(E_f - E_i \pm \hbar\omega)$$

$$(3.97)$$

with the minus sign corresponding to absorption and the plus sign to stimulated emission.

In order to assign a meaning to the expression (3.97), we first write, following Lippmann and Schwinger [11],

$$\delta^2(E_f - E_i \pm \hbar\omega) \to \delta(0)\delta(E_f - E_i \pm \hbar\omega).$$

$$(3.98)$$

Next, while we have formally taken the interaction time to be infinite, we must take into account the fact that the atom interacts with the laser field for a long, but finite, time. For an interaction time comprising a large number of optical cycles $(t \gg 2\pi/\omega)$, we therefore modify the right-hand side of the relation (3.98) as follows:

$$\delta(0)\delta(E_f - E_i \pm \hbar\omega) \rightarrow (2\pi\hbar)^{-1}\left(\int_{-t/2}^{t/2} \exp[i(E_f - E_i \pm \hbar\omega)t_1/\hbar]dt_1\right)$$

$$\times \delta(E_f - E_i \pm \hbar\omega)$$

$$= (2\pi\hbar)^{-1} t\, \delta(E_f - E_i \pm \hbar\omega). \tag{3.99}$$

As a result, the first-order transition probabilities (3.97) become

$$P_{fi}^{(1)} = \frac{2\pi}{\hbar}\frac{\mathcal{E}_0^2}{4}|M_{fi}|^2\, t\, \delta(E_f - E_i \pm \hbar\omega), \tag{3.100}$$

in agreement with the results (3.81) and (3.89) obtained for one-photon absorption and stimulated emission, respectively, when the interaction is turned on suddenly and the limit of large interaction times ($t \rightarrow \infty$) is taken. In the same way, the first-order transition rates obtained from the transition probabilities (3.100) are given by the time-independent expressions

$$W_{fi}^{(1)} = \frac{2\pi}{\hbar}\frac{\mathcal{E}_0^2}{4}|M_{fi}|^2\delta(E_f - E_i \pm \hbar\omega), \tag{3.101}$$

which are in agreement with our previous results (3.82) and (3.90) obtained in the limit $t \rightarrow \infty$ when the interaction is turned on suddenly.

As explained above, the results (3.100) and (3.101) are unphysical because of the presence of the Dirac delta functions $\delta(E_f - E_i \pm \hbar\omega)$ imposing energy conservation between an initial state i and a final state f having infinitely "sharp" energies E_i and E_f, respectively. Again, one should consider transitions to a group of final states. Following a procedure similar to that used for the case of a sudden "turn on" of the interaction, one finds that the first-order transition rate for an adiabatic switching of the interaction is time-independent and is given by

$$W_{fi}^{(1)} = \frac{2\pi}{\hbar}\frac{\mathcal{E}_0^2}{4}|M_{fi}|^2\rho_f(E_f), \tag{3.102}$$

with $E_f \simeq E_i + \hbar\omega$ for absorption and $E_f \simeq E_i - \hbar\omega$ for stimulated emission of one photon.

3.1.7 Two-photon absorption and stimulated emission. AC Stark shift

Using Equation (3.72), in which we set $n = 2$, and assuming that the interaction is switched on adiabatically at $t_0 = -\infty$, the second-order contribution to the probability amplitude for the transition $i \rightarrow f$ is given by

$$\bar{a}_{fi}^{(2)}(t) = \left(\frac{i\mathcal{E}_0}{2\hbar}\right)^2 \lim_{\epsilon \rightarrow 0^+} \sum_k M_{fk} M_{ki}$$

$$\times \left[\int_{-\infty}^t dt_1 \exp(-\epsilon|t_1|) \exp(i(\omega_{fk}t_1))\{\exp[i(\omega t_1 + \varphi)]\right.$$

$$+ \exp[-i(\omega t_1 + \varphi)]\}\Bigg]$$

$$\times \left[\int_{-\infty}^{t_1} dt_2 \exp(-\epsilon|t_2|) \exp(i(\omega_{ki}t_2))\{\exp[i(\omega t_2 + \varphi)]\right.$$

$$+ \exp[-i(\omega t_2 + \varphi)]\}\Bigg]. \tag{3.103}$$

By letting $t \rightarrow \infty$ in this expression and using Equation (3.95), we find that

$$\bar{a}_{fi}^{(2)} = 2\pi i \left(\frac{\mathcal{E}_0}{2}\right)^2 \left[\delta(E_f - E_i)\sum_k \left(\frac{M_{fk}M_{ki}}{E_k - E_i + \hbar\omega} + \frac{M_{fk}M_{ki}}{E_k - E_i - \hbar\omega}\right)\right.$$

$$+ \delta(E_f - E_i - 2\hbar\omega) \exp(2i\varphi) \sum_k \frac{M_{fk}M_{ki}}{E_k - E_i - \hbar\omega}$$

$$+ \delta(E_f - E_i + 2\hbar\omega) \exp(-2i\varphi) \sum_k \frac{M_{fk}M_{ki}}{E_k - E_i + \hbar\omega}\Bigg]. \tag{3.104}$$

The second term on the right-hand side of this equation corresponds to the absorption of two photons ($E_f = E_i + 2\hbar\omega$) and the third one refers to the stimulated emission of two photons ($E_f = E_i - 2\hbar\omega$). As in the case of the one-photon transitions discussed in the preceding section, the presence of the Dirac delta functions $\delta(E_f - E_i \pm 2\hbar\omega)$ leads to unphysical transition probabilities and transition rates. By considering transitions to a group of final states and following a procedure similar to that used for one-photon transitions, we find that the second-order (LOPT) transition rate for two-photon absorption is given by

$$W_{fi}^{(2)} = \frac{2\pi}{\hbar} \left(\frac{\mathcal{E}_0^2}{4}\right)^2 |T_{fi}^{(2)}|^2 \rho_f(E_f), \tag{3.105}$$

with $E_f \simeq E_i + 2\hbar\omega$ and

$$T_{fi}^{(2)} = \sum_k \frac{\langle\psi_f|\hat{\boldsymbol{\epsilon}} \cdot \mathbf{D}|\psi_k\rangle\langle\psi_k|\hat{\boldsymbol{\epsilon}} \cdot \mathbf{D}|\psi_i\rangle}{E_i + \hbar\omega - E_k}, \tag{3.106}$$

where we have used Equation (3.70).

Proceeding in a similar way with the third term on the right-hand side of Equation (3.104), we find that the second-order (LOPT) transition rate for stimulated emission of two photons is given by Equation (3.105), where now $E_f \simeq E_i - 2\hbar\omega$ and

$$T_{fi}^{(2)} = \sum_k \frac{\langle \psi_f | \hat{\boldsymbol{\epsilon}} \cdot \mathbf{D} | \psi_k \rangle \langle \psi_k | \hat{\boldsymbol{\epsilon}} \cdot \mathbf{D} | \psi_i \rangle}{E_i - \hbar\omega - E_k}. \tag{3.107}$$

The first term on the right-hand side of Equation (3.104) is non-vanishing only when $E_i = E_f$, and two cases must be considered. The first one is such that the state $|\psi_f\rangle$ is *different* from the state $|\psi_i\rangle$, and hence must be *degenerate* with $|\psi_i\rangle$. The first term on the right-hand side of Equation (3.104) then gives the *mixing amplitude* between this degenerate pair of states. The second case arises if the state $|\psi_f\rangle$ is *identical* to the state $|\psi_i\rangle$. The first term on the right-hand side of Equation (3.104) then yields the second-order contribution $\Delta E_i^{(2)}$ to a shift of the energy level E_i called *dynamic* or *alternating current* (AC) *Stark shift*. Indeed, by keeping the time t finite in Equation (3.103), we see that this term contributes to $\bar{a}_{fi}^{(2)}(t)$ by an amount

$$\mathrm{i}\frac{t}{\hbar}\left(\frac{\mathcal{E}_0}{2}\right)^2 \sum_{k\neq i} |M_{ki}|^2 \left(\frac{1}{E_k - E_i + \hbar\omega} + \frac{1}{E_k - E_i - \hbar\omega}\right). \tag{3.108}$$

Using Equation (3.53), in which we set $f = i$, we see that the state vector $|\Psi_i(t)\rangle$ is modified to second order in $\mathcal{E}_0$ to read

$$|\Psi_i(t)\rangle = |\psi_i\rangle \exp(-\mathrm{i}E_i t/\hbar)\left[1 + \mathrm{i}\frac{t}{\hbar}\left(\frac{\mathcal{E}_0}{2}\right)^2 \sum_{k\neq i} |M_{ki}|^2 \right.$$

$$\left. \times \left(\frac{1}{E_k - E_i + \hbar\omega} + \frac{1}{E_k - E_i - \hbar\omega}\right)\right]$$

$$\simeq |\psi_i\rangle \exp[-\mathrm{i}(E_i + \Delta E_i^{(2)})t/\hbar], \tag{3.109}$$

where

$$\Delta E_i^{(2)} = -\left(\frac{\mathcal{E}_0}{2}\right)^2 \sum_{k\neq i} |M_{ki}|^2 \left(\frac{1}{E_k - E_i + \hbar\omega} + \frac{1}{E_k - E_i - \hbar\omega}\right). \tag{3.110}$$

We can also write Equation (3.110) of the second-order (LOPT) contribution to the AC Stark shift of the energy level E_i as

$$\Delta E_i^{(2)} = -\bar{\mathcal{E}}^2 \sum_{k\neq i} |\langle \psi_k | \hat{\boldsymbol{\epsilon}} \cdot \mathbf{D} | \psi_i \rangle|^2 \frac{E_k - E_i}{(E_k - E_i)^2 - \hbar^2\omega^2}, \tag{3.111}$$

where $\bar{\mathcal{E}}^2$ is the average of $\mathcal{E}^2(t)$ over an optical period $T = 2\pi/\omega$. For linear polarization,

$$\bar{\mathcal{E}}^2 = \frac{1}{T} \int_0^T \mathcal{E}^2(t)\, dt$$

$$= \frac{\mathcal{E}_0^2}{T} \int_0^T \cos^2(\omega t + \varphi)\, dt = \frac{\mathcal{E}_0^2}{2}. \tag{3.112}$$

Introducing the dynamic (or frequency-dependent) dipole polarizability of the atom in the state $|\psi_i\rangle$, namely

$$\alpha(i,\omega) = 2 \sum_{k \neq i} |\langle \psi_k | \hat{\boldsymbol{\epsilon}} \cdot \mathbf{D} | \psi_i \rangle|^2 \frac{E_k - E_i}{(E_k - E_i)^2 - \hbar^2 \omega^2}, \tag{3.113}$$

we see that the second-order contribution to the AC Stark shift of the energy level E_i can be written in the form

$$\Delta E_i^{(2)} = -\frac{1}{2}\alpha(i,\omega)\bar{\mathcal{E}}^2. \tag{3.114}$$

We also note that when $\omega \to 0$ Equation (3.111) reduces to the level shift corresponding to the quadratic Stark effect for a static electric field of strength $\mathcal{E} = \bar{\mathcal{E}}$ directed parallel to the polarization vector $\hat{\boldsymbol{\epsilon}}$. That is,

$$\left(\Delta E_i^{(2)}\right)_S = -\mathcal{E}^2 \sum_{k \neq i} \frac{|\langle \psi_k | \hat{\boldsymbol{\epsilon}} \cdot \mathbf{D} | \psi_i \rangle|^2}{E_k - E_i}$$

$$= -\frac{1}{2}\bar{\alpha}(i)\mathcal{E}^2, \tag{3.115}$$

where

$$\bar{\alpha}(i) = 2 \sum_{k \neq i} \frac{|\langle \psi_k | \hat{\boldsymbol{\epsilon}} \cdot \mathbf{D} | \psi_i \rangle|^2}{E_k - E_i} \tag{3.116}$$

is the *static dipole polarizability* of the atom in the state $|\psi_i\rangle$.

3.1.8 Transitions involving n photons

The LOPT expression (3.105) for the transition rate corresponding to two-photon absorption is readily generalized to the LOPT transition rate for the absorption of n photons. We find that

$$W_{fi}^{(n)} = \frac{2\pi}{\hbar} \left(\frac{\mathcal{E}_0^2}{4}\right)^n |T_{fi}^{(n)}|^2 \rho_f(E_f), \tag{3.117}$$

where $E_f \simeq E_i + n\hbar\omega$ and $T_{fi}^{(n)}$ is the LOPT transition matrix element for the absorption of n photons, given in the length gauge by

$$T_{fi}^{(n)} = \sum_{k_1}\sum_{k_2}\cdots\sum_{k_{n-1}} \frac{\langle\psi_f|\hat{\boldsymbol{\epsilon}}\cdot\mathbf{D}|\psi_{k_{n-1}}\rangle\ldots\langle\psi_{k_2}|\hat{\boldsymbol{\epsilon}}\cdot\mathbf{D}|\psi_{k_1}\rangle\langle\psi_{k_1}|\hat{\boldsymbol{\epsilon}}\cdot\mathbf{D}|\psi_i\rangle}{(E_i + (n-1)\hbar\omega - E_{k_{n-1}})\ldots(E_i + 2\hbar\omega - E_{k_2})(E_i + \hbar\omega - E_{k_1})}.$$

(3.118)

Using Equation (2.39), remembering that $\mathbf{D} = -e\mathbf{R}$ and introducing the fine-structure constant $\alpha = e^2/(4\pi\epsilon_0\hbar c)$, we can also rewrite Equations (3.117) and (3.118) as

$$W_{fi}^{(n)} = \frac{2\pi}{\hbar}(2\pi\alpha\hbar)^n I^n |\tilde{T}_{fi}^{(n)}|^2 \rho_f(E_f),$$

(3.119)

where $I = \epsilon_0 c \mathcal{E}_0^2/2$ denotes the intensity, and the LOPT transition matrix element $\tilde{T}_{fi}^{(n)}$ is given by

$$\tilde{T}_{fi}^{(n)} = \sum_{k_1}\sum_{k_2}\cdots\sum_{k_{n-1}} \frac{\langle\psi_f|\hat{\boldsymbol{\epsilon}}\cdot\mathbf{R}|\psi_{k_{n-1}}\rangle\ldots\langle\psi_{k_2}|\hat{\boldsymbol{\epsilon}}\cdot\mathbf{R}|\psi_{k_1}\rangle\langle\psi_{k_1}|\hat{\boldsymbol{\epsilon}}\cdot\mathbf{R}|\psi_{k_i}\rangle}{(E_i + (n-1)\hbar\omega - E_{k_{n-1}})\ldots(E_i + 2\hbar\omega - E_{k_2})(E_i + \hbar\omega - E_{k_1})}.$$

(3.120)

We remark that LOPT expressions for the stimulated emission of n photons can be obtained in a similar manner, as well as higher-order contributions to the AC Stark shift.

Formally equivalent expressions for the LOPT transition rates $W_{fi}^{(n)}$ can be written down in other gauges, such as the velocity gauge. Since gauge transformations are particular cases of unitary transformations, which leave invariant the transition probabilities, it follows that transition rates are *formally* gauge invariant, and this property also holds for any order of perturbation theory, as expected on physical grounds. However, in nearly all *practical* cases the transition probabilities can only be calculated approximately, so that differences between results obtained by using different gauges will inevitably occur. Of course, these differences will be small if accurate calculations are performed.

The calculation of LOPT transition matrix elements is in general a difficult task, particularly for higher-order multiphoton processes and (or) for complex atoms. The simplest case is that of non-resonant multiphoton ionization in one-electron atoms for which LOPT has been applied successfully for intensities $I < 10^{13}\,\mathrm{W\,cm^{-2}}$ and angular frequencies ω such that $\omega \gg U_p$ [3, 12–15].

Discrepancies from the I^n power law of Equation (3.119) for an n-photon process, which are found at higher intensities, signal the breakdown of the LOPT approach. These discrepancies are due to the fact that the contributions of higher-order matrix elements become dominant, a behavior typical of asymptotic series. This behavior has been studied in detail for the case of the DC Stark effect [16].

In the case of AC fields of interest here, various resummation schemes, based on matrix continued-fraction expansions, have been considered [17, 18], but they are difficult to implement in applications. The breakdown of perturbation theory is also clearly exhibited in other strong field phenomena such as the "peak suppression" in ATI spectra (see Fig. 1.4), the existence of a plateau in ATI spectra (see Fig. 1.9) or in high-order harmonic generation (see Fig. 1.12), and in successive "free–free transition" peaks of comparable height in laser-assisted electron–atom collisions (see Fig. 1.16).

It is also clear from Equations (3.118) or (3.120) that the perturbative approach always fails for *resonant* multiphoton processes, such that one or several factors in the denominator vanish, that is if

$$E_i + r\hbar\omega = E_{k_r} \qquad (3.121)$$

for a particular $r \in \{1, 2, \ldots, n-1\}$. In this case, modifications of the theory are required, in which the resonantly coupled states are treated in a non-perturbative way, whereas the other states are treated by perturbation theory [1,2]. This approach, called the *method of essential states*, belongs to the category of semi-perturbative methods, and will be discussed in Section 3.3.1.

3.1.9 Selection rules

In Equations (3.118) or (3.120) the summations over the intermediate states $|\psi_{k_1}\rangle$, $|\psi_{k_2}\rangle, \ldots, |\psi_{k_{n-1}}\rangle$ include integrations over the continuum for all sequences of one-photon electric dipole (E1) transitions allowed by angular momentum and parity selection rules [10].

As a simple example, let us consider a one-electron atom or ion, which is initially in an unperturbed energy eigenstate $|\psi_i\rangle$. Neglecting spin-orbit coupling, we assume that this state is characterized by the orbital angular momentum quantum number l_i and the magnetic quantum number m_i. Denoting by l_f the final state orbital angular momentum quantum number, and by m_f the final state magnetic quantum number, the electric dipole (E1) selection rules for *one-photon absorption* are given in this case by

$$\Delta l = l_f - l_i = \pm 1 \qquad (3.122)$$

with

$$
\begin{aligned}
\Delta m = m_f - m_i &= 0 && \text{for linear polarization,} \\
\Delta m = m_f - m_i &= +1 && \text{for left circular polarization,} \\
\Delta m = m_f - m_i &= -1 && \text{for right circular polarization.}
\end{aligned}
\qquad (3.123)
$$

For *one-photon emission*, the orbital angular momentum quantum number selection rule is given by Equation (3.122), while the magnetic quantum number selection rules are given by

$$
\begin{aligned}
\Delta m = m_f - m_i = 0 \qquad & \text{for linear polarization,} \\
\Delta m = m_f - m_i = -1 \qquad & \text{for left circular polarization,} \\
\Delta m = m_f - m_i = +1 \qquad & \text{for right circular polarization.}
\end{aligned} \qquad (3.124)
$$

We also note that for one-photon electric dipole (E1) transitions the initial and final atomic states have opposite parity.

The multiphoton selection rules are obtained by applying successively the electric dipole selection rules to obtain the angular momentum and parity quantum numbers of the intermediate states and of the final state. Thus, in the example considered above, the selection rules for *two-photon absorption* can be obtained by applying twice the one-photon absorption selection rules (3.122) and (3.123). We find in this way that the orbital angular momentum and magnetic quantum numbers of the intermediate and final state are given in the following way.

(a) For linear polarization

$$
\begin{array}{ccc}
 & & (l_i + 2, m_i) \\
 & \nearrow & \\
 & (l_i + 1, m_i) & \\
\nearrow & & \searrow \\
(l_i, m_i) & & (l_i, m_i) \\
\searrow & & \nearrow \\
 & (l_i - 1, m_i) & \\
 & & \searrow \\
 & & (l_i - 2, m_i)
\end{array} \qquad (3.125)
$$

$$
\text{Initial state} \qquad \text{Intermediate state} \qquad \text{Final state}
$$

(b) For left circular polarization

$$
\begin{array}{ccc}
 & & (l_i + 2, m_i + 2) \\
 & \nearrow & \\
 & (l_i + 1, m_i + 1) & \\
\nearrow & & \searrow \\
(l_i, m_i) & & (l_i, m_i + 2) \\
\searrow & & \nearrow \\
 & (l_i - 1, m_i + 1) & \\
 & & \searrow \\
 & & (l_i - 2, m_i + 2)
\end{array} \qquad (3.126)
$$

$$
\text{Initial state} \qquad \text{Intermediate state} \qquad \text{Final state}
$$

(c) For right circular polarization

$$
\begin{array}{ccc}
 & & (l_i+2, m_i-2) \\
 & \nearrow & \\
(l_i+1, m_i-1) & & \\
\nearrow & \searrow & \\
 & & (l_i, m_i-2) \\
(l_i, m_i) & \nearrow & \\
\searrow & & \\
(l_i-1, m_i-1) & & \\
 & \searrow & \\
 & & (l_i-2, m_i-2)
\end{array}
\tag{3.127}
$$

Initial state   Intermediate state   Final state

It is clear from the foregoing discussion that the spectroscopy of multiphoton processes is much richer than that of single-photon transitions. For example, the $1s - 2s$ transition in atomic hydrogen, which is "forbidden" by the electric dipole one-photon selection rules, can be induced by two-photon absorption.

3.1.10 Green's operators and functions

The transition matrix elements (3.118) or (3.120) can be expressed in a more compact way by introducing the *Green's operator*

$$
G_0(E) = \frac{1}{E - H_0},
\tag{3.128}
$$

which has the spectral representation

$$
G_0(E) = \sum_k \frac{|\psi_k\rangle\langle\psi_k|}{E - E_k}.
\tag{3.129}
$$

As usual, the sum over the index k includes a summation over the discrete part and an integration over the continuum part of the set of energy eigenstates $|\psi_k\rangle$ of H_0. The states belonging to the continuum part of the spectrum can correspond in the position representation to either outgoing, incoming or stationary waves, as each form a complete set of continuum states. The object

$$
G_0(z) = \frac{1}{z - H_0}
\tag{3.130}
$$

considered as a function of the complex variable z is called the *resolvent* of the operator H_0. It is a bounded operator in Hilbert space for every value of the variable z except at the eigenvalues of H_0. The Green's operator (3.128) can therefore be defined as the limit of the resolvent (3.130) when z approaches the real axis.

Using Equation (3.128), we can rewrite the transition matrix element (3.118) as

$$T_{fi}^{(n)} = \langle \psi_f | \hat{\boldsymbol{\epsilon}} \cdot \mathbf{D} \, G_0 [E_i + (n-1)\hbar\omega] \dots \hat{\boldsymbol{\epsilon}} \cdot \mathbf{D} \, G_0(E_i + \hbar\omega) \, \hat{\boldsymbol{\epsilon}} \cdot \mathbf{D} | \psi_i \rangle . \quad (3.131)$$

Let us now elect to work in the position representation. We then obtain the *Green's function* corresponding to the Green's operator $G_0(E)$, namely

$$G_0(E, X, X') \equiv \langle X | \frac{1}{E - H_0} | X' \rangle$$

$$= \sum_k \frac{\langle X | \psi_k \rangle \langle \psi_k | X' \rangle}{E - E_k}$$

$$= \sum_k \frac{\psi_k(X) \psi_k^*(X')}{E - E_k}, \quad (3.132)$$

where $\psi_k(X) \equiv \langle X | \psi_k \rangle$ is the wave function (in the position representation) corresponding to the unperturbed energy eigenstate $|\psi_k\rangle$.

3.1.11 The Coulomb Green's function

As an example, let us consider the simple case of one-electron (hydrogenic) atoms and ions in the infinite nuclear mass approximation, for which

$$H_0 \equiv H_c = \frac{\mathbf{p}^2}{2} - \frac{Z}{r} \quad (3.133)$$

is the non-relativistic Coulomb Hamiltonian. Here and in the remaining part of this sub-section we shall use atomic units (a.u.). The corresponding *Coulomb Green's function* is given by

$$G_c(E, \mathbf{r}, \mathbf{r}') = \langle \mathbf{r} | \frac{1}{E - H_c} | \mathbf{r}' \rangle$$

$$= \sum_{n=1}^{\infty} \sum_{l=0}^{n-1} \sum_{m=-l}^{l} \frac{\psi_{nlm}^c(Z, \mathbf{r}) \psi_{nlm}^{c*}(Z, \mathbf{r}')}{E - E_n}$$

$$+ \lim_{\epsilon \to 0^+} \int d\kappa \frac{\psi_\kappa^{c(+)}(Z, \mathbf{r}) \psi_\kappa^{c(+)*}(Z, \mathbf{r}')}{E - E_\kappa + i\epsilon}. \quad (3.134)$$

In this equation,

$$E_n = -\frac{Z^2}{2n^2}, \quad n = 1, 2, \dots, \quad (3.135)$$

are the energy eigenvalues of the bound states, and the summation on the indices (n, l, m) runs only over the *discrete* hydrogenic energy eigenfunctions

$$\psi_{nlm}^c(Z, \mathbf{r}) = R_{nl}(Z, r) Y_{lm}(\theta, \phi), \quad (3.136)$$

where $Y_{lm}(\theta, \phi)$ are spherical harmonics with $l = 0, 1, \ldots, n-1$ and $m = -l, -l+1, \ldots, l$. The radial functions $R_{nl}(Z, r)$ are given by [10]

$$R_{nl}(Z, r) = \frac{1}{(2l+1)!} \left\{ \left(\frac{2Z}{n} \right)^3 \frac{(n+l)!}{2n(n-l-1)!} \right\}^{1/2}$$

$$\times \exp(-\rho/2)\rho^l \, _1F_1(l+1-n, 2l+2, \rho), \tag{3.137}$$

where $\rho = 2Zr/n$ and $_1F_1(l+1-n, 2l+2, \rho)$ is a confluent hypergeometric function which, apart from a multiplicative constant, reduces to an associated Laguerre polynomial of order $n-l-1$ [10]. In Equation (3.134) the integral runs over the continuum energy eigenstates, with $E_\kappa = \hbar^2\kappa^2/(2m)$. In writing this equation, we have adopted the continuum Coulomb energy eigenfunctions $\psi_\kappa^{c(+)}(Z, \mathbf{r})$ describing the motion of an electron of wave vector κ and (positive) energy E_κ moving in the Coulomb potential

$$V_c(r) = -\frac{Z}{r} \tag{3.138}$$

and exhibiting an outgoing (+) spherical wave asymptotic behavior. However, any complete set for the continuum part of the spectrum, for example the continuum Coulomb energy eigenfunctions $\psi_\kappa^{c(-)}(Z, \mathbf{r})$, have an incoming (−) spherical wave asymptotic behavior is equally suitable. We shall normalize these Coulomb wave functions to a delta function in wave vector space, namely

$$\langle \psi_{\kappa'}^{c(\pm)}(Z, \mathbf{r}) | \psi_\kappa^{c(\pm)}(Z, \mathbf{r}) \rangle = \delta(\kappa - \kappa'), \tag{3.139}$$

so that they are given by [10]

$$\psi_\kappa^{c(\pm)}(Z, \mathbf{r}) = (2\pi)^{-3/2} \exp(-\pi\eta_\kappa/2)\Gamma(1 \pm i\eta_\kappa)$$

$$\times \exp(i\kappa \cdot \mathbf{r})_1F_1[\mp i\eta_\kappa, 1, \pm i(\kappa r \mp \kappa \cdot \mathbf{r})], \tag{3.140}$$

where Γ is Euler's gamma function and η_κ is the Sommerfeld parameter, which is given by $\eta_\kappa = -Z/\kappa$.

Using the fact that the Coulomb potential (3.138) is spherically symmetric, we can separate the radial part of the Coulomb Green's function from its angular part by expanding it in spherical harmonics:

$$G_c(E, \mathbf{r}, \mathbf{r}') = \sum_{l=0}^{\infty} g_l^c(E, r, r') \sum_{m=-l}^{l} Y_{lm}(\hat{\mathbf{r}}) Y_{lm}^*(\hat{\mathbf{r}}'). \tag{3.141}$$

A number of representations of the radial Coulomb Green's functions $g_l^c(E, r, r')$ are available [2]. The first one is the closed-form representation

$$g_l^c(E, r, r') = -\frac{v/Z}{rr'} \frac{\Gamma(l+1-v)}{\Gamma(2l+2)} M_{v,l+1/2}\left(\frac{2Z}{v} r_<\right) W_{v,l+1/2}\left(\frac{2Z}{v} r_>\right),$$

(3.142)

where

$$v = \frac{Z}{(-2E)^{1/2}},$$

(3.143)

$r_<$ is the lesser and $r_>$ the greater of r and r'. In Equation (3.142), $M_{v,l+1/2}$ and $W_{v,l+1/2}$ denote the regular and irregular Whittaker functions [19], respectively.

A second representation of the radial Coulomb Green's function is the integral representation [20]

$$g_l^c(E, r, r') = -2(rr')^{1/2} \int_0^\infty dx \, \exp\left\{\left[-\frac{Z}{v}(r+r')\right] \cosh x\right\}$$

$$\times (\coth \frac{x}{2})^{2v} I_{2l+1}\left[\frac{2Z}{v}(rr')^{1/2} \sinh x\right],$$

(3.144)

where I_{2l+1} is a modified Bessel function [19].

A third representation of $g_l^c(E, r, r')$ is the Sturmian representation [21, 22]

$$g_l^c(E, r, r') = \sum_{n=l+1}^\infty \frac{S_{nl}^\kappa(r) S_{nl}^\kappa(r')}{Z + in\kappa},$$

(3.145)

where the functions S_{nl}^κ are *Sturmian functions* [23, 24] defined as

$$S_{nl}^\kappa(r) = \frac{1}{(2l+1)!}\left[\frac{(n+l)!}{(n-l-1)!}\right]^{1/2} (-2i\kappa r)^{l+1}$$

$$\times \exp(i\kappa r) \, {}_1F_1(l+1-n, 2l+2, -2i\kappa r),$$

(3.146)

where the "wave number" κ is in general complex. The confluent hypergeometric function ${}_1F_1(l+1-n, 2l+2, -2i\kappa r)$ is a polynomial of order $n-l-1$. It should be noted that the variable r appears in the combination κr and that the number of nodes of $S_{nl}^\kappa(r)$ in the variable κr is $n-l$. For large r, one has

$$S_{nl}^\kappa(r) \simeq a_{nl} \, r^n \exp(i\kappa r),$$

(3.147)

where the quantities a_{nl} are constants. The Sturmian functions (3.146) are orthonormal with respect to the weight function $1/r$:

$$\int_0^\infty S_{n'l'}^\kappa(r) \frac{1}{r} S_{nl}^\kappa(r) dr = \delta_{n'n},$$

(3.148)

and this orthonormality property can be analytically continued throughout the complex κ plane. The Sturmian representation (3.145) provides a completely separable and denumerable (purely "discrete") representation of the radial Coulomb Green's function $g_l^c(E, r, r')$. It has been very useful in the calculation of multiphoton processes in hydrogenic systems.

Finally, we note that representations of the full Coulomb Green's function (3.134) have been given in three-dimensional space [25, 26] and in hyperspherical space [27]. Representations of the corresponding Coulomb Green's function in momentum space

$$G_c(E, \mathbf{p}, \mathbf{p}') = \langle \mathbf{p} | \frac{1}{E - H_c} | \mathbf{p}' \rangle \tag{3.149}$$

are also available [28–30].

A detailed review of the Coulomb Green's function has been given by Maquet, Véniard and Marian [31]. We also remark that the usefulness of the Coulomb Green's function can be extended to alkali-like quasi-one-electron atoms and ions by using quantum defect theory [32].

3.1.12 Inhomogeneous differential equations

Another useful scheme for performing the summations on the intermediate states appearing in the expressions (3.118) or (3.120) is the method of *inhomogeneous differential equations* (IDEs) [33–35]. This method has been applied first to calculate second-order processes in atomic hydrogen [35–38] and was then generalized [17, 39–41] to treat many-photon ionization calculations. An important advantage of the IDE method is that it can be applied to systems which are non-hydrogenic and therefore cannot be treated analytically.

Let us work in the length gauge and write the transition matrix element (3.120) in the form

$$\tilde{T}_{fi}^{(n)} = \langle \psi_f | \hat{\boldsymbol{\epsilon}} \cdot \mathbf{R} | \bar{\psi}_i^{(n-1)} \rangle, \tag{3.150}$$

where

$$|\bar{\psi}_i^{(n-1)}\rangle = \sum_{k_1} \sum_{k_2} \cdots \sum_{k_{n-2}} \sum_{k_{n-1}} |\psi_{k_{n-1}}\rangle$$

$$\times \frac{\langle \psi_{k_{n-1}} | \hat{\boldsymbol{\epsilon}} \cdot \mathbf{R} | \psi_{k_{n-2}} \rangle \cdots \langle \psi_{k_2} | \hat{\boldsymbol{\epsilon}} \cdot \mathbf{R} | \psi_{k_1} \rangle \langle \psi_{k_1} | \hat{\boldsymbol{\epsilon}} \cdot \mathbf{R} | \psi_i \rangle}{(E_i + (n-1)\hbar\omega - E_{k_{n-1}}) \cdots (E_i + 2\hbar\omega - E_{k_2})(E_i + \hbar\omega - E_{k_1})} \tag{3.151}$$

is the part of the $(n-1)$th-order perturbed wave function, corresponding to the absorption of $n - 1$ photons, that has evolved from the unperturbed initial state

$|\psi_i\rangle$. Starting with the sum on k_1, we write, for $n = 2$,

$$|\bar{\psi}_i^{(1)}\rangle = \sum_{k_1} |\psi_{k_1}\rangle \frac{\langle \psi_{k_1} | \hat{\epsilon} \cdot \mathbf{R} | \psi_i \rangle}{E_i + \hbar\omega - E_{k_1}}. \tag{3.152}$$

Premultiplying both sides of this equation by the operator $(E_i + \hbar\omega - H_0)$ and remembering that $H_0 |\psi_k\rangle = E_k |\psi_k\rangle$, we find that $|\bar{\psi}_i^{(1)}\rangle$ satisfies the inhomogeneous differential equation

$$(E_i + \hbar\omega - H_0)|\bar{\psi}_i^{(1)}\rangle = \sum_{k_1} |\psi_{k_1}\rangle \langle \psi_{k_1} | \hat{\epsilon} \cdot \mathbf{R} | \psi_i \rangle$$

$$= \hat{\epsilon} \cdot \mathbf{R} | \psi_i \rangle, \tag{3.153}$$

where we have used the closure property (3.59) of the unperturbed atomic states. The inhomogeneous differential equation (3.153) can then be solved to obtain $|\bar{\psi}_i^{(1)}\rangle$. This procedure can then be repeated to obtain $|\bar{\psi}_i^{(2)}\rangle$, $|\bar{\psi}_i^{(3)}\rangle$, ..., and finally $|\bar{\psi}_i^{(n-1)}\rangle$ by solving successively a sequence of IDEs.

3.1.13 Multiphoton transition rates and cross sections for excitation and ionization. Generalized cross sections

Let us return to Equation (3.119) giving the LOPT transition rate for the absorption of n photons. The following three cases of interest arise,

(1) "Bound–bound" transitions from a discrete initial state to discrete final states with a line shape function $S_f(E_f)$. In this case, the LOPT transition rate for an n-photon excitation transition is given by

$$W_{\text{exc}}^{(n)} = \frac{2\pi}{\hbar} (2\pi\hbar\alpha)^n I^n |\tilde{T}_{fi}^{(n)}|^2 S_f(E_f). \tag{3.154}$$

We can also express this transition rate in terms of the incident photon flux $\Phi = I/(\hbar\omega)$, namely

$$W_{\text{exc}}^{(n)} = \frac{2\pi}{\hbar} (2\pi\hbar^2\alpha\omega)^n \Phi^n |\tilde{T}_{fi}^{(n)}|^2 S_f(E_f). \tag{3.155}$$

The corresponding LOPT excitation cross section is obtained by dividing the transition rate (3.155) by the incident photon flux. That is,

$$\sigma_{\text{exc}}^{(n)} = \Phi^{-1} W_{\text{exc}}^{(n)}. \tag{3.156}$$

Similar expressions can be written down for the LOPT transition rate and cross section corresponding to stimulated emission of n photons between two bound states.

(2) "Bound–free" transitions from a discrete initial state to final continuum simply ionized states with a density of final states $\rho_f(E_f)$, where E_f is the energy of the ejected electron. Assuming the motion of the ejected electron to be non-relativistic, and denoting by $\mathbf{k}_f$ its wave vector, we have

$$E_f = \frac{\hbar^2 k_f^2}{2m}, \tag{3.157}$$

and, by energy conservation,

$$E_f = E_i + n\hbar\omega, \tag{3.158}$$

where E_i is the (negative) energy of the initial *bound* state. Let $\tilde{\rho}_f(E_f)\mathrm{d}\Omega\mathrm{d}E_f$ be the number of final continuum states whose wave vector $\mathbf{k}_f$ lies within the solid angle $\mathrm{d}\Omega$ about the direction $\Omega \equiv (\theta, \phi)$ and whose energy is in the interval $(E_f, E_f + \mathrm{d}E_f)$. Assuming that the final continuum wave function of the ejected electron is normalized to a Dirac delta function in wave vector space, as in Equation (3.139), we have

$$\tilde{\rho}_f(E_f)\mathrm{d}\Omega\mathrm{d}E_f = \mathrm{d}\mathbf{k}_f = k_f^2\,\mathrm{d}k_f\,\mathrm{d}\Omega, \tag{3.159}$$

so that the density of levels within the particular subgroup of final states whose wave vectors $\mathbf{k}_f$ lie within $\mathrm{d}\Omega$ is given by

$$\tilde{\rho}_f(E_f) = k_f^2 \frac{\mathrm{d}k_f}{\mathrm{d}E_f} = \frac{mk_f}{\hbar^2}. \tag{3.160}$$

It is worth stressing that $\tilde{\rho}_f(E_f)$ depends on the normalization adopted for the free states $|\psi_f\rangle$. The LOPT *differential* multiphoton single ionization rate corresponding to the absorption by an atom of n photons with the ejection of an electron of wave vector $\mathbf{k}_f$ within a solid angle $\mathrm{d}\Omega$ about the direction $\Omega \equiv (\theta, \Phi)$ is then given by

$$\frac{\mathrm{d}W_{\mathrm{ion}}^{(n)}}{\mathrm{d}\Omega} = \frac{2\pi}{\hbar}(2\pi\hbar\alpha)^n I^n |\tilde{T}_{fi}^{(n)}|^2 \tilde{\rho}_f(E_f). \tag{3.161}$$

This transition rate can also be expressed in terms of the incident photon flux Φ as

$$\frac{\mathrm{d}W_{\mathrm{ion}}^{(n)}}{\mathrm{d}\Omega} = \frac{2\pi}{\hbar}(2\pi\hbar^2\alpha\omega)^n \Phi^n |\tilde{T}_{fi}^{(n)}|^2 \tilde{\rho}_f(E_f). \tag{3.162}$$

The LOPT total n-photon ionization rate, $W_{\mathrm{ion}}^{(n)}$, is obtained by integrating the differential rate with respect to the angular variables of the ejected electron.

Thus,

$$
\begin{aligned}
W_{\text{ion}}^{(n)} &= \int \frac{\mathrm{d} W_{\text{ion}}^{(n)}}{\mathrm{d}\Omega} \, \mathrm{d}\Omega \\
&= \int_0^{2\pi} \mathrm{d}\phi \int_0^{\pi} \mathrm{d}\theta \, \sin\theta \, \frac{\mathrm{d} W_{\text{ion}}^{(n)}}{\mathrm{d}\Omega}.
\end{aligned}
\tag{3.163}
$$

The corresponding cross sections are obtained by dividing the transition rates by the incident photon flux Φ. Thus, the LOPT differential n-photon ionization cross section is given by

$$
\frac{\mathrm{d}\sigma_{\text{ion}}^{(n)}}{\mathrm{d}\Omega} = \Phi^{-1} \frac{\mathrm{d} W_{\text{ion}}^{(n)}}{\mathrm{d}\Omega},
\tag{3.164}
$$

and the LOPT total n-photon ionization cross section is given by

$$
\sigma_{\text{ion}}^{(n)} = \int \frac{\mathrm{d}\sigma_{\text{ion}}^{(n)}}{\mathrm{d}\Omega} \mathrm{d}\Omega = \Phi^{-1} W_{\text{ion}}^{(n)}.
\tag{3.165}
$$

We remark that the excitation cross section (3.156) and the ionization cross section (3.165) have the usual dimensions of an area. However, they are proportional to the $(n-1)$th power of the incident photon flux Φ, and hence do not represent a pure target (atomic) characteristic at a given photon energy, except for $n = 1$. This difficulty can be avoided by defining LOPT *generalized* excitation and ionization total cross sections by the relation

$$
\tilde{\sigma}_{fi}^{(n)} = \Phi^{-n} W_{fi}^{(n)}
\tag{3.166}
$$

and adopting a similar definition for the LOPT differential n-photon ionization cross section. The generalized cross sections $\tilde{\sigma}_{fi}^{(n)}$ are *independent of the incident photon flux*, like the single-photon excitation or ionization cross sections, and they coincide with these when $n = 1$. However, in contrast to the LOPT total excitation cross section (3.156) and ionization cross section (3.165), which have the usual dimensions of an area, the LOPT generalized cross sections (3.166) have the dimensions of $(\text{time})^{n-1} \times (\text{area})^n$, which is awkward when $n \geq 2$.

(3) "Free–free" transitions between continuum states. Transitions of this type will be studied in Chapter 10, which is devoted to laser-assisted electron–atom collisions.

3.2 Method of variation of the constants

3.2.1 Basic equations

An alternative formulation of time-dependent perturbation theory, proposed by Dirac [42, 43], is the *method of variation of the constants*. In this approach, the state vector of the system, $|\Psi(t)\rangle$, satisfying the TDSE (3.1), is expanded in the complete set of energy eigenstates of the unperturbed Hamiltonian H_0. That is,

$$|\Psi(t)\rangle = \sum_k c_k(t)|\phi_k(t)\rangle$$

$$= \sum_k c_k(t)|\psi_k\rangle \exp(-iE_k t/\hbar), \qquad (3.167)$$

where the summation symbol implies a sum over the entire set (discrete plus continuum) of energy eigenstates $|\psi_k\rangle$ of H_0.

In contrast to the coefficients $c_k^{(0)}$ of the expansion (3.6) of the unperturbed state vector $|\Psi_0(t)\rangle$, which are *constants*, the unknown coefficients $c_k(t)$ are *functions of time*. Assuming that $|\Psi(t)\rangle$ is normalized to unity, we see that

$$|c_k(t)|^2 = |\langle\psi_k|\Psi(t)\rangle|^2 \qquad (3.168)$$

is the probability of finding the system in the unperturbed energy eigenstate $|\psi_k\rangle$ at time t, and $c_k(t)$ is the corresponding probability amplitude. Moreover, since the unperturbed energy eigenstates $|\psi_k\rangle$ are orthonormal and form a complete set, we have

$$\sum_k |c_k(t)|^2 = 1. \qquad (3.169)$$

Upon comparison of Equations (3.6) and (3.167), we also see that if there is no perturbation, the coefficients $c_k(t)$ reduce to the constants $c_k^{(0)}$, which are therefore the initial values of the $c_k(t)$, specifying the state of the system at times $t \leq t_0$, before the perturbation $H_{int}(t)$ is applied.

In order to obtain equations for the unknown coefficients $c_k(t)$, we substitute the expansion (3.167) into the TDSE (3.1) and use Equations (3.2) and (3.3), together with the fact that the unperturbed energy eigenstates $|\psi_k\rangle$ are orthonormal. We then find that

$$\frac{d}{dt}c_k(t) = -\frac{i}{\hbar}\sum_j \langle\psi_k|H_{int}(t)|\psi_j\rangle \exp(i\omega_{kj}t)c_j(t), \qquad (3.170)$$

where $\omega_{kj} = (E_k - E_j)/\hbar$ is a Bohr angular frequency.

The set of equations (3.170) for all values of the index k constitutes a system of first-order coupled differential equations which is strictly equivalent to the TDSE (3.1), and therefore to Equation (3.8), which is the starting point of the time-dependent perturbation theory for the evolution operator $U(t, t_0)$ developed in Section 3.1.

In Dirac's method of variation of the constants, the system (3.170) is solved perturbatively, assuming that $H_{int}(t)$ is a small perturbation. It is convenient for this purpose to replace formally $H_{int}(t)$ by $\lambda H_{int}(t)$, where the parameter λ is used to identify the different orders of the perturbation calculation. This parameter has the value $\lambda = 1$ for the actual physical problem. We note that one can pass smoothly from the physical problem to the unperturbed one by letting λ tend to zero.

Let us now expand the coefficients $c_k(t)$ in powers of λ as

$$c_k(t) = \sum_{n=0}^{\infty} \lambda^n \bar{c}_k^{(n)}(t), \tag{3.171}$$

where $\bar{c}_k^{(n)}(t)$ denotes the contribution of order n to $c_k(t)$. We shall also write

$$c_k^{(N)}(t) = \sum_{n=0}^{N} \lambda^n \bar{c}_k^{(n)}(t) \tag{3.172}$$

to denote the coefficient $c_k(t)$ calculated through order N. Substituting the expansion (3.171) into the system (3.170), we find by equating the coefficients of equal powers of λ that

$$\frac{d}{dt} \bar{c}_k^{(0)}(t) = 0, \tag{3.173}$$

$$\frac{d}{dt} \bar{c}_k^{(1)}(t) = -\frac{i}{\hbar} \sum_j \langle \psi_k | H_{int}(t) | \psi_j \rangle \exp(i\omega_{kj}t) \bar{c}_j^{(0)}(t), \tag{3.174}$$

$$\frac{d}{dt} \bar{c}_k^{(n)}(t) = -\frac{i}{\hbar} \sum_j \langle \psi_k | H_{int}(t) | \psi_j \rangle \exp(i\omega_{kj}t) \bar{c}_j^{(n-1)}(t), \quad n = 2, 3, \dots \tag{3.175}$$

Thus the original system (3.170) has been decoupled in such a way that Equations (3.173)–(3.175) can now in principle be integrated successively to any given order in the perturbation.

3.2.2 Probability amplitudes, transition probabilities and transition rates

The first equation (3.173) confirms that the coefficients $\bar{c}_k^{(0)}$ are time-independent. Indeed, we have $\bar{c}_k^{(0)} \equiv c_k^{(0)}$, where the constants $c_k^{(0)}$ are the coefficients of the expansion (3.6) of the unperturbed state vector $|\Psi_0(t)\rangle$. These constants provide the initial conditions of the problem, so that

$$c_k(t \leq t_0) = c_k^{(0)}. \tag{3.176}$$

In what follows we shall assume, as in Section 3.1, that the system is initially (for $t \leq t_0$) in a particular unperturbed energy eigenstate $|\psi_i\rangle$ of H_0. The constants $c_k^{(0)}$ are then given by Equation (3.37) and the probability amplitude for the transition

$i \to f$ is simply

$$a_{fi}(t) = c_f(t). \tag{3.177}$$

The corresponding transition probability and transition rate are then given by Equations (3.54) and (3.55), respectively. By using Equation (3.177), together with Equations (3.56) and (3.171), in which we set $\lambda = 1$, we also find that

$$\bar{a}_{fi}^{(n)}(t) = \bar{c}_f^{(n)}(t), \quad n = 1, 2, \ldots \tag{3.178}$$

We can readily verify Equation (3.178) by integrating successively Equations (3.174) and (3.175). Thus, upon substituting Equation (3.37) into Equation (3.174) we have, to first order in the perturbation $H_{\text{int}}(t)$, the following:

$$\frac{d}{dt}\bar{c}_k^{(1)}(t) = -\frac{i}{k}\langle\psi_k|H_{\text{int}}(t)|\psi_i\rangle \exp(i\omega_{ki}t). \tag{3.179}$$

The solution of this first-order differential equation is readily obtained. Taking the state k to be the final state f of the transition $i \to f$, and choosing the integration constant in such a way that $\bar{c}_f^{(1)}(t_0) = 0$, we find that

$$\bar{c}_f^{(1)}(t) = -\frac{i}{\hbar}\int_{t_0}^{t}\langle\psi_f|H_{\text{int}}(t_1)|\psi_i\rangle\exp(i\omega_{fi}t_1)dt_1$$

$$= \bar{a}_{fi}^{(1)}(t), \tag{3.180}$$

where we have used Equations (3.61) and (3.62). It is also worth noting that for $f = i$ we have

$$\bar{c}_i^{(1)}(t) = -\frac{i}{\hbar}\int_{t_0}^{t}\langle\psi_i|H_{\text{int}}(t_1)|\psi_i\rangle\,dt_1, \tag{3.181}$$

so that for $t > t_0$ the coefficient $c_i(t)$ of the state i is given to first order in the perturbation by

$$c_i(t) \simeq c_i^{(0)} + \bar{c}_i^{(1)}(t)$$

$$\simeq 1 - \frac{i}{\hbar}\int_{t_0}^{t}\langle\psi_i|H_{\text{int}}(t_1)|\psi_i\rangle\,dt_1$$

$$\simeq \exp\left[-\frac{i}{\hbar}\int_{t_0}^{t}\langle\psi_i|H_{\text{int}}(t_1)|\psi_i\rangle\,dt_1\right]. \tag{3.182}$$

Thus

$$|c_i(t)|^2 \simeq 1, \tag{3.183}$$

and the main effect of the perturbation on the initial state i is to change its phase.

To second order in the interaction we obtain, from Equation (3.175) with $n = 2$, for the transition $i \to f$

$$\bar{c}_f^{(2)}(t) = \left(-\frac{i}{\hbar}\right)^2 \sum_{k_1} \int_{t_0}^{t} dt_1 \int_{t_0}^{t_1} dt_2 \langle \psi_f | H_{int}(t_1) | \psi_{k_1} \rangle$$

$$\times \exp(i\omega_{fk_1}t_1) \langle \psi_{k_1} | H_{int}(t_2) | \psi_i \rangle \exp(i\omega_{k_1 i}t_2)$$

$$= \bar{a}_{fi}^{(2)}(t), \tag{3.184}$$

where we have used Equation (3.63). Proceeding by iteration, one can readily generalize this result, and it is found that the contribution of order n to the probability amplitude for the transition $i \to f$ is given by

$$\bar{c}_f^{(n)}(t) = \bar{a}_{fi}^{(n)}(t). \tag{3.185}$$

3.2.3 Dressed wave functions

As an example, we shall obtain to first order in the electric-field strength $\mathcal{E}_0$ the "dressed" wave functions for an atom in a linearly polarized monochromatic laser field described in the dipole approximation by the electric field (3.68). It is assumed that the laser frequency is not close to resonance with an atomic transition frequency.

We shall first perform the calculation in the *length gauge*, where the interaction Hamiltonian is given by Equation (3.69). We suppose that, before the laser field is turned on, the atom is in a particular unperturbed stationary state $|\phi_j(t)\rangle = |\psi_j\rangle \exp(-iE_j t/\hbar)$. The corresponding unperturbed wave function in the position representation is

$$\phi_j(X, t) \equiv \langle X | \phi_j(t) \rangle = \psi_j(X) \exp(-iE_j t/\hbar). \tag{3.186}$$

We shall denote by $\Phi_j^L(X, t)$ the corresponding dressed atomic wave function in the length gauge. Through first order in the electric field strength $\mathcal{E}_0$, this dressed wave function is given by

$$\Phi_j^L(X, t) = \sum_{j'} c_{j'}^{L(1)}(t) \psi_{j'}(X) \exp(-iE_{j'} t/\hbar) \tag{3.187}$$

with

$$c_{j'}^{L(1)}(t) = c_{j'}^{(0)} + \bar{c}_{j'}^{L(1)}(t). \tag{3.188}$$

We assume that the interaction is switched on adiabatically at $t_0 = -\infty$. The initial condition is therefore

$$c_{j'}^{L(1)}(t_0 = -\infty) = c_{j'}^{(0)} = \delta_{j'j}. \tag{3.189}$$

From Equations (3.178) and (3.94), we have

$$\bar{c}_{j'}^{L(1)}(t) = \frac{\mathcal{E}_0}{2\hbar} M_{j'j} \left\{ \frac{\exp[i(\omega_{j'j} + \omega)t]}{\omega_{j'j} + \omega} \exp(i\varphi) + \frac{\exp[i(\omega_{j'j} - \omega)t]}{\omega_{j'j} - \omega} \exp(-i\varphi) \right\}.$$

(3.190)

Using Equations (3.187) – (3.190), we then find that the dressed atomic wave function in the length gauge is given to first order in $\mathcal{E}_0$ by the following expression:

$$\Phi_j^L(X, t) = \exp(-iE_j t/\hbar) \left\{ \psi_j(X) \right.$$

$$\left. + \frac{\mathcal{E}_0}{2\hbar} \sum_{j'} \left[\frac{\exp(i\omega t)}{\omega_{j'j} + \omega} \exp(i\varphi) + \frac{\exp(-i\omega t)}{\omega_{j'j} - \omega} \exp(-i\varphi) \right] M_{j'j} \psi_{j'}(X) \right\}.$$

(3.191)

In particular, for a linearly polarized, spatially homogeneous and monochromatic laser field described by the electric field

$$\mathcal{E}(t) = \hat{\epsilon} \mathcal{E}_0 \sin(\omega t),$$

(3.192)

we find by using Equation (3.68) and setting $\varphi = -\pi/2$ that the dressed atomic wave function in the length gauge is given to first order in $\mathcal{E}_0$ by

$$\Phi_j^L(X, t) = \exp(-iE_j t/\hbar) \left\{ \psi_j(X) \right.$$

$$\left. - \frac{i\mathcal{E}_0}{2\hbar} \sum_{j'} \left[\frac{\exp(i\omega t)}{\omega_{j'j} + \omega} - \frac{\exp(-i\omega t)}{\omega_{j'j} - \omega} \right] M_{j'j} \psi_{j'}(X) \right\}. \quad (3.193)$$

Knowing the dressed atomic wave function in the length gauge, we can readily obtain it in the velocity gauge by using the gauge transformations discussed in Section 2.4. Thus, by performing the gauge transformation

$$\Phi_j(X, t) = \exp\left[-\frac{ie}{\hbar} \mathbf{A}(t) \cdot \mathbf{R} \right] \Phi_j^L(X, t),$$

(3.194)

we see by using Equation (2.132) that the wave function $\Phi_j(X, t)$ satisfies a TDSE of the form (2.121) in the original (untransformed) gauge, with an interaction Hamiltonian containing a term in $\mathbf{A}^2$. This term can be eliminated by performing a second gauge transformation which yields the dressed atomic wave function in the velocity gauge. Denoting this wave function by $\Phi_j^V(X, t)$ and using Equation (2.124), we have

$$\Phi_j^V(X, t) = \exp\left[\frac{i}{\hbar} \frac{e^2 N}{2m} \int_{-\infty}^t \mathbf{A}^2(t') dt' \right] \Phi_j(X, t),$$

(3.195)

where N denotes the number of atomic electrons. It follows from Equations (3.194) and (3.195) that we can obtain the dressed atomic wave functions in the velocity gauge from those in the length gauge by performing two successive gauge transformations. Thus we have

$$\Phi_j^V(X,t) = \exp\left[\frac{\mathrm{i}}{\hbar}\frac{e^2 N}{2m}\int_{-\infty}^t \mathbf{A}^2(t')\mathrm{d}t'\right]\exp[-\mathrm{i}\mathbf{a}(t)\cdot\mathbf{R}]\Phi_j^L(X,t), \qquad (3.196)$$

where

$$\mathbf{a}(t) = \frac{e}{\hbar}\mathbf{A}(t). \qquad (3.197)$$

The gauge factor $\exp[-\mathrm{i}\mathbf{a}(t)\cdot\mathbf{R}]$ which appears in this context is often called the *Göppert-Mayer gauge factor* [44].

As an example, let us consider a laser field described by the electric field (3.192), so that the corresponding vector potential is

$$\mathbf{A}(t) = \hat{\boldsymbol{\epsilon}} A_0 \cos(\omega t), \qquad (3.198)$$

with $A_0 = \mathcal{E}_0/\omega$. Using Equations (3.193) and (3.196), we find that the dressed atomic wave function in the velocity gauge corresponding to the length gauge expression (3.193) is given to first order in $\mathcal{E}_0$ (and therefore to first order in A_0) by

$$\Phi_j^V(X,t) = \exp(-\mathrm{i}E_j t/\hbar)\exp\left[\frac{\mathrm{i}}{\hbar}\frac{e^2 N}{2m}\int_{-\infty}^t \mathbf{A}^2(t')\mathrm{d}t'\right]\exp[-\mathrm{i}\mathbf{a}(t)\cdot\mathbf{R}]$$
$$\times\left\{\psi_j(X) - \frac{\mathrm{i}\mathcal{E}_0}{2\hbar}\sum_{j'}\left[\frac{\exp(\mathrm{i}\omega t)}{\omega_{j'j}+\omega} - \frac{\exp(-\mathrm{i}\omega t)}{\omega_{j'j}-\omega}\right]M_{j'j}\psi_{j'}(X)\right\}.$$
$$(3.199)$$

This result will be used in Section 10.3 to study target-dressing effects in laser-assisted electron–atom collisions.

It is interesting to compare the expression (3.199) for the dressed atomic wave function in the velocity gauge, derived above from the length gauge result (3.193) by performing two successive gauge transformations, with that obtained by solving through first order in $\mathcal{E}_0$ the system of equations (3.170) directly in the velocity gauge. The interaction Hamiltonian is then given by

$$H_{\mathrm{int}}(t) = \frac{e}{m}\mathbf{A}(t)\cdot\mathbf{P}$$
$$= \frac{e}{2m}A_0[\exp(\mathrm{i}\omega t)+\exp(-\mathrm{i}\omega t)]\hat{\boldsymbol{\epsilon}}\cdot\mathbf{P}, \qquad (3.200)$$

where we have used Equation (3.198) and $\mathbf{P}$ is the total momentum operator given by Equation (2.120).

Let us denote by

$$\tilde{\Phi}_j^V(X,t) = \sum_{j'} c_{j'}^{V(1)}(t)\psi_{j'}(X)\exp(-iE_{j'}t/\hbar) \tag{3.201}$$

the dressed atomic wave function obtained by determining the coefficients $c_{j'}^{V(1)}(t)$ through first order in $\mathcal{E}_0$. Thus we have

$$c_{j'}^{V(1)}(t) = c_{j'}^{(0)} + \bar{c}_{j'}^{V(1)}(t), \tag{3.202}$$

with $c_{j'}^{(0)} = \delta_{j'j}$. Assuming that the interaction is turned on adiabatically at $t_0 = -\infty$, we find by using Equations (3.180) and (3.200) that

$$\bar{c}_{j'}^{V(1)}(t) = -\frac{A_0}{2\hbar}\frac{e}{m}\langle\psi_{j'}|\hat{\boldsymbol{\epsilon}}\cdot\mathbf{P}\psi_j\rangle\left[\frac{\exp[i(\omega_{j'j}+\omega)t]}{\omega_{j'j}+\omega} + \frac{\exp[i(\omega_{j'j}-\omega)t]}{\omega_{j'j}-\omega}\right] \tag{3.203}$$

and therefore

$$\tilde{\Phi}_j^V(X,t) = \exp(-iE_jt/\hbar)$$

$$\times\left\{\psi_j(X) - \frac{A_0}{2\hbar}\frac{e}{m}\sum_{j'}\left[\frac{\exp(i\omega t)}{\omega_{j'j}+\omega} + \frac{\exp(-i\omega t)}{\omega_{j'j}-\omega}\right]\langle\psi_{j'}|\hat{\boldsymbol{\epsilon}}\cdot\mathbf{P}|\psi_j\rangle\psi_{j'}(X)\right\}. \tag{3.204}$$

Now, since

$$\langle\psi_{j'}|\mathbf{P}|\psi_j\rangle = \sum_{i=1}^{N}\langle\psi_{j'}|\mathbf{p}_i|\psi_j\rangle$$

$$= im\omega_{j'j}\sum_{i=1}^{N}\langle\psi_{j'}|\mathbf{r}_i|\psi_j\rangle$$

$$= im\omega_{j'j}\langle\psi_{j'}|\mathbf{R}|\psi_j\rangle \tag{3.205}$$

we have, using Equations (2.135) and (3.70),

$$\langle\psi_{j'}|\hat{\boldsymbol{\epsilon}}\cdot\mathbf{P}|\psi_j\rangle = -\frac{im\omega_{j'j}}{e}\langle\psi_{j'}|\hat{\boldsymbol{\epsilon}}\cdot\mathbf{D}|\psi_j\rangle$$

$$= -\frac{im\omega_{j'j}}{e}M_{j'j}. \tag{3.206}$$

Using this relation and remembering that $A_0 = \mathcal{E}_0/\omega$, we can rewrite the expression (3.204) for $\tilde{\Phi}_j^V(X,t)$ in the form

$$\tilde{\Phi}_j^V(X,t) = \exp(-iE_j t/\hbar)$$

$$\times \left\{ \psi_j(X) + \frac{i\mathcal{E}_0}{2\hbar} \sum_{j'} \frac{\omega_{j'j}}{\omega} \left[\frac{\exp(i\omega t)}{\omega_{j'j}+\omega} + \frac{\exp(-i\omega t)}{\omega_{j'j}-\omega} \right] M_{j'j}\psi_{j'}(X) \right\}. \tag{3.207}$$

At first sight, this result appears to be strange because, in contrast to the expression (3.193) obtained in the length gauge and the result (3.199) found in the velocity gauge by performing gauge transformations, Equation (3.207) *diverges* when $\omega \to 0$. In order to elucidate this point, we rewrite Equation (3.207) in the following form:

$$\tilde{\Phi}_j^V(X,t) = \exp(-iE_j t/\hbar)\left\{ \psi_j(X) + \frac{i\mathcal{E}_0}{2\hbar} \sum_{j'} \left[\frac{\exp(i\omega t)}{\omega_{j'j}+\omega}\left(-1+\frac{\omega_{j'j}+\omega}{\omega}\right) \right.\right.$$

$$\left.\left. + \frac{\exp(-i\omega t)}{\omega_{j'j}-\omega}\left(1+\frac{\omega_{j'j}-\omega}{\omega}\right) \right] M_{j'j}\psi_{j'}(X) \right\}$$

$$= \exp(-iE_j t/\hbar)\left\{ \psi_j(X) - \frac{i\mathcal{E}_0}{2\hbar} \sum_{j'} \left[\frac{\exp(i\omega t)}{\omega_{j'j}+\omega} - \frac{\exp(-i\omega t)}{\omega_{j'j}-\omega} \right] M_{j'j}\psi_{j'}(X) \right.$$

$$\left. + \frac{i\mathcal{E}_0}{\hbar\omega}\cos(\omega t)\sum_{j'} M_{j'j}\psi_{j'}(X) \right\}. \tag{3.208}$$

The summation appearing in the final term on the right-hand side of Equation (3.208) can be performed by using Equation (3.70) and the closure relation (3.59). That is,

$$\sum_{j'} M_{j'j}\psi_{j'}(X) = -e\sum_{j'}\langle\psi_{j'}|\hat{\epsilon}\cdot\mathbf{R}|\psi_j\rangle\psi_{j'}(X)$$

$$= -e\sum_{j'}\langle X|\psi_{j'}\rangle\langle\psi_{j'}|\hat{\epsilon}\cdot\mathbf{R}|\psi_j\rangle$$

$$= -e\,\hat{\epsilon}\cdot\mathbf{R}\,\psi_j(X). \tag{3.209}$$

Hence, using Equations (3.197) and (3.198) and the fact that $\mathcal{E}_0 = \omega A_0$, we find that

$$
\tilde{\Phi}_j^{\mathrm{V}}(X, t) = \exp(-\mathrm{i}E_j t/\hbar) \left\{ \psi_j(X) \right.
$$

$$
\left. -\frac{\mathrm{i}\mathcal{E}_0}{2\hbar} \sum_{j'} \left[\frac{\exp(\mathrm{i}\omega t)}{\omega_{j'j} + \omega} - \frac{\exp(-\mathrm{i}\omega t)}{\omega_{j'j} - \omega} \right] M_{j'j} \psi_{j'}(X) - \mathrm{ia}(t) \cdot \mathbf{R} \, \psi_j(X) \right\}.
$$

$$(3.210)$$

Let us now compare this result with the dressed wave function $\Phi_j^{\mathrm{V}}(X, t)$, given by Equation (3.199), which we obtained by starting from the length gauge result (3.193) and performing two gauge transformations. First of all, we note that the gauge factor $\exp[(\mathrm{i}/\hbar)(e^2 N/(2m)) \int_{-\infty}^{t} \mathbf{A}^2(t')\mathrm{d}t']$ appearing in Equation (3.199) is independent of the electron coordinates and hence does not affect any probability amplitude calculated using the dressed wave function. Secondly, if we expand the Göppert-Mayer gauge factor $\exp[-\mathrm{ia}(t) \cdot \mathbf{R}]$ to first order in $\mathcal{E}_0$ (or A_0), we find that

$$
\exp[-\mathrm{ia}(t) \cdot \mathbf{R}] = 1 - \mathrm{ia}(t) \cdot \mathbf{R} + \mathcal{O}(\mathcal{E}_0^2). \tag{3.211}
$$

As a result, the probability amplitudes calculated to first order in $\mathcal{E}_0$ by using either the velocity gauge dressed atomic wave functions $\Phi_j^{\mathrm{V}}(X, t)$ given by Equation (3.199) or $\tilde{\Phi}_j^{\mathrm{V}}(X, t)$ given by Equation (3.210) are the same. The spurious divergence in ω^{-1} is entirely contained in the gauge term $-\mathrm{ia}(t) \cdot \mathbf{R} \psi_j(X)$.

It is apparent from the preceding discussion that the transition probabilities calculated to first order of perturbation theory are formally gauge-invariant. As mentioned in Section 3.1.8, this property is true for any order of perturbation theory.

3.2.4 Two-level atomic system with harmonic perturbation

As a second application of Dirac's method of variation of the constants, let us consider a two-level atomic system with discrete unperturbed energies $E_a < E_b$ and corresponding unperturbed energy eigenstates $|\psi_a\rangle$ and $|\psi_b\rangle$, respectively. We assume that the system is initially (for times $t \leq t_0$) in the state a. At time $t_0 = 0$, a perturbation is switched on which has the form

$$
H_{\mathrm{int}}(t) = H_+ \exp(-\mathrm{i}\omega t) + H_- \exp(\mathrm{i}\omega t), \tag{3.212}
$$

where H_+ and $H_- = (H_+)^\dagger$ are time-independent operators. Using Equation (3.170), we obtain the system of two coupled first-order differential equations,

$$i\hbar \frac{d}{dt} c_a(t) = \{(H_-)_{aa} \exp[i\omega t] + (H_+)_{aa} \exp[-i\omega t]\} c_a(t)$$
$$+ \{(H_-)_{ab} \exp[i(\Delta\omega)t] + (H_+)_{ab} \exp[-i(\omega + \omega_{ba})t]\} c_b(t) \quad (3.213)$$

and

$$i\hbar \frac{d}{dt} c_b(t) = \{(H_-)_{ba} \exp[i(\omega_{ba} + \omega)t] + (H_+)_{ba} \exp[-i(\Delta\omega)t]\} c_a(t)$$
$$+ \{(H_-)_{bb} \exp[i\omega t] + (H_+)_{bb} \exp[-i\omega t]\} c_b(t), \quad (3.214)$$

where $\omega_{ba} = (E_b - E_a)/\hbar$, and we have introduced the "detuning" angular frequency

$$\Delta\omega = \omega - \omega_{ba}. \quad (3.215)$$

The system of equations (3.213)–(3.214) must be solved subject to the initial conditions

$$c_a(t \le 0) = 1, \quad c_b(t \le 0) = 0. \quad (3.216)$$

This system cannot be solved exactly, but if it is assumed that $|\Delta\omega| \ll \omega$ (so that the angular frequency ω is always close to its resonant value $\omega = \omega_{ba}$), then the terms in $\exp[\pm i(\Delta\omega)t]$ will be much more important than those in $\exp[\pm i(\omega + \omega_{ba})t]$ and $\exp(\pm i\omega t)$. This is due to the fact that the latter terms oscillate much more rapidly and, on average, make little contribution to dc_a/dt or dc_b/dt. It is therefore reasonable to neglect the higher-frequency terms. This is known as the *rotating wave approximation* (RWA) because the only terms which are kept are those in which the time-dependence of the system and of the perturbation are in phase. In this approximation, the system of equations (3.213)–(3.214) reduces to

$$i\hbar \frac{d}{dt} c_a(t) = (H_-)_{ab} \exp[i(\Delta\omega)t] c_b(t) \quad (3.217)$$

and

$$i\hbar \frac{d}{dt} c_b(t) = (H_+)_{ba} \exp[-i(\Delta\omega)t] c_a(t). \quad (3.218)$$

This system, which is much simpler than the system (3.213)–(3.214), can be solved exactly. The solutions $c_a(t)$ and $c_b(t)$ satisfying the initial conditions (3.216) are given by

$$c_a(t) = \exp[i(\Delta\omega)t/2] \left[\cos(\omega_R t/2) - i \left(\frac{\Delta\omega}{\omega_R} \right) \sin(\omega_R t/2) \right] \quad (3.219)$$

and

$$c_b(t) = -i\frac{2(H_+)_{ba}}{\hbar\omega_R}\exp[-i(\Delta\omega)t/2]\sin(\omega_R t/2), \tag{3.220}$$

where

$$\omega_R = \left[(\Delta\omega)^2 + \frac{4|(H_+)_{ba}|^2}{\hbar^2}\right]^{1/2} \tag{3.221}$$

is called the *Rabi flopping frequency*. Using Equation (3.219), the probability of finding the system at time $t > 0$ in the state a is given by

$$|c_a(t)|^2 = \cos^2(\omega_R t/2) + \frac{(\Delta\omega)^2}{\omega_R^2}\sin^2(\omega_R t/2). \tag{3.222}$$

On the other hand, it follows from Equation (3.220) that the probability of finding the system at time $t > 0$ in the state b (that is, the probability that the excitation transition $a \to b$ will occur) is

$$P_{ba}(t) = |c_b(t)|^2 = \frac{4|(H_+)_{ba}|^2}{\hbar^2\omega_R^2}\sin^2(\omega_R t/2). \tag{3.223}$$

As expected, the excitation $a \to b$ is a typical *resonant* process, since the transition probability (3.223) rapidly decreases when the absolute value $|\Delta\omega|$ of the detuning angular frequency increases. Using Equations (3.219)–(3.223) it is also easily verified that

$$|c_a(t)|^2 + |c_b(t)|^2 = 1 \tag{3.224}$$

and that the system oscillates between the two levels with a period $T = 2\pi/\omega_R$.

Having obtained exact results for this problem within the framework of the RWA, we can compare them with those arising from first-order perturbation theory. Using Equation (3.222), we see that when $\omega_R t \ll 1$ we have $|c_a(t)|^2 \simeq 1$, in agreement with the result (3.183) derived by using first-order perturbation theory. From Equations (3.74), (3.75), (3.69), (3.70) and (3.212), it follows that the first-order transition probability is given by

$$P_{ba}^{(1)}(t) = \frac{2}{\hbar^2}|(H_+)_{ba}|^2 F(t, \Delta\omega)$$

$$= \frac{4|(H_+)_{ba}|^2}{\hbar^2(\Delta\omega)^2}\sin^2[(\Delta\omega)t/2]. \tag{3.225}$$

If $\Delta\omega \neq 0$, this formula agrees with the RWA expression (3.223) provided that the perturbation is weak enough so that one can write $\omega_R \simeq |\Delta\omega|$. At resonance ($\Delta\omega = 0$) we see that $P_{ba}^{(1)}(t)$ increases quadratically with time. This result is only in agreement with the RWA expression (3.223) for small enough times and small enough perturbations, in agreement with the discussion following Equation (3.77).

3.3 Semi-perturbative methods

We shall now describe two semi-perturbative approaches, in which some of the interactions are treated in a non-perturbative way, and the remaining ones are treated by using perturbation theory. The first one is the method of *essential states*, which can be applied to analyze resonantly multiphoton ionization (REMPI) at moderate laser intensities. The second one is a semi-perturbative approach which allows to take into account *target-dressing effects* for laser fields having moderate intensities; this method will be used in Chapter 10 to analyze such target-dressing effects in laser-assisted electron–atom collisions.

3.3.1 Essential states

When resonances are present, and for laser fields which are not too intense, the multiphoton processes are usually dominated by the resonant contributions. A convenient way of accounting for this fact is to use an approach based on the Feshbach projection operator formalism [45, 46]. Two projection operators P and Q are defined as follows. The first one, P, projects onto the space of the field-free states that are resonantly coupled by the laser field. The second one,

$$Q = I - P, \tag{3.226}$$

projects onto the complement space. Since P and Q are projection operators, the following relations hold:

$$P^2 = P, \quad Q^2 = Q, \quad PQ = QP = 0. \tag{3.227}$$

Writing the state vector as

$$|\Psi(t)\rangle = P|\Psi(t)\rangle + Q|\Psi(t)\rangle, \tag{3.228}$$

and inserting this expression into the TDSE (3.1), the following two coupled equations are obtained:

$$P\left(i\hbar\frac{\partial}{\partial t} - H(t)\right)P|\Psi(t)\rangle = PH(t)Q|\Psi(t)\rangle \tag{3.229}$$

and

$$Q\left(i\hbar\frac{\partial}{\partial t} - H(t)\right)Q|\Psi(t)\rangle = QH(t)P|\Psi(t)\rangle. \tag{3.230}$$

Defining the Green's operator in the Q space,

$$G_Q(t) = Q\left(i\hbar\frac{\partial}{\partial t} - H(t)\right)^{-1}Q, \tag{3.231}$$

Equation (3.230) can be formally solved for $Q|\Psi(t)\rangle$. Upon substitution in Equation (3.229), an effective time-dependent Schrödinger equation in the P space is found. That is,

$$\left(i\hbar\frac{\partial}{\partial t} - PH(t)P - PH(t)QG_Q(t)QH(t)P\right)P|\Psi(t)\rangle = 0. \quad (3.232)$$

The Green's operator G_Q can be approximated by accounting for the non-resonant part of the interaction using perturbation theory. By expanding the state vector $P|\Psi(t)\rangle$ on the basis of the resonantly coupled, unperturbed states as

$$P|\Psi(t)\rangle = \sum_{k\in P} c_k(t)|\phi_k(t)\rangle, \quad (3.233)$$

one obtains a system of coupled first-order differential equations for the resonantly coupled, or "essential," states of the system. That is,

$$i\hbar\frac{d}{dt}\mathbf{c}(t) = \mathbf{H}_{\text{eff}}(t)\mathbf{c}(t), \quad (3.234)$$

where $\mathbf{c}(t)$ is a column vector containing the expansion coefficients $c_k(t)$, and the matrix $\mathbf{H}_{\text{eff}}(t)$ represents the effective Hamiltonian which incorporates the resonant couplings exactly and takes into account the contribution of the remaining states in a perturbative way.

Various formulations of essential states methods have been widely employed to study a range of resonant, multiphoton phenomena. Examples include resonantly enhanced multiphoton ionization (REMPI), the formation and evolution of Rydberg wave packets [47,48] and ionization suppression by quantum interference [49,50]. Further applications of the method of essential states are discussed in the review articles of Lambropoulos and Tang [4] and Burnett *et al.* [51].

3.3.2 Semi-perturbative method for laser-assisted electron–atom collisions

Let us consider electron–atom collisions in the presence of a laser field, such as those described in Section 1.5. The theoretical analysis of these processes implies that three types of interactions must be taken into account. The first one, which is also present in the absence of the laser field, is the interaction between the projectile electron and the target atom. The second one is the interaction between the laser field and the unbound electron(s). The third one is the interaction between the laser field and the target atom, which is responsible for the "dressing" of the atomic target states.

In order to deal with this problem, Byron and Joachain [52] have proposed a semi-perturbative theory which we shall now briefly outline. We consider the case for which the incident electron is fast, so that the electron–atom interaction can be treated perturbatively by using the Born series [53]. Assuming that the laser field can be described as a monochromatic, spatially homogeneous electric field, and that the incident electron is non-relativistic, the interaction between the laser field and the projectile electron can be treated exactly by using a Gordon–Volkov wave function. Finally, the interaction between the laser field and the target atom can be treated by using first-order time-dependent perturbation theory, provided that the electric-field strength $\mathcal{E}_0$ remains small with respect to the atomic unit of electric field strength, $\mathcal{E}_a \simeq 5.1 \times 10^9 \, \mathrm{V \, cm^{-1}}$. We remark that this condition is required for laser-assisted collisions, since otherwise the target atom would be ionized by the laser field. The "dressed" atomic wave functions can then be obtained to first order in $\mathcal{E}_0$ by following the treatment of Section 3.2.3. A detailed account of this semi-perturbative theory and of its applications to the three kinds of laser-assisted electron–atom collisions described in Section 1.5 ("elastic" collisions, inelastic collisions and (e, 2e) reactions) will be given in Section 10.3.

References

[1] P. Lambropoulos, *Adv. At. Mol. Phys.* **12**, 87 (1976).

[2] F. H. M. Faisal, *Theory of Multiphoton Processes* (New York: Plenum Press, 1987).

[3] M. Crance, *Phys. Rep.* **114**, 117 (1987).

[4] P. Lambropoulos and X. Tang, *Adv. At. Mol. Opt. Phys. Suppl.* **1**, 335 (1992).

[5] C. J. Joachain, M. Dörr, and N. J. Kylstra, *Adv. At. Mol. Opt. Phys.* **42**, 225 (2000).

[6] S. Tomonaga, *Progr. Theor. Phys. (Kyoto)* **1**, 27 (1946).

[7] J. Schwinger, *Phys. Rev.* **74**, 1439 (1948).

[8] B. H. Bransden and C. J. Joachain, *Quantum Mechanics*, 2nd edn (Harlow, UK: Prentice Hall-Pearson, 2000).

[9] P. A. M. Dirac, *The Principles of Quantum Mechanics*, 4th edn (Oxford: Oxford University Press, 1958).

[10] B. H. Bransden and C. J. Joachain, *Physics of Atoms and Molecules*, 2nd edn (Harlow, UK: Prentice Hall-Pearson, 2003).

[11] B. A. Lippmann and J. Schwinger, *Phys. Rev.* **79**, 469 (1950).

[12] H. Bebb and A. Gold, *Phys. Rev.* **143**, 1 (1966).

[13] H. Bebb, *Phys. Rev.* **149**, 25 (1966).

[14] H. Bebb, *Phys. Rev.* **153**, 23 (1967).

[15] Y. Gontier and M. Trahin, *J. Phys. B* **13**, 4383 (1980).

[16] H. Silverstone, B. Adams, J. Cizek and P. Otto, *Phys. Rev. Lett.* **43**, 1498 (1979).

[17] Y. Gontier, N. Rahman and M. Trahin, *Phys. Rev. A* **14**, 2109 (1976).

[18] A. Maquet, S. I. Chu and W. P. Reinhardt, *Phys. Rev. A* **27**, 2946 (1983).

[19] M. Abramowitz and I. A. Stegun, *Handbook of Mathematical Functions* (New York: Dover, 1970).

[20] L. Hostler, *J. Math. Phys.* **5**, 591 (1964).

[21] L. Hostler, *J. Math. Phys.* **11**, 2966 (1970).

[22] S. Khristenko and S. Vetchinkin, *Opt. Spectrosc. (USSR)* **31**, 269 (1971).

[23] M. Rotenberg, *Ann. Phys. NY* **19**, 262 (1962).

[24] M. Rotenberg, *Adv. At. Mol. Phys.* **6**, 233 (1970).

[25] L. Hostler, *J. Math. Phys.* **8**, 642 (1967).

[26] B. Laurenzi, *J. Chem. Phys.* **52**, 3049 (1970).

[27] J. Schwinger, *J. Math. Phys.* **5**, 1606 (1964).
[28] S. Okubo and D. Feldman, *Phys. Rev.* **117**, 292 (1960).
[29] L. Hostler, *J. Math. Phys.* **5**, 1235 (1964).
[30] V. Gorshkov, *Sov. Phys. JETP* **20**, 234 (1965).
[31] A. Maquet, V. Véniard and T. Marian, *J. Phys. B* **31**, R3743 (1988).
[32] M. Seaton, *Rep. Prog. Phys.* **46**, 167 (1983).
[33] R. M. Sternheimer, *Phys. Rev.* **84**, 244 (1951).
[34] A. Dalgarno and J. T. Lewis, *Proc. Roy. Soc. London A* **233**, 70 (1955).
[35] C. Schwartz and J. Tiemann, *Ann. Phys. NY* **6**, 178 (1959).
[36] M. H. Mittleman and F. Wolf, *Phys. Rev.* **128**, 2686 (1962).
[37] W. Zernik, *Phys. Rev.* **132**, 320 (1963).
[38] W. Zernik, *Phys. Rev.* **135**, A51 (1964).
[39] Y. Gontier and M. Trahin, *Phys. Rev.* **172**, 83 (1968).
[40] Y. Gontier and M. Trahin, *Phys. Rev. A* **4**, 1896 (1971).
[41] M. Aymar and M. Crance, *J. Phys. B* **14**, 3585 (1981).
[42] P. A. M. Dirac, *Proc. Roy. Soc. London A* **112**, 661 (1926).
[43] P. A. M. Dirac, *Proc. Roy. Soc. London A* **114**, 243 (1927).
[44] M. Göppert-Mayer, *Ann. Phys.* **9**, 273 (1931).
[45] H. Feshbach, *Ann. Phys. NY* **5**, 357 (1958).
[46] H. Feshbach, *Ann. Phys. NY* **19**, 287 (1962).
[47] O. Zobay and G. Alber, *Phys. Rev. A* **54**, 5361 (1996).
[48] B. S. Mecking and P. Lambropoulos, *J. Phys. B* **31**, 3353 (1998).
[49] M. V. Fedorov and A. M. Movsesian, *J. Phys. B* **21**, L155 (1988).
[50] M. V. Fedorov, M. Y. Ivanov and A. M. Movsesian, *J. Phys. B* **23**, 2245 (1990).
[51] K. Burnett, V. C. Reed and P. L. Knight, *J. Phys. B* **26**, 561 (1993).
[52] F. W. Byron and C. J. Joachain, *J. Phys. B* **17**, L295 (1984).
[53] C. J. Joachain, *Quantum Collision Theory*, 3rd edn (Amsterdam: North Holland, 1983).

4

Floquet theory

In this chapter, we shall analyze the particular case of an atom interacting with a laser pulse whose duration is sufficiently long, so that the evolution of the atom in the laser field is adiabatic. When this condition is fulfilled, the atom can be considered to interact with a monochromatic laser field. As a consequence, the Hamiltonian of the system is periodic in time, and the Floquet theory [1] can be used to solve the time-dependent Schrödinger equation (TDSE) non-perturbatively.

We begin in Section 4.1 by considering the Hermitian Floquet theory. We first derive the Floquet theorem for a monochromatic, spatially homogeneous laser field and show that the solutions of the TDSE correspond to *dressed states* having *real quasi-energies*, which can be obtained by solving an infinite system of *time-independent* coupled equations. We then generalize the Floquet theory to multicolor laser fields and to "non-dipole" laser fields which are not spatially homogeneous. In Section 4.2, the Floquet theory is applied to study the dynamics of a model atom having M discrete levels interacting with a monochromatic laser field. In this case, the coupling between the bound and continuum atomic states is neglected. We analyze the relationship between the Floquet theory and the rotating wave approximation, and examine the perturbative limit of the Floquet theory. We also consider the population transfer between Floquet dressed states. In Section 4.3, we use the Floquet theory to obtain dressed continuum states for laser-assisted scattering. Section 4.4 is devoted to the non-Hermitian Floquet theory, which is used to study decaying dressed states having *complex quasi-energies*. In Section 4.5, we formulate the Floquet theory in the Kramers–Henneberger frame, where the boundary conditions are easier to express since the channels decouple asymptotically. In the remaining two sections, we describe two methods which have been used to implement the Floquet theory to study multiphoton processes in atoms: the Sturmian-Floquet theory and the R-matrix–Floquet theory. Reviews of the application of the Floquet theory to atomic multiphoton processes have been given by Chu [2], Potvliege and Shakeshaft [3] and Joachain, Dörr and Kylstra [4].

4.1 Hermitian Floquet theory

4.1.1 The Floquet theorem. Dressed states and quasi-energies

The TDSE describing the atom in a laser field is as follows:

$$i\hbar\frac{\partial}{\partial t}|\Psi(t)\rangle = H(t)|\Psi(t)\rangle, \tag{4.1}$$

where the time-dependent Hamiltonian $H(t)$ of the system is given by the field-free atomic Hamiltonian H_0 plus the atom–field interaction Hamiltonian $H_{int}(t)$, namely

$$H(t) = H_0 + H_{int}(t). \tag{4.2}$$

We seek a method of solution that does not resort to a perturbative expansion in the laser–atom interaction. The laser pulse duration is assumed to be sufficiently long so that the corresponding laser field can be approximated by a monochromatic field of angular frequency ω. In this case, the key feature of the laser–atom interaction Hamiltonian is that it is periodic in time, so that from Equation (4.2) the Hamiltonian $H(t)$ is also periodic in time:

$$H(t+T) = H(t), \tag{4.3}$$

where $T = 2\pi/\omega$ is the optical period of the laser field.

The TDSE is then a first-order partial differential equation in the time variable, with periodic coefficients. The general form of the solutions of differential equations with periodic coefficients was obtained in 1883 by Floquet [1]. Within the context of laser–atom interactions, the Floquet result can also be derived using the formalism of quantum mechanics, as we shall now show.

In our discussion, we will assume that the Hamiltonian $H(t)$ is Hermitian. However, in Section 4.4 it will become clear that the description of the multiphoton ionization of bound states in a monochromatic laser field requires the introduction of a non-Hermitian Hamiltonian. As in Chapters 2 and 3, we introduce the evolution operator $U(t, t_0)$ that transforms the state vector at time t_0 into the state vector at time t:

$$|\Psi(t)\rangle = U(t, t_0)|\Psi(t_0)\rangle. \tag{4.4}$$

We recall that this operator satisfies Equation (2.150), namely

$$i\hbar\frac{\partial}{\partial t}U(t, t_0) = H(t)U(t, t_0), \tag{4.5}$$

subject to the initial condition $U(t_0, t_0) = I$. This equation is valid for any times t_0 and t. In particular we can write

$$i\hbar\frac{\partial}{\partial t}U(t+T, t_0+T) = H(t+T)U(t+T, t_0+T), \tag{4.6}$$

with $U(t_0 + T, t_0 + T) = I$. Using the fact that $H(t)$ is periodic in time, we also have

$$i\hbar \frac{\partial}{\partial t} U(t + T, t_0 + T) = H(t)U(t + T, t_0 + T). \tag{4.7}$$

Since $U(t + T, t_0 + T)$ and $U(t, t_0)$ satisfy the same linear differential equation with the same initial condition, we conclude that

$$U(t + T, t_0 + T) = U(t, t_0). \tag{4.8}$$

We now concentrate on the evolution of the atom over one cycle of the laser field. The corresponding evolution operator is $U(t + T, t)$. Being a unitary operator, $U(t + T, t)$ can be diagonalized. Let us denote by $|\Psi_T(t)\rangle$ an eigenstate of $U(t + T, t)$ and by λ_T the corresponding eigenvalue, so that

$$U(t + T, t)|\Psi_T(t)\rangle = \lambda_T(t)|\Psi_T(t)\rangle. \tag{4.9}$$

It follows from Equation (2.145) that we also have, for any time t',

$$U(t + T, t' + T)U(t' + T, t')U(t', t)|\Psi_T(t)\rangle = \lambda_T(t)|\Psi_T(t)\rangle. \tag{4.10}$$

Using Equation (4.8), this relation can be written as

$$U(t, t')U(t' + T, t')U(t', t)|\Psi_T(t)\rangle = \lambda_T(t)|\Psi_T(t)\rangle, \tag{4.11}$$

which in turn, using Equation (2.146), can be recast into the following form:

$$U(t' + T, t')U(t', t)|\Psi_T(t)\rangle = \lambda_T(t)U(t', t)|\Psi_T(t)\rangle. \tag{4.12}$$

Therefore, if $|\Psi_T(t)\rangle$ is an eigenstate of $U(t + T, t)$ and $\lambda_T(t)$ is the corresponding eigenvalue, then $U(t', t)|\Psi_T(t)\rangle$ is an eigenstate of $U(t' + T, t')$ at any time t', and the corresponding eigenvalue is the same $\lambda_T(t)$. As a result, the eigenvalues of $U(t + T, t)$ are constant and its eigenvectors evolve in time according to the evolution operator $U(t, t_0)$.

It follows from Equation (4.8) that

$$U(t + nT, t) = [U(t + T, t)]^n \tag{4.13}$$

for any integer n. In particular,

$$U(t - T, t)U(t + T, t) = U(t + T, t)U(t - T, t) = I. \tag{4.14}$$

This implies that any eigenvalue λ_{nT} of $U(t + nT, t)$ can be written in terms of an eigenvalue λ_T of $U(t + T, t)$ as $\lambda_{nT} = (\lambda_T)^n$, and also that $\lambda_{-T}\lambda_T = 1$. As the evolution operator is unitary, λ_T is a complex number of unit modulus. It is convenient to write it in the form

$$\lambda_T = \exp(-iET/\hbar), \tag{4.15}$$

where E is a real quantity called the Floquet *quasi-energy* of the state. Like λ_T, the quasi-energy E is a constant of the motion. Henceforth the corresponding eigenstate $|\Psi_T(t)\rangle$ of $U(t+T,t)$ will be denoted by $|\Psi_\mathsf{E}(t)\rangle$.

From Equations (4.9) and (4.15) we see that

$$U(t+T,t)|\Psi_\mathsf{E}(t)\rangle = |\Psi_\mathsf{E}(t+T)\rangle$$

$$= \exp(-i\mathsf{E}T/\hbar)|\Psi_\mathsf{E}(t)\rangle. \qquad (4.16)$$

This result has an important consequence. If we write $|\Psi_\mathsf{E}(t)\rangle$ in the form

$$|\Psi_\mathsf{E}(t)\rangle = \exp(-i\mathsf{E}t/\hbar)|P_\mathsf{E}(t)\rangle, \qquad (4.17)$$

which defines the ket $|P_\mathsf{E}(t)\rangle$, then, in order for Equation (4.16) to be fulfilled for any t, $|P_\mathsf{E}(t)\rangle$ must be periodic in time with the same period as the laser field, namely

$$|P_\mathsf{E}(t+T)\rangle = |P_\mathsf{E}(t)\rangle. \qquad (4.18)$$

The existence of such solutions of the TDSE is a particular case of the Floquet theorem concerning the solutions of linear differential equations with periodic coefficients [1]. Another important particular case arises in the study of the time-independent Schrödinger equation describing a particle in a potential that is periodic in space. Examples include atoms in an optical lattice and electrons in solids. In these contexts, the Floquet theorem is usually referred to as the Bloch theorem [5]. It leads directly to Bloch waves and conduction bands [6–8].

It is important to remark that solutions of the time-dependent Schrödinger equation of the form (4.17), with $|P_\mathsf{E}(t)\rangle$ being periodic in time, can still be found for the non-Hermitian Hamiltonians considered in Section 4.4, where the extension of the theory to the case of decaying bound states is discussed. The major difference with the Hermitian case discussed so far is that the evolution operator is not unitary in the non-Hermitian case, and that the quasi-energies are in general complex rather than purely real. Unless otherwise indicated, the results obtained in the remainder of this section are applicable to both Hermitian and non-Hermitian Hamiltonians.

We now show that both the state vector $|P_\mathsf{E}(t)\rangle$ and the quasi-energy E can be obtained by recasting the problem in the form of a time-independent eigensystem, as was recognized by Shirley[9]. Since $|P_\mathsf{E}(t)\rangle$ is periodic in t, it can be expanded in a Fourier series, with harmonic components $|F_n(\mathsf{E})\rangle$. That is,

$$|P_\mathsf{E}(t)\rangle = \sum_{n=-\infty}^{\infty} \exp(-in\omega t)|F_n(\mathsf{E})\rangle, \qquad (4.19)$$

so that the state vector (4.17) assumes the "Floquet–Fourier" form

$$|\Psi_E(t)\rangle = \exp(-iEt/\hbar) \sum_{n=-\infty}^{\infty} \exp(-in\omega t)|F_n(E)\rangle. \qquad (4.20)$$

If we also expand in a Fourier series the interaction Hamiltonian,

$$H_{int}(t) = \sum_{n=-\infty}^{\infty} (H_{int})_n \exp(-in\omega t), \qquad (4.21)$$

and substitute both Equations (4.20) and (4.21) into Equation (4.1), we obtain for the harmonic components the infinite system of time-independent coupled equations

$$(H_0 - n\hbar\omega)|F_n(E)\rangle + \sum_{k=-\infty}^{+\infty} (H_{int})_{n-k}|F_k(E)\rangle = E|F_n(E)\rangle, \quad n = 0, \pm 1, \pm 2, \ldots$$

$$(4.22)$$

It is convenient to group the harmonic components $|F_n(E)\rangle$ into a column vector $|F(E)\rangle$ and write Equation (4.22) in the more compact form

$$H_F|F(E)\rangle = E|F(E)\rangle. \qquad (4.23)$$

Here, the Hamiltonian H_F, which we shall refer to as the *Floquet Hamiltonian*, is an infinite matrix of time-independent operators. The column vector $|F(E)\rangle$ will be called the *Floquet vector*.

Of particular interest is the case of a laser field that is monochromatic and spatially homogeneous, so that the dipole approximation can be made. As we have seen in Section 2.4, the interaction Hamiltonian can then be written in the velocity gauge or in the length gauge in the form of Equation (2.137). That is:

$$H_{int}(t) = H_+ \exp(-i\omega t) + H_- \exp(i\omega t), \qquad (4.24)$$

where H_+ and H_- are time-independent operators and, in the Hermitian case, $H_-^\dagger = H_+$. As a result, the coupled equations (4.22) take the simpler form

$$(H_0 - n\hbar\omega)|F_n(E)\rangle + H_+|F_{n-1}(E)\rangle + H_-|F_{n+1}(E)\rangle = E|F_n(E)\rangle, \quad (4.25)$$

with $n = 0, \pm 1, \pm 2, \ldots$, and the Floquet Hamiltonian is an infinite tridiagonal matrix of operators of the form

$$
H_F = \begin{pmatrix}
\ddots & & & & & \\
& H_+ & H_0 - (n-1)\hbar\omega & H_- & 0 & \\
& 0 & H_+ & H_0 - n\hbar\omega & H_- & 0 \\
& & 0 & H_+ & H_0 - (n+1)\hbar\omega & H_- \\
& & & & & \ddots
\end{pmatrix}.
$$

$$(4.26)$$

The structure of the Hamiltonian H_F has a simple interpretation. The harmonic component $|F_n(\mathsf{E})\rangle$ can be seen as describing the state that the atom reaches from the state of energy E by exchanging $|n|$ photons with the laser field, namely absorbing n photons if $n > 0$ or emitting $|n|$ photons by stimulated emission if $n < 0$. The elementary processes of absorption and stimulated emission of a single photon change n by one unit. They are effected, respectively, by the dipole operators H_+ and H_-. In the limit of zero intensity, these operators vanish and the system (4.25) decouples into the equations

$$
(H_0 - n\hbar\omega)|F_n(\mathsf{E})\rangle = \mathsf{E}|F_n(\mathsf{E})\rangle, \qquad n = 0, \pm 1, \pm 2, \ldots, \qquad (4.27)
$$

which can be solved, for example, by taking $|F_0(\mathsf{E})\rangle$ to be an eigenstate of the field-free Hamiltonian H_0 and setting all the remaining harmonic components to be zero. The laser field is said to "dress" the atom through absorption and stimulated emission of photons, and Floquet states are referred to as *dressed states*.

Such an interpretation can be more rigorously founded using a quantum electrodynamical description of the laser field. The Fourier decomposition (4.19) can then be obtained in the limit that the number of photons in the laser mode becomes very large so that the Fourier expansion then simply corresponds to the photon states of the field relative to the mean number of photons in the laser mode. For this reason, one can still speak in terms of photons of the laser field, despite the fact that the field is treated classically. The correspondence between this classical description and the quantized description of the laser field has been discussed by Shirley [9] and by Mittleman [10].

It is worth noting that $|n|$ is the *net* number of photons that the atom must absorb (if $n > 0$) or emit (if $n < 0$) to couple the harmonic component $|F_0(\mathsf{E})\rangle$ to $|F_n(\mathsf{E})\rangle$. The system of equations (4.25) couple $|F_n(\mathsf{E})\rangle$ to $|F_0(\mathsf{E})\rangle$ in multiple ways, differing in the *total* number of photons exchanged with the field. For example, if $n > 0$, then $|F_n(\mathsf{E})\rangle$ is coupled to $|F_0(\mathsf{E})\rangle$ not only by the absorption of n photons, but also by the absorption of $n + k$ photons and the stimulated emission of k photons for any positive integer k.

Equation (4.15) does not define the quasi-energy E uniquely, since changing E into $\mathsf{E} + p\hbar\omega$, where p is an arbitrary integer, leaves λ_T invariant. The same redundancy also exists in the non-Hermitian case. Using Equation (4.20) we may write

$$|\Psi_\mathsf{E}(t)\rangle = \exp[-\mathrm{i}(\mathsf{E} + p\hbar\omega)t/\hbar] \sum_{n=-\infty}^{\infty} \exp[-\mathrm{i}(n-p)\omega t]|F_n(\mathsf{E})\rangle$$

$$= \exp[-\mathrm{i}(\mathsf{E} + p\hbar\omega)t/\hbar] \sum_{n=-\infty}^{\infty} \exp(-\mathrm{i}n\omega t)|F_{n+p}(\mathsf{E})\rangle \quad (4.28)$$

and also

$$|\Psi_{\mathsf{E}+p\hbar\omega}(t)\rangle = \exp[-\mathrm{i}(\mathsf{E} + p\hbar\omega)t/\hbar] \sum_{n=-\infty}^{\infty} \exp(-\mathrm{i}n\omega t)|F_n(\mathsf{E} + p\hbar\omega)\rangle. \quad (4.29)$$

If $|\Psi_{\mathsf{E}+p\hbar\omega}(t)\rangle$ and $|\Psi_\mathsf{E}(t)\rangle$ denote the same eigenstate of $U(t+T,t)$, then, on comparing Equations (4.28) and (4.29), we see that

$$|F_n(\mathsf{E} + p\hbar\omega)\rangle = |F_{n+p}(\mathsf{E})\rangle \quad (4.30)$$

for all n. Therefore if we find an eigenstate of the Floquet Hamiltonian $|F(\mathsf{E})\rangle$ of quasi-energy E, then we know infinitely many other eigenstates $|F(\mathsf{E} + p\hbar\omega)\rangle$ with quasi-energies $\mathsf{E} + p\hbar\omega$, $p = 0, \pm 1, \ldots$, which are related to $|F(\mathsf{E})\rangle$ by Equation (4.30) and which describe the same physical state of the atom. Each dressed state is thus associated with an infinite periodic comb of quasi-energies.

We emphasize that Equation (4.30) applies only if the states $|\Psi_\mathsf{E}(t)\rangle$ and $|\Psi_{\mathsf{E}+p\hbar\omega}(t)\rangle$ refer, respectively, to the same physical state. This is always the case if the quasi-energy spectrum is non-degenerate. However, if it is degenerate then there may exist two (or more) physically different Floquet states with the same quasi-energy $\mathsf{E} + p\hbar\omega$. For example, let us introduce the state $|\Psi'_{\mathsf{E}+p\hbar\omega}(t)\rangle$ having quasi-energy $\mathsf{E} + p\hbar\omega$ which does not correspond to the same physical state as $|\Psi_\mathsf{E}(t)\rangle$. In this case, the Floquet–Fourier components of the two states are not related to each other by Equation (4.30). Nevertheless, Equation (4.30) can be used to express the state $|\Psi'_{\mathsf{E}+p\hbar\omega}(t)\rangle$ as

$$|\Psi'_{\mathsf{E}+p\hbar\omega}(t)\rangle = \exp[-\mathrm{i}(\mathsf{E} + p\hbar\omega)t/\hbar] \sum_{n=-\infty}^{\infty} \exp(-\mathrm{i}n\omega t)|F'_n(\mathsf{E} + p\hbar\omega)\rangle$$

$$= \exp(-\mathrm{i}\mathsf{E}t/\hbar) \sum_{n=-\infty}^{\infty} \exp(-\mathrm{i}n\omega t)|F'_{n-p}(\mathsf{E} + p\hbar\omega)\rangle$$

$$= |\Psi'_\mathsf{E}(t)\rangle \quad (4.31)$$

so that the state $|\Psi_E(t)\rangle$ and the state $|\Psi'_E(t)\rangle$ have the same quasi-energy E.

We also note that when carrying out numerical calculations the Fourier expansion (4.20) must be truncated to a finite number of terms. As a result, the comb of quasi-energies associated with each dressed state is finite in extension and not strictly periodic, and Equation (4.30) holds only approximately.

The equivalence of the Floquet states of quasi-energies differing by an integer multiple of $\hbar\omega$ allows us to restrict ourselves to states belonging to a particular interval of the quasi-energy spectrum of size $\hbar\omega$, since any state having a quasi-energy outside this interval will always be equivalent to one lying inside this interval. An analogous situation is found when describing the electronic band structure in crystals: only wave numbers in the first Brillouin zone are required as the wave numbers are defined only up to integer multiples of reciprocal lattice vectors.

Before closing this section, we make three observations on the solutions of the coupled equations (4.25). The first one concerns the symmetry of the Floquet states. When the dipole operators H_+ and H_- act on a wave function which is either even or odd under a reflection about the origin, the result is a wave function of opposite parity. Therefore, if H_0 does not change the parity of the wave function, the solutions of the system can be divided into two classes: those for which the harmonic components $|F_n(E)\rangle$ are of even parity for even values of n and of odd parity for odd values of n, and those for which they are of odd parity for even values of n and of even parity for odd values of n. Depending on the polarization of the laser field, the solutions may also possess other symmetries. For example, if the laser field is linearly polarized, the projection of the orbital angular momentum operator on the polarization axis commutes with $H(t)$ (neglecting spin-orbit coupling), and therefore the solutions of the equations (4.25) decouple into eigenstates of this operator.

The second observation concerns the role of the phase, φ, of the laser field defined by Equation (2.114). This phase enters in the coupled equations (4.25) through the dipole operators. Since the laser field is monochromatic, however, this phase can be cancelled by a shift of the origin of the time axis. By introducing the time variable $t' = t - \varphi/\omega$, we can eliminate the phase φ from the TDSE. Hence, for any dressed state the state vector

$$|\Psi_E(t')\rangle = \exp(-iEt'/\hbar) \sum_{n=-\infty}^{\infty} \exp(-in\omega t')|F_n(E)\rangle \qquad (4.32)$$

does not depend on φ. Going back to the original time variable t, we see that the dressed states depend on φ only through phase factors. That is,

$$|\Psi_E(t - \varphi/\omega)\rangle = \exp[-iEt/\hbar + iE\varphi/(\hbar\omega)] \sum_{n=-\infty}^{\infty} \exp(-in\omega t)\exp(in\varphi)|F_n(E)\rangle.$$

$$(4.33)$$

Thus, varying φ only changes the relative phases of the harmonic components and the (irrelevant) overall phase of $|\Psi_E(t)\rangle$, and does not affect the quasi-energy.

Finally, we note that transforming the state vector between the length gauge and the velocity gauge does not alter the form of the Floquet states. Let us assume that the wave function

$$\Psi_{EL}^L(X, t) = \exp(-iE^L t/\hbar) \sum_{n=-\infty}^{\infty} \exp(-in\omega t) F_n^L \left(E^L; X \right) \qquad (4.34)$$

is a solution of the TDSE (2.133) for an N-electron atom (or ion) in a monochromatic laser field described in the length gauge. Then, it follows from Equation (2.132) that the wave function

$$\Psi_E(X, t) = \exp\left[-\frac{ie}{\hbar} \mathbf{A}(t) \cdot \mathbf{R} \right] \Psi_{EL}^L(X, t), \qquad (4.35)$$

with $\mathbf{R}$ defined by Equation (2.130), is a solution of the original TDSE (2.121). Since the phase factor appearing on the right-hand side of this equation is periodic in time, with the same period as the laser field, $\Psi_E(X, t)$ has the same form as $\Psi_{EL}^L(X, t)$, although with different harmonic components:

$$\Psi_E(X, t) = \exp(-iE^L t/\hbar) \sum_{n=-\infty}^{\infty} \exp(-in\omega t) F_n (E; X). \qquad (4.36)$$

The quasi-energy is therefore unaffected by this gauge transformation. However, the harmonic components $F_n (E; X)$ satisfy a five-term recursion relation rather than the simpler three-term recursion relation satisfied by the harmonic components $F_n^L (E; X)$. We can now go to the velocity gauge by introducing the wave function

$$\Psi_{EV}^V(X, t) = \exp\left[\frac{i}{\hbar} \frac{e^2 N}{2m} \int_{-\infty}^t \mathbf{A}^2(t')dt' \right] \Psi_E(X, t), \qquad (4.37)$$

which satisfies the TDSE (2.127). For a monochromatic laser field of electric field strength $\mathcal{E}_0$, the integral of $\mathbf{A}^2(t)$ over t is the sum of $\mathcal{E}_0^2 t/(2\omega^2)$ and a term oscillating in time at twice the frequency of the laser field. Therefore, using Equations (4.35) and (4.37), $\Psi_{EV}^V(X, t)$ can be written in the form

$$\Psi_{EV}^V(X, t) = \exp(-iE^V t/\hbar) \sum_{n=-\infty}^{\infty} \exp(-in\omega t) F_n^V \left(E^V; X \right), \qquad (4.38)$$

and the quasi-energy in the velocity gauge, E^V, is related to the quasi-energy in the length gauge, E^L, by the equation

$$E^V = E^L - N U_p, \qquad (4.39)$$

where U_p is the electron ponderomotive energy defined in Equation (1.5). Like their length-gauge counterparts, the harmonic components $F_n^V(E^V; X)$ satisfy a three-term recursion relation. However, in the velocity gauge the quasi-energies are lower by NU_p compared to the quasi-energies in the length gauge. This shift affects the *whole* quasi-energy spectrum uniformly and does not modify the separation between the different quasi-energies. We stress that the relationship between the Floquet states in the velocity gauge and those in the length gauge is a property of the *exact* solutions of the coupled equations (4.25) and may not hold for approximate solutions.

4.1.2 Multicolor laser fields

We now consider the application of the Floquet theory to the case in which the atom interacts with several laser fields having different frequencies. We assume that these different laser fields act coherently, or in other words that the total electric field to which the atom is exposed is given by the sum of the electric fields of each laser field. We begin by considering the case for which the atom interacts with a laser field of angular frequency $\omega_1 = \omega$ and with one of its harmonics, of angular frequency $\omega_2 = q\omega$, $q = 2, 3, 4, \ldots$. The corresponding generalization of Equation (4.24) is

$$H_{\text{int}}(t) = H_+^{(1)} \exp(-i\omega t) + H_-^{(1)} \exp(i\omega t) + H_+^{(2)} \exp(-iq\omega t) + H_-^{(2)} \exp(iq\omega t).$$
$$(4.40)$$

The general Floquet theory developed above can be applied as the Hamiltonian $H(t)$ is still periodic in time. However, the harmonic components of the Floquet states are now related to each other by the following system of coupled equations:

$$(H_0 - n\hbar\omega)| F_n(E) \rangle + H_+^{(1)}| F_{n-1}(E) \rangle + H_-^{(1)}| F_{n+1}(E) \rangle$$
$$+ H_+^{(2)}| F_{n-q}(E) \rangle + H_-^{(2)}| F_{n+q}(E) \rangle = E| F_n(E) \rangle, \quad (4.41)$$

rather than by the system of coupled equations (4.25), and the Floquet Hamiltonian is no longer a tridiagonal matrix of operators. The case of an atom exposed to a more general coherent superposition of several harmonics of the same laser field can be treated along similar lines.

Returning to the interpretation of the dipole operators in terms of the absorption and the emission of photons, we see that any harmonic component can now be reached from another one by infinitely many interfering pathways which differ from each other by the net number of photons involved in the coupling. For instance, if we take a superposition of a laser field and its third harmonic ($q = 3$), then the harmonic component $| F_n(E) \rangle$ may be reached from the harmonic component $| F_{n-3}(E) \rangle$ by the absorption of three photons of angular frequency ω, or by the absorption of one photon of angular frequency 3ω, or by the absorption of six photons of angular frequency ω and the emission of one photon of angular frequency 3ω, etc. Because these different pathways interfere with each other, the dressed states depend on the relative phases of the laser fields. This dependence cannot be removed by a translation of the time variable.

Another difference with respect to the monochromatic case is that the solutions of the system of equations (4.41) no longer separate into two distinct parity classes when the second laser field is an even harmonic of the first one. Let us take, for example, $q = 2$. Since the harmonic component $|F_n(\mathrm{E})\rangle$ may be reached from the harmonic component $|F_{n-2}(\mathrm{E})\rangle$ by the absorption of two photons of angular frequency ω as well as by the absorption of one photon of angular frequency 2ω, $|F_{n-2}(\mathrm{E})\rangle$ and $|F_n(\mathrm{E})\rangle$ cannot both have a well defined parity with respect to a reflection about the origin.

In principle, phase-dependent interferences exist whenever the atom is exposed to two laser fields with commensurable frequencies, namely laser fields whose angular frequencies ω_1 and ω_2 are multiples of a common fundamental angular frequency ω. Therefore, the angular frequencies are commensurable if $\omega_1 = p\omega$ and $\omega_2 = q\omega$, where p and q are two integers. The Floquet states can then be written in the form (4.20) and the harmonic components satisfy the following system of equations:

$$(H_0 - n\hbar\omega)|F_n(\mathrm{E})\rangle + H_+^{(1)}|F_{n-p}(\mathrm{E})\rangle + H_-^{(1)}|F_{n+p}(\mathrm{E})\rangle$$
$$+ H_+^{(2)}|F_{n-q}(\mathrm{E})\rangle + H_-^{(2)}|F_{n+q}(\mathrm{E})\rangle = \mathrm{E}|F_n(\mathrm{E})\rangle. \quad (4.42)$$

These equations imply that if two harmonic components are coupled, for example by the absorption of n_1 photons from the first laser field and n_2 photons from the second laser field, then they are also coupled by the absorption of $n_1 + kq$ photons of the first laser field and $n_2 - kp$ photons of the second laser field for infinitely many values of the integer k. The solutions of these equations depend on the phases of the two laser fields in a non-trivial way because these multiple interfering couplings all have a different dependence in these phases. Unless p and q are sufficiently small, however, the interference effects depend on the exchange of many photons, and are therefore likely to be small. It should be noted that p and q are usually large for two arbitrary angular frequencies. For example, for a superposition of a Nd:YAG laser field ($\lambda = 1064$ nm) and a KrF laser field ($\lambda = 248$ nm), $\omega_1/\omega_2 = 31/133$ and thus $p = 31$ and $q = 133$.

Equations (4.20) and (4.42) cannot be used if the ratio of the angular frequencies of the two laser fields is not commensurable. From the mathematical point of view, these equations are formally valid when ω_1/ω_2 is a rational number and are invalid when ω_1/ω_2 is irrational. However, it is clear that distinguishing between these two cases cannot have any physical relevance. Moreover, varying one of the angular frequencies by a small amount will completely change the values of p, q and ω, even when the variation has no physical significance. As we shall now show, it is possible to avoid these problems, when p and q are large, by modifying Equations (4.20) and (4.42).

Let us focus our attention on the solutions for which $|F_{n=0}(E)\rangle \neq 0$. We can safely assume that the probability that the atom absorbs more than a net number n_1 of photons from the first laser field becomes negligible when $|n_1|$ exceeds a certain upper bound N_1, and similarly for the second laser field. Therefore, only a finite number of harmonic components must be included in the Fourier expansion (4.20) for representing the dressed states accurately, and we can restrict the system (4.42) to coupling only those harmonic components that can be reached from $|F_{n=0}(E)\rangle$ by the *net* absorption (or emission) of at most N_1 photons of the first laser field and N_2 photons of the second laser field. Given these restrictions, the system couples $|F_{n=0}(E)\rangle$ only to a subset of the other harmonic components $|F_n(E)\rangle$. The harmonic components that do not belong to this subset can then be set equal to zero without affecting the solutions of interest. The others are each coupled to $|F_{n=0}(E)\rangle$ by absorption of a certain net number n_1 of photons of angular frequency ω_1 and of a certain net number n_2 photons of angular frequency ω_2. For the harmonic component $|F_n(E)\rangle$, the coupling is possible if, and only if, $n_1\omega_1 + n_2\omega_2 = n\omega$. If p and q are both small, there might be more than one possible combination of values of n_1 and n_2 fulfilling this condition, with $|n_1| \leq N_1$ and $|n_2| \leq N_2$. But when p or q are sufficiently large, there will be at most one. In this case, we can replace the expansion (4.20) by

$$|\Psi_E(t)\rangle = \exp(-iEt/\hbar) \sum_{|n_1| \leq N_1} \sum_{|n_2| \leq N_2} \exp[-i(n_1\omega_1 + n_2\omega_2)t] |F_{n_1,n_2}(E)\rangle.$$

$$(4.43)$$

Writing

$$H_{\text{int}}(t) = H_+^{(1)} \exp(-i\omega_1 t) + H_-^{(1)} \exp(i\omega_1 t) + H_+^{(2)} \exp(-i\omega_2 t) + H_-^{(2)} \exp(i\omega_2 t),$$

$$(4.44)$$

we then have

$$(H_0 - n_1\hbar\omega_1 - n_2\hbar\omega_2)|F_{n_1,n_2}(E)\rangle + H_+^{(1)}|F_{n_1-1,n_2}(E)\rangle + H_-^{(1)}|F_{n_1+1,n_2}(E)\rangle$$

$$+ H_+^{(2)}|F_{n_1,n_2-1}(E)\rangle + H_-^{(2)}|F_{n_1,n_2+1}(E)\rangle = E|F_{n_1,n_2}(E)\rangle. \qquad (4.45)$$

We note that the expansion (4.43) and the system of equations (4.45) do not depend on the values of p and q, and hence do not depend on the rational number used to approximate the ratio ω_1/ω_2. These equations can also be extended to more general coherent superpositions of laser fields of several angular frequencies.

We have seen above that it is possible to eliminate the phase of a monochromatic laser field by a shift of the origin of the time axis. The phases φ_1 and φ_2 of the two components of a two-color laser field cannot both be removed in general by a similar operation. However, the solutions of the coupled equations (4.45) still depend on these phases in a simple way. Indeed, when the two laser fields are defined as in Equation (2.114), we can write $H_\pm^{(1)} = \exp(\mp i\varphi_1)\bar{H}_\pm^{(1)}$ and $H_\pm^{(2)} = \exp(\mp i\varphi_2)\bar{H}_\pm^{(2)}$, where $\bar{H}_\pm^{(1)}$ and $\bar{H}_\pm^{(2)}$ do not depend on φ_1 and φ_2. If we now substitute the harmonic components $|F_{n_1,n_2}(\mathsf{E})\rangle$ by $\exp(-in_1\varphi_1 - in_2\varphi_2)|\bar{F}_{n_1,n_2}(\mathsf{E})\rangle$, we see that the harmonic components $|\bar{F}_{n_1,n_2}(\mathsf{E})\rangle$ satisfy the phase-independent equations

$$(H_0 - n_1\hbar\omega_1 - n_2\hbar\omega_2)|\bar{F}_{n_1,n_2}(\mathsf{E})\rangle + \bar{H}_+^{(1)}|\bar{F}_{n_1-1,n_2}(\mathsf{E})\rangle + \bar{H}_-^{(1)}|\bar{F}_{n_1+1,n_2}(\mathsf{E})\rangle$$
$$+ \bar{H}_+^{(2)}|\bar{F}_{n_1,n_2-1}(\mathsf{E})\rangle + \bar{H}_-^{(2)}|\bar{F}_{n_1,n_2+1}(\mathsf{E})\rangle = \mathsf{E}|\bar{F}_{n_1,n_2}(\mathsf{E})\rangle. \tag{4.46}$$

Thus, the harmonic components $|F_{n_1,n_2}(\mathsf{E})\rangle$ depend on φ_1 and φ_2 in a similar way as in the case of a monochromatic field, and the quasi-energy does not depend on these phases.

4.1.3 Non-dipole laser fields

Until now, we have assumed that the laser field can be taken to be spatially homogeneous, or in other words that the dipole approximation can be made. If this is not the case, the interaction Hamiltonian is given in the Coulomb gauge by Equation (2.108),

$$H_{\text{int}}(t) = \frac{e}{m}\sum_{i=1}^{N}\mathbf{A}(\mathbf{r}_i, t)\cdot\mathbf{p}_i + \frac{e^2}{2m}\sum_{i=1}^{N}\mathbf{A}^2(\mathbf{r}_i, t). \tag{4.47}$$

Let us assume that the laser field is monochromatic and linearly polarized. Using Equation (2.10) with $\varphi = 0$, the vector potential can be written in the form

$$\mathbf{A}(\mathbf{r}, t) = \hat{\boldsymbol{\epsilon}}\frac{A_0}{2i}\{\exp[i(\mathbf{k}_{\mathrm{L}}\cdot\mathbf{r} - \omega t)] - \exp[-i(\mathbf{k}_{\mathrm{L}}\cdot\mathbf{r} - \omega t)]\}. \tag{4.48}$$

The two terms on the right-hand side of Equation (4.47) can be expressed as

$$\frac{e}{m}\sum_{i=1}^{N}\mathbf{A}(\mathbf{r}_i, t)\cdot\mathbf{p}_i = H_{\text{int}}^{(1)}\exp(-i\omega t) + H_{\text{int}}^{(-1)}\exp(i\omega t) \tag{4.49}$$

and

$$\frac{e^2}{2m}\sum_{i=1}^{N}\mathbf{A}^2(\mathbf{r}_i,t) = H_{\text{int}}^{(0)} + H_{\text{int}}^{(2)}\exp(-2i\omega t) + H_{\text{int}}^{(-2)}\exp(2i\omega t), \qquad (4.50)$$

where we have introduced the time-independent quantities

$$H_{\text{int}}^{(0)} = N\frac{e^2 A_0^2}{4m}, \qquad (4.51)$$

$$H_{\text{int}}^{(\pm 1)} = \pm\frac{eA_0}{2mi}\sum_{i=1}^{N}\exp(\pm i\mathbf{k}_{\text{L}}\cdot\mathbf{r}_i)\hat{\boldsymbol{\epsilon}}\cdot\mathbf{p}_i \qquad (4.52)$$

and

$$H_{\text{int}}^{(\pm 2)} = \frac{e^2 A_0^2}{8m}\sum_{i=1}^{N}\exp(\pm 2i\mathbf{k}_{\text{L}}\cdot\mathbf{r}_i). \qquad (4.53)$$

The TDSE (4.1) then takes the form

$$i\hbar\frac{\partial}{\partial t}|\Psi(t)\rangle = \Big[H_0 + H_{\text{int}}^{(0)} + H_{\text{int}}^{(1)}\exp(-i\omega t) + H_{\text{int}}^{(-1)}\exp(i\omega t)$$

$$+ H_{\text{int}}^{(2)}\exp(-2i\omega t) + H_{\text{int}}^{(-2)}\exp(2i\omega t)\Big]|\Psi(t)\rangle. \qquad (4.54)$$

We note that $H_{\text{int}}^{(0)}$ is a constant term that modifies the energy of the system. Remembering that $A_0 = \mathcal{E}_0/\omega$, we see that

$$H_{\text{int}}^{(0)} = N\frac{e^2\mathcal{E}_0^2}{4m\omega^2} = NU_{\text{p}}, \qquad (4.55)$$

where U_{p} is the ponderomotive energy given by Equation (1.5). Since the Hamiltonian in the TDSE (4.54) is periodic in time, we can use for the state vector $|\Psi(t)\rangle$ the Floquet–Fourier expansion (4.20). We then obtain for the harmonic components $|F_n(E)\rangle$ the infinite system of time-independent coupled equations

$$(H_0 - n\hbar\omega + NU_{\text{p}})|F_n(E)\rangle + H_{\text{int}}^{(1)}|F_{n-1}(E)\rangle + H_{\text{int}}^{(-1)}|F_{n+1}(E)\rangle$$

$$+ H_{\text{int}}^{(2)}|F_{n-2}(E)\rangle + H_{\text{int}}^{(-2)}|F_{n+2}(E)\rangle = E|F_n(E)\rangle. \qquad (4.56)$$

In contrast to the dipole approximation case considered above, we see that the Floquet Hamiltonian is now an infinite pentadiagonal matrix of operators.

While non-dipole Floquet calculations have been carried out along these lines [11], going beyond the dipole approximation makes the computations considerably more demanding because of the complicated structure of the system of equations (4.56) and the difficulty of calculating the relevant matrix elements. However, it should be noted that taking into account the spatial variation of the vector potential to all orders in $1/c$, as done here, leads to solutions of Equations (4.56) for a *free* electron in a plane wave electromagnetic field that has spurious features [12, 13]. This difficulty does not affect the non-dipole Gordon–Volkov wave functions introduced in Section 2.8, since in their derivation the non-dipole corrections to the vector potential are limited to the first order in $1/c$.

4.2 Bound–bound Floquet dynamics

In this section, we shall consider the interaction of a model atom having M bound states with a laser field. We shall assume that the coupling of the atomic states to the continuum can be neglected, which is the case when photoionization is sufficiently weak. This situation arises, for instance, when a few bound states are strongly coupled to each other by a weak near-resonant laser field. In this approximation, the dressed bound states cannot decay by ionization and their quasi-energies are therefore real. The atom can then be described in terms of a basis of dressed states constructed from the field-free eigenstates of H_0. For simplicity, we will first take the laser field to be monochromatic and spatially homogeneous. The case of a slow temporal variation of the intensity of the laser pulse will be considered at the end of this section.

4.2.1 Bases of dressed states

We express the harmonic component $|F_n\rangle$ as a sum of field-free, orthonormal, bound eigenstates $|\psi_k\rangle$ of the atomic Hamiltonian H_0, and we restrict the dipole operators H_+ and H_-, the Hamiltonian $H(t)$ and the evolution operator $U(t, t_0)$ to the subspace spanned by M bound atomic states. Thus, we write

$$|F_n(\mathsf{E})\rangle = \sum_{k=1}^{M} c_{nk}(\mathsf{E})|\psi_k\rangle, \tag{4.57}$$

with

$$H_0|\psi_k\rangle = E_k|\psi_k\rangle. \tag{4.58}$$

Inserting this expansion in Equation (4.23) and projecting on the eigenstate $|\psi_j\rangle$, one finds that the coefficients c_{nk} satisfy the equation

$$(E_j - \mathsf{E} - n\hbar\omega)\, c_{nj} + \sum_{k=-\infty}^{+\infty} \left[(H_+)_{jk}\, c_{n-1,k} + (H_-)_{jk}\, c_{n+1,k}\right] = 0, \tag{4.59}$$

where

$$(H_\pm)_{jk} = \langle \psi_j | H_\pm | \psi_k \rangle. \tag{4.60}$$

Equation (4.59) can be rewritten as the matrix eigenvalue equation

$$(\mathbf{H}_F - \mathsf{E})\mathbf{c} = 0, \tag{4.61}$$

where $\mathbf{H}_F$ is the matrix representation of the Floquet Hamiltonian H_F. Introducing the matrix $\mathbf{H}_0$ with elements

$$(\mathbf{H}_0)_{jk} = E_j \delta_{jk} \tag{4.62}$$

and the matrices $\mathbf{H}_\pm$ having elements $(H_\pm)_{jk}$, the Floquet matrix $\mathbf{H}_F$ can be written in the form of the infinite block-tridiagonal matrix

$$(\mathbf{H}_F)_{nn'} = (\mathbf{H}_0 - n\hbar\omega\mathbf{I})\delta_{nn'} + \mathbf{H}_+\delta_{n,n'+1} + \mathbf{H}_-\delta_{n,n'-1}, \tag{4.63}$$

which has the explicit form

$$\mathbf{H}_F = \begin{pmatrix} \ddots & & & & & \\ & \mathbf{H}_+ & \mathbf{H}_0 - (n-1)\hbar\omega\mathbf{I} & \mathbf{H}_- & 0 & \\ & 0 & \mathbf{H}_+ & \mathbf{H}_0 - n\hbar\omega\mathbf{I} & \mathbf{H}_- & 0 \\ & & & \mathbf{H}_+ & \mathbf{H}_0 - (n+1)\hbar\omega\mathbf{I} & \mathbf{H}_- \\ & & & & & \ddots \end{pmatrix}. \tag{4.64}$$

Here, $\mathbf{I}$ is the $M \times M$ unit matrix. The diagonal matrices $\mathbf{H}_0 - n\hbar\omega$ are often referred to as Floquet blocks. The solution of the TDSE has thus been transformed into a time-independent infinite matrix eigenvalue problem, which yields the possible values of the quasi-energy E. However, in view of the above discussion, only M eigenvalues and eigenvectors of the matrix are required.

Since the Floquet Hamiltonian H_F is Hermitian, all the quasi-energies are real. Each one reduces to one of the field-free eigenenergies E_j in the zero-field limit, modulo $\hbar\omega$. Any quasi-energy E_j can therefore be written in the form

$$\mathsf{E}_j = E_j + \Delta_j + n_j\hbar\omega, \tag{4.65}$$

where n_j is an integer, E_j is the energy of the initial field-free state and Δ_j is the AC Stark shift of that state. The corresponding Floquet vectors can be taken to be orthonormal:

$$\langle F(\mathsf{E}_i) | F(\mathsf{E}_j) \rangle = \sum_{n=-\infty}^{\infty} \langle F_n(\mathsf{E}_i) | F_n(\mathsf{E}_j) \rangle = \delta_{ij}. \tag{4.66}$$

Moreover, since the norm of the state vector $|\Psi_E(t)\rangle$ is conserved, we can normalize the dressed states such that

$$\langle \Psi_E(t) | \Psi_E(t) \rangle = 1 \tag{4.67}$$

for all times. From the Floquet–Fourier expansion (4.20) of the state vector, this implies that

$$\sum_{m,n} \exp[-i(n-m)\omega t] \langle F_m(E) | F_n(E) \rangle = \sum_{p,n} \exp(ip\omega t) \langle F_{n+p}(E) | F_n(E) \rangle$$

$$= 1. \tag{4.68}$$

In obtaining this result, we have used the fact that

$$\sum_n \langle F_{n+p}(E) | F_n(E) \rangle = \delta_{p0} \tag{4.69}$$

for all values of p, which follows from Equations (4.30) and (4.66). We also have

$$\langle \Psi_{E_i}(0) | \Psi_{E_j}(0) \rangle = \sum_{m,n} \langle F_m(E_i) | F_n(E_j) \rangle = 0 \tag{4.70}$$

if $|\Psi_{E_i}(0)\rangle$ and $|\Psi_{E_j}(0)\rangle$ correspond to two different physical states. These solutions then remain orthogonal at all times, since, from Equation (4.70),

$$\langle \Psi_{E_i}(t) | \Psi_{E_j}(t) \rangle = \langle \Psi_{E_i}(0) | U^\dagger(t,0) U(t,0) | \Psi_{E_j}(0) \rangle = \langle \Psi_{E_i}(0) | \Psi_{E_j}(0) \rangle = 0. \tag{4.71}$$

Thus, when the atom is in a particular dressed state, it remains in that dressed state. We recall that spontaneous emission is neglected.

Until now, we have examined particular solutions of the TDSE. It is clear that the M orthonormal dressed states $|\Psi_{E_i}(t)\rangle$ span the same subspace as the M eigenstates of H_0 from which they are constructed. A general state of the system, if restricted to this subspace, can thus be expanded in terms of these dressed states. For instance, for any time t_0, we may always write

$$|\Psi(t_0)\rangle = \sum_{i=1}^{M} a_i |\Psi_{E_i}(t_0)\rangle \tag{4.72}$$

if $|\Psi(t_0)\rangle$ is a linear superposition of the M eigenstates $|\psi_k\rangle$. Since $|\Psi(t)\rangle = U(t,t_0)|\Psi(t_0)\rangle$ and $|\Psi_{E_i}(t)\rangle = U(t,t_0)|\Psi_{E_i}(t_0)\rangle$, we also have

$$|\Psi(t)\rangle = \sum_{i=1}^{M} a_i |\Psi_{E_i}(t)\rangle \tag{4.73}$$

at all times. The expansion coefficients a_i are time-independent and are given by

$$a_i = \langle \Psi_{E_i}(t_0)|\Psi(t_0)\rangle = \langle \Psi_{E_i}(t)|\Psi(t)\rangle. \tag{4.74}$$

In Section 4.2.4 below, we will consider the more general case in which the expansion coefficients are time-dependent.

The Floquet evolution operator can also be expressed in terms of the M states $|\Psi_{E_i}(t)\rangle$. Indeed, we have:

$$U(t, t_0) = \sum_{i=1}^{M} |\Psi_{E_i}(t)\rangle \langle \Psi_{E_i}(t_0)|$$

$$= \sum_{i=1}^{M} \exp[-iE_i(t - t_0)/\hbar] \sum_{n,m=-\infty}^{\infty} \exp[-i\omega(nt - mt_0)]|F_n(E_i)\rangle \langle F_m(E_i)|$$

$$= \sum_{i=1}^{M} \exp[-iE_i(t - t_0)/\hbar]$$

$$\times \sum_{n,m=-\infty}^{\infty} \exp[-i\omega n(t - t_0) + i\omega(m - n)t_0]|F_n(E_i)\rangle \langle F_m(E_i)|. \tag{4.75}$$

As in the case of the evolution operator corresponding to a system described by a time-independent Hamiltonian, the Floquet evolution operator $U(t, t_0)$ depends on t only through the time difference $(t - t_0)$. Using Equation (4.75), and recalling that $T = 2\pi/\omega$, one can verify that Equation (4.8) is satisfied.

4.2.2 *Rotating wave approximation in the Floquet approach*

To illustrate the theory developed above, we now rederive the probability for making a transition between an initial state a and a final state b obtained using the rotating wave approximation in Section 3.2. To this end, we consider only two bound atomic states, a and b, strongly coupled by a near-resonant laser field. We assume that $E_b > E_a$ and that $\hbar\omega \simeq E_b - E_a$. Within this two-state model, the harmonic components can be written in the form

$$|F_n(E)\rangle = c_{na}|\psi_a\rangle + c_{nb}|\psi_b\rangle. \tag{4.76}$$

However, the states a and b must have opposite parities under a reflection about the origin in order for them to be coupled by the laser field. Since the harmonic components can be taken to be either even or odd under this operation, as was mentioned in Section 4.1.1, we can restrict ourselves to the solutions of Equations (4.59) such that $c_{na} = 0$ for n odd and $c_{nb} = 0$ for n even. Moreover, as seen in

Chapter 3, we can assume that the transitions from state a to state b will be far more probable when accompanied by the absorption of a photon rather than by the emission of a photon, and inversely for the transitions from b to a. The rotating wave approximation introduced in Section 3.2.4 consists in neglecting emission in the transitions from a to b and absorption in the transitions from b to a. In the Floquet calculation developed here, it amounts to setting $|F_n(\mathsf{E})\rangle = 0$ for all values of n except 0 and 1.

Within this approximation, the matrix eigenvalue equation (4.61) reduces to the system of homogeneous linear equations

$$\begin{pmatrix} E_a - \mathsf{E} & (H_-)_{ab} \\ (H_+)_{ba} & E_b - \hbar\omega - \mathsf{E} \end{pmatrix} \begin{pmatrix} c_{0a} \\ c_{1b} \end{pmatrix} = 0. \tag{4.77}$$

This system has non-trivial solutions if its determinant vanishes, so that

$$(E_a - \mathsf{E})(E_b - \hbar\omega - \mathsf{E}) - (H_-)_{ab}(H_+)_{ba} = 0. \tag{4.78}$$

Since $H_- = H_+^\dagger$, we note that $(H_-)_{ab}(H_+)_{ba} = |(H_+)_{ba}|^2$. In terms of the detuning angular frequency (3.215), the condition (4.78) can also be written as

$$(E_a - \mathsf{E})^2 - \hbar\Delta\omega(E_a - \mathsf{E}) - |(H_+)_{ba}|^2 = 0. \tag{4.79}$$

Hence, the two possible values of the quasi-energies are

$$\mathsf{E}_\pm = E_a - \frac{\hbar}{2}(\Delta\omega \pm \omega_R), \tag{4.80}$$

where ω_R is the Rabi flopping frequency defined by Equation (3.221). We find only two quasi-energies rather than multiple periodic combs of quasi-energies because of the approximation made in replacing the infinite system (4.25) by just two coupled equations.

The two resulting Floquet states, normalized according to Equation (4.67), are

$$|\Psi_{\mathsf{E}_+}(t)\rangle = \exp(-i\mathsf{E}_+ t/\hbar)\, C\,[(H_-)_{ab}|\psi_a\rangle - \exp(-i\omega t)(\hbar/2)(\Delta\omega + \omega_R)|\psi_b\rangle]$$

$$\tag{4.81}$$

and

$$|\Psi_{\mathsf{E}_-}(t)\rangle = \exp(-i\mathsf{E}_- t/\hbar)\, C\,[(\hbar/2)(\Delta\omega + \omega_R)|\psi_a\rangle + \exp(-i\omega t)(H_+)_{ba}|\psi_b\rangle],$$

$$\tag{4.82}$$

with

$$C = \left[\hbar^2(\Delta\omega + \omega_R)^2/4 + |(H_+)_{ba}|^2\right]^{-1/2}. \tag{4.83}$$

In the zero-field limit, $|\Psi_{\mathsf{E}_+}(t)\rangle \rightarrow \exp(-iE_b t/\hbar)|\psi_b\rangle$ and $|\Psi_{\mathsf{E}_-}(t)\rangle \rightarrow \exp(-iE_a t/\hbar)|\psi_a\rangle$.

When the atom is either in the dressed state $|\Psi_{E_+}(t)\rangle$ or in the dressed state $|\Psi_{E_-}(t)\rangle$, the probabilities of finding it either in the field-free state $|\psi_a\rangle$ or in the field-free state $|\psi_b\rangle$ are the same at all times. In these dressed states, there is therefore no "Rabi flopping" between the populations in the two field-free states. To retrieve the oscillating solution found in Section 3.2, we look for a state $|\Psi(t)\rangle$ in the form of a linear combination of $|\Psi_{E_+}(t)\rangle$ and $|\Psi_{E_-}(t)\rangle$ such that $|\Psi(t)\rangle = |\psi_a\rangle$ at time $t = 0$. Writing

$$|\Psi(t)\rangle = a_+|\Psi_{E_+}(t)\rangle + a_-|\Psi_{E_-}(t)\rangle \tag{4.84}$$

and requiring that

$$a_+|\Psi_{E_+}(0)\rangle + a_-|\Psi_{E_-}(0)\rangle = |\psi_a\rangle, \tag{4.85}$$

we find that

$$a_+ = C(H_+)_{ba}, \qquad a_- = C\hbar(\Delta\omega + \omega_R)/2. \tag{4.86}$$

The probability that the atom is in state b at time t,

$$P_{ba}(t) = |\langle\psi_b|\Psi(t)\rangle|^2, \tag{4.87}$$

can then be calculated easily. The result is

$$P_{ba}(t) = \frac{4|(H_+)_{ba}|^2}{\hbar^2\omega_R^2}\sin^2(\omega_R t/2), \tag{4.88}$$

in agreement with Equation (3.223).

The fact that the probability $P_{ba}(t)$ may be non-zero, even when $E_b - E_a \neq \hbar\omega$ is not in contradiction with the photon picture of the Floquet theory. Indeed, the atom is in the same superposition of two dressed states at all times, and both superpositions include the state b. If instead of a monochromatic laser field we consider a laser field suddenly turned on to a constant intensity at time $t_0 = 0$ and suddenly turned off at a time t_1, then the atom, initially in the state a for $t \leq 0$, will be in the state (4.84) for $0 < t < t_1$ with the same amplitudes a_+ and a_- as above. It will thus be left in the state b at $t = t_1$ with the probability $P_{ba}(t_1)$. There is no violation of energy conservation since a laser field suddenly turned on and off has an infinite bandwidth and the turn-on and turn-off may transfer an arbitrary amount of energy to the atom.

The Floquet approach is a convenient framework for developing the theory beyond the rotating wave approximation. The simplest extension is to allow the two states a and b to be coupled to each other by both the emission and the absorption of a single photon, while still neglecting the possibility that the atom absorbs or emits two or more photons in succession. Within this approximation, the dressed states satisfy the finite system of coupled equations

$$
\begin{pmatrix}
E_b + \hbar\omega - \mathsf{E} & (H_-)_{ba} & 0 & 0 \\
(H_+)_{ab} & E_a - \mathsf{E} & (H_-)_{ab} & 0 \\
0 & (H_+)_{ba} & E_b - \hbar\omega - \mathsf{E} & (H_-)_{ba} \\
0 & 0 & (H_+)_{ab} & E_b - 2\hbar\omega - \mathsf{E}
\end{pmatrix}
\begin{pmatrix}
c_{-1b} \\
c_{0a} \\
c_{1b} \\
c_{2a}
\end{pmatrix}
= 0. \quad (4.89)
$$

Setting the determinant of this system equal to zero yields for the quasi-energy the quartic equation

$$
|(H_+)_{ba}|^4 - |(H_+)_{ba}|^2 \left[(E_b - \hbar\omega - \mathsf{E})(E_a - 2\hbar\omega - \mathsf{E}) + (E_a - \mathsf{E})(E_b + \hbar\omega - \mathsf{E}) \right]
$$

$$
+ (E_b + \hbar\omega - \mathsf{E})(E_a - 2\hbar\omega) \left[(E_a - \mathsf{E})(E_b - \hbar\omega - \mathsf{E}) - |(H_+)_{ba}|^2 \right] = 0.
$$

$$
(4.90)
$$

The resulting changes in the quasi-energies and in the probability $P_{ba}(t)$ have been given by Shirley [9]. The most important effect of the "counter-rotating" components of the laser field can be readily obtained if one assumes that the intensity is sufficiently weak, and the detuning sufficiently small, so that $\hbar\omega \gg |(H_+)_{ba}|$ and $\hbar\omega \gg \Delta\omega$. Then $|E_a - \mathsf{E}_\pm| \ll \hbar\omega$ for the two quasi-energies E_+ and E_- found in the rotating wave approximation, and one can expect that the same inequality also holds for the two perturbed quasi-energies that E_+ and E_- transform into when the counter-rotating components of the laser field are taken into account. We therefore look for solutions of Equation (4.90) for which $E_a - 2\hbar\omega - \mathsf{E} \simeq -2\hbar\omega$ and $E_b + \hbar\omega - \mathsf{E} \simeq 2\hbar\omega$. In this approximation,

$$
|(H_+)_{ba}|^4 + 2\hbar\omega |(H_+)_{ba}|^2 [(E_b - \hbar\omega - \mathsf{E}) - (E_a - \mathsf{E})]
$$

$$
- 4\hbar\omega^2 [(E_a - \mathsf{E})(E_b - \hbar\omega - \mathsf{E}) - |(H_+)_{ba}|^2] = 0. \quad (4.91)
$$

This equation can also be written in the following form:

$$
\left(E_a - \frac{|(H_+)_{ba}|^2}{2\hbar\omega} - \mathsf{E} \right) \left(E_b + \frac{|(H_+)_{ba}|^2}{2\hbar\omega} - \hbar\omega - \mathsf{E} \right) - |(H_+)_{ba}|^2 = 0. \quad (4.92)
$$

The counter-rotating couplings thus shift the resonance energy from $E_b - E_a$ to $E_b - E_a + |(H_+)_{ba}|^2/(\hbar\omega)$, in a first approximation. This displacement, called the *Bloch–Siegert shift* [14, 15], is usually negligible for an atom exposed to a near-resonant laser field. However, large departures from the predictions of the rotating wave approximation occur at high intensities.

4.2.3 Floquet theory and perturbation theory

Let us now return to Equation (4.64) and express the block-tridiagonal matrix $\mathbf{H}_F$ as

$$\mathbf{H}_F = \mathbf{H}_F^{(0)} + \mathbf{H}_F^{int}. \tag{4.93}$$

The first term on the right-hand side of this equation contains the block-diagonal matrices $\mathbf{H}_0 + n\hbar\omega\mathbf{I}$, $n = \ldots, -2, -1, 0, 1, 2, \ldots$, while the second term is a block-banded matrix composed of the matrix $\mathbf{H}_-$ along the upper diagonal and the matrix $\mathbf{H}_+$ along the lower diagonal. Whereas the quasi-energies and the corresponding Floquet state vectors can be obtained directly by solving Equation (4.61) as an eigenvalue equation, it is sometimes preferable to obtain them by treating $\mathbf{H}_F^{int}$ as a perturbation and using Rayleigh–Schrödinger or Brillouin–Wigner perturbation theory.

To be more specific, let us consider a particular Floquet quasi-energy, E_i, and focus on the case of a monochromatic, spatially homogeneous laser field. We use Equation (4.57) and express the Floquet–Fourier components of the corresponding Floquet–Fourier state, $|F(\mathsf{E}_i)\rangle$, as

$$|F_n(\mathsf{E}_i)\rangle = \sum_{k=1}^{M} c_{nk}(\mathsf{E}_i)|\psi_k\rangle, \tag{4.94}$$

where the coefficients $c_{nk}(\mathsf{E}_i)$ are determined by Equation (4.61). That is,

$$(\mathbf{H}_F - \mathsf{E}_i)\mathbf{c}(\mathsf{E}_i) = 0. \tag{4.95}$$

Making use of Equation (4.93), we can also write

$$(\mathbf{H}_F^{(0)} + \mathbf{H}_F^{int} - \mathsf{E}_i)\mathbf{c}(\mathsf{E}_i) = 0. \tag{4.96}$$

In the limit that the intensity of the laser field goes to zero, we have

$$\lim_{\mathcal{E}_0 \to 0} \mathbf{H}_F = \mathbf{H}_F^{(0)}, \tag{4.97}$$

and it follows that, with the appropriate choice of the energy scale, the Floquet quasi-energy E_i reduces to the unperturbed energy E_i of one of the M discrete atomic eigenstates of H_0, namely

$$\lim_{\mathcal{E}_0 \to 0} \mathsf{E}_i \equiv \mathsf{E}_i^{(0)} = E_i. \tag{4.98}$$

In this limit, the components of the Floquet–Fourier state reduce to

$$\lim_{\mathcal{E}_0 \to 0} |F_n(\mathsf{E}_i)\rangle \equiv |F_n^{(0)}(\mathsf{E}_i^{(0)})\rangle = |\psi_i\rangle\delta_{n0}, \tag{4.99}$$

where $|\psi_i\rangle$ is an unperturbed eigenstate of H_0 having energy E_i. This last relation allows us to introduce the vector $\mathbf{c}^{(0)}$ as

$$\lim_{\mathcal{E}_0 \to 0} \mathbf{c}(\mathsf{E}_i) = \mathbf{c}^{(0)}. \tag{4.100}$$

This vector has only one non-zero component, namely $c_{0i}^{(0)} = 1$, and is an eigenvector of $\mathbf{H}_F^{(0)}$ with eigenvalue E_i. The state $|\psi_i\rangle$ can be considered to be the "initial" state of the atom that evolves adiabatically into the dressed state described by the quasi-energy E_i and the Floquet–Fourier state $|F(E_i)\rangle$.

It is useful to introduce the block-diagonal Floquet matrices $\mathbf{P}_F$ and $\mathbf{Q}_F$ such that

$$\mathbf{P}_F = \mathbf{c}_i^{(0)} \mathbf{c}_i^{(0)\mathsf{T}} \tag{4.101}$$

and $\mathbf{Q}_F = \mathbf{I}_F - \mathbf{P}_F$, where $\mathbf{I}_F$ is the unit Floquet matrix. The dyad $\mathbf{P}_F$ is the matrix representation of the operator that projects onto the zero-field Fourier–Floquet state $|F^{(0)}(E_i^{(0)})\rangle$. In particular, multiplying the vector $\mathbf{c}(E_i)$ by $\mathbf{P}_F$ yields

$$\mathbf{P}_F \mathbf{c}(E_i) = \beta \mathbf{c}_i^{(0)}, \tag{4.102}$$

where β is a constant depending on the normalization of $\mathbf{c}(E_i)$. Rayleigh–Schrödinger perturbation theory is easiest to formulate when the normalization is chosen such that $\beta = 1$ (with $\mathbf{c}_i^{(0)\mathsf{T}} \mathbf{c}_i^{(0)} = 1$). However, as shown by Equation (4.115) below, the quasi-energy does not depend on the value of β.

To proceed, we formally expand E_i and $\mathbf{c}(E_i)$ as

$$E_i = \sum_{k=0}^{\infty} E_i^{(k)}, \qquad \mathbf{c}(E_i) = \sum_{k=0}^{\infty} \mathbf{c}^{(k)}, \tag{4.103}$$

where $\mathbf{c}^{(0)}$ is defined by Equation (4.100), $E_i^{(0)} \equiv E_i$ and the terms $E_i^{(k)}$ and $\mathbf{c}^{(k)}$ are of kth order in $\mathbf{H}_F^{\text{int}}$. These terms can be obtained order by order, starting with the unperturbed quantities $E_i^{(0)}$ and $\mathbf{c}^{(0)}$, by writing Equation (4.96) in the form

$$(\mathbf{H}_F^{(0)} - E_i)\mathbf{c}(E_i) = (E_i - E_i - \mathbf{H}_F^{\text{int}})\mathbf{c}(E_i), \tag{4.104}$$

expanding E_i and $\mathbf{c}(E_i)$ as in Equations (4.103) and identifying the contributions of the same order in $\mathbf{H}_F^{\text{int}}$ in the left-hand and right-hand sides of Equation (4.104). With $\beta = 1$, acting on the left with $\mathbf{P}_F$ yields the following expression for the energy coefficients:

$$E_i^{(k)} = \mathbf{c}^{(0)\mathsf{T}} \mathbf{H}_F^{\text{int}} \mathbf{c}^{(k-1)}, \qquad k > 0. \tag{4.105}$$

Acting on the left with $\mathbf{Q}_F$ results in an algorithm for calculating the corresponding eigenvectors:

$$\mathbf{Q}_F(\mathbf{H}_F^{(0)} - E_i)\mathbf{c}^{(k)} = \mathbf{Q}_F \sum_{0<p<k} E_i^{(p)} \mathbf{c}^{(k-p)} - \mathbf{Q}_F \mathbf{H}_F^{\text{int}} \mathbf{c}^{(k-1)}.$$

$$\tag{4.106}$$

In favorable cases, these equations can been solved to high orders in H_F^{int} [16–18]. The radius of convergence of the series (4.103) is limited by singularities arising from multiphoton resonances between dressed states [19].

Another way of proceeding is to introduce a non-zero complex parameter z and write Equation (4.96) in the form

$$\mathbf{c}(E_i) = \frac{1}{H_F^{(0)} - z}(E_i - H_F^{int} - z)\mathbf{c}(E_i). \tag{4.107}$$

Since the matrices $\mathbf{P}_F$ and $\mathbf{Q}_F$ satisfy the relation $\mathbf{P}_F + \mathbf{Q}_F = \mathbf{I}_F$, we can express Equation (4.107) as

$$\mathbf{c}(E_i) = (\mathbf{P}_F + \mathbf{Q}_F)\frac{1}{H_F^{(0)} - z}(E_i - H_F^{int} - z)\mathbf{c}(E_i), \tag{4.108}$$

which gives

$$\mathbf{c}(E_i) = \beta\mathbf{c}^{(0)} + \mathbf{Q}_F\frac{1}{H_F^{(0)} - z}(E_i - H_F^{int} - z)\mathbf{c}(E_i). \tag{4.109}$$

Iterating once yields

$$\mathbf{c}(E_i) = \beta\mathbf{c}^{(0)} + \beta\mathbf{Q}_F\frac{1}{H_F^{(0)} - z}(E_i - H_F^{int} - z)\mathbf{c}^{(0)}$$

$$+ \left[\mathbf{Q}_F\frac{1}{H_F^{(0)} - z}(E_i - H_F^{int} - z)\right]^2\mathbf{c}(E_i). \tag{4.110}$$

Continuing this iteration process, we obtain a formal solution of Equation (4.96) in the form of an infinite series. Introducing the resolvent matrix

$$G_F^{(0)}(z) \equiv \mathbf{Q}_F\frac{1}{z - H_F^{(0)}}\mathbf{Q}_F, \tag{4.111}$$

this solution can be expressed as follows:

$$\mathbf{c}(E_i) = \beta\sum_{l=0}^{\infty}\left[G_F^{(0)}(z)(z - E_i + H_F^{int})\right]^l\mathbf{c}^{(0)}. \tag{4.112}$$

Next, pre-multiplying Equation (4.96) by the transpose vector $\mathbf{c}^{(0)T}$ allows us to express the energy shift of the state $|\psi_i\rangle$ in the laser field as

$$E_i - E_i = \frac{\mathbf{c}^{(0)T}H_F^{int}\mathbf{c}(E_i)}{\mathbf{c}^{(0)T}\mathbf{c}(E_i)} \tag{4.113}$$

or

$$E_i - E_i = \mathbf{c}^{(0)T}H_F^{int}\mathbf{c}(E_i)/\beta. \tag{4.114}$$

Using Equation (4.112), we find that

$$\mathsf{E}_i - E_i = \sum_{l=0}^{\infty} \mathbf{c}^{(0)\mathsf{T}} \mathbf{H}_F^{int} \left[\mathbf{G}_F^{(0)}(z)(z - \mathsf{E}_i + \mathbf{H}_F^{int}) \right]^l \mathbf{c}^{(0)}. \qquad (4.115)$$

The preceding relation shows that, as expected, the quasi-energy does not depend on the constant β (which merely defines the normalization of the Floquet state vector). Setting $z = \mathsf{E}_i$ in Equations (4.112) and (4.115) gives the Brillouin–Wigner formulation of perturbation theory: Equation (4.112) can be used to obtain $\mathbf{c}(\mathsf{E}_i)$ if the quasi-energy is known, while Equation (4.115) can be solved iteratively for E_i if the quasi-energy is not known.

Whether in the Rayleigh–Schrödinger case or in the Brillouin–Wigner case, truncating the Floquet–Fourier expansion (4.20) so that $-N \leq n \leq N$ implies that only processes involving the emission or absorption of a net number of photons not larger than N are taken into account. Thus, the value of N determines the maximum order of perturbative calculations which can be consistently carried out. For instance, let us consider the Floquet dressed state which, in the field-free limit, tends to the unperturbed ground state of the atom, and let us assume that the Floquet–Fourier expansion is truncated to $|n| \leq 1$. The second-order term in the expansion of the quasi-energy $\mathsf{E}_i^{(2)}$, can then be calculated consistently. However, the next non-vanishing term, $\mathsf{E}_i^{(4)}$, will miss the contributions of the processes in which the initial state emits or absorbs two photons in succession.

As an illustrative example, let us re-derive the result (3.110) for the leading-order contribution to the AC Stark shift. Using Equations (4.105) and (4.106), we obtain, to first order in $\mathbf{H}_F^{int}$, $\mathsf{E}_i - E_i = \mathbf{c}^{(0)\mathsf{T}} \mathbf{H}_F^{int} \mathbf{c}^{(0)}$, which vanishes since the diagonal blocks of $\mathbf{H}_F^{int}$ are zero. To second order we have that $\mathsf{E}_i - E_i = \Delta E_i^{(2)}$, with

$$\Delta E_i^{(2)} = \mathbf{c}^{(0)\mathsf{T}} \mathbf{H}_F^{int} \mathbf{G}_F^{(0)}(E_i) \mathbf{H}_F^{int} \mathbf{c}^{(0)}, \qquad (4.116)$$

where the Green's matrix $\mathbf{G}_F^{(0)}(E_i)$ is obtained from the resolvent matrix $\mathbf{G}_F^{(0)}(z)$ by letting $z \to E_i$. Explicitly,

$$\Delta E_i^{(2)} = \sum_{k \neq i} \frac{|\langle \psi_i | H_- | \psi_k \rangle|^2}{E_i - E_k + \hbar\omega} + \sum_{k \neq i} \frac{|\langle \psi_i | H_+ | \psi_k \rangle|^2}{E_i - E_k - \hbar\omega}. \qquad (4.117)$$

Working in the length gauge, and referring to Equation (2.138), we have $H_+ = H_-^\dagger = (e\mathcal{E}_0/2)\hat{\boldsymbol{\epsilon}} \cdot \mathbf{R} \exp(-i\varphi)$. Recalling that $\mathbf{D} = -e\mathbf{R}$, Equation (4.117) yields the same result for the AC Stark shift as given by Equation (3.110).

In Section 4.2.3 we shall show how this perturbative approach can be applied in the framework of the non-Hermitian Floquet theory.

4.2.4 Population transfer between dressed states

Thus far we have considered atoms interacting with a periodic laser field and we have seen how the Floquet theory can be applied to obtain quasi-stationary solutions of the TDSE. Implicit in our analysis has been the assumption that the laser field is applied adiabatically so that no population transfer between dressed Floquet states occurs. In other words, the atomic system is described as a single dressed Floquet state that evolves in the laser field. Here we will develop a multistate Floquet theory that can be used to describe the more general case of population transfer between dressed states. As we shall see, population transfer can take place when the atom interacts with a short laser pulse, in particular when resonances occur.

For a monochromatic laser field the Floquet quasi-energies E_i and the corresponding Floquet states $|\Psi_{E_i}(t)\rangle$ are determined for a particular set of values of the laser electric field strength $\mathcal{E}_0$ and angular frequency ω. Hence the Floquet quasi-energies, Floquet states and Floquet Hamiltonian can all be considered as functions of $\mathcal{E}_0$ and ω. In the present discussion, we will focus on laser pulses whose carrier angular frequency ω is fixed, as this is the case encountered most often in practice. The Floquet quasi-energies, the Floquet–Fourier states and the Floquet Hamiltonian will be expressed as functions only of the electric field strength $\mathcal{E}_0$ of the laser. We therefore write

$$E_i = E_i(\mathcal{E}_0), \tag{4.118}$$

$$|\Psi_{E_i}(t)\rangle = \exp[-iE_i(\mathcal{E}_0)t/\hbar] \sum_{n=-\infty}^{\infty} \exp(-in\omega t)|F_n[E_i(\mathcal{E}_0)]\rangle \tag{4.119}$$

and

$$H_F = H_F(\mathcal{E}_0). \tag{4.120}$$

We now turn our attention to the more general case of a laser pulse and take the electric field strength to be time-dependent. Making the dipole approximation and letting $\mathcal{E}_0 \rightarrow \mathcal{E}_0(t) \equiv \mathcal{E}_0 F(t)$, where $F(t)$ is the laser pulse envelope function, we assume that the laser electric-field component is linearly polarized and can be expressed as

$$\mathcal{E}(t) = \hat{\epsilon}\mathcal{E}_0(t)\sin(\omega t). \tag{4.121}$$

If the electric-field strength $\mathcal{E}_0(t)$ varies very slowly in time, so that the cycle-averaged intensity is nearly constant from one laser cycle to the next, and if there are no resonances, then at any point in time during the laser pulse, the state of the atom can be approximately described by a single Floquet state. That is,

$$|\Psi(t)\rangle = |\Psi_{E_i[\mathcal{E}_0(t)]}(t)\rangle$$

$$= \exp\left(-\frac{i}{\hbar}\int_{-\infty}^{t}dt'\,E_i[\mathcal{E}_0(t')]\right)\sum_{n=-\infty}^{\infty}\exp(-in\omega t)|F_n\left(E_i[\mathcal{E}_0(t)]\right)\rangle.$$

(4.122)

The Floquet quasi-energy and Floquet vector of this state are given by the "instantaneous" solutions of the eigensystem:

$$H_F[\mathcal{E}_0(t)]|F\left(E_i[\mathcal{E}_0(t)]\right)\rangle = E_i[\mathcal{E}_0(t)]|F\left(E_i[\mathcal{E}_0(t)]\right)\rangle.\qquad (4.123)$$

The Floquet quasi-energy and Floquet vector are thus determined using the instantaneous electric-field strength of the laser pulse $\mathcal{E}_0(t)$.

Let us now consider the more general case for which the approximation (4.122) does not hold. In order to simplify the notation, in what follows we will write the quasi-energies simply as $E_i(t)$, with the understanding that these quantities depend on time via $\mathcal{E}_0(t)$.

At the initial time $t = t_0$, we take the atomic system to be in the ground state $|\psi_0\rangle$, which is an eigenstate of the field-free Hamiltonian. The components of the Floquet vector $|F[E(t)]\rangle$ at the time $t = t_0$ are then given by

$$|\Psi(t_0)\rangle = |F_n[E(t_0)]\rangle = |\psi_0\rangle\delta_{n0},\qquad (4.124)$$

where $E(t_0) = E_0$. At time t_0, a laser pulse is applied. We will again assume that the atom can be described in terms of M bound states. An atomic state in the laser pulse can then be expanded in terms of the complete set of M instantaneous Floquet solutions of the TDSE (4.1). In particular, using Equation (4.18) we may write

$$|\Psi(t)\rangle = \sum_{j=1}^{M}c_j(t)\exp\left[-\frac{i}{\hbar}\int_{t_0}^{t}E_j(t')dt'\right]|P_{E_j(t)}(t)\rangle.\qquad (4.125)$$

We require that the integer n_j appearing in Equation (4.65) be zero for all the quasi-energy states included in this expansion. Indeed, Equation (4.17) alone does not define the kets $|P_{E_j(t)}(t)\rangle$ uniquely, since $|\Psi_{E_j(t)}(t)\rangle \equiv |\Psi_{E_j(t)+p\hbar\omega}(t)\rangle$ for any integer p whereas $|P_{E_j(t)}(t)\rangle$ and $|P_{E_j(t)+p\hbar\omega}(t)\rangle$ differ by a time-dependent phase factor. Inserting the expansion (4.125) into the TDSE (4.1), pre-multiplying by $\langle P_{E_i(t)}(t)|$ and using the property

$$\langle P_{E_i(t)}(t)|P_{E_j(t)}(t)\rangle = \delta_{ij},\qquad (4.126)$$

which follows from Equations (4.67) and (4.71), we obtain the following set of coupled ordinary differential equations:

$$i\hbar\frac{d}{dt}\mathbf{c} = \mathbf{Hc} \tag{4.127}$$

for the expansion coefficients $c_j(t)$ with the initial condition $c_j(t_0) = \delta_{j0}$. The elements of the time-dependent matrix $\mathbf{H}$ are given by

$$(\mathbf{H})_{ij} = -i\hbar\frac{d\mathcal{E}_0}{dt}\exp\left(\frac{i}{\hbar}\int_{t_0}^{t}[\mathsf{E}_i(t') - \mathsf{E}_j(t')]dt'\right)\langle P_{\mathsf{E}_i(t)}(t)\,|\,P'_{\mathsf{E}_j(t)}(t)\rangle, \tag{4.128}$$

where

$$|P'_{\mathsf{E}_j(t)}(t)\rangle \equiv \frac{d}{d\mathcal{E}_0}|P_{\mathsf{E}_j(t)}(t)\rangle. \tag{4.129}$$

From Equation (4.126), we have

$$\frac{d}{d\mathcal{E}_0}\langle P_{\mathsf{E}_i(t)}(t)\,|\,P_{\mathsf{E}_j(t)}(t)\rangle = 0, \tag{4.130}$$

which implies that

$$\langle P_{\mathsf{E}_i(t)}(t)\,|\,P'_{\mathsf{E}_j(t)}(t)\rangle = -\langle P'_{\mathsf{E}_i(t)}(t)\,|\,P_{\mathsf{E}_j(t)}(t)\rangle$$

$$= -\langle P_{\mathsf{E}_j(t)}(t)\,|\,P'_{\mathsf{E}_i(t)}(t)\rangle^*. \tag{4.131}$$

It follows that the matrix $\mathbf{H}$ is Hermitian and that for all times t the quantities $\langle P_{\mathsf{E}_j(t)}(t)\,|\,P'_{\mathsf{E}_j(t)}(t)\rangle$ are purely imaginary (or zero).

We now show how Equation (4.128) can be written entirely in terms of the instantaneous Floquet eigenstates and quasi-energies, which in principle are known. Making use of the Fourier expansion (4.19), we write the matrix elements (4.131) as

$$\langle P_{\mathsf{E}_i(t)}(t)\,|\,P'_{\mathsf{E}_j(t)}(t)\rangle = \sum_{p=-\infty}^{\infty}\exp[ip\omega(t - t_0)]\left(\sum_{n=-\infty}^{\infty}\langle F_{n+p}[\mathsf{E}_i(t)]\,|\,F'_n[\mathsf{E}_j(t)]\rangle\right), \tag{4.132}$$

with

$$|F'_n[\mathsf{E}_j(t)]\rangle \equiv \frac{d}{d\mathcal{E}_0}|F_n[\mathsf{E}_j(t)]\rangle. \tag{4.133}$$

Using Equation (4.30), we also have

$$\langle P_{\mathsf{E}_i(t)}(t)\,|\,P'_{\mathsf{E}_j(t)}(t)\rangle = \sum_{p=-\infty}^{\infty}\exp[ip\omega(t - t_0)]$$

$$\times\left(\sum_{n=-\infty}^{\infty}\langle F_n[\mathsf{E}_i(t) + p\hbar\omega]\,|\,F'_n[\mathsf{E}_j(t)]\rangle\right), \tag{4.134}$$

from which we obtain

$$(\mathbf{H})_{ij} = -i\hbar \frac{d\mathcal{E}_0}{dt} \sum_{p=-\infty}^{\infty} \exp\left(\frac{i}{\hbar} \int_{t_0}^{t} [\mathsf{E}_i(t') + p\hbar\omega - \mathsf{E}_j(t')]dt'\right)$$

$$\times \langle F[\mathsf{E}_i(t) + p\hbar\omega] | F'[\mathsf{E}_j(t)]\rangle, \tag{4.135}$$

with $|F'[\mathsf{E}_j(t)]\rangle$ denoting the column vector of components $|F'_n[\mathsf{E}_j(t)]\rangle$. Formally,

$$|F'[\mathsf{E}_j(\mathcal{E}_0)]\rangle = \lim_{\delta\mathcal{E}_0 \to 0} \frac{1}{\delta\mathcal{E}_0}\left(|F[\mathsf{E}_j(\mathcal{E}_0 + \delta\mathcal{E}_0)]\rangle - |F[\mathsf{E}_j(\mathcal{E}_0)]\rangle\right), \tag{4.136}$$

where the state $|F[\mathsf{E}_j(\mathcal{E}_0 + \delta\mathcal{E}_0)]\rangle$ is an eigenstate of the "perturbed" Floquet Hamiltonian

$$H_\mathrm{F}(\mathcal{E}_0 + \delta\mathcal{E}_0) = H_\mathrm{F}(\mathcal{E}_0) + \delta\mathcal{E}_0 \frac{d}{d\mathcal{E}_0} H_\mathrm{F}(\mathcal{E}_0). \tag{4.137}$$

We note that we can write the Floquet Hamiltonian $H_\mathrm{F}(\mathcal{E}_0)$, in the dipole approximation, as the sum of a field-free Hamiltonian $H_\mathrm{F}^{(0)}$ and an interaction Hamiltonian linear in $\mathcal{E}_0$, whether the interaction of the atom with the field is described in the length gauge or in the velocity gauge. Thus, we have

$$H_\mathrm{F}(\mathcal{E}_0) = H_\mathrm{F}^{(0)} + \mathcal{E}_0 D_\mathrm{F}, \tag{4.138}$$

where D_F is a matrix operator independent of $\mathcal{E}_0$. As a consequence, we may write

$$H_\mathrm{F}(\mathcal{E}_0 + \delta\mathcal{E}_0) = H_\mathrm{F}(\mathcal{E}_0) + \delta\mathcal{E}_0 D_\mathrm{F}. \tag{4.139}$$

Treating the second term on the right-hand side as a perturbation, we can obtain the vector $|F[\mathsf{E}_j(\mathcal{E}_0 + \delta\mathcal{E}_0)]\rangle$ to first order in $\delta\mathcal{E}_0$. The result is

$$|F[\mathsf{E}_j(\mathcal{E}_0 + \delta\mathcal{E}_0)]\rangle = |F[\mathsf{E}_j(\mathcal{E}_0)]\rangle$$

$$+ \delta\mathcal{E}_0 \sum_{n=-\infty}^{\infty} \sum_{k \neq j} |F[\mathsf{E}_k(\mathcal{E}_0) + n\hbar\omega]\rangle \frac{\langle F[\mathsf{E}_k(\mathcal{E}_0) + n\hbar\omega] | D_\mathrm{F} | F[\mathsf{E}_j(\mathcal{E}_0)]\rangle}{\mathsf{E}_j(\mathcal{E}_0) + n\hbar\omega - \mathsf{E}_k(\mathcal{E}_0)}.$$

$$\tag{4.140}$$

The sum over the intermediate states runs over all the eigenstates of $H_\mathrm{F}(\mathcal{E}_0)$, with the exception of j when $n = 0$. Inserting this expression of $|F[\mathsf{E}_j(\mathcal{E}_0 + \delta\mathcal{E}_0)]\rangle$ into Equation (4.136) and using the property (4.66), we find that

$$\langle F[\mathsf{E}_i(t) + p\hbar\omega] | F'[\mathsf{E}_j(t)]\rangle = 0 \tag{4.141}$$

when $i = j$ and $p = 0$, and that

$$\langle F[\mathsf{E}_i(t) + p\hbar\omega] | F'[\mathsf{E}_j(t)]\rangle = -\frac{\langle F(\mathsf{E}_i[\mathcal{E}_0(t)] + p\hbar\omega) | D_\mathrm{F} | F(\mathsf{E}_j[\mathcal{E}_0(t)])\rangle}{\mathsf{E}_i[\mathcal{E}_0(t)] + p\hbar\omega - \mathsf{E}_j[\mathcal{E}_0(t)]}$$

$$\tag{4.142}$$

otherwise. The diagonal matrix elements $\langle P_{\mathsf{E}_i(t)}(t) | P'_{\mathsf{E}_i(t)}(t) \rangle$ are imaginary, as mentioned above, because

$$\langle F[\mathsf{E}_i(t) + p\hbar\omega] | F'[\mathsf{E}_i(t)] \rangle = -\langle F[\mathsf{E}_i(t) - p\hbar\omega] | F'[\mathsf{E}_i(t)] \rangle^*. \quad (4.143)$$

It is worth emphasizing that the state $|\Psi(t)\rangle$ given in Equation (4.125) with the coefficients $c_j(t)$ satisfying Equations (4.127) constitutes an *exact* solution of the TDSE (4.1). These equations are cumbersome to solve since this would require the knowledge of a complete set of Floquet states and quasi-energies over a range of electric-field strengths. However, Equations (4.125) and (4.127) are a convenient starting point for studying systems whose evolution in time is nearly adiabatic, or when the system can be reduced to a few resonantly coupled states.

In general, the adiabatic approximation will hold when the quasi-energies of the relevant dressed states are separated by large gaps. Indeed, the relevant matrix elements $(\mathbf{H})_{ij}$ are then both small in magnitude (because the Floquet states vary slowly with $\mathcal{E}_0$ away from resonances) and rapidly oscillating in time. However, a non-adiabatic evolution is likely at multiphoton resonances, where one of the quasi-energies $\mathsf{E}_i(t)$ becomes approximately equal to one of the other ones, say, $\mathsf{E}_j(t)$, modulo $\hbar\omega$. The Floquet state vectors $|F(\mathsf{E}_j[\mathcal{E}_0(t)])\rangle$ and $|F(\mathsf{E}_i[\mathcal{E}_0(t) + p\hbar\omega])\rangle$ vary rapidly with $\mathcal{E}_0$, where $\mathsf{E}_j(t) \simeq \mathsf{E}_i(t) + p\hbar\omega$ for a certain value of p, unless their interaction is forbidden by selection rules. Referring to Equation (4.135), we observe that the time-dependent matrix elements coupling the two Floquet states can then become large and oscillate slowly in time. As in the Landau–Zener model [20–23], the probability of population transfer between resonant dressed states can become significant if $\mathcal{E}_0(t)$ varies rapidly enough. The non-adiabatic evolution of the atomic system is a complex process which depends on the angular frequency, the envelope and peak intensity of the laser pulse, as well as the specific system under consideration. In particular, population transfer can occur between *non*-resonant dressed states in the case of ultra-short, intense laser pulses.

The Hermitian theory outlined above can be generalized to a multistate non-Hermitian Floquet theory applicable to the case where dressed states decaying by multiphoton ionization interact with each other [24].

4.3 Floquet dressed continuum states

As a further illustration of the Floquet theory, we shall now solve the coupled equations (4.25) for the case of a non-relativistic electron interacting only with a monochromatic laser field and prove that the resulting dressed continuum states are the Gordon–Volkov wave functions introduced in Section 2.7. We will then study the boundary conditions which must be satisfied by dressed continuum states for laser-assisted scattering. Initially, scattering from a spherical square-well potential will be considered. This analysis will allow us to obtain and interpret the general asymptotic form of the laser-assisted scattering wave function.

4.3.1 Gordon–Volkov waves as dressed states

We will work within the dipole approximation and in the velocity gauge. For simplicity, we assume that the laser electric field is of the form (2.118) with $\varphi = -\pi/2$, so that

$$\mathcal{E}(t) = \hat{\epsilon}\mathcal{E}_0 \sin(\omega t). \tag{4.144}$$

In the velocity gauge, the dipole operators H_+ and H_- are given by Equation (2.139) and the system of coupled equations (4.25) becomes, in the position representation,

$$\left[-\frac{\hbar^2}{2m}\nabla^2 - (E^V + n\hbar\omega)\right] F_n^V(E^V; \mathbf{r})$$
$$-\left(\frac{ie\hbar\mathcal{E}_0}{2m\omega}\right)\hat{\epsilon}\cdot\nabla\left[F_{n+1}^V(E^V; \mathbf{r}) + F_{n-1}^V(E^V; \mathbf{r})\right] = 0. \tag{4.145}$$

To solve this system of equations we make the ansatz

$$F_n^V(E^V; \mathbf{r}) = (2\pi)^{-3/2} f_n \exp(i\mathbf{k}\cdot\mathbf{r}), \tag{4.146}$$

where the coefficients f_n are independent of $\mathbf{r}$. Hence,

$$\left[E_k - (E^V + n\hbar\omega)\right] f_n + \hbar\omega(x/2)(f_{n+1} + f_{n-1}) = 0, \quad n = 0, \pm 1, \pm 2, \ldots, \tag{4.147}$$

with $E_k = \hbar^2 k^2/(2m)$ and $x = \mathbf{k}\cdot\boldsymbol{\alpha}_0$, where we recall that $\boldsymbol{\alpha}_0 = \alpha_0\hat{\epsilon}$ and $\alpha_0 = e\mathcal{E}_0/(m\omega^2)$ is the excursion amplitude (2.79) of the electron in the laser field. Defining $\nu = (E^V - E_k)/(\hbar\omega)$, we can recast the equations (4.147) into the three-term recursion relation

$$xf_{n-1} + xf_{n+1} = 2(n + \nu)f_n. \tag{4.148}$$

This relation implies that each f_n is a linear superposition of the Bessel function of the first kind $J_{n+\nu}(x)$ and of the Bessel function of the second kind $Y_{n+\nu}(x)$. However, the coefficients f_n would not vanish in the limits $|\pm n| \to \infty$ if they contained any admixture of the latter, or if ν was not an integer, which would make it impossible for the functions $F_n^V(E^V; \mathbf{r})$ to be the harmonic components of a convergent Floquet–Fourier expansion. We can thus set $E^V = E_k + p\hbar\omega$, with p an integer, and take $f_n = J_{n+p}(x)$ for all n. The functions $F_n^V(E^V; \mathbf{r})$ are then the harmonic components of the wave function

$$\chi_{\mathbf{k}}^{V}(\mathbf{r}, t) = (2\pi)^{-3/2} \exp(-iE_k t/\hbar) \sum_{n=-\infty}^{\infty} \exp[-i(n+p)\omega t] J_{n+p}(\mathbf{k} \cdot \boldsymbol{\alpha}_0)$$

$$\times \exp(i\mathbf{k} \cdot \mathbf{r}), \tag{4.149}$$

or, by shifting the index n,

$$\chi_{\mathbf{k}}^{V}(\mathbf{r}, t) = (2\pi)^{-3/2} \exp(-iE_k t/\hbar) \sum_{n=-\infty}^{\infty} \exp(-in\omega t) J_n(\mathbf{k} \cdot \boldsymbol{\alpha}_0) \exp(i\mathbf{k} \cdot \mathbf{r}). \tag{4.150}$$

As we shall show in Section 10.2.1, this solution is the non-relativistic Gordon–Volkov wave function (2.167) in the velocity gauge, expanded in Floquet–Fourier form, the Floquet quasi-energy being $\mathsf{E}^V = E_k$. It reduces to a simple plane wave in the zero-field limit:

$$\chi_{\mathbf{k}}^{V}(\mathbf{r}, t) \underset{\varepsilon_0 \to 0}{\longrightarrow} (2\pi)^{-3/2} \exp(-iE_k t/\hbar) \exp(i\mathbf{k} \cdot \mathbf{r}). \tag{4.151}$$

The wave function $\chi_{\mathbf{k}}^{V}(\mathbf{r}, t)$ is regular at the origin. It is also useful to obtain solutions of the system of coupled equations (4.145) diverging for $r \to 0$. To this end, we expand the plane wave $\exp(i\mathbf{k} \cdot \mathbf{r})$ appearing in Equation (4.150) in products of spherical harmonics and spherical Bessel functions,

$$\exp(i\mathbf{k} \cdot \mathbf{r}) = 4\pi \sum_{\lambda=0}^{\infty} \sum_{\mu=-\lambda}^{\lambda} i^\lambda j_\lambda(kr) Y_{\lambda\mu}^*(\hat{\mathbf{k}}) Y_{\lambda\mu}(\hat{\mathbf{r}}). \tag{4.152}$$

Similarly, we perform a partial wave decomposition of the Bessel function $J_n(\mathbf{k} \cdot \boldsymbol{\alpha}_0)$, which yields

$$J_n(\mathbf{k} \cdot \boldsymbol{\alpha}_0) = \sum_{\lambda'=0}^{\infty} \sum_{\mu'=-\lambda'}^{\lambda'} J_{\lambda'}^n(k\alpha_0) Y_{\lambda'\mu'}^*(\hat{\mathbf{k}}) Y_{\lambda'\mu'}(\hat{\boldsymbol{\alpha}}_0), \tag{4.153}$$

where

$$J_l^k(x) = 2\pi \int_{-1}^{1} da\, J_k(ax) P_l(a) \tag{4.154}$$

and P_l is a Legendre polynomial. Properties of the functions $J_l^k(x)$ have been discussed by Kylstra and Joachain [25]. We now form the following linear superposition of Gordon–Volkov wave functions:

$$\mathcal{J}_{lm}(k; \mathbf{r}, t) = \int d\hat{\mathbf{k}}\, Y_{lm}(\hat{\mathbf{k}}) \chi_{\mathbf{k}}^{V}(\mathbf{r}, t). \tag{4.155}$$

From Equation (4.150), and making use of the expansions (4.152) and (4.153), this superposition is given explicitly by

$$\mathcal{J}_{lm}(k; \mathbf{r}, t) = (2\pi)^{-3/2} \exp(-iE_k t/\hbar) \sum_{n=-\infty}^{\infty} \exp(-in\omega t)$$

$$\times \sum_{\lambda=0}^{\infty} \sum_{\mu=-\lambda}^{\lambda} i^\lambda \, \Upsilon_{lm,\lambda\mu}^n(k, \boldsymbol{\alpha}_0) j_\lambda(kr) Y_{\lambda\mu}(\hat{\mathbf{r}}), \qquad (4.156)$$

with

$$\Upsilon_{lm,\lambda\mu}^n(k, \boldsymbol{\alpha}_0) = 4\pi \left[\int d\hat{\mathbf{k}} \, Y_{lm}(\hat{\mathbf{k}}) J_n(\mathbf{k} \cdot \boldsymbol{\alpha}_0) Y_{\lambda\mu}^*(\hat{\mathbf{k}}) \right]$$

$$= (-1)^m \sum_{\lambda'=0}^{\infty} \sum_{\mu'=-\lambda'}^{\lambda'} [4\pi(2l+1)(2\lambda'+1)(2\lambda+1)]^{1/2}$$

$$\times \begin{pmatrix} l & \lambda' & \lambda \\ 0 & 0 & 0 \end{pmatrix} \begin{pmatrix} l & \lambda' & \lambda \\ -m & \mu' & \mu \end{pmatrix} J_{\lambda'}^n(k\alpha_0) Y_{\lambda'\mu'}(\hat{\boldsymbol{\alpha}}_0), \quad (4.157)$$

where we have used Wigner 3-j symbols. As they are linear combinations of solutions of the TDSE, the functions $\mathcal{J}_{lm}(k; \mathbf{r}, t)$ are also solutions of this equation. Irregular Floquet solutions of the TDSE are obtained by replacing the (regular) spherical Bessel functions $j_\lambda(kr)$ by the (irregular) spherical Neumann functions $n_\lambda(kr)$. In this way, we obtain the functions

$$\mathcal{N}_{lm}(k; \mathbf{r}, t) = (2\pi)^{-3/2} \exp(-iE_k t/\hbar) \sum_{n=-\infty}^{\infty} \exp(-in\omega t)$$

$$\times \sum_{\lambda=0}^{\infty} \sum_{\mu=-\lambda}^{\lambda} i^\lambda \, \Upsilon_{lm,\lambda\mu}^n(k, \boldsymbol{\alpha}_0) n_\lambda(kr) Y_{\lambda\mu}(\hat{\mathbf{r}}). \qquad (4.158)$$

The substitution leading to Equation (4.158) is possible because the spherical Neumann functions $n_\lambda(kr)$ satisfy the same differential equations as the spherical Bessel functions $j_\lambda(kr)$. Although the former diverge for $\lambda \to \infty$, one has

$$\lim_{\lambda \to \infty} \Upsilon_{lm,\lambda\mu}^n(k, \boldsymbol{\alpha}_0) n_\lambda(kr) = 0, \qquad (4.159)$$

which ensures the convergence of the partial wave expansion.

Any linear combination of the functions $\mathcal{J}_{lm}(k; \mathbf{r}, t)$ and $\mathcal{N}_{lm}(k; \mathbf{r}, t)$ is also a Floquet solution of the TDSE. In Section 4.3.2, we will use the linear combinations

$$\mathcal{H}_{lm}^{(1)}(k; \mathbf{r}, t) = \mathcal{J}_{lm}(k; \mathbf{r}, t) + i\mathcal{N}_{lm}(k; \mathbf{r}, t), \qquad (4.160)$$

which make use of the spherical Hankel functions of the first kind $h_\lambda^{(1)}(kr) = j_\lambda(kr) + i n_\lambda(kr)$. Thus,

$$\mathcal{H}_{lm}^{(1)}(k; \mathbf{r}, t) = (2\pi)^{-3/2} \exp(-iE_k t/\hbar) \sum_{n=-\infty}^{\infty} \exp(-in\omega t)$$

$$\times \sum_{\lambda=0}^{\infty} \sum_{\mu=-\lambda}^{\lambda} i^\lambda \, \Upsilon_{lm,\lambda\mu}^n(k, \boldsymbol{\alpha}_0) h_\lambda^{(1)}(kr) Y_{\lambda\mu}(\hat{\mathbf{r}}). \qquad (4.161)$$

These solutions behave asymptotically as outgoing spherical waves. This follows from the fact that

$$h_\lambda^{(1)}(r) \underset{r\to\infty}{\longrightarrow} -i \frac{\exp[i(kr - \lambda\pi/2)]}{kr}. \qquad (4.162)$$

We also note that the functions $\mathcal{J}_{lm}(k; \mathbf{r}, t)$, $\mathcal{N}_{lm}(k; \mathbf{r}, t)$ and $\mathcal{H}_{lm}^{(1)}(k; \mathbf{r}, t)$ have simple forms in the zero-field limit, $\mathcal{E}_0 \to 0$, or equivalently in the limit $\alpha_0 \to 0$ with ω fixed. Using $J_n(0) = \delta_{n0}$ and the orthonormality property of the spherical harmonics, we see that

$$\Upsilon_{lm,\lambda\mu}^n(k, \boldsymbol{\alpha}_0) \underset{\alpha_0\to 0}{\longrightarrow} 4\pi \, \delta_{l\lambda} \, \delta_{m\mu} \, \delta_{n0}, \qquad (4.163)$$

from which we obtain

$$\mathcal{J}_{lm}(k; \mathbf{r}, t) \underset{\mathcal{E}_0\to 0}{\longrightarrow} (2\pi)^{-3/2} \exp(-iE_k t/\hbar) 4\pi \, i^l \, j_l(kr) \, Y_{lm}(\hat{\mathbf{r}}), \qquad (4.164)$$

$$\mathcal{N}_{lm}(k; \mathbf{r}, t) \underset{\mathcal{E}_0\to 0}{\longrightarrow} (2\pi)^{-3/2} \exp(-iE_k t/\hbar) 4\pi \, i^l \, n_l(kr) \, Y_{lm}(\hat{\mathbf{r}}) \qquad (4.165)$$

and

$$\mathcal{H}_{lm}^{(1)}(k; \mathbf{r}, t) \underset{\mathcal{E}_0\to 0}{\longrightarrow} (2\pi)^{-3/2} \exp(-iE_k t/\hbar) 4\pi \, i^l \, h_l^{(1)}(kr) \, Y_{lm}(\hat{\mathbf{r}}). \qquad (4.166)$$

We conclude this discussion by pointing out that the Floquet quasi-energy spectrum of the Gordon–Volkov states is infinitely degenerate. This can be seen by considering an electron having a kinetic energy $E_k + N\hbar\omega$, where N is an integer, and thus a wave number

$$k_N = [(2m/\hbar^2)(E_k + N\hbar\omega)]^{1/2}. \qquad (4.167)$$

From Equation (4.150), the Gordon–Volkov wave describing this electron in the laser field is given by

$$\chi_{\mathbf{k}_N}^V(\mathbf{r}, t) = (2\pi)^{-3/2} \exp[-i(E_k + N\hbar\omega)t/\hbar]$$

$$\times \sum_{n=-\infty}^{\infty} \exp(-in\omega t) J_n(\mathbf{k}_N \cdot \boldsymbol{\alpha}_0) \exp(i\mathbf{k}_N \cdot \mathbf{r}), \qquad (4.168)$$

with $\mathbf{k}_N = k_N \hat{\mathbf{k}}$. As shown in Section 4.1.1, this Floquet wave function can also be expressed as

$$\chi_{\mathbf{k}_N}^{\text{V}}(\mathbf{r}, t) = (2\pi)^{-3/2} \exp(-iE_k t/\hbar) \sum_{n=-\infty}^{\infty} \exp(-in\omega t) J_{n-N}(\mathbf{k}_N \cdot \boldsymbol{\alpha}_0) \exp(i\mathbf{k}_N \cdot \mathbf{r}),$$

(4.169)

so that the Floquet quasi-energy of the state is now E_k, just as in Equation (4.150). Although both states have the same quasi-energy, Equation (4.169) describes a different physical state of the electron. In the zero-field limit, it corresponds to a plane wave state of an electron having wave vector $\mathbf{k}_N$. In exactly the same way, the Gordon–Volkov waves $\mathcal{J}_{lm}(k_N; \mathbf{r}, t)$, $\mathcal{N}_{lm}(k_N; \mathbf{r}, t)$ and $\mathcal{H}_{lm}^{(1)}(k_N; \mathbf{r}, t)$, which describe spherical waves with wave number k_N and Floquet quasi-energy $E_k + N\hbar\omega$, can also be expressed as Floquet solutions of the TDSE with Floquet quasi-energy E_k. This degeneracy will be exploited in Section 4.3.2 when constructing Floquet scattering solutions of the TDSE.

4.3.2 Dressed continuum states

It is instructive to begin our discussion of dressed continuum states by comparing these states with field-free continuum states. As laser-assisted electron scattering will be studied in detail in Chapter 10, we limit ourselves here to examining the asymptotic form of the Floquet wave function, which represents an electron scattered by a short-range central potential $V(r)$ in the presence of a linearly polarized monochromatic field. This asymptotic form is more complicated than for field-free scattering, due to the fact that the electron can absorb or emit both real and virtual photons during the collision. In the absence of the laser field, the scattering of an electron by a time-independent potential is necessarily elastic. The electron has the same energy and the same wave number before and after the collision. However, when the collision happens in the presence of a monochromatic laser field of angular frequency ω, its final energy may differ from its initial one by an integer multiple of $\hbar\omega$. Laser-assisted scattering by a potential is therefore a multichannel problem, in which the elastic channel is coupled to (infinitely many) inelastic channels differing by the net number of photons exchanged with the laser field. Correspondingly, one should expect that at asymptotic distances from the center of force, the scattering wave function reduces to the sum of an incident Gordon–Volkov wave and of a superposition of outgoing spherical waves, each one of the latter having a wave number given by

$$k_N = \left[(2m/\hbar^2)(E_{k_i} + N\hbar\omega) \right]^{1/2},$$

(4.170)

where $E_{k_i} = \hbar^2 k_i^2/(2m)$ is the initial electron energy.

To see how this asymptotic behavior arises from the Floquet equations, let us take the potential to be an attractive spherical square well of range a. Thus

$$V(r) = \begin{cases} -V_0 & r < a, \\ 0 & r > a, \end{cases} \tag{4.171}$$

with $V_0 > 0$.

In the absence of a laser field, the wave function describing the scattering of an electron of initial wave vector $\mathbf{k}_i$ by the potential (4.171) would take on the following form for $r > a$:

$$\Psi_{\mathbf{k}_i}^{(0)}(\mathbf{r}, t) = (2\pi)^{-3/2} \exp(-iE_{k_i}t/\hbar)$$

$$\times \left[\exp(i\mathbf{k}_i \cdot \mathbf{r}) + 4\pi \sum_{l=0}^{\infty} \sum_{m=-l}^{l} c_{lm}^{(0)} i^l h_l^{(1)}(k_i r) Y_{lm}(\hat{\mathbf{r}}) \right]. \tag{4.172}$$

In this equation, the coefficients $c_{lm}^{(0)}$ are constants. Here we use the superscript (0) to denote the fact that this wave function describes the field-free case. Given the asymptotic form of the spherical Hankel functions of the first kind, we see that $\Psi_{\mathbf{k}_i}^{(0)}(\mathbf{r}, t)$ reduces at large distances to the sum of an incident plane wave and of a scattered outgoing spherical wave, namely

$$\Psi_{\mathbf{k}_i}^{(0)}(\mathbf{r}, t) \xrightarrow[r \to \infty]{} (2\pi)^{-3/2} \exp(-iE_{k_i}t/\hbar) \left[\exp(i\mathbf{k}_i \cdot \mathbf{r}) + f(\hat{\mathbf{r}}) \exp(ik_i r)/r \right], \tag{4.173}$$

where $f(\hat{\mathbf{r}})$ is the scattering amplitude.

For $r < a$, only spherical Bessel functions are allowed in the partial wave expansion of $\Psi_{\mathbf{k}_i}^{(0)}(\mathbf{r}, t)$, since this wave function must be regular at $r = 0$. Therefore, for $r < a$, $\Psi_{\mathbf{k}_i}^{(0)}(\mathbf{r}, t)$ must have the form

$$\Psi_{\mathbf{k}_i}^{(0)}(\mathbf{r}, t) = (2\pi)^{-3/2} \exp(-iE_{k_i}t/\hbar) 4\pi \sum_{l=0}^{\infty} \sum_{m=-l}^{l} C_{lm}^{(0)} i^l j_l(Kr) Y_{lm}(\hat{\mathbf{r}}), \tag{4.174}$$

where $K = (k_i^2 + 2mV_0/\hbar^2)^{1/2}$ and the coefficients $C_{lm}^{(0)}$ are constants. These coefficients and the coefficients $c_{lm}^{(0)}$ appearing in Equation (4.172), and therefore the scattering amplitude $f(\hat{\mathbf{r}})$, can be obtained by matching the right-hand sides of Equations (4.172) and (4.174) at $r = a$. This procedure gives a non-trivial solution because the number of free parameters each partial wave depends on – the constant $C_{lm}^{(0)}$ and the constant $c_{lm}^{(0)}$ – is the same as the number of conditions to be fulfilled – the continuity of the partial wave and the continuity of its first derivative at $r = a$.

We now turn our attention to the wave function of the dressed continuum state $\Psi_{\mathbf{k}_i}(\mathbf{r}, t)$. As noted above, for $r \to \infty$, the wave function must have the form of an incident Gordon–Volkov plane wave plus a superposition of outgoing spherical waves. The electron can emit or absorb photons when scattering from the potential. As a consequence, the outgoing spherical wave will have wave numbers given by Equation (4.167). We therefore construct linear combinations of these functions to obtain an outer solution of the form

$$\Psi_{\mathbf{k}_i}^{V}(\mathbf{r}, t) = \left[\chi_{\mathbf{k}_i}^{V}(\mathbf{r}, t) + \sum_{N=-\infty}^{\infty} \sum_{l=0}^{\infty} \sum_{m=-l}^{l} c_{Nlm} \, \mathcal{H}_{lm}^{(1)}(k_N; \mathbf{r}, t) \right], \qquad r > a.$$

$$(4.175)$$

In a similar manner, we find that the inner solution has the form

$$\Psi_{\mathbf{k}_i}^{V}(\mathbf{r}, t) = \sum_{N=-\infty}^{\infty} \sum_{l=0}^{\infty} \sum_{m=-l}^{l} C_{Nlm} \, \mathcal{J}_{lm}(K_N; \mathbf{r}, t), \qquad r < a, \qquad (4.176)$$

with

$$K_N = \left[(2m/\hbar^2)(E_{k_i} + N\hbar\omega + V_0) \right]^{1/2}. \qquad (4.177)$$

Both of the wave functions (4.175) and (4.176) are linear combinations of Floquet solutions of the TDSE, each having quasi-energy E_{k_i}. Indeed, recalling the discussion at the end of Section 4.3.1, and using Equations (4.156) and (4.161), we can express the functions $\mathcal{J}_{lm}(K_N; \mathbf{r}, t)$ and $\mathcal{H}_{lm}^{(1)}(k_N; \mathbf{r}, t)$ as

$$\mathcal{J}_{lm}(k_N; \mathbf{r}, t) = (2\pi)^{-3/2} \exp(-iE_{k_i}t/\hbar) \sum_{n=-\infty}^{\infty} \exp(-in\omega t)$$

$$\times \sum_{\lambda=0}^{\infty} \sum_{\mu=-\lambda}^{\lambda} i^{\lambda} \Upsilon_{lm,\lambda\mu}^{n-N}(k_N, \alpha_0) j_{\lambda}(k_N r) Y_{\lambda\mu}(\hat{\mathbf{r}}) \qquad (4.178)$$

and

$$\mathcal{H}_{lm}^{(1)}(k_N; \mathbf{r}, t) = (2\pi)^{-3/2} \exp(-iE_{k_i}t/\hbar) \sum_{n=-\infty}^{\infty} \exp(-in\omega t)$$

$$\times \sum_{\lambda=0}^{\infty} \sum_{\mu=-\lambda}^{\lambda} i^{\lambda} \Upsilon_{lm,\lambda\mu}^{n-N}(k_N, \alpha_0) h_{\lambda}^{(1)}(k_N r) Y_{\lambda\mu}(\hat{\mathbf{r}}). \qquad (4.179)$$

In order to determine the inner- and outer-region expansion coefficients C_{Nlm} and c_{Nlm}, the right-hand sides of Equations (4.175) and (4.176), respectively, are first expanded in partial waves. The partial wave expansion of the Gordon–Volkov wave function $\chi_{\mathbf{k}_i}^V(\mathbf{r}, t)$ appearing in Equation (4.175) is obtained by inserting the expansions (4.152) and (4.153) in Equation (4.150), while the partial wave expansions of the functions $\mathcal{J}_{lm}(K_N; \mathbf{r}, t)$ and $\mathcal{H}_{lm}^{(1)}(k_N; \mathbf{r}, t)$ follow directly from Equations (4.178) and (4.179). The partial wave expansions are characterized by a *triplet* of indexes (n, λ, μ), and the coefficients C_{Nlm} and c_{Nlm} are determined from the condition that each of the partial waves contributions to the wave function is continuous and has a continuous first derivative at $r = a$.

The resulting Floquet wave function describes the scattering of an electron by the potential (4.171) in the presence of a homogeneous monochromatic laser field. Asymptotically, this wave function reduces to the superposition of an incident Gordon–Volkov wave and outgoing spherical waves, namely

$$\Psi_{\mathbf{k}_i}^V(\mathbf{r}, t) \underset{r \to \infty}{\longrightarrow} (2\pi)^{-3/2} \exp(-iE_{k_i}t/\hbar) \sum_{n=-\infty}^{\infty} \exp(-in\omega t)$$

$$\times \left[J_n(\mathbf{k}_i \cdot \boldsymbol{\alpha}_0) \exp(i\mathbf{k}_i \cdot \mathbf{r}) + \sum_{N=-\infty}^{\infty} f_{Nn}(E_{k_i}; \hat{\mathbf{r}}) \exp(ik_N r)/r \right],$$

$$(4.180)$$

where the coefficients $f_{Nn}(E_{k_i}, \hat{\mathbf{r}})$ are scattering amplitudes. This general asymptotic form of the wave function is valid for laser-assisted electron scattering by any short-range potential.

Equation (4.180) shows that every harmonic component of the Floquet wave function $\Psi_{\mathbf{k}_i}^V(\mathbf{r}, t)$ contains contributions from every scattering channel and that the two photon indices, n and N, play different roles. Their interpretation becomes clear if, making use of Equation (4.166), one notes that

$$\mathcal{H}_{lm}^{(1)}(k_N; \mathbf{r}, t) \underset{\varepsilon_0 \to 0}{\longrightarrow} (2\pi)^{-3/2} \exp[-i(E_{k_i} + N\hbar\omega)t/\hbar] 4\pi \, i^l \, h_l^{(1)}(k_N r) \, Y_{lm}(\hat{\mathbf{r}}).$$

$$(4.181)$$

If the laser field is adiabatically turned off far from the center of force, the dressed continuum wave function reduces to the superposition of an incident plane wave and outgoing spherical waves describing electrons moving with momentum $\hbar k_N$. For $N = 0$, we have $k_{N=0} = k_i$, and the photon index N can be understood as representing the net number of photons absorbed or emitted during the collision.

When the laser field is adiabatically turned off, all the harmonic components of the Gordon–Volkov wave function vanish, apart from the terms with $n = 0$. Thus n can be viewed as denoting the number of virtual photons exchanged with the laser field. In particular, when the wave number k_N appears in the nth harmonic component of the wave function, we can interpret this feature as describing the absorption by the electron of n photons, of which N are "real photons" and $n - N$

are "virtual photons." The fact that the absorption of real photons is entirely due to the collision can be seen from the observation that the Gordon–Volkov waves in Equation (4.150) have no components with $N \neq 0$, which is consistent with the well known result that a non-relativistic electron cannot absorb or emit photons in the dipole approximation.

The wave numbers k_N are purely imaginary for negative values of N such that $E_{k_i} + N\hbar\omega < 0$. In this case, the square root function in Equation (4.170) should be evaluated so that $\exp(ik_N r)$ decreases when $r \to \infty$, to prevent the wave function from diverging exponentially at large distances. Thus, if $E_{k_i} + N\hbar\omega < 0$,

$$k_N = i\left[(2m/\hbar^2)|E_{k_i} + N\hbar\omega|\right]^{1/2} \tag{4.182}$$

and the probability that the electron emerges at asymptotically large distances after having emitted $|N|$ real photons is equal to zero. Clearly, the scattering amplitudes $f_{Nn}(E_{k_i}; \hat{\mathbf{r}})$ vanish in the zero-field limit when $E_{k_i} + N\hbar\omega$ is negative and not equal to the energy of a field-free bound state.

We also note that the asymptotic form of the Floquet wave function given by Equation (4.180) is gauge-dependent. For example, we have seen in Section 4.1.1 that transforming $\Psi_{\mathbf{k}_i}^V(\mathbf{r}, t)$ to the length gauge would shift the quasi-energy upwards by a constant amount (the ponderomotive energy U_p) and multiply the wave function by a position-dependent time-periodic phase factor. The latter gives rise to a more complicated asymptotic behavior of each harmonic component of the wave function.

4.4 Non-Hermitian Floquet theory

In this section, we shall first examine the boundary conditions which must be satisfied by dressed bound states that are coupled to the continuum. We have seen above that for the case of laser-assisted electron scattering, involving dressed continuum states, one can impose boundary conditions characterized by the presence of both incoming and outgoing waves. In this case, the outgoing probability flux is compensated by the incident probability flux due to the incoming wave, so that the Hermitian Floquet theory (with real quasi-energies E) can be used. On the other hand, in the case of multiphoton ionization of a bound state, we shall show that *Siegert boundary conditions* [26] must be imposed. The corresponding *Siegert states* are complex energy states with purely outgoing waves. As a result, no incoming wave can compensate for the loss of probability flux due to the outgoing waves, and the electron probability density decreases as a function of time. We shall see that the Floquet quasi-energy E must then be allowed to become complex (with a negative imaginary part), so that the theory becomes *non-Hermitian*.

4.4.1 Decaying dressed states

The relationship between dressed bound states that are coupled to the continuum (or decaying dressed states) and dressed continuum states can be easily understood by considering the laser-assisted scattering of an electron by a potential or by an atomic system. Let us first assume that the laser field is weak and that the energy of the incident electron, E_{k_i}, differs from the energy of a bound state of the field-free atom, E_B, by an integer multiple of the photon energy, $N\hbar\omega$, where N is a positive integer. That is,

$$E_{k_i} = E_B + N\hbar\omega. \qquad (4.183)$$

During the scattering process, the electron may emit N photons by stimulated emission and be captured into the bound state. However, the electron will not stay indefinitely in that bound state, even if spontaneous emission is neglected, as it can always re-absorb enough photons from the laser field to become unbound again. The process of capture followed by re-emission leads to a resonance (capture–escape resonance) in the scattering amplitudes at the incident energy at which Equation (4.183) is satisfied. Such resonances are thus similar to shape resonances in the scattering by a potential, in which the incident particle is temporarily trapped into an unstable bound state by an effective potential barrier [27].

Dressed atomic bound states in a laser field can ionize and therefore have a finite lifetime. As a consequence, their energies are not sharply defined and have a non-zero width. Contrary to the field-free case, the wave function of a decaying bound state does not vanish at infinity because such a state is coupled to the continuum. Instead, asymptotically, the wave function must reduce to a superposition of outgoing spherical waves describing the ionized electron. The appropriate boundary conditions for decaying bound states differ from those discussed in Section 4.3.2 for dressed continuum states by the absence of an incident plane wave. For example, in the velocity gauge,

$$\Psi_{E^V}^V(\mathbf{r}, t) \underset{r\to\infty}{\longrightarrow} (2\pi)^{-3/2} \exp(-iE^V t/\hbar) \sum_{n=-\infty}^{\infty} \exp(-in\omega t)$$

$$\times \sum_{N=-\infty}^{\infty} F_{Nn}^V(E^V; \hat{\mathbf{r}}) \exp(ik_N r)/r, \qquad (4.184)$$

where E^V is the quasi-energy of the decaying bound state, the functions $F_{Nn}^V(E^V; \hat{\mathbf{r}})$ are multiphoton ionization amplitudes and

$$k_N = \left[(2m/\hbar^2)(E^V + N\hbar\omega)\right]^{1/2}. \qquad (4.185)$$

The boundary conditions (4.184) are a generalized form of the Siegert boundary conditions often used in the theory of single-channel scattering resonances [26]. If the binding potential has a Coulomb tail, so that

$$V(r) = V_s(r) - \frac{Ze^2}{(4\pi\epsilon_0)r}, \tag{4.186}$$

where $V_s(r)$ is a short-range potential, Equation (4.184) must be corrected for the long-range effect of the Coulombic part of $V(r)$. We have then

$$\Psi_{E^V}^V(\mathbf{r}, t) \underset{r\to\infty}{\longrightarrow} (2\pi)^{-3/2} \exp(-iE^V t/\hbar) \sum_{n=-\infty}^{\infty} \exp(-in\omega t)$$

$$\times \sum_{N=-\infty}^{\infty} F_{Nn}^V(E^V; \hat{\mathbf{r}}) \exp[ik_N r - \eta_N \log(2k_N r)]/r, \tag{4.187}$$

where

$$\eta_N = -\frac{Ze^2 m}{(4\pi\epsilon_0)\hbar^2 k_N} \tag{4.188}$$

is the Sommerfeld parameter.

As we have seen in Section 4.3.2, the boundary conditions relevant for dressed continuum states are such that it is possible to find Floquet solutions of the TDSE for any incident electron energy. The corresponding quasi-energies are real, being equal to the incident energy of the electron within an integer multiple of the photon energy. By contrast, finding solutions of the Floquet equations that are regular at $r = 0$ and satisfy the boundary conditions (4.184) or (4.187) is an eigenvalue problem: a non-trivial solution exists only for certain discrete values of the quasi-energy E. These values are in general complex, and can be expressed modulo $n\hbar\omega$ as

$$E = E_i + \Delta - i\Gamma/2, \tag{4.189}$$

where E_i is the energy of the initial field-free state, Δ is the AC Stark shift of that state and Γ is its energy width. This interpretation of Γ can be understood by noting that the relative position probability density of the state is given by

$$|\Psi_E(\mathbf{r}, t)|^2 = \exp(-\Gamma t/\hbar)|P(\mathbf{r}, t)|^2, \tag{4.190}$$

where $P(\mathbf{r}, t)$ is a periodic function of t having the same period as the laser field. Thus $\Gamma/\hbar$ is the decay rate, or ionization rate, of the atomic state, and $\hbar/\Gamma$ is the lifetime of that state, and therefore, from the energy–time uncertainty relation, Γ can be identified with its energy uncertainty or "width." It is clear, on physical grounds, that Γ cannot depend on the choice of gauge and must be positive. On the other hand, the real part of E is gauge-dependent.

The decaying bound states are not the only discrete dressed states that may be found by solving the Floquet equations subject to the boundary conditions (4.184) or (4.187). Other solutions include dressed resonance states, which reduce to bare resonance states in the zero-field limit, and, as we shall see later, to *light-induced states* (LIS), which in some cases have no clear counterpart in the field-free spectrum. As discussed, these discrete decaying states are all related to the dressed continuum states. Let us return to Equations (4.180)–(4.184) and multiply the Bessel function appearing in Equation (4.180) by an arbitrary factor ϵ, the same for every harmonic component, and let us allow E_{k_i} to take on complex values. Since the TDSE is linear and homogeneous, the scattering amplitudes $f_{Nn}(E_{k_i}; \hat{\mathbf{r}})$ are proportional to ϵ. Hence, if we let ϵ approach zero, each $f_{Nn}(E_{k_i}; \hat{\mathbf{r}})$ vanishes, except at those values of E_{k_i} at which these scattering amplitudes have a pole. In the limit $\epsilon \to 0$, the scattering Floquet wave function, at these values of E_{k_i}, acquires the purely outgoing wave asymptotic behavior given by Equation (4.184). Therefore, as functions of the energy, the amplitudes for laser-assisted electron scattering have poles in the lower half of the complex plane at the decaying dressed state quasi-energies. These decaying bound states play the same role as scattering resonances in the field-free case. By turning on the laser field, all of the bound states of the field-free system become resonances. In other words, the laser field embeds the bound states into the continuum, thereby creating *laser-induced continuum structures* (or LICS). These states are similar to auto-ionizing states in the field-free case.

Because the quasi-energy E is complex, the wave numbers k_N, defined by Equation (4.185), are also complex. Care should be exercised in evaluating the square root function: $\mathrm{Re}\, k_N$ must be positive when N is such that $\mathrm{Re}\, \mathsf{E} + N\hbar\omega > 0$ (the "open channels"), so that the function $\exp(ik_N r)/r$ has the character of an outgoing spherical wave, whereas $\mathrm{Im}\, k_N$ must be positive when N is such that $\mathrm{Re}\, \mathsf{E} + N\hbar\omega < 0$ (the "closed channels"), so that $\exp(ik_N r)/r$ vanishes for $r \to \infty$. Because $\mathrm{Im}\, \mathsf{E} < 0$, it follows that $\mathrm{Im}\, k_N$ is negative in the open channels, which means that the corresponding functions $\exp(ik_N r)/r$ diverge exponentially at large distances.

This divergence is an inescapable consequence of imposing the appropriate boundary conditions on the Floquet wave function. From the physical point of view, it can be understood as arising from the assumption that the laser field is monochromatic. The Floquet wave functions thus describe states which have been continuously exposed to the laser field from times infinitely remote in the past, and have continuously decayed at the rate $\Gamma/\hbar$. For example, let us consider an electron escaping to infinity with a wave number k_N, thus with a velocity of magnitude

$v_N = \mathrm{Re}\,\hbar k_N/m$, and let us treat this electron classically. If it is emitted at a certain time t_1, it will be at a certain distance r_1 from the origin at a later time t. If emitted at a certain time $t_2 > t_1$ it will be at a certain distance $r_2 < r_1$ at time t, with $r_1 - r_2 = (t_2 - t_1)v_N$. Quantum mechanically, the probability density of finding the electron at r_1 at time t is larger than that of finding it at r_2 at time t by a factor $\exp[-2\,\mathrm{Im}\,k_N(r_1 - r_2)]$. If we assume that $\Gamma \ll mv_N^2/2$, then $\mathrm{Im}\,k_N \simeq -\Gamma/(2\hbar v_N)$, so that $\exp[-2\,\mathrm{Im}\,k_N(r_1 - r_2)] \simeq \exp[(\Gamma/\hbar)(r_1 - r_2)/v_N] = \exp[-\Gamma(t_2 - t_1)/\hbar]$. Therefore the exponential increase of the wave function at large distances reflects the temporal decrease of the electronic density close to the origin.

Decaying bound states are thus associated with complex energies and divergent wave functions instead of real energies and square-integrable wave functions. These features, which are common to all resonance states, cannot be described within the same mathematical framework as the ordinary, field-free bound states. A rigorous mathematical theory of resonance states exists nonetheless, and is based on the concept of *complex scaling* introduced by Aguilar and Combes [28] and Balslev and Combes [29] and applied in the context of Floquet theory by Howland [30], Yajima [31] and Graffi, Grecchi and Silverstone [32]. This theory relies on the fact that the divergent asymptotic wave functions of the decaying bound states can be transformed into square-integrable wave functions by transforming the radial variable r into $r\exp(i\theta_r)$ for suitably chosen values of the rotation angle θ_r. Indeed, under this transformation,

$$\exp(ik_N r) \to \exp[i(\mathrm{Re}\,k_N + i\,\mathrm{Im}\,k_N)(r\cos\theta_r + ir\sin\theta_r)]. \qquad (4.191)$$

The result is, for $\mathrm{Re}\,E + N\hbar\omega > 0$, that the exponential divergence of $\exp(ik_N r)$ at large values of r becomes instead an exponential decrease provided that $\theta_r > -\arctan(\mathrm{Im}\,k_N/\mathrm{Re}\,k_N)$. We recall that, in the open channels, $\mathrm{Re}\,k_N < 0$ and $\mathrm{Im}\,k_N < 0$. As long as θ_r is chosen to be sufficiently large, but not too large so as to make $\exp[ik_N r\exp(i\theta_r)]$ asymptotically diverge in the closed channels, the complex scaling transforms the Floquet state $\Psi_E(\mathbf{r}, t)$ describing the decaying bound state into an L^2 wave function. It follows that by replacing r by $r\exp(i\theta_r)$ in the Floquet Hamiltonian H_F, one transforms it into a non-Hermitian Hamiltonian, which we denote by $H_F(\theta_r)$. The complex Floquet quasi-energies of the decaying bound states and their corresponding L^2 Floquet state vectors are obtained as the eigenvalues and eigenstates, respectively, of $H_F(\theta_r)$.

Complex scaling allows, in many respects, this eigenvalue problem to be treated like an ordinary bound-state problem by carrying out a partial or full diagonalization of the matrix representing $H_F(\theta_r)$ in an L^2 basis. It should be stressed that, in principle, the complex Floquet quasi-energies and corresponding Floquet states do not depend on θ_r, provided that its value is chosen within the bounds mentioned above. We also note that accidental degeneracies between quasi-energies at particular values of θ_r can occur. This can lead to difficulties as non-Hermitian operators that have degenerate eigenvalues may not possess a complete set of eigenstates. However, this situation rarely arises in applications.

4.4.2 Non-Hermitian perturbation theory

As noted above, the complex Floquet quasi-energy and the corresponding L^2 representation of a decaying bound state in a monochromatic laser field can be obtained by solving the time-independent Schrödinger equation for the scaled Hamiltonian,

$$H_F(\theta_r)|F(\mathsf{E})\rangle = \mathsf{E}\,|\,F(\mathsf{E})\rangle. \tag{4.192}$$

In this section we will examine how to solve this equation using time-independent perturbation theory. As in Section 4.2.3 we will work in the dipole approximation.

We start by expanding the state $|F(\mathsf{E})\rangle$ on a set of L^2 basis states $|\phi_k\rangle$, with each of the harmonic components $|F_n(\mathsf{E})\rangle$ being written as

$$|F_n(\mathsf{E})\rangle = \sum_k c_{nk}(\mathsf{E})|\phi_k\rangle. \tag{4.193}$$

We shall assume that this sum is limited to a finite number of basis states. Likewise, we also limit the Fourier–Floquet expansion (4.20) to a finite number of harmonic components. The accuracy of the calculation improves as the number of basis states and harmonic components is increased.

Using Equation (4.193) and pre-multiplying Equation (4.192) by $\langle\phi_k|$ yields a matrix eigenvalue problem for the quasi-energy. For an orthonormal basis, we have

$$\mathbf{H}_F(\theta_r)\,\mathbf{c}(\mathsf{E}) = \mathsf{E}\,\mathbf{c}(\mathsf{E}), \tag{4.194}$$

where $\mathbf{H}_F(\theta_r)$ is the non-Hermitian matrix representation of the Hamiltonian $H_F(\theta_r)$ on the chosen basis and $\mathbf{c}(\mathsf{E})$ is a column vector formed by the coefficients $c_{nk}(\mathsf{E})$. The matrix $\mathbf{H}_F(\theta_r)$ can be expressed as the sum of a field-free part and an interaction part as

$$\mathbf{H}_F(\theta_r) = \mathbf{H}_F^{(0)}(\theta_r) + \mathbf{H}_F^{\text{int}}(\theta_r) \tag{4.195}$$

with

$$\mathbf{H}_F^{(0)}(\theta_r) = \lim_{\mathcal{E}_0\to 0} \mathbf{H}_F(\theta_r). \tag{4.196}$$

We assume that the eigenvalues and eigenvectors of $\mathbf{H}_F^{(0)}(\theta_r)$ are known, and treat $\mathbf{H}_F^{\text{int}}(\theta_r)$ as a perturbation.

The matrix $\mathbf{H}_F^{(0)}(\theta_r)$ is block-diagonal, with the diagonal matrices given by non-Hermitian square matrices corresponding to a particular photon number n and to a particular combination Λ of the quantum numbers which characterize the symmetry of the field-free states. Let us take $\mathbf{H}_{n\Lambda}$ to be the non-Hermitian square matrix forming the (n, Λ) block, with its dimension being given by $\mathcal{N}_{n\Lambda}$, where $\mathcal{N}_{n\Lambda}$ may vary from block to block. The eigenvalues of $\mathbf{H}_{n\Lambda}$ are denoted by $w_{n\Lambda,q}$, $q = 1, 2, \ldots, \mathcal{N}_{n\Lambda}$. Some of these eigenvalues have a small imaginary part, with their real parts being approximately equal to the energies of the field-free bound states of the atom, modulo $\hbar\omega$, within the accuracy with which these can be represented on the finite basis set used in the calculation. The remaining eigenvalues may have a large imaginary part and either correspond to field-free resonant states of the atom or represent the discretized continuum spectrum of the field-free atom.

The eigenvalues of a given block can be assumed to be distinct. Barring accidental degeneracies, the following four properties hold [33]. Firstly, each of the eigenvalues $w_{n\Lambda,q}$ of $\mathbf{H}_{n\Lambda}$ is associated with two eigenvectors, namely a *right-hand eigenvector* $\mathbf{c}_{n\Lambda,q}$ that satisfies the equation

$$\mathbf{H}_{n\Lambda}\mathbf{c}_{n\Lambda,q} = w_{n\Lambda,q}\mathbf{c}_{n\Lambda,q} \tag{4.197}$$

and a *left-hand eigenvector* $\mathbf{d}_{n\Lambda,q}^{\mathsf{T}}$, which is a solution of the equation

$$\mathbf{d}_{n\Lambda,q}^{\mathsf{T}} \mathbf{H}_{n\Lambda} = w_{n\Lambda,q} \mathbf{d}_{n\Lambda,q}^{\mathsf{T}}. \tag{4.198}$$

Secondly, the left-hand and right-hand eigenvectors are bi-orthogonal to each other, in the sense that $\mathbf{d}_{n\Lambda,q}^{\mathsf{T}} \mathbf{c}_{n\Lambda,r} = 0$ if $q \neq r$. Thirdly, $\mathbf{d}_{n\Lambda,q}^{\mathsf{T}} \mathbf{c}_{n\Lambda,r} \neq 0$ if $q = r$. Fourthly, any $\mathcal{N}_{n\Lambda}$-component column vector can be written as a linear superposition of the right-hand eigenvectors $\mathbf{c}_{n\Lambda,q}$. In what follows, we will choose the basis states $|\phi_k\rangle$ such that the matrices $\mathbf{H}_{n\Lambda}$ are symmetric. In this case, the left-hand eigenvectors are the transpose of the right-hand eigenvectors (within an overall constant factor). We adopt the normalization $\mathbf{c}_{n\Lambda,q}^{\mathsf{T}} \mathbf{c}_{n\Lambda,q} = 1$.

Let us now introduce the eigenvectors $\mathbf{c}_j^{(0)}$ satisfying the equation

$$\mathbf{H}_{\mathrm{F}}^{(0)}(\theta_r)\mathbf{c}_j^{(0)} = \mathsf{E}_j^{(0)}\mathbf{c}_j^{(0)}. \tag{4.199}$$

These eigenvectors are orthogonal, and are normalized so that $\mathbf{c}_j^{(0)\mathsf{T}} \mathbf{c}_j^{(0)} = 1$. It follows from the above discussion that each zero-field quasi-energy $\mathsf{E}_j^{(0)}$ is equal to one of the eigenvalues $w_{n\Lambda,q}$ and conversely. We note that an eigenvalue $\mathsf{E}_i^{(0)}$ may be identical to an eigenvalue $\mathsf{E}_k^{(0)}$ belonging to a *different* diagonal block. However, because they belong to different diagonal blocks, the two eigenvalues are necessarily associated with two linearly independent eigenvectors $\mathbf{c}_i^{(0)}$ and $\mathbf{c}_k^{(0)}$. As long as there are no accidental degeneracies within a diagonal block, the eigenvectors $\mathbf{c}_j^{(0)}$ form a complete orthonormal set.

A particular solution of Equation (4.192) corresponding to the complex Floquet quasi-energy E_i and eigenvector $c_i(E_i)$, which in the limit $\mathcal{E}_0 \to 0$ reduce to $E_i^{(0)}$ and $c_i^{(0)}$, respectively, can now be found following the procedure discussed in Section 4.2.3. We shall assume that $E_i^{(0)}$ is not degenerate. One obtains

$$c(E_i) = \sum_{l=0}^{\infty} \left(G_F^{(0)}(\theta_r; z)[z - E_i + H_F^{int}(\theta_r)] \right)^l c^{(0)}, \qquad (4.200)$$

where

$$G_F^{(0)}(\theta_r; z) \equiv Q_F \frac{1}{z - H_F^{(0)}(\theta_r)} Q_F. \qquad (4.201)$$

It is again straightforward to re-derive the result (3.110) for the leading-order contribution to the AC Stark shift of a decaying bound state in a monochromatic laser field. If the one-photon ionization channel is closed, then ionization can be neglected to lowest order in the laser field–atom interaction and we may set $\theta_r = 0$. However, if the absorption of one photon is energetically allowed, then a non-zero value of θ_r must be chosen and the lowest-order correction to the field-free energy E_i of the state will contain a complex contribution, $-i\Gamma_i^{(2)}/2$, where $\Gamma_i^{(2)}$ is the LOPT contribution to the energy width of the state.

4.4.3 *Quasi-energy maps and multiphoton resonances*

Quasi-energy maps provide a useful method for analyzing the interaction of an atomic system with a laser field within the Floquet theory. These are plots of the real and/or imaginary parts of the complex Floquet quasi-energies (4.189) as a function of the laser-field parameters. For example, for a monochromatic laser-field, the quasi-energies can be plotted as a function of the electric field strength $\mathcal{E}_0$. Each quasi-energy then follows a trajectory in the complex plane that is parameterized by $\mathcal{E}_0$, or equivalently by the laser field intensity I.

In Fig. 4.1 the real part of the Floquet quasi-energy for a number of states of atomic hydrogen is shown as a function of the laser-field intensity. Results were obtained in the dipole approximation for a linearly polarized monochromatic field of wavelength 616 nm using the Sturmian-Floquet approach approach [34]. Quasi-energy trajectories for all of the Floquet states that have the same symmetry as the dressed ground state and reduce in the zero-field limit to bound states with principal quantum number $n \leq 6$, plus the 7g, 7h, 8g and 8h states, are given. States within the same Rydberg manifold are labeled by their dominant orbital angular momentum component. The quasi-energies of the excited dressed states are shifted down by $N_j \hbar \omega$, where N_j has been chosen so that the crossing of the jth quasi-energy is in

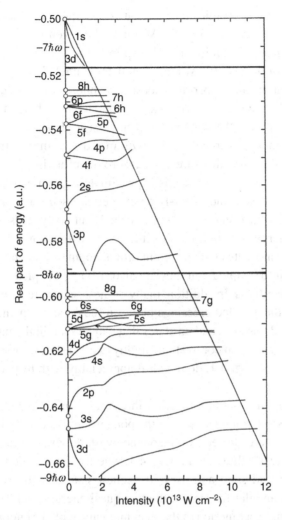

Figure 4.1. Real parts of Floquet quasi-energies for various levels of atomic hydrogen, each of which in the low-intensity limit corresponds predominantly to the atomic configuration indicated next to the curve. The laser field is linearly polarized and its wavelength is 616 nm. (From M. Dörr, R.M. Potvliege and R. Shakeshaft, *Phys. Rev. A* **41**, 558 (1990).)

N_j-photon resonance with the dressed ground state. The values of N_j are $N_{1s} = 0$, $N_{2s} = 0$, $N_{2p} = 0$ and $N_{3d} = 0$, while $N_k = 7$ and $N_k = 8$ for all levels above the seven- and eight-photon thresholds, respectively.

The ionization threshold is fixed at zero on the energy scale, and it is seen that the energy of the dressed ground state exhibits a dramatic downward shift as the intensity increases. At a wavelength of 616 nm, the absorption of seven photons is required to ionize atomic hydrogen at low laser intensities. However, as the intensity is increased to 1.35×10^{13} W cm^{-2}, the seven-photon ionization channel closes, and ionization requires the absorption of at least eight photons. Similarly, at an intensity of 6.88×10^{13} W cm^{-2} another channel closing appears. Above this intensity, the atom must absorb at least nine photons to ionize. As discussed in Section 1.3, when interacting with a low-frequency laser field of angular frequency ω, ionization requires the absorption of at least n photons, with $n\hbar\omega \geq I_P + U_p$, where I_P is the ionization potential of the atom and U_p is the electron ponderomotive energy. Fixing the ionization threshold at zero therefore implies that the downward AC Stark shift of the ground state is approximately given by $-U_p$.

The energy levels of the lower-lying excited dressed states in Fig. 4.1 shift in a complicated way as a function of the laser-field intensity. This is a general feature of the quasi-energy spectra for atoms interacting with intense, low-frequency laser fields. Not only are states with different orbital angular momentum within the same Rydberg manifold coupled, but they are also strongly coupled to a large number of neighboring manifolds. In addition, many of these states are shifted into resonance. Examples are the avoided crossings between the dressed 2p and 3s states as well as between the dressed 4s and 4d states that can be seen in the intensity range from 2 to 3×10^{13} W cm^{-2}. In contrast, the highly excited states, for example the 7g, 7h, 8g and 8h states in Fig. 4.1, do not shift appreciably with respect to the ionization threshold.

The upward shift of the ionization threshold – or equivalently the downward shift of the dressed ground state – by the ponderomotive energy implies that a large number of states are shifted into resonance with the ground state during the laser pulse. This is clearly illustrated in Fig. 4.1, where many avoided crossings between the dressed ground state and dressed excited state are observed. At resonance, and for a fixed angular frequency, the two quasi-energies will typically exhibit an avoided crossing as a function of the laser intensity, with the energy gap between the levels being approximately equal to $\hbar\omega_R$, where ω_R is the Rabi flopping frequency (3.221). If the variation of the intensity is rapid enough, such that the passage through the resonance occurs on a time scale that is short compared to ω_R^{-1}, then the Floquet state will follow the diabatic quasi-energy curve. If, on the other hand, the passage through resonance occurs on a time scale that is very long compared with ω_R^{-1}, the state will remain on the adiabatic curve, and its character will have changed upon emerging from the resonance. For these two limiting cases, the atom will remain in a single Floquet dressed state. However, in general, resonances will involve the transfer of population between the two dressed Floquet states, as discussed in Section 4.2.4.

The real parts of the Floquet quasi-energies corresponding to two resonantly coupled Floquet states can also undergo a true crossing at resonance [35]. Let us denote by Γ_a and Γ_b the energy widths of the two resonant Floquet states. When analyzing the behavior of these quasi-energies at a resonance, we can distinguish between two limiting cases [35–37]. When $|\Gamma_a - \Gamma_b| \ll \hbar\omega_R$, the real parts of the two Floquet quasi-energies undergo an avoided crossing, with the minimum distance between the real parts of the energies given approximately by $\hbar\omega_R$. The imaginary parts, on the other hand, exhibit a true crossing. In contrast, when $|\Gamma_a - \Gamma_b| \gg \hbar\omega_R$ the opposite occurs: the real parts of the two Floquet quasi-energies cross while the imaginary parts do not as the laser intensity increases and then decreases through the resonance. In fact, as we shall see in Section 4.7.5, for particular cases and for specific laser angular frequencies and intensities, both the real and the imaginary parts of the quasi-energies of the two resonantly coupled dressed Floquet states can cross, so that *laser-induced degenerate states* (LIDS) occur.

4.4.4 Shadow states and light-induced states

Let us return to the boundary conditions (4.184) satisfied by the decaying bound states $\Psi_E^V(\mathbf{r}, t)$ in the velocity gauge. As pointed out above, two possible choices can be made for the sign of each of the infinitely many channel wave numbers k_N, due to the square root appearing in Equation (4.185). Each combination of signs leads to a different eigenvalue problem. The choice leading to *physical dressed states* is to take $\mathrm{Re}(k_N) > 0$ in the open channels (when $\mathrm{Re}(E^V + N\hbar\omega) > 0$) and $\mathrm{Im}(k_N) > 0$ in the closed channels (when $\mathrm{Re}(E^V + N\hbar\omega) < 0$)). This choice ensures that $\exp(ik_N r)/r$ is an outgoing spherical wave if channel N is open, and an exponentially decaying wave if it is closed. The solutions corresponding to physical dressed states are called *dominant*. Other choices lead to wave functions having an unphysical asymptotic behavior. For example, they may have a non-decreasing closed-channel component, or an open-channel component with ingoing wave behavior. Thus, associated with each dominant solution are infinitely many other solutions such that the eigenvalues reduce to the same unperturbed energy level in the field-free limit $\mathcal{E}_0 \to 0$, but for $\mathcal{E}_0 > 0$ at least one of the k_N has the "wrong" sign. These solutions are called *shadow solutions* [38] for the following reason. The S-matrix for an atom A in the absence of the laser field can have poles on the negative real axis of the "physical" sheet $\mathrm{Im}\,k > 0$ of the complex-energy plane, corresponding to bound-state energies [39]. It can also have poles on the "unphysical" sheet ($\mathrm{Im}\,k < 0$) of the complex-energy plane, at complex values of the energy, $E = E_r - i\Gamma/2$, corresponding to resonances of energy E_r and width Γ in the scattering of an electron by the ion A^+. Due to the dependence of the dressed states on the channel wave numbers, the S-matrix has a branch point at $E = 0$ as well as branch points at the other multiphoton ionization thresholds. Since the ionization amplitudes vanish in the zero-field limit, if the field-free S-matrix has a pole on a certain sheet, it also has a pole at the same location on every sheet that can be reached without crossing the cut starting at $E = 0$. Thus, the dominant poles "cast shadows" on infinitely many unphysical sheets; for this reason these poles are referred to as *shadow poles* [38]. When the laser field is turned on, all the channels become coupled. The dominant and shadow poles then move on the Riemann energy surface as the parameters of the laser field vary, each pole following its own trajectory. For example, the poles lying close to the real axis of the physical sheet induce resonances in laser-assisted electron–atom scattering. As we have seen above, these resonances can be interpreted as capture–escape resonances arising from the temporary capture of the projectile electron into a decaying state, by absorption or stimulated emission of a certain number of photons having the angular frequency of the laser field [40]. In most cases, the shadow poles lie farther away from the real axis of the physical sheet than do the dominant ones, so that the dominant poles play the prominent role in determining the observable quantities. However, the shadow poles are sometimes important. In particular, they may give rise to threshold effects, or, as will be discussed below, they may become dominant for certain ranges of the laser-field parameters.

As illustrated in Fig. 4.1, the Floquet quasi-energy spectrum of an atom can display complicated interactions between a large number of Floquet states. Therefore, in order to determine, for example, the multiphoton ionization rate of the ground state of atomic hydrogen at a particular laser intensity and angular frequency, it is necessary to find the Floquet quasi-energy that is adiabatically connected to the ground state. In other words, to ensure that we have the "correct" quasi-energy, we must verify that as the laser intensity goes to zero the quasi-energy tends to the field-free ground-state energy (modulo $\hbar\omega$) of atomic hydrogen. Following this procedure, the quasi-energies shown in Fig. 4.1 can be identified.

Remarkably, there exist decaying bound Floquet states that occur in strong fields but cannot be connected adiabatically to a field-free state in the limit of zero intensity. These states are generally referred to as *light-induced states* (LIS) of atoms in intense laser fields. These states appear at certain laser intensities and frequencies, and, as we shall now see, can arise in a number of different situations [41].

We will first consider the case for which a multiphoton ionization threshold is crossed as the laser intensity increases. For the ground state of an atomic system in a low-frequency laser field, the crossing of a multiphoton ionization threshold corresponds to the closing of a multiphoton ionization channel. As an example, in Fig. 4.1 the ponderomotive energy shifts the position of the dressed hydrogenic ground state through the seven-photon ionization and the eight-photon ionization thresholds, respectively, as the laser intensity is increased. It is important to understand how k_N behaves across these thresholds.

In fact, upon reaching a threshold, a dominant state normally swaps character with a shadow state, so that the dominant state becomes a shadow state and the shadow state becomes a dominant state [42]. This dominant state is a LIS as it is not adiabatically connected to a field-free state of the atom. We note that the diabatic quasi-energy trajectory connecting the two dominant states across the threshold does not have to be smooth. For short-range potentials, large discontinuities in the imaginary part of the quasi-energy can occur [41]. However, typically the interchange of dominant and shadow states results in a nearly continuous quasi-energy trajectory. The designation as a LIS of a state that is a diabatic continuation of a decaying Floquet state across a threshold is largely formal, in the sense that its existence can be viewed as a consequence of the multisheeted structure of the complex-energy plane on which the Floquet quasi-energies are defined [43].

More interesting LIS occur at single- or multi-photon ionization thresholds when a shadow pole has a quasi-energy trajectory that brings it onto the physical sheet. This happens when a multiphoton ionization threshold is crossed as the laser intensity increases above a certain appearance intensity. The resulting LIS is a new dominant state whose properties cannot be distinguished from those of decaying bound states [41]. Light-induced states have been found in one-dimensional model atom calculations [44,45] and have been analyzed in terms of shadow poles [46–48]. In Section 7.3, we will discuss examples of LIS that have been found in Floquet calculations at high laser frequencies.

4.5 Floquet theory in the Kramers–Henneberger frame

Let us consider the simple case of an atom with one active electron in a spatially homogeneous monochromatic laser field of arbitrary polarization, described by the electric field $\mathcal{E}(t)$ of Equation (2.114). The corresponding vector potential $\mathbf{A}(t)$ is given by Equation (2.115). The TDSE (2.141) for the wave function of the system $\Psi^A(\mathbf{r}, t)$ in the accelerated, or space-translated, Kramers–Henneberger (K–H) frame then reduces to

$$i\hbar\frac{\partial}{\partial t}\Psi^A(\mathbf{r}, t) = \left(-\frac{\hbar^2}{2m}\nabla^2 + V\left[\mathbf{r} + \boldsymbol{\alpha}(t)\right]\right)\Psi^A(\mathbf{r}, t), \qquad (4.202)$$

where $\boldsymbol{\alpha}(t)$ is the displacement vector (2.116) corresponding to the "quiver" motion of a classical electron in the laser field.

Since the space-translated TDSE (4.202) has periodic coefficients, one can seek solutions $\Psi_E^A(\mathbf{r}, t)$ having the Floquet–Fourier form

$$\Psi_{E^A}^A(\mathbf{r}, t) = \exp(-iE^A t/\hbar)\sum_{n=-\infty}^{\infty}\exp(-in\omega t)F_n^A(E^A; \mathbf{r}). \qquad (4.203)$$

Making also a Fourier analysis of the potential, we write

$$V[\mathbf{r} + \boldsymbol{\alpha}(t)] = \sum_{n=-\infty}^{\infty} \exp(-in\omega t) V_n(\alpha_0, \mathbf{r}), \qquad (4.204)$$

where

$$V_n(\alpha_0, \mathbf{r}) = (T)^{-1} \int_0^T \exp(in\omega t) V[\mathbf{r} + \boldsymbol{\alpha}(t)] dt \qquad (4.205)$$

and $T = 2\pi/\omega$ is the optical period. We remark that the Fourier components $V_n(\alpha_0, \mathbf{r})$ depend on the laser field only through the excursion amplitude α_0, and not separately on the angular frequency ω and the intensity I. Moreover, since $V[\mathbf{r} + \boldsymbol{\alpha}(t)]$ is real,

$$V_{-n}(\alpha_0, \mathbf{r}) = V_n^*(\alpha_0, \mathbf{r}), \qquad (4.206)$$

and if the potential is symmetric, so that $V(-\mathbf{r}) = V(\mathbf{r})$, then V_n has the parity of n:

$$V_n(\alpha_0, -\mathbf{r}) = (-1)^n V_n(\alpha_0, \mathbf{r}). \qquad (4.207)$$

Substituting Equations (4.203) and (4.204) into the TDSE (4.202), one obtains for the harmonic components $F_n^A(E^A; \mathbf{r})$ of the wave function $\Psi_{E^A}^A(\mathbf{r}, t)$ the following infinite system of time-independent coupled equations:

$$\left(-\frac{\hbar^2}{2m} \nabla^2 - n\hbar\omega \right) F_n^A(E^A; \mathbf{r}) + \sum_{\nu=-\infty}^{\infty} V_{n-\nu}(\alpha_0, \mathbf{r}) F_\nu^A(E^A; \mathbf{r}) = E^A F_n^A(E^A; \mathbf{r}),$$

$$n = 0, \pm 1, \pm 2, \ldots$$
$$(4.208)$$

In order to define a solution of these equations, we must specify the boundary conditions to be satisfied by the harmonic components $F_n^A(E^A; \mathbf{r})$. It is important to note that in the Kramers–Henneberger frame, each $F_n^A(E^A; \mathbf{r})$ corresponds to a *single* wave number k_n, thus to a *definite* channel n. This is due to the fact that in the limit $r \to \infty$ the displacement vector $\boldsymbol{\alpha}(t)$ becomes negligible with respect to $\mathbf{r}$ and the potential $V[\mathbf{r} - \boldsymbol{\alpha}(t)]$ appearing in Equation (4.202) reduces to $V(\mathbf{r})$, which tends to zero when $r \to \infty$. Thus, in the limit $r \to \infty$, the channels are *decoupled* in the Kramers–Henneberger frame. As we have seen above, this is not the case in the laboratory frame.

To examine this point in more detail, let us first focus on the case of laser-assisted scattering of a non-relativistic electron of initial wave vector $\mathbf{k}_i$ and energy $E_{k_i} = \hbar^2 k_i^2/(2m)$ by a potential $V(\mathbf{r})$. In the Kramers–Henneberger frame, the Gordon–Volkov wave function describing a free electron in a laser field with wave vector $\mathbf{k}_i$ is a plane wave of the form (2.174), namely

$$\chi_{\mathbf{k}_i}^A(\mathbf{r}, t) = (2\pi)^{-3/2} \exp[i(\mathbf{k}_i \cdot \mathbf{r} - E_k t/\hbar)]. \qquad (4.209)$$

We denote the Floquet wave function describing the corresponding scattering state by $\Psi_{\mathbf{k}_i}^A(\mathbf{r}, t)$. We take the quasi-energy of this dressed continuum state, E^A, to be equal to the energy E_{k_i}. Assuming that the potential $V(\mathbf{r})$ is of short-range, Equations (4.208) become, in the limit $r \to \infty$,

$$\left(-\frac{\hbar^2}{2m}\nabla^2 - n\hbar\omega\right) F_n^A(E_{k_i}; \mathbf{r}) = E_k F_n^A(E_{k_i}; \mathbf{r}), \qquad n = 0, \pm 1, \pm 2, \dots \quad (4.210)$$

These equations show that asymptotically the radial wave function of each harmonic component $F_n^A(E_{k_i}; \mathbf{r})$ reduces to an outgoing spherical wave $\exp(ik_n r)/r$, or to an ingoing spherical wave $\exp(-ik_n r)/r$, or to a linear combination of those. The wave numbers k_n are related to the energy E_{k_i} as in the laboratory frame, namely

$$k_n = \begin{cases} \left[(2m/\hbar^2)(E_{k_i} + n\hbar\omega)\right]^{1/2} & E_{k_i} + n\hbar\omega \geq 0, \\ i\left[(2m/\hbar^2)|E_{k_i} + n\hbar\omega|\right]^{1/2} & E_{k_i} + n\hbar\omega \leq 0. \end{cases} \quad (4.211)$$

Hence, the boundary conditions to be imposed on the wave function of a dressed continuum state are given by

$$\Psi_{\mathbf{k}_i}^A(\mathbf{r}, t) \underset{r \to \infty}{\longrightarrow} (2\pi)^{-3/2} \exp(-iE_{k_i}t/\hbar) \sum_{n=-\infty}^{\infty} \exp(-in\omega t)$$

$$\times \left[\exp(i\mathbf{k}_i \cdot \mathbf{r}) + f_n(E_{k_i}; \hat{\mathbf{r}})\exp(ik_n r)/r\right]. \quad (4.212)$$

The asymptotic behavior of $\Psi_{\mathbf{k}_i}^A(\mathbf{r}, t)$ is therefore simpler than that of its velocity-gauge counterpart, which is given by Equation (4.180).

We can now obtain the boundary conditions satisfied by the wave functions $\Psi_{E^A}^A(\mathbf{r}, t)$ of the corresponding decaying dressed states. Denoting by E^A a complex pole of the scattering amplitude $f_n(E_k; \hat{\mathbf{r}})$ and setting

$$k_n = \left[(2m/\hbar^2)(E^A + n\hbar\omega)\right]^{1/2}, \quad (4.213)$$

we have

$$\Psi_{E^A}^A(\mathbf{r}, t) \underset{r \to \infty}{\longrightarrow} (2\pi)^{-3/2} \exp(-iE^A t/\hbar) \sum_{n=-\infty}^{\infty} \exp(-in\omega t) F_n(E^A; \hat{\mathbf{r}})\exp(ik_n r)/r$$

$$(4.214)$$

if $V(\mathbf{r})$ is a short-range potential. If $V(\mathbf{r})$ has a Coulomb tail of the form $-Ze^2/(4\pi\epsilon_0 r)$, then the potentials $V_n(\alpha_0, \mathbf{r})$ decrease as $r^{-|n|-1}$ asymptotically, and Equation (4.214) must be replaced by the Siegert boundary condition

$$\Psi_{\mathrm{EA}}^{\mathrm{A}}(\mathbf{r},t) \underset{r\to\infty}{\longrightarrow} (2\pi)^{-3/2}\exp(-i\mathsf{E}^{\mathrm{A}}t/\hbar)$$

$$\times \sum_{n=-\infty}^{\infty} \exp(-in\omega t)F_n(\mathsf{E}^{\mathrm{A}};\hat{\mathbf{r}})\exp\left[ik_n r - \eta_n \log(2k_n r)\right]/r, \quad (4.215)$$

where the Sommerfeld parameter η_n is given by Equation (4.188) with $N=n$. As in the laboratory frame, the decaying dressed states have complex quasi-energies and the determination of the square root functions in Equation (4.213) must be chosen such that $\exp(ik_n r)/r$ is an outgoing wave in the open channels (for n such that $\mathrm{Re}\,\mathsf{E}^{\mathrm{A}} + n\hbar\omega \geq 0$) and a damped exponential in the closed channels (for n such that $\mathrm{Re}\,\mathsf{E}^{\mathrm{A}} + n\hbar\omega < 0$).

4.6 Sturmian-Floquet theory

We shall now discuss a method in which a *discrete* basis set of Sturmian functions [49] is used to obtain a solution of the system of coupled Floquet equations (4.25). This Sturmian basis set is particularly appropriate for obtaining accurate solutions to these equations for atoms with one active electron moving either in a Coulomb potential (as in the case of hydrogenic atoms) or in a modified Coulomb effective potential. In this context, Sturmian basis sets were first used by Maquet, Chu and Reinhardt [50], who carried out their calculations for atomic hydrogen using complex scaling. We will outline a formulation of this method that makes use of complex wave numbers [3].

Denoting by $\mathbf{r}$ the position vector of the active electron and letting $F_n(\mathbf{r}) \equiv \langle\mathbf{r}|F_n\rangle$, we expand each harmonic component as

$$F_n(\mathbf{r}) = \sum_{NLM} c_{NLM}^{(n)} r^{-1} S_{NL}^{\kappa}(r)Y_{LM}(\hat{\mathbf{r}}), \quad (4.216)$$

where the $Y_{LM}(\hat{\mathbf{r}})$ are spherical harmonics and the radial functions $S_{NL}^{\kappa}(r)$ are the complex Sturmian functions defined in Equation (3.146). The parameter κ is chosen to be complex, allowing the computation of quasi-bound (Siegert) states, in analogy to the complex-rotation transformation [2].

Any well-behaved function $f(r)$ that vanishes as r^{L+1} at the origin and behaves as an outgoing wave $\exp(ikr)$ at large r can be expanded in the form

$$f(r) = \sum_{N} f_N S_{NL}^{\kappa}(r), \quad (4.217)$$

with coefficients f_N that decrease as N increases provided that [51, 52]

$$|\arg(\kappa) - \arg(k_N)| < \frac{\pi}{2}. \quad (4.218)$$

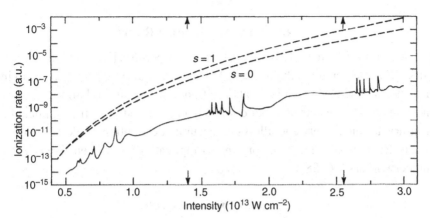

Figure 4.2. Total ionization rate of atomic hydrogen as a function of the intensity of a monochromatic, linearly polarized laser field of wavelength 1064 nm. Dashed curves are partial rates for $(12 + s)$-photon ionization obtained using LOPT. The arrows indicate the intensities at which the real part of the 1s Floquet eigenvalue crosses the 13- and 14-photon ionization thresholds. (From R. M. Potvliege and R. Shakeshaft, *Phys. Rev. A* **40**, 3061 (1989).)

The Sturmian-Floquet method has been applied extensively [3] to study multiphoton ionization and harmonic generation in atomic hydrogen and other atoms modeled as effective one-electron systems. As an example, we show in Fig. 4.2 the total ionization rate as a function of the intensity for H(1s) interacting with a monochromatic, linearly polarized laser field of wavelength $\lambda = 1064$ nm, as calculated by Potvliege and Shakeshaft [35] using the Sturmian-Floquet approach. Also shown in Fig. 4.2 are the partial rates for $(12 + s)$-photon ionization obtained by using lowest-order perturbation theory (LOPT). It is apparent from Fig. 4.2 that LOPT is inadequate to calculate the ionization rate of H(1s) by infra-red radiation. More generally, at low frequencies, the total rate for multiphoton ionization from the ground state, as obtained from LOPT, increases much too fast with increasing intensity.

The Sturmian-Floquet results displayed in Fig. 4.2 also illustrate the role of the Stark shift-induced resonances, which were briefly described in Chapter 1. For example, the first group of peaks in Fig. 4.2 corresponds to intermediate 11- or 12-photon transitions to Rydberg sublevels associated, in the intensity range from 6×10^{12} W cm^{-2} to 10^{13} W cm^{-2}, with the crossings of the 1s level with a group of levels. Since the ionization potential of the atom is different at each crossing, the Stark shift-induced resonances manifest themselves prominently in the energy spectrum of the ejected electrons [53].

4.7 *R*-matrix–Floquet theory

The *R*-matrix–Floquet (RMF) theory is a non-perturbative method proposed by Burke, Francken and Joachain [54, 55] to analyze multiphoton processes in intense laser fields. The RMF theory treats multiphoton ionization, harmonic generation and laser-assisted electron–atom collisions in a unified way. It is applicable to an arbitrary atomic system and allows an accurate description of electron correlation effects. Reviews of the RMF theory and its applications have been given by Joachain and co-workers [56–58].

4.7.1 *Basic concepts*

Let us consider an atomic system (atom or ion) composed of a nucleus of atomic number Z and N electrons, interacting with a laser field which is treated as a spatially homogeneous electric field $\mathcal{E}(t)$. Neglecting relativistic effects, the atomic system in the presence of this laser field is described by the TDSE (2.121). Using atomic units (a.u.) we have

$$i\frac{\partial}{\partial t}\Psi(X,t) = \left[H_0 + \mathbf{A}(t)\cdot\mathbf{P} + \frac{N}{2}\mathbf{A}^2(t)\right]\Psi(X,t), \qquad (4.219)$$

where we recall that X denotes the ensemble of the coordinates of the N electrons, H_0 is the non-relativistic field-free Hamiltonian of the N-electron atomic system, $\mathbf{A}(t)$ is the vector potential and $\mathbf{P}$ is the total momentum operator (2.120).

According to the *R*-matrix method [59–61], configuration space is subdivided into two regions as shown in Fig. 4.3. The internal region is defined by the condition that the radial coordinates r_i of all N electrons are such that $r_i \leq a$ $(i = 1, 2, ..., N)$, where the sphere of radius a envelops the charge distribution of the target atom states retained in the calculation. In this region, exchange and correlation effects involving all N electrons are important. The external region is defined so that one of the electrons (say electron N) has a radial coordinate $r_N \geq a$, and the remaining

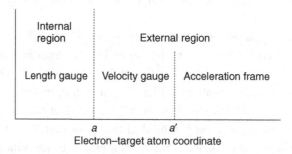

Figure 4.3. Partitioning of configuration space in the *R*-matrix–Floquet theory.

$N-1$ electrons are confined within the sphere of radius a. Hence, in this region, exchange and correlation effects between the "external" electron and the remaining $N-1$ electrons can be neglected. We remark that by allowing only one electron to be present in the external region, we restrict here our analysis to processes involving at most one unbound electron, namely multiphoton single ionization, harmonic generation and laser-assisted electron–atom (ion) elastic and inelastic collisions.

Having divided configuration space into an internal and an external region, we must solve the TDSE (4.219) in these two regions separately. This will be done by using the Floquet method, which, as we have seen above, reduces the problem to that of solving an infinite set of coupled time-independent equations for the harmonic components $F_n(X) \equiv \langle X | F_n \rangle$ of the wave function $\Psi(X, t)$. The solutions obtained in the internal and external regions will then be matched on the boundary at $r = a$.

4.7.2 Internal region solution

In the internal region, it is convenient to use the length gauge, because in this gauge the laser–atom coupling tends to zero at the origin. Indeed, if we start from the original TDSE (4.219) and use Equation (2.132) to perform the gauge transformation (in a.u.)

$$\Psi^{L}(X, t) = \exp[i\mathbf{A}(t) \cdot \mathbf{R}] \Psi(X, t), \tag{4.220}$$

we find that the wave function in the length gauge, $\Psi^{L}(X, t)$, satisfies the TDSE (2.133), namely

$$i \frac{\partial}{\partial t} \Psi^{L}(X, t) = \left[H_0 + H_{\text{int}}^{L}(t) \right] \Psi^{L}(X, t), \tag{4.221}$$

where, according to Equations (2.134) and (2.136),

$$H_{\text{int}}^{L}(t) = \mathcal{E}(t) \cdot \mathbf{R} = -\mathcal{E}(t) \cdot \mathbf{D}. \tag{4.222}$$

Although, as we have seen in Section 4.1, the Floquet method can be applied to polychromatic fields, we shall assume for the moment that the laser field is not only homogeneous (dipole approximation) but also monochromatic. For simplicity, we shall also suppose that it is linearly polarized, so that we can write it in the form

$$\mathcal{E}(t) = \hat{\boldsymbol{\epsilon}} \mathcal{E}_0 \cos(\omega t). \tag{4.223}$$

The interaction Hamiltonian $H_{\text{int}}^{L}(t)$ can then be written in the form of Equation (4.24), where $H_+ = H_- = (\mathcal{E}_0/2)\hat{\boldsymbol{\epsilon}} \cdot \mathbf{R}$. To solve Equation (4.221), we then use the Floquet–Fourier expansion

$$\Psi^{L}(X, t) = \exp(-iE^{L}t) \sum_{n=-\infty}^{+\infty} \exp(-in\omega t) F_n^{L}(X), \tag{4.224}$$

where the explicit dependence of the wave function and its harmonic components on E^L has been omitted. Upon substitution of Equation (4.224) into Equation (4.221), we find that the harmonic components F_n^L satisfy a set of coupled equations of the form (4.25), namely

$$(H_0 - n\omega)F_n^L(X) + H_+ F_{n-1}^L(X) + H_- F_{n+1}^L(X) = E^L F_n^L(X). \quad (4.225)$$

The functions F_n^L can be regarded as the components of a vector $|F^L\rangle$ in Fourier space. This allows us to express Equation (4.225) in a form similar to that of Equation (4.23). That is

$$H_F |F^L\rangle = E^L |F^L\rangle, \quad (4.226)$$

where the Floquet Hamiltonian H_F is an infinite tridiagonal matrix of operators having the form (4.26).

It is important to note that the Hamiltonian H_F is not Hermitian in the internal region due to surface terms at $r = a$ arising from the kinetic energy operator in H_0. These surface terms can be eliminated using a procedure introduced by Bloch [62], who defined an operator L_B that cancels those terms, so that $H_F + L_B$ is Hermitian in the internal region. One can then rewrite Equation (4.226) in the form

$$(H_F + L_B - E^L) |F^L\rangle = L_B |F^L\rangle, \quad (4.227)$$

which has the advantage over Equation (4.226) that the operator on the left-hand side can be inverted in a straightforward way, as we shall now show. The Bloch operator L_B is diagonal in Fourier space, being given by

$$L_B = \frac{1}{2} \sum_{\gamma ij} |\phi_i^\gamma (r_j^{-1})\rangle \delta(r_j - a) \left[\frac{d}{dr_j} - \frac{b-1}{r_j} \right] \langle \phi_i^\gamma (r_j^{-1})|, \quad (4.228)$$

where b is an arbitrary constant and the channel functions

$$\phi_i^\gamma (r_j^{-1}) \equiv \phi_i^\gamma (q_1, ..., q_{j-1}, q_{j+1}, ..., q_N, \hat{r}_j, \sigma_j) \quad (4.229)$$

are formed by coupling the atomic target states $\psi_i(q_1, ..., q_{j-1}, q_{j+1}, ..., q_N)$ and possibly the pseudo-states $\bar{\psi}_i(q_1, ..., q_{j-1}, q_{j+1}, ..., q_N)$ retained in the calculation with the spin-angle functions of the scattered or ejected electron (j) to give a state whose quantum numbers are collectively denoted by γ. We recall that $q_j \equiv (\mathbf{r}_j, \sigma_j)$ denote the space and spin coordinates of electron j, and that the symbol $\hat{r}_j$ means the angular variables (θ_j, ϕ_j) of the vector $\mathbf{r}_j$.

Equation (4.227) can then be solved using the spectral representation of the operator $(H_F + L_B - E^L)$. In analogy with the R-matrix method in the absence of an external field [63], we introduce the functions

$$F_{kn}^L(X) = A \sum_{\gamma il} \phi_i^\gamma (r_j^{-1}) r_j^{-1} u_l^\gamma (r_j) a_{ilkn}^\gamma + \sum_{\gamma i} \chi_i^\gamma (X) b_{ikn}^\gamma, \quad (4.230)$$

where $\mathcal{A}$ is the antisymmetrization operator, the u_i^γ are radial basis functions that are non-vanishing on the boundary of the internal region, the χ_i^γ are quadratically integrable antisymmetric functions that vanish by the boundary of the internal region and a_{ilkn}^γ, as well as b_{ikn}^γ, are coefficients obtained by diagonalizing the operator $(H_F + L_B)$ in the internal region. In practice, only a finite number of low-lying bound-target eigenstates can be included in the first expansion on the right-hand side of Equation (4.230). Hence, additional pseudo-states are often used in that expansion and quadratically integrable functions are retained in the second expansion on the right-hand side of Equation (4.230) to ensure that the total expansion represents the essential physics.

Let $|F_k^L\rangle$ be the vector in Fourier space whose components are F_{kn}^L. We have

$$\langle F_{k'}^L | H_F + L_B | F_k^L \rangle = E_k \delta_{kk'}. \tag{4.231}$$

The matrix representation of $(H_F + L_B)$ in the basis defined by Equation (4.230) is block-tridiagonal. By retaining a finite number of basis functions F_{kn}^L, this matrix is first truncated to become finite. It can then be diagonalized to obtain the eigenvalues E_k and the corresponding eigenvectors. Making use of the spectral representation of the operator $(H_F + L_B - \mathsf{E}^L)$ and of Equation (4.231), we can rewrite Equation (4.227) as follows:

$$|F^L\rangle = \sum_k |F_k^L\rangle \frac{1}{E_k - \mathsf{E}^L} \langle F_k^L | L_B | F^L \rangle. \tag{4.232}$$

Projecting this equation onto the channel functions $\phi_i^\gamma(r_j^{-1})$ and onto the nth component in Fourier space, and evaluating on the boundary $r = a$, yields

$$F_{in}^\gamma(a) = \sum_{\gamma'i'n'} {}^L R_{ini'n'}^{\gamma\gamma'}(\mathsf{E}^L) \left(r \frac{\mathrm{d} F_{i'n'}^{\gamma'}}{\mathrm{d} r} - b F_{i'n'}^{\gamma'} \right)_{r=a}, \tag{4.233}$$

where we have introduced the R-matrix calculated in the length gauge,

$$^L R_{ini'n'}^{\gamma\gamma'}(\mathsf{E}^L) = \frac{1}{2a} \sum_k \frac{w_{ink}^\gamma w_{i'n'k'}^{\gamma'}}{E_k - \mathsf{E}^L}, \tag{4.234}$$

the reduced radial functions,

$$F_{in}^\gamma(r_j) = r_j \langle \phi_i^\gamma(r_j^{-1}) | F_n^L \rangle, \tag{4.235}$$

and the surface amplitudes,

$$w_{ikn}^\gamma = a \langle \phi_i^\gamma(r_j^{-1}) | F_{kn}^L \rangle_{r_j=a}. \tag{4.236}$$

Equations (4.233) and (4.234) are the basic equations describing the solution of the TDSE (4.219) in the internal region. We note that Equation (4.233) can be written in matrix form as

$$\mathbf{F}(a) = {}^{L}\mathbf{R}(\mathbf{E}^{L}) \left[r\frac{d\mathbf{F}}{dr} - b\mathbf{F} \right]_{r=a}, \tag{4.237}$$

where $\mathbf{F}$ denotes the set of reduced radial functions defined by Equation (4.235).

The R-matrix obtained from Equation (4.234) with a finite number of basis states can be improved by including a Buttle correction [64, 65], which corrects for the high-lying omitted poles. The logarithmic derivative of the reduced radial functions (4.235) on the boundary $r = a$, which provides the boundary condition for obtaining the solution of the TDSE (4.219) in the external region, is then given by Equation (4.237). In practice, the constant b arising from the Bloch operator (4.228) is often taken to be zero, so that Equation (4.237) reduces to

$$\mathbf{F}(a) = {}^{L}\mathbf{R}(\mathbf{E}^{L}) \left[r\frac{d\mathbf{F}}{dr} \right]_{r=a}. \tag{4.238}$$

4.7.3 External region solution

In the external region, we have only one electron ($r_N \geq a$), which we describe by using the velocity (V) gauge, while the remaining ($N-1$) electrons are confined to lie within the sphere of radius a ($r_i \leq a, i = 1, 2, ..., N-1$) and are still treated by using the length gauge. Thus, starting from the original TDSE (4.219), we perform the unitary transformation

$$\Psi^{V}(X, t) = \exp\left[i\mathbf{A}(t) \cdot \mathbf{R}' + \frac{i}{2} \int_{-\infty}^{t} \mathbf{A}^2(t')dt' \right] \Psi(X, t),$$
$$r_i \leq a \ (i = 1, 2, ..., N-1), \quad r_N \geq a, \tag{4.239}$$

where

$$\mathbf{R}' = \sum_{i=1}^{N-1} \mathbf{r}_i. \tag{4.240}$$

Substituting Equation (4.239) into Equation (4.219), we find that the wave function $\Psi^{V}(X, t)$ must satisfy the TDSE

$$i\frac{\partial}{\partial t}\Psi^{V}(X, t) = [H^{V} + \mathcal{E}(t) \cdot \mathbf{R}']\Psi^{V}(X, t), \tag{4.241}$$

where

$$H^{V} = H_0 + \mathbf{A}(t) \cdot \mathbf{p}_N \tag{4.242}$$

and $\mathbf{p}_N = -i\nabla_{\mathbf{r}_N}$.

In order to solve Equation (4.241), we introduce, as in the internal region, the Floquet–Fourier expansion

$$\Psi^V(X,t) = \exp(-iE^V t) \sum_{n=-\infty}^{+\infty} \exp(-in\omega t) F_n^V(X), \qquad (4.243)$$

and upon substitution into Equation (4.241), we obtain for the harmonic components F_n^V the system of coupled equations

$$(H^V - n\omega) F_n^V(X) + H'_+ F_{n-1}^V(X) + H'_- F_{n+1}^V(X) = E^V F_n^V(X), \quad (4.244)$$

where $H'_+ = H'_- = (\mathcal{E}_0/2)\hat{\epsilon} \cdot \mathbf{R}'$.

The functions F_n^V can be considered as the components of a vector $|F^V\rangle$ in Fourier space, and Equation (4.244) can be written in the form

$$H_F^V |F^V\rangle = E^V |F^V\rangle. \qquad (4.245)$$

To solve this system of equations, we introduce for the harmonic components F_n^V the close-coupling expansion

$$F_n^V(X) = \sum_{\gamma i} \phi_i^\gamma(r_N^{-1}) r_N^{-1} \, {}^V G_{in}^\gamma(r_N), \quad r_a \geq a, \qquad (4.246)$$

where the expansion over γ and i goes over the same range as in the expansion (4.230) adopted in the internal region. However, we note that the expansion (4.246), unlike the expansion (4.230), is not antisymmetrized with respect to the Nth electron. This is due to the fact that this electron now lies outside the sphere of radius a, while the remaining $(N-1)$ electrons are confined within this sphere. Hence, exchange effects between the Nth electron and the remaining electrons are negligible.

Substituting the expansion (4.246) into Equation (4.244), projecting onto the channel functions $\phi_i^\gamma(r_N^{-1})$ and onto the nth component in Fourier space yields for the reduced radial functions ${}^V G_{in}^\gamma$ a system of coupled differential equations describing the motion of the Nth electron in the external region. These equations are of the form

$$\left[\frac{d^2}{dr^2} - \frac{l_i(l_i+1)}{r^2} + \frac{2(Z-N+1)}{r} + k_{in}^2\right] {}^V G_{in}^\gamma(r)$$

$$= 2 \sum_{\gamma' i' n'} {}^V W_{in i' n'}^{\gamma\gamma'}(r) \, {}^V G_{i'n'}^\gamma(r), \quad r \geq a, \qquad (4.247)$$

where we have written $r \equiv r_N$ to simplify the notation. In these equations l_i is the orbital angular momentum of the Nth electron in the ith channel,

$$k_{in}^2 = 2(E^V - E_i + n\omega), \qquad (4.248)$$

where E_i is the energy of the atomic target state defined by

$$\langle \psi_i | H_0 | \psi_j \rangle = E_i \delta_{ij}, \tag{4.249}$$

and $^{\mathrm{V}} W_{ini'n'}^{\gamma\gamma'}(r)$ is a long-range potential coupling the channels.

Equations (4.247) must be solved subject to boundary conditions at $r = a$ and $r \to \infty$. At $r = a$, the matching of the internal and external region wave functions yields the relation

$$\mathsf{E}^{\mathrm{L}} = \mathsf{E}^{\mathrm{V}} + U_{\mathrm{p}}, \tag{4.250}$$

where U_{p} is the electron ponderomotive energy. This result agrees with Equation (4.39) as there is only one electron in the external region. This matching also provides $^{\mathrm{V}}\mathbf{R}(\mathsf{E}^{\mathrm{V}})$, the R-matrix in the velocity gauge which relates the reduced radial functions $^{\mathrm{V}} G_{in}^{\gamma}$ and their derivative $\mathrm{d}^{\mathrm{V}} G_{in}^{\gamma}/\mathrm{d}r$ on the boundary $r = a$, and provides the boundary condition satisfied by the solutions of Equation (4.247).

The coupled equations (4.247) can be solved [66] from $r = a$ to a sufficiently large distance $r = \bar{a}$ by propagating the R-matrix $^{\mathrm{V}}\mathbf{R}(\mathsf{E}^{\mathrm{V}})$ using standard propagators [67,68]. The solutions obtained in this way are matched at $r = \bar{a}$ with those satisfying given boundary conditions for $r \to \infty$ calculated using asymptotic expansions.

The boundary conditions for $r \to \infty$ are conveniently formulated in the Kramers–Henneberger (K–H) accelerated frame because in this frame the channels decouple asymptotically, as we have seen in Section 4.5. The transformation to the K–H frame can be made at any large radius $r = a'$ such that $a' \geq \bar{a}$. It is accomplished by performing the unitary transformation

$$\Psi^{\mathrm{A}}(X,t) = \exp[-\mathrm{i}\mathbf{p}_N \cdot \boldsymbol{\alpha}(t)]\Psi^{\mathrm{V}}(X,t), \quad r_i \leq a \quad (i = 1, 2, \ldots, N-1), \; r_N \geq a', \tag{4.251}$$

where, according to Equations (2.113) and (4.223),

$$\boldsymbol{\alpha}(t) = \int_{-\infty}^{t} \mathbf{A}(t')\mathrm{d}t'$$
$$= \hat{\boldsymbol{\epsilon}}\alpha_0 \cos(\omega t). \tag{4.252}$$

The TDSE (4.241) is then replaced by

$$\mathrm{i}\frac{\partial}{\partial t}\Psi^{\mathrm{A}}(X,t) = [H^{\mathrm{A}} + \boldsymbol{\mathcal{E}}(t) \cdot \mathbf{R}']\Psi^{\mathrm{A}}(X,t), \tag{4.253}$$

where

$$H^{\mathrm{A}} = H_0^{N-1} - \frac{1}{2}\nabla_{\mathbf{r}_N}^2 - \frac{Z}{|\mathbf{r}_N + \boldsymbol{\alpha}(t)|} + \sum_{i=1}^{N-1}\frac{1}{|\mathbf{r}_N + \boldsymbol{\alpha}(t) - \mathbf{r}_i|}. \tag{4.254}$$

In this equation, H_0^{N-1} is the field-free Hamiltonian of the nucleus and the $(N-1)$ inner electrons. As seen from Equation (4.254), the Hamiltonian H^{A} reduces to the field-free Hamiltonian H_0 when the radial coordinate $r_N \to \infty$. Thus, as we have seen in Section 4.5, the channels decouple asymptotically in the K–H frame, so that the asymptotic boundary conditions can be easily formulated.

In order to solve Equation (4.253), we introduce, as above, the Floquet–Fourier expansion

$$\Psi^A(X, t) = \exp(-iE^A t) \sum_{n=-\infty}^{+\infty} \exp(-in\omega t) F_n^A(X), \qquad (4.255)$$

and, upon substitution into Equation (4.253), we obtain for the harmonic components F_n^A the system of coupled equations

$$(H^A - n\omega) F_n^A(X) + H'_+ F_{n-1}^A(X) + H'_- F_{n+1}^A(X) = E^A F_n^A(X). \qquad (4.256)$$

Again, the functions F_n^A can be regarded as the components of a vector $|F^A\rangle$ in Fourier space. The system of equations (4.256) can be solved by introducing for the harmonic components F_n^A a close-coupling expansion similar to that of Equation (4.246), namely

$$F_n^A(X) = \sum_{\gamma i} \phi_i^\gamma(r_N^{-1}) r_N^{-1} {}^A G_{in}^\gamma(r_N), \quad r_N \ge a'. \qquad (4.257)$$

It is found that the reduced radial functions ${}^A G_{in}^\gamma$ must satisfy a system of equations of the form

$$\left[\frac{d^2}{dr^2} - \frac{l(l_i + 1)}{r^2} + \frac{2(Z - N + 1)}{r} + \bar{k}_{in}^2 \right] {}^A G_{in}^\gamma(r)$$
$$= 2 \sum_{\gamma' i' n'} {}^A W_{ini'n'}^{\gamma\gamma'}(r) \, {}^A G_{i'n'}^\gamma(r), \quad r \ge a', \qquad (4.258)$$

where

$$\bar{k}_{in}^2 = 2(E^A - E_i + n\omega) \qquad (4.259)$$

and ${}^A W_{ini'n'}^{\gamma\gamma'}(r)$ is a long-range potential coupling the channels.

In order to solve Equations (4.258) in the region $r \ge a'$, we need to determine the boundary conditions satisfied by the functions ${}^A G_{in}^\gamma$ at $r = a'$. These boundary conditions can be obtained by propagating the solutions ${}^V G_{in}^\gamma$ of Equations (4.247) in the velocity gauge from $r = a$ to $r = a'$ and then transforming the resulting solutions from the velocity gauge to the accelerated frame. The matching of the external region solutions at $r = a'$ yields the relation

$$E^A = E^V \qquad (4.260)$$

and also gives ${}^A R(E^A)$, the R-matrix in the accelerated frame.

Although the unitary transformation (4.251) to the accelerated frame can, in principle, be performed at any radius $r = a'$ such that $a' \geq \bar{a}$, it turns out that in practice this transformation is hard to implement accurately at finite r. It is therefore usually performed at $r = \infty$ and incorporated into the asymptotic expansion of the external region solution [66].

4.7.4 Asymptotic boundary conditions

We shall now discuss the boundary conditions for $r \to \infty$ which must be satisfied by the external region solution. We assume that we are working in the K–H frame where the channels decouple asymptotically.

These asymptotic boundary conditions differ according to the process considered. In the case of laser-assisted electron–atom collisions, the solutions of Equations (4.258) satisfy the asymptotic boundary conditions

$$G_{\nu\nu'}(r) \underset{r \to \infty}{\to} k_\nu^{-1/2}(\sin\theta_\nu \delta_{\nu\nu'} + \cos\theta_\nu K_{\nu\nu'}), \quad k_\nu^2 \geq 0,$$

$$\underset{r \to \infty}{\to} \exp(-|k_\nu|r)N_{\nu\nu'}, \quad k_\nu^2 < 0, \tag{4.261}$$

where we assume that there are m coupled channels of which m_0 are open ($k_\nu^2 > 0$) and $m - m_0$ are closed ($k_\nu^2 < 0$), and we have dropped the superscript A of the functions $G_{\nu\nu'}$ for notational simplicity. For convenience, the open channels are ordered first. We have also combined the channel indices (γin) in Equations (4.258) into a single index ν in Equation (4.261) and we have introduced a second index ν' to denote the m_0 linearly independent solutions. In addition, we have

$$\theta_\nu = k_\nu r - l_i \pi/2 - \eta_\nu \log(2k_\nu r) + \sigma_\nu, \tag{4.262}$$

where

$$\eta_\nu = -\frac{Z - N + 1}{k_\nu} \tag{4.263}$$

is the corresponding Sommerfeld parameter (in a.u.) and

$$\sigma_\nu = \arg \Gamma(l_i + 1 + i\eta_\nu). \tag{4.264}$$

Equation (4.261) then defines an ($m_0 \times m_0$)-dimensional K-matrix from which the ($m_0 \times m_0$)-dimensional S-matrix can be obtained from the matrix equation

$$\mathbf{S} = \frac{\mathbf{I} + i\mathbf{K}}{\mathbf{I} - i\mathbf{K}}. \tag{4.265}$$

The cross sections are then given in terms of the matrix elements of $\mathbf{S}$.

In order to relate the ($m_0 \times m_0$)-dimensional matrix $\mathbf{K}$ of Equation (4.261) to the ($m \times m$)-dimensional R-matrix $^A\mathbf{R}$ obtained in the accelerated frame, we introduce, following Burke, Hibbert and Robb [69], $m + m_0$ linearly independent solutions $v_{\nu\nu'}$ of Equations (4.258) satisfying the boundary conditions

$$v_{\nu\nu'}(r) \underset{r\to\infty}{\to} \sin\theta_\nu \delta_{\nu\nu'}, \quad \nu = 1, ..., m, \quad \nu' = 1, ..., m_0,$$

$$\underset{r\to\infty}{\to} \cos\theta_\nu \delta_{\nu\nu'-m_0}, \quad \nu = 1, ..., m, \quad \nu' = m_0+1, ..., 2m_0,$$

$$\underset{r\to\infty}{\to} \exp(-|k_\nu|r)\delta_{\nu\nu'-m_0}, \quad \nu = 1, ..., m, \quad \nu' = 2m_0+1, ..., m+m_0.$$

$$(4.266)$$

We can then write

$$G_{\nu\nu'}(r) = \sum_{\nu''=1}^{m+m_0} v_{\nu\nu''}(r) x_{\nu''\nu'}, \quad \nu = 1, ..., m, \quad \nu' = 1, ..., m_0, \quad (4.267)$$

where the coefficients $x_{\nu''\nu'}$ satisfy the linear simultaneous equations

$$x_{\nu\nu'} = k_\nu^{-1/2}\delta_{\nu\nu'}, \quad \nu = 1, ..., m_0,$$

$$\sum_{\nu''=1}^{m+m_0} \left[v_{\nu\nu''}(a') - \sum_{\nu'''=1}^{m} R_{\nu\nu'''} \left(a'\frac{\mathrm{d}v_{\nu'''\nu''}}{\mathrm{d}r} - bv_{\nu'''\nu''} \right)_{r=a'} \right] x_{\nu''\nu'} = 0, \quad \nu = 1, ..., m,$$

$$(4.268)$$

which must be solved for $\nu' = 1, ..., m_0$. It follows from Equation (4.261) that

$$K_{\nu\nu'} = k_\nu^{1/2} x_{\nu+m_0\nu'}, \quad \nu, \nu' = 1, ..., m_0. \quad (4.269)$$

In the case of multiphoton ionization and harmonic generation the above procedure must be modified since there are only outgoing waves corresponding to Siegert boundary conditions. We introduce m linearly independent solutions $w_{\nu\nu'}$ of Equations (4.258) satisfying the boundary conditions

$$w_{\nu\nu'}(r) \underset{r\to\infty}{\to} \exp(ik_\nu r)\delta_{\nu\nu'}, \quad \nu, \nu' = 1, ..., m. \quad (4.270)$$

Equations (4.267) and (4.268) are then replaced by the equations

$$G_\nu(r) = \sum_{\nu'=1}^{m} w_{\nu\nu'}(r) x_{\nu'}, \quad \nu = 1, ..., m \quad (4.271)$$

and

$$\sum_{\nu'=1}^{m} \left[w_{\nu\nu'}(a') - \sum_{\nu''=1}^{m} R_{\nu\nu''} \left(a'\frac{\mathrm{d}w_{\nu''\nu'}}{\mathrm{d}r} - bw_{\nu''\nu'} \right)_{r=a'} \right] x_{\nu'} = 0, \quad (4.272)$$

respectively, where we have dropped the second index on G_ν and $x_{\nu'}$ since there is now only one solution. This solution must be found by an iterative procedure since Equation (4.272) is homogeneous and hence only has solutions when the determinant of the matrix in square brackets vanishes. This occurs for complex values of the quasi-energy, which we write in the form of Equation (4.189). From our knowledge of the coefficients $x_{v'}$, that is of the eigenvectors, one can obtain all the other physical observables, such as the branching ratios into the channels and the angular distribution of the ejected electrons.

The RMF theory, discussed above for the case of a linearly polarized, monochromatic field has been generalized [70] to incommensurate bichromatic laser fields of the form

$$\mathcal{E}(t) = \hat{\boldsymbol{\epsilon}}\left[\mathcal{E}_1 \cos(\omega_1 t) + \mathcal{E}_2 \cos(\omega_2 t + \phi)\right], \qquad (4.273)$$

where $\hat{\boldsymbol{\epsilon}}$ is the common unit polarization vector, $\mathcal{E}_1$ and $\mathcal{E}_2$ are the amplitudes of the electric fields oscillating with the angular frequencies ω_1 and ω_2, respectively, and ϕ is a phase. This extension of the RMF theory has led to the study of a variety of two-color multiphoton ionization processes.

Another generalization of the RMF theory [71] has been to allow *two* electrons to be present in the external region. In this way, processes such as multiphoton double ionization of atoms and ions, and laser-assisted electron–atom (ion) single-ionization collision processes can be studied within the framework of the RMF theory.

4.7.5 Applications

The RMF theory has been used to analyze multiphoton processes in a number of atomic systems. We shall consider here a few examples of multiphoton ionization calculations. Applications to harmonic generation and laser-assisted electron–atom collisions will be considered in Chapters 9 and 10, respectively.

The results discussed below have all been obtained for linearly polarized, monochromatic and homogeneous laser fields as described by Equation (4.223). We begin by considering the multiphoton ionization of H(1s) at an angular frequency $\omega = 0.184$ a.u. corresponding to a KrF laser. At low intensities, three-photon absorption is required for ionization to occur. The RMF results of Dörr *et al.* [72] for the total ionization rate are shown in Fig. 4.4(a); they are in excellent agreement with the values obtained by using the Sturmian-Floquet method [3]. In Fig. 4.4(b), we display the RMF branching ratios into the dominant ionization channels for the absorption of three photons; they are seen to agree with the results calculated at the intensity of 10^{14} W cm^{-2} by using the Sturmian-Floquet method [3]. In both Figs. 4.4(a) and (b) we note the striking difference between the lowest-order perturbation theory (LOPT) results and the non-perturbative RMF values.

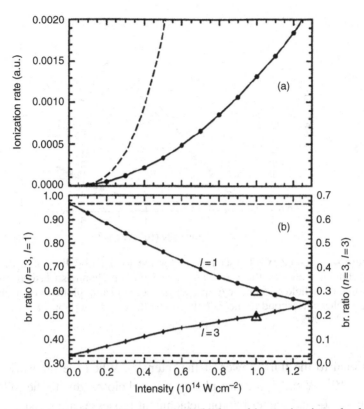

Figure 4.4. (a) Total ionization rate and (b) branching ratios into the lowest ionization channel where $n = 3$ photons have been absorbed, resolved into the angular momentum components, as a function of intensity, for H(1s) in a laser field of angular frequency $\omega = 0.184$ a.u. The solid lines correspond to the RMF calculations. The results of lowest-order perturbation theory (LOPT) are given by the broken lines. The triangles in plot (b) indicate the results obtained using the Sturmian-Floquet method. (From M. Dörr *et al.*, *J. Phys. B* **26**, L275 (1993).)

We now consider two-electron atoms and ions, for which the RMF theory has been applied to calculate non-perturbative multiphoton ionization rates taking into account electron correlation and exchange effects. As a first example, we show in Fig. 4.5 the RMF results of Dörr *et al.* [73] for the multiphoton detachment of H$^-$, obtained at an angular frequency $\omega = 0.0149$ a.u. such that at least two photons are necessary to detach an electron from H$^-$. The total detachment rate is shown, as are the partial rates into the two- and three-photon detachment channels. At low intensities, the total detachment rate increases perturbatively, with the second power of the intensity, since the dominant detachment channel is the two-photon channel. At these low intensities, the three-photon partial rate is very small, being

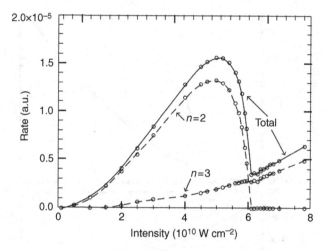

Figure 4.5. *R*-matrix–Floquet results for the total (solid line) and the partial detachment rates into the $n = 2$ and $n = 3$ photon channels (dashed lines), for H⁻ in a laser field of angular frequency $\omega = 0.0149$ a.u., plotted versus intensity. The circles correspond to the calculated results. (From M. Dörr *et al.*, *J. Phys. B* **28**, 4481 (1995).)

proportional to the third power of the intensity. When the intensity reaches the value $6 \times 10^{10}\,\mathrm{W\,cm^{-2}}$, the two-photon channel closes, due to the AC Stark shift, and only three- and higher photon detachment processes are possible. Above this intensity, the difference between the total detachment rate and the partial rate into the three-photon detachment channel is due to higher-order processes.

As a second example of two-electron systems in a laser field, we show in Fig. 4.6 the RMF ionization rate into the two-photon channel, versus photoelectron energy, for He in a laser field of intensity $10^{12}\,\mathrm{W\,cm^{-2}}$, obtained by Purvis *et al.* [74]. The Rydberg series of resonance-enhanced multiphoton ionization (REMPI) peaks visible below the $n = 1$ threshold corresponds to one-photon resonances due to intermediate ¹P bound states, while the REMPI peaks below the $n = 2$ threshold are two-photon resonances due to ¹S and ¹D auto-ionizing states.

R-matrix–Floquet calculations for helium have also been carried out by Glass and Burke [75] at the longer wavelength, $\lambda = 248\,\mathrm{nm}$, corresponding to the KrF laser, where at low intensities at least five photons are required to ionize the helium atom. As illustrated in Fig. 4.7, at low intensities non-resonant five-photon ionization occurs and the ionization rate follows closely the perturbative power law I^5 for the dependence on intensity. Channel closing occurs at an intensity of $6.7 \times 10^{13}\,\mathrm{W\,cm^{-2}}$, while at higher intensities six-photon ionization is the dominant process and REMPI peaks due to the influence of intermediate resonances on the total ionization rate are seen. Just above the intensity of $6.7 \times 10^{13}\,\mathrm{W\,cm^{-2}}$ at which

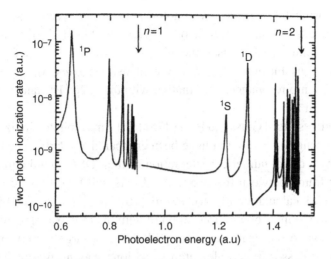

Figure 4.6. *R*-matrix–Floquet results for the two-photon ionization rate of He versus photoelectron energy at a laser intensity of $10^{12}\,\mathrm{W\,cm^{-2}}$. The positions of the $\mathrm{He^+}(n=1)$ and $\mathrm{He^+}(n=2)$ thresholds are indicated by the arrows. Below the $\mathrm{He^+}(n=1)$ threshold, the resonances are due to intermediate 1snp ^{1}P bound states. Below the $\mathrm{He^+}(n=2)$ threshold, the ^{1}S and ^{1}D auto-ionizing resonances appear. (From J. Purvis *et al.*, *Phys. Rev. Lett.* **71**, 3943 (1993).)

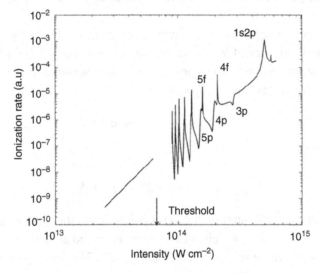

Figure 4.7. Total *R*-matrix–Floquet ionization rate of He versus intensity for a laser field of wavelength $\lambda = 248$ nm. The arrow marks the intensity where a channel closing occurs. (From D. H. Glass and P. G. Burke, *J. Phys. B* **33**, 407 (2000).)

channel closing takes place, five-photon resonances occur between the ground state and excited bound states of odd parity, and the sixth photon ionizes these states. As the intensity increases, the fifth photon sweeps through Rydberg series of these resonances. Further increases of the intensity progressively bring lower-lying bound states into resonance, culminating with the $1s2p$ $^1P^o$ state at an intensity of 5×10^{14} W cm^{-2}.

The RMF results of Glass and Burke [75] for the multiphoton ionization of helium at the wavelength $\lambda = 248$ nm have been compared by Parker *et al.* [76] to those obtained by a direct numerical integration of the TDSE. Good agreement (within 10%) was found between the two methods. As will be discussed in Section 5.5, the benchmark calculation of Parker *et al.* [76] has been extended by van der Hart *et al.* [77] to the multiphoton ionization of helium at the wavelength $\lambda = 390$ nm.

In addition to the examples discussed above, a large variety of multiphoton ionization processes in complex atoms and ions (having two or more electrons) have been studied by using the RMF theory. These include resonant multiphoton ionization of the noble gases He, Ne and Ar [78–85], of Ca [86] and Sr [87], multiphoton detachment of the negative ions H$^-$, Li$^-$, Na$^-$, K$^-$, F$^-$ and Cl$^-$ [88–93], multiphoton ionization of the highly charged ion Ar^{7+} [94], of interest for the development of X-ray lasers, and multiphoton emission of inner electrons from Li$^-$ [95] as well as from excited ^1S states of He [96, 97] subjected to VUV radiation.

A particularly interesting effect which has been predicted by using the RMF theory is the occurrence of *laser-induced degenerate states* (LIDS) involving auto-ionizing states in complex atoms or ions [79]. To understand the LIDS mechanism, we consider the following two processes:

$$e^- + A^+ \rightarrow A_1^{**} \rightarrow e^- + A^+ \tag{4.274}$$

and

$$n\hbar\omega + A \rightarrow A_2^{**} \rightarrow e^- + A^+. \tag{4.275}$$

The upper process of Equation (4.274) corresponds to the scattering of an electron e^- by a positive ion A$^+$, which occurs via an intermediate auto-ionizing state A$_1^{**}$ lying in the continuum with a complex energy E_1. The lower process of Equation (4.275) corresponds to multiphoton ionization of the corresponding atom A which takes place through a *laser-induced continuum structure* (LICS) state A$_2^{**}$ [98, 99], which is also embedded in the continuum with a complex energy E_2. By varying the laser angular frequency ω and intensity I, both the real and imaginary parts of the complex energies E_1 and E_2 can be made degenerate, giving rise to a double pole in the corresponding laser-assisted electron–ion scattering S-matrix. Double

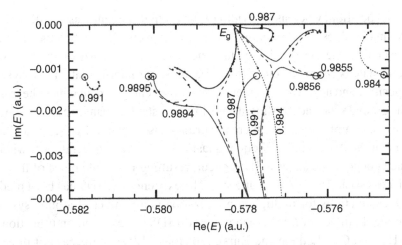

Figure 4.8. Trajectories of the complex quasi-energies for the ground state and the $3s3p^6\,4p\,^1P^o$ auto-ionizing state of argon, calculated by using the RMF theory, for laser intensities varying from zero to $5 \times 10^{13}\,\mathrm{W\,cm^{-2}}$. The values of the laser angular frequency ω are indicated next to the trajectories. For each angular frequency ω, there are two trajectory curves: one corresponding to the ground state and the other to the auto-ionizing state. (From O. Latinne *et al.*, *Phys. Rev. Lett.* **74**, 46 (1995).)

poles in laser-assisted electron–atom (ion) collisions will be discussed in Section 10.3.4.

As an example, we show in Fig. 4.8 the results of the first RMF calculation, performed by Latinne *et al.* [79], in which the existence of LIDS was reported. In this work, the interaction between the $3s3p^64p\,^1P^o$ auto-ionizing state of Ar and the $3s^23p^6\,^1S^e$ ground state of Ar dressed by one photon was analyzed. The resulting trajectories of the complex Floquet quasi-energies E_1 and $E_2 = E_2$ (both shifted down by the laser angular frequency ω) are displayed in Fig. 4.8, for laser intensities I ranging from zero to $5 \times 10^{13}\,\mathrm{W\,cm^{-2}}$, and for several fixed values of the angular frequency ω, chosen in the vicinity of 0.99 a.u., which corresponds to a one-photon resonance between the $3s^23p^6\,^1S^e$ ground state and the $3s3p^64p\,^1P^o$ auto-ionizing state of Ar. The zero-field position of the Ar ground state lies on the real axis at $E_g = -0.57816$ a.u., while the shifted zero-field position of the auto-ionizing state (denoted by circles in Fig. 4.8) changes with ω and lies at a complex energy of $(0.40936 - i0.00119 - \omega)$ a.u. $= E_a - i\Gamma_a/2 - \omega$, where $\Gamma_a = 2 \times 0.00119$ a.u. is the zero-field width. For each angular frequency there are two trajectories: one connected adiabatically to the zero-field position of the ground state, and the other with the shifted zero-field position of the auto-ionizing state. The detuning from resonance is defined as $\delta = E_a - E_g - \omega$. At large values of $|\delta|$ (e.g. $\omega = 0.984$ a.u. or $\omega = 0.991$ a.u.), the auto-ionizing state does not move much from its position, whereas the width of the ground state increases with intensity. At very small values of $|\delta|$ (e.g. $\omega = 0.987$ a.u.), just the opposite happens: the curve connected to the auto-ionizing state increases in width with intensity, while the ground state is "trapped" close to the real axis. For intermediate detunings, both on the positive and negative side, there is a complex energy about which the trajectories of the ground state and the shifted auto-ionizing state exchange their roles. At each of these two complex energies, there is a critical laser angular frequency and intensity such that the two complex Floquet quasi-energies are degenerate, i.e. where laser-induced degenerate states (LIDS) occur, resulting in double poles of the S-matrix for laser-assisted $e^- - Ar^+$ scattering. The existence of LIDS has been predicted in RMF calculations for multiphoton ionization in a number of atomic systems. For example, Latinne *et al.* [79] also reported LIDS for two-photon transitions in H^- and He, and Cyr, Latinne and Burke [80] found LIDS in the case of three-photon ionization of Ar. Laser-induced degenerate states have also been found in two-color multiphoton ionization, for example in Sturmian-Floquet calculations for H [100, 101] and in RMF calculations for Mg [102].

The LIDS phenomenon is therefore a general one, which can be understood [79] by constructing models that retain the essential ingredients of the full RMF calculations. We can also infer from Fig. 4.8 that it is possible in principle to use the existence of LIDS for achieving population transfer from the ground state to an auto-ionizing state by adiabatic adjustments of the laser angular frequency ω and intensity I. Looking at Fig. 4.8, we see that this could be achieved by (i) choosing $\omega = 0.9856$ a.u. and increasing I from zero to 2×10^{13} W cm^{-2}, (ii) changing ω slightly, from 0.9856 a.u. to 0.9855 a.u. and (iii) decreasing I back to zero. These operations should be timed precisely due on one side to the constraint of adiabaticity (the changes in angular frequency and intensity must not be too fast) and on the other side by the fact that the atomic system is decaying in the laser field. Instead of step (iii), further adjustments of ω and I could allow a circuit to be completed around the degeneracy by following an adiabatic path. In this way, LIDS can be viewed [79] as an extension of the work of Berry [103], where the adiabatic passage around degeneracies in a parameter space was described. In the case of the LIDS illustrated in Fig. 4.8, the parameter space is two-dimensional and is characterized by the laser angular frequency and intensity.

Let us now consider two-color calculations of multiphoton ionization of atoms and ions with more than one electron. The RMF theory has been used to perform such calculations for He [70, 104], Mg [102] and Ar^{7+} [105]. In particular, LICS in He have been studied by van der Hart [70] and by Kylstra, Paspalakis and Knight [104]. The latter analyzed the case in which a Nd:YAG laser is used to embed the $1s4s$ $^1S^e$ Rydberg state into the continuum. Their RMF results are in good agreement with the experimental and theoretical values of Halfmann *et al.* [106].

Another interesting application of two-color ionization processes is that of multiply resonant multiphoton ionization involving auto-ionizing states of complex atoms. For example, Kylstra *et al.* [102] used the RMF theory to study the two-color doubly resonant multiphoton ionization process depicted in Fig. 4.9(a). A high-frequency laser field of angular frequency ω_H couples the $3s^2\,^1S^e$ ground state of Mg and the $3p^2\,^1S^e$ auto-ionizing state by a two-photon process which is nearly resonant with the intermediate bound state $3s3p\,^1P^o$. A second, low-frequency laser field of angular frequency ω_L couples the $3p^2\,^1S^e$ and the $3p3d\,^1P^o$ auto-ionizing states. The RMF ionization rates calculated by Kylstra *et al.* [102] are shown in Fig. 4.9(b) as a function of ω_L, for three different values of ω_H, and for fixed intensities $I_L = 2 \times 10^9\,\mathrm{W\,cm^{-2}}$ and $I_H = 5 \times 10^9\,\mathrm{W\,cm^{-2}}$ of the two laser fields. They are seen to be in good agreement with the experimental data of Karapanagioti *et al.*

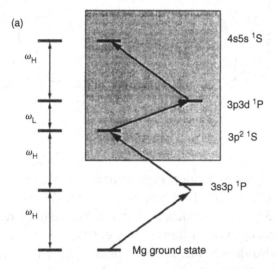

Figure 4.9. (a) A high-frequency laser field of angular frequency ω_H couples the $3s^2\,^1S^e$ ground state of Mg and the $3p^2\,^1S^e$ auto-ionizing state by a two-photon process which is nearly one-photon resonant with the intermediate $3s3p\,^1P^o$ bound state. A low-frequency laser field of angular frequency ω_L couples the $3p^2\,^1S^e$ auto-ionizing state and the $3p3d\,^1P^o$ auto-ionizing state which, in turn, induces a resonant coupling between the $3p3d\,^1P^o$ and the $4s5s\,^1S^e$ auto-ionizing states. (b) The experimental ionization yields of Karapanagioti *et al.* [107–109] (left) and the RMF calculated values of Kylstra *et al.* [102] (right) in the presence of the two laser fields, as a function of ω_L for three different values of ω_H. The quantity $\delta_H = E_g - E_a + 2\omega_H$ denotes the detuning from resonance, where E_g and E_a refer to the field-free energies of the ground state and the $3p^2\,^1S^e$ auto-ionizing state, respectively. (From N. J. Kylstra, H. W. van der Hart, P. G. Burke and C. J. Joachain, *J. Phys. B* **31**, 3089 (1998).)

(b)

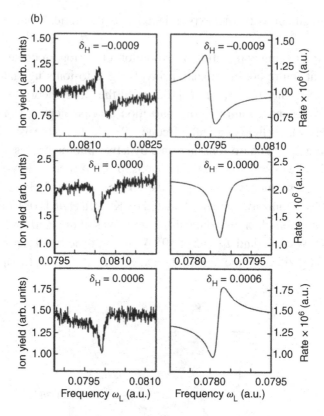

Figure 4.9. (*cont.*)

[107–109]. At higher intensities, higher-order processes become non-negligible. In particular, Kylstra *et al.* [102] have performed RMF calculations including the effects of the $4s5s\ ^1S^e$ state, which, as seen from Fig. 4.9(a), is resonantly coupled to the $3p3d\ ^1P^o$ state by the high-frequency laser field, thus inducing a triply resonant process. They also showed that LIDS occur among the three auto-ionizing states, and studied their effect on the decay rate of the ground state.

References

[1] G. Floquet, *Ann. Ec. Norm.* (2) **13**, 47 (1883).
[2] S. I. Chu, *Adv. At. Mol. Phys.* **21**, 197 (1985).
[3] R. M. Potvliege and R. Shakeshaft, *Adv. At. Mol. Opt. Phys. Suppl.* **1**, 373 (1992).
[4] C. J. Joachain, M. Dörr and N. J. Kylstra, *Adv. At. Mol. Opt. Phys.* **42**, 225 (2000).
[5] F. Bloch, *Z. Physik* **52**, 5055 (1928).
[6] N. W. Ashcroft and N. D. Mermin, *Solid State Physics* (Orlando, Fl.: Harcourt, 1976).
[7] C. Kittel, *Introduction to Solid State Physics* (New York: Wiley, 1996).
[8] B. H. Bransden and C. J. Joachain, *Quantum Mechanics* 2nd edn, (Harlow, UK: Prentice Hall-Pearson, 2000).

[9] J. H. Shirley, *Phys. Rev. B* **138**, 979 (1965).
[10] M. H. Mittleman, *Introduction to the Theory of Laser–Atom Interactions*, 2nd edn (New York: Plenum Press, 1993).
[11] R. M. Potvliege, *Laser Phys.* **10**, 143 (2000).
[12] J. Bergou and S. Varró, *J. Phys. A* **13**, 3553 (1980).
[13] K. Drühl and J. K. McIver, *J. Math. Phys.* **24**, 705 (1983).
[14] F. Bloch and A. Siegert, *Phys. Rev.* **57**, 522 (1940).
[15] L. Allen and J. H. Eberly, *Optical Resonance and Two-level Atoms* (New York: Dover, 1987).
[16] B. Gao and A. F. Starace, *Phys. Rev. Lett.* **61**, 404 (1988).
[17] R. M. Potvliege and R. Shakeshaft, *Phys. Rev. A* **39**, 1545 (1989).
[18] R. M. Potvliege and R. Shakeshaft, *Z. Phys. D* **11**, 93 (1989).
[19] M. Pont and R. Shakeshaft, *Phys. Rev. A* **43**, 3764 (1991).
[20] L. D. Landau, *Phys. Z. Soviet Union* **2**, 46 (1932).
[21] C. Zener, *Proc Roy. Soc. A* **137**, 696 (1932).
[22] E. C. G. Stueckelberg, *Helv. Phys. Acta* **5**, 369 (1932).
[23] B. H. Bransden and M. R. C. McDowell, *Charge Exchange and the Theory of Ion-Atom Collisions* (Oxford: Oxford University Press, 1992).
[24] H. C. Day, B. Piraux and R. M. Potvliege, *Phys. Rev. A* **61**, 031402(R) (2000).
[25] N. J. Kylstra and C. J. Joachain, *Phys. Rev. A* **60**, 2255 (1999).
[26] A. J. F. Siegert, *Phys. Rev. A* **56**, 750 (1939).
[27] B. H. Bransden and C. J. Joachain, *Physics of Atoms and Molecules*, 2nd edn (Harlow, UK: Prentice Hall-Pearson, 2003).
[28] J. Aguilar and J. M. Combes, *Commun. Math. Phys.* **22**, 269 (1971).
[29] E. Balslev and J. M. Combes, *Commun. Math. Phys.* **22**, 280 (1971).
[30] J. S. Howland, in J. A. DeSanto, A. W. Sáenz and W. W. Zachary, eds. *Mathematical Methods and Applications of Scattering Theory*, Lecture Notes in Physics (Berlin: Springer-Verlag, 1980).
[31] K. Yajima, *Commun. Math. Phys.* **87**, 331 (1982).
[32] S. Graffi, V. Grecchi and H. Silverstone, *Ann. Inst. Henri Poincaré – Phys. Théor.* **42**, 215 (1985).
[33] J. H. Wilkinson, *The Algebraic Eigenvalue Problem* (Oxford: Clarendon Press, 1965).
[34] M. Dörr, R. M. Potvliege and R. Shakeshaft, *Phys. Rev. A* **41**, 558 (1990).
[35] R. M. Potvliege and R. Shakeshaft, *Phys. Rev. A* **40**, 3061 (1989).
[36] Y. Gontier and M. Trahin, *Phys. Rev. A* **19**, 264 (1979).
[37] C. R. Holt, M. G. Raymer and W. P. Reinhardt, *Phys. Rev. A* **27**, 2971 (1983).
[38] R. J. Eden and J. R. Taylor, *Phys. Rev. B* **133**, 1575 (1964).
[39] C. J. Joachain, *Quantum Collision Theory*, 3rd edn (Amsterdam: North Holland, 1983).
[40] I. J. Berson, *J. Phys. B* **8**, 3078 (1975).
[41] R. M. Potvliege, *Phys. Scripta* **68**, C18 (2003).
[42] R. M. Potvliege and R. Shakeshaft, *Phys. Rev. A* **38**, 6190 (1988).
[43] P. Schlagheck, K. Hornberger and A. Buchleitner, *Phys. Rev. Lett.* **82**, 664 (1999).
[44] R. Bhatt, B. Piraux and K. Burnett, *Phys. Rev. A* **37**, 98 (1988).
[45] J. N. Bardsley, A. Szöke and M. Comella, *J. Phys. B* **21**, 3899 (1988).
[46] M. Dörr and R. M. Potvliege, *Phys. Rev. A* **41**, 1472 (1990).
[47] M. Dörr, R. M. Potvliege, D. Proulx and R. Shakeshaft, *Phys. Rev. A* **43**, 3729 (1991).
[48] A. S. Fearnside, R. M. Potvliege and R. Shakeshaft, *Phys. Rev. A* **51**, 1471 (1995).
[49] M. Rotenberg, *Adv. At. Mol. Phys.* **6**, 233 (1970).
[50] A. Maquet, S. I. Chu and W. P. Reinhardt, *Phys. Rev. A* **27**, 2946 (1983).

[51] R. Shakeshaft, *Phys. Rev. A* **34**, 244, 5119 (1986).

[52] R. M. Potvliege and R. Shakeshaft, *Phys. Rev. A* **38**, 1098 (1988).

[53] R. R. Freeman, P. H. Bucksbaum, H. Milchberg, S. Darack, D. Schumacher and M. E. Geusic, *Phys. Rev. Lett.* **59**, 1092 (1987).

[54] P. G. Burke, P. Francken and C. J. Joachain, *Europhys. Lett.* **13**, 617 (1990).

[55] P. G. Burke, P. Francken and C. J. Joachain, *J. Phys. B* **24**, 761 (1991).

[56] C. J. Joachain, In P. Lambropoulos and H. Walther, eds., *Multiphoton Processes 1996* (Bristol: Institute of Physics, 1997), p. 46.

[57] C. J. Joachain, M. Dörr and N. J. Kylstra, *Comments At. Mol. Phys.* **33**, 247 (1997).

[58] C. J. Joachain, *J. Mod. Opt. Suppl. 1* **54**, 15 (2007).

[59] E. P. Wigner, *Phys. Rev.* **70**, 15 (1946).

[60] E. P. Wigner and L. Eisenbud, *Phys. Rev.* **72**, 29 (1947).

[61] P. G. Burke, *R-Matrix Theory of Atomic Collisions* (Heideberg: Springer, 2011).

[62] C. Bloch, *Nucl. Phys.* **4**, 503 (1957).

[63] P. G. Burke and K. A. Berrington, eds., *Atomic and Molecular Processes: An R-matrix Approach* (Bristol: Institute of Physics, 1993).

[64] P. J. A. Buttle, *Phys. Rev.* **160**, 719 (1967).

[65] P. G. Burke and C. J. Joachain, *Theory of Electron–Atom Collisions. Part.1: Potential Scattering* (New York: Plenum Press, 1995).

[66] M. Dörr, M. Terao-Dunseath, J. Purvis, C. J. Noble, P. G. Burke and C. J. Joachain, *J. Phys. B* **25**, 2809 (1992).

[67] J. C. Light and R. B. Walker, *J. Chem. Phys.* **65**, 4272 (1976).

[68] K. L. Baluja, P. G. Burke and L. A. Morgan, *Comput. Phys. Commun.* **27**, 299 (1982).

[69] P. G. Burke, A. Hibbert and W. D. Robb, *J. Phys. B: At. Mol. Phys.* **4**, 153 (1971).

[70] H. W. van der Hart, *J. Phys. B* **29**, 2217 (1996).

[71] L. Feng and H. W. van der Hart, *Phys. Rev. A* **66**, 031402 (2002).

[72] M. Dörr, P. G. Burke, C. J. Joachain, C. J. Noble, J. Purvis and M. Terao-Dunseath, *J. Phys. B* **26**, L275 (1993).

[73] M. Dörr, J. Purvis, M. Terao-Dunseath, P. G. Burke, C. J. Joachain and C. J. Noble, *J. Phys. B* **28**, 4481 (1995).

[74] J. Purvis, M. Dörr, M. Terao-Dunseath, C. J. Joachain, P. G. Burke and C. J. Noble, *Phys. Rev. Lett.* **71**, 3943 (1993).

[75] D. H. Glass and P. G. Burke, *J. Phys. B* **33**, 407 (2000).

[76] J. S. Parker, D. H. Glass, L. R. Moore, E. S. Smyth, K. T. Taylor and P. G. Burke, *J. Phys. B* **33**, L239 (2000).

[77] H. W. van der Hart, B. J. S. Doherty, J. S. Parker and K. T. Taylor, *J. Phys. B* **38**, L207 (2005).

[78] N. J. Kylstra, M. Dörr, C. J. Joachain and P. G. Burke, *J. Phys. B* **28**, L685 (1995).

[79] O. Latinne, N. J. Kylstra, M. Dörr, J. Purvis, M. Terao-Dunseath, C. J. Joachain, P. G. Burke and C. J. Noble, *Phys. Rev. Lett.* **74**, 46 (1995).

[80] A. Cyr, O. Latinne and P. G. Burke, *J. Phys. B* **30**, 659 (1997).

[81] D. H. Glass, P. G. Burke, H. W. van der Hart and C. J. Noble, *J. Phys. B: At. Mol. Opt. Phys.* **30**, 3801 (1997).

[82] M. Plummer and C. J. Noble, *J. Phys. B* **33**, L807 (2000).

[83] C. McKenna and H. W. van der Hart, *J. Phys. B* **37**, 457 (2004).

[84] H. W. van der Hart and P. Bingham, *J. Phys. B* **38**, 207 (2005).

[85] H. W. van der Hart, *Phys. Rev. A* **73**, 023417 (2006).

[86] C. McKenna and H. W. van der Hart, *J. Phys. B* **36**, 1627 (2003).

[87] M. Madine and H. W. van der Hart, *J. Phys. B* **38**, 1895 (2005).

[88] H. W. van der Hart, *J. Phys. B* **29**, 3059 (1996).

[89] H. W. van der Hart and A. S. Fearnside, *J. Phys. B* **30**, 5657 (1997).

[90] D. H. Glass, P. G. Burke, C. J. Noble and G. B. Wöste, *J. Phys. B* **31**, L667 (1998).

[91] H. W. van der Hart, *J. Phys. B* **33**, 1789 (2000).

[92] N. Vinci, D. H. Glass, K. T. Taylor and P. G. Burke, *J. Phys. B* **33**, 4799 (2000).

[93] N. Vinci, D. H. Glass, H. W. van der Hart, K. T. Taylor and P. Burke, *J. Phys. B* **36**, 1795 (2003).

[94] H. W. van der Hart and L. Feng, *J. Phys. B* **35**, 1185 (2002).

[95] H. W. van der Hart, *Phys. Rev. Lett.* **95**, 153001 (2005).

[96] M. Madine and H. W. van der Hart, *J. Phys. B* **38**, 3963 (2005).

[97] M. Madine and H. W. van der Hart, *J. Phys. B* **39**, 4040 (2006).

[98] P. L. Knight, *Commun. At. Mol. Phys.* **15**, 193 (1984).

[99] P. L. Knight, M. A. Lauder and B. J. Dalton, *Phys. Rep.* **190**, 1 (1990).

[100] R. M. Potvliege and P. H. G. Smith, *J. Phys. B* **24**, L641 (1991).

[101] M. Pont, R. M. Potvliege, R. Shakeshaft and P. H. G. Smith, *Phys. Rev. A* **46**, 555 (1992).

[102] N. J. Kylstra, H. W. van der Hart, P. G. Burke and C. J. Joachain, *J. Phys. B* **31**, 3089 (1998).

[103] M. V. Berry, *Proc. Roy. Soc. London A* **392**, 45 (1984).

[104] N. J. Kylstra, E. Paspalakis and P. L. Knight, *J. Phys. B* **31**, L719 (1998).

[105] E. Costa i Bricha, C. L. S. Lewis and H. W. van der Hart, *J. Phys. B* **37**, 2755 (2004).

[106] T. Halfmann, L. P. Yatsenko, M. Shapiro, B. W. Shore and K. Bergmann, *Phys. Rev. A* **58**, R46 (1998).

[107] N. E. Karapanagioti, O. Faucher, Y. L. Shao, D. Charalambidis, H. Bachau and E. Cormier, *Phys. Rev. Lett.* **74**, 2431 (1995).

[108] N. E. Karapanagioti, D. Charalambidis, C. J. G. Uiterwaal, C. Fotakis, H. Bachau, I. Sanchez and E. Cormier, *Phys. Rev. A* **53**, 2587 (1996).

[109] N. E. Karapanagioti, O. Faucher, C. J. G. Uiterwaal, D. Charalambidis, H. Bachau, I. Sanchez and E. Cormier, *J. Mod. Opt.* **43**, 953 (1996).

5

Numerical integration of the wave equations

One of the major non-perturbative methods used to study atoms in intense laser fields is the direct numerical integration of the wave equations describing atoms interacting with laser fields [1]. This is an attractive alternative to the methods discussed in the two preceding chapters, since solutions of the wave equations can be obtained by numerical integration for a wide range of laser intensities and frequencies. In addition, no restrictions need to be imposed on the type of laser pulses which are used, making the numerical integration of wave equations particularly useful for the study of interaction of atoms with short laser pulses.

However, the numerical integration of the wave equations is computationally very intensive, for the following reasons. Firstly, at high laser intensities, and especially for low frequencies, the ionized electrons can acquire quite high velocities and their quiver motion becomes much larger than the size of the initial atomic orbit. The corresponding wave packets can therefore travel large distances in short time intervals. As a result, the spatial grids used to follow the motion of these wave packets must be large and have small spatial separations. Secondly, the discretization of time, used in all numerical integration schemes, requires a large number of small steps in order to obtain accurate results. Thirdly, the direct numerical integration of the wave equations becomes extremely demanding for atoms with more than one active electron. In recent years, advances in computer technology have allowed the numerical solution of the time-dependent Schrödinger equation (TDSE) for two-electron atoms to be obtained, but the generalization of this approach to the solution of the TDSE for multielectron atoms is beyond the present computer capabilities, although approaches such as the time-dependent R-matrix (TDRM) theory, which will also be discussed in this chapter, have been proposed. Fourthly, once a numerical solution has been obtained, it must be analyzed in order to obtain physical observables. Finally, for super-intense, ultra-short laser pulses, non-dipole and eventually relativistic effects become important. In this case, the TDSE in the dipole approximation must be replaced by other wave equations, such as the non-relativistic, non-dipole TDSE if relativistic effects are negligible, or the time-dependent relativistic Schrödinger equation (TDRSE) or the time-dependent Klein–Gordon equation (TDKGE) if the relativistic motion of the electrons must be accounted for, or the time-dependent Dirac equation (TDDE) if both relativistic and spin effects must be taken into account. The numerical solution of the TDDE is a particularly difficult task, even for one-electron atoms or ions.

In this chapter, we shall describe some of the methods which have been used to solve numerically the wave equations describing atoms in intense laser pulses. We begin in Section 5.1 by discussing time propagation schemes, grids and basis sets, and the choice of gauge used for solving the TDSE. The calculation of observables such as ionization rates and harmonic spectra is considered in Section 5.2. The following three sections are devoted to the numerical solution of the TDSE in the dipole approximation. They deal successively with one- and two-dimensional model atoms, atomic hydrogen and complex atoms. Non-dipole effects are considered in Section 5.6, and the numerical solution of relativistic wave equations is discussed in Section 5.7. Atomic units will be used, unless otherwise stated.

5.1 Discrete-time propagators, representations of the wave function and choice of gauge

5.1.1 Discrete-time propagators

Let us consider an atom interacting with a laser pulse of duration τ_p. We assume that the system is in the state $|\Psi(t_0)\rangle$ at time t_0 when the pulse is applied, and that the angular frequency and peak intensity of the pulse are such that the atom in the laser pulse can be described by the TDSE (2.121) in the dipole approximation. As discussed in Section 2.6, the state of the system at any time t can be expressed as

$$|\Psi(t)\rangle = U(t, t_0)|\Psi(t_0)\rangle, \tag{5.1}$$

where $U(t, t_0)$ is the evolution operator. Obtaining a numerical solution of the TDSE therefore requires that a numerical representation of the evolution operator be constructed. Let us examine how this can be done.

We first divide the interval $[t_0, t_0 + \tau_p]$ into N equal sub-intervals. Defining $\Delta t = \tau_p/N$ and writing $t_i = t_0 + i\Delta t$, $i = 0, 1, \ldots, N$, so that $t_N = t_0 + \tau_P$, the evolution operator can be expressed as the product of N evolution operators:

$$U(t_0 + \tau_p, t_0) = U(t_N, t_{N-1})U(t_{N-1}, t_{N-2})\ldots U(t_1, t_0), \tag{5.2}$$

where we have made successive use of the group property (2.145).

Let us now concentrate on obtaining an approximation for the evolution operator over the time interval $[t_i, t_{i+1}]$ with $0 \leq i \leq N - 1$. Using Equation (2.151) we have

$$U(t_{i+1}, t_i) = I - i \int_{t_i}^{t_{i+1}} H(t_1) U(t_1, t_i) dt_1. \tag{5.3}$$

This equation can be formally solved for $U(t_{i+1}, t_i)$ in exactly the same way that the integral equation (3.19) satisfied by the evolution operator $U_I(t, t_0)$ was solved in Section 3.1. The result is given by the series

$$U(t_{i+1}, t_i) = \sum_{n=0}^{\infty} \overline{U}^{(n)}(t_{i+1}, t_i), \tag{5.4}$$

where

$$\overline{U}^{(0)}(t_{i+1}, t_i) = I, \tag{5.5}$$

while, for $n \geq 1$,

$$\overline{U}^{(n)}(t_{i+1}, t_i) = (-i)^n \int_{t_i}^{t_{i+1}} dt_1 \int_{t_i}^{t_1} dt_2 \cdots \int_{t_i}^{t_{n-1}} dt_n \, H(t_1) H(t_2) \cdots H(t_n). \tag{5.6}$$

We now consider the case for which the number of intervals N is taken to be sufficiently large so that the Hamiltonian operator $H(t)$ can be assumed to be constant during each of the time intervals $[t_i, t_{i+1}]$. This allows us to make the approximation that

$$H(t) = H(t_i) \quad \text{for } t_i \leq t \leq t_{i+1}. \tag{5.7}$$

An important consequence of this approximation is that the time integrals in Equation (5.6) can now be carried out, with the following result:

$$\overline{U}^{(n)}(t_{i+1}, t_i) = (-i)^n \frac{[(\Delta t) H(t_i)]^n}{n!}. \tag{5.8}$$

The series (5.4) can then be formally summed, yielding for the evolution operator $U(t_{i+1}, t_i)$ the approximation

$$U(t_{i+1}, t_i) = \exp[-i\Delta t H(t_i)]. \tag{5.9}$$

The integral equation (2.152) for the evolution operator can be used instead as the starting point for our iteration scheme. In this case, one is led to make the approximation

$$H(t) = H(t_{i+1}) \quad \text{for } t_i \leq t \leq t_{i+1}. \tag{5.10}$$

As before, we take the Hamiltonian operator to be constant during the time interval $[t_i, t_{i+1}]$. However, the time integrals are now evaluated using the value of $H(t)$

at the *end* of the intervals instead of its value at the *beginning*. Again the series (5.4) can be formally summed, yielding for the evolution operator the alternative approximation

$$U(t_{i+1}, t_i) = \exp[-i\Delta t H(t_{i+1})].$$ (5.11)

We remark that if the approximation (5.7) is made during the first time interval, when $i = 0$, the operator $U(t_1, t_0)$ of Equation (5.9) is simply the field-free evolution operator. Similarly, for $i = N - 1$ the operator $U(t_N, t_{N-1})$ of Equation (5.11) is also the field-free evolution operator. Therefore, in both of these approximation schemes, the total atom–laser field interaction time is only $\tau_p - \Delta t$.

It is possible to improve the foregoing approach by evaluating the Hamiltonian operator at the *mid-points* of the time intervals:

$$U(t_{i+1}, t_i) = \exp\left[-i\Delta t H(t_{i+1/2})\right].$$ (5.12)

For Hamiltonian operators that vary sufficiently smoothly over time, the leading corrections to this formula are of order $(\Delta t)^2$. Because the laser pulses typically have smooth temporal profiles, these corrections are small as long as the number of time steps per optical cycle is sufficiently large.

We have thus obtained three approximate evolution operators $U(t_{i+1}, t_i)$ that can be applied to evolve the state vector over each time interval. However, these operators are defined via a power series in the Hamiltonian operator, and we are therefore required to make further approximations in order to cast these evolution operators into a form that will make them useful when performing numerical calculations. We will now consider some well known approaches. No analysis of the stability of the approximation schemes will be given here. A more detailed discussion of the methods that can be used for obtaining numerical solutions of partial differential equations can be found in Press *et al.* [2].

5.1.2 Explicit and implicit schemes

The simplest approximation for the evolution operator is the *explicit* Euler scheme

$$U(t_{i+1}, t_i) = I - i\Delta t H(t_{i+1/2}),$$ (5.13)

which consists of retaining the first two terms of the power series expansion of the exponential operator on the right-hand side of Equation (5.12). Hence, the state $|\Psi(t_{i+1})\rangle$ is obtained from the state $|\Psi(t_i)\rangle$ according to

$$|\Psi(t_{i+1})\rangle = \left[I - i\Delta t H(t_{i+1/2})\right]|\Psi(t_i)\rangle.$$ (5.14)

In this scheme, the evolved state vector at time t_{i+1} depends explicitly on the state vector at the previous time t_i. Assuming that the operation $H(t_{i+1/2})|\Psi(t_i)\rangle$

can be performed exactly for every i, the error in the state vector $|\Psi(t_{i+1})\rangle$ is of order $(\Delta t)^2$. In practice, this approach requires that N be unacceptably large. For example, let us consider an atom interacting with a Nd:YAG laser pulse ($\omega = 0.043$ a.u.). If the laser pulse intensity is sufficiently high so that the absorption of 50 ATI photons can occur with a non-negligible probability, then Δt must be chosen such that

$$(\Delta t)(50\omega) \ll 1. \tag{5.15}$$

As a result, more than 100π time intervals per optical period $T = 2\pi/\omega$ are required.

It is tempting to reduce the number of time intervals by including an additional term in the expansion of the exponential operator in Equation (5.12), so that

$$U(t_{i+1}, t_i) = I - i\Delta t H(t_{i+1/2}) + \frac{1}{2}(-i\Delta t)^2 \left[H(t_{i+1/2}) \right]^2. \tag{5.16}$$

This scheme does not always constitute an improvement over the Euler scheme (5.13). Indeed, explicit schemes for the evolution operator are, in general, not unitary, and therefore the norm of the wave function is not conserved. It is easy to show that – in principle – the error in the norm will be of order $(\Delta t)^2$ and $(\Delta t)^3$, respectively, for the schemes (5.13) and (5.16). However, in practice, the errors due to unphysical states with imaginary energies that are introduced in the numerical calculation of $H(t_{i+1/2})|\Psi(t_i)\rangle$ will grow exponentially over time.

This defect can be remedied by using the *implicit* Euler scheme. Recalling that the evolution operator (5.9) is unitary, we have

$$U(t_i, t_{i+1}) = U^{-1}(t_{i+1}, t_i) = U^\dagger(t_{i+1}, t_i)$$

$$= \exp[i\Delta t H(t_{i+1})], \tag{5.17}$$

so that we obtain for $|\Psi(t_{i+1})\rangle$ the implicit equation

$$U(t_i, t_{i+1})|\Psi(t_{i+1})\rangle = |\Psi(t_i)\rangle. \tag{5.18}$$

Making use of the explicit Euler scheme, we have that

$$U(t_i, t_{i+1}) = I + i\Delta t H(t_{i+1/2}) \tag{5.19}$$

and hence

$$|\Psi(t_{i+1})\rangle = \left[I + i\Delta t H(t_{i+1/2}) \right]^{-1} |\Psi(t_i)\rangle. \tag{5.20}$$

This implicit Euler scheme is only correct to first order in Δt. However it can be shown [2] that unphysical components of the state vector do not grow exponentially with time.

5.1.3 Crank–Nicolson and split-operator schemes

By combining the explicit and implicit Euler schemes one obtains the Crank–Nicolson scheme, which has the important properties of exactly conserving the norm of the state vector and being correct to second order in Δt. Let us use the explicit scheme (5.14) to evolve the state vector from time t_i to time $t_{i+1/2}$, with the Hamiltonian operator evaluated at the end of the time interval. That is,

$$|\Psi(t_{i+1/2})\rangle = \left[I - \frac{i\Delta t}{2}H(t_{i+1/2})\right]|\Psi(t_i)\rangle. \tag{5.21}$$

An implicit Euler step can then be used to evolve the state vector from time $t_{i+1/2}$ to time t_{i+1}, namely

$$|\Psi(t_{i+1})\rangle = \left[I + \frac{i\Delta t}{2}H(t_{i+1/2})\right]^{-1}|\Psi(t_{i+1/2})\rangle. \tag{5.22}$$

Combining these two steps gives for the evolution operator the Cayley–Klein formula

$$U(t_{i+1}, t_i) = \left[I + \frac{i\Delta t}{2}H(t_{i+1/2})\right]^{-1}\left[I - \frac{i\Delta t}{2}H(t_{i+1/2})\right], \tag{5.23}$$

which yields for the state vector $|\Psi(t_{i+1})\rangle$ the Crank–Nicolson scheme

$$|\Psi(t_{i+1})\rangle = \left[I + \frac{i\Delta t}{2}H(t_{i+1/2})\right]^{-1}\left[I - \frac{i\Delta t}{2}H(t_{i+1/2})\right]|\Psi(t_i)\rangle. \tag{5.24}$$

Since

$$\left[I - \frac{i\Delta t}{2}H(t_{i+1/2})\right]^{\dagger} = \left[I + \frac{i\Delta t}{2}H(t_{i+1/2})\right], \tag{5.25}$$

if follows that the Cayley–Klein operator is unitary. Moreover, by expanding this operator in powers of Δt, it is seen that it is correct to order $(\Delta t)^2$. A potential difficulty with this scheme stems from the requirement that an inverse operator, $\left[I + i(\Delta t/2)H(t_{i+1/2})\right]^{-1}$, must be formed at each time step. We will see below that in many cases this inverse operator can be readily evaluated numerically. It is also interesting to point out that the Cayley–Klein form of the evolution operator can be viewed as a Padé approximant [2] of the evolution operator (5.12), so that higher-order unitary schemes can be derived.

We shall now consider two unitary schemes that fall into a large class of so-called split-operator schemes. The basic idea behind split-operator schemes is to partition the Hamiltonian into a number of "sub-Hamiltonians." If this partitioning is carried out judiciously, based on some property or characteristic of the Hamiltonian, the problem of evolving the state vector over a time interval can be broken down into a number of sub-problems, each of which are more amenable to numerical calculations than the entire problem. We begin by considering the split-operator "fast Fourier transform" (FFT) scheme and then briefly discuss the "alternating direction implicit" (ADI) scheme for solving problems in two or more dimensions.

Let us consider the TDSE (2.127) in the velocity gauge. We will restrict our attention to the case of a one-electron atom. The Hamiltonian $H(t)$ can be split into two parts as follows:

$$H(t) = H_{\mathrm{p}}(t) + H_{\mathrm{r}}, \tag{5.26}$$

where (in a.u.)

$$H_{\mathrm{p}}(t) = \frac{\mathbf{p}^2}{2} + \mathbf{A}(t) \cdot \mathbf{p} \tag{5.27}$$

and

$$H_{\mathrm{r}} = V(\mathbf{r}). \tag{5.28}$$

The first term on the right-hand side of Equation (5.26) depends on the momentum operator $\mathbf{p}$, while the second term depends on the position operator $\mathbf{r}$. Using the Zassenhaus formula for the two operators A and B, and the parameter μ, namely

$$\exp(\mu A)\exp(\mu B) = \exp(\mu A + \mu B)\exp\left(\frac{\mu^2}{2}[A, B]\right) + \mathcal{O}(\mu^3), \tag{5.29}$$

where $[A, B] = AB - BA$ is the commutator of A and B, allows us to write the evolution operator (5.12), correct to order $(\Delta t)^2$, as

$$U(t_{i+1}, t_i) = \exp\left(-\mathrm{i}\Delta t[H_{\mathrm{p}}(t_{i+1/2}) + H_{\mathrm{r}}]\right)$$

$$= \exp\left(-\mathrm{i}\Delta t\, H_{\mathrm{p}}(t_{i+1/2})\right)\exp\left(-\mathrm{i}\Delta t\, H_{\mathrm{r}}\right)\exp\left(\frac{1}{2}(\mathrm{i}\Delta t)^2[H_{\mathrm{p}}(t_{i+1/2}), H_{\mathrm{r}}]\right).$$

$$\tag{5.30}$$

Ignoring terms of order $(\Delta t)^2$ and higher in the right-most operator leads to the following approximate evolution operator:

$$U(t_{i+1}, t_i) = \exp\left[-\mathrm{i}\Delta t\, H_{\mathrm{p}}(t_{i+1/2})\right]\exp\left[-\mathrm{i}\Delta t\, H_{\mathrm{r}}\right]. \tag{5.31}$$

This results in a scheme that is unitary and correct to order Δt. Its accuracy can be improved in the following way. We start by writing

$$\exp\left(-\mathrm{i}\Delta t[H_{\mathrm{p}}(t_{i+1/2}) + H_{\mathrm{r}}]\right) = \exp\left(-\frac{\mathrm{i}\Delta t}{2}[H_{\mathrm{r}} + H_{\mathrm{p}}(t_{i+1/2})]\right)$$

$$\times \exp\left(-\frac{\mathrm{i}\Delta t}{2}[H_{\mathrm{p}}(t_{i+1/2}) + H_{\mathrm{r}}]\right). \tag{5.32}$$

Successive applications of the Zassenhaus formula then yield, correct to order $(\Delta t)^2$,

$$U(t_{i+1}, t_i) = \exp\left(-i\Delta t[H_p(t_{i+1/2}) + H_r]\right)$$

$$= \exp\left(-\frac{i\Delta t}{2}H_r\right)\exp\left[-i\Delta t\,H_p(t_{i+1/2})\right]\exp\left(-\frac{i\Delta t}{2}H_r\right). \quad (5.33)$$

At this point we remark that the chosen partition of the Hamiltonian has resulted in an operator that is diagonal (i.e. a multiplicative operator) in configuration space and an operator that is diagonal in momentum space. To see how this fact can be exploited, let us write the wave function at time t_i in configuration space as $\Psi(\mathbf{r}, t_i)$. Using Equation (5.33) together with the closure relation, the wave function at time t_{i+1} is then given by

$$\Psi(\mathbf{r}, t_{i+1}) = \int d\mathbf{r}' \int d\mathbf{p} \int d\mathbf{r}''$$

$$\times \langle \mathbf{r}|\exp\left(-\frac{i\Delta t}{2}H_{r'}\right)|\mathbf{r}'\rangle\langle \mathbf{r}'|\exp\left[-i\Delta t\,H_p(t_{i+1/2})\right]$$

$$\times |\mathbf{p}\rangle\langle \mathbf{p}|\exp\left(-\frac{i\Delta t}{2}H_{r''}\right)|\mathbf{r}''\rangle\Psi(\mathbf{r}'', t_i). \quad (5.34)$$

Carrying out the integration over $\mathbf{r}'$ and using the fact that the operators are diagonal, we can rewrite the above equation in the following form:

$$\Psi(\mathbf{r}, t_{i+1}) = \exp\left(-\frac{i\Delta t}{2}H_r\right)$$

$$\times \left\{\int d\mathbf{p}\langle \mathbf{r}|\mathbf{p}\rangle\exp\left[-i\Delta t\,H_p(t_{i+1/2})\right]\right.$$

$$\times \left.\left[\int d\mathbf{r}''\langle \mathbf{p}|\mathbf{r}''\rangle\exp\left(-\frac{i\Delta t}{2}H_{r''}\right)\Psi(\mathbf{r}'', t_i)\right]\right\}, \quad (5.35)$$

where

$$\langle \mathbf{r}|\mathbf{p}\rangle = (2\pi)^{-3/2}\exp(i\mathbf{p}\cdot\mathbf{r}). \quad (5.36)$$

We see that five steps are required to evaluate the right-hand side of Equation (5.35). The wave function is first multiplied by a diagonal operator. A Fourier transform is then performed on the resulting wave function, yielding the corresponding wave function in momentum space. Next, this momentum space wave function is multiplied by a diagonal operator, and is transformed back to configuration space via an inverse Fourier transform, where it is finally multiplied by a diagonal operator.

The result of these five steps – three multiplications, one Fourier transform and one inverse Fourier transform – yields the wave function $\Psi(\mathbf{r}, t_{i+1})$. In practice, the Fourier transforms can be carried out very efficiently using the fast Fourier transform (FFT) algorithm [2]. Indeed, if the wave function is represented on a grid of M points, the computational size of the problem only scales as $M \log M$. If the wave function at intermediate times during the laser pulse is not explicitly required, then the last step can be combined with the first step of the calculation of the evolution operator over the subsequent time interval.

In order to apply the split-operator FFT scheme, the key requirement is that it must be possible to partition the Hamiltonian exactly into a sub-Hamiltonian that is diagonal in momentum space and a sub-Hamiltonian that is diagonal in configuration space. This scheme is therefore of particular interest when numerically solving problems using coordinates such that the Hamiltonian does not contain terms involving both position and momentum operators. This is the case, for example, for the TDSE in the dipole approximation expressed in Cartesian coordinates.

We will now discuss the "alternating direction implicit," or ADI, scheme. This scheme requires the partition of the Hamiltonian into sub-Hamiltonians in such a way that each sub-Hamiltonian contains differential operators that act on at most a single spatial variable. To illustrate this approach, we consider the case of a single-electron atom that is described by the TDSE (2.133) in the length gauge. Assuming a linearly polarized laser field, the system exhibits rotational symmetry about the polarization vector $\hat{\boldsymbol{\epsilon}}$, which we choose to be the quantization axis. As a result, using cylindrical polar coordinates (ρ, ϕ, z), the wave function does not depend on ϕ. The relevant part of the Hamiltonian operator, $\tilde{H}(t)$, can therefore be written as the sum of two operators,

$$\tilde{H}(t) = H_\rho + H_z(t). \tag{5.37}$$

The first operator is given by

$$H_\rho = -\frac{1}{2\rho} \frac{\partial}{\partial \rho} \left(\rho \frac{\partial}{\partial \rho} \right) + V[(\rho^2 + z^2)^{1/2}], \tag{5.38}$$

and contains the contribution to the Laplacian involving the coordinate ρ. The second operator,

$$H_z(t) = -\frac{1}{2} \frac{\partial^2}{\partial z^2} + \mathcal{E}(t)z, \tag{5.39}$$

contains the contribution to the Laplacian involving the coordinate z.

Making use of the above partition, the explicit Euler scheme which is correct to first order in Δt can be expressed as

$$|\Psi(t_{i+1})\rangle = \left[I - \mathrm{i}\Delta t\, H_z(t_{i+1/2}) \right] \left[I - \mathrm{i}\Delta t\, H_\rho \right] |\Psi(t_i)\rangle. \tag{5.40}$$

The Euler scheme now involves the product of two operators, one involving H_ρ and the other $H_z(t)$. The implicit Euler scheme (5.20) can also be written in a similar form. Combining the implicit and explicit steps, respectively, as in Equation (5.24) then yields the ADI scheme

$$|\Psi(t_{i+1})\rangle = \left[I + i\frac{\Delta t}{2}H_\rho\right]^{-1}\left[I + i\frac{\Delta t}{2}H_z(t_{i+1/2})\right]^{-1}\left[I - i\frac{\Delta t}{2}H_z(t_{i+1/2})\right]$$
$$\times \left[I - i\frac{\Delta t}{2}H_\rho\right]|\Psi(t_i)\rangle. \tag{5.41}$$

The ADI scheme exploits the fact that, in practice, the operators $[I + i(\Delta t/2)H_\rho]$ and $[I + i(\Delta t/2)H_z(t_{i+1/2})]$ can both be readily inverted, while this is not the case for the operator $[I + i(\Delta t/2)\tilde{H}(t_{i+1/2})]$. Despite being correct to second order in Δt and stable, it is not unitary as the operators H_ρ and $H_z(t)$ do not commute. The ADI scheme will be encountered again in our discussion of the numerical solution of the TDSE for atomic hydrogen in Section 5.4.

5.1.4 Representation of the wave function

The wave function of the system can be represented in space by points on a *spatial grid* or in terms of an expansion in *basis functions*, or by employing a *hybrid* representation involving both a grid and basis functions.

5.1.4.1 Spatial grids

Since the wave function is of finite extent, it is possible, in principle, to choose for any laser pulse duration a large enough grid to contain the entire wave function, including the part of it that belongs to the continuum. However, as explained above, at high laser intensities, and especially at low frequencies, the ionized electrons can acquire high kinetic energies, and their quiver motion can become large, which implies that large grids would be required. For this reason, absorbing potentials or mask functions must be incorporated near the end of realistically sized grids in order to remove the probability flux that reaches the boundaries, and avoid in this way spurious reflections of the wave function. We also observe that, as the TDSE is an initial value problem, the initial (often the ground state) wave function must be computed first on the grid (in the absence of the laser field) and therefore depends on the grid spacing. This finite-difference ground-state wave function can be generated either by diagonalizing the field-free atomic Hamiltonian H_0 directly [3] or by integrating the field-free TDSE in imaginary time as a diffusion equation [4].

Using the finite difference method, a solution of the TDSE is sought at a set of discrete points in space. If the grid point density is sufficiently high and the grid boundaries encompass a sufficiently large volume, then the values of the wave function at this set of points can give an accurate representation of the exact wave function. The required grid density must be determined by the highest momentum components in the wave function. For example, if the largest electron energy that is expected to occur is $10U_p$, the corresponding maximum momentum is $p_{max} = (20U_p)^{1/2}$. The grid spacing must then be chosen to be smaller than $1/p_{max}$.

To illustrate the method, we will set up the finite difference equations in the Crank–Nicolson scheme for a one-dimensional model atom interacting with a laser pulse. Our starting point is the TDSE in the length gauge,

$$i\frac{\partial}{\partial t}\Psi(x,t) = \left[-\frac{1}{2}\frac{\partial^2}{\partial x^2} + V(x) + \mathcal{E}(t)x\right]\Psi(x,t), \tag{5.42}$$

where $\mathcal{E}(t) = \mathcal{E}_0 F(t)\cos(\omega t + \varphi)$. The spatial variable x is discretized by setting $x_j = (j - M/2)\Delta x$, $j = 0, 1, 2, \ldots, M$, so that the grid extends from $-M\Delta x/2$ to $M\Delta x/2$. Using the second-order central difference formula,

$$\frac{\partial^2}{\partial x^2}\Psi(x_j, t) = \frac{\Psi(x_{j+1}, t) - 2\Psi(x_j, t) + \Psi(x_{j-1}, t)}{(\Delta x)^2}, \tag{5.43}$$

we denote the finite difference representation of the evolution operators (5.13) and (5.19), respectively, by the matrices $\mathbf{U}^+(t_{i+1/2})$ and $\mathbf{U}^-(t_{i+1/2})$, whose elements are given by

$$U_{j,j'}^\pm(t_{i+1/2}) = \begin{cases} \pm i\Delta t/[2(\Delta x)^2], & j = j'+1, j'-1, \\ 1 \mp i\Delta t\left[1/(\Delta x)^2 + V(x_j) + \mathcal{E}(t_{i+1/2})x_j\right], & j = j', \\ 0, & \text{otherwise.} \end{cases} \tag{5.44}$$

We also denote by $\mathbf{\Psi}(t_i)$ the finite difference representation of the wave function at time t_i. This vector has $M+1$ components, with $\Psi_j(t_i) = \Psi(x_j, t_i)$. Given the finite difference representation of the wave function at time t_i, $\mathbf{\Psi}(t_i)$, the wave function at time t_{i+1} is obtained by solving the following set of linear equations:

$$\mathbf{U}^-(t_{i+1/2})\mathbf{\Psi}(t_{i+1}) = \mathbf{U}^+(t_{i+1/2})\mathbf{\Psi}(t_i). \tag{5.45}$$

From Equation (5.44), it is seen that the matrix $\mathbf{U}^-$ is tridiagonal, allowing $\mathbf{\Psi}(t_{i+1})$ to be obtained efficiently [2], with the computational effort scaling linearly with M.

5.1.4.2 Basis-states methods

The wave function of the system can also be represented in terms of an expansion in a set of basis states. To this end, we return to the Dirac method of variation of the constants, discussed in Section 3.2. Taking a hydrogenic system as an example, a complete set of energy eigenfunctions $\psi_k(\mathbf{r})$ of the field-free Hamiltonian H_0, satisfying the equation

$$H_0\psi_k(\mathbf{r}) = E_k\psi_k(\mathbf{r}), \tag{5.46}$$

is introduced. The wave function of the system is then expanded in terms of the set of discrete eigenfunctions and a set of suitably discretized continuum eigenfunctions as

$$\Psi(\mathbf{r}, t) = \sum_k a_k(t)\psi_k(\mathbf{r}). \tag{5.47}$$

Inserting this expansion into the TDSE (2.101), we obtain for the expansion coefficients $a_k(t)$ the following set of ordinary coupled differential equations:

$$i\frac{d}{dt}\mathbf{a}(t) = \mathbf{H}(t)\mathbf{a}(t), \tag{5.48}$$

where the elements of the matrix $\mathbf{H}(t)$ are given by

$$H_{jk}(t) = \langle \psi_j | H(t) | \psi_k \rangle = E_k\delta_{jk} + \langle \psi_j | H_{\text{int}}(t) | \psi_k \rangle \tag{5.49}$$

and $a_k(t_0) = \delta_{ki}$, so that the system is initially in the state $| \psi_i \rangle$. In the dipole approximation, $\mathbf{H}(t)$ can be cast into a banded matrix.

A formal solution of Equation (5.48) can be obtained using Equation (5.1) together with the expansion (5.47). We find that

$$\mathbf{a}(t) = \mathbf{U}(t, t_0)\mathbf{a}(t_0), \tag{5.50}$$

where the matrix representation of the evolution operator is given by

$$U_{jl}(t, t_0) = \langle \psi_j | U(t, t_0) | \psi_l \rangle. \tag{5.51}$$

The methods outlined above for obtaining approximations to the evolution operator $U(t, t_0)$ over small time intervals can be directly applied to derive similar approximations to the evolution matrix $\mathbf{U}(t, t_0)$. For example, in the explicit Euler scheme, we have, to order Δt,

$$\begin{aligned}
\mathbf{a}(t_{i+1}) &= \mathbf{U}(t_{i+1}, t_i)\mathbf{a}(t_i) \\
&= \left[\mathbf{I} - i\Delta t\mathbf{H}(t_{i+1/2})\right]\mathbf{a}(t_i), \tag{5.52}
\end{aligned}$$

where $\mathbf{I}$ is the identity matrix. Higher-order explicit schemes are usually preferred over implicit schemes due to the computational effort required to invert banded matrices of the form $\mathbf{I} + i\Delta t\mathbf{H}$. Higher-order explicit schemes for solving the system of Equations (5.48) are discussed by Press *et al.* [2]. These schemes are typically found to be stable for even relatively large values of Δt since the state vector is expressed directly in terms of the eigenstates of the field-free Hamiltonian.

In order to limit the size of the system of ordinary differential equations to be solved, the expansion of the wave function in terms of field-free hydrogenic basis states must be truncated at some maximum principal quantum number for the bound states and at some maximum kinetic energy for the continuum states. In addition, states having orbital angular momentum quantum numbers greater than some maximum cut-off value must also be excluded. In general, the success of basis-states approaches depends critically on finding a balance between retaining a sufficiently large enough set of field-free eigenstates of the atom to ensure an accurate representation of the wave function while avoiding computational issues associated with obtaining numerical solutions of very large systems of ordinary differential equations. The number of states required to span the space of physically relevant bound and continuum states can be reduced by expanding the radial part of the wave function in terms of the Sturmian functions introduced in Section 3.1 or B-spline functions. The latter are piecewise polynomials defined with respect to a knot sequence over a finite interval [5]. Both the Sturmian and B-spline functions are L^2 functions, and thereby provide a discrete representation of the continuum states.

A basis-states approach has the advantage that it is straightforward to obtain observable quantities from the expansion coefficients $a_k(t)$. For example, since the probability of finding the atom in the state $|\psi_k\rangle$ at the end of the pulse is simply $|a_k(t_0 + \tau_p)|^2$, photoelectron energy spectra can be readily calculated. In addition, as discussed by Lambropoulos, Maragakis and Zhang [6], insight into the photoionization dynamics can be obtained by monitoring populations in ground and excited states while the atom is interacting with the laser pulse.

5.1.4.3 Hybrid methods

Hybrid methods consist of expanding a subset of the wave function spatial coordinates in terms of a set of basis states, while the remaining coordinates are represented on a grid. To understand why such an approach can be appropriate, we recall that when solving the TDSE numerically, it is important to choose a coordinate system that exploits the exact or approximate symmetries of the system. For example, as noted above, when solving the TDSE for a hydrogenic atom interacting with a linearly polarized laser pulse, the Hamiltonian is invariant under rotations about the polarization vector $\hat{\epsilon}$, which we take as the z-axis. The system has therefore cylindrical symmetry about that axis, so that the magnetic quantum number m_l of the electron is conserved; its initial value may be chosen arbitrarily and will be taken to be zero. Thus, the number of effective spatial dimensions is reduced to two, which significantly decreases the computational requirements. In addition, for not too intense pulses, the departure of the system from spherical symmetry is not too great. This means that an expansion of the wave function in terms of partial waves is appropriate. In a hybrid approach, the remaining radial components of the wave function are then represented on a grid, so that the wave function $\Psi(\mathbf{r}, t)$ is expanded in spherical harmonics as

$$\Psi(r, \theta, t) = \sum_{l=0}^{l_{max}} r^{-1} u_l(r, t) Y_{l,0}(\theta)$$

$$= (4\pi)^{-1/2} \sum_{l=0}^{l_{max}} (2l+1)^{1/2} r^{-1} u_l(r, t) P_l(\cos \theta), \qquad (5.53)$$

where l_{max} is the maximum orbital angular momentum quantum number included in the expansion, θ is the angle between the vector $\mathbf{r}$ and the z-axis and P_l are Legendre polynomials. We note that all the time dependence is included in the reduced radial functions $u_l(r, t)$. The computational task is to determine the reduced radial functions, and it will be shown how this can be done in our discussion of the TDSE for atomic hydrogen in Section 5.4

5.1.5 Choice of gauge

As noted in Section 2.4, observable quantities computed from the wave function of the system are gauge-independent. However, in practice the choice of the gauge is important as it affects various aspects of the numerical calculations. Cormier and Lambropoulos [7] have studied the convergence of numerical solutions of the TDSE for atomic hydrogen interacting with intense, low-frequency laser pulses. They employed a partial wave expansion of the form (5.53), in which the radial part of the wave function was represented by B-spline basis functions. It was found that nearly an order of magnitude fewer partial waves were required in the velocity gauge than in the length gauge. These results show that it can be advantageous to work in the velocity gauge when solving the TDSE numerically.

In order to illustrate why this is the case, let us consider the one-electron wave function $\Psi^V(\mathbf{r}, t)$, which is a solution of the TDSE (2.127) in the velocity gauge. Referring to Equation (5.53), for an electron moving in a spherically symmetric potential and interacting with a laser pulse that is linearly polarized, the wave function in spherical polar coordinates can be expressed as

$$\Psi^V(r, \theta, t) = (4\pi)^{-1/2} \sum_{l=0}^{l_{max}} (2l+1)^{1/2} r^{-1} u_l^V(r, t) P_l(\cos \theta). \qquad (5.54)$$

It has been found that, for laser pulses with carrier frequencies in the infra-red and intensities of the order of $10^{14}\,\mathrm{W\,cm^{-2}}$, l_{max} is typically between 20 and 30 [8].

Recalling Equations (2.124) and (2.132), the wave function in the length gauge is obtained from the wave function in the velocity gauge by performing the following unitary transformation (in a.u.):

$$\Psi^L(r,\theta,t) = \exp[i\mathbf{A}(t)\cdot\mathbf{r}]\exp\left[-\frac{i}{2}\int_{-\infty}^{t}\mathbf{A}^2(t')dt'\right]\Psi^V(r,\theta,t). \quad (5.55)$$

The operator $\exp[i\mathbf{A}(t)\cdot\mathbf{r}]$ can be expanded in partial waves as follows:

$$\exp[i\mathbf{A}(t)\cdot\mathbf{r}] = \exp[iA(t)z]$$

$$= \sum_{l'=0}^{\infty}(2l'+1)i^{l'}\,j_{l'}[A(t)r]\,P_{l'}(\cos\theta), \quad (5.56)$$

where we have written $\mathbf{A}(t) = \hat{\mathbf{z}}A(t)$ and the functions $j_{l'}$ are the spherical Bessel functions. For an intense laser pulse of angular frequency ω and electric-field strength $\mathcal{E}_0$, the "size" of the atomic system can be estimated to be $\alpha_0 = \mathcal{E}_0/\omega^2$. It follows that the expansion (5.56) must include at least

$$l'_{\max} \simeq A_0\alpha_0 = \mathcal{E}_0^2/\omega^3 \quad (5.57)$$

partial waves before it can be truncated. We note that $l'_{\max}$ increases approximately linearly with the laser intensity.

As an example, let us consider a Nd:YAG laser pulse ($\omega = 0.043$ a.u.) having a peak intensity of $10^{14}\,\mathrm{W\,cm^{-2}}$. The estimate given by Equation (5.57) leads to a cut-off value of $l'_{\max} \simeq 125$ for the partial wave expansion of the wave function in the length gauge, $\Psi^L(r,\theta,t)$. This cut-off value is much larger than the cut-off value $l_{\max}$ that is typically required for the wave function expansion (5.54) in the velocity gauge. However, we note that is has been found that the advantage of working in the velocity gauge can be offset when dealing with complex atoms, as we shall see in Section 5.5.5.

5.2 Calculation of the observables

In this section, we shall examine how the most important properties of the system can be extracted from the knowledge of the wave function.

5.2.1 Total ionization probabilities and rates

The ionization probability, P_{ion}, is defined as the fraction of the atomic population which is in the continuum at the end of the pulse. From a computational point of view it is more efficient to calculate the atomic population left in bound states, and obtain the ionization probability by subtraction. The ionization probability is then given by

$$P_{\text{ion}} = 1 - \sum_{j \in \text{bound states}} |\langle \psi_j | \Psi(t_0 + \tau_{\text{p}}) \rangle|^2. \tag{5.58}$$

For *non-resonant* ionization in not too short laser pulses, a simple estimate of the ionization probability can be obtained by projecting onto the field-free initial state of the atom.

If absorbing boundary conditions or mask functions are employed, then the ionization probability can be obtained by calculating the norm of the state vector.

Ionization *rates* can also be readily extracted from the numerical solution of the TDSE. These are obtained using a laser pulse whose envelope rises to a maximum over several optical cycles and then remains constant over a time interval that is long compared to the lifetime of the atom in the laser field. First, one calculates the "time-dependent ionization probability" given by

$$P_{\text{ion}}(t) = 1 - \sum_{j \in \text{bound states}} |\langle \psi_j | \Psi(t) \rangle|^2 \tag{5.59}$$

over the constant part of the pulse. The ionization rate is then estimated from the slope of the log of $P_{\text{ion}}(t)$. When using this approach, it should be borne in mind that for some angular frequencies and intensities resonantly enhanced multiphoton ionization (REMPI) can occur. This complicates the ionization process so that a single multiphoton ionization rate cannot be defined [3].

5.2.2 Photoelectron energy and angular distributions

Final-state *electron energy distributions* can be obtained from $| \Psi(t_0 + \tau_{\text{p}}) \rangle$ in several ways. The most obvious method involves generating the eigenfunctions of the field-free Hamiltonian H_0 for a large number of energies. The probability distributions are then calculated by projecting $| \Psi(t_0 + \tau_{\text{p}}) \rangle$ on these field-free eigenstates [9, 10]. This approach requires that a large number of field-free eigenfunctions be generated. It is difficult to implement for complex atoms.

A second method [11] involves propagating the state vector $| \Psi(t) \rangle$ in time in the absence of the laser field (for $t > t_0 + \tau_{\text{p}}$) and calculating its overlap with the state vector $| \Psi(t_0 + \tau_{\text{p}}) \rangle$ at the end of the laser pulse. This yields the function

$$\langle \Psi(t_0 + \tau_{\text{p}}) | \Psi(t) \rangle = \sum_j \exp[-iE_j(t - t_0 - \tau_{\text{p}})] |a_j|^2, \tag{5.60}$$

where the sum includes all of the bound and continuum eigenstates of the field-free Hamiltonian H_0. The Fourier transform of this correlation function yields the atomic population, $|a_j|^2$, in the energy eigenstate $|\psi_j\rangle$. The energy resolution that is obtained is inversely proportional to the propagation time. In order to calculate an ATI spectrum with an energy resolution of better than, for example, $0.1\hbar\omega$ requires that the wave function be propagated over an additional time of about $20\pi/\omega$, or ten optical cycles.

A third approach relies on a spectral analysis of the state vector $|\Psi(t_0 + \tau_p)\rangle$ based on the use of a window operator. In particular, Schafer and Kulander [12] have used the operator

$$W(E_k, n, \gamma) = \frac{\gamma^{2^n}}{(H_0 - E_k)^{2^n} + \gamma^{2^n}}, \tag{5.61}$$

where the parameter γ fixes the width of the spectral window, and therefore determines the energy resolution of the analysis. To a good approximation, the expectation value

$$P(E_k, n, \gamma) = \langle \Psi | W(E_k, n, \gamma) | \Psi \rangle \tag{5.62}$$

is proportional to the total probability of finding the energy of the photoelectron within the interval $E_k \pm \gamma$ at the end of the laser pulse. This result can be obtained by expressing the expectation value on the right-hand side of Equation (5.62) in terms of the eigenfunctions ψ_j of H_0 as

$$P(E_k, n, \gamma) = \sum_j |\langle \Psi | \psi_j \rangle|^2 \left[\frac{\gamma^{2^n}}{(E_j - E_k)^{2^n} + \gamma^{2^n}} \right], \tag{5.63}$$

where the weight function or "energy bin" in the brackets on the right-hand side of this equation has a width of 2γ. Setting E_k to be equal to the eigenenergy E_s of the eigenstate $|\psi_s\rangle$, we have

$$P(E_s, n, \gamma) = |\langle \Psi | \psi_s \rangle|^2 + \sum_{j \neq s} |\langle \Psi | \psi_j \rangle|^2 \left[\frac{\gamma^{2^n}}{(E_j - E_s)^{2^n} + \gamma^{2^n}} \right]. \tag{5.64}$$

Because as n is increased the window function becomes increasingly rectangular, the weight function ensures that the second term will be approximately zero, provided that the density of eigenstates is small compared to γ^{-1}. As a result, the function (5.63) evaluated for a set of energies E_k, with $E_{k+1} = E_k + 2\gamma$, is independent of γ and gives an energy spectrum with a resolution of approximately 2γ. In practice, the choice $n = 2$ has been found to give good results [8, 12].

Let us now consider how the wave function spectral energy density (5.62) can be calculated numerically. Schafer and Kulander [12] have pointed out that this quantity can be expressed as

$$P(E_k, n, \gamma) = \gamma^{2^n} \langle \zeta_k | \zeta_k \rangle, \tag{5.65}$$

where

$$\left[(H_0 - E_k)^{2^{n-1}} + i\gamma^{2^{n-1}}\right]|\zeta_k\rangle = |\Psi\rangle. \tag{5.66}$$

The operator acting on $|\zeta_k\rangle$ can be decomposed in terms of simple powers of H_0. For $n = 2$, we have

$$(H_0 - E_k + \sqrt{i}\gamma)(H_0 - E_k - \sqrt{i}\gamma)|\zeta_k\rangle = |\Psi\rangle. \tag{5.67}$$

The state $|\zeta_k\rangle$ can now be obtained in two steps. First, we introduce the state $|\zeta_k^0\rangle$ as a solution of the inhomogeneous equation

$$(H_0 - E_k + \sqrt{i}\gamma)|\zeta_k^0\rangle = |\Psi\rangle. \tag{5.68}$$

Since the matrix representation of H_0 is both sparse and banded, efficient algorithms are available for solving this linear system. Next we determine $|\zeta_k\rangle$ from the equation

$$(H_0 - E_k - \sqrt{i}\gamma)|\zeta_k\rangle = |\zeta_k^0\rangle. \tag{5.69}$$

The spectral energy density at E_k is then obtained from Equation (5.65).

Photoelectron *angular distributions* can also be calculated using the window operator technique. For a linearly polarized laser pulse, expressing the function $\zeta_k(\mathbf{r})$ in terms of partial waves as

$$\zeta_k(r, \theta) = \sum_{l=0}^{\infty} r^{-1} v_l(k, r) P_l(\cos\theta), \tag{5.70}$$

we see that the probability density of an electron being ejected at an angle θ with respect to the polarization direction $\hat{\mathbf{z}}$ and having energy within the range $[E_k - \gamma, E_k + \gamma]$ is given by

$$P(E_k, \theta, n, \gamma) = \gamma^{2^n} \sum_{l=0}^{l_{max}} \sum_{l'=0}^{l_{max}} \left[\int_0^{\infty} v_l^*(k, r) v_{l'}(k, r) dr\right] P_l(\cos\theta) P_{l'}(\cos\theta). \tag{5.71}$$

An alternative method for obtaining angular distributions has been proposed by Guyétand *et al.* [13,14], who calculated the electron probability flux within a given energy bin through a surface element located far from the nucleus.

5.2.3 Photon emission spectra

As we have seen in Section 1.4, the spectrum of radiation emitted by an atom in an intense laser field exhibits peaks at the odd harmonics of the laser carrier angular frequency ω. The power spectrum of the emitted radiation is proportional to the modulus squared of the Fourier transform of the dipole acceleration, $\mathbf{a}(t) = (\mathrm{d}^2/\mathrm{d}t^2)\mathbf{d}(t)$, where $\mathbf{d}(t)$ is the laser-induced atomic dipole moment, given by Equation (1.12). In order to determine the power spectrum, the dipole acceleration must be calculated at equally spaced time intervals during the laser pulse. In order to obtain an estimate of the required time interval Δt, let us assume that ω_c is the angular frequency of the highest harmonic emitted by the atom in the laser field. We recall that, for a low-frequency laser field, one has $\omega_c \simeq (I_p + 3U_p)/\hbar$. As a result, the dipole acceleration must be obtained on time intervals Δt during the laser pulse that are smaller than π/ω_c in order to ensure that the highest emitted harmonics are correctly resolved. Hence, to obtain an accurate photon emission spectrum, the numerical integration scheme must be chosen such that at least the components of the wave function with energy $\hbar\omega_c$ are correctly represented in both space and time.

After the end of the laser pulse, fluorescence can also occur due to atomic population left in excited states. To determine the fluorescence spectrum, one first projects the state vector at the end of the laser pulse, $|\Psi(t_0 + \tau_p)\rangle$, onto the field-free excited states of the atom to obtain the atomic populations in these states. Using these populations, together with the respective radiative lifetimes and branching ratios, the fluorescence spectrum can be calculated.

5.3 One- and two-dimensional studies of the time-dependent Schrödinger equation

In this section, we begin our discussion of the numerical solution of the TDSE in the dipole approximation by considering one- and two-dimensional model atoms, for which the computational effort is significantly reduced.

One-dimensional model atoms interacting with intense laser fields were first considered by Geltman [15], Goldberg and Shore [16] and Austin [17]. They have been investigated extensively by several authors [9,18–34] and reviewed by Eberly, Javanainen and Rząžewski [35] and Eberly *et al.* [36].

Because one-dimensional calculations are relatively easy to perform, it is possible to conduct "numerical experiments" by investigating a large range of parameters. However, one-dimensional calculations present a number of disadvantages due to the oversimplification of the problem. In particular, effects related to the transverse direction, such as those arising from the use of elliptically (and in particular circularly) polarized light, are neglected. Moreover, in one dimension, the singularity of the Coulomb potential at the origin is overemphasized, and leads to numerical problems. To avoid this difficulty, a "soft-core Coulomb" potential given by the expression

$$V(x) = -\frac{1}{\sqrt{a^2 + x^2}}, \tag{5.72}$$

where a is a positive constant, has been adopted to represent the electron–nucleus interaction of a one-dimensional model of the hydrogen atom [9, 36] in a large number of studies of multiphoton processes. Analytic expressions for the bound-state energies and eigenfunctions of this potential have been given by Liu and Clark [37].

An important advantage of this potential is that the electron motion is continuous across $x = 0$. The softening in the vicinity of $x = 0$ corrects for the unphysical increase of the electron–nucleus interaction in one dimension. Potentials like the soft-core Coulomb interaction (5.72), which behave as $1/|x|$ when $x \to \pm\infty$ have an infinite range and support an infinite number of bound states. For $a = 1$, the ionization potential of the soft-Coulomb potential (5.72) is $I_P = 0.67$ a.u. $= 18$ eV. The higher lying bound energy levels of the potential (5.72) give rise to a Rydberg-like series.

The soft-core Coulomb potential (5.72) has been used to carry out a number of strong-field numerical experiments, which have been reviewed by Eberly *et al.* [36]. The response of the one-dimensional model atom has been analyzed for bound–bound, bound–free and free–free transitions. As an illustration, we show in Fig. 5.1 the results of a simulation performed by Eberly *et al.* [36] for $a = 1$, displaying the threshold shifting of ATI peaks due to the AC Stark effect, for 4.25-cycle square laser pulses, and peak electric-field strengths $\mathcal{E}_0 = 0.05$ a.u., 0.07071 a.u. and 0.085 a.u. The ATI peaks are seen to shift to lower energies with increasing laser intensity. The bottom graph, corresponding to the highest electric-field strength $\mathcal{E}_0 = 0.085$ a.u., shows that the lowest ionization channel is closed. It is interesting to note the presence, in this graph, of a peak below the threshold. This is possible for a short-pulse interaction because the soft-core Coulomb potential (5.72) has Rydberg-type levels which are populated as a result of resonantly enhanced multiphoton ionization (REMPI).

The soft-core Coulomb potential (5.72) can also be scaled to model other atomic species by writing [38]

$$V(x) = -\frac{Z}{\sqrt{(a/Z)^2 + x^2}}. \tag{5.73}$$

For example, in a one-dimensional model of He^+, the interaction of the electron with the helium nucleus can be represented by the above potential with $Z = 2$ and $a = \sqrt{2}$ a.u., giving an ionization potential $I_P = 2$ a.u. [39]. As an example, we show in Fig. 5.2 the photon emission spectra of He^+ calculated by Potvliege,

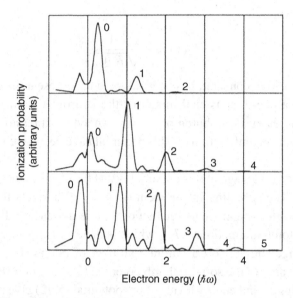

Figure 5.1. Ionization probability (in arbitrary units) as a function of the electron energy (in units of the photon energy $\hbar\omega$) for a one-dimensional model of the hydrogen atom interacting with 4.25-cycle square laser pulses having peak electric-field strengths $\mathcal{E}_0 = 0.05$ a.u., 0.07071 a.u. and 0.085 a.u. (from top to bottom). The numbers refer to the corresponding peaks at different intensities. (From J. H. Eberly, *Adv. At. Mol. Opt. Phys. Suppl.* **1**, 301 (1992).)

Kylstra and Joachain [39] using this one-dimensional electron–nucleus potential. The calculations were carried out for two-cycle laser pulses having a carrier wavelength of 800 nm (corresponding to a Ti:sapphire laser) and a peak electric-field strength $\mathcal{E}_0 = 0.4$ a.u., corresponding to a peak intensity of 5.6×10^{15} W cm^{-2}. The photon emission spectra of Fig. 5.2 exhibit interesting features which depend on the carrier-envelope phase (CEP) φ of the laser pulse. Indeed, as seen from Fig. 5.2, depending on the value of φ, either one or two plateaus are present. In addition, the position of the cut-off of the plateaus depends strongly on φ.

One-dimensional studies of the TDSE have also been carried out for potentials of *finite* range. These include the potential

$$V(x) = -Z\frac{\exp(-\epsilon|x|) - \exp(-|x|)}{|x|} \tag{5.74}$$

employed by Dörr and Shakeshaft [23]. This potential is finite at the origin. For large values of $|x|$, it behaves as $-Z\exp(-\epsilon|x|)/|x|$, where ϵ is a small parameter that determines the range of the potential. The interaction potential (5.74) supports a finite number of bound states; this number increases as ϵ decreases.

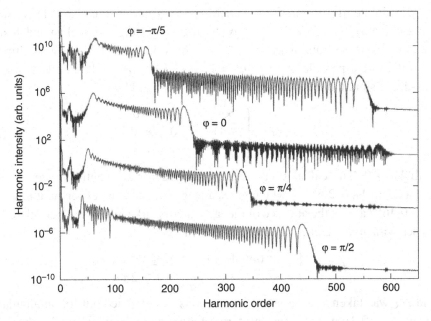

Figure 5.2. Magnitude squared of the Fourier transform of the dipole acceleration as a function of the photon energy (in units of $\hbar\omega$) for four different values of the carrier-envelope phase φ. The results were obtained using a one-dimensional model He$^+$ ion interacting with a two-cycle pulse with $\mathcal{E}_0 = 0.4$ a.u. and carrier wavelength $\lambda = 800$ nm. For clarity, the curves have been offset along the vertical axis. (From R. M. Potvliege, N. J. Kylstra and C. J. Joachain, *J. Phys.* B **33**, L743 (2000).)

Another one-dimensional potential of finite range, used by Patel, Kylstra and Knight [40] is given by

$$V(x) = -\frac{1}{\sqrt{2+x^2}}\exp(-x^2/\beta^2). \qquad (5.75)$$

This potential supports only a finite number of bound states. As β increases this number grows, and their energies approach those of the soft-core Coulomb potential (5.73), with $Z = 1$ and $a = \sqrt{2}$ a.u.

Before leaving one-dimensional studies of the TDSE, we mention the time-dependent R-matrix (TDRM) calculation of Burke and Burke [41], who studied a model multiphoton ionization problem. The TDRM approach is complementary to the R-matrix–Floquet (RMF) theory of Burke, Francken and Joachain [42, 43] discussed in Section 4.7, in that it can be applied to shorter laser pulses, where the use of the Floquet method is questionable. In the TDRM approach, the TDSE is solved within a box using R-matrix-based inner-region basis sets. The wave function of the system is propagated in time using the Crank–Nicolson scheme.

As a simple illustration of the TDRM theory, Burke and Burke [41] considered a one-dimensional model previously discussed by Burke, Francken and Joachain [42,43] within the framework of the RMF theory. It corresponds to the multiphoton ionization of a particle of mass $m = 1$ and charge $q = -1$ initially bound in a one-dimensional, one-sided square well $V(x)$ defined by

$$V(x) = \begin{cases} \infty, & x \leq 0, \\ -V_0, & 0 < x < a, \\ 0, & x \geq a, \end{cases} \tag{5.76}$$

with $V_0 > 0$. The calculations were performed for a potential range $a = 1$ a.u. and a depth $V_0 = 2.5$ a.u., which gives a single bound state of ionization potential $I_P = 0.4657$ a.u. in the absence of the laser field. Burke and Burke considered laser pulses with an envelope function $F(t)$ given by

$$F(t) = \begin{cases} \sin^2(\omega t/4N_R), & 0 \leq t \leq 2\pi N_R/\omega, \\ 1, & t \geq 2\pi N_R/\omega, \end{cases} \tag{5.77}$$

and N_R was taken to be 5. Most calculations were carried out for an angular frequency $\omega = 0.2$ a.u., so that at least three photons were required to ionize the system. A few calculations were also performed for $\omega = 0.3$ a.u. and 0.6 a.u. The TDRM results were checked by comparing them to those obtained by using the RMF theory.

In Fig. 5.3, the logarithm of the total ionization rate obtained by Burke and Burke [41] is plotted as a function of the logarithm of the peak (final) electric-field strength $\mathcal{E}_0$ for $\omega = 0.2$ a.u., 0.3 a.u. and 0.6 a.u. Results for both TDRM calculations (represented by points) and for RMF calculations (full curves) are displayed and are seen to be in excellent agreement. Also shown are LOPT (third-order) perturbation theory results for $\omega = 0.2$ a.u., giving the broken line with slope 6, which corresponds to three-photon ionization. At low electric-field strengths the RMF and TDRM calculations agree with the LOPT results, but at larger electric-field strengths they deviate from the LOPT values, as expected. For $\omega = 0.2$ a.u. the calculations were extended to higher electric-field strengths, where discontinuities in the curve are seen to appear. These discontinuities can be understood as being due to quasi-energy trajectories in the Floquet solution moving onto unphysical sheets and being replaced by new trajectories originating as shadow poles on other unphysical sheets, as discussed in Section 4.4.

We now consider briefly two-dimensional studies of the TDSE in the dipole approximation. The dependence of multiphoton ionization and harmonic generation on the ellipticity of the laser field has been studied by Protopapas, Lappas and Knight [44] and by Patel *et al.* [45] using a soft-core Coulomb potential of the form

$$V(\rho) = -\frac{Z}{\sqrt{(a/Z)^2 + \rho^2}}, \tag{5.78}$$

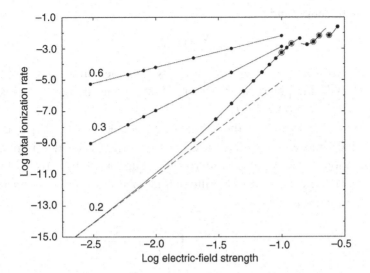

Figure 5.3. Logarithm of the total ionization rate (in a.u.) as a function of the logarithm of the peak (final) electric-field strength $\mathcal{E}_0$ (in a.u.). Full curves: RMF results for $\omega = 0.2$ a.u., 0.3 a.u. and 0.6 a.u. Points: TDRM results for peak electric-field strengths $\mathcal{E}_0 = 0.10$ a.u., 0.12 a.u., 0.18 a.u., 0.20 a.u. and 0.24 a.u. Broken curve: third-order perturbation theory for $\omega = 0.2$ a.u. (From P. G. Burke and V. M. Burke, *J. Phys. B* **30**, L383 (1997).)

with $\rho^2 = x^2 + y^2$, $Z = 1$ and $a = 0.8$ a.u., giving the ionization potential $I_P = 0.5$ a.u. of the hydrogen atom. Using this model potential with $Z = 2$ and $a = 0.798$ a.u. (which yields an ionization potential $I_P = 2.0$ a.u.), Potvliege, Kylstra and Joachain [39] have calculated the photon emission by He$^+$ in intense, ultra-short pulses of carrier wavelength $\lambda = 800$ nm and peak intensity $I = 5.6 \times 10^{15}$ W cm^{-2}.

5.4 Time-dependent Schrödinger equation for atomic hydrogen

We shall now discuss the numerical integration, in three dimensions, of the TDSE in the dipole approximation for the hydrogen atom. The TDSE to be solved is

$$i\frac{\partial}{\partial t}\Psi(\mathbf{r}, t) = [H_0 + H_{\text{int}}(t)]\,\Psi(\mathbf{r}, t), \tag{5.79}$$

where

$$H_0 = -\frac{1}{2}\nabla^2 - \frac{1}{r} \tag{5.80}$$

is the field-free Hamiltonian of the hydrogen atom. The interaction Hamiltonian $H_{\text{int}}(t)$ is given by

$$H_{\text{int}}^{\text{L}}(t) = \mathcal{E}(t) \cdot \mathbf{r}, \tag{5.81}$$

in the length gauge, and by

$$H_{\text{int}}^{\text{V}}(t) = \mathbf{A}(t) \cdot \mathbf{p} \tag{5.82}$$

in the velocity gauge. As in Section 5.1.5, we assume that the laser field is linearly polarized, with the polarization vector $\hat{\boldsymbol{\epsilon}}$ lying along the positive z-axis, so that $\mathbf{A}(t) = \hat{\mathbf{z}}A(t)$ and $\mathcal{E}(t) = \hat{\mathbf{z}}\mathcal{E}(t)$.

Following the hybrid method of Schafer and Kulander [12] and Kulander, Schafer and Krause [3, 46], the wave function is first expanded in partial waves, as in Equation (5.53). In order to derive the equations to be satisfied by the functions $u_l(r, t)$, we express the Hamiltonian in spherical polar coordinates. For the field-free Hamiltonian we have

$$H_0 = -\frac{1}{2}\left[\frac{1}{r^2}\frac{\partial}{\partial r}\left(r^2\frac{\partial}{\partial r}\right) - \frac{\mathbf{L}^2}{r^2}\right] - \frac{1}{r}, \tag{5.83}$$

where $\mathbf{L} = \mathbf{r} \times \mathbf{p}$ is the orbital angular momentum operator, while the interaction Hamiltonian $H_{\text{int}}(t)$ is given by

$$H_{\text{int}}^{\text{L}}(t) = \mathcal{E}(t)r\cos\theta, \tag{5.84}$$

in the length gauge, and

$$H_{\text{int}}^{\text{V}}(t) = -\mathrm{i}A(t)\left(\cos\theta\frac{\partial}{\partial r} - \frac{\sin\theta}{r}\frac{\partial}{\partial\theta}\right) \tag{5.85}$$

in the velocity gauge.

Inserting the expansion (5.53) into the TDSE (5.79), using the fact that

$$\mathbf{L}^2 Y_{l,m} = l(l+1)Y_{l,m} \tag{5.86}$$

and the orthonormality property of the spherical harmonics, we find that the reduced radial functions must satisfy the following coupled equations:

$$\mathrm{i}\frac{\partial}{\partial t}u_l(r, t) = \left[-\frac{1}{2}\frac{\partial^2}{\partial r^2} + \frac{l(l+1)}{2r^2} - \frac{1}{r}\right]u_l(r, t)$$

$$+ r\sum_{l'=0}^{\infty}\langle l, 0|H_{\text{int}}(t)|l', 0\rangle r^{-1}u_{l'}(r, t). \tag{5.87}$$

The matrix elements $\langle l, 0|\cos\theta|l', 0\rangle$ and $\langle l, 0|\sin\theta(\partial/\partial\theta)|l', 0\rangle$ can be written in terms of Clebsch–Gordan coefficients [47]. In the length gauge, the resulting system of coupled equations is found to be

$$\mathrm{i}\frac{\partial}{\partial t}u_0^{\text{L}}(r, t) = \left[-\frac{1}{2}\frac{\partial^2}{\partial r^2} - \frac{1}{r}\right]u_0^{\text{L}}(r, t) + \mathcal{E}(t)r c_0 u_1^{\text{L}}(r, t) \tag{5.88}$$

for $l = 0$, while for $l > 0$ we have

$$i\frac{\partial}{\partial t}u_l^L(r,t) = \left[-\frac{1}{2}\frac{\partial^2}{\partial r^2} + \frac{l(l+1)}{2r^2} - \frac{1}{r}\right]u_l^L(r,t)$$
$$+ \mathcal{E}(t)r\left[c_l u_{l+1}^L(r,t) + c_{l-1}u_{l-1}^L(r,t)\right], \tag{5.89}$$

where the coefficients c_l are given by

$$c_l = \frac{l+1}{\sqrt{(2l+1)(2l+3)}}. \tag{5.90}$$

In the velocity gauge, for $l = 0$, we have

$$i\frac{\partial}{\partial t}u_0^V(r,t) = \left[-\frac{1}{2}\frac{\partial^2}{\partial r^2} - \frac{1}{r}\right]u_0^V(r,t) - iA(t)c_0\left(\frac{\partial}{\partial r} + \frac{2}{r}\right)u_1^V(r,t) \tag{5.91}$$

and, for $l > 0$,

$$i\frac{\partial}{\partial t}u_l^V(r,t) = \left[-\frac{1}{2}\frac{\partial^2}{\partial r^2} + \frac{l(l+1)}{2r^2} - \frac{1}{r}\right]u_l^V(r,t)$$
$$- iA(t)\left[c_l\left(\frac{\partial}{\partial r} + \frac{l+1}{r}\right)u_{l+1}^V(r,t)\right.$$
$$\left. + c_{l-1}\left(\frac{\partial}{\partial r} - \frac{l}{r}\right)u_{l-1}^V(r,t)\right]. \tag{5.92}$$

The problem has therefore been transformed from a two-dimensional partial differential equation to an infinite set of coupled one-dimensional partial differential equations involving the radial variable r and the time t. In practice, the system of coupled equations is solved up to a maximum value, l_{max}, of the orbital angular momentum quantum number.

In the second step of this hybrid approach, Equations (5.88) and (5.89) in the length gauge and Equations (5.91) and (5.92) in the velocity gauge are solved numerically. Firstly, a set of equidistant grid points $r_j = j\Delta r$, with $j = 1, 2, \ldots, M$, are defined for the radial coordinate. Next, we introduce the vectors $\mathbf{u}_l(t)$, with components $u_{l,j}(t) = u_l(r_j, t)$, and the matrix $\mathbf{u}(t)$ having columns $\mathbf{u}_l(t)$, where $l = 0, 1, \ldots, l_{max}$.

Using the second order central difference formula (5.43), the non-zero elements of the partial wave field-free finite difference Hamiltonian matrix are given by

$$H_{0,l,l'}^{j,j'} = \delta_{l',l}\begin{cases} -1/[2(\Delta r)^2], & j' = j \pm 1, \\ 1/(\Delta r)^2 + l(l+1)/2r_j^2 - 1/r_j, & j' = j. \end{cases} \tag{5.93}$$

We note that this matrix is diagonal in the angular momentum quantum number and is tridiagonal in the grid index j. Turning our attention first to the length gauge, the non-zero matrix elements of the interaction Hamiltonian are

$$H_{\text{int},l,l'}^{j,j'}(t) = \delta_{j'j}\mathcal{E}(t)r_j \begin{cases} c_l, & l' = l+1, \\ c_{l-1}, & l' = l-1. \end{cases} \tag{5.94}$$

This matrix is tridiagonal in the angular momentum quantum number and diagonal in the grid index.

The fact that both the matrix $\mathbf{H}_0$ and the matrix $\mathbf{H}_{\text{int}}(t)$ are tridiagonal suggests that an ADI integration scheme be employed. Introducing the matrix $\mathbf{H} = \mathbf{H}_0 + \mathbf{H}_{\text{int}}(t)$, in analogy with Equation (5.40) we can express the explicit Euler scheme as

$$\mathbf{I} + \mathrm{i}\Delta t\mathbf{H}(t) = [\mathbf{I} + \mathrm{i}\Delta t\mathbf{H}_{\text{int}}(t)][\mathbf{I} + \mathrm{i}\Delta t\mathbf{H}_0], \tag{5.95}$$

which is correct to first order in Δt. Following Equation (5.41), we combine this explicit Euler scheme with the implicit Euler scheme to obtain the ADI scheme

$$\left[\mathbf{I} - \mathrm{i}\frac{\Delta t}{2}\mathbf{H}_0\right]\left[\mathbf{I} - \mathrm{i}\frac{\Delta t}{2}\mathbf{H}_{\text{int}}(t_{i+1/2})\right]\mathbf{u}(t_{i+1})$$

$$= \left[\mathbf{I} + \mathrm{i}\frac{\Delta t}{2}\mathbf{H}_{\text{int}}(t_{i+1/2})\right]\left[\mathbf{I} + \mathrm{i}\frac{\Delta t}{2}\mathbf{H}_0\right]\mathbf{u}(t_i). \tag{5.96}$$

Solving for $\mathbf{u}(t_{i+1})$ now requires that two tridiagonal matrices be inverted. As noted above, the inversion of tridiagonal matrices can be performed very efficiently so that the computation effort of the method increases approximately linearly in $M \times l_{\text{max}}$.

The numerical solution in the velocity gauge proceeds in a similar manner. In this case the interaction Hamiltonian contains a radial derivative. Therefore, in order to express the finite difference representation of the interaction Hamiltonian matrix in terms of tridiagonal matrices, the matrix must be decomposed into four terms. The increase in computational time as compared with the length gauge is a factor two. However, as discussed above, typically fewer partial waves are required to achieve convergence in the velocity gauge.

In Fig. 5.4 we show the photoelectron energy spectrum calculated by Schafer and Kulander [12] for a hydrogen atom in a laser pulse of carrier wavelength $\lambda = 532\,\text{nm}$ (corresponding to a photon energy of 2.33 eV) and peak intensity $I = 2 \times 10^{13}\,\text{W cm}^{-2}$. The laser pulse shape was taken to be trapezoidal, with a linear turn-on of two optical cycles, followed by ten cycles of constant intensity, and a two-cycle turn-off. The peaks for which $E < 0$ correspond to atomic population remaining in the bound excited states of the atom at the end of the pulse. The ATI peaks are labeled by the number s of excess photons (above the weak-field limit of six) absorbed by the photoelectron. For the short laser pulse considered in this calculation, the expected (shifted) photoelectron energies, $\tilde{E}_s$, are given by Equation (1.6). The ATI peaks in Fig. 5.4 are separated by one photon energy and

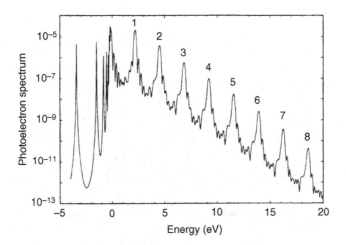

Figure 5.4. Photoelectron energy spectrum for a hydrogen atom in a linearly polarized trapezoidal laser pulse having a carrier wavelength of 532 nm and a peak intensity of 2×10^{13} W cm^{-2}. The labels above the peaks refer to the number s of excess photons absorbed by the ionizing electron. The negative-energy peaks correspond to the atomic population left in bound excited states at the end of the laser pulse (From K. J. Schafer and K. C. Kulander, *Phys. Rev. A* **42**, 5794 (1990).)

indeed appear at the shifted energies $\tilde{E}_s$. The widths of these peaks are consistent with the Fourier transform of the pulse shape, as is expected in the absence of REMPI. At the peak intensity of 2×10^{13} W cm^{-2} chosen for the computation, the ponderomotive shift of the ionization potential is responsible for the fact that the lowest peak ($s = 0$) now lies below the field-free ionization potential.

Following the work of Kulander, Schafer and Krause, the TDSE in the dipole approximation has been solved numerically for the case of atomic hydrogen by several authors [48–56]. As a first example, we show in Fig. 5.5 the results of Dörr, Latinne and Joachain [54], who integrated numerically the TDSE in the velocity gauge using the ADI method in order to study in the high-frequency regime the time evolution of the ionization probability for a hydrogen atom initially in the ground state. The laser pulse was assumed to be linearly polarized and described by the vector potential $\mathbf{A}(t) = \hat{\boldsymbol{\epsilon}} A(t)$, where the polarization vector $\hat{\boldsymbol{\epsilon}}$ was taken along the z-axis and $A(t)$ had a Gaussian envelope, namely

$$A(t) = A_0 \exp\left[-(t - t_0)^2/\beta^2\right]\cos(\omega t), \qquad (5.97)$$

with $\omega = 2$ a.u., $\beta = 9.43$ a.u. and $\mathcal{E}_0 = \omega A_0 = 10$ a.u. (corresponding to a peak intensity of 3.5×10^{18} W cm^{-2}). The line labeled "A^2" in Fig. 5.5 shows $A^2(t)$ in arbitrary units. The laser pulse comprised approximately ten optical cycles. The wavy line in Fig. 5.5 corresponds to the quantity $1 - P_{1s}$ obtained from the numerical

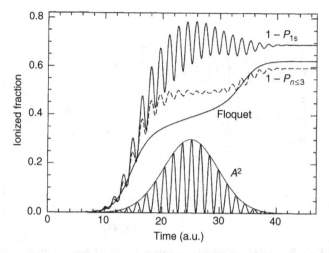

Figure 5.5. Time evolution of the ionization probability in a short, linearly polarized laser pulse of carrier angular frequency $\omega = 2.0$ a.u. and peak electric-field strength $\mathcal{E}_0 = 10.0$ a.u. with a Gaussian envelope, for which the quantity $A^2(t)$ is shown in arbitrary units. The curves corresponding to the Floquet adiabatic ionization probability and the ionization probabilities $1 - P_{1s}(t)$ and $1 - P_{n \leq 3}(t)$ obtained from the numerical integration of the TDSE are discussed in the text. (From M. Dörr, O. Latinne and C. J. Joachain, *Phys. Rev. A* **52**, 4289 (1995).)

solution of the TDSE, while the smooth line corresponds to the Floquet adiabatic ionization probability

$$P_F(t) = 1 - \exp\left(-\int_0^t \Gamma[A(t')]dt'\right). \tag{5.98}$$

We note that at the end of the laser pulse the Floquet result $P_F = 0.625$ is quite close to the result $1 - P_{1s} = 0.692$. By including the atomic population left in bound excited states, which are populated due to the frequency spread of the pulse, one obtains the ionization probability as follows:

$$1 - P_{n \leq 3} = 1 - \sum_j P_j$$

$$= 1 - \sum_j |\langle \psi_j | \Psi(t_0 + \tau)\rangle|^2, \quad j = 1s, 2s, 2p, 3s, 3p, 3d. \tag{5.99}$$

This probability is represented in Fig. 5.5 by the wavy dashed line. At the end of the pulse, this quantity was found to be 0.595, with the largest excited state population being $P_{2s} = 0.090$.

As a second example, we show in Fig. 5.6 a high-order ATI photoelectron spectrum of atomic hydrogen in a 25 fs linearly polarized laser pulse of carrier angular

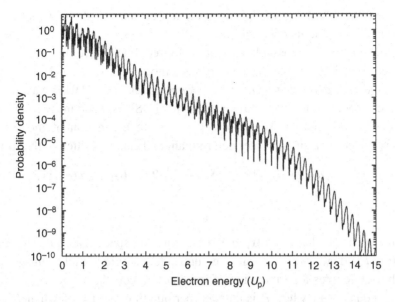

Figure 5.6. Photoelectron ATI spectrum of atomic hydrogen in a linearly polarized laser pulse of 25 fs FWHM, photon energy $\hbar\omega = 2$ eV and peak intensity of 2×10^{14} W cm^{-2}. (From E. Cormier and P. Lambropoulos, *J. Phys. B* **30**, 77 (1997).)

frequency $\omega = 0.073$ a.u. (corresponding to a photon energy of 2 eV) and peak intensity $I = 2 \times 10^{14}$ W cm^{-2} calculated by Cormier and Lambropoulos [56], who solved the TDSE in the velocity gauge by expanding the angular part of the wave function in spherical harmonics and the radial part in B-spline basis functions. These calculations exhibit a plateau in the ATI spectra, as observed in experiments [57, 58]. The discovery of this plateau was mentioned in Section 1.3 (see Fig. 1.9).

5.5 Time-dependent Schrödinger equation for complex atoms

5.5.1 Time-dependent Hartree–Fock approximation

Let us consider the numerical integration of the TDSE, in the dipole approximation, for an atomic system composed of a nucleus of atomic number Z and N electrons. A standard approximation used to perform time-independent calculations for many-electron atomic systems is the *Hartree–Fock method*, based on the *independent particle model*, in which each of the atomic electrons moves in a *self-consistent field* that takes into account the attraction of the nucleus and the average effect of the repulsive interactions due to the other electrons [59]. In the Hartree–Fock approximation, each electron in a multi-electron system is therefore described by its own spin-orbital. Although a certain amount of electron correlation is already included in the Hartree–Fock wave function because of the fact that it is totally antisymmetric, one usually denotes by *electron correlation effects* the remaining electron correlations not included in the Hartree–Fock wave function.

The extension of the Hartree–Fock method to time-dependent calculations, called the *time-dependent Hartree–Fock approximation* (TDHF) was first used by Kulander [60, 61] to solve numerically the TDSE for helium and xenon atoms in intense laser pulses. The time-dependent wave function of the system was approximated by a single Slater determinant and can be written in the form

$$\Psi_{HF}(q_1, q_2, \ldots, q_N, t) = (N!)^{-1/2} \sum_P (-1)^P P \{\varphi_\alpha(q_1, t)\varphi_\beta(q_2, t) \ldots \varphi_\nu(q_N, t)\},$$

$$(5.100)$$

where we recall that $q_i \equiv (\mathbf{r}_i, \sigma_i)$ represents the space and spin coordinates of electron i and $\varphi_\alpha, \varphi_\beta, \ldots, \varphi_\nu$ are time-dependent single-particle spin-orbitals. The symbol P denotes a permutation of the electron coordinates $q_1, q_2, \ldots, q_N$, with $(-1)^P$ equal to $+1$ when P is an even permutation and to -1 when P is an odd permutation; the sum is over all permutations P. Equation (5.100) can be expressed in the more compact form

$$\Psi_{HF}(q_1, q_2, \ldots, q_N, t) = (N!)^{1/2} \mathcal{A}\{\varphi_\alpha(q_1, t)\varphi_\beta(q_2, t) \ldots \varphi_\nu(q_N, t)\}, \quad (5.101)$$

where

$$\mathcal{A} = \frac{1}{N!} \sum_P (-1)^P P \qquad (5.102)$$

is an *antisymmetrization* operator, which is also a projection operator since $\mathcal{A}^2 = \mathcal{A}$.

Substituting the ansatz (5.101) into the TDSE (2.121), we obtain for a laser pulse linearly polarized along the z-axis, the system of coupled integro-partial differential equations

$$i\frac{\partial}{\partial t}\varphi_\lambda(q_i, t) = \left\{ -\frac{1}{2}\nabla_{r_i}^2 - \frac{Z}{r_i} + \sum_\mu \int dq_j \frac{|\varphi_\mu(q_j, t)|^2}{r_{ij}} + H_{int}^i(t) \right\} \varphi_\lambda(q_i, t)$$

$$- \sum_\mu \int dq_j\, \varphi_\mu^*(q_j, t) \frac{1}{r_{ij}} \varphi_\lambda(q_j, t)\varphi_\mu(q_i, t), \qquad (5.103)$$

where $r_{ij} = |\mathbf{r}_i - \mathbf{r}_j|$, $H_{int}^i(t) = \mathcal{E}(t)z_i$ in the length gauge and $H_{int}^i(t) = -iA(t)\partial/\partial z_i$ in the velocity gauge. The symbol $\int dq_j$ means an integration over the spatial coordinates $\mathbf{r}_j$ and a summation over the spin coordinate σ_j of electron j, and the summation over μ extends over the N occupied spin-orbitals. The last term on the right-hand side of Equation (5.103) is due to electron exchange. It is seen that both the direct and the exchange terms on the right-hand side of Equation (5.103) depend on the spin-orbitals, so that these equations are *non-linear*. Moreover, the exchange term is non-local so that its evaluation is computationally intensive. It is often replaced by a local approximation that depends only on the electron density [62]. This replacement, known as the *local density approximation* (LDA), does not modify the dynamics significantly since the electron exchange term itself is of relatively small importance.

Although the TDHF method has the advantage of reducing considerably the computational effort at the expense of neglecting electron correlation effects (as defined above), it suffers from important shortcomings. This can be illustrated for the case of helium, initially in the ground state, by writing the doubly occupied spatial orbital as

$$\varphi_\lambda(\mathbf{r}, t) = c_{1s}(t)\psi_{1s}(\mathbf{r}) + c_k(t)\psi_k(\mathbf{r}), \tag{5.104}$$

where ψ_{1s} and ψ_k are, respectively, the helium ground-state orbital and a continuum orbital. The time-dependent expansion coefficients $c_{1s}(t)$ and $c_k(t)$ are determined by inserting the expansion (5.104) into the TDHF equation (5.103). From the resulting set of non-linear coupled equations, if follows that the states ψ_{1s} and ψ_k are still coupled at time $t_0 + \tau$ at the end of the pulse. As a result, stable transition probabilities cannot be obtained from TDHF wave functions. Indeed, projecting a TDHF wave function onto eigenstates gives probabilities that oscillate in time even after the laser–atom interaction has been turned off [3, 63]. In addition, the TDHF ionization potential decreases approximately linearly with the population $|c_{1s}|^2$ in the ψ_{1s} orbital. The TDHF approach therefore gives rise to a system that evolves in a continuous manner from one partially ionized state to another. This is in contrast to a real system which, on the time scale of an optical cycle, will either ionize or not. Moreover, because the emission process is extremely sensitive to the instantaneous ionization potential, unphysical population trapping in the ground state occurs, making it impossible to extract ionization rates from the ground-state occupation probabilities [63].

Finally, the double ionization probability depends in an unphysical way on the single ionization probability. Indeed, using the simple expansion (5.104), the single ionization probability is given by $2\text{Re}[c_{1s}^*(t_0 + \tau)c_k(t_0 + \tau)]$, while the double ionization probability is $|c_k(t_0 + \tau)|^2$. It should not be expected that such a relationship would exist. In fact, this relationship between single and double ionization is not observed experimentally. Multiphoton double ionization of atoms will be discussed in Section 8.2.

5.5.2 *Time-dependent density functional theory*

Density functional theory (DFT) was first introduced by Hohenberg and Kohn [64] to describe a system of interacting particles in terms of its *number density* $\rho(\mathbf{r})$ in the ground state. Indeed, according to a theorem proved by Hohenberg and Kohn [64], every observable of a stationary system of interacting particles is uniquely determined by the number density $\rho(\mathbf{r})$. The theory is based on an exact mapping between densities and external potentials, which implies that the density of the interacting system can be obtained from the density of an auxiliary system of non-interacting particles moving in an *effective local single-particle potential* $V_{eff}[\rho](\mathbf{r})$ which is a functional of the number density $\rho(\mathbf{r})$ and includes an exchange-correlation contribution $V_{xc}[\rho](\mathbf{r})$. The success or failure of the DFT depends on finding an adequate approximation for V_{xc}. Kohn and Sham [65] introduced for this purpose a LDA, which has proved to be good for systems having a large number of electrons [66, 67].

A time-dependent DFT (TDDFT) has been proposed by Runge and Gross [68]. They extended the Hohenberg–Kohn result by proving that, in a time-dependent problem, every observable can be obtained from the *time-dependent number density* $\rho(\mathbf{r}, t)$. For the case of an N-electron atom in a laser field, we have

$$\rho(\mathbf{r}, t) = \sum_{j=1}^{N} |\phi_j(\mathbf{r}, t)|^2, \tag{5.105}$$

where the single-electron orbitals $\phi_j(\mathbf{r}, t)$ satisfy the time-dependent Kohn–Sham equations

$$i\frac{\partial}{\partial t}\phi_j(\mathbf{r}, t) = \left[-\frac{1}{2}\nabla^2 + V_{eff}[\rho](\mathbf{r}, t) \right]\phi_j(\mathbf{r}, t), \tag{5.106}$$

with

$$V_{eff}[\rho](\mathbf{r}, t) = V_{ext}(\mathbf{r}, t) + \int \frac{\rho(\mathbf{r}', t)}{|\mathbf{r} - \mathbf{r}'|} \, d\mathbf{r}' + V_{xc}[\rho](\mathbf{r}, t). \tag{5.107}$$

In the above equation,

$$V_{ext}(\mathbf{r}, t) = -\frac{Z}{r} + H_{int}(t), \tag{5.108}$$

where $H_{int}(t)$ is the Hamiltonian describing the laser–atom interaction. The exchange-correlation potential $V_{xc}[\rho](\mathbf{r}, t)$ is a functional of the time-dependent number density $\rho(\mathbf{r}, t)$ [69].

In practice, the exchange-correlation effects present in $V_{xc}[\rho](\mathbf{r}, t)$ must be approximated. Two approximation schemes have generally been used. The first one is the LDA described above, and the second one is a time-dependent version of the optimized potential method discussed by Ullrich, Grossman and Gross [70].

The TDDFT approach is structurally similar to the TDHF method described above, and therefore suffers from the same drawbacks. However, it is computationally less intensive since it does not include non-local potentials. Using the TDDFT, the interaction of various multi-electron atoms with intense laser pulses has been studied by Gross, Dobson and Petersilka [69]. In general, the time evolution of several multiply occupied orbitals has been calculated. For example, Gross, Dobson and Petersilka have determined the time evolution of the 2s, $2p_0$ and $2p_1$ orbitals in neon for a laser pulse of carrier wavelength $\lambda = 248$ nm, ramped up linearly over ten optical cycles to a peak intensity of 3×10^{15} W cm^{-2}. From this time evolution, they estimated atomic populations for the first three charged states of neon.

5.5.3 Single-active-electron model

In order to avoid problems associated with the non-linear Hartree–Fock equations (5.103) when calculating multiphoton transitions, Kulander [61] and Krause, Schafer and Kulander [71] considered a simpler approximation in which the atomic system is treated by using a *single-active-electron* (SAE) model, such that the atomic electron undergoing the transition moves in a mean field potential. This potential can be generated, for example, from a Hartree–Slater calculation [72] for the ground state or an excited state, with different potentials for different orbital angular momentum quantum numbers l.

As an example, let us consider the numerical simulations of high-order ATI spectra in argon performed by Muller [8] using an SAE model argon atom interacting with a linearly polarized laser pulse in the velocity gauge. Muller constructed a spherically symmetric mean field potential of the form

$$V(r) = V_1(r) + V_2(r), \qquad (5.109)$$

where the potential well V_1 is given by

$$V_1 = -\frac{1}{r}[1 + A\exp(-Br) + (17 - A)\exp(-Cr)], \qquad (5.110)$$

with $A = 5.4$, $B = 1$ and $C = 3.682$. The eigenenergies of an electron bound in this potential reproduce accurately the configuration averages of the energies of the singly excited states. In addition, the ionization potentials of the K-shell, L-shell and 3s sub-shell are well reproduced, and $V_1(r)$ has the correct behavior for $r \to 0$ and $r \to \infty$. Thus, the potential $V_1(r)$ is an excellent approximation to the electron–ion interaction (including exchange) provided that core excitations are absent. In addition, to account for the presence of inner shells, Muller [8] corrected the potential $V_1(r)$ by a soft repulsive core $V_2(r)$. The TDSE was solved numerically in the velocity gauge using the hybrid method discussed in Section 5.4 for a laser pulse linearly polarized along the z-axis. At the outer grid boundary, an absorbing wall removed the ejected electrons by multiplying, after each time step, the reduced radial functions with a mask function $1 - (d/60)^4$, where d is the penetration depth into the absorber.

The calculations of Muller [8] were performed for argon atoms, initially in the ground 3p state oriented along the laser polarization vector $\hat{\boldsymbol{\epsilon}}$, subjected to an N-optical-cycle flat-top laser pulse described by the vector potential

$$\mathbf{A}(t) = \hat{\boldsymbol{\epsilon}} A_0 \cos(\omega t), \quad 0 < t < NT, \tag{5.111}$$

preceded by a half-cycle turn-on

$$\mathbf{A}(t) = \hat{\boldsymbol{\epsilon}} A_0 [0.5 \cos(\omega t) + 0.125 \cos(2\omega t) + 0.375], \quad -T/2 < t < 0, \tag{5.112}$$

and followed by a similar turn-off, where $T = 2\pi/\omega$ is the optical period. The form of the turn-on and turn-off was chosen so that the vector potential is four times differentiable at $t = -T/2$, $t = 0$, $t = NT$ and $t = (N+1/2)T$, thus minimizing non-adiabatic shake-up transitions from the ground state. After the interaction with the laser pulse, the wave function was energy-analyzed for each orbital angular momentum l by using the energy window operator described in Section 5.2.

As an illustration, we show in Fig. 5.7 a photoelectron ATI spectrum calculated by Muller [8] from a 19-cycle flat-top linearly polarized laser pulse of the type described by Equations (5.111) and (5.112), with carrier wavelength $\lambda = 800$ nm and peak electric-field strength $\mathcal{E}_0 = 0.044$ a.u., corresponding to a peak intensity

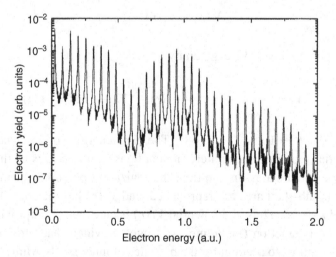

Figure 5.7. Calculated photoelectron ATI spectrum of argon from the final ten cycles of a 19-cycle flat-top laser pulse with a carrier wavelength of 800 nm and a peak intensity of 6.8×10^{13} W cm^{-2}, along the polarization vector. (From H. G. Muller, *Phys. Rev. A* **60**, 1341 (1999).)

of 6.8×10^{13} W cm^{-2}. By varying the position of the absorber, only photoelectrons produced during the final ten cycles of the laser pulse were retained. This spectrum contains peaks up to 54 eV. Faster electrons are at least partly absorbed at the outer boundary of the grid. For the chosen peak intensity, the spectrum exhibits a pronounced plateau around 25 eV. At intensities 5% lower or higher, the corresponding peaks are reduced by an order of magnitude, suggesting that the plateau is due to a resonance. This behavior is very similar to high-order enhancements observed in experiments [73–75]. We shall return to this point in Section 8.1, where we will see that the SAE model is capable of describing the physics of multiphoton single ionization – and more generally multiphoton processes involving single one-electron transitions – very accurately.

5.5.4 Time-dependent Schrödinger equation for two-electron atoms

We now turn to the numerical solution of the TDSE, in the dipole approximation, for two-electron atoms. This equation can be written in the form

$$i\frac{\partial}{\partial t}\Psi(q_1, q_2) = [H_0 + H_{int}(t)]\Psi(q_1, q_2), \tag{5.113}$$

where we recall that $q_i \equiv (\mathbf{r}_i, \sigma_i)$ represents the space and spin coordinates of electron i. The field-free Hamiltonian of the two-electron atom is given by

$$H_0 = -\frac{1}{2}\nabla_{\mathbf{r}_1}^2 - \frac{1}{2}\nabla_{\mathbf{r}_2}^2 - \frac{Z}{r_1} - \frac{Z}{r_2} + \frac{1}{r_{12}}, \tag{5.114}$$

with $r_{12} = |\mathbf{r}_1 - \mathbf{r}_2|$. The interaction Hamiltonian is given in the length gauge by

$$H_{int}^{L}(t) = \mathcal{E}(t) \cdot \mathbf{R}, \tag{5.115}$$

where $\mathbf{R} = \mathbf{r}_1 + \mathbf{r}_2$. In the velocity gauge, we have

$$H_{int}^{V}(t) = \mathbf{A}(t) \cdot \mathbf{P} \tag{5.116}$$

with $\mathbf{P} = \mathbf{p}_1 + \mathbf{p}_2$. We shall assume that the laser pulse is linearly polarized, so that the number of effective spatial dimensions in Equation (5.113) is five.

A first approach to the solution of the TDSE (5.113) has been proposed by Tang, Rudolph and Lambropoulos [76], who used an atomic basis set in a box. This type of calculation consists of two steps. In the first one, the field-free states of the atom are represented, using B-spline basis functions, by a set of eigenstates calculated inside the box, so that the complete set of field-free eigenstates (bound and continuum) is approximated by an orthonormal set of discretized states [77]. An important advantage of using discretized atomic basis sets is that they are relatively easy to construct, since one diagonalization is required to obtain a set of states for each total angular momentum symmetry. In the second step, the time-dependent wave function is expanded in this atomic basis set. Substituting this expansion into the TDSE leads to a set of coupled ordinary differentially equations for the expansion coefficients, as was shown in Section 5.1.4. Zhang and Lambropoulos [78] have carried out extensions of this type of calculations to larger ranges of angular frequencies and intensities, and have calculated photoelectron spectra for helium [78] and the quasi-two-electron magnesium atom [79]. Typically, their calculations were performed using a box with a radius of 1000 a.u. and 902 B-splines for each subset of single-electron orbitals with angular momentum quantum numbers from $l = 0$ to $l = 20$. The influence of open channels associated with excited states of the remaining ion (corresponding to multiple continua) was also investigated. Similar to the inner-region basis used in R-matrix–Floquet calculations [42, 43], a two-electron basis consisting of suitably chosen orbitals for one of the electrons and a set of basis functions for the second (ionizing) electron was employed. A detailed account of these calculations can be found in the review article of Lambropoulos, Maragakis and Zhang [6].

Another approach has been proposed by Scrinzi and Piraux [80,81]. They solved the TDSE for helium using a discretized atomic basis constructed in terms of two-electron Hylleraas-type functions [59], together with complex scaling. They obtained results for single ionization and for atomic populations in doubly excited states, for laser pulses of relatively short carrier wavelength ($\lambda = 39.6$ nm and 151.8 nm) and short duration (3.8 to 15 fs). At the intensity considered (about 10^{14} W cm^{-2}) departures from the LOPT power-law dependence I^n of multiphoton ionization were found.

Mercouris, Dionissopoulou and Nicolaides [82] have used a basis of field-free atomic states obtained using a multi-configuration Hartree–Fock approach to calculate ionization rates for helium, which they have compared with experimental and theoretical results of Charalambidis *et al.* [83] for the generalized cross section corresponding to ionization from the helium ground state interacting with a laser pulse of carrier wavelength $\lambda = 248$ nm. Multiphoton ionization from the metastable 2^1S state of helium has also been investigated [84].

Using a massively parallel computer (Cray T3D), Taylor *et al.* [85–89] have performed a direct numerical integration of the TDSE (5.113) for laser-driven helium, initially in the ground state. They used a hybrid basis set in which the radial coordinates r_1 and r_2 were discretized on a grid, and the four angular coordinates $\theta_1, \theta_2, \phi_1, \phi_2$ were handled by expanding the wave function on a basis set of coupled spherical harmonics. Assuming that the laser pulse is linearly polarized, with the polarization vector $\hat{\epsilon}$ along the direction of the z-axis, the total magnetic quantum number M is constant so that, as noted above, the number of effective spatial dimensions is five. The total electron spin quantum number S is conserved and is equal to $S = 0$, so that the atom remains in a singlet state at all times. The spatial part of the time-dependent wave function $\Psi(r_1, r_2, \theta_1, \theta_2, \phi_1, \phi_2, t)$ must therefore be symmetric under the interchange of the spatial coordinates of the two electrons. Hence, this wave function can be written in the form

$$\Psi(r_1, r_2, \theta_1, \theta_2, \phi_1 - \phi_2, t) = S \sum_{l_1, l_2, L} f_{l_1, l_2, L}(r_1, r_2, t) \mathcal{Y}_{l_1, l_2}^{L0}(\theta_1, \theta_2, \phi_1 - \phi_2).$$

(5.117)

The coupled spherical harmonics $\mathcal{Y}_{l_1, l_2}^{LM}$ are given by

$$\mathcal{Y}_{l_1, l_2}^{LM}(\theta_1, \phi_1, \theta_2, \phi_2) = \sum_{m_1, m_2} \langle l_1, l_2, m_1, m_2 | LM \rangle Y_{l_1, m_1}(\theta_1, \phi_1) Y_{l_2, m_2}(\theta_2, \phi_2),$$

(5.118)

where $\langle l_1, l_2, m_1, m_2 | LM \rangle$ are Clebsch–Gordan coefficients. Since the initial state of the system is chosen to be the field-free helium atom in the ground state 1^1S with $M = 0$, the quantum number M can be omitted so that

$$\mathcal{Y}_{l_1, l_2}^{L0}(\theta_1, \theta_2, \phi_1 - \phi_2) = \sum_{m_1} \langle l_1, l_2, m_1, -m_1 | L0 \rangle Y_{l_1, m_1}(\theta_1, \phi_1) Y_{l_2, -m_1}(\theta_2, \phi_2).$$

(5.119)

The operator S in Equation (5.117) enforces even symmetry of the wave function under interchange of the electron spatial coordinates.

The calculations were performed in the velocity gauge. Upon substitution of the expansion (5.117) into the TDSE (5.113), one obtains for the radial functions $f_{l_1, l_2, L}(r_1, r_2, t)$ a system of coupled two-dimensional time-dependent partial differential equations, each of which is solved by finite difference techniques. The electron–electron interaction term $1/r_{12}$ is expanded in Legendre polynomials as

$$\frac{1}{r_{12}} = \sum_{l=0}^{\infty} \frac{(r_<)^l}{(r_>)^{l+1}} P_l(\cos\theta_{12}),$$

(5.120)

where $r_<$ is the smaller and $r_>$ the larger of r_1 and r_2, and θ_{12} is the angle between the vectors $\mathbf{r}_1$ and $\mathbf{r}_2$. This expansion was truncated to a limited number of terms, up to a maximum value l_{max}. The ground state of the finite difference grid was found by using the Lanczos method [2] to obtain an initial estimate of the eigenvector and subsequently integrating the TDSE in imaginary time. These calculations have also been used to assess the accuracy of SAE models [87, 90].

As an example, we show in Fig. 5.8 the results of a benchmark calculation performed by van der Hart *et al.* [91] for the multiphoton ionization of helium at the wavelength $\lambda = 390\,$nm, which is accessible to Ti:sapphire lasers

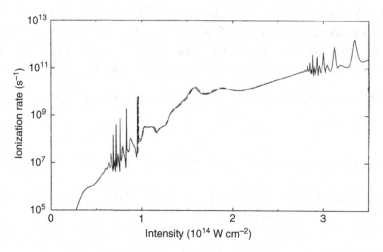

Figure 5.8. Total ionization rate of helium versus intensity for a laser field of wavelength $\lambda = 390$ nm. The solid line corresponds to the results obtained from the numerical integration of the TDSE and the dashed line to the RMF results. (From H. W. van der Hart *et al.*, *J. Phys. B* **38**, L207 (2005).)

by means of frequency doubling. The total ionization rates obtained from the direct numerical integration of the TDSE are compared with those obtained using the *R*-matrix–Floquet (RMF) method. The figure shows two occurrences of channel closing. Below an intensity of 0.58×10^{14} W cm^{-2} at least eight photons are required to ionize He. Between the intensities of 0.58×10^{14} W cm^{-2} and 2.75×10^{14} W cm^{-2} at least nine photons are necessary for ionization, while between the intensities of 2.75×10^{14} W cm^{-2} and 5.0×10^{14} W cm^{-2} at least ten photons are required for ionization. At energies just above the threshold at 0.58×10^{14} W cm^{-2}, where at least nine photons are required to ionize the atom, the first eight photons excite one of the $1snl$ Rydberg bound states of the helium atom, and the ninth photon ionizes this state. This gives rise to REMPI peaks, as seen in Fig. 5.8. In a similar way, REMPI peaks appear just above the threshold at 2.75×10^{14} W cm^{-2}.

A prominent feature of the results displayed in Fig. 5.8 is the excellent agreement between the values obtained by using the RMF theory and those computed by numerical integration of the TDSE. This agreement between two different methods, which complement each other by emphasizing the energy and time domains, respectively, is gratifying.

Finally, we remark that a stringent test for the various approaches to the non-perturbative solution of the TDSE for He in a laser field is to account correctly for multiphoton *double ionization*. We shall return to this question in Section 8.2.

5.5.5 Time-dependent R-matrix method

In order to study the dynamics of complex atoms ($N \geq 2$) in intense laser pulses beyond the SAE approximation, an approach is needed which can accurately describe the multielectron atomic structure (including electron correlations) and the multielectron response to the laser field. For atoms interacting with relatively long pulses, such an approach is provided by the R-matrix–Floquet (RMF) theory [42,43]. Since the RMF theory assumes that the laser field has a constant intensity, it cannot be applied to atoms interacting with ultra-short laser pulses. However, we have seen in Section 5.3 that Burke and Burke [41] proposed an extension of the RMF theory, called the time-dependent R-matrix (TDRM) theory, which they used to study a model one-dimensional multiphoton ionization problem.

More recently, the TDRM theory has been used by van der Hart, Lysaght and Burke [92] to analyze the time-dependent, multielectron atomic dynamics in intense, short laser pulses. This approach solves the TDSE within a box using R-matrix-based inner-region basis sets. The time-dependent wave function of the system is propagated in time using the Crank–Nicolson scheme discussed in Section 5.1.

The wave function is expanded in terms of eigenstates of the field-free Hamiltonian H_0 for each accessible total orbital angular momentum quantum number $L \leq L_{\max}$. In their calculations, van der Hart, Lysaght and Burke have used the length gauge. This is in contrast with intense-field calculations for one- and two-electron atoms, where the velocity gauge is, in general, preferable, as we have seen above. However, the choice of the velocity gauge emphasizes short-range electron excitations. In particular, the neglect of energetic three- and four-electron excitations would lead to significant loss of accuracy. As a result, the use of the velocity gauge requires a better description of the atomic structure than that of the length gauge. It is worth noting that, in the RMF approach, the length gauge is also used in the inner region. While the length gauge allows a simpler description of the atomic structure in the calculations, this choice has it own drawbacks since the description of the system must include many angular momenta, leading to large values of $L_{\max}$.

To demonstrate the accuracy of the TDRM approach, van der Hart, Lysaght and Burke [92] studied the multiphoton ionization of argon interacting with linearly polarized laser pulses of carrier wavelength $\lambda = 390$ nm, with a three-cycle $\sin^2$ turn-on, followed by N (between 0 and 14) optical cycles of constant electric-field strength $\mathcal{E}_0$ and a three-cycle $\sin^2$ turn-off. The calculations were performed in a box of radius 100 a.u., taking $L_{\max} = 18$ and typically 2000 time steps per optical cycle.

As as illustration, the intensity dependence of the ionization rates for argon are displayed in Fig. 5.9, where they are compared with the rates obtained by van der Hart [93] using the RMF theory. Also shown are rates computed using

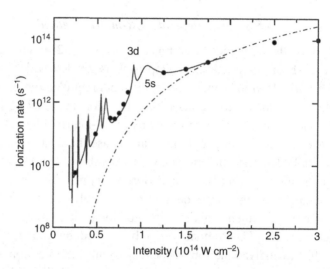

Figure 5.9. Comparison of ionization rates for argon interacting with a linearly polarized 390 nm laser field as a function of intensity. Solid circles: TDRM results. Solid line: RMF results. Dashed line: ADK results. The intermediate $3s^2 3p^5 3d$ ^1P, ^1F and $3s^2 3p^5 5s$ ^1P resonances are indicated (From H. W. van der Hart, M. A. Lysaght and P. G. Burke, *Phys. Rev. A* **76**, 043405 (2007).)

the Ammosov–Delone–Krainov (ADK) tunneling formula [94], which will be discussed in Section 6.1. The agreement between the TDRM and RMF ionization rates is excellent, typically well within 10%. At the higher intensities, the rates are in close agreement with those obtained from the ADK approximation, although the ADK rates tend to increase faster with intensity. At lower intensities, the differences between the TDRM or RMF rates and the ADK tunneling rates increase, as the ADK method becomes less appropriate. We also note in Fig. 5.9 that near an intensity of 0.75×10^{14} W cm^{-2} the ionization rates in the RMF calculations are enhanced due to $3s^2 3p^5 3d$ ^1P and ^1F resonances [93]. This REMPI phenomenon can also been seen in the TDRM results.

5.6 Non-dipole effects

As discussed in Section 2.8, the dipole approximation becomes inadequate to treat the interaction of an atom with electromagnetic radiation at sufficiently high-intensities. We shall consider in this section the numerical solution of the non-relativistic, non-dipole TDSE, which includes *non-dipole* terms arising from the fact that the vector potential $\mathbf{A}(\mathbf{r}, t)$ depends not only on the time t, but also on the position $\mathbf{r}$. As a result, the magnetic-field component of the laser field,

$\mathcal{B}(\mathbf{r}, t) = \nabla \times \mathbf{A}(\mathbf{r}, t)$, no longer vanishes (as it does in the dipole approximation) and *magnetic-field* effects come into play.

In what follows, we shall consider the case of one-electron atoms and ions interacting with a laser pulse linearly polarized and with wave vector $\mathbf{k}_L$, where $k_L = \omega/c$. Using Equation (2.193) and assuming that the laser pulse is described by the vector potential $\mathbf{A}(\mathbf{r}, t) = \mathbf{A}(\eta) = \hat{\boldsymbol{\epsilon}} A(\eta)$, where $\eta = \omega t - \mathbf{k}_L \cdot \mathbf{r}$, the corresponding non-relativistic, non-dipole TDSE is given by

$$i\frac{\partial}{\partial t}\Psi(\mathbf{r}, t) = \left(\frac{1}{2}[-i\nabla + \mathbf{A}(\eta)]^2 + V(r)\right)\Psi(\mathbf{r}, t). \tag{5.121}$$

Let us first consider the case of atomic hydrogen. The above equation with $V(r) = -1/r$ was solved numerically by Latinne, Joachain and Dörr [52] using the ADI method in three spatial dimensions. Their calculations were performed for laser pulses turned on very rapidly with a two-cycle ramp. They found only small differences from the corresponding dipole approximation results for a carrier angular frequency $\omega = 2$ a.u. and a peak electric field strength $\mathcal{E}_0 = 16$ a.u. Similar results were obtained for $\omega = 5$ a.u. and $\mathcal{E}_0 = 40$ a.u.

Bugacov, Pont and Shakeshaft [95] calculated the effects of the lowest-order correction to the dipole approximation by solving the corresponding TDSE for atomic hydrogen, using Sturmian basis functions. They showed that the corrections are comparatively larger for excited states. In particular, they calculated ionization yields for atomic hydrogen initially in the 2p state, interacting with laser pulses having a Gaussian profile with a width of five optical cycles, a carrier angular frequency $\omega = 0.5$ a.u. and peak intensities up to $4 \times 10^{17}\,\mathrm{W\,cm^{-2}}$. Calculations have also been carried out by Meharg, Parker and Taylor [96] using a hybrid method similar to that of Kulander, Schafer and Krause [3] discussed in Sections 5.1 and 5.4. As seen from Fig. 5.10, the results of Meharg, Parker and Taylor [96] are in good agreement with those of Bugacov, Pont and Shakeshaft [95].

As a further step, Latinne, Joachain and Dörr [52] included the coupling of the magnetic-field component of the laser pulse to the spin of the electron within a non-relativistic treatment. To this end, they solved the time-dependent Pauli equation (2.269) for a two-component time-dependent wave function describing the two spin orientations of the electron. Only small differences in the ionization probability were found in comparison with the result obtained from the solution of the non-dipole TDSE (5.121).

The non-dipole TDSE (5.121) has also been integrated numerically for two-dimensional model atoms [39, 97–102]. In particular, Vázquez de Aldana and Roso [98], Kylstra *et al.* [99, 101] and Vázquez de Aldana *et al.* [100] considered a model two-dimensional hydrogen atom described by the soft-core Coulomb

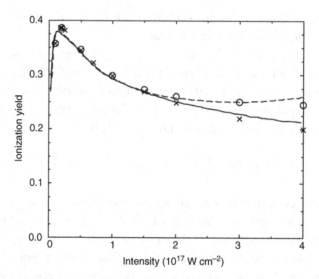

Figure 5.10. Ionization yield as a function of peak laser intensity for atomic hydrogen initially in the $2p_x$ state interacting with a Gaussian profile laser pulse having a five-cycle width and carrier angular frequency of 0.5 a.u. The results of Bugacov, Pont and Shakeshaft [95] within the dipole approximation are displayed as a solid line, while those including non-dipole corrections are given by the dashed line. The results of Meharg, Parker and Taylor [96] obtained in the dipole approximation correspond to the crosses, while those incorporating non-dipole corrections are given by circles. (From K. J. Meharg, J. S. Parker and K. T. Taylor, *J. Phys. B* **38**, 237 (2005).)

potential (5.78) with $Z = 1$ and $a = 0.80$ a.u. They studied the influence of non-dipole effects on the ionization probability in the high-frequency regime, a problem to which we shall return in Section 7.3.

Potvliege, Kylstra and Joachain [39] have calculated photon emission spectra beyond the dipole approximation by solving Equation (5.121) for the two-dimensional model of the He$^+$ ion described by the soft-core Coulomb potential (5.78) with $Z = 2$ and $a = 0.798$ a.u. These calculations were performed for few-cycle laser pulses with carrier wavelength $\lambda = 800$ nm and a peak electric-field strength $\mathcal{E}_0 = 0.16$ a.u., corresponding to an intensity of 5.6×10^{15} W cm^{-2}. They were compared with calculations using the dipole approximation. At the intensity of 5.6×10^{15} W cm^{-2}, the non-dipole effects were found to be small. The calculated photoemission spectra were also used to assess the accuracy of the *non-dipole, strong-field approximation*, which will be discussed in Section 6.6.

5.7 Relativistic effects

We have seen in Section 2.9 that relativistic effects arise when atoms interact with ultra-strong laser fields. Such effects are expected to become important when the quiver velocity of the electron becomes comparable to the velocity of light c, or in other words when its ponderomotive energy U_p is of the order of its rest mass energy mc^2. As noted in Section 2.9, this means that the quantity $q = U_p/(mc^2) = e^2\mathcal{E}_0^2/(4m^2\omega^2c^2)$ must then be of the order of unity.

In what follows, we shall restrict ourselves to one-electron atoms. A full quantum-mechanical treatment of relativistic effects in the interaction of such atoms with ultra-intense laser pulses requires the numerical solution of the time-dependent Dirac equation (TDDE). This is a very difficult task since the computational problem scales approximately as $\mathcal{E}_0^4/\omega^4$. For this reason, nearly all numerical calculations have been restricted to angular frequencies of the order of the atomic unit or higher. Even in this case the calculations are extremely time consuming, and have been restricted to one- and two-dimensional model atoms. In addition, several of the calculations have been performed using the time-dependent relativistic Schrödinger equation (TDRSE) or the time-dependent Klein–Gordon equation (TDKGE). We shall consider these calculations first, following the review of Joachain and Kylstra [103].

Protopapas, Keitel and Knight [104] have solved numerically the TDRSE in the K-H frame for a model one-dimensional atom described by the soft-core Coulomb potential (5.73) with $Z = 10$ and $a = 1$ a.u. interacting with laser pulses of carrier angular frequency $\omega = 1$ a.u. and peak electric-field strength $\mathcal{E}_0 = 200$ a.u. Magnetic-field effects are not included in one-dimensional model calculations. On the other hand, we have seen in Section 2.9 that a relativistic electron propagates in an electromagnetic field like a particle having a variable, "dressed" mass $m^* = m(1 + 2q)^{1/2}$. This relativistic effect, due to the dressing of the electron mass by the laser field, was investigated by Protopapas, Keitel and Knight [104]. They found differences between the relativistic and non-relativistic treatments, the model atom being slightly more stable against ionization when solving the TDRSE rather than its non-relativistic counterpart.

Taïeb, Véniard and Maquet [105] have solved the TDRSE for a one-dimensional atom, described by the soft-core Coulomb potential (5.73) with $Z = 1$ and $a = \sqrt{2}$ a.u., in a *two-color* laser field. A high-frequency ($\omega_H = 10$ a.u.), ultra-intense ($I_H = 5 \times 10^{21}$ W cm^{-2}) laser pulse was used to dress the atom. A lower-frequency ($\omega_L = 0.2$ a.u.), less intense ($I_L = 10^{15}$ W cm^{-2}) pulse was used to probe the wave function. They found that the ATI peaks in the photoelectron energy spectrum are shifted with respect to the corresponding non-relativistic calculations.

Let us now consider the TDDE. Kylstra, Ermolaev and Joachain [106] solved this equation numerically for a model one-dimensional atom. In one dimension, the four-component Dirac equation reduces to two uncoupled two-component equations, corresponding to the two possible orientations of the electron spin. The calculations were carried out in momentum space, the potential in momentum space, $\hat{V}(q)$

being obtained from that in configuration space, $V(x)$, by performing the Fourier transform

$$\hat{V}(q) = (2\pi)^{-1/2} \int \exp(-iqx) V(x) dx. \tag{5.122}$$

The potential $V(x)$ was chosen to be given by Equation (5.74) with $Z = 1$ and $\epsilon = 10^{-3}$. The corresponding ionization potential of the ground state is $I_P = 0.60$ a.u. The exponential damping of $V(x)$ ensures that $\hat{V}(q)$ is regular at $q = 0$, which is convenient for numerical calculations. In a momentum space formulation, no absorbing boundary conditions need be imposed and the "fermion doubling" problem [107, 108] is not encountered.

The calculations of Kylstra, Ermolaev and Joachain were performed for laser pulses having a peak electric-field strength $\mathcal{E}_0 = 175$ a.u. and carrier angular frequencies $\omega = 1$ a.u. and $\omega = 2$ a.u. The two-component TDDE was solved using a basis of field-free (positive- and negative-energy) eigenstates obtained by diagonalizing the field-free Dirac Hamiltonian using a basis of B-spline functions. They found that for $\omega = 1$ a.u. (such that $q = 0.4$), relativistic effects become apparent. We show in Fig. 5.11 the Dirac and non-relativistic Schrödinger probability densities at the end of the ninth cycle, when the electric field is maximum, for a laser pulse with a four-cycle $\sin^2$ turn-on. As will be discussed in Chapter 7, in this high-frequency regime the wave function remains essentially localized in a superposition of field-free bound states and low-energy continuum states. The peak in the Dirac probability density corresponds to the relativistic classical excursion amplitude, which occurs at $\alpha_0^* = 124$ a.u. The peak in the Schrödinger probability

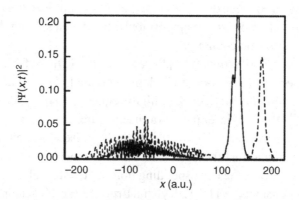

Figure 5.11. Probability densities of the Dirac (solid line) and non-relativistic Schrödinger (dashed line) wave packets at the end of the ninth cycle for a laser pulse with a four-cycle $\sin^2$ turn-on, a carrier angular frequency $\omega = 1$ a.u. and a peak electric-field strength $\mathcal{E}_0 = 175$ a.u. (From N. J. Kylstra, A. M. Ermolaev and C. J. Joachain, *J. Phys. B* **30**, L449 (1997).)

density is at $\alpha_0 = 175$ a.u., which is the value of the non-relativistic classical excursion amplitude. Kylstra, Ermolaev and Joachain found that the coupling by the laser field of the field-free positive and negative energy states is responsible for the relativistic corrections to the quiver motion. At the end of the laser pulse, the ionization probability is 0.52 for the Dirac wave function and 0.58 for the Schrödinger wave function, showing that the Dirac wave function is slightly more stable against ionization.

In Section 2.9, it was shown that solutions of the single-electron TDRSE are expected to be a good approximation to the solutions of the single-electron TDKGE. By considering a model one-dimensional atom, Lenz, Dörr and Sandner [109] have demonstrated that solutions of the TDRSE can also be approximate solutions of the TDDE. In their investigation, they compared the numerical solution of the TDRSE and TDDE for a deeply bound electron having an ionization potential of 2.2×10^3 a.u. interacting with a single-cycle laser pulse of angular frequency $\omega = 6.3 \times 10^2$ a.u. and peak electric-field strength $\mathcal{E}_0 = 6.3 \times 10^5$ a.u. They found only very small differences between the probability densities obtained from the TDRSE and those obtained from the TDDE. Insight into the origin of this result can be obtained by referring to the second-order Dirac equation, (2.253). We note that if the third and fourth terms on the left-hand side of this equation can be neglected, then each component of the second-order Dirac spinor is a solution of the TDKGE. When this is the case, and provided that the initial Dirac ground-state density of the system can be well approximated by the Klein–Gordon ground-state density, then the probability densities calculated using the TDRSE, TDKGE and the TDDE will be approximately the same.

Rathe *et al.* [110] have solved numerically the TDDE for a model two-dimensional atom interacting with an intense laser pulse of carrier angular frequency $\omega = 10$ a.u. and peak electric-field strengths up to $\mathcal{E}_0 = 411$ a.u. The atom was modeled using the soft-core coulomb potential (5.78) with $Z = 1$ and $a = 0.775$ a.u. The effects of the magnetic-field component of the laser pulse on the wave packet and spin dynamics of the system were demonstrated. In particular, the spin-flip probability was found to be substantial.

References

[1] K. J. Schafer, Numerical methods in strong field physics. In T. Brabec, ed., *Strong Field Laser Physics*, Springer Series in Optical Sciences 134 (New York: Springer, 2009), p. 111.

[2] W. H. Press, S. A. Teukolsky, W. T. Vetterling and B. P. Flannery, *Numerical Recipes* (Cambridge: Cambridge University Press, 2007).

[3] K. C. Kulander, K. J. Schafer and J. C. Krause, *Adv. At. Mol. Opt. Phys. Suppl.* **1**, 247 (1992).

[4] H. Flocard, S. E. Koonin and M. S. Weiss, *Phys. Rev. C* **17**, 1682 (1978).

[5] C. de Boor, *A Practical Guide to Splines* (New York: Springer-Verlag, 1978).

[6] P. Lambropoulos, P. Maragakis and J. Zhang, *Phys. Rep.* **305**, 203 (1998).
[7] E. Cormier and P. Lambropoulos, *J. Phys. B* **29**, 1667 (1996).
[8] H. G. Muller, *Phys. Rev. A* **60**, 1341 (1999).
[9] J. Javanainen, J. Eberly and Q. Su, *Phys. Rev. A* **38**, 3430 (1988).
[10] K. J. La Gattuta, *Phys. Rev. A* **41**, 5110 (1990).
[11] M. D. Feit, J. A. Fleck and A. Steiger, *J. Comp. Phys.* **47**, 412 (1982).
[12] K. J. Schafer and K. C. Kulander, *Phys. Rev. A* **42**, 5794 (1990).
[13] O. Guyétand, M. Gisselbrecht, A. Huetz, *et al.*, *J. Phys. B* **38**, L357 (2005).
[14] O. Guyétand, M. Gisselbrecht, A. Huetz, *et al.*, *J Phys. B* **41**, 051002 (2008).
[15] S. Geltman, *J. Phys. B* **10**, 831 (1977).
[16] A. Goldberg and B. W. Shore, *J. Phys. B* **11**, 3339 (1978).
[17] E. J. Austin, *J. Phys. B* **12**, 4045 (1979).
[18] J. H. Eberly, *Hyperfine Int.* **37**, 33 (1987).
[19] C. Cerjan and R. Kosloff, *J. Phys. B* **20**, 4441 (1987).
[20] M. S. Pindzola and C. Bottcher, *J. Opt. Soc. Am. B* **4**, 752 (1987).
[21] S. M. Susskind and R. V. Jensen, *Phys. Rev. A* **38**, 711 (1988).
[22] B. Sundaram and L. Armstrong, *Phys. Rev. A* **38**, 152 (1988).
[23] M. Dörr and R. Shakeshaft, *Phys. Rev. A* **38**, 543 (1988).
[24] J. N. Bardsley and M. J. Comella, *Phys. Rev. A* **39**, 2252 (1989).
[25] J. H. Eberly, Q. Su and J. Javanainen, *Phys. Rev. Lett.* **62**, 881 (1989).
[26] J. H. Eberly, Q. Su and J. Javanainen, *J. Opt. Soc. Am. B* **6**, 1289 (1989).
[27] J. H. Eberly, Q. Su, J. Javanainen, K. C. Kulander, B. W. Shore and L. Roso-Franco, *J. Mod. Opt.* **36**, 829 (1989).
[28] R. A. Sacks and A. Szüke, *Phys. Rev. A* **40**, 5614 (1989).
[29] B. Ritchie, C. M. Bowden, C. C. Sung and Y. Q. Li, *Phys. Rev. A* **41**, 6114 (1990).
[30] V. C. Reed and K. Burnett, *Phys. Rev. A* **42**, 3152 (1990).
[31] V. C. Reed, P. L. Knight and K. Burnett, *Phys. Rev. Lett.* **67**, 1415 (1991).
[32] V. C. Reed and K. Burnett, *Phys. Rev. A* **46**, 424 (1992).
[33] L. A. Bloomfield, *J. Opt. Soc. Am. B* **7**, 472 (1990).
[34] R. Blümel and U. Smilansky, *J. Opt. Soc. Am. B* **7**, 664 (1990).
[35] J. H. Eberly, J. Javanainen and K. Rzążewski, *Phys. Rep.* **204**, 331 (1991).
[36] J. H. Eberly, R. Grobe, C. K. Law and Q. Su, *Adv. At. Mol. Opt. Phys. Suppl.* **1**, 301 (1992).
[37] W. C. Liu and C. W. Clark, *J. Phys. B* **25**, L517 (1992).
[38] J. H. Eberly, *Phys. Rev. A* **42**, 5750 (1990).
[39] R. M. Potvliege, N. J. Kylstra and C. J. Joachain, *J. Phys. B* **33**, L743 (2000).
[40] A. Patel, N. J. Kylstra and P. L. Knight, *J. Phys. B* **32**, 5759 (1999).
[41] P. G. Burke and V. M. Burke, *J. Phys. B* **30**, L383 (1997).
[42] P. G. Burke, P. Francken and C. J. Joachain, *Europhys. Lett.* **13**, 617 (1990).
[43] P. G. Burke, P. Francken and C. J. Joachain, *J. Phys. B* **24**, 761 (1991).
[44] M. Protopapas, D. G. Lappas and P. L. Knight, *Phys. Rev. Lett.* **79**, 4550 (1997).
[45] A. Patel, M. Protopapas, D. G. Lappas and P. L. Knight, *Phys. Rev. A* **58**, R2652 (1998).
[46] K. C. Kulander, K. J. Schafer and J. L. Krause, *Phys. Rev. Lett.* **66**, 2601 (1991).
[47] A. R. Edmonds, *Angular Momentum in Quantum Mechanics* (Princeton, N.J.: Princeton University Press, 1957).
[48] M. S. Pindzola and M. Dörr, *Phys. Rev. A* **43**, 439 (1991).
[49] M. Pont, D. Proulx and R. Shakeshaft, *Phys. Rev. A* **44**, 4486 (1991).
[50] K. J. LaGattuta, *Phys. Rev. A* **43**, 5157 (1991).
[51] M. Horbatsch, *J. Phys. B* **24**, 4149 (1991).

[52] O. Latinne, C. J. Joachain and M. Dörr, *Europhys. Lett.* **26**, 333 (1994).
[53] P. Antoine, B. Piraux and A. Maquet, *Phys. Rev. A* **51**, R1750 (1995).
[54] M. Dörr, O. Latinne and C. J. Joachain, *Phys. Rev. A* **52**, 4289 (1995).
[55] M. Dörr, O. Latinne and C. J. Joachain, *Phys. Rev. A* **55**, 3697 (1997).
[56] E. Cormier and P. Lambropoulos, *J. Phys. B* **30**, 77 (1997).
[57] G. G. Paulus, W. Nicklich, H. Xu, P. Lambropoulos and H. Walther, *Phys. Rev. Lett.* **72**, 2851 (1994).
[58] G. G. Paulus, W. Nicklich, F. Zacher, P. Lambropoulos and H. Walther, *J. Phys. B* **29**, L249 (1996).
[59] B. H. Bransden and C. J. Joachain, *Physics of Atoms and Molecules*, 2nd edn (Harlow, UK: Prentice Hall-Pearson, 2003).
[60] K. C. Kulander, *Phys. Rev. A* **36**, 2726 (1987).
[61] K. C. Kulander, *Phys. Rev. A* **38**, 778 (1988).
[62] F. Herman and S. Skillman, *Atomic Structure Calculations* (Englewood Cliffs, N.J.: Prentice Hall, 1963).
[63] M. S. Pindzola, P. Gavras and T. W. Gorczyca, *Phys. Rev. A* **51**, 3999 (1995).
[64] P. Hohenberg and W. Kohn, *Phys. Rev.* **136**, B864 (1964).
[65] W. Kohn and L. J. Sham, *Phys. Rev. A* **140**, 1133 (1965).
[66] R. G. Parr and W. Yang, *Density Functional Theory of Atoms and Molecules* (New York: Oxford University Press, 1989).
[67] R. M. Dreizler and E. K. U. Gross, *Density Functional Theory* (Berlin: Springer-Verlag, 1990).
[68] E. Runge and E. K. U. Gross, *Phys. Rev. Lett.* **52**, 997 (1984).
[69] E. K. U. Gross, J. F. Dobson and M. Petersilka, *Density Functional Theory II* (Berlin: Springer-Verlag, 1996), p. 81.
[70] C. A. Ullrich, U. J. Gossmann and E. K. U. Gross, *Phys. Rev. Lett.* **74**, 872 (1995).
[71] J. L. Krause, K. J. Schafer and K. C. Kulander, *Phys. Rev. A* **45**, 4998 (1992).
[72] R. D. Cowan, *The Theory of Atomic Structure and Spectra* (Berkeley, Calif.: University of California Press, 1981).
[73] M. P. Hertlein, P. H. Bucksbaum and H. G. Muller, *J. Phys. B* **30**, L197 (1997).
[74] H. G. Muller and F. C. Kooiman, *Phys. Rev. Lett.* **81**, 1207 (1998).
[75] M. J. Nandor, M. A. Walker, L. D. van Woerkom and H. G. Muller, *Phys. Rev. A* **60**, R1771 (1999).
[76] X. Tang, H. Rudolph and P. Lambropoulos, *Phys. Rev. A* **44**, R6994 (1991).
[77] H. Bachau, P. Lambropoulos and X. Tang, *Phys. Rev. A* **42**, R5801 (1997).
[78] J. Zhang and P. Lambropoulos, *J. Phys. B* **28**, L101 (1995).
[79] J. Zhang and P. Lambropoulos, *Phys. Rev. Lett.* **77**, 2186 (1996).
[80] A. Scrinzi and B. Piraux, *Phys. Rev. A* **56**, R13 (1997).
[81] A. Scrinzi and B. Piraux, *Phys. Rev. A* **58**, 1310 (1998).
[82] T. Mercouris, S. Dionissopoulou and C. A. Nicolaides, *J. Phys. B* **30**, 4751 (1997).
[83] D. Charalambidis, D. Xenakis, C. J. G. J. Uiterwaal *et al.*, *J. Phys. B* **30**, 1467 (1997).
[84] C. A. Nicolaides, S. Dionissopoulou and T. Mercouris, *J. Phys. B* **29**, 231 (1996).
[85] J. Parker, K. T. Taylor, C. W. Clark and S. Blodgett-Ford, *J. Phys. B* **29**, L33 (1996).
[86] K. T. Taylor, J. S. Parker, D. Dundas, E. Smyth and S. Vivirito. In P. Lambropoulos and H. Walther, eds., *Multiphoton Processes 1996* (Bristol: Institute of Physics, 1997), p. 56.
[87] J. S. Parker, E. S. Smyth and K. T. Taylor, *J. Phys. B* **31**, L571 (1998).
[88] E. Smyth, J. S. Parker and K. T. Taylor, *Comp. Phys. Commun.* **114**, 1 (1998).
[89] K. T. Taylor, J. S. Parker, D. Dundas, E. Smyth and S. Vivirito, *Laser Phys.* **9**, 98 (1999).

[90] J. S. Parker, L. R. Moore, E. Smyth and K. T. Taylor, *J. Phys. B* **33**, 1057 (2000).
[91] H. W. van der Hart, B. J. S. Doherty, J. S. Parker and K. T. Taylor, *J. Phys. B* **38**, L207 (2005).
[92] H. W. van der Hart, M. A. Lysaght and P. G. Burke, *Phys. Rev. A* **76**, 0430405 (2007).
[93] H. W. van der Hart, *Phys. Rev. A* **73**, 023417 (2006).
[94] M. V. Ammosov, N. B. Delone and V. P. Krainov, *Sov. Phys. JETP* **64**, 1191 (1986).
[95] A. Bugacov, M. Pont and R. Shakeshaft, *Phys. Rev. A* **48**, R4027 (1993).
[96] K. J. Meharg, J. S. Parker and K. T. Taylor, *J. Phys. B* **38**, 237 (2005).
[97] A. V. Kim, M. Y. Ryabikin and A. M. Sergeev, *Sov. Phys.-Usp.* **42**, 54 (1999).
[98] J. R. Vázquez de Aldana and L. Roso, *Opt. Exp.* **5**, 144 (1999).
[99] N. J. Kylstra, R. A. Worthington, A. Patel, P. L. Knight, J. R. Vázquez de Aldana and L. Roso, *Phys. Rev. Lett.* **85**, 1835 (2000).
[100] J. R. Vázquez de Aldana, N. J. Kylstra, L. Roso, P. L. Knight, A. S. Patel, and R. A. Worthington, *Phys. Rev. A* **64**, 013411 (2001).
[101] N. J. Kylstra, R. M. Potvliege, R. A. Worthington *et al.*, In B. Piraux and K. Rząźewski, eds., *Super-Intense Laser-Atom Physics* (Dordrecht: Kluwer Academic, 2001), p. 345.
[102] N. J. Kylstra, R. M. Potvliege and C. J. Joachain, *J. Phys. B* **34**, L55 (2001).
[103] C. J. Joachain and N. J. Kylstra, *Laser Phys.* **11**, 212 (2001).
[104] M. Protopapas, C. H. Keitel and P. L. Knight, *J. Phys. B* **29**, L591 (1996).
[105] R. Taïeb, V. Véniard and A. Maquet, *Phys. Rev. Lett.* **81**, 2882 (1998).
[106] N. J. Kylstra, A. M. Ermolaev and C. J. Joachain, *J. Phys. B* **30**, L449 (1997).
[107] C. Bottcher and M. R. Strayer, *Ann. Phys. NY* **175**, 64 (1987).
[108] J. C. Wells, V. E. Oberacker, A. S. Umar, C. Bottcher, M. R. Strayer, J. S. Wu and G. Plunien, *Phys. Rev. A* **45**, 6296 (1992).
[109] E. Lenz, M. Dörr and W. Sandner, *Laser Phys.* **11**, 216 (2001).
[110] U. W. Rathe, C. H. Keitel, M. Protopapas and P. L. Knight, *J. Phys. B* **30**, L531 (1997).

6

The low-frequency regime

In this chapter we turn to the formulation of the theory of the interaction of intense laser fields with atoms in the important case where the laser photon energy is much smaller than the ionization potential of the initial atomic state. When the intensity is sufficiently high and the frequency sufficiently low, ionization proceeds as if the laser electric field were quasi-static. In this regime, it is appropriate to make the "strong-field approximation," or SFA, in which one assumes that an active electron, after having been ionized, interacts only with the laser field and not with its parent core. Using this approximation, Keldysh [1] showed that analytical expressions for the rate of ionization can be obtained when the electric-field amplitude, the laser frequency and the binding energy of the initial state are such that the Keldysh parameter γ_K defined by Equation (1.8) is much less than unity and the photoelectron does not escape by over-the-barrier ionization (OBI). However, the applicability of the SFA extends beyond this regime and, more importantly, it can be used to investigate high-order ATI and high-order harmonic generation. The SFA also provides a framework in which the physical origin of these processes, embodied in the semi-classical three-step recollision model introduced in Section 1.3, can be understood.

We begin in Section 6.1 by examining the low-frequency limit of the Floquet theory and showing how the total ionization rate of the atom can be obtained using the adiabatic approximation. In Section 6.2 we introduce the SFA, which we apply in Section 6.3 to describe direct multiphoton single ionization. In this process, the ionized electrons do not recollide with their parent core. In Section 6.4, we consider high harmonic generation (HHG). This important process occurs when ionized electrons are driven back to their parent ion by the laser field, where they radiatively recombine. By analyzing the laser-induced atomic dipole moment in the SFA, we show that HHG (at the single-atom level) can be understood in terms of the trajectories of classical unbound electrons oscillating in the laser field. The case in which ionized electrons return to their parent core and collide with it, leading to multiphoton single or multiple ionization, is discussed in Section 6.5. We will see that the same classical electron trajectories that were used in our analysis of HHG play an important role in the description of these recollision processes. Finally, using the HHG process as an example, we examine in Section 6.6 how the SFA can be generalized to account for non-dipole effects, due to the magnetic component of the laser field, which arise when the intensity becomes sufficiently high.

Reviews of the SFA and its application to the study of multiphoton processes have been given by Salières *et al.* [2], Becker *et al.* [3] , Becker and Faisal [4] and Lewenstein and L'Huillier [5]. Atomic units (a.u.) will be used, unless otherwise indicated.

6.1 Quasi-static ionization

6.1.1 Ionization in the adiabatic approximation

We start our analysis by obtaining expressions for the probability that an atom is ionized by a laser pulse in the limit where the laser electric field varies slowly in time. As in Chapter 2, we denote by X the ensemble of the coordinates of all the atomic electrons, by $\mathbf{R}$ the sum of their position vectors and by H_0 the field-free Hamiltonian. We assume that the atom is initially in the ground state $\phi_0(X, t) = \psi_0(X) \exp(-iE_0 t)$ of H_0, with

$$H_0 \psi_0(X) = E_0 \psi_0(X) \tag{6.1}$$

and $E_0 = -I_P$, where I_P is the field-free ionization potential of the atom. In the presence of the laser field, the Hamiltonian governing the atomic dynamics is given in the dipole approximation and the length gauge by

$$H(t) = H_0 + \mathcal{E}(t) \cdot \mathbf{R} \tag{6.2}$$

and the atomic wave function evolves according to the time-dependent Schrödinger equation (TDSE)

$$i \frac{\partial}{\partial t} \Psi(X, t) = H(t) \Psi(X, t). \tag{6.3}$$

We describe the laser electric field $\mathcal{E}(t)$ by Equation (2.111). The ionization probability depends on the orientation of the atom with respect to $\mathcal{E}(t)$ if the initial state is not spherically symmetric. We assume that the laser pulse envelope function $F(t) = 0$ for $t \leq 0$, so that $\mathcal{E}(t) = 0$ and $\Psi(X, t) = \phi_0(X, t)$ for $t \leq 0$.

If the eigenvalue E_0 is not degenerate and $\mathcal{E}(t) \cdot \mathbf{R}$ varies sufficiently slowly, the adiabatic theorem [6] ensures that the atom will remain in the eigenstate of the

Hamiltonian $H(t)$ that evolves from the initial state $\psi_0(X)$. In this limit, the atom is described for all times by an eigenstate of the Stark Hamiltonian $H_0 + \mathcal{E} \cdot \mathbf{R}$, where $\mathcal{E}$ is a static electric field. The relevant eigenfunction of this Hamiltonian is the solution of the time-independent Schrödinger equation

$$(H_0 + \mathcal{E} \cdot \mathbf{R}) \psi_{DC}(\mathcal{E}; X) = E_{DC}(\mathcal{E}) \psi_{DC}(\mathcal{E}; X) \tag{6.4}$$

that reduces to $\psi_0(X)$ for $\mathcal{E} \to 0$. Solving this equation is not straightforward because the electric field, no matter how weak, couples the bound states to the continuum states and turns all the bound states into resonances of finite lifetime. As a consequence, none of the solutions of Equation (6.4) are square integrable. However, it is possible to give a rigorous meaning to the eigenenergies $E_{DC}(\mathcal{E})$ within the theory of complex scaling mentioned in Section 4.4 [7–9]. As in the case of the decaying bound states studied in Section 4.4, this leads to complex energies of the form

$$E_{DC}(\mathcal{E}) = E_0 + \Delta_{DC}(\mathcal{E}) - \frac{i}{2}\Gamma_{DC}(\mathcal{E}), \tag{6.5}$$

where $\Delta(\mathcal{E})$ is the DC Stark shift of the initial state and the quantity $\Gamma_{DC}(\mathcal{E})$ is the ionization width of the Stark state. In atomic units, the latter is also the ionization rate of the atom in the static electric field $\mathcal{E}$. The corresponding wave function $\psi_{DC}(\mathcal{E}; X)$ can be obtained either by solving Equation (6.4) subject to appropriate outgoing wave boundary conditions or by transforming the Hamiltonian by complex scaling [10, 11].

If $\mathcal{E}(t)$ varies slowly enough, we can make the adiabatic approximation [6] and take $\Psi(X, t) \simeq \Psi_{ad}(X, t)$ with

$$\Psi_{ad}(X, t) \equiv \exp\left(-i \int_0^t E_{DC}[\mathcal{E}(t')] dt'\right) \psi_{DC}[\mathcal{E}(t); X]. \tag{6.6}$$

The energy $E_{DC}[\mathcal{E}(t)]$ and the wave function $\psi_{DC}[\mathcal{E}(t); X]$ satisfy Equation (6.4) for a static electric field $\mathcal{E}$ equal to the instantaneous electric field of the laser $\mathcal{E}(t)$. As $t \to \infty$ and the laser pulse envelope function $F(t)$ goes back to zero at the end of the pulse, the wave function $\Psi_{ad}(X, t)$ returns to the initial unperturbed wave function $\psi_0(X) \exp(-iE_0t)$, but only within an overall complex factor of norm less than unity since $\text{Im}\, E_{DC}[\mathcal{E}(t')] < 0$ when $|\mathcal{E}(t')| \neq 0$. The probability that the atom is still in the initial state at the end of the pulse is therefore not unity, and its ionization probability P_{ion} is given by

$$P_{ion} = 1 - \exp\left(-\int_0^\infty \Gamma_{DC}[\mathcal{E}(t)] dt\right), \tag{6.7}$$

where we have assumed that population transfer does not occur at crossings between Stark states.

The ionization rate of an atom interacting with a monochromatic laser field can also be calculated in the adiabatic approximation. To this end, we use Equation (2.114) for the laser electric field and look for an approximate quasi-stationary solution of Equation (6.3) which is valid for sufficiently small values of ω and reduces to the unperturbed wave function $\phi_0(X, t)$ for $\mathcal{E}_0 \to 0$. Thus, following Equation (4.17), we write the wave function in the Floquet form of a product of a time-dependent phase factor and a function $P(\mathsf{E}; X, t)$ oscillating at the frequency of $\mathcal{E}(t)$, namely

$$\Psi(\mathsf{E}; X, t) = \exp(-\mathrm{i}\mathsf{E}t)\, P(\mathsf{E}; X, t). \tag{6.8}$$

The quasi-energy E is a function of the amplitude $\mathcal{E}_0$, the ellipticity parameter ξ and the angular frequency ω of the laser field. As noted in Section 4.1, the quasi-energy E does not depend on the phase φ if the laser field is monochromatic. The adiabatic wave function satisfying $\mathsf{E} \to E_0$ for $\mathcal{E}_0 \to 0$ can be found by writing $\Psi(\mathsf{E}; X, t)$ in the same form as the wave function (6.6) obtained for a laser pulse. Introducing the function $\widetilde{E}(\omega t)$ with the dimension of energy and a wave function $\widetilde{\Psi}(X, \omega t)$, we make the ansatz

$$\Psi(\mathsf{E}; X, t) = \exp\left[-\frac{\mathrm{i}}{\omega}\int_0^{\omega t}\widetilde{E}(\omega t')\,\mathrm{d}(\omega t')\right]\widetilde{\Psi}(X, \omega t). \tag{6.9}$$

Like the monochromatic laser electric field described by Equation (2.114), both $\widetilde{\Psi}(X, \omega t)$ and $\widetilde{E}(\omega t)$ are taken to depend on time only through the product ωt. It follows from the TDSE (6.3) that

$$\widetilde{E}(\omega t)\widetilde{\Psi}(X, \omega t) + \mathrm{i}\omega\frac{\partial\widetilde{\Psi}}{\partial(\omega t)} = [H_0 + \mathcal{E}(t)\cdot\mathbf{R}]\,\widetilde{\Psi}(X, \omega t). \tag{6.10}$$

This equation shows that when $\omega \simeq 0$ we can replace $\widetilde{E}(\omega t)$ by $E_{\mathrm{DC}}[\mathcal{E}(t)]$ and $\widetilde{\Psi}(X, \omega t)$ by $\psi_{\mathrm{DC}}[\mathcal{E}(t); X]$, thereby giving the Floquet wave function in the adiabatic approximation. By comparing the resulting expression of the wave function to Equation (6.8), we obtain the quasi-energy in the adiabatic approximation. That is,

$$\mathsf{E}(\mathcal{E}_0, \xi, \omega) \simeq E_{\mathrm{ad}}(\mathcal{E}_0, \xi) = \frac{\omega}{2\pi}\int_0^{2\pi/\omega}E_{\mathrm{DC}}[\mathcal{E}(t)]\,\mathrm{d}t. \tag{6.11}$$

The corresponding Floquet rate of ionization by an oscillating monochromatic electric field is the cycle-average of the static rate of ionization $\Gamma_{\mathrm{DC}}[\mathcal{E}(t)]$ of the atom in the instantaneous static electric field $\mathcal{E}(t)$, namely

$$\Gamma(\mathcal{E}_0, \xi\,\omega) \simeq \Gamma_{\mathrm{ad}}(\mathcal{E}_0, \xi) = \frac{\omega}{2\pi}\int_0^{2\pi/\omega}\Gamma_{\mathrm{DC}}[\mathcal{E}(t)]\,\mathrm{d}t. \tag{6.12}$$

The integral on the right-hand side of this equation must usually be evaluated numerically, although we shall see below that approximate analytical forms can be derived. It reduces to a simple form in the case of ionization from an S-state by a circularly polarized monochromatic laser field ($\xi = \pm 1$). In this case the rate of ionization by a static electric field does not depend on the direction of the field. Since $\xi = \pm 1$, the laser electric field defined by Equation (2.114) has a constant magnitude $\mathcal{E}_0/\sqrt{2}$, and we have

$$E_{\text{ad}}(\mathcal{E}_0, \xi = \pm 1) = E_{\text{DC}}[\mathcal{E} = (\mathcal{E}_0/\sqrt{2})\hat{\epsilon}]. \tag{6.13}$$

In the adiabatic approximation, neither the AC Stark shift nor the ionization rate depend on the angular frequency of the laser field. The fact that these quantities are independent of ω arises from the assumption that the atom behaves at every instant of time as if the electric-field component of the laser field were static. From a mathematical point of view, it is a consequence of neglecting the term that contains the derivative of $\widetilde{\Psi}(X, \omega t)$ in the left-hand side of Equation (6.10). By taking this term into account, it is possible to include non-adiabatic corrections. In particular, treating this term as a small perturbation leads to an asymptotic expansion of the exact quasi-energy in even powers of ω, of which the adiabatic quasi-energy is the leading term [12, 13]. Thus, for small values of ω,

$$\mathsf{E}(\mathcal{E}_0, \xi, \omega) \simeq E_{\text{ad}}(\mathcal{E}_0, \xi) + \sum_{m=1}^{\infty} E^{(2m)}(\mathcal{E}_0, \xi)\omega^{2m}. \tag{6.14}$$

This expansion can be used to estimate the rate of ionization at frequencies that are too low for carrying out a full Floquet calculation and too high for the adiabatic approximation to be accurate.

The convergence of the adiabatic ionization rate to the exact rate for long wavelengths is illustrated in Fig. 6.1, where the rate resulting from an "exact" Floquet calculation is compared to $\Gamma_{\text{ad}}(\mathcal{E}_0, \xi)$ for two different wavelengths, 616 nm and 1064 nm, for a circularly polarized laser field interacting with atomic hydrogen. The ground state can be resonantly coupled only to excited states of large angular momentum by such laser fields, since each absorption of a circularly polarized photon increases the absolute value of the magnetic quantum number by one unit. Such resonances hardly affect the ground state at long wavelengths, which is why the Floquet results vary smoothly with intensity. As seen from Fig. 6.1, the adiabatic rate $\Gamma_{\text{ad}}(\mathcal{E}_0, \xi)$ is too low by orders of magnitude in weak fields. However, the adiabatic approximation improves when the wavelength and the intensity increase. For example, at an intensity of 8×10^{13} W cm^{-2}, the difference between the exact rate and the adiabatic rate is a factor five at 616 nm and only a factor two at 1064 nm.

A simple expression of the AC Stark shift of a low-lying non-degenerate S bound state in a low-frequency monochromatic laser field can be derived from

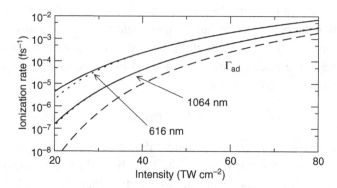

Figure 6.1. Ionization rate of the ground state of atomic hydrogen in a circularly polarized monochromatic laser-field. Results are shown for laser-field wavelengths of 616 nm and 1064 nm, and intensities ranging from 2×10^{13} to $8 \times 10^{13}\,\mathrm{W\,cm^{-2}}$. Solid curves: results of a full Floquet calculation. Dotted curves: rate predicted by Equation (6.14) summed up to the term of order ω^6. Dashed curve: rate in the adiabatic approximation.

Equation (6.11). For a static electric field that is not too strong, we have

$$\Delta_{\mathrm{DC}} \simeq -\frac{1}{2}\bar{\alpha}\,\mathcal{E}^2, \tag{6.15}$$

where $\bar{\alpha}$ is the static dipole polarizability of the state, defined by Equation (3.116). Hence, within the adiabatic approximation,

$$\Delta_{\mathrm{AC}} \simeq -\frac{1}{4}\bar{\alpha}\,\mathcal{E}_0^2 \tag{6.16}$$

for a linearly polarized monochromatic laser field of amplitude $\mathcal{E}_0$ described in the length gauge. The AC Stark shift given by Equation (6.16) is proportional to the intensity, like the ponderomotive energy U_p, but typically amounts only to a small fraction of the latter. As an example, for a laser wavelength of $\lambda = 800\,\mathrm{nm}$, this shift amounts to 0.4% of U_p for He and 9% of U_p for Xe.

6.1.2 Tunneling formulae

As shown by Equation (6.12), the adiabatic approximation gives the rate of ionization in a low-frequency monochromatic laser field in terms of the cycle-average of the rate of ionization $\Gamma_{\mathrm{DC}}[\mathcal{E}(t)]$ by a static electric field $\mathcal{E}$ whose instantaneous value at time t is $\mathcal{E}(t)$. Insight can be obtained by deriving an approximate analytical expression for the quantity $\Gamma_{\mathrm{DC}}(\mathcal{E})$.

Let us suppose that only one of the atomic electrons is active, and that the other electrons and the nucleus form an inert core of total charge Z_c, with $Z_\mathrm{c} = 0$ for a negative ion. Hence, the active electron interacts with the rest of the atom only via the Coulomb potential $-Z_c/r$ when it is far away from the core. It is convenient to assume here that the static electric field is oriented in the positive z-direction, so that $\boldsymbol{\mathcal{E}} = \mathcal{E}\hat{\mathbf{z}}$. The potential energy of the active electron in this static electric field is then given approximately by

$$V(\rho, z) = -\frac{Z_c}{\sqrt{\rho^2 + z^2}} + \mathcal{E}z, \tag{6.17}$$

where ρ is the perpendicular distance of the electron from the z-axis. Classically, the total energy of this electron is

$$E_{\text{tot}} = \frac{1}{2}\left(\dot{z}^2 + \dot{\rho}^2 + \rho^2\dot{\phi}^2\right) + V(\rho, z). \tag{6.18}$$

Since $V(\rho, z)$ does not depend on ϕ, the angular momentum about the z-axis, $M_z = \rho^2\dot{\phi}$, is a constant of the motion. We may therefore write

$$E_{\text{tot}} = \frac{1}{2}\left(\dot{z}^2 + \dot{\rho}^2\right) + \frac{M_z^2}{2\rho^2} + V(\rho, z). \tag{6.19}$$

The classical motion of this electron is therefore limited to the region of space where $E_{\text{tot}} \geq V_{\text{eff}}(\rho, z, M_z)$, with the effective potential

$$V_{\text{eff}}(\rho, z, M_z) = \frac{M_z^2}{2\rho^2} - \frac{Z_c}{\sqrt{\rho^2 + z^2}} + \mathcal{E}z. \tag{6.20}$$

As can be seen in Fig. 1.7, the effective potential has a barrier which may prevent the classical electron from leaving the vicinity of the core. We remark that in the present case the electric field pulls the electron in the negative z-direction.

When $M_z = 0$, the function $V_{\text{eff}}(\rho, z, M_z)$ has a minimum with respect to ρ and a maximum with respect to z when $\rho = \rho_s = 0$ and $z = z_s = -(Z_c/\mathcal{E})^{1/2}$. At this saddle point, $V_{\text{eff}}(\rho_s, z_s, 0) = -2(Z_c\mathcal{E})^{1/2}$. Let us suppose that the electron has an ionization potential I_P and that the static electric field does not change this energy significantly. Then, if $\mathcal{E} > \mathcal{E}_{\text{OBI}}$, where $\mathcal{E}_{\text{OBI}} = I_P^2/(4Z_c)$ is the critical electric-field strength at which $-2(Z_c\mathcal{E})^{1/2} = -I_P$, the electron may pass "over the barrier" and escape. Conversely, if $\mathcal{E} \leq \mathcal{E}_{\text{OBI}}$ its classical motion is confined to the vicinity of the core. It should be noted that for all neutral atoms the critical electric-field strength for over-the-barrier ionization from the ground state, while very large by laboratory standards for static electric fields, is always somewhat less than one atomic unit and can be easily achieved in the focus of a very intense laser pulse. For example, in the case of helium, $\mathcal{E}_{\text{OBI}} = 0.204$ a.u., which is the electric-field strength of a linearly polarized laser field with an intensity of $1.46 \times 10^{15}\,\text{W cm}^{-2}$.

When $M_z \neq 0$, the angular momentum barrier $M_z^2/(2\rho^2)$ repels the electron from the z-axis. The effective potential $V_{\text{eff}}(\rho, z, M_z)$ still has a saddle point, but this saddle point is displaced to a non-zero value of ρ and to a smaller value of $|z|$ compared to the case for which $M_z = 0$ [14]. The resulting change in the value of $V_{\text{eff}}(\rho, z, M_z)$ makes the corresponding critical electric-field strength for over-the-barrier ionization M_z-dependent and higher for $M_z \neq 0$ than for $M_z = 0$.

Whereas the classical motion is unbound only above the critical electric-field strength for over-the-barrier ionization, this is not the case quantum mechanically as the electron can always escape by tunneling through the effective potential barrier. Since the z-component of the orbital angular momentum operator L_z commutes with the Stark Hamiltonian for a static electric field oriented in the z-direction, the corresponding Stark states $|\psi\rangle_{\text{DC}}$ can be taken to be eigenstates of L_z. Writing $L_z|\psi\rangle_{\text{DC}} = m|\psi\rangle_{\text{DC}}$, we can identify the classical angular momentum M_z with the corresponding eigenvalues m of L_z. The analysis of the classical motion thus suggests that states with $m \neq 0$ are more difficult to field-ionize than $m = 0$ states with the same binding energy.

Since the motion of the electron in the z-direction is coupled to its motion in the ρ-variable, it may seem that the barrier through which the outgoing electron tunnels is two-dimensional. However, for the potential given by Equation (6.17), the Schrödinger equation can be separated by using parabolic coordinates ($r+z$, $r-z, \phi$), which decouple the motion in the $r+z$ coordinate from that in the $r-z$ coordinate. Tunneling occurs only along the coordinate $r-z$ for the static field considered here. The tunneling process is thus effectively one-dimensional, and the rate of tunneling ionization can be calculated using the WKB method [6]. The result is particularly simple for the ground state of atomic hydrogen [15]. For a static electric field of strength $\mathcal{E}$,

$$\Gamma_{\text{DC}}(\mathcal{E}) \simeq \frac{4}{\mathcal{E}} \exp\left(-\frac{2}{3\mathcal{E}}\right). \tag{6.21}$$

More generally, for a hydrogenic positive ion of nuclear charge Z initially in its ground state,

$$\Gamma_{\text{DC}}(\mathcal{E}) \simeq \frac{4Z^5}{\mathcal{E}} \exp\left(-\frac{2Z^3}{3\mathcal{E}}\right). \tag{6.22}$$

These rates do not depend on the direction of the electric field since the initial state is spherically symmetric.

For atomic hydrogen and hydrogenic ions, Equation (6.17) is exact and the Schrödinger equation is separable in parabolic coordinates everywhere in space. For other species, Equation (6.17) holds only sufficiently far from the core (within the single-active-electron approximation). Tunneling formulae generalizing Equations (6.21) and (6.22) to arbitrary bound states and to non-hydrogenic systems can nonetheless be obtained. However, it is necessary to postulate (i) the existence of a region of space in which the interaction of the active electron with the core can be represented by a pure Coulomb potential and (ii) that, at the same time, the potential energy $\mathcal{E}z$ of the electron in the electric field is too small to perturb the wave function significantly. Such a region always exists when the electric field is weak enough. The WKB calculation can then be carried out, as the assumption that the electron–core interaction is Coulombic in the vicinity of the inner turning point and beyond makes it possible to calculate the wave function inside the barrier and to connect it to the unperturbed wave function of the initial bound state. In the single-active-electron approximation, the latter can be written in spherical polar coordinates as

$$\phi_{n^*lm}(\mathbf{r}, t) = R_{n^*l}(r) Y_{lm}(\hat{\mathbf{r}}) \exp(iI_\mathrm{P}t), \qquad (6.23)$$

where n^* denotes the effective principal quantum number and $I_\mathrm{P} = Z_\mathrm{c}^2/(2n^{*2})$. Although $\phi_{n^*lm}(\mathbf{r}, t)$ can be a complicated function close to or inside the core, its asymptotic form is simple. Indeed, the requirement that $\phi_{n^*lm}(\mathbf{r}, t)$ be a solution of the field-free TDSE far from the core, where the potential energy is $-Z_\mathrm{c}/r$, implies that

$$R_{n^*l}(r) \underset{r \to \infty}{\to} C_\mathrm{as} \kappa^{3/2}(\kappa r)^{Z_\mathrm{c}/\kappa - 1} \exp(-\kappa r), \qquad (6.24)$$

with $\kappa = (2I_\mathrm{P})^{1/2}$. The coefficient C_as is obtained from the initial field-free bound state; it is defined apart from an arbitrary complex multiplicative factor of modulus one, and can be taken to be positive. It is important to stress that for weak enough electric fields, the rate of tunneling ionization does not depend on how $R_{n^*l}(r)$ varies in or close to the core. Only the asymptotic form of the wave function, as given by Equation (6.24), is required for carrying out the WKB calculation.

The general expression of the tunnel ionization rate was first obtained by Smirnov and Chibisov [16]. A detailed derivation of this result has been given by Bisgaard and Madsen [17]. For an electric field of strength $\mathcal{E}$ oriented either in the positive or the negative z-direction, the resulting expression for the rate of ionization is

$$\Gamma_\mathrm{DC}(\mathcal{E}) \simeq \Gamma_\mathrm{WKB}(\mathcal{E}) = C_\mathrm{as}^2 A(l, m) \frac{\kappa^2}{2} \left(\frac{2\kappa^3}{\mathcal{E}}\right)^{2Z_\mathrm{c}/\kappa - |m| - 1} \exp\left(-\frac{2\kappa^3}{3\mathcal{E}}\right), \qquad (6.25)$$

with

$$A(l, m) = \frac{2l+1}{2^{|m|}|m|!} \frac{(l+|m|)!}{(l-|m|)!}. \qquad (6.26)$$

The accuracy of the rate $\Gamma_\mathrm{DC}(\mathcal{E})$ given by Equation (6.25) improves as $\mathcal{E}$ becomes small. We note that it depends on Z_c and $\mathcal{E}$ only through the scaled variables Z_c/κ and $\mathcal{E}/\kappa^3$. This can be understood by observing that, in the weak-field limit and far from the core, the wave function of the ejected electron is a solution of the

Schrödinger equation,

$$\left[-\frac{1}{2}\nabla_{\mathbf{r}}^2 - \frac{Z_c}{r} + \mathcal{E}\cdot\mathbf{r}\right]\psi(\mathbf{r}) = -\frac{\kappa^2}{2}\psi(\mathbf{r}), \qquad (6.27)$$

which can be transformed into the equation

$$\left[-\frac{1}{2}\nabla_{\mathbf{r}'}^2 - \frac{Z_c}{\kappa}\frac{1}{r'} + \frac{1}{\kappa^3}\mathcal{E}\cdot\mathbf{r}'\right]\psi(\mathbf{r}') = -\frac{1}{2}\psi(\mathbf{r}') \qquad (6.28)$$

by introducing scaled coordinates $\mathbf{r}' = \kappa\,\mathbf{r}$. For hydrogenic systems initially in their ground state, $C_{as} = 2$, $Z_c = \kappa = Z$, $l = m = 0$ and $A(l,m) = 1$, so that Equation (6.25) reduces to Equation (6.22), as expected.

While the exponential factor in Equation (6.25) does not depend on the atomic species, the pre-exponential factor, in particular its functional dependence in the strength of the electric field, does. When $\mathcal{E} \ll \kappa^3$, which is typically the case for the systems for which Equation (6.25) applies, $\Gamma_{DC}(\mathcal{E})$ is smaller for ionization from a state with $m \neq 0$ than for ionization from a state of the same energy aligned with the electric field. This difference is consistent with the observation made above that the classical critical electric-field strength for over-the-barrier ionization is higher for $M_z \neq 0$ than for $M_z = 0$.

As noted above, the value of the asymptotic coefficient C_{as} appearing in the tunneling formula is a property of the initial field-free bound state. This coefficient is known exactly for atomic hydrogen and one-electron ions. For an initial state with principal quantum number n and orbital quantum number l, it is given by

$$C_{as} = \left[\frac{2^{2n}}{n(n+l)!(n-l-1)!}\right]^{1/2}. \qquad (6.29)$$

This coefficient has been estimated for many non-hydrogenic species by fitting the right-hand side of Equation (6.24) to a numerical wave function or by using approximation methods [18, 19]. Hartree [20] noticed that the general properties of the radial wave function suggest that

$$C_{as} \simeq \left[\frac{2^{2n^*}}{n^*\,\Gamma(n^*+l+1)\Gamma(n^*-l)}\right]^{1/2}, \qquad (6.30)$$

where n^* is the effective principal quantum number and Γ denotes Euler's gamma function. Moreover, if the field-free atom is described as a one-electron system in which the electron interacts with the core through an l-dependent model potential of the form

$$W_l(r) = \frac{l^*(l^*+1) - l(l+1)}{2r^2} - \frac{Z_c}{r}, \qquad (6.31)$$

then Equation (6.30) with l replaced by l^* is exact for the bound-state wave functions obtained for this model potential provided that $n^* - l^*$ is an integer, $l^* > -3/2$ and $l^* \leq n^* - 1$ [21]. Ammosov, Delone and Krainov [22] proposed to evaluate C_{as} within this model, with l^* taken to be equal to $n_0^* - 1$, where n_0^* is the effective principal quantum number of the lowest term of the series to which the initial state belongs. In order to facilitate the calculation, they also evaluated the resulting Euler's gamma functions by using the Stirling approximation. This yields the approximate asymptotic coefficient

$$C_{\mathrm{ADK}} = \left[\frac{1}{2\pi n^*} \left(\frac{4\mathrm{e}^2}{n^{*2} - l^{*2}} \right)^{n^*} \left(\frac{n^* - l^*}{n^* + l^*} \right)^{l^* + 1/2} \right]^{1/2}, \qquad (6.32)$$

where $\mathrm{e} = 2.718\ldots$ For $n^* \gg l^*$,

$$C_{\mathrm{ADK}} \simeq \left[\frac{1}{2\pi n^*} \left(\frac{4\mathrm{e}^2}{n^{*2}} \right)^{n^*} \right]^{1/2}, \qquad (6.33)$$

which is also the asymptotic coefficient one would obtain by using the WKB approximation. Although the approximations made in the derivation of Equation (6.33) do not appear to be justified for low quantum numbers, the resulting value of the asymptotic coefficient for the ground state of atoms or ions is often in remarkably good agreement with the asymptotic behavior of the Hartree–Fock wave function [18, 19, 22, 23].

The good accuracy of the tunneling formulae for sufficiently weak fields is illustrated by Fig. 6.2. It predicts total ionization rates that are in close agreement with the results of exact calculations up to $\mathcal{E} \simeq \mathcal{E}_{\mathrm{OBI}}$ [24], the critical electric-field strength at which the classical effective potential barrier vanishes so that OBI occurs. However, these simple formulae overestimate the ionization rate significantly for $\mathcal{E} \gg \mathcal{E}_{\mathrm{OBI}}$. As seen in Fig. 6.2, the breakdown of the tunneling formulae when $\mathcal{E}$ approaches and exceeds $\mathcal{E}_{\mathrm{OBI}}$ varies from species to species.

Three important observations can be made from the examination of Fig. 6.2. Firstly, and contrary to what the classical model might suggest, the rate of ionization varies smoothly, without any sudden increase, at the critical electric-field strength for over-the-barrier ionization. Secondly, the ionization rate increases rapidly as $\mathcal{E}$ approaches $\mathcal{E}_{\mathrm{OBI}}$. Thus, reaching the barrier-suppression regime is generally precluded since the atom will ionize with almost unit probability during the laser pulse turn-on, unless the pulse is very short. Thirdly, the rate of ionization is a very rapidly varying function in the tunneling regime, which means that small changes in the strength of the electric field induce large changes in the ionization yield. This last fact plays a crucial role in the dynamics of atoms in intense few-cycle pulses.

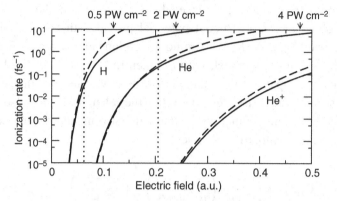

Figure 6.2. Rates of ionization from the ground state of a hydrogen atom, a helium atom and a He$^+$ ion in a static electric field. Solid curves: exact rate; the results for He have been obtained by Scrinzi, Geissler and Brabec [24]. Dashed curves: rate predicted by the tunneling formulae, Equation (6.21) for H, Equation (6.22) for He$^+$ and Equation (6.25) with $C_{as} = 1.99$ for He [18]. The dotted lines indicate the critical fields for over-the-barrier ionization of H and He. ($\mathcal{E}_{OBI} = 0.5$ a.u. for He$^+$.) The three arrows at the top of the diagram point to the values of $\mathcal{E}$ at which a linearly polarized laser field with $\mathcal{E}_0 = \mathcal{E}$ would have an intensity of 0.5, 2 or 4×10^{15} W cm^{-2}.

These results for ionization in a static electric field can be readily extended to the case of a slowly oscillating laser electric field using Equation (6.12). In particular, one can obtain a rate of tunneling ionization in the adiabatic approximation by calculating the cycle-average of the approximate DC rate given by Equation (6.25). For a linearly polarized monochromatic laser field described by the electric field $\mathcal{E}(t) = \hat{\epsilon}\mathcal{E}_0 \cos(\omega t)$, this procedure yields

$$\Gamma_{ad}(\mathcal{E}_0, 0) \simeq \frac{2\omega}{\pi} \int_0^{\pi/(2\omega)} \Gamma_{WKB}[|\mathcal{E}(t)|]\,dt. \tag{6.34}$$

The right-hand side of this equation involves an integral of the form

$$I(a) = \frac{2}{\pi} \int_0^{\pi/2} (\cos\theta)^\nu \exp(-a/\cos\theta)\,d\theta, \tag{6.35}$$

where a and ν are real numbers. The case of interest here is one for which $\kappa^3/\mathcal{E}_0 \gg 1$, so that a is large and positive. The integral $I(a)$ is a well-behaved function of a for any value of ν. Although $\cos\theta$ vanishes at $\theta = \pi/2$, when $a > 0$ the exponential function goes to zero faster than $(\cos\theta)^\nu$ diverges. Upon changing the integration variable from θ to η, where $\cosh\eta = 1/\cos\theta$, Equation (6.35) becomes

$$I(a) = \frac{2}{\pi} \int_0^\infty (\cosh\eta)^{-\nu-1} \exp(-a\cosh\eta)\,d\eta. \tag{6.36}$$

The behavior of this integral for large values of a can be obtained by the method of Laplace [25]. The result is

$$I(a) \simeq \left(\frac{2}{\pi a}\right)^{1/2} \exp(-a) \tag{6.37}$$

as $a \to \infty$. Hence, for a linearly polarized monochromatic laser field,

$$\Gamma_{\text{ad}}(\mathcal{E}_0, 0) \simeq \left(\frac{3\mathcal{E}_0}{\pi\kappa^3}\right)^{1/2} \Gamma_{\text{WKB}}(\mathcal{E} = \mathcal{E}_0)$$

$$= C_{\text{as}}^2 A(l,m) \frac{\kappa^2}{2} \left(\frac{3\mathcal{E}_0}{\pi\kappa^3}\right)^{1/2} \left(\frac{2\kappa^3}{\mathcal{E}_0}\right)^{2Z_c/\kappa - |m| - 1} \exp\left(-\frac{2\kappa^3}{3\mathcal{E}_0}\right).$$
$$\tag{6.38}$$

Apart from an overall factor that varies slowly with $\mathcal{E}_0$, the cycle-average rate of tunneling ionization in a linearly polarized laser field of electric-field strength $\mathcal{E}_0$ is identical to the rate of ionization in a static electric field of strength $\mathcal{E}_0$. Indeed, due to the strongly non-linear variation of $\Gamma_{\text{DC}}(\mathcal{E})$ with $\mathcal{E}$, ionization occurs predominantly at the maxima of $|\mathcal{E}(t)|$.

Equation (6.38) is due to Perelomov, Popov and Terent'ev [26, 27]. It is often used with the asymptotic coefficient C_{as} evaluated in the approximation (6.32) proposed by Ammosov, Delone and Krainov [22]. The resulting expression for the total rate of ionization in a linearly polarized monochromatic laser field is referred to as the "ADK formula," and is given by

$$\Gamma_{\text{ad}}^{\text{ADK}}(\mathcal{E}_0) = C_{\text{ADK}}^2 A(l,m) \frac{\kappa^2}{2} \left(\frac{3\mathcal{E}_0}{\pi\kappa^3}\right)^{1/2} \left(\frac{2\kappa^3}{\mathcal{E}_0}\right)^{2Z_c/\kappa - |m| - 1} \exp\left(-\frac{2\kappa^3}{3\mathcal{E}_0}\right).$$
$$\tag{6.39}$$

6.1.3 The Keldysh adiabaticity parameter

Let us now return to the question of how small the angular frequency ω of the laser field must be for the adiabatic approximation to remain valid. We have seen above that the exact rate of ionization can be formally expanded in powers of ω^2. Since the zeroth-order term of this series is the adiabatic rate, in order for the adiabatic approximation to hold, the remaining terms of the series should be sufficiently small. However, this requirement cannot be expressed in a simple way in terms of the parameters of the laser field, and the tunneling theory introduced in Section 6.1.2 does not lend itself to an analysis along these lines. We will see in Section 6.3 that a very simple adiabaticity criterion emerges from the asymptotic analysis of the ionization amplitude within the strong-field approximation (SFA). Indeed, the rate of ionization in a laser field reduces to the adiabatic ionization rate when $\gamma_K \ll 1$, where γ_K is the dimensionless Keldysh parameter [1] defined by Equation (1.8). For a laser field with electric-field strength $\mathcal{E}_0$, this parameter can also be expressed as

$$\gamma_K = \frac{\omega (2I_P)^{1/2}}{\mathcal{E}_0}. \tag{6.40}$$

At intensities below the barrier-suppression regime, ionization can be described as occurring via tunneling through a static barrier when $\gamma_K \lesssim 1$. For $\gamma_K \gtrsim 1$, the multi-photon character of the ionization process dominates. These dynamical regimes are usually referred to as the "tunneling regime" and the "multiphoton regime," respectively.

While the adiabaticity condition $\gamma_K \ll 1$ is justified by a mathematical analysis of the ionization amplitude, it can also be understood from a simple physical picture. This condition expresses the requirement that the laser field remains essentially static while the electron tunnels through the effective potential barrier V_{eff} defined by Equation (6.20). Thus, ionization is adiabatic if the time taken by the electron to tunnel, T_t, is short compared to the time during which the electric field changes significantly. The electron tunneling time is not a well defined concept, as tunneling is not amenable to a classical description. Nevertheless, let us define an effective tunneling time by treating the sub-barrier motion classically, but with an imaginary velocity. When the laser field is described by the electric-field vector $\mathcal{E}(t) = \mathcal{E}_0 \hat{z} \cos(\omega t)$ and the electron has zero angular momentum about the z-axis, the effective potential is given by Equation (6.20). Setting $\rho = 0$ and $M_z = 0$, the problem is one-dimensional. One can assign an instantaneous velocity

$$v(z) = \{2[I_P - V_{\text{eff}}(0, z, 0)]\}^{1/2} \tag{6.41}$$

to the tunneling electron. Letting $z_{\min}$ and $z_{\max}$ be the two roots of the equation $v(z) = 0$, we have that under the barrier, namely between $z = z_{\min}$ and $z = z_{\max}$, the function $v(z)$ is imaginary. The tunneling time is then the corresponding characteristic time for the sub-barrier motion, and is given by

$$T_t = \int_{z_{\min}}^{z_{\max}} \frac{dz}{|v(z)|}. \tag{6.42}$$

The right-hand side of Equation (6.42) can be expressed in terms of a complete elliptic integral of the second kind, which reduces to unity for $(Z_c\mathcal{E})^{1/2}/I_P \to 0$. The result is that $T_t \simeq (2I_P)^{1/2}/\mathcal{E}$ for $I_P \gg 2(Z_c\mathcal{E})^{1/2}$, and hence for $\mathcal{E} \ll \mathcal{E}_{\text{OBI}}$. The barrier changes little over a time interval of the order of T_t if T_t is short compared to about one-quarter of an optical period of the laser field, namely $T_t \ll \pi/(2\omega)$. Replacing $\pi/2$ by 1, this condition can also be expressed as $\omega (2I_P)^{1/2}/\mathcal{E}(t) \ll 1$, where $\mathcal{E}(t) = |\mathcal{E}(t)|$. Irrespective of the value of the electric-field strength $\mathcal{E}_0$, this inequality is violated if $\mathcal{E}(t)$ vanishes during the course of the oscillation of the laser field. However, because of the sharp increase of $\Gamma_{DC}(\mathcal{E})$ with $\mathcal{E}$ in the tunneling regime, ionization occurs predominantly when $\mathcal{E}(t) \simeq \mathcal{E}_0$. As long as $\mathcal{E}_0 < \mathcal{E}_{OBI}$, it follows that ionization can be described as occurring through a quasi-static barrier when $\omega(2I_P)^{1/2}/\mathcal{E}_0 \ll 1$, or equivalently when $\gamma_K \ll 1$.

It is interesting to note that the characteristic time T_t associated with the tunneling process in the present problem is identical in form to the Büttiker and Landauer traversal time through a one-dimensional oscillating potential barrier [28]. In both cases the transition between the regimes where tunneling proceeds as if the barrier were static and if it were not occurs when $\omega T_t \simeq 1$.

It follows from Equation (1.8) that the inequality $\gamma_K \ll 1$ implies that the ponderomotive energy U_p is large compared to the ionization potential I_P. Even for the case of helium, this condition can be satisfied for frequencies and intensities at which the motion of the electron is non-relativistic and the dipole approximation is completely justified. However, the double requirement that γ_K be sufficiently small and that $\mathcal{E}_0 < \mathcal{E}_{OBI}$ is difficult or impossible to satisfy for most elements at wavelengths in the visible part of the spectrum [29]. In contrast, ionization deep in the tunneling regime is possible experimentally at wavelengths in the mid- or far-infra-red.

6.2 Multiphoton processes in the strong-field approximation

In Section 6.1 we showed, within an adiabatic approximation, how DC tunneling ionization rates can be used to calculate total ionization rates for atoms interacting with intense, low-frequency laser fields. It what follows we shall describe a low-frequency theory that is based on the strong-field approximation (SFA). In the SFA, the influence of the atomic potential on an ionized electron is either neglected entirely, so that the electron moves "freely" in the laser field, or treated as a perturbation [1, 26, 30]. We will see that this approach leads to relatively simple expressions for the ionization probability amplitudes. The SFA theory, in contrast to the adiabatic tunneling ionization description introduced above, yields partial and total ionization rates that depend explicitly on the angular frequency of the laser field. We will begin our discussion by developing a strong-field perturbation theory in which the interaction between the electron and the atomic core is treated as a perturbation while the interaction of the electron with the laser field is treated exactly. This leads to a series expansion for the ionization amplitude, and we will see that the SFA amplitude corresponding to direct ionization is given by the leading term of this expansion. The SFA amplitude corresponding to recollision ionization, in turn, is given by the second term of the expansion.

6.2.1 Basic equations

We focus on the case of an atom (or ion) described within the single-active-electron approximation. The Hamiltonian of the system is given by the equation

$$H(t) = H_0 + H_{\text{int}}(t), \tag{6.43}$$

where H_0 is the field-free atomic Hamiltonian and $H_{\text{int}}(t)$ is the Hamiltonian describing the interaction of the laser field with the active electron. It is convenient to introduce the Hamiltonian of a "free" electron in the laser field, namely

$$H_{\text{F}}(t) = H(t) - V = K + H_{\text{int}}(t). \tag{6.44}$$

Here V is the potential describing the interaction of the active electron with the core and $K = \mathbf{p}^2/2$ is the electron kinetic energy operator.

At the time t_0 when the laser pulse is applied, the initial state of the atom is taken to be the ground state $|\phi_0(t_0)\rangle = |\psi_0\rangle \exp(-iE_0 t_0)$, where $|\psi_0\rangle$ is an eigenstate of H_0 and $E_0 = -I_{\text{P}}$. Using Equation (3.25), a formal solution of the TDSE (2.149) is found to be

$$|\Psi(t)\rangle = |\phi_0(t)\rangle - i \int_{t_0}^{t} dt' \, U(t,t') H_{\text{int}}(t') |\phi_0(t')\rangle, \tag{6.45}$$

where $U(t,t')$ is the evolution operator introduced in Section 2.6.

Let us now obtain an expression for the ionization amplitude at the end of the laser pulse. We denote by $|\phi_f(t)\rangle$ a continuum eigenstate of H_0 with energy $E_f = k_f^2/2$ and a drift wave vector $\mathbf{k}_f$ that is normalized to a delta function in momentum space, as specified by Equation (2.168) for Gordon–Volkov states. Taking the laser pulse duration to be τ_{p}, and letting $t_1 = t_0 + \tau_{\text{p}}$, the amplitude corresponding to a transition to the final state $|\phi_f(t)\rangle$ is

$$T_{f0} = \langle \phi_f(t_1) | \Psi(t_1) \rangle$$

$$= -i \int_{t_0}^{t_1} dt' \langle \phi_f(t_1) | U(t_1,t') H_{\text{int}}(t') | \phi_0(t') \rangle. \tag{6.46}$$

The probability that the atom will emit an electron having an energy within the interval $(E_f, E_f + dE_f)$ and into the solid angle $d\hat{\mathbf{k}}_f$ centered about $\hat{\mathbf{k}}_f$ is then given by

$$P(E_f, \hat{\mathbf{k}}_f) = (2E_f)^{1/2} |T_{f0}|^2 \, dE_f \, d\hat{\mathbf{k}}_f. \tag{6.47}$$

Correspondingly,

$$P_{\text{tot}} = \int_0^{\infty} dE_f \int d\hat{\mathbf{k}}_f \, (2E_f)^{1/2} |T_{f0}|^2 \tag{6.48}$$

is the total probability that the atom will be ionized. The factor $(2E_f)^{1/2}$ appearing in Equations (6.47) and (6.48) is not present when the energy normalization is adopted for the continuum states.

6.2.2 Strong-field perturbation theory

The solution of the TDSE (2.149), given by Equation (6.45), expresses the state vector of the system in terms of an integral equation involving the evolution operator $U(t, t')$. As it is in general not possible to obtain exact expressions for this operator, we must make use of a method of approximation. In Section 3.1, we developed a series expansion for $U(t, t')$ in which each term of the series was expressed in terms of the laser–atom interaction Hamiltonian $H_{int}(t)$ and the evolution operator $U_0(t, t')$ corresponding to the field-free Hamiltonian H_0. This expansion of the evolution operator is appropriate when describing the evolution of an atom interacting with a weak laser field. Let us now consider an expansion that is appropriate for intense, low-frequency laser fields.

In order to develop a framework for approximating $U(t, t')$ in this regime, we follow a procedure similar to the one discussed in Section 3.1. However, the atomic interaction potential V will now be taken to be the perturbation. Using the Gordon–Volkov evolution operator $U_F(t, t')$, we find, in analogy with Equations (3.24) and (3.25), that the evolution operator $U(t, t')$ satisfies the following two integral equations:

$$U(t, t') = U_F(t, t') - i \int_{t'}^{t} U_F(t, t'_1) V U(t'_1, t') dt'_1 \qquad (6.49)$$

and

$$U(t, t') = U_F(t, t') - i \int_{t'}^{t} U(t, t'_1) V U_F(t'_1, t') dt'_1. \qquad (6.50)$$

A formal solution of these equations, in the form of an infinite series, can be obtained by iteration. The result is

$$U(t, t') = \sum_{n=0}^{\infty} \overline{U}^{(n)}(t, t'). \qquad (6.51)$$

The first term is simply the Gordon–Volkov evolution operator,

$$\overline{U}^{(0)}(t, t') = U_F(t, t'), \qquad (6.52)$$

while the remaining terms are given by

$$\overline{U}^{(n)}(t,t') = (-\mathrm{i})^n \int_{t'}^{t} \mathrm{d}t_1' \int_{t'}^{t_1'} \mathrm{d}t_2' \dots \int_{t'}^{t_{n-1}'} \mathrm{d}t_n'$$
$$\times U_{\mathrm{F}}(t,t_1')V U_{\mathrm{F}}(t_1',t_2')V \dots U_{\mathrm{F}}(t_{n-1}',t_n')V U_{\mathrm{F}}(t_n',t'), \quad n \geq 1.$$

$$(6.53)$$

Inserting Equation (6.51) into Equation (6.45) yields

$$|\Psi(t)\rangle = |\phi_0(t)\rangle + \sum_{n=0}^{\infty} |\overline{\Psi}^{(n)}(t)\rangle, \tag{6.54}$$

where

$$|\overline{\Psi}^{(n)}(t)\rangle = -\mathrm{i} \int_{t_0}^{t} \mathrm{d}t' \, \overline{U}^{(n)}(t,t') H_{\mathrm{int}}(t') |\phi_0(t')\rangle. \tag{6.55}$$

Referring to Equations (6.52), (6.53) and (6.55), the terms on the right-hand side of Equation (6.54) can be interpreted as follows. The first term is simply the time-evolved initial state of the atom. The second term with $n = 0$ gives the direct ionization contribution to the state vector. From Equations (6.52) and (6.55), this term represents the process whereby the initial state interacts with the laser field at time t' and forms a superposition of Gordon–Volkov states. The terms in Equation (6.54) with $n \geq 1$ contain the recollision contributions to the state vector. The superposition of Gordon–Volkov states formed by the interaction of the initial state with the laser field at time t' now evolves to time t according to the operator $\overline{U}^{(n)}(t,t')$. From Equation (6.53), we see that this superposition first evolves in time as a free electron in the laser field according to the Gordon–Volkov evolution operator $U_{\mathrm{F}}(t_n',t')$. The electron wave packet then interacts with its parent core at time t_n'. The resulting electron wave packet subsequently evolves in time according to the Gordon–Volkov evolution operator and interacts with its parent core at time t_{n-1}', and so on, so that the electron interacts n times with its parent core. Each interaction can be interpreted as a recollision event. For a single-electron atom, this event will be an "elastic" process. For complex atoms, the recollision event can be elastic or inelastic as defined in Equations (1.20) and (1.21), respectively, or a reaction such as the (e, 2e) collision defined by Equation (1.22). The contribution to the state vector is then obtained by integrating over all the possible recollision times $t_n', t_{n-1}', \dots, t_1'$ that satisfy the time ordering

$$t \geq t_1' \geq t_2' \dots \geq t_{n-1}' \geq t_n' \geq t'. \tag{6.56}$$

As in Equation (6.51) for the evolution operator, the transition amplitude (6.46) can be written as an infinite series of the form

$$T_{f0} = \sum_{n=0}^{\infty} \overline{T}_{f0}^{(n)}. \qquad (6.57)$$

Using Equations (6.52), (6.53) and (6.55), we see that the first term is given by

$$\overline{T}_{f0}^{(0)} = \langle \phi_f(t_1) | \overline{\Psi}^{(0)}(t_1) \rangle$$

$$= -i \int_{t_0}^{t_1} dt' \langle \chi_{\mathbf{k}_f}(t') | H_{\text{int}}(t') | \phi_0(t') \rangle, \qquad (6.58)$$

where $|\chi_{\mathbf{k}_f}(t')\rangle$ is a Gordon–Volkov state with wave vector $\mathbf{k}_f$. The term (6.58) does not contain the atomic interaction potential V, while the terms

$$\overline{T}_{f0}^{(n)} = \langle \phi_f(t_1) | \overline{\Psi}^{(n)}(t_1) \rangle$$

$$= (-i)^{n+1} \int_{t_0}^{t_1} dt' \int_{t'}^{t_1} dt'_1 \int_{t'}^{t'_1} dt'_2 \ldots \int_{t'}^{t'_{n-1}} dt'_n$$

$$\times \langle \chi_{\mathbf{k}_f}(t'_1) | V U_{\text{F}}(t'_1, t'_2) V \ldots U_{\text{F}}(t'_{n-1}, t'_n) V U_{\text{F}}(t'_n, t') H_{\text{int}}(t') | \phi_0(t') \rangle,$$

$$n \geq 1, \qquad (6.59)$$

contain n "interactions" of the electron with the potential V. It is important to bear in mind that the initial state $|\phi(t_0)\rangle$ is an eigenstate of H_0, and therefore takes into account the influence of V exactly.

In analogy with the weak-field perturbation theory discussed in Chapter 3, we can formulate a strong-field perturbation theory by successively evaluating the terms in the series (6.57) up to some maximum order in V. We will denote by $|\Psi^{(N)}(t)\rangle$ and $T_{f0}^{(N)}$, respectively, the approximate state vector and transition amplitude in which only the N leading terms in V are retained, so that

$$T_{f0}^{(N)} = \langle \phi_f(t_1) | \Psi^{(N)}(t_1) \rangle$$

$$= \sum_{n=0}^{N} \langle \phi_f(t_1) | \overline{\Psi}^{(n)}(t_1) \rangle$$

$$= \sum_{n=0}^{N} \overline{T}_{f0}^{(n)}, \qquad (6.60)$$

where we have used the fact that the initial and final states are orthogonal.

6.2.3 The strong-field approximation

The SFA transition amplitude corresponding to direct ionization is given by Equation (6.60) with $N = 0$. Therefore, we have

$$T_{f0}^{\text{SFA}} = \overline{T}_{f0}^{(0)} = T_{f0}^{(0)} = \langle \phi_f(t_1) | \Psi^{(0)}(t_1) \rangle, \tag{6.61}$$

where

$$| \Psi^{(0)}(t) \rangle = | \phi_0(t) \rangle - i \int_{t_0}^{t} dt' \, U_{\text{F}}(t, t') H_{\text{int}}(t') | \phi_0(t') \rangle \tag{6.62}$$

so that T_{f0}^{SFA} is given by Equation (6.58). The ionization process is described by a transition from the initial ground state of the atom to a final Gordon–Volkov state. The SFA direct ionization amplitude will be investigated in detail in Section 6.3. At this level of approximation, the time-dependent laser-induced atomic dipole moment is given by

$$\begin{aligned}
\mathbf{d}(t) &= -\langle \Psi^{(0)}(t) | \mathbf{r} | \Psi^{(0)}(t) \rangle \\
&= -\Bigg[\langle \phi_0(t) | \mathbf{r} | \phi_0(t) \rangle + 2 \, \text{Im} \int_{t_0}^{t} dt' \langle \phi_0(t) | \mathbf{r} U_{\text{F}}(t, t') H_{\text{int}}(t) | \phi_0(t') \rangle \\
&\quad + \int_{t_0}^{t} dt' \int_{t_0}^{t} dt'' \langle \phi_0(t') | H_{\text{int}}^{\dagger}(t') U_{\text{F}}(t', t) \mathbf{r} U_{\text{F}}(t, t'') H_{\text{int}}(t'') | \phi_0(t'') \rangle \Bigg].
\end{aligned} \tag{6.63}$$

Section 6.4 is devoted to an analysis of this quantity.

If we now incorporate the atomic interaction potential V to first order, and use the state vector $| \Psi^{(1)}(t) \rangle$ to calculate the ionization transition amplitude, we obtain

$$\begin{aligned}
T_{f0}^{(1)} &= \langle \phi_f(t_1) | \Psi^{(1)}(t_1) \rangle \\
&= \overline{T}_{f0}^{(0)} + \overline{T}_{f0}^{(1)},
\end{aligned} \tag{6.64}$$

where, from Equation (6.59),

$$\overline{T}_{f0}^{(1)} = -\int_{t_0}^{t_1} dt' \int_{t'}^{t_1} dt_1' \langle \chi_{\mathbf{k}_f}(t_1') | V U_{\text{F}}(t_1', t') H_{\text{int}}(t') | \phi_0(t') \rangle. \tag{6.65}$$

In addition to the direct ionization transition amplitude, the total ionization transition amplitude now contains a contribution from electrons that rescatter from the parent core. We will discuss the rescattering transition amplitude $\overline{T}_{f0}^{(1)}$ in Section 6.5.

Let us now comment on the validity of the SFA. In contrast to the expansion (3.33), which is a power series in the perturbation $H_{\text{int}}(t)$ and therefore converges rapidly in the weak-field limit, the convergence properties of the series (6.51) are difficult to determine. Nevertheless, it can be expected that the SFA is applicable when (i) the laser angular frequency is low ($\omega \ll I_P$) and (ii) the laser intensity is sufficiently high so that $\gamma_K \lesssim 1$. We recall that in Section 6.1 it was shown that when the above two criteria are met, ionization in a linearly polarized laser field occurs predominantly twice per laser cycle at times near the maxima of $|\mathcal{E}(t)|$. Since in this frequency and intensity regime the excursion amplitude of the ionized electron wave packet, given by $\alpha_0 = \mathcal{E}_0/\omega^2 = 2(U_p)^{1/2}/\omega$, is large, the wave packet will be driven away from its parent ion by the laser field. As a result, the influence of the potential V on the electron dynamics will be small. We will see in Section 6.3 that when $\gamma_K \ll 1$ the transition amplitude $T_{f0}^{\text{SFA}} = \overline{T}_{f0}^{(0)}$ given by Equation (6.58) describes the ionization of the atom in a manner similar to the quasi-static approximation while including the multiphoton character of the ionization process.

If the electron is driven back to its parent core by the laser field, it can then rescatter. Referring to the interpretation of the operators $\overline{U}^{(n)}(t, t_0)$ given above, we expect that the expansion (6.51) will be appropriate when no, or very few, recollision events contribute to the multiphoton process being studied. This is typically the case for atoms interacting with intense, low-frequency laser fields. Let us consider the case of a hydrogenic system initially in its ground state. The ionization potential is then given by $I_P = Z^2/2$ and the width of the initial wave function in momentum space is approximately given by $\Delta p = Z = (2I_P)^{1/2}$ [6]. An electron wave packet that is formed at the time t_d when $|\mathcal{E}(t)|$ is near a maximum will then also have a width in momentum space given by $\Delta p \simeq (2I_P)^{1/2}$. If $T = 2\pi/\omega \gg 1/(\Delta p)^2$, or $\omega \ll I_P$, then this wave packet will have spread appreciably at the later time $t_d + T$. As a consequence, the first recollision of the wave packet with its parent core, represented by the term with $n = 1$ in Equation (6.54), will give the dominant recollision contribution to the evolution of the state vector $|\Psi(t)\rangle$. Subsequent recollisions, described by the terms with $n > 1$ in Equation (6.54), will give rise to progressively smaller contributions.

The expansion (6.51) is in general not applicable when the atom interacts with an intense, high-frequency laser pulse whose duration is longer than a few optical cycles. Insight into why this is the case can be gained by considering a very intense laser pulse with angular frequency such that $\omega \gg I_P$. We will also assume that the pulse is applied instantaneously. Under these conditions, the initial dynamics of the electron wave packet will be similar to that of a Gordon–Volkov wave packet. Because now $|\langle \Psi(t)|\phi_0(t)\rangle|^2 \simeq 0$, instead of using Equations (6.54) and (6.55) it is more appropriate to apply the evolution operator $U(t, t_0)$ expressed in terms of the series expansion (6.51) directly to the initial state at time t_0. This yields

$$|\Psi(t)\rangle = U_F(t, t_0)|\phi_0(t_0)\rangle + \sum_{n=1}^{\infty} \overline{U}^{(n)}(t, t_0)|\phi_0(t_0)\rangle. \tag{6.66}$$

For times $t \ll T = 2\pi/\omega$, only the first term on the right-hand side of this equation, which describes the evolution of the system as a Gordon–Volkov wave packet, needs to be retained. However, when $t \gg T$, the electron wave packet will recollide with its parent core many times before it spreads appreciably. In contrast with the low-frequency regime, it is now necessary to retain a large number of terms in the expansion (6.51) to ensure convergence. In Chapter 7, we will develop a theory appropriate to the high-frequency regime.

As will be shown below, the integrals over the intermediate times and momenta that appear in the SFA ionization transition amplitudes and laser-induced atomic dipole moment can in general be evaluated by the application of the saddle-point method [31]. For the three multiphoton processes considered in this chapter, namely direct ionization, harmonic generation and recollision ionization, we will show how the corresponding amplitudes can be written succinctly as a sum over contributions arising from particular complex electron trajectories. Each contribution can, in turn, be expressed as an amplitude multiplied by a phase factor, with the phase being given by a modified classical action corresponding to the process. The value of this action is determined from the saddle-point equations that define the complex electron trajectories. The saddle-point equations have a straightforward physical interpretation, thereby allowing the corresponding classical trajectories to be readily determined from the complex classical trajectories. In this way, we recover the semi-classical model of multiphoton process in the low-frequency regime, which allows the physical origin of many features of multiphoton processes to be uncovered. This will be illustrated by considering, for example, the $2U_p$ "low-energy" ATI cut-off, the $3.17U_p$ harmonic generation cut-off and the $10U_p$ "high-energy" ATI cut-off.

It is worth noting that the SFA direct ionization transition amplitude (6.61) can be related to the alternative SFA direct ionization transition amplitude [32]

$$\tilde{T}_{f0}^{\mathrm{SFA}} = \langle \phi_f(t_1) | U_{\mathrm{F}}(t_1, t_0) | \phi_0(t_0) \rangle, \tag{6.67}$$

which is obtained by retaining only the first term on the right-hand side of Equation (6.66), by using the fact that $V = H_0 + H_{\mathrm{int}}(t) - H_{\mathrm{F}}(t)$ and recalling that the evolution operator $U_{\mathrm{F}}(t, t_0)$ is a solution of the TDSE

$$\mathrm{i}\frac{\partial}{\partial t} U_{\mathrm{F}}(t, t_0) = H_{\mathrm{F}}(t) U_{\mathrm{F}}(t, t_0). \tag{6.68}$$

6.3 Multiphoton direct ionization

In this section we will consider in detail the SFA direct ionization transition amplitude $T_{f0}^{\mathrm{SFA}} = \overline{T}_{f0}^{(0)}$, where $\overline{T}_{f0}^{(0)}$ is given by Equation (6.58). We recall that the SFA ionization transition amplitude describes an electron which is in the field-free ground state $|\psi_0\rangle$ before the laser pulse arrives, is detached from the atom by the pulse, subsequently propagates as a "free" particle in the laser field without any further interaction with its parent core, and is finally detected, after the end of the pulse, with a drift momentum $\mathbf{k}_f$. This amplitude describes only direct ionization and does not include the process whereby the ionized electron recollides with the parent core. Likewise, it does not take into account the atomic bound states other than the initial state; hence, it cannot account for resonant multiphoton processes, and in particular REMPI. Also neglected in this approximation is the non-ponderomotive AC Stark shift of the initial bound state with respect to the continuum, which arises from the dressing of this bound state by the laser field. In contrast, the ponderomotive part of the relative AC Stark shift that originates from the quivering motion of the unbound electron is taken into account in the transition amplitude T_{f0}^{SFA}. The absence of a non-ponderomotive shift is not important for ionization in low-frequency fields, since, as we have seen in Section 6.1.1, this non-ponderomotive shift is much smaller than U_{p} at long wavelengths. In high-frequency laser fields, with $\omega > I_{\text{P}}$, the non-ponderomotive shift often exceeds U_{p} so that, as noted above, the validity of the SFA becomes questionable.

A final point worth noting concerning the SFA ionization transition amplitude is that it is not gauge-invariant, contrary to the exact ionization amplitude. As a consequence, the results obtained by expressing the interaction Hamiltonian $H_{\text{int}}(t)$ and the Gordon–Volkov state $|\chi_{\mathbf{k}_f}(t)\rangle$ in the length gauge will not be the same as those obtained using either the velocity gauge or the original, untransformed, TDSE. The length-gauge formulation, which as we have seen in Section 6.1 is the most natural one for low-frequency fields, was introduced by Keldysh [1], while Faisal [33] and Reiss [34] have considered the SFA ionization transition amplitude using the untransformed TDSE. We shall discuss the former in Sections 6.3.1 and 6.3.2 and the latter in Section 6.3.3.

6.3.1 The Keldysh theory for an ultra-short laser pulse

The right-hand side of Equation (6.61) is simple enough to be calculated numerically without further approximations. However, the analytical approach given in this section provides considerable insight into the multiphoton direct ionization process.

It is convenient to express the SFA transition amplitude (6.61) in a somewhat different form. To this end, we first note that we can formally write the Gordon–Volkov state $|\chi_{\mathbf{k}_f}(t)\rangle$ as

$$|\chi_{\mathbf{k}_f}(t)\rangle = \exp\left[-\mathrm{i}\int_{t_0}^{t} \mathrm{d}t' H_{\text{F}}(t')\right]|\mathbf{k}_f\rangle, \qquad (6.69)$$

where $|\mathbf{k}_f\rangle$ stands for the field-free plane wave state of wave vector (or momentum in a.u.) $\mathbf{k}_f$ satisfying the Schrödinger equation $K|\mathbf{k}_f\rangle = (k_f^2/2)|\mathbf{k}_f\rangle$. Taking this result into account and expressing the interaction Hamiltonian as

$$H_{\text{int}}(t) = H_F(t) - K, \tag{6.70}$$

we integrate Equation (6.61) by parts. The result is:

$$
\begin{aligned}
T_{f0}^{\text{SFA}} &= -i \int_{t_0}^{t_1} dt' \langle \mathbf{k}_f | \exp\left[i \int_{t_0}^{t'} dt'' H_F(t'')\right] [H_F(t') - K] | \phi_0(t') \rangle \\
&= -\int_{t_0}^{t_1} dt' \left[\frac{d}{dt'} \langle \chi_{\mathbf{k}_f}(t') | \right] | \phi_0(t') \rangle + i \int_{t_0}^{t_1} dt' \langle \chi_{\mathbf{k}_f}(t') | K | \phi_0(t') \rangle \\
&= -\langle \chi_{\mathbf{k}_f}(t') | \phi_0(t') \rangle \Big|_{t_0}^{t_1} + i \int_{t_0}^{t_1} dt' \langle \chi_{\mathbf{k}_f}(t') | [K + I_P] | \phi_0(t') \rangle. \tag{6.71}
\end{aligned}
$$

This equation can be simplified further by using the fact that the Gordon–Volkov state $|\chi_{\mathbf{k}_f}(t)\rangle$ is an eigenstate of the momentum operator and of the kinetic-energy operator K. From Equation (2.172) we have, in the length gauge,

$$\langle \chi_{\mathbf{k}_f}^{\text{L}}(t) | K | \phi_0(t) \rangle = \frac{1}{2}\pi^2(\mathbf{k}_f, t) \langle \chi_{\mathbf{k}_f}^{\text{L}}(t) | \phi_0(t) \rangle, \tag{6.72}$$

where

$$\pi(\mathbf{k}_f, t) = \mathbf{k}_f + \mathbf{A}(t). \tag{6.73}$$

Classically, the vector $\pi(\mathbf{k}_f, t)$ is the kinetic momentum of the electron, while $\mathbf{k}_f$ is its canonical momentum. Excluding an irrelevant phase factor, we also have that

$$\langle \chi_{\mathbf{k}_f}^{\text{L}}(t) \phi_0(t) \rangle = \exp[iS(\mathbf{k}_f, t, t_0)] \langle \pi(\mathbf{k}_f, t) | \psi_0 \rangle \tag{6.74}$$

with

$$S(\mathbf{k}_f, t, t_0) = \frac{1}{2} \int_{t_0}^{t} dt' \left[\pi(\mathbf{k}_f, t')\right]^2 + I_P(t - t_0). \tag{6.75}$$

The quantity $S(\mathbf{k}_f, t, t_0)$ is of central importance in the theory. It is usually referred to as the "modified action." We note that neither S, as it is defined here, nor $S - I_P(t - t_0)$ is the integral of the classical Lagrangian of the system. Let us also introduce the quantity

$$\tilde{S}(\mathbf{k}_f, t_1, t) = -\frac{1}{2} \int_{t}^{t_1} dt' \left[\pi(\mathbf{k}_f, t')\right]^2 + I_P t. \tag{6.76}$$

Since $S - \tilde{S}$ does not depend on t, we may also write, apart from an overall phase factor,

$$\langle \chi_{\mathbf{k}_f}^{\text{L}}(t) | \phi_0(t) \rangle = \exp[i\tilde{S}(\mathbf{k}_f, t_1, t)] \langle \pi(\mathbf{k}_f, t) | \psi_0 \rangle. \tag{6.77}$$

As we will see below, the use of $\tilde{S}$ will lead to an intuitive interpretation of the ionization amplitude.

We now express the ionization transition amplitude, in the length gauge, as

$$T_{f0}^{\mathrm{SFA}} = -\exp[\mathrm{i}\tilde{S}(\mathbf{k}_f, t_1, t')] \langle \boldsymbol{\pi}(\mathbf{k}_f, t') | \psi_0 \rangle \Big|_{t_0}^{t_1}$$

$$+\mathrm{i} \int_{t_0}^{t_1} \mathrm{d}t' \left(\frac{\boldsymbol{\pi}^2(\mathbf{k}_f, t')}{2} + I_{\mathrm{P}} \right) \langle \boldsymbol{\pi}(\mathbf{k}_f, t') | \psi_0 \rangle \exp[\mathrm{i}\tilde{S}(\mathbf{k}_f, t_1, t')]. \quad (6.78)$$

The integral appearing in Equation (6.78) can be performed using the saddle-point method, since the modified action $\tilde{S}(\mathbf{k}_f, t_1, t')$ varies rapidly with t' at the frequencies and field strengths for which the strong-field approximation applies. In this way, the amplitude reduces to a sum of contributions from the end points t_0 and t_1, and of contributions at the saddle points of the modified action, namely, at the complex times t_s at which

$$\frac{\partial}{\partial t'} \tilde{S}(\mathbf{k}_f, t_1, t') \Big|_{t'=t_s} \equiv \frac{\boldsymbol{\pi}^2(\mathbf{k}_f, t_s)}{2} + I_{\mathrm{P}} = 0. \quad (6.79)$$

The analysis of the asymptotic expansion involved in the calculation shows that the first term on the right-hand side of Equation (6.78) cancels contributions of the end points of the integral. The result is that the total contribution of the end points is negligible if the laser field varies sufficiently smoothly at t_0 and t_1. Indeed, as one would expect on physical grounds, the ionization probability cannot depend on the details of the form of the laser field at the very beginning and the very end of the pulse.

It is worth noticing that the integrand in Equation (6.78) does not vanish at the saddle times t_s. The reason is that $\langle \boldsymbol{\pi}(\mathbf{k}_f, t_s) | \psi_0 \rangle$ is singular when $\boldsymbol{\pi}^2(\mathbf{k}_f, t_s) = -2I_{\mathrm{P}}$. Indeed, making use of the quantity

$$\kappa = (2I_{\mathrm{P}})^{1/2}, \quad (6.80)$$

and using the fact that the radial part of the wave function of the initial state behaves asymptotically as defined by Equation (6.24), one has [30]

$$\langle \mathbf{k} | \psi_0 \rangle \simeq \xi_l C_{\mathrm{as}} \frac{\Gamma(1 + Z_{\mathrm{c}}/\kappa)}{(2\pi\kappa^3)^{1/2}} \left(\frac{2\kappa^2}{k^2 + \kappa^2} \right)^{Z_{\mathrm{c}}/\kappa + 1} Y_{lm}(\hat{\mathbf{k}}) \quad (6.81)$$

for an initial state of orbital angular momentum l and magnetic quantum number m. Here $\xi_l = 1$ for $k \to \mathrm{i}\kappa$ and $\xi_l = (-1)^l$ for $k \to -\mathrm{i}\kappa$. For the ground state of atomic hydrogen or the ground state of a one-electron ion, one has, exactly,

$$\langle \mathbf{k} | \psi_0 \rangle = \frac{1}{(2\pi^2\kappa^3)^{1/2}} \left(\frac{2\kappa^2}{k^2 + \kappa^2} \right)^2. \quad (6.82)$$

Taking $\mathbf{k} = \boldsymbol{\pi}(\mathbf{k}_f, t')$, we obtain the result

$$\left(\frac{\boldsymbol{\pi}^2(\mathbf{k}_f, t')}{2} + I_P\right) \langle \boldsymbol{\pi}(\mathbf{k}_f, t') | \psi_0 \rangle \bigg|_{t' \to t_s}$$

$$\simeq \xi_l(t_s) C_{as} \frac{\kappa^{2\nu+2}}{(2\pi\kappa^3)^{1/2}} \Gamma(1+\nu) \left[S''(t_s)(t'-t_s)\right]^{-\nu} Y_{lm}(\hat{\boldsymbol{\pi}}).$$

(6.83)

In this equation, $\nu \equiv n^* = Z_c/\kappa$,

$$S''(t_s) = \frac{\partial^2}{\partial t'^2} \tilde{S}(\mathbf{k}_f, t_1, t') \bigg|_{t'=t_s}$$

(6.84)

and $\hat{\boldsymbol{\pi}}$ denotes the polar angles of the vector $\boldsymbol{\pi}(\mathbf{k}_f, t')$. In addition, we have $\xi_l(t_s) = (\sigma)^l$, where σ is the sign of the imaginary part of $\boldsymbol{\pi}(\mathbf{k}_f, t')$ for $t' \to t_s$. Assuming that there is no confluence of the complex saddle times t_s, the ionization amplitude can thus be reduced to a sum of integrals of the form

$$J = g_s \int dt \, \frac{\exp[-\lambda f(t)]}{(t-t_s)^\nu},$$

(6.85)

where λ is large, $f(t)$ is a slowly varying function with $f'(t_s) = 0$ and $f''(t_s) \neq 0$, and g_s is a constant. The contribution of each saddle time can then be obtained by using the fact that [35]

$$J \underset{\lambda \to \infty}{\longrightarrow} i^\nu g_s \pi^{1/2} \frac{\Gamma(\nu/2)}{\Gamma(\nu)} [2\lambda f''(t_s)]^{(\nu-1)/2} \exp[-\lambda f(t_s)].$$

(6.86)

The SFA ionization transition amplitude obtained using the saddle-point integration is therefore given by

$$T_{f0}^{\text{SFA}} = i C_{as} 2^{\nu/2} \kappa^{2\nu+1/2} \Gamma(1+\nu/2) \sum_{t_s} \frac{\xi_l(t_s)}{[-iS''(t_s)]^{(\nu+1)/2}} \exp[iS(t_s)] Y_{lm}(\hat{\boldsymbol{\pi}}),$$

(6.87)

where

$$S(t_s) = \tilde{S}(\mathbf{k}_f, t_1, t_s).$$

(6.88)

We note that $\hat{\boldsymbol{\pi}}$ is complex, and that the spherical harmonics Y_{lm} with complex arguments can be defined by analytical continuation.

For the particular case of ionization from an S-state, we have

$$T_{f0}^{\text{SFA}} = i(4\pi)^{-1/2} C_{as} 2^{\nu/2} \kappa^{2\nu+1/2} \Gamma(1+\nu/2) \sum_{t_s} \frac{\exp[iS(t_s)]}{[-iS''(t_s)]^{(\nu+1)/2}}.$$

(6.89)

It is worth pointing out that in this length-gauge formulation of the SFA, the ionization amplitude depends on the initial state only via its binding energy and the asymptotic behavior of its wave function. The form of the wave function close to the nucleus is immaterial.

Equation (6.87) shows that, within the SFA, the amplitude for the production of a photoelectron with a given drift momentum $\mathbf{k}_f$ is the sum of discrete contributions, each one corresponding to a particular "complex trajectory" in which the electron is detached at a complex time t_s. The resulting interferences between different trajectories in the ionization probability $P(E_f, \hat{\mathbf{k}}_f)$ give rise to oscillations in the energy and angular distributions of the photoelectrons.

As we shall now see, these complex trajectories have a very simple interpretation in terms of real, classical trajectories when the complex saddle times t_s are sufficiently close to the real axis. We first express Equation (6.88) as the sum of two terms: the contribution of the integral in the complex plane from t_s to $t_s^R = \mathrm{Re}\, t_s$ and the contribution from t_s^R to t_1. Thus, we write

$$S(t_s) = -\int_{t_s}^{t_s^R} dt' \left(\frac{1}{2}\pi^2(\mathbf{k}_f, t') + I_P \right) + S(t_s^R). \tag{6.90}$$

We now assume that the imaginary part of t_s is small enough such that one can approximate $\pi^2(\mathbf{k}_f, t')$ by the first three terms of its Taylor expansion about t_s^R. In this approximation,

$$\pi^2(\mathbf{k}_f, t') \simeq \pi^2(\mathbf{k}_f, t_s^R) + \pi^{2\prime}(\mathbf{k}_f, t_s^R)(t' - t_s^R) + \frac{1}{2}\pi^{2\prime\prime}(\mathbf{k}_f, t_s^R)(t' - t_s^R)^2, \tag{6.91}$$

with

$$\pi^{2\prime}(\mathbf{k}_f, t_s^R) = \frac{\partial}{\partial t}\pi^2(\mathbf{k}_f, t)\Big|_{t=t_s^R} \tag{6.92}$$

and

$$\pi^{2\prime\prime}(\mathbf{k}_f, t_s^R) = \frac{\partial^2}{\partial t^2}\pi^2(\mathbf{k}_f, t)\Big|_{t=t_s^R}. \tag{6.93}$$

We will assume that Equation (6.91) holds at any point along the path in the complex plane running from t_s^R to t_s parallel to the imaginary axis. Setting $t' = t_s$ and separating Equation (6.79) into its real and imaginary parts yields two conditions on t_s^R and $t_s^I = \mathrm{Im}\, t_s$, namely

$$\frac{1}{2}\pi^2(\mathbf{k}_f, t_s^R) - \frac{1}{4}\pi^{2\prime\prime}(\mathbf{k}_f, t_s^R)(t_s^I)^2 + I_P \simeq 0 \tag{6.94}$$

and

$$\pi^{2\prime}(\mathbf{k}_f, t_{\mathrm{s}}^{\mathrm{R}}) \simeq 0. \tag{6.95}$$

It can be shown that $\pi^{2\prime\prime}(\mathbf{k}_f, t_{\mathrm{s}}^{\mathrm{R}}) > 0$ and that the only physically relevant saddle times are those for which $t_{\mathrm{s}}^{\mathrm{I}} > 0$. Combining Equations (6.90), (6.91), (6.94) and (6.95) yields

$$S(t_{\mathrm{s}}) \simeq S(t_{\mathrm{s}}^{\mathrm{R}}) + \mathrm{i} \frac{\zeta^3(\mathbf{k}_f, t_{\mathrm{s}}^{\mathrm{R}})}{3\varepsilon(\mathbf{k}_f, t_{\mathrm{s}}^{\mathrm{R}})} \tag{6.96}$$

and

$$S''(t_{\mathrm{s}}) \simeq \mathrm{i}\,\zeta(\mathbf{k}_f, t_{\mathrm{s}}^{\mathrm{R}})\,\varepsilon(\mathbf{k}_f, t_{\mathrm{s}}^{\mathrm{R}}), \tag{6.97}$$

with $\zeta(\mathbf{k}_f, t_{\mathrm{s}}^{\mathrm{R}}) = \{[\mathbf{k}_f + \mathbf{A}(t_{\mathrm{s}}^{\mathrm{R}})]^2 + \kappa^2\}^{1/2}$ and $\varepsilon(\mathbf{k}_f, t_{\mathrm{s}}^{\mathrm{R}}) = [\pi^{2\prime\prime}(\mathbf{k}_f, t_{\mathrm{s}}^{\mathrm{R}})/2]^{1/2}$. When Equation (6.91) holds, Equation (6.89) can thus be entirely written in terms of the real times $t_{\mathrm{s}}^{\mathrm{R}}$. For ionization from an S-state,

$$T_{f0}^{\mathrm{SFA}} \simeq \mathrm{i} \sum_{t_{\mathrm{s}}} \frac{C_{\mathrm{as}}}{(4\pi)^{1/2}} \frac{2^{\nu/2}\kappa^{2\nu+1/2}\Gamma(1+\nu/2)\exp[\mathrm{i}S(t_{\mathrm{s}}^{\mathrm{R}})]}{[\zeta(\mathbf{k}_f, t_{\mathrm{s}}^{\mathrm{R}})\,\varepsilon(\mathbf{k}_f, t_{\mathrm{s}}^{\mathrm{R}})]^{(\nu+1)/2}} \exp\left[-\frac{\zeta^3(\mathbf{k}_f, t_{\mathrm{s}}^{\mathrm{R}})}{3\varepsilon(\mathbf{k}_f, t_{\mathrm{s}}^{\mathrm{R}})}\right]. \tag{6.98}$$

This analysis associates each saddle time t_{s} with a complex trajectory in which the photoelectron first propagates in imaginary time from t_{s} to $t_{\mathrm{s}}^{\mathrm{R}}$, emerges as an unbound electron at the detachment time $t_{\mathrm{s}}^{\mathrm{R}}$, and then follows the real trajectory of a quivering electron drifting in the laser field with drift momentum $\mathbf{k}_f$. The motion in imaginary time can be identified with the tunneling through the effective potential barrier discussed in Sections 6.1.2 and 6.1.3. It gives rise to the decreasing exponential appearing in the right-hand side of Equation (6.98). The motion in real time results in the factor $\exp[\mathrm{i}S(t_{\mathrm{s}}^{\mathrm{R}})]$, where $S(t_{\mathrm{s}}^{\mathrm{R}})$ is the phase accumulated along the trajectory of the photoelectron between the detachment time $t_{\mathrm{s}}^{\mathrm{R}}$ and the end of the pulse. We remark that $S(t_{\mathrm{s}}^{\mathrm{R}}) - I_{\mathrm{P}}t_{\mathrm{s}}^{\mathrm{R}}$ is not the classical action for the motion in real time. The classical action for the real part of the trajectory is the integral from $t_{\mathrm{s}}^{\mathrm{R}}$ to t_1 of the corresponding classical Lagrangian of an electron quivering in the field with drift momentum $\mathbf{k}_f$.

Let us now turn our attention to the particular case of a linearly polarized laser field described by the vector potential $\mathbf{A}(t) = \hat{\boldsymbol{\epsilon}} A(t)$. Equation (6.79) yields

$$A(t_{\mathrm{s}}) = -k_{\parallel} \pm \mathrm{i}(k_{\perp}^2 + \kappa^2)^{1/2}, \tag{6.99}$$

with $k_{\parallel} = \hat{\boldsymbol{\epsilon}} \cdot \mathbf{k}_f$ and $k_{\perp}^2 = k_f^2 - k_{\parallel}^2$. Thus, a pair of conjugate saddle times exist for each half-cycle of the laser field. As noted above, only the values of t_{s} for which

$\text{Im} \, t_s > 0$ are physically acceptable. The sign of the imaginary part of $A(t_s)$ should be chosen accordingly. If $k_\perp$ and κ are sufficiently small, it is clear that

$$A(t_s^R) \simeq -k_\parallel, \tag{6.100}$$

in agreement with Equation (6.95). Equation (6.100) has a simple interpretation. At its detachment time t_s^R – the time at which it starts its classical trajectory – the photoelectron has zero velocity in the direction of polarization and its drift momentum in that direction is $-A(t_s^R)$. Let us therefore approximate t_s^R by the "detachment time" t_d, where t_d is determined from

$$A(t_d) = -k_\parallel. \tag{6.101}$$

We remark that this equation can be satisfied only if k_f is not larger than the maximum value A_{max} attained by $|A(t)|$ during the pulse. Detachment with larger drift momenta is still possible; however, the corresponding complex saddle times are further away from the real axis. This makes the probability of this process exponentially small. If the laser field is strong enough for the expansion (6.91) to be applicable, then the energy spectrum of the direct photoelectrons extends up to a cut-off energy $E_{max} = A_{max}^2/2$. Below this cut-off energy, we approximate the real parts of the saddle times by the detachment times t_d defined by Equation (6.101), in which case Equation (6.98) gives the ionization transition amplitude for an S-state as

$$T_{f0}^{SFA} \simeq i \sum_{t_d} a_{free}(t_d) a_{ion}(t_d), \tag{6.102}$$

where the "free"-motion and ionization amplitudes, respectively, are

$$a_{free}(t_d) = \exp[iS(t_d)] \tag{6.103}$$

and

$$a_{ion}(t_d) = \frac{C_{as}}{(4\pi)^{1/2}} \frac{2^{\nu/2} \kappa^{2\nu+1/2} \Gamma(1+\nu/2)}{[(k_\perp^2 + \kappa^2)^{1/2} |\mathcal{E}(t_d)|]^{(\nu+1)/2}} \exp\left[-\frac{(k_\perp^2 + \kappa^2)^{3/2}}{3|\mathcal{E}(t_d)|} \right], \tag{6.104}$$

with

$$\mathcal{E}(t_d) = -\frac{\partial}{\partial t} A(t) \bigg|_{t=t_d}. \tag{6.105}$$

We see that the corresponding angular distribution of the photoelectrons has a peak in the polarization direction ($k_\perp = 0$) and decreases rapidly as $k_\perp$ increases. Moreover, under the conditions where Equation (6.102) applies, the emission of slow electrons is more probable than the emission of fast electrons, since the electrons with the smallest drift momentum are detached near the zeros of $A(t)$, in other words near the maxima of $|\mathcal{E}(t)|$.

We note that there is a clear similarity between Equation (6.25) and the ionization transition amplitude (6.104). We shall come back to this point in Section 6.3.2. We shall also see how the theory must be modified to take into account the effect of the interaction between the photoelectron and the core during the tunneling stage of the ionization process when $Z_c \neq 0$.

6.3.2 The Keldysh theory for a monochromatic laser field

The ionization transition amplitude derived in Section 6.3.1 for an atomic system interacting with a laser pulse of finite duration can be readily used to obtain the ionization rate in a monochromatic laser field. To this end, we will take the pulse to have a square intensity profile and determine how $|T_{f0}^{\mathrm{SFA}}|^2$ varies for increasingly long pulse durations. Thus, we assume that the laser field is turned on and off abruptly, at times t_0 and t_1, and that it is periodic between these two instants in time. In addition, we take the pulse duration τ_p to encompass exactly N optical cycles. Choosing $t_0 = 0$, it follows that $t_1 = \tau_p = NT = 2N\pi/\omega$. The periodicity of the laser field implies that each optical cycle produces a set of saddle times identical to those produced by the preceding optical cycle, apart from a shift in the real part of the saddle times by $T = 2\pi/\omega$. The saddle times t_s can thus be divided into the saddle times produced by the first cycle, which we denote by t_μ, and the N copies of these. For any t_s there is an integer q between 0 and $N - 1$ such that $S(t_s) = S_0(t_\mu) + q\,S_c + S_p$, with

$$S_0(t_\mu) = \frac{1}{2}\int_0^{t_\mu} \mathrm{d}t'\left[\boldsymbol{\pi}(\mathbf{k}_f, t')\right]^2 + I_P t_\mu, \tag{6.106}$$

$$S_c = \frac{1}{2}\int_0^{T} \mathrm{d}t'\left[\boldsymbol{\pi}(\mathbf{k}_f, t')\right]^2 + I_P T \tag{6.107}$$

and

$$S_p = -\frac{1}{2}\int_0^{NT} \mathrm{d}t'\left[\boldsymbol{\pi}(\mathbf{k}_f, t')\right]^2. \tag{6.108}$$

Contrary to $S_0(t_\mu)$, S_c and S_p do not depend on t_s. For a square pulse, the right-hand side of Equation (6.87) can thus be written as the sum of a geometric progression, namely

$$T_{f0}^{\mathrm{SFA}} \simeq M_{f0}^{\mathrm{SFA}}\exp(\mathrm{i}S_p)\sum_{q=0}^{N-1}\exp(\mathrm{i}q\,S_c). \tag{6.109}$$

The amplitude M_{f0}^{SFA} does not depend on the duration of the laser pulse and is defined as

$$M_{f0}^{\text{SFA}} = i C_{\text{as}} 2^{\nu/2} \kappa^{2\nu+1/2} \Gamma(1+\nu/2) \sum_{t_\mu} \frac{\xi_l(t_\mu)}{[-iS_0''(t_\mu)]^{(\nu+1)/2}} \exp[iS_0(t_\mu)] Y_{lm}(\hat{\pi}_\mu),$$

(6.110)

where $\hat{\pi}_\mu$ denotes the polar angles of the vector $\pi(\mathbf{k}_f, t_\mu)$. Using Equation (6.47), the ionization probability per optical cycle for an N-cycle square pulse, in the SFA, is

$$P_N^{\text{SFA}}(E_f, \hat{\mathbf{k}}_f)/N = (2E_f)^{1/2} |M_{f0}^{\text{SFA}}|^2 \frac{\sin^2(NS_c/2)}{N \sin^2(S_c/2)} dE_f d\hat{\mathbf{k}}_f. \qquad (6.111)$$

In general, as N becomes large, the contribution from boundary terms appearing in Equation (6.78) tends to zero and can be neglected. In our case, this contribution is, in fact, exactly zero for all $N > 0$, as we will show below.

The right-hand side of Equation (6.111) vanishes for $N \to \infty$, unless $S_c/2$ is an integer multiple of π. Indeed, one has

$$\lim_{N \to \infty} \frac{\sin^2(NS_c/2)}{N \sin^2(S_c/2)} = \pi \delta(S_c/2 - n\pi). \qquad (6.112)$$

Using the fact that the laser field is periodic and that the average of $A^2(t)/2$ over one optical cycle yields the ponderomotive energy U_{p}, we find that

$$S_c = (k_f^2/2 + U_{\text{p}} + I_{\text{P}})(2\pi/\omega). \qquad (6.113)$$

Hence,

$$\pi \delta(S_c/2 - n\pi) = \omega \delta(k_f^2/2 + U_{\text{p}} + I_{\text{P}} - n\omega), \qquad (6.114)$$

and we see that when $N \to \infty$, the ionization probability per optical cycle, $P_N^{\text{SFA}}(E_f, \hat{\mathbf{k}}_f)/N$, vanishes unless E_f is equal to one of the energies E_n defined by the equation

$$E_n = n\omega - I_{\text{P}} - U_{\text{p}}, \qquad (6.115)$$

where $n \geq n_0$ is an integer and n_0 is the smallest integer for which $E_n \geq 0$.

Integrating $P_N^{\text{SFA}}(E_f, \hat{\mathbf{k}}_f)/N$ over a narrow energy interval centered about E_n, dividing the result by T and taking the limit $N \to \infty$ yields the SFA differential rate for electron emission at an energy E_n, namely

$$\frac{d\Gamma_n^{\text{SFA}}}{d\hat{\mathbf{k}}_f} = \frac{\omega^2}{2\pi} (2E_n)^{1/2} |M_{f0}^{\text{SFA}}|^2. \qquad (6.116)$$

By integrating this differential ionization rate over the angles of the ejected electron, we find the rate of electron emission at an energy E_n. That is,

$$\Gamma_n^{\mathrm{SFA}} = \int d\hat{\mathbf{k}}_f \, \frac{d\Gamma_n^{\mathrm{SFA}}}{d\hat{\mathbf{k}}_f}. \tag{6.117}$$

The total ionization rate, Γ^{SFA}, can then be obtained by summing the partial rates Γ_n^{SFA} over all the possible values of n:

$$\Gamma^{\mathrm{SFA}} = \sum_{n \geq n_0} \Gamma_n^{\mathrm{SFA}}. \tag{6.118}$$

The restriction of E_f to the energies E_n is exactly what is expected from conservation of energy. In the limit of infinite interaction times with a monochromatic laser field, the difference between the energy of the final state of the atom and that of its initial state must be an integer multiple of ω (in a.u.). In the length gauge, the former is the sum of the drift kinetic energy E_f and the ponderomotive energy U_{p}, while the latter is simply the field-free ground-state energy $-I_{\mathrm{P}}$, where we have used the fact that the non-ponderomotive AC Stark shift of the initial state is neglected in the SFA. The final-state energies $E_f = E_n$ are therefore the only ones permitted in a monochromatic laser field. We also have

$$\exp[i\tilde{S}(\mathbf{k}_f, t_1, t')]\langle \boldsymbol{\pi}(\mathbf{k}_f, t')|\psi_0\rangle \Big|_{t_0}^{t_1} = 0 \tag{6.119}$$

when $k_f^2/2 = E_n$ and $(t_1 - t_0)/T$ is an integer, which explains why no boundary terms appear in Equation (6.111).

Whereas finding the saddle times for a pulsed laser field typically involves a numerical search in the complex plane, their values are given by simple analytical expressions for monochromatic laser fields that are either linearly or circularly polarized. For a linearly polarized laser field whose electric-field component is given by $\boldsymbol{\mathcal{E}}(t) = \hat{\boldsymbol{\epsilon}}\mathcal{E}_0\cos(\omega t)$, Equation (6.99) reduces to

$$\sin(\omega t_\mu) = \tilde{k}_\parallel \pm i\left(\tilde{k}_\perp^2 + \gamma_{\mathrm{K}}^2\right)^{1/2}, \tag{6.120}$$

where $\tilde{k}_\parallel = \omega k_\parallel/\mathcal{E}_0$, $\tilde{k}_\perp = \omega k_\perp/\mathcal{E}_0$ and γ_{K} is the Keldysh parameter. Equation (6.120) has two relevant solutions, namely

$$t_1 = \frac{1}{\omega}\arcsin\left[\tilde{k}_\parallel + i\left(\tilde{k}_\perp^2 + \gamma_{\mathrm{K}}^2\right)^{1/2}\right] \tag{6.121}$$

and $t_2 = \pi - t_1^*$. Given these results, the modified action $S_0(t_\mu)$ appearing in the ionization amplitude M_{f0}^{SFA} can be expressed in closed form, and therefore the energy and angular distributions of the photoelectrons can be obtained directly from Equations (6.110) and (6.116) without further approximations. It is instructive, however, to compare the exact saddle times t_μ to the real semi-classical detachment times $t_{d\mu}$ that are given by the solutions of Equation (6.101). We find that $t_{d1} = (1/\omega)\arcsin(\tilde{k}_\parallel)$ and $t_{d2} = \pi - t_{d1}$. These solutions exist provided $|\tilde{k}_\parallel| \leq 1$. Expanding the exact saddle time t_1 in a Taylor series about the approximate saddle time t_{d1} yields

$$t_1 = t_{d1} + i\frac{1}{\omega}\frac{(\tilde{k}_\perp^2 + \gamma_K^2)^{1/2}}{(1 - \tilde{k}_\parallel^2)^{1/2}} - \frac{1}{2\omega}\frac{\tilde{k}_\parallel(\tilde{k}_\perp^2 + \gamma_K^2)}{(1 - \tilde{k}_\parallel^2)^{3/2}} + \cdots . \tag{6.122}$$

This series converges provided $(\tilde{k}_\perp^2 + \gamma_K^2)^{1/2} < \min(1 - \tilde{k}_\parallel, 1 + \tilde{k}_\parallel)$. Thus, we see that $t_1 \simeq t_{d1}$ and $t_2 \simeq t_{d2}$ when $|\tilde{k}_\parallel| \ll 1$ and $|\tilde{k}_\perp| \ll 1$ together with $\gamma_K \ll 1$. That is, $A(\mathrm{Re}\,t_\mu) \simeq -k_\parallel$ for emission of photoelectrons with drift momentum much smaller than $\mathcal{E}_0/\omega$ in the tunneling regime. The classical solutions t_{d1} and t_{d2} do not exist for drift momenta larger than $\mathcal{E}_0/\omega$, which corresponds to a classical cut-off energy of $2U_p$. The exact saddle times move away from the real axis when the photoelectron drift energy approaches $2U_p$ from below, which for $\gamma_K \ll 1$ leads to an exponential suppression of direct ionization at and above this cut-off energy. As will be discussed in Section 6.5, photoelectrons may acquire much larger drift energies upon recolliding with the core.

The dependence of $S_0(t_\mu)$ on $k_\parallel$ and $k_\perp$ simplifies considerably in the limit where $k_\parallel^2 + k_\perp^2 = k_f^2 \ll \kappa^2$. Neglecting terms cubic in k_f/κ in the modified action $S_0(t_\mu)$, one finds that

$$|\exp[iS_0(t_1)]| = |\exp[iS_0(t_2)]|$$

$$\simeq \exp\left\{-\frac{\kappa^3}{2\mathcal{E}_0}\frac{1}{\gamma_K}\left[f(\gamma_K) + c_\parallel(\gamma_K)\left(\frac{k_\parallel}{\kappa}\right)^2 + c_\perp(\gamma_K)\left(\frac{k_\perp}{\kappa}\right)^2\right]\right\}, \tag{6.123}$$

where

$$f(\gamma_K) = \left(1 + \frac{1}{2\gamma_K^2}\right)\mathrm{arcsinh}\,\gamma_K - \frac{1}{2\beta_K}, \tag{6.124}$$

$$c_\parallel(\gamma_K) = \mathrm{arcsinh}\,\gamma_K - \beta_K \tag{6.125}$$

and

$$c_\perp(\gamma_K) = \mathrm{arcsinh}\,\gamma_K \tag{6.126}$$

with $\beta_K = \gamma_K/(1 + \gamma_K^2)^{1/2}$. A relatively simple analytical expression of the differential ionization rate is then obtained.

Let us now consider the case for which $\nu = Z_c/\kappa = 0$, which corresponds to the case of detachment from a negative ion. Multiphoton detachment from negative ïons has been considered by Gribakin and Kuchiev [35] and Frolov *et al.* [36]. Using Equations (6.110), (6.116) and (6.123), one finds that

$$\frac{d\Gamma_n^{SFA}}{d\hat{k}_f} \simeq (2E_n)^{1/2} \frac{\omega}{\kappa} \frac{C_{as}^2}{4\pi^2} \frac{A(l,m)}{2^{|m|}|m|!} \left(\frac{k_\perp}{\kappa}\right)^{2|m|} \left[1 + (-1)^{n+l+m} \cos\left(\frac{2\kappa k_\parallel}{\omega\beta_K}\right)\right]$$

$$\times \beta_K \exp\left\{-\frac{\kappa^3}{\mathcal{E}_0} \frac{1}{\gamma_K}\left[f(\gamma_K) + c_\parallel(\gamma_K)\left(\frac{k_\parallel}{\kappa}\right)^2 + c_\perp(\gamma_K)\left(\frac{k_\perp}{\kappa}\right)^2\right]\right\},$$

$$(6.127)$$

where the coefficient $A(l,m)$ is defined by Equation (6.26). One also finds that

$$\Gamma_n^{SFA} \simeq C_{as}^2 \frac{A(l,m)}{|m|!} \frac{4\beta_K^2}{(3\pi)^{1/2}} \exp[-2c_\parallel(\gamma_K)E_n/\omega] F_m\left(\sqrt{2\beta_K E_n/\omega}\right)$$

$$\times \frac{\kappa^2}{2}\left(\frac{6}{\pi}\right)^{1/2}\left[\frac{\mathcal{E}_0(1+\gamma_K)^{1/2}}{2\kappa^3}\right]^{|m|+3/2} \exp\left[-\frac{\kappa^3}{\mathcal{E}_0}\frac{f(\gamma_K)}{\gamma_K}\right], \quad (6.128)$$

with

$$F_m(x) = \exp(-x^2)\int_0^x dy\,(x^2 - y^2)^{|m|}\exp(y^2). \quad (6.129)$$

For $m = 0$, $F_m(x)$ is the Dawson integral [37]. Finally, summing the partial rates Γ_n^{SFA} given by Equation (6.128) yields the total ionization rate when $\gamma_K \ll 1$:

$$\Gamma^{SFA} \simeq C_{as}^2 A(l,m) \frac{\kappa^2}{2}\sqrt{\frac{3\mathcal{E}_0}{\pi\kappa^3}}\left(\frac{2\kappa^3}{\mathcal{E}_0}\right)^{-|m|-1}\exp\left[-\frac{2\kappa^3}{3\mathcal{E}_0}\left(1 - \frac{1}{10}\gamma_K^2\right)\right]. \quad (6.130)$$

This analysis shows that when $\gamma_K \to 0$ the SFA total ionization rate reduces to the adiabatic tunneling rate Γ_{ad} given by Equation (6.38) with $Z_c = 0$. We therefore have the important result that in this limit, the ionization process can be viewed as proceeding as if the electric-field component of the laser pulse were static, in agreement with the discussion in Section 6.1.3. We also see that the adiabatic rate is modified by non-adiabatic corrections for non-vanishing values of γ_K. In particular, from Equation (6.130) we have that for $\gamma_K \ll 1$ there is a correction of order γ_K^2 in the exponential factor. This correction makes Γ^{SFA} larger than Γ_{ad}, and the relative difference increases as the laser field becomes weaker.

While the SFA detachment rate for negative ions, Γ^{SFA}, calculated as described above reduces to the tunneling formula in the adiabatic limit, this is not the case when $Z_c \neq 0$. As an example, for ionization from the ground state of atomic hydrogen ($C_{as} = 2$, $Z_c = \kappa = 1$, $l = m = 0$, $A(l,m) = 1$), one finds

$$\Gamma^{SFA} \simeq \frac{(3\pi)^{1/2}}{4}\mathcal{E}_0^{1/2}\exp\left[-\frac{2}{3\mathcal{E}_0}\left(1 - \frac{1}{10}\gamma_K^2\right)\right], \quad (6.131)$$

while, from Equation (6.38),

$$\Gamma_{\mathrm{ad}}(\mathcal{E}_0) \simeq 4 \left(\frac{3}{\pi}\right)^{1/2} \mathcal{E}_0^{-1/2} \exp\left(-\frac{2}{3\mathcal{E}_0}\right). \tag{6.132}$$

As is always the case when $Z_c \neq 0$, in the limit $\gamma_K \to 0$ the exponentials are the same in Γ^{SFA} and in $\Gamma_{\mathrm{ad}}(\mathcal{E}_0)$, but not the pre-exponential factors. The origin of the difference is the neglect of the Coulomb interaction between the core and the outgoing electron. As a consequence, the total ionization rates for atoms and positive ions predicted by the SFA are too low, often by over an order of magnitude.

A number of approximations have been proposed to incorporate the Coulomb interaction in the SFA ionization transition amplitude [27, 38–41]. An overview of these and earlier work on SFA Coulomb corrections has been given by Chirilă and Potvliege [42]. The effect of the Coulomb interaction during tunneling can be approximately taken into account by multiplying the final-state plane wave function by the factor [39, 43]

$$I(r) = \left(\frac{2\kappa^2}{\mathcal{E}_0 r}\right)^{Z/\kappa}, \tag{6.133}$$

where here Z is the charge of the residual ion. Within the SFA, this yields a "tunneling-corrected" direct ionization amplitude that is obtained by multiplying the final Gordon–Volkov wave function by the factor $I(r)$ given by Equation (6.133). An additional correction, proposed by Gordienko and Meyer-ter-Vehn [44], can be used to include the effect of the Coulomb interaction on the motion of the ejected electron. This leads to a "continuum-corrected" SFA ionization amplitude.

To illustrate the accuracy of the Coulomb-corrected SFA, we show in Fig. 6.3 results obtained by Chirilă and Potvliege [42]. They calculated the probability density $P(E_f, \theta)$ that an electron is ejected with energy E_f from an He$^+$ ion interacting with a linearly polarized four-cycle laser pulse of 400 nm carrier wavelength and peak intensity 1×10^{16} W cm^{-2}. In the figure, the results are shown for an ejection angle $\theta = 10°$, which is detected with respect to the laser polarization vector $\hat{\epsilon}$. The thin solid curve shows the results obtained from the direct numerical integration of the TDSE. The remaining two curves correspond to SFA results obtained using the tunneling-corrected ionization amplitude (dotted curves) and the continuum- and tunneling-corrected ionization amplitude (thick solid curves). In this low-energy part of the electron emission spectrum, it is expected that the ejected electrons will be affected by the Coulomb interaction. We see that by including both the continuum and tunneling corrections to the SFA ionization amplitude, good agreement is found with the results obtained by numerically integrating the TDSE. The influence of the Coulomb interaction on multiphoton ionization processes has also been investigated by de Bohan *et al.* [45].

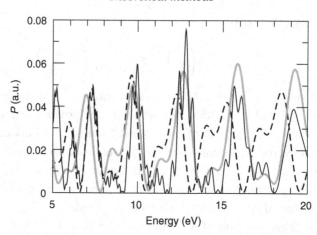

Figure 6.3. The probability density $P(E_f, \theta)$ that an electron is ejected with energy E_f from an He^+ ion interacting with a linearly polarized four-cycle laser pulse of 400 nm carrier wavelength and peak intensity $1 \times 10^{16}\,W\,cm^{-2}$. The ejection angle θ, which is detected with respect to the laser polarization vector, is $10°$. Thin solid curves: TDSE results. Dashed curves: tunneling-corrected SFA results. Thick solid curves: continuum- and tunneling-corrected SFA results. (From C. Chirilă and R. M. Potvliege, *Phys. Rev. A* **71**, 021402(R) (2005).)

6.3.3 The Faisal–Reiss theory

Let us consider again the case of a hydrogenic atom. We shall now calculate the SFA ionization transition amplitude using the interaction Hamiltonian

$$H_{\text{int}}(t) = \mathbf{A}(t) \cdot \mathbf{p} + \frac{1}{2}\mathbf{A}^2(t) \tag{6.134}$$

corresponding to the untransformed TDSE (2.121). We will restrict our attention to the particular case of a spatially homogeneous, monochromatic laser field, for which analytical results can be obtained readily.

This amplitude has been investigated by Faisal [33] and Reiss [34]. As noted above, the choice of the length gauge is most appropriate for calculations in the low-frequency regime. However, we will see that the Faisal–Reiss SFA amplitude can be evaluated analytically in terms of ordinary Bessel functions, in contrast with the length-gauge formulation of Keldysh.

The starting point of our analysis is the SFA ionization amplitude (6.71). Using Equation (2.171), we have

$$\langle \chi_{\mathbf{k}_f}(t) | \phi_0(t) \rangle = \exp[iS(\mathbf{k}_f, t, t_0)]\langle \mathbf{k}_f | \psi_i \rangle, \tag{6.135}$$

where the modified action $S(\mathbf{k}_f, t, t_0)$ is given by Equation (6.75). As in Section 6.3.2, we will set $t_0 = 0$ and $t_1 = \tau_p = 2N\pi/\omega = NT$, so that the laser pulse duration encompasses N optical cycles. The Faisal–Reiss SFA amplitude is obtained by substituting Equation (6.135) into the expression (6.71) for the SFA amplitude, thereby yielding

$$T_{fi}^{\mathrm{FR}} = -\langle \mathbf{k}_f | \psi_i \rangle \left[\exp[\mathrm{i} S(\mathbf{k}_f, t', 0)] \Big|_0^{NT} - \mathrm{i} \left(\frac{k_f^2}{2} + I_{\mathrm{P}} \right) \right.$$

$$\left. \times \int_0^{NT} \mathrm{d}t' \exp[\mathrm{i} S(\mathbf{k}_f, t', 0)] \right]. \tag{6.136}$$

From Equations (6.75) and (2.115), the quasi-classical action is seen to be composed of a function that is periodic in time and of a term that is linear in time. In particular, we have

$$S(\mathbf{k}_f, t + T, 0) = S(\mathbf{k}_f, t, 0) + S_{\mathrm{c}}, \tag{6.137}$$

where the quantity S_{c}, given by Equation (6.107), does not depend on t. We can now express Equation (6.136) as

$$T_{f0}^{\mathrm{FR}} = -\langle \mathbf{k}_f | \psi_i \rangle [\exp(\mathrm{i} S_{\mathrm{c}}) - 1] + M_{f0}^{\mathrm{FR}} \sum_{q=0}^{N-1} \exp(\mathrm{i} S_{\mathrm{c}} q)$$

$$= -\langle \mathbf{k}_f | \psi_i \rangle [\exp(\mathrm{i} S_{\mathrm{c}}) - 1] + M_{f0}^{\mathrm{FR}} \frac{\sin(N S_{\mathrm{c}}/2)}{\sin(S_{\mathrm{c}}/2)}, \tag{6.138}$$

where

$$M_{f0}^{\mathrm{FR}} = \mathrm{i} \langle \mathbf{k}_f | \psi_i \rangle \left(\frac{k_f^2}{2} + I_{\mathrm{P}} \right) \int_0^T \mathrm{d}t' \exp[\mathrm{i} S(\mathbf{k}_f, t', 0)]. \tag{6.139}$$

Recalling our discussion in Section 6.3.2, energy conservation requires that, for N large, $S_{\mathrm{c}}/2 = n\pi$, where $n \geq n_0$ and n_0 is the minimum number of photons needed to ionize the atom. We can therefore drop the boundary term in Equation (6.138). The Faisal–Reiss SFA ionization probability per optical cycle for an N-cycle square laser pulse is then given by

$$P_N^{\mathrm{FR}}(E_f, \hat{\mathbf{k}}_f)/N = (2E_f)^{1/2} |M_{f0}^{\mathrm{FR}}|^2 \frac{\sin^2(N S_{\mathrm{c}}/2)}{N \sin^2(S_{\mathrm{c}}/2)} \, \mathrm{d}E_f \, \mathrm{d}\hat{\mathbf{k}}_f. \tag{6.140}$$

Let us now evaluate the quantity M_{f0}^{FR}. To this end, we return to Equation (6.75) for the quasi-classical action and Equation (2.115) describing the vector potential of the laser field. For an elliptically polarized laser field with $\varphi = -\pi/2$, we find that

$$S(\mathbf{k}_f, t, 0) = (S_{\mathrm{c}}/T)t + a \sin(2\omega t) + b \sin(\omega t + \zeta), \tag{6.141}$$

where $\tan \zeta = \xi k_{f,2}/k_{f,1}$ with $k_{f,j} = \mathbf{k}_f \cdot \hat{\boldsymbol{\epsilon}}_j$, $j = 1, 2$. In this equation, we have introduced the quantities

$$a = \frac{U_p}{2\omega}\left(\frac{1-\xi^2}{1+\xi^2}\right) \tag{6.142}$$

and

$$b = \mathrm{sgn}(k_{f,1})\frac{A_0}{\omega\sqrt{1+\xi^2}}\left(k_{f,1}^2 + \xi^2 k_{f,2}^2\right)^{1/2}. \tag{6.143}$$

Using the generating function of the ordinary Bessel functions [37],

$$\exp(ix\sin\gamma) = \sum_{m=-\infty}^{\infty}\exp(im\gamma)J_m(x), \tag{6.144}$$

and the energy conservation condition (6.115), we can now write

$$\exp[S(\mathbf{k}_f,t,0)] = \sum_{l=-\infty}^{\infty}\sum_{m=-\infty}^{\infty}J_l(a)J_m(b)\exp[i(2l+m+n)\omega t]\exp(im\zeta). \tag{6.145}$$

This equation is readily integrated over one optical cycle, with the result

$$\int_0^T dt'\exp[iS(\mathbf{k}_f,t',0)] = (-1)^n T\sum_{l=-\infty}^{\infty}J_l(a)J_{2l+n}(b)\exp[-i(2l+n)\zeta]. \tag{6.146}$$

As in Section 6.3.2, we now integrate $P_N^{FR}(E_f,\hat{\mathbf{k}}_f)/N$ over a narrow energy interval centered about E_n, divide the result by T and take the limit $N \to \infty$. We then obtain the Faisal–Reiss SFA differential rate for electron emission at an energy E_n, namely

$$\frac{d\Gamma_n^{FR}}{d\hat{\mathbf{k}}_f} = \frac{\omega^2}{2\pi}(2E_n)^{1/2}|M_{f0}^{FR}|^2, \tag{6.147}$$

where M_{f0}^{FR} is given by using Equations (6.139) and (6.146).

For the particular case of a hydrogenic atom in the ground state interacting with a linearly polarized laser field ($\xi = 0$), Equation (6.147) reduces to

$$\frac{d\Gamma_n^{FR}}{d\hat{\mathbf{k}}_f} = \frac{(2I_p)^{5/2}}{\pi}\frac{k_f}{(n\omega - U_p)^2}\left|\sum_{l=-\infty}^{\infty}J_l(U_p/2\omega)J_{2l+n}(A_0 k_{f,1}/\omega)\right|^2, \tag{6.148}$$

while for a left circularly polarized laser field ($\xi = -1$) we have

$$\frac{d\Gamma_n^{\text{FR}}}{d\hat{\mathbf{k}}_f} = \frac{(2I_{\text{P}})^{5/2}}{\pi} \frac{k_f}{(n\omega - U_{\text{p}})^2} \left| J_n[(A_0/2)(k_{f,\perp}^2/2)^{1/2}] \right|^2, \qquad (6.149)$$

where $k_{f,\perp} = (k_{f,1}^2 + k_{f,2}^2)^{1/2}$. This result differs from the expression (6.148) for linear polarization by the absence of summation over Bessel functions. As a consequence, one expects less interference effects in the ionization spectrum for circularly polarized radiation.

For linear polarization, comparisons of the Keldysh and the Faisal–Reiss SFA ionization rates have indicated that the two are typically in good agreement [46,47]. The relationship between the Keldysh and Faisal–Reiss SFA amplitudes has also been analyzed by Faisal [48].

6.4 High harmonic generation

In our discussion of the Keldysh theory of multiphoton direct ionization of atoms interacting with intense, low-frequency laser fields, we applied the saddle-point method to evaluate the Keldysh ionization amplitude. This led to an expression involving a sum over contributions from complex electron trajectories. A close correspondence was found to exist between each complex trajectory and the trajectory of a classical electron in the laser field. As a consequence, a simple semi-classical picture of the ionization process emerged: the electron is ejected with a near-zero velocity at times when the magnitude of the electric-field component of the laser pulse is near its maximum and subsequently oscillates "freely" in the laser field.

In this section we will apply the SFA to the study of high harmonic generation (HHG). In our analysis of the laser-induced atomic dipole moment, we will also make use of the saddle-point method. We will see once again that this approach leads to a simple semi-classical interpretation of the HHG processes whereby the dominant contributions to photon emission are those coming from ionized electrons that return to their parent core and recombine radiatively.

6.4.1 Laser-induced atomic dipole moment

We shall focus our attention on the case of a single-electron atomic system interacting with a linearly polarized laser pulse. Working in the dipole approximation and adopting the length gauge, the interaction Hamiltonian is given by

$$H_{\text{int}}^{\text{L}}(t) = \mathcal{E}(t) \cdot \mathbf{r}. \qquad (6.150)$$

The starting point of our analysis is Equation (6.63) for the laser-induced atomic dipole moment in the SFA. The first term on the right-hand of this equation vanishes because **r** is an operator of odd parity. The third term contains the contributions from continuum–continuum transitions, which can be neglected [49]. The laser-induced dipole moment of the atom is then given by

$$\mathbf{d}(t) = -2\mathrm{Im} \int_{t_0}^{t} dt' \langle \phi_0(t) | \mathbf{r} U_{\mathrm{F}}^{\mathrm{L}}(t, t') H_{\mathrm{int}}^{\mathrm{L}}(t) | \phi_0(t') \rangle, \qquad (6.151)$$

where the Gordon–Volkov evolution operator in the length gauge $U_{\mathrm{F}}^{\mathrm{L}}(t, t')$ is given by Equation (2.180). The only bound atomic state contributing to $\mathbf{d}(t)$ is the initial state of the atom. As a consequence, resonances between bound states are not taken into account.

The single-atom spectrum of the emitted photons is obtained by calculating $|\hat{\boldsymbol{\epsilon}} \cdot \mathbf{a}(\Omega)|^2$, for emission polarized parallel to the polarization direction of the incident laser pulse, where Ω is the angular frequency of the emitted photon and $\mathbf{a}(\Omega)$ is the Fourier transform of $\ddot{\mathbf{d}}(t)$. The ratio Ω/ω is an effective "harmonic order." As we shall discuss below, the SFA is readily modified to include the depletion of the ground state [2, 50]. We will assume here that the laser parameters and the atomic system have been chosen in such a way that the ionization probability per laser optical cycle is much less than unity, and hence depletion can be neglected.

Following Lewenstein *et al.* [50], we will now obtain an approximate expression for the laser-induced dipole moment given by Equation (6.151). First, we write this equation in the form

$$\mathbf{d}(t) = 2\mathrm{Im} \int_{t_0}^{t} dt' \int d\mathbf{k} \, \mathbf{d}_{\mathrm{rec}}^{*}[\boldsymbol{\pi}(\mathbf{k}, t)] \exp[-iS(\mathbf{k}, t, t')] d_{\mathrm{ion}}[\boldsymbol{\pi}(\mathbf{k}, t'), t'],$$

$$(6.152)$$

where $\boldsymbol{\pi}(\mathbf{k}, t)$ and $S(\mathbf{k}, t, t')$ are given by Equations (6.73) and (6.75), respectively. In addition, we have introduced the dipole recombination amplitude

$$\mathbf{d}_{\mathrm{rec}}(\mathbf{q}) = -\langle \mathbf{q} | \mathbf{r} | \psi_0 \rangle \qquad (6.153)$$

and the ionization amplitude

$$d_{\mathrm{ion}}(\mathbf{q}, t) = \langle \mathbf{q} | H_{\mathrm{int}}^{\mathrm{L}}(t) | \psi_0 \rangle. \qquad (6.154)$$

The modified action $S(\mathbf{k}, t, t')$ is a rapidly varying function of $\mathbf{k}$, t and t', and therefore we will again make use of the saddle-point method to carry out the integrations in Equation (6.152). To this end, let us express the laser-induced atomic dipole moment as

$$d_{\mathrm{ion}}[\boldsymbol{\pi}(\mathbf{k}, t'), t'] = -i \frac{d}{dt'} \langle \boldsymbol{\pi}(\mathbf{k}, t') | \psi_0 \rangle. \qquad (6.155)$$

We will assume that $|\psi_0\rangle$ is the ground state of a hydrogenic atomic system with nuclear charge $Z = (2I_{\mathrm{P}})^{1/2}$. Referring to Equation (6.82), and recalling that

$\kappa = (2I_P)^{1/2}$, we have

$$\langle \pi(\mathbf{k}, t') | \psi_0 \rangle = \frac{(8I_P)^{5/4}}{8\pi} \frac{1}{\left[\pi^2(p, t')/2 + I_P \right]^2}$$

$$= \frac{(8I_P)^{5/4}}{8\pi} \left[-\frac{\partial}{\partial t'} S(\mathbf{k}, t, t') \right]^{-2} \tag{6.156}$$

and

$$\mathbf{d}_{\mathrm{rec}}(\mathbf{q}) = \mathrm{i} \frac{(8I_P)^{5/4}}{4\pi} \frac{\mathbf{q}}{(q^2/2 + I_P)^3}. \tag{6.157}$$

Making use of Equations (6.155) and (6.156), we now integrate Equation (6.152) by parts, with the result

$$\mathbf{d}(t) = 2\,\mathrm{Im} \frac{(8I_P)^{5/4}}{8\pi} \int_{t_0}^{t} \mathrm{d}t' \int \mathrm{d}\mathbf{k}\, \mathbf{d}_{\mathrm{rec}}^*[\pi(\mathbf{k}, t)] \exp[-\mathrm{i}S(\mathbf{k}, t, t')] \left[\frac{\partial}{\partial t'} S(\mathbf{k}, t, t') \right]^{-1}. \tag{6.158}$$

Since $H_{\mathrm{int}}^{\mathrm{L}}(t' = 0) = 0$, the boundary term at $t' = 0$ vanishes while the boundary term at $t' = t$ can be ignored. It corresponds to the process whereby the electron both ionizes and recombines at time t.

Next, the integral over $\mathbf{k}$ is approximated using the saddle-point method. The saddle momentum $\mathbf{k}_s$ depends on t and t' and is obtained by solving the equation

$$\nabla_{\mathbf{k}} S(\mathbf{k}, t, t')|_{\mathbf{k}=\mathbf{k}_s} = 0. \tag{6.159}$$

In order to understand the significance of the saddle momentum, we point out that

$$\nabla_{\mathbf{k}} S(\mathbf{k}, t, t') = \int_{t'}^{t} \mathrm{d}t'' \left[\pi(\mathbf{k}, t'') \right] = \mathbf{r}(t), \tag{6.160}$$

where $\mathbf{r}(t)$, given by Equation (2.76), is the displacement of a classical electron interacting with the laser field. Therefore, the saddle momentum corresponds to the initial drift momentum that a classical electron would be required to have at time t' to ensure that its displacement in the laser field is zero at the later time t. In particular, an electron that is ionized at time t' with an initial momentum $\mathbf{k}_s$ would return to its parent ion at time t.

Solving Equation (6.159) yields for the saddle momentum the following expression:

$$\mathbf{k}_s(t, t') = -\frac{1}{t - t'} \int_{t'}^{t} \mathrm{d}t'' \mathbf{A}(t'')$$

$$= -\frac{1}{t - t'} \boldsymbol{\alpha}(t, t'). \tag{6.161}$$

Applying the saddle-point method, the laser-induced atomic dipole moment is approximated as

$$\mathbf{d}(t) \simeq 2\mathrm{Im} \frac{(8I_\mathrm{P})^{5/4}}{8\pi} \int_{t_0}^{t} \mathrm{d}t' \, C(t-t') \mathbf{d}_\mathrm{rec}^{*} [\boldsymbol{\pi}(\mathbf{k}_\mathrm{s}, t)] \exp[-\mathrm{i}S(\mathbf{k}_\mathrm{s}, t, t')]$$

$$\times \left[\frac{\partial}{\partial t'} S(\mathbf{k}_\mathrm{s}, t, t') \right]^{-1}. \tag{6.162}$$

The factor

$$C(t-t') = (2\pi)^{3/2} [\varepsilon + \mathrm{i}(t-t')]^{-3/2}, \tag{6.163}$$

where ε is a small positive parameter, accounts for the spreading of the wave packet of the ionized electron.

The integral over t' in Equation (6.162) will now be evaluated also using the saddle-point method. We see that, as in Equation (6.79), the saddle times t_s are determined from the solutions of

$$-\frac{\partial}{\partial t'} S(\mathbf{k}_\mathrm{s}, t, t') \bigg|_{t'=t_\mathrm{s}} = [\boldsymbol{\pi}^2(\mathbf{k}_\mathrm{s}, t_\mathrm{s})/2 + I_\mathrm{P}] = 0. \tag{6.164}$$

In order to give a physical interpretation of the saddle times t_s, let us first neglect I_P in Equation (6.164). Denoting the saddle times by t_d in this case, the values of t_d are determined from the solutions of

$$\boldsymbol{\pi}(\mathbf{k}_\mathrm{s}, t_\mathrm{d}) = 0. \tag{6.165}$$

As we saw in Section 6.3.1, this condition simply imposes the requirement that the initial velocity of the electron be zero. Using Equation (6.161) for the saddle momentum, Equation (6.165) becomes

$$-\frac{1}{t-t_\mathrm{d}} \int_{t_\mathrm{d}}^{t} \mathrm{d}t'' \mathbf{A}(t'') + \mathbf{A}(t_\mathrm{d}) = 0. \tag{6.166}$$

For a linearly polarized laser pulse, we shall show in Section 6.4.2 that this is equivalent to finding the particular values of t_d such that an electron initially at rest at the origin at time t_d returns to the origin at the later time t. In terms of the semi-classical recollision model, this corresponds to an electron that is detached at time t_d, subsequently oscillates in the laser field as a "free" particle, and then returns to its parent ion at time t.

Turning now to the realistic case for which $I_\mathrm{P} > 0$ in Equation (6.164), we see that this equation cannot be satisfied by a purely real t_s. Instead, as in Section 6.3, one obtains complex saddle times. An electron tunneling through the effective Coulomb barrier has a negative kinetic energy, and as in Section 6.4.2 we can interpret the imaginary part of t_s as corresponding to the electron tunneling time [22, 50].

While the saddle times can be obtained numerically, we will now consider the semi-analytical approximation of Ivanov, Brabec and Burnett [38]. When the Keldysh parameter $\gamma_K \ll 1$, the tunneling rates at the peak laser pulse intensities are large. Hence the tunneling times are small, and we can take t_d as a zeroth-order approximation to the saddle times. In this approximation, we therefore express the saddle times as

$$t_s = t_d + \Delta. \tag{6.167}$$

We then expand Equation (6.164) in powers of Δ:

$$-\frac{\partial}{\partial t'} S(\mathbf{k}_s, t, t')\bigg|_{t'=t_s} = s_0 + s_1 \Delta + \tfrac{1}{2} s_2 \Delta^2 + O(\Delta^3). \tag{6.168}$$

Making use of Equation (6.165), we readily obtain for the first coefficient

$$s_0 = I_P, \tag{6.169}$$

while the second coefficient is found to be zero. The third coefficient is given by

$$s_2 = \mathcal{E}^2(t_d). \tag{6.170}$$

Using Equations (6.164) and (6.168) and solving for Δ yields

$$\Delta = \pm i \frac{(2 I_P)^{1/2}}{|\mathcal{E}(t_d)|}. \tag{6.171}$$

For detachment times t_d occurring at the laser pulse peak, when $|\mathcal{E}(t)|$ is maximum, we see that $\omega \Delta \simeq \pm i \gamma_K$. On physical grounds, the root that lies on the negative imaginary axis must be chosen. The approximate expression for the modified action at the saddle time is then

$$S(\mathbf{k}_s, t, t_s) = S(\mathbf{k}_s, t, t_d) - s_0 \Delta - \frac{1}{6} s_2 \Delta^3$$

$$= S(\mathbf{k}_s, t, t_d) - \frac{(2 I_P)^{3/2}}{3 |\mathcal{E}(t_d)|}. \tag{6.172}$$

Moreover, we have

$$-\frac{\partial^2}{\partial t_s^2} S(\mathbf{k}_s, t, t')\bigg|_{t'=t_s} = \left[\boldsymbol{\pi}(\mathbf{k}_s, t') \cdot \frac{\partial}{\partial t'} \boldsymbol{\pi}(\mathbf{k}_s, t') \right]_{t'=t_s}$$

$$= \Delta s_2$$

$$= -i (2 I_P)^{1/2} |\mathcal{E}(t_d)|. \tag{6.173}$$

Referring to the saddle condition given by Equation (6.164), it is seen that the laser-induced atomic dipole moment (6.162) is singular at the saddle times. This situation was also encountered in Section 6.4.2, and in order to deal with this singularity we again expand the denominator in Equation (6.162) in a Taylor series about $t' = t_s$, namely

$$\frac{\partial}{\partial t'} S(\mathbf{k}_s, t, t') = \frac{\partial}{\partial t'} S(\mathbf{k}_s, t, t')\Big|_{t'=t_s} + (t' - t_s)\frac{\partial^2}{\partial t'^2} S(\mathbf{k}_s, t, t')\Big|_{t'=t_s} + \cdots \quad (6.174)$$

Retaining only the linear term in the Taylor expansion (6.174) so that the integrand has a first-order pole at the saddle times, we must evaluate an integral of the form (6.85). Using Equation (6.86), the laser-induced atomic dipole moment is given by

$$\mathbf{d}(t) \simeq -2\mathrm{Re}\frac{(8I_P)^{5/4}}{8}\sum_{t_s} C(t - t_s)\mathbf{d}^*_{\mathrm{rec}}[\boldsymbol{\pi}(\mathbf{k}_s, t)]$$

$$\times \exp[-iS(\mathbf{k}_s, t, t_s)]\left[\boldsymbol{\pi}(\mathbf{k}_s, t')\cdot\frac{\partial}{\partial t'}\boldsymbol{\pi}(\mathbf{k}_s, t')\right]^{-1}_{t'=t_s}. \quad (6.175)$$

Using the approximations (6.172) and (6.173), the resulting laser-induced atomic dipole moment can be expressed as [38]

$$\mathbf{d}(t) = 2\mathrm{Im}\sum_{t_d} \mathbf{a}_{\mathrm{rec}}(t, t_d)a^*_{\mathrm{free}}(t, t_d)a^*_{\mathrm{ion}}(t_d), \quad (6.176)$$

with the ionization, "free"-motion and recombination amplitudes, respectively, given by

$$a_{\mathrm{ion}}(t_d) = \frac{(8I_P)^{3/4}}{4|\mathcal{E}(t_d)|}\exp\left[-\frac{1}{3}\frac{(2I_P)^{3/2}}{|\mathcal{E}(t_d)|}\right], \quad (6.177)$$

$$a_{\mathrm{free}}(t, t_d) = C(t - t_d)\exp[-iS(\mathbf{k}_s, t, t_d)] \quad (6.178)$$

and

$$\mathbf{a}_{\mathrm{rec}}(t, t_d) = \mathbf{d}_{\mathrm{rec}}[\boldsymbol{\pi}(\mathbf{k}_s, t)]. \quad (6.179)$$

At each time, $\mathbf{d}(t)$ is given by a sum over contributions from the set of electron trajectories that correspond to an electron that has been detached at time t_d and returns to its parent ion at time t.

The final expression (6.176) for the laser-induced atomic dipole moment is a remarkably simple formula that can be readily evaluated numerically. As a consequence, it has been applied to investigate a wide range of problems involving HHG and, in particular, issues concerning how the generated harmonics propagate

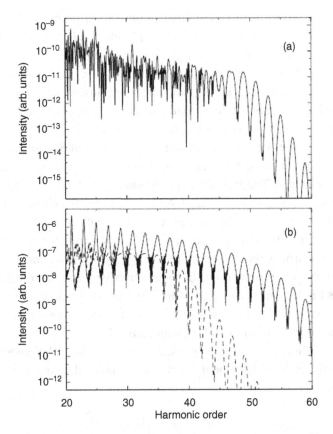

Figure 6.4. (a) Calculated SFA photon emission spectrum generated by an argon atom interacting with a 27 fs laser pulse of peak intensity $3 \times 10^{14} \, \mathrm{W \, cm^{-2}}$ and carrier wavelength 810 nm. (b) Propagated harmonic spectrum with the atomic jet located 2 mm after the focus (solid line) and at the focus (dashed line). (From P. Salières *et al.*, *Phys. Rev. Lett.* **81**, 5544 (1998).)

through a macroscopic medium [51, 52]. In Fig. 6.4, we give an example of the results of such a study carried out by Salières *et al.* [53]. Figure 6.4(a) shows the calculated photon emission spectrum obtained by using the laser-induced atomic moment in the SFA for an argon atom interacting with a 27 fs laser pulse of peak intensity $3 \times 10^{14} \, \mathrm{W \, cm^{-2}}$ and carrier wavelength 810 nm (corresponding to a Ti:sapphire laser). We note the cut-off in the emission spectrum located at approximately the harmonic order $q = 51$. Figure 6.4(b) illustrates the influence of the spatial intensity profile of the laser pulse in the focal region on the propagation of HHG through the macroscopic gas medium, a subject to which we will return in Chapter 9. Shown are the propagated harmonic spectra with the atomic jet located 2 mm after the focus (solid line) and at the focus (dashed line).

Let us now discuss each of the three amplitudes appearing in Equation (6.176). We start by comparing the ionization amplitude $a_{\text{ion}}(t, t_{\text{d}})$ defined by Equation (6.177) with the tunneling formula given by Equation (6.38). Recalling that, for a hydrogenic system in its ground state, $C_{\text{as}} = 2$, $\kappa = Z_{\text{c}} = Z = (2I_{\text{P}})^{1/2}$, $l = m = 0$ and $A(l, m) = 1$, we see that the ionization amplitude (6.177) can be viewed as corresponding to a quasi-static tunneling ionization amplitude. The pre-exponential factor appearing in $a_{\text{ion}}(t_{\text{d}})$ is not, however, correct. This was also found to be the case in our analysis of the SFA ionization amplitude in Section 6.3.2, and can be again attributed to the neglect of the Coulomb interaction between the core and the ejected electron. The free-motion amplitude $a_{\text{free}}(t, t_{\text{d}})$ consists of two factors. One is a phase factor, with $-iS(\mathbf{k}_{\text{s}}, t, t_{\text{d}})$ being simply the phase acquired by a free electron, with zero initial velocity, as it interacts with the laser field from time t_{d} to time t. As noted above, the factor $C(t - t_{\text{d}})$ accounts for the spreading of the initial free electron wave packet. Finally, $\mathbf{a}_{\text{rec}}(t, t_{\text{d}})$ is the recombination amplitude in the first Born approximation for an electron with momentum $\boldsymbol{\pi}(\mathbf{k}_{\text{s}}, t)$, where $\boldsymbol{\pi}(\mathbf{k}_{\text{s}}, t)$ is the kinetic momentum of the electron as it returns to the core. We can therefore conclude that Equation (6.176) is a quantum-mechanical version of the semi-classical recollision model introduced in Chapter 1.

Let us continue this analysis by establishing another relationship between the quantum-mechanical SFA model and the semi-classical model. First, we introduce the quantity

$$E_{\text{kin}}(t, t_{\text{d}}) = \frac{1}{2}[\boldsymbol{\pi}(\mathbf{k}_{\text{s}}, t)]^2, \tag{6.180}$$

which is the kinetic energy of an electron that is detached at time t_{d} and returns to its parent core at time t. We recall that $\mathbf{k}_{\text{s}}$, which is defined by Equation (6.161), is a function of both t and t_{d}. According to the semi-classical model, an electron returning to its parent core at time t with kinetic energy $E_{\text{kin}}(t, t_{\text{d}})$ can recombine by emitting a photon of energy $\Omega = E_{\text{kin}}(t, t_{\text{d}}) + I_{\text{P}}$. As will be shown in Section 9.2, the photon emission spectrum of an atom interacting with a *long* laser pulse can be obtained from the Fourier transform of the laser-induced atomic dipole moment, as given by Equation (1.15). In order to calculate this Fourier transform, it is convenient to express the laser-induced atomic dipole moment as

$$\mathbf{d}(t) = -i\left[\tilde{\mathbf{d}}(t) - \tilde{\mathbf{d}}^*(t)\right]. \tag{6.181}$$

Using Equations (6.152) and (6.176), we see that

$$\tilde{\mathbf{d}}(t) = \int_{t_0}^{t} dt' \int d\mathbf{k}\, \mathbf{d}_{\text{rec}}^*[\boldsymbol{\pi}(\mathbf{k}, t)] \exp[-iS(\mathbf{k}, t, t')] d_{\text{ion}}[\boldsymbol{\pi}(\mathbf{k}, t'), t']$$

$$\simeq \sum_{t_{\text{d}}} \mathbf{a}_{\text{rec}}(t, t_{\text{d}}) a_{\text{free}}^*(t, t_{\text{d}}) a_{\text{ion}}^*(t_{\text{d}}). \tag{6.182}$$

Introducing the Fourier transform of $\tilde{\mathbf{d}}(t)$, namely

$$\tilde{\mathbf{d}}(\Omega) = (2\pi)^{-1/2} \int_{-\infty}^{\infty} \exp(-\mathrm{i}\Omega t)\,\tilde{\mathbf{d}}(t)\,\mathrm{d}t, \tag{6.183}$$

the Fourier transform of the laser-induced atomic dipole moment can be written as

$$\mathbf{d}(\Omega) = -\mathrm{i}\left[\tilde{\mathbf{d}}(\Omega) - [\tilde{\mathbf{d}}(-\Omega)]^*\right]. \tag{6.184}$$

We will now carry out the integral over t in Equation (6.183) using the saddle-point method. The saddle times, which we denote by t_r, are determined from the solutions of the equation

$$\left[\frac{\partial}{\partial t}S(\mathbf{k}_\mathrm{s}, t, t_\mathrm{d}) - \Omega\right]_{t=t_\mathrm{r}} = E_\mathrm{kin}(t_\mathrm{r}, t_\mathrm{d}) + I_\mathrm{P} - \Omega = 0. \tag{6.185}$$

This equation determines the kinetic energy that the returning electron must have in order to generate the emission of a photon of energy Ω, and expresses the same energy conservation law as the semi-classical recollision model, namely that the energy of the emitted photon is equal to the kinetic energy of the returning electron plus the ionization potential of the atom or ion.

The saddle-point approximation yields

$$\tilde{\mathbf{d}}(\Omega) = \sum_{\{t_\mathrm{r}, t_\mathrm{d}\}} \frac{\exp(-\mathrm{i}\Omega t_\mathrm{r})}{[2\pi\mathrm{i}E'_\mathrm{kin}(t_\mathrm{r}, t_\mathrm{d})]^{1/2}} \mathbf{a}_\mathrm{rec}(t_\mathrm{r}, t_\mathrm{d})a^*_\mathrm{free}(t_\mathrm{r}, t_\mathrm{d})a^*_\mathrm{ion}(t_\mathrm{d}), \tag{6.186}$$

where

$$E'_\mathrm{kin}(t_\mathrm{r}, t_\mathrm{d}) = \left.\frac{\partial^2}{\partial t^2}S(\mathbf{k}_\mathrm{s}, t, t_\mathrm{d})\right|_{t=t_\mathrm{r}}, \tag{6.187}$$

and the times t_r are calculated using Equation (6.185). Within the SFA, the amplitude corresponding to the emission of a photon with energy Ω is therefore given by a sum over contributions from the classical trajectories of the electrons that return to their parent core with kinetic energy $\Omega - I_\mathrm{P}$. In fact, Equation (6.186) suggests that photon emission into some range of angular frequencies $\Omega + \delta\Omega$ will occur at well defined moments in time during the laser pulse. In Chapter 9 we will see that this is indeed the case.

As noted in Section 1.4, for sufficiently long laser pulses, photons are emitted at angular frequencies that are odd multiples of the driving laser angular frequency ω. Describing HHG in this long-pulse regime by calculating the laser-induced atomic dipole moment using a monochromatic driving field, one finds that Ω is non-zero only for discrete values, with $\Omega = q\omega$ and $q = 1, 3, 5, \ldots$ This case has been investigated in detail by Lewenstein *et al.* [50]. For a laser field described by a spatially homogeneous, linearly polarized and monochromatic electric field of the form

$$\mathcal{E}(t) = \hat{\boldsymbol{\epsilon}}\mathcal{E}_0\cos(\omega t), \tag{6.188}$$

and using Equations (6.161) and (6.166), the modified action takes the form

$$S(\mathbf{k}_s, t_r, t_d) = U_p \left(\cos(2\omega t_d)(t_r - t_d) - \frac{1}{2\omega} [\sin(2\omega t_r) - \sin(2\omega t_d)] \right) + I_P(t_r - t_d).$$

(6.189)

Because the dominant contributions to HHG come from electrons that are ionized when $|\mathcal{E}(t)|$ is maximum and return approximately half an optical period later, we see that the electrons acquire an overall phase that is approximately equal to $U_p \tau$, where $\tau = t_r - t_d$ is the time that the electron spends in the continuum. This phase depends linearly on the intensity of the laser field. We recall that, in a typical experiment, HHG is produced by focusing a laser beam onto a gas jet. The resulting position-dependent intensity profile in the jet gives rise to a spatial dependence of the phases of the laser-induced atomic dipole moments of the atoms. This has important consequences for phase matching, and hence HHG conversion efficiency, as will be discussed in Chapter 9.

To conclude this section, we note that the saddle-point approximation fails when $|E'_{\text{kin}}(t_r, t_d)| \to 0$. This occurs when Ω approaches the cut-off frequency ω_c in the HHG spectrum whereby photon emission with $\Omega > \omega_c$ is classically forbidden. For $\Omega < \omega_c$, there exist two dominant classical trajectories with return times $\tau < T$ and energies satisfying $E_{\text{kin}}(t_r, t_d) + I_P = \Omega < \omega_c$. However, as $\Omega \to \omega_c$, these two trajectories coalesce. The requirement that the saddle times be sufficiently isolated is therefore violated near a cut-off, and a uniform approximation [54, 55] can instead be used to evaluate the integral in Equation (6.183). Insight into this and other aspects regarding the relationship between the semi-classical recollision model and Equation (6.176) for the laser-induced atomic dipole moment can be gained by analyzing classical electron trajectories in the laser field.

6.4.2 Classical electron trajectories

Let us begin by calculating the trajectory of a classical electron that, at time t_d, is located at the origin and has zero drift velocity. From Equations (2.72) and (2.76), the velocity and position of the electron, respectively, at the later time t are

$$\mathbf{v}(t) = \mathbf{A}(t) - \mathbf{A}(t_d) \tag{6.190}$$

and

$$\mathbf{r}(t) = \int_{t_d}^{t} \mathbf{A}(t')dt' - (t - t_d)\mathbf{A}(t_d). \tag{6.191}$$

The requirement that the detachment time t_d corresponds to an electron trajectory with zero net displacement at time t implies that t_d is determined by setting

$\mathbf{r}(t) = 0$ in Equation (6.191). As expected, this is equivalent to the condition (6.166) introduced above.

Equation (6.190) gives the kinetic energy of the returning electron as

$$E_{\text{kin}}(t, t_{\text{d}}) = \frac{1}{2} [\mathbf{A}(t) - \mathbf{A}(t_{\text{d}})]^2. \tag{6.192}$$

The equivalence of this equation and Equation (6.180) is readily verified using Equation (6.190) together with Equations (6.161) and (6.166). It is worth bearing in mind that setting $\mathbf{r}(t) = 0$ in Equation (6.191) gives the detachment time t_{d} as an implicit function of t.

Of particular interest is the maximum value that $E_{\text{kin}}(t, t_{\text{d}})$ can attain for a given laser pulse. We will denote this value by $E_{\text{kin}}^{\text{max}}$, so that

$$E_{\text{kin}}^{\text{max}} = E_{\text{kin}}(t_r^{\text{max}}, t_{\text{d}}), \tag{6.193}$$

where the return times t_r^{max} are determined by solving the equation

$$\frac{\partial}{\partial t} E_{\text{kin}}(t, t_{\text{d}}) \bigg|_{t = t_r^{\text{max}}} = E'_{\text{kin}}(t, t_{\text{d}}) \big|_{t = t_r^{\text{max}}} = 0. \tag{6.194}$$

The maximum electron kinetic energy of a returning electron determines the corresponding cut-off frequency ω_c in the HHG spectrum. That is,

$$\omega_c = I_P + E_{\text{kin}}^{\text{max}}. \tag{6.195}$$

It is instructive to calculate ω_c for the particular case of a linearly polarized monochromatic laser field described by the electric field (6.188). Taking the polarization vector $\hat{\boldsymbol{\epsilon}}$ to lie along the x-axis, the velocity of the electron as it returns to the core is found to be

$$v_x(t) = -\frac{\mathcal{E}_0}{\omega} [\sin(\omega t) - \sin(\omega t_{\text{d}})], \tag{6.196}$$

where the detachment time t_{d} is obtained by solving the equation

$$x(t) = \frac{\mathcal{E}_0}{\omega^2} [\cos(\omega t) - \cos(\omega t_{\text{d}}) + \omega(t - t_{\text{d}}) \sin(\omega t_{\text{d}})] = 0. \tag{6.197}$$

The kinetic energy of the electron is then given by

$$E_{\text{kin}}(t, t_{\text{d}}) = 2U_{\text{p}} [\sin(\omega t) - \sin(\omega t_{\text{d}})]^2. \tag{6.198}$$

The maximum kinetic energy of a returning electron can now be determined. Using Equation (6.198) to solve Equation (6.194) yields the result $3.17U_{\text{p}}$ [56,57]. These electrons are detached at the time $t_{\text{d}} = 0.10\pi/\omega$ ($\omega t_{\text{d}} = 18°$) and return to the parent core at the time $t_r^{\text{max}} = 1.94\pi/\omega$ ($\omega t_r^{\text{max}} = 252°$). Within this model, the cut-off harmonic energy is therefore $\omega_c = I_P + 3.17U_{\text{p}}$. As mentioned in Chapter 1, the predicted harmonic cut-off obtained from this simple analysis is in good agreement with experiment.

Let us now look at the classical trajectories in more detail, with the aim of showing how they can be used to understand the essential features of HHG spectra within the semi-classical model [56,57]. We first recall that in this model it is assumed that in the first step an electron is detached from its parent core by tunneling ionization. In the second step, the electron motion is dominated by the laser field, so that the influence of the core potential is neglected and the motion of the electron in the laser field is treated classically. Finally, an electron that returns to its parent core with kinetic energy E_{kin} can recombine radiatively and emit a photon of energy $\Omega = E_{kin} + I_P$.

A number of features of the classical trajectories that return to the core are illustrated in Fig. 6.5 for a linearly polarized, monochromatic laser field. The detachment times t_d as a function of the trajectory duration τ are shown in Fig. 6.5(a), while Fig. 6.5(b) depicts the corresponding kinetic energy of the electron when it returns. The size of the dots is proportional to the strength of the electric field (6.188) at the time of detachment: the larger the dot, the stronger the electric field. Each dot is associated with a trajectory, with the trajectory detachment times chosen on an equidistant

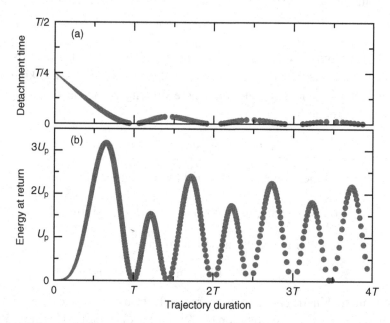

Figure 6.5. (a) Detachment times t_d as a function of the trajectory duration τ. The trajectories are those of a classical electron that is initially at rest at the origin. From time t_d onwards, the electron interacts with the electric field (6.188), oscillates in the field and is finally driven back to the origin at time t. (b) Kinetic energy of the electron when it returns to the origin.

grid between $t_d = 0$ and $t_d = T/4$, where $T = 2\pi/\omega$ is a laser period or optical cycle. Electrons detached in the time interval $(T/4, T)$ do not return to the core.

From the examination of Fig. 6.5, we can make the following observations. (i) The shortest trajectories correspond to electrons that are detached when the electric field is weak and the tunneling ionization probability is extremely small. As a result, their contribution to photon emission spectra is small. (ii) There exist trajectories that return to the core many times. (iii) As noted above, the $3.17U_P$ cut-off is due to electrons that detach when the electric field is close to its maximum and return to the core just over half a laser period later. (iv) The trajectories whose durations are approximately equal to $T + T(2n-1)/4$, with $n = 1, 2, 3, \ldots$, give rise to additional cut-offs. The corresponding energies of these cut-offs are less than $3.17U_p$. (v) For each energy below $3.17U_p$ there is a short and a long trajectory whose duration is less than one laser period. These are the "short" and "long" trajectories introduced in Section 1.4, and these trajectories typically provide the dominant contributions to photon emission spectra.

We will now consider an example that illustrates how the main features of HHG spectra can be understood in terms of the semi-classical model. In Fig. 6.6(a) we show the HHG spectra obtained by Potvliege, Kylstra and Joachain [58] for a one-dimensional (1D) and two-dimensional model (2Dx) He$^+$ ion ($I_P = 2$ a.u.) interacting with an intense, two-cycle pulse. The spectra were obtained from the "exact" numerical solutions of the TDSE, with the interaction of the electron with the nucleus being represented by the soft-core potentials (5.73) and (5.78), respectively, with $a = \sqrt{2}$ in 1D and $a = 0.798$ in 2D. The laser pulse was described by the vector potential

$$\mathbf{A}(t) = \hat{\boldsymbol{\epsilon}}(\mathcal{E}_0/\omega) F(t) \sin(\omega t), \tag{6.199}$$

where the envelope function was taken to be $F(t) = \sin^2(\pi t/\tau_p)$ and $\tau_p = 2T$. The peak electric-field strength is $\mathcal{E}_0 = 0.4$ a.u. (corresponding to a peak intensity of 5.6×10^{15} W cm^{-2}) and $\omega = 0.057$ a.u. (corresponding to a Ti:sapphire laser of wavelength 800 nm).

Two distinct plateaus can be seen in the figure. The first one extends from approximately harmonic order $q = \Omega/\omega = 55$ to $q = 220$, while the second continues to about $q = 570$. Since in this case $\omega/I_P \ll 1$ and $\gamma_K \ll 1$, the HHG process can be analyzed in terms of the semi-classical model. To this end, we show in Fig. 6.6(b) the 1D static tunneling rate at the detachment time t_d as a function of the kinetic energy of the returning electrons expressed as the corresponding harmonic order. The similarity with the HHG spectra of Fig. 6.6(a) is remarkable. The two plateaus end at the same harmonic order as the classical trajectories, and their relative height is comparable to the difference between the rates at the cut-offs. This shows that the first plateau is due to a distinct set of trajectories whose main contribution to

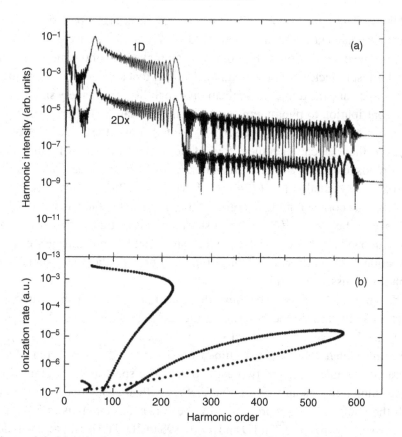

Figure 6.6. Magnitude squared of the Fourier transform of the dipole acceleration as a function of the photon energy (in units of $\hbar\omega$). The two-cycle laser pulse is described by the vector potential (6.199) with peak electric-field strength $\mathcal{E}_0 = 0.4$ a.u. and a wavelength of 800 nm. The emission spectra obtained from a one-dimensional model (1D) and a two-dimensional model (2Dx) are shown. (b) Instantaneous tunneling ionization rate at the time at which the electron ionizes shown as a function of the energy of the photon emitted when it recombines with the core. (From R. M. Potvliege, N. J. Kylstra and C. J. Joachain, *J. Phys. B* **33**, L743 (2000).)

photon emission comes from electrons that are ionized at a rate of about 10^{-3} a.u. Similarly, the second plateau arises from another set of trajectories that correspond to electrons that are ionized at a rate of about 10^{-5} a.u.

6.4.3 Extensions of the theory

The analysis of the SFA laser-induced atomic dipole moment can be generalized to the case of an atom interacting with an elliptically polarized laser field [32,59–61].

Within the semi-classical recollision model, it is clear that an electron ejected with zero velocity into an elliptically polarized laser field will never return to its parent core. However, if the electron is ejected with a zero velocity component along the direction opposite to $\mathcal{E}(t)$ and a sufficiently large velocity component along the tangential component of $\mathcal{E}(t)$, the electron can return to the core. In the regime where $\gamma_K \ll 1$, this implies that $k_\perp$, the component of the drift momentum perpendicular to $\mathcal{E}(t)$, must be non-zero. As the magnitude of the ellipticity of the laser field increases, the corresponding values of $k_\perp$ must increase. The ionization amplitudes decrease exponentially as $k_\perp$ increases, as can be seen, for example, by generalizing Equation (6.104), which gives the SFA ionization amplitude for a linearly polarized laser pulse. Within the SFA, HHG is therefore suppressed as the magnitude of the ellipticity of the laser field increases, which has been confirmed experimentally [62–64].

In Section 6.3 it was shown that, in the limit that $\gamma_K \ll 1$, the neglect of the Coulomb correction between the ejected electron and the ionic core leads to SFA ionization rates that did not agree with the adiabatic tunneling ionization rates. Let us return to the ionization amplitude $a_{\text{ion}}(t)$, given by Equation (6.177), that appears in Equation (6.176) for the laser-induced atomic dipole moment. Assuming that $\gamma_K \ll 1$, the quantity $|a_{\text{ion}}(t)|^2$ can be interpreted as the instantaneous quasi-static ionization rate of the atom in the laser field. Comparing this quantity with Equation (6.22), it is seen that in this limit the pre-exponential factor is not correct. As a consequence, the total HHG emission yields calculated by using the SFA are typically too low, thereby requiring that a Coulomb correction be applied [65] to the ionization amplitude $a_{\text{ion}}(t)$. The influence of the Coulomb potential on the recombination amplitude (6.179) can also be included in an approximate way. An illustration of the accuracy that can be attained using the resulting Coulomb-corrected SFA laser-induced atomic dipole moment will be given in Section 9.2.1.3.

For a sufficiently intense laser pulse, there is a significant probability that the atoms will be ionized by the laser pulse. This is particularly true for long laser pulses. As the ground-state population of the atoms is depleted, photon emission by the atoms will be suppressed. The SFA does not correctly account for the depletion of the ground-state atomic population. This has been discussed by Lewenstein *et al.* [50], who introduced a general procedure for including depletion in the calculation of the SFA laser-induced atomic dipole moment $\mathbf{d}(t)$. When $\gamma_K \ll 1$, a simple approximation can be applied. If the evolution of the atom in the laser pulse is assumed to be adiabatic, one can multiply $\mathbf{d}(t)$ by the atomic ground-state population, which, using Equation (6.7), can be approximated at time t by

$$P(t) = \exp\left(-\int_{t_0}^{t} \Gamma_{DC}[\mathcal{E}(t')]\,dt'\right), \tag{6.200}$$

where Γ_{DC} is given by Equation (6.22) for a hydrogenic system.

6.5 Recollision ionization

In this section we will discuss the SFA recollision ionization transition amplitude (6.65). Within the semi-classical model, this amplitude describes the process whereby a detached electron is driven back to its parent core by the laser field and then rescatters elastically. The recollision ionization process is therefore similar to the process leading to harmonic generation. Only the final state is different. For the emission of a photon, in the SFA the final state of the system is the atomic ground state. If the returning electron rescatters, the final state describes a free electron with kinetic energy E_{k_f} and drift momentum $\mathbf{k}_f$. As we shall see, the similarity between the HHG and recollision ionization processes will allow us to apply directly a number of the results obtained in Section 6.4.1 to our analysis of the recollision amplitude.

Using the Gordon–Volkov evolution operator (2.181), we first write the transition amplitude (6.65) as

$$\overline{T}_{f0}^{(1)} = -\int_{t_0}^{t_1} dt' \int_{t'}^{t_1} dt \int d\mathbf{k} \langle \chi_{\mathbf{k}_f}(t)|V|\chi_{\mathbf{k}}(t)\rangle \langle \chi_{\mathbf{k}}(t')|H_{\mathrm{int}}(t')|\phi_0(t')\rangle. \tag{6.201}$$

We now focus on the case in which the interaction between the atom and the laser field is described in the length gauge by the interaction Hamiltonian (6.150). An overall constant phase factor aside, Equation (6.201), can be expressed as

$$\overline{T}_{f0}^{(1)} = -\int_{t_0}^{t_1} dt' \int_{t'}^{t_1} dt \int d\mathbf{k}\, a_{\mathrm{scat}}(\mathbf{k}_f, \mathbf{k}, t)]\exp[iS_{\mathbf{k}_f}(\mathbf{k}, t, t')]d_{\mathrm{ion}}[\boldsymbol{\pi}(\mathbf{k}, t'), t'], \tag{6.202}$$

where, using Equation (6.75), we have introduced the modified action

$$S_{\mathbf{k}_f}(\mathbf{k}, t, t') = S(\mathbf{k}_f, t, t_0) - S(\mathbf{k}, t, t'). \tag{6.203}$$

The quantity d_{ion} is the ionization amplitude (6.154) and the scattering amplitude is given by

$$a_{\mathrm{scat}}(\mathbf{k}_f, \mathbf{k}, t) = \langle \boldsymbol{\pi}(\mathbf{k}_f, t)|V|\boldsymbol{\pi}(\mathbf{k}, t)\rangle. \tag{6.204}$$

Comparing Equation (6.202) for the recollision transition amplitude $\overline{T}_{f0}^{(1)}$ with the equations (6.182) and (6.183) that define the Fourier transform of the laser-induced atomic dipole moment, $\tilde{\mathbf{d}}(\Omega)$, we see that the expression for $\tilde{\mathbf{d}}(\Omega)$ has exactly the same form as that of $\overline{T}_{f0}^{(1)}$. As a consequence, we can directly apply the analysis of Section 6.4.1 to the evaluation of $\overline{T}_{f0}^{(1)}$. Carrying out the integrals over $\mathbf{k}$, t' and t using the saddle-point method, we find that, for a hydrogenic atomic system, the recollision transition amplitude takes the form

$$\overline{T}_{f0}^{(1)} = \sum_{\{t_r, t_d\}} \frac{1}{\left[2\pi i S_{\mathbf{k}_f}''(\mathbf{k}_s, t_r, t_d)\right]^{1/2}} \exp\left[iS(\mathbf{k}_f, t_r, t_0)\right] a_{\text{scat}}(\mathbf{k}_f, \mathbf{k}_s, t_r)$$

$$\times\, a_{\text{free}}(t_r, t_d) a_{\text{ion}}(t_d), \tag{6.205}$$

where the ionization amplitude a_{ion} and the free-electron amplitude a_{free}, given by Equations (6.177) and (6.178), respectively, are the same quantities that appear in Equation (6.176) for the laser-induced atomic dipole moment. The scattering amplitude a_{scat} is defined by Equation (6.204), and

$$S_{\mathbf{k}_f}''(\mathbf{k}_s, t_r, t_d) = \left.\frac{\partial^2}{\partial t^2} S_{\mathbf{k}_f}(\mathbf{k}_s, t, t_d)\right|_{t=t_r}, \tag{6.206}$$

where the set of return and detachment times, $\{t_r, t_d\}$, are obtained from the solutions of Equation (6.166) and

$$\left.\frac{\partial}{\partial t} S_{\mathbf{k}_f}(\mathbf{k}_s, t, t_d)\right|_{t=t_r} = 0. \tag{6.207}$$

The saddle momentum $\mathbf{k}_s = \mathbf{k}_s(t_r, t_d)$ is given by Equation (6.161). In obtaining Equation (6.205), the boundary contributions to the recollision amplitude at times $t = t_0$ and $t = t_1$ have been neglected.

Equation (6.207) plays the same role in the recollision process as Equation (6.185) does for photon emission. Within the semi-classical recollision model, it expresses energy conservation of the elastically scattered electron. Thus,

$$\left.\frac{\partial}{\partial t} S_{\mathbf{k}_f}(\mathbf{k}_s, t, t_d)\right|_{t=t_r} = \frac{1}{2}[\mathbf{k}_f + \mathbf{A}(t_r)]^2 - E_{\text{kin}}(t_r, t_d) = 0, \tag{6.208}$$

where E_{kin}, given by Equation (6.180), is the kinetic energy of the returning electron.

The similarity between Equation (6.186) for the SFA quantity $\tilde{\mathbf{d}}(\Omega)$ and Equation (6.205) for the SFA recollision ionization amplitude is remarkable, as has been pointed out by Salières *el al.* [66]. In both formulae, a five-dimensional integral over intermediate states has been reduced to a coherent sum over contributions from a small set of relevant classical trajectories. Each contribution has an associated amplitude and a phase, so that the contributions from different trajectories can interfere.

As for the HHG process, it is interesting to look at the recollision ionization from the point of view of the semi-classical model. We start by noting that in order for a rescattered electron to attain a high final kinetic energy, the electron must, at the instant that it returns to the core, also have a high kinetic energy. From our discussion in Section 6.2, we recall that these high-energy electrons are ionized just after the magnitude of the electric-field component of the laser field $|\mathcal{E}(t)|$ is maximum and return to the core when it is close to a minimum. If the returning electron scatters in the forward direction, then the electric-field component of the laser field will decelerate the electron during the following half of the optical cycle, and the drift energy of the electron will never exceed $2U_p$. If, on the other hand, the electron returns to the core and is back-scattered, then the electric field can continue to accelerate the electron. Let us focus on these back-scattered electrons and assume that at time t_r the electron returns to the core with velocity

$$\mathbf{v}(t_r) = \mathbf{A}(t_r) - \mathbf{A}(t_d), \tag{6.209}$$

and then back-scatters. Using Equation (2.72), and neglecting the recoil momentum of the ion, the velocity of the electron at time $t > t_r$ is then given by

$$\mathbf{v}(t) = \mathbf{A}(t) - \mathbf{A}(t_r) + \mathbf{v}_0$$
$$= \mathbf{A}(t) - [2\mathbf{A}(t_r) - \mathbf{A}(t_d)]. \tag{6.210}$$

The term between the brackets is the drift velocity of the electron in the laser field. For sufficiently short laser pulses, this drift velocity determines the final kinetic energy of the scattered electron at the end of the laser pulse, so that

$$E_{\text{drift}}(t_r, t_d) = \frac{1}{2}[2\mathbf{A}(t_r) - \mathbf{A}(t_d)]^2. \tag{6.211}$$

It is of interest to determine the maximum value that $E_{\text{drift}}(t_r, t_d)$ can attain for a given laser pulse. Denoting this value by $E_{\text{drift}}^{\text{max}}$, we must determine the return times t_r^{max} such that

$$E_{\text{drift}}^{\text{max}} = E_{\text{drift}}(t_r^{\text{max}}, t_d), \tag{6.212}$$

where the values of t_r^{max} are obtained by solving the following equation:

$$\frac{\partial}{\partial t}E_{\text{drift}}(t, t_d)\bigg|_{t=t_r^{\text{max}}} = 0. \tag{6.213}$$

For the particular case of a spatially homogeneous, linearly polarized and monochromatic laser field described by the electric field (6.188), the drift energy of the electron is

$$E_{\text{drift}}(t, t_d) = 2U_p [2\sin(\omega t) - \sin(\omega t_d)]^2, \qquad (6.214)$$

where t_d is calculated using Equation (6.197). Solving Equation (6.213), one obtains $t_r^{\text{max}} = 1.45\pi/\omega (\omega t_r^{\text{max}} = 262°)$. The corresponding detachment time is $t_d = 0.08\pi/\omega (\omega t_d = 18°)$, yielding $E_{\text{drift}}^{\text{max}} = 10.01U_p$ [67]. Within the semi-classical model, this is the maximum kinetic energy that a rescattered electron can have at the end of the laser pulse. As will be discussed in Section 8.1, both ATI electron spectra obtained from *ab initio* calculations and experimental measurements exhibit a high-energy cut-off around $10U_p$.

If the returning electron is scattered in the forward direction, the maximum drift energy the electron can attain is $2U_p$, as in the case of direct ionization. Introducing the observed emission angle θ_f, where θ_f is defined to be the angle between $\mathbf{k}_f$ and the polarization vector $\hat{\epsilon}$, it follows that for a given emission angle θ_f there exists a corresponding maximum classically allowed drift energy $E_{\text{drift}}^{\text{max}}(\theta_f)$. We will analyze this in more detail in Section 8.1.

Let us return to Equation (6.206) defining $S_{\mathbf{k}_f}''(\mathbf{k}_s, t_r, t_d)$. This quantity vanishes when $\mathbf{k}_f = -[2\mathbf{A}(t_r^{\text{max}}) - \mathbf{A}(t_d)]$, so that $E_{k_f} = E_{\text{drift}}^{\text{max}}$. As a result, Equation (6.205) cannot be applied to calculate the contribution of this trajectory to the recollision amplitude. For the general case of recollision leading to electron emission into the angle θ_f, Equation (6.205) can only be applied for energies E_{k_f} less than the corresponding classical cut-off energy $E_{\text{drift}}^{\text{max}}(\theta_f)$. As in the case of direct ionization of electrons with drift energies close to the $2U_p$ cut-off and the emission of photons with angular frequencies close to the HHG cut-off angular frequency $\omega_c = I_p + 3.17U_p$, alternative methods must be applied to evaluate the contributions to the recollision amplitude of the electrons having energies $E_{k_f} \simeq E_{\text{drift}}^{\text{max}}(\theta_f)$ [54, 55].

We conclude this section by noting that the SFA recollision amplitude discussed here can be generalized to incorporate recollision processes such as the inelastic collision (1.21) and the (e, 2e) reaction (1.22) [4, 68, 69].

6.6 Non-dipole effects

In this section we will show how the SFA can be generalized to account for non-dipole effects that can arise when the laser-field intensity becomes sufficiently high [70–73]. We will limit our discussion to the dynamical regime in which spin and relativistic effects are small and can be therefore neglected.

As discussed in Section 2.8, a classical electron is accelerated not only by the electric-field component of the laser, but also by the magnetic-field component. The latter introduces a drift motion along the laser propagation direction. For not too high intensities, typically 10^{14} to 10^{15} W cm^{-2} at visible or near infra-red wavelengths, this drift is small. As a result, the magnetic force does not need to be taken into account and the coupling of the electron with the laser field is accurately described within the dipole approximation. However, as the laser pulse peak intensity increases, this drift motion can no longer be neglected and the dipole approximation ceases to be valid.

In the semi-classical model, the magnetic-field-induced drift motion of an ionized electron along the laser propagation direction will reduce the probability that it will return to the vicinity of the parent ion, thereby making the radiative recombination and recollision processes less likely. In fact, when taken literally, the semi-classical model predicts that HHG and high-order ATI will be completely suppressed if an ionized electron, initially having zero velocity, acquires a drift velocity in the propagation direction because the electron will never return to its parent core. However, there is a non-zero probability that the electron will ionize with an initial velocity in the direction *opposite* to the propagation direction of the incident laser pulse. When this initial velocity exactly compensates the acquired drift velocity, the electron will return to its parent core. We will show below that it is precisely these electron trajectories that provide the dominant contributions to the SFA non-dipole laser-induced atomic dipole moment.

The starting point of our analysis is the non-dipole (ND), non-relativistic Hamiltonian for the single-electron atomic system that was introduced in Section 2.8, namely

$$H(t) = H_0 + H_{\text{int}}^{\text{ND}}(t), \tag{6.215}$$

where H_0 is the field-free Hamiltonian and

$$H_{\text{int}}^{\text{ND}}(t) = \left[\mathbf{r} - i\frac{z}{c}\nabla\right] \cdot \boldsymbol{\mathcal{E}}(t) \tag{6.216}$$

is the laser–atom interaction Hamiltonian in the "length gauge." The laser pulse propagation direction is taken to be along the z-axis.

We note that, using Equations (2.199) and (2.200), the non-dipole, non-relativistic Gordon–Volkov states in the length gauge can be expressed as

$$|\chi_{\mathbf{k}}^{\text{L,ND}}(t)\rangle = |\boldsymbol{\pi}^{\text{ND}}(\mathbf{k}, t)\rangle \exp\left(-\frac{i}{2}\int_{-\infty}^{t} [\boldsymbol{\pi}^{\text{ND}}(\mathbf{k}, t')]^2 dt'\right), \tag{6.217}$$

where

$$\boldsymbol{\pi}^{\text{ND}}(\mathbf{k}, t) = \boldsymbol{\pi}(\mathbf{k}, t) + \frac{1}{c}\left[\mathbf{k} \cdot \mathbf{A}(t) + \frac{1}{2}\mathbf{A}^2(t)\right]\hat{\mathbf{z}}$$

$$= \mathbf{k} + \mathbf{A}(t) + \frac{1}{c}\left[\mathbf{k} \cdot \mathbf{A}(t) + \frac{1}{2}\mathbf{A}^2(t)\right]\hat{\mathbf{z}}. \tag{6.218}$$

We also introduce the modified action

$$S^{\text{ND}}(\mathbf{k}, t, t') = \frac{1}{2} \int_{t'}^{t} dt'' \left[\pi^{\text{ND}}(\mathbf{k}, t'') \right]^2 + I_{\text{P}}(t - t'). \qquad (6.219)$$

As a first illustration of the application of this non-relativistic, non-dipole theory within the SFA, we will briefly consider the SFA direct ionization amplitude. Using Equations (6.216)–(6.219), it is straightforward to show that Equation (6.78), which gives the ionization amplitude in the dipole approximation, generalizes to

$$
\begin{aligned}
T_{f0}^{\text{SFA,ND}} &= -i \int_{t_0}^{t_1} dt' \langle \chi_{\mathbf{k}_f}^{\text{L,ND}}(t') | H_{\text{int}}^{\text{ND}}(t') | \phi_0(t') \rangle \\
&= - \exp[-iS^{\text{ND}}(\mathbf{k}_f, t', t_0)] \langle \pi^{\text{ND}}(\mathbf{k}_f, t') | \psi_0 \rangle \Big|_{t_0}^{t_1} \\
&\quad + i \int_{t_0}^{t_1} dt' \left(\frac{[\pi^{\text{ND}}(\mathbf{k}_f, t')]^2}{2} + I_{\text{P}} \right) \langle \pi^{\text{ND}}(\mathbf{k}_f, t') | \psi_0 \rangle \\
&\quad \times \exp[-iS^{\text{ND}}(\mathbf{k}_f, t', t_0)].
\end{aligned}
$$

$$(6.220)$$

This expression can be evaluated using the same methods introduced in Section 6.3.1.

Let us now turn our attention to investigating how non-dipole effects modify the laser-induced atomic dipole moment, and hence harmonic generation, in the SFA. As in Section 6.4, we will carry out our analysis for a hydrogenic atomic system. In the non-relativistic, non-dipole theory, the induced atomic dipole moment is given by

$$\mathbf{d}^{\text{ND}}(t) = 2\text{Im} \int_{0}^{t} dt' \int d\mathbf{k} \, \mathbf{d}_{\text{rec}}^{*}[\pi^{\text{ND}}(\mathbf{k}, t)] \exp[-iS^{\text{ND}}(\mathbf{k}, t, t')] d_{\text{ion}}^{\text{ND}}[\pi^{\text{ND}}(\mathbf{k}, t'), t'],$$

$$(6.221)$$

where

$$d_{\text{ion}}^{\text{ND}}(\mathbf{q}, t) = \langle \mathbf{q} | H_{\text{int}}^{\text{ND}}(t) | \psi_0 \rangle, \qquad (6.222)$$

and $\pi^{\text{ND}}(\mathbf{k}, t)$ and $S^{\text{ND}}(\mathbf{k}, t, t')$ are defined by Equations (6.218) and (6.219), respectively. The recombination amplitude $\mathbf{d}_{\text{rec}}(\mathbf{q})$ is the same quantity appearing in the expression (6.152) for the SFA laser-induced atomic dipole moment in the dipole approximation, and is given by Equation (6.153).

We shall, by following exactly the same procedure as in Section 6.4, obtain a semi-analytic expression for the laser-induced atomic dipole moment. We first carry out the integral over $\mathbf{k}$ using the saddle-point method. The result is as follows:

$$\mathbf{d}(t) \simeq 2\mathrm{Im} \frac{(8I_{\mathrm{P}})^{5/4}}{8\pi} \int_0^t dt' \, C^{\mathrm{ND}}(t-t') \mathbf{d}_{\mathrm{rec}}^*[\boldsymbol{\pi}^{\mathrm{ND}}(\mathbf{k}_{\mathrm{s}}^{\mathrm{ND}},t)]$$

$$\times \exp[-iS^{\mathrm{ND}}(\mathbf{k}_{\mathrm{s}}^{\mathrm{ND}}(t,t'))] \left[\frac{\partial}{\partial t'} S^{\mathrm{ND}}(\mathbf{k}_{\mathrm{s}}^{\mathrm{ND}},t,t')\right]^{-1}. \qquad (6.223)$$

In this equation, the factor $C^{\mathrm{ND}}(t-t')$ is given by

$$C^{\mathrm{ND}}(t-t') = (2\pi)^{3/2} \left([\varepsilon + i(t-t')]^3 \left[1 - \frac{1}{c^2} \left(\hat{\epsilon} \cdot \mathbf{k}_{\mathrm{s}}^{\mathrm{ND}}\right)^2\right]\right)^{-1/2} \qquad (6.224)$$

and the saddle momentum $\mathbf{k}_{\mathrm{s}}^{\mathrm{ND}}(t,t')$ is obtained by solving the equation

$$\nabla_{\mathbf{k}} S^{\mathrm{ND}}(\mathbf{k},t,t')|_{\mathbf{k}=\mathbf{k}_{\mathrm{s}}^{\mathrm{ND}}} = 0. \qquad (6.225)$$

To order $1/c$, we find that

$$\mathbf{k}_{\mathrm{s}}^{\mathrm{ND}}(t,t') = \mathbf{k}_{\mathrm{s}}(t,t') + \frac{1}{c}\left[\mathbf{k}_{\mathrm{s}}^2(t,t') - \frac{1}{2(t-t')}\int_{t'}^t dt'' \mathbf{A}^2(t'')\right]\hat{\mathbf{z}}. \qquad (6.226)$$

The quantity $\mathbf{k}_{\mathrm{s}}(t,t')$ is the saddle momentum in the dipole approximation given by Equation (6.161). As discussed above, the saddle momentum has a simple classical interpretation: it is the momentum that an electron in the laser field at time t' would be required to have in order that at the later time t its net displacement is zero.

Let us now evaluate the integral over t' in Equation (6.223) by making use of the same saddle-point approximation that was employed in Section 6.4.1. In particular, we express the saddle times as

$$t_{\mathrm{s}} = t_{\mathrm{d}} + \Delta^{\mathrm{ND}}, \qquad (6.227)$$

where we recall that t_{d} is the detachment time of a classical electron, initially at rest, that interacts with the laser pulse and then returns to its original position at time t. Next, we make the expansion

$$-\frac{\partial}{\partial t'} S(\mathbf{k}_{\mathrm{s}}^{\mathrm{ND}},t,t')\bigg|_{t'=t_{\mathrm{s}}} = s_0^{\mathrm{ND}} + s_1^{\mathrm{ND}}\Delta^{\mathrm{ND}} + \frac{1}{2}s_2^{\mathrm{ND}}(\Delta^{\mathrm{ND}})^2 + \cdots. \qquad (6.228)$$

Retaining the dominant contributions yields

$$s_0^{\mathrm{ND}} = s_0 + \frac{1}{2}[\pi_z^{\mathrm{ND}}(\mathbf{k}_{\mathrm{s}}^{\mathrm{ND}},t_{\mathrm{d}})]^2, \qquad (6.229)$$

$$s_1^{\mathrm{ND}} = 0, \qquad (6.230)$$

$$s_2^{\mathrm{ND}} = s_2, \qquad (6.231)$$

and

$$\Delta^{\text{ND}} = -i \left(\frac{2s_0^{\text{ND}}}{s_2^{\text{ND}}} \right)^{1/2}, \tag{6.232}$$

where s_0 and s_2 are defined by Equations (6.169) and (6.170), respectively, and

$$\pi_z^{\text{ND}}(\mathbf{k}_s^{\text{ND}}, t_d) = \hat{\mathbf{z}} \cdot \boldsymbol{\pi}^{\text{ND}}(\mathbf{k}_s^{\text{ND}}, t_d)$$

$$= -\frac{1}{2c(t - t_d)} \int_{t_d}^t dt' \left| \mathbf{A}(t') - \mathbf{A}(t_d) \right|^2 \tag{6.233}$$

is the component of $\boldsymbol{\pi}^{\text{ND}}(\mathbf{k}_s, t_d)$ in the propagation direction. This key quantity will be discussed in more detail below.

With the above approximations, the laser-induced atomic dipole moment in the non-dipole SFA can be expressed as

$$\mathbf{d}^{\text{ND}}(t) = 2\text{Im} \sum_{t_d} \mathbf{a}_{\text{rec}}^{\text{ND}}(t, t_d)[a_{\text{free}}^{\text{ND}}(t, t_d)a_{\text{ion}}^{\text{ND}}(t, t_d)]^*, \tag{6.234}$$

with the ionization, "free" motion and recombination amplitudes, respectively, given by

$$a_{\text{ion}}^{\text{ND}}(t, t_d) = \frac{(8I_P)^{5/4}}{8 \left(2s_0^{\text{ND}} s_2^{\text{ND}} \right)^{1/2}} \exp\left[-\frac{1}{3} \left(\frac{8(s_0^{\text{ND}})^3}{s_2^{\text{ND}}} \right)^{1/2} \right], \tag{6.235}$$

$$a_{\text{free}}^{\text{ND}}(t, t_d) = C^{\text{ND}}(t - t_s) \exp[-iS^{\text{ND}}(\mathbf{k}_s^{\text{ND}}, t, t_d)] \tag{6.236}$$

and

$$\mathbf{a}_{\text{rec}}^{\text{ND}}(t, t_d) = \mathbf{d}_{\text{rec}}[\boldsymbol{\pi}(\mathbf{k}_s^{\text{ND}}, t)]. \tag{6.237}$$

These amplitudes are very similar to the amplitudes (6.177)–(6.179) obtained in the dipole approximation. Aside from the prefactor in the ionization amplitude (6.235), the non-dipole corrections appearing in the amplitudes are of order $1/c^2$. However, the most important non-dipole correction is due to the quantity s_0^{ND} appearing in the argument of the exponential function in the ionization amplitude (6.235).

In order to understand the significance of this correction, let us generalize the discussion of Section 2.8 on the non-dipole dynamics of a classical electron in a laser field. We start with the classical analog of the Hamiltonian operator in Equation (6.215), namely

$$H_{\text{cl}} = \frac{1}{2} \left[\mathbf{p}_{\text{cl}}(t) + \mathbf{A}(t) \right]^2 + \frac{1}{c} [\hat{\mathbf{z}} \cdot \mathbf{r}(t)] \left[\mathbf{p}_{\text{cl}}(t) + \mathbf{A}(t) \right] \cdot \boldsymbol{\mathcal{E}}(t), \tag{6.238}$$

obtained by replacing the operator $-i\nabla$ with the classical canonical momentum $\mathbf{p}_{\mathrm{cl}}(t)$. Solving the classical equations of motion, one finds that if, at time t_0, the electron is located at the origin and has a velocity $\mathbf{v}_0$, then at time t its position and velocity are, respectively,

$$\mathbf{r}(t) = \int_{t_0}^{t} dt' \, \mathbf{v}(t') \tag{6.239}$$

and

$$\mathbf{v}(t) = v_\epsilon(t)\hat{\boldsymbol{\epsilon}} + v_z(t)\hat{\mathbf{z}}, \tag{6.240}$$

with

$$v_\epsilon(t) = \hat{\boldsymbol{\epsilon}} \cdot \mathbf{v}_0 + [\mathbf{A}(t) - \mathbf{A}(t_0)] \cdot \hat{\mathbf{z}} + \frac{1}{c} E(t) \int_{t_0}^{t} dt' v_k(t') \tag{6.241}$$

and

$$v_z(t) = \hat{\mathbf{k}} \cdot \mathbf{v}_0 + \frac{1}{2c} \left(v_\epsilon(t_0) + [\mathbf{A}(t) - \mathbf{A}(t_0)] \cdot \hat{\mathbf{z}} \right)^2 - \frac{1}{2c} v_\epsilon^2(t_0). \tag{6.242}$$

A wave packet formed by a linear superposition of non-dipole Volkov waves (2.199) that is, at time t_0, localized at the origin and has velocity $\mathbf{v}_0$ follows the classical trajectory (6.239) [73].

Let us consider an electron that is detached from the core at time t_0. From Equations (6.241) and (6.242), we see that to order $1/c$ two conditions must be fulfilled to ensure that the electron will return to the core at time t. Firstly, t_0 must be equal to one of the saddle times t_d defined by Equation (6.165). Secondly, the initial velocity of the electron at time t_d should be $\mathbf{v}_0 = \pi_z^{\mathrm{ND}}(\mathbf{k}_s^{\mathrm{ND}}, t_d)\hat{\mathbf{z}}$, where $\pi_z^{\mathrm{ND}}(\mathbf{k}_s^{\mathrm{ND}}, t_d)$ is defined by Equation (6.233). In the classical model, $\pi_z^{\mathrm{ND}}(\mathbf{k}_s^{\mathrm{ND}}, t_d)$ is thus the initial velocity that the electron must have along the laser-field propagation direction $\hat{\mathbf{z}}$ at the detachment time t_d in order to return to the core at time t. As such, this initial velocity compensates exactly the displacement imparted by the magnetic field component of the laser field on the free electron as it oscillates in the laser field.

We now return to the ionization amplitude (6.235). Delone and Krainov [74] have derived an expression giving the tunneling ionization rate of a hydrogenic system for which the electron exits the barrier with a momentum $k_\perp$ perpendicular to the electric field direction. This rate can be expressed in the form

$$\Gamma_{\mathrm{DC}}(\mathcal{E}_0, k_\perp) = \frac{4}{\pi} \frac{(2I_{\mathrm{P}})^3}{\mathcal{E}_0^2} \exp\left[-\frac{2}{3} \frac{(2I_{\mathrm{P}} + k_\perp^2)^{3/2}}{\mathcal{E}_0} \right], \tag{6.243}$$

where the prefactor has been determined to ensure that the integral of Γ_{DC} over $k_\perp$ in the plane normal to the electric-field direction yields the total ionization rate of a hydrogenic system in a static electric field, as given by Equation (6.22). We note that the magnitude squared of the the SFA ionization amplitude (6.104) has the same exponential dependence on $k_\perp$ as the Delone and Krainov tunneling ionization rate (6.243). In the quasi-static tunneling approximation, we can replace $\mathcal{E}_0$ by the instantaneous electric-field amplitude $|\mathcal{E}(t)|$. Setting $k_\perp = \pi_z^{ND}(\mathbf{k}_s^{ND}, t)$ and $t = t_d$, and comparing Equation (6.243) to the non-dipole ionization amplitude (6.235), we see that the ionization rate (6.243) varies with the electric-field amplitude in the same way as the magnitude squared of the ionization amplitude. Moreover, it has the same exponential dependence on I_P and $k_\perp$. The amplitude (6.235) can therefore be interpreted as the quasi-static tunneling amplitude for an electron exiting the barrier with initial momentum $\pi_z^{ND}(\mathbf{k}_s^{ND}, t)$. As long as $[\pi_z^{ND}(\mathbf{k}_s^{ND}, t)]^2 \ll 2I_P$, the magnitude of the ionization amplitude increases as the laser electric-field strength increases. It reaches a maximum when $[\pi_z^{ND}(\mathbf{k}_s^{ND}, t)]^2$ approaches $2I_P$ and subsequently decreases as the electric-field strength increases. Hence, as expected, for sufficiently high laser intensities, harmonic emission will be suppressed due to non-dipole effects.

This suppression, already mentioned in Section 1.4, is illustrated in Fig. 6.7, which shows the results of calculations performed by Chirila *et al.* [72]. The upper plots show the magnitude squared of the Fourier transform of the SFA dipole moment acceleration along the polarization direction as a function of the harmonic order for the ions Li^{2+} ($I_P = 122.5$ eV) and Be^{3+} ($I_P = 217.7$ eV) in the dipole approximation (gray curves) and in the non-dipole, non-relativistic approximation (black curves). The ions were taken to interact with the same linearly polarized and monochromatic laser field of intensity 1.83×10^{17} W cm^{-2} and wavelength 800 nm. As can be seen in Fig. 6.7 the results obtained in the dipole approximation clearly overestimate photon emission rates, with the effect being more pronounced in Be^{3+}. We point out that it is necessary to consider positive ions having large ionization potentials when investigating non-dipole effects in HHG. Neutral species interacting with a Ti:sapphire laser pulse will ionize at intensities that are lower than those required to give rise to non-dipole effects.

The strong dependence on I_P, and in fact all of the main differences between the dipole and non-dipole spectra, can be understood within the framework of the recollision model. As in our discussion in Section 6.4.2, we need only consider two key quantities, namely the effective tunneling rate at the electron detachment time t_d and its kinetic energy when it returns to the core. In the lower plots of Fig. 6.7, the tunneling rate

$$W_{ion}(t, t_d) = \Gamma_{DC}[|\mathcal{E}(t_d)|, \pi_z^{ND}(\mathbf{p}_s^{ND}, t_d)] \tag{6.244}$$

is given for electron trajectories that return to the core as a function of its kinetic energy expressed as the harmonic order of the photon emitted at recombination.

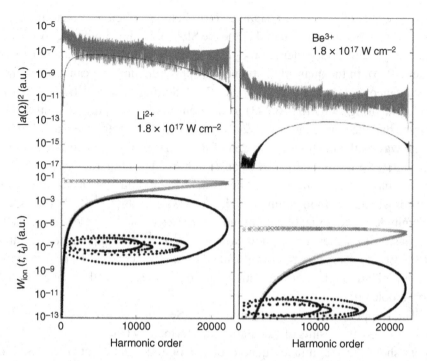

Figure 6.7. Upper plots: Magnitude squared of the Fourier transform of the dipole moment acceleration (in a.u.) as a function of the harmonic order. Results for the ions Li^{2+} and Be^{3+} in the dipole approximation (gray curves) and in the non-dipole, non-relativistic approximation (black curves) are shown. The ions are irradiated by a monochromatic laser field of intensity $1.83 \times 10^{17}\,W\,cm^{-2}$ and wavelength 800 nm. Lower plots: ionization rate $W_{ion}(t, t_d)$, defined by Equation (6.244), plotted for electron trajectories that return to the core as a function of the harmonic order of the photon emitted. The dipole and non-dipole values of $W_{ion}(t, t_d)$ are indicated by crosses and circles, respectively. (From C. C. Chirilă, N. J. Kylstra, R. M. Potvliege and C. J. Joachain, *Phys. Rev. A* **66**, 063411 (2002).)

The dipole and non-dipole values of $W_{ion}(t, t_d)$ are indicated by crosses and circles, respectively.

In the dipole approximation, the ionization amplitude, given by Equation (6.177), is largest for the long trajectories, since electrons having short trajectories are ionized at lower electric fields. Therefore, short trajectories tend to contribute less to the harmonic spectrum. However, when non-dipole effects are taken into account, the opposite is true. This is due to the fact that $\pi_z^{ND}(\mathbf{k}_s, t_d)$ is larger for the long trajectories than for the short ones. As a consequence, the exponential dependence of the ionization amplitude on the initial transverse momentum $k_\perp = \pi_z^{ND}(\mathbf{k}_s, t_d)$ means that this amplitude is typically smaller for the longer trajectories than for the short ones. Therefore, the latter end up dominating the spectrum. We also note that the oscillations in the dipole HHG spectrum largely disappear at most emission frequencies because only one set of trajectories (the short ones) contribute to emission. Finally, the reduction in HHG is proportionally larger for Be^{3+} than for Li^{2+} because the ionization amplitude varies faster with $k_\perp$ when I_P is larger.

We conclude this section by noting that harmonic generation has also been investigated using relativistic versions of the SFA [23, 75].

References

[1] L. V. Keldysh, *Sov. Phys. JETP* **20**, 1307 (1965).

[2] P. Salières, A. L'Huillier, P. Antoine and M. Lewenstein, *Adv. At. Mol. Opt. Phys.* **41**, 83 (1999).

[3] W. Becker, F. Grasbon, R. Kopold, D. B. Milosevic, G. G. Paulus and H. Walther, *Adv. At. Mol. Opt. Phys.* **48**, 35 (2002).

[4] A. Becker and F. H. M. Faisal, *J. Phys. B* **38**, R1 (2005).

[5] M. Lewenstein and A. L'Huillier, Principles of single atom physics: high-order harmonic generation, above-threshold ionization and non-sequential ionization. In T. Brabec, ed., *Strong Field Laser Physics*, Springer Series in Optical Sciences 134 (New York: Springer, 2009), p. 147.

[6] B. H. Bransden and C. J. Joachain, *Quantum Mechanics*, 2nd edn (Harlow, UK: Prentice Hall-Pearson, 2000).

[7] I. W. Herbst and B. Simon, *Phys. Rev. Lett.* **41**, 67 (1978).

[8] I. W. Herbst and B. Simon, *Phys. Rev. Lett.* **41**, 1759 (1978).

[9] S. Graffi and V. Grecchi, *Commun. Math. Phys.* **79**, 91 (1981).

[10] A. Maquet, S. I. Chu and W. P. Reinhardt, *Phys. Rev. A* **27**, 2946 (1983).

[11] C. A. Nicolaides and S. I. Themelis, *Phys. Rev. A* **45**, 349 (1992).

[12] M. Pont, R. Shakeshaft and R. M. Potvliege, *Phys. Rev. A* **42**, 6969 (1990).

[13] M. Pont, R. M. Potvliege, R. Shakeshaft and Z. J. Teng, *Phys. Rev. A* **45**, 8235 (1992).

[14] R. Shakeshaft, R. M. Potvliege, M. Dörr and W. E. Cooke, *Phys. Rev. A* **42**, 1656 (1990).

[15] L. D. Landau and E. M. Lifshitz, *Quantum Mechanics* (Reading, Mass.: Addison Wesley, 1958).

[16] B. M. Smirnov and M. I. Chibisov, *Sov. Phys. JETP* **22**, 585 (1966).

[17] C. Z. Bisgaard and L. B. Madsen, *Am. J. Phys.* **72**, 249 (2004).

[18] A. A. Radzig and B. M. Smirnov, *References Data on Atoms, Molecules and Ions* (Berlin: Springer-Verlag, 1985).

[19] V. D. Mur, B. M. Karnakov and V. S. Popov, *Sov. Phys. JETP* **88**, 286 (1999).

[20] D. R. Hartree, *Proc. Camb. Phil. Soc.* **24**, 89 (1927).

[21] I. Martin and G. Simons, *J. Chem. Phys.* **62**, 4799 (1975).

[22] M. V. Ammosov, N. B. Delone and V. P. Krainov, *Sov. Phys. JETP* **64**, 1191 (1986).

[23] V. S. Popov, *Phys. Usp.* **47**, 855 (2004).

[24] A. Scrinzi, M. Geissler and T. Brabec, *Phys. Rev. Lett.* **83**, 706 (1999).

[25] E. T. Copson, *Asymptotic Expansions* (Cambridge: Cambridge University Press, 1965).

[26] A. M. Perelomov, V. S. Popov and M. V. Terent'ev, *Sov. Phys. JETP* **23**, 924 (1966).

[27] A. M. Perelomov and V. S. Popov, *Sov. Phys. JETP* **25**, 336 (1967).

[28] M. Büttiker and R. Landauer, *Phys. Rev. Lett.* **49**, 1739 (1982).

[29] F. A. Ilkov, J. E. Decker and S. L. Chin, *J. Phys. B* **25**, 4005 (1992).

[30] A. M. Perelomov, V. S. Popov and M. V. Terent'ev, *Sov. Phys. JETP* **24**, 207 (1967).

[31] A. Erdélyi, *Asymptotic Expansions* (New York: Dover, 1956).

[32] W. Becker, A. Lohr, M. Kleber and M. Lewenstein, *Phys. Rev. A* **56**, 645 (1997).

[33] F. H. M. Faisal, *J. Phys. B* **6**, L89 (1973).

[34] H. R. Reiss, *Phys. Rev. A* **22**, 1786 (1980).

[35] G. F. Gribakin and M. Y. Kuchiev, *Phys. Rev. A* **55**, 3760 (1997).

[36] M. V. Frolov, N. L. Manakov, E. A. Pronin and A. F. Starace, *Phys. Rev. Lett.* **91**, 053003 (2003).

[37] M. Abramowitz and I. A. Stegun, *Handbook of Mathematical Functions* (New York: Dover, 1970).

[38] M. Y. Ivanov, T. Brabec and N. Burnett, *Phys. Rev. A* **54**, 742 (1996).

[39] V. P. Krainov, *J. Opt. Soc. Am. B* **14**, 425 (1997).

[40] A. Lohr, M. Kleber, R. Kopold and W. Becker, *Phys. Rev. A* **55**, R4003 (1997).

[41] D. B. Milosevic and F. Ehlotzky, *Phys. Rev. A* **57**, 5002 (1998).

[42] C. C. Chirilă and R. M. Potvliege, *Phys. Rev. A* **71**, 021402(R) (2005).

[43] V. P. Krainov and B. Shokri, *Sov. Phys. JETP* **80**, 657 (1995).

[44] S. Gordienko and J. Meyer-ter-Vehn. In O. N. Krokhin, S. Y. Gus'kov and Y. A. Merkul'ev, eds., *Laser Interaction with Matter*, Proceedings of SPIE 5228 (Bellingham: Wash.: SPIE, 2003), p. 416.

[45] A. de Bohan, B. Piraux, L. Ponce, R. Taïeb, V. Véniard and A. Maquet, *Phys. Rev. Lett.* **89**, 113002 (2002).

[46] M. Dörr, R. M. Potvliege, D. Proulx and R. Shakeshaft, *Phys. Rev. A* **42**, 4138 (1990).

[47] D. Bauer, D. B. Milosevic and W. Becker, *Phys. Rev. A* **72**, 023415 (2005).

[48] F. H. M. Faisal, *J. Phys. B* **40**, F145 (2007).

[49] D. B. Milosevic. In B. Piraux and K. Rząžewski eds., *Super-Intense Laser-Atom Physics* (Dordrecht: Kluwer Academic, 2001), p. 229.

[50] M. Lewenstein, P. Balcou, M. Y. Ivanov, A. L'Huillier and P. B. Corkum, *Phys. Rev. A* **49**, 2117 (1994).

[51] T. Brabec and F. Krausz, *Rev. Mod. Phys.* **72**, 545 (2000).

[52] F. Krausz and M. Ivanov, *Rev. Mod. Phys.* **81**, 163 (2009).

[53] P. Salières, P. Antoine, A. de Bohan and M. Lewenstein, *Phys. Rev. Lett.* **81**, 5544 (1998).

[54] C. F. D. Faria, H. Schomerus and W. Becker, *Phys. Rev. A* **66**, 043413 (2002).

[55] D. B. Milosevic and W. Becker, *Phys. Rev. A* **66**, 063417 (2002).

[56] P. B. Corkum, *Phys. Rev. Lett.* **71**, 1994 (1993).

[57] K. C. Kulander, K. J. Schafer and J. L. Krause. In B. Piraux, A. L'Huillier and K. Rząžewski, eds., *Super-Intense Laser-Atom Physics* (New York: Plenum Press, 1993), p. 95.

[58] R. M. Potvliege, N. J. Kylstra and C. J. Joachain, *J. Phys. B* **33**, L743 (2000).

[59] W. Becker, S. Long and J. K. McIver, *Phys. Rev. A* **50**, 1540 (1994).

[60] P. Antoine, A. L'Huillier, M. Lewenstein, P. Salières and B. Carré, *Phys. Rev. A* **53**, 1725 (1996).

[61] D. B. Milosevic, *J. Phys. B* **33**, 2479 (2000).

[62] K. S. Budil, P. Salières, A. L'Huillier, T. Ditmire and M. D. Perry, *Phys. Rev. A* **48**, R3437 (1993).

[63] P. Dietrich, N. H. Burnett, M. Ivanov and P. B. Corkum, *Phys. Rev. A* **50**, R3585 (1994).

[64] Y. Liang, M. V. Ammosov and S. L. Chin, *J. Phys. B* **27**, 1269 (1994).

[65] C. C. Chirilă, C. J. Joachain, N. J. Kylstra and R. M. Potvliege, *Phys. Rev. Lett.* **93**, 243603 (2004).

[66] P. Salières, B. Carré, L. Le Déroff *et al.*, *Science* **292**, 902 (2001).

[67] G. Paulus, W. Becker, W. Nicklich and H. Walther, *J. Phys. B* **27**, L703 (1994).

[68] A. Becker and F. H. M. Faisal, *J. Phys. B* **29**, L197 (1996).

[69] S. Popruzhenko and S. Goreslavskii, *J. Phys. B* **34**, L329 (2001).

[70] M. W. Walser, C. H. Keitel, A. Scrinzi and T. Brabec, *Phys. Rev. Lett.* **85**, 5082 (2000).

[71] N. J. Kylstra, R. M. Potvliege and C. J. Joachain, *J. Phys. B* **34**, L55 (2001).

[72] C. C. Chirilă, N. J. Kylstra, R. M. Potvliege and C. J. Joachain, *Phys. Rev. A* **66**, 063411 (2002).

[73] N. J. Kylstra, C. J. Joachain and M. Dörr. In D. Batani, C. J. Joachain, S. Martellucci and A. N. Chester, eds., *Atoms, Solids and Plasmas in Super-Intense Laser Fields* (New York: Kluwer Academic-Plenum Publishers, 2001), p. 15.

[74] N. B. Delone and V. P. Krainov, *J. Opt. Soc. Am. B* **8**, 1207 (1991).

[75] D. B. Milosevic, S. X. Hu and W. Becker, *Phys. Rev. A* **63**, R011403 (2001).

7

The high-frequency regime

In this chapter, we shall analyze the interaction of atoms with intense laser fields whose frequency is much larger than the threshold frequency for one-photon ionization. We begin in Section 7.1 by discussing the high-frequency Floquet theory (HFFT) within the framework of the non-relativistic theory of laser–atom interactions in the dipole approximation. In Section 7.2, the HFFT is applied to study the structure of atomic hydrogen in intense, high-frequency laser fields. An interesting prediction of the HFFT is *atomic stabilization*, whereby the ionization rate of an atom interacting with an intense, high-frequency laser field decreases as the laser intensity increases. This phenomenon is analyzed in Section 7.3, where we discuss ionization rates obtained within the HFFT as well as from *ab initio* Floquet calculations. We then consider investigations of stabilization based on the direct numerical integration of the time-dependent Schrödinger equation (TDSE). Finally, we examine the influence of non-dipole and relativistic effects on atomic stabilization. Detailed reviews of the HFFT and stabilization have been given by Gavrila [1–3].

7.1 High-frequency Floquet theory

The HFFT is based on analyzing the atom–laser field interaction in the accelerated, or Kramers–Henneberger (K–H), frame [4, 5]. It was developed by Gavrila and Kaminski [6] to study electron scattering by a potential in the presence of a high-frequency laser field and generalized by Gavrila [7] to investigate the atomic structure and ionization of decaying dressed states. The HFFT was applied to study laser-assisted electron scattering by a Yukawa (screened Coulomb) potential [8], a Coulomb potential [9, 10] and to analyze the structure [11–14] and ionization [13, 14] of atomic hydrogen in an intense high-frequency laser field. Atomic units will be used, unless otherwise stated.

Let us consider an atom with one active electron in a spatially homogeneous, monochromatic laser field of arbitrary polarization. The corresponding electric field $\mathcal{E}(t)$ is given by Equation (2.114) and the associated vector potential $\mathbf{A}(t)$ by Equation (2.115). We recall that a classical electron interacting with this laser field has a "quiver" motion characterized by the displacement vector $\boldsymbol{\alpha}(t)$, given in the present case by Equation (2.116).

In Section 4.5 we formulated for this one-electron problem the Floquet theory in the accelerated (K–H) frame. The wave function of the system in this frame, $\Psi^A(\mathbf{r}, t)$, is a solution of the TDSE (4.202). Expanding the wave function in the Floquet–Fourier form (4.203), its harmonic components $F_n^A(\mathsf{E}^A; \mathbf{r})$ are found to satisfy the infinite system of time-independent coupled equations (4.208), which we rewrite in the form

$$\left[\left(\frac{\mathbf{p}^2}{2} + V_0(\alpha_0, \mathbf{r})\right) - \mathsf{E}^A - n\omega\right] F_n^A(\mathsf{E}^A; \mathbf{r}) + \sum_{\substack{s=-\infty \\ s \neq n}}^{\infty} V_{n-s}(\alpha_0, \mathbf{r}) F_s^A(\mathsf{E}^A; \mathbf{r}) = 0,$$

(7.1)

where $\alpha_0 = \mathcal{E}_0/\omega^2$ is the excursion amplitude of the electron in the laser field and $\mathcal{E}_0$ is the electric-field strength. The quantities $V_n(\alpha_0, \mathbf{r})$, given by Equation (4.205), are the Fourier components associated with the oscillating interaction potential $V[\mathbf{r} + \boldsymbol{\alpha}(t)]$ in the K–H frame. As we shall see, of particular importance is the potential $V_0(\alpha_0, \mathbf{r})$, which appears on the left-hand side of Equation (7.1). This is the static (time-averaged) "dressed" potential

$$V_0(\alpha_0, \mathbf{r}) = (T)^{-1} \int_0^T V[\mathbf{r} + \boldsymbol{\alpha}(t)] \mathrm{d}t,$$

(7.2)

where $T = 2\pi/\omega$ is the laser field period.

In what follows, we shall solve Equation (7.1) by using an iteration scheme. It is convenient to rewrite this equation in terms of Dirac kets as follows:

$$\left(H_n^A - \mathsf{E}^A\right) | F_n^A \rangle + \sum_{\substack{s=-\infty \\ (s \neq n)}}^{\infty} V_{n-s}(\alpha_0) | F_s^A \rangle = 0,$$

(7.3)

where $\langle \mathbf{r} | F_n^A \rangle \equiv F_n^A(\mathsf{E}^A; \mathbf{r})$ and

$$H_n^A = \frac{\mathbf{p}^2}{2} + V_0(\alpha_0, \mathbf{r}) - n\omega.$$

(7.4)

Defining the Floquet–Fourier matrix operators H^A and $V^A(\alpha_0)$, whose matrix elements are given by

$$[H^A]_{ns} = H_n^A \delta_{ns}$$

(7.5)

and

$$[V^A(\alpha_0)]_{ns} = V_{n-s}(\alpha_0)(1 - \delta_{ns}),$$

(7.6)

respectively, Equation (7.3) can be now written in the form

$$\left[H^A + V^A(\alpha_0) - \mathsf{E}^A \right] | F^A \rangle = 0. \tag{7.7}$$

Let us obtain a formal solution of this equation, in the form of an infinite series, by applying the same technique used in Section 4.2.3. We start by introducing a "zeroth-order" Floquet vector $| F^{A(0)} \rangle$, solution of the equation

$$H^A | F^{A(0)} \rangle = \mathsf{E}^{(0)} | F^{A(0)} \rangle \tag{7.8}$$

with components

$$[| F^{A(0)} \rangle]_n = | F_0^{A(0)} \rangle \delta_{n0}, \tag{7.9}$$

so that only the $n = 0$ component is non-zero. This state will be taken to be an eigenstate of $H_0^A = \mathbf{p}^2/2 + V_0(\alpha_0, \mathbf{r})$, so that

$$H_0^A | F_0^{A(0)} \rangle = \mathsf{E}^{(0)} | F_0^{A(0)} \rangle. \tag{7.10}$$

As in Section 4.2.3, the projection operators

$$P = | F^{A(0)} \rangle \langle F^{A(0)} | \tag{7.11}$$

and

$$Q = 1 - P \tag{7.12}$$

that project onto the zeroth-order solution and its complement space, respectively, are defined. The projection of the zeroth-order solution onto the full solution $| F^A \rangle$ is denoted by

$$\beta = \langle F^{A(0)} | F^A \rangle. \tag{7.13}$$

The solution of Equation (7.7) can then be expressed as

$$| F^A \rangle = \beta \sum_{l=0}^{\infty} \left(Q \, G^A(z) \, Q \, [z - \mathsf{E}^A + V^A(\alpha_0)] \right)^l | F^{A(0)} \rangle, \tag{7.14}$$

where the resolvent operator is given by

$$G^A(z) = \frac{1}{z - H^A}. \tag{7.15}$$

Let us investigate Equation (7.14) in the intense-field, high-frequency limit. We define this limit as the one in which $\omega \to \infty$ while the electric-field strength $\mathcal{E}_0$ is increased in such a way that α_0 remains finite. Referring to Equations (7.4) and (7.15), we note that in Equation (7.14) the only explicit dependence on ω occurs via the matrix operator $G^A(z)$. Focusing our attention on the components of this operator with $n \neq 0$, it is seen that as $\omega \to \infty$ the dominant behavior is given by

$$[G^A(z)]_{nn} \to (n\omega)^{-1}I, \quad n \neq 0, \tag{7.16}$$

where I is the unit operator in the atomic subspace. The $n = 0$ component of $G^A(z)$ does not depend on ω, and therefore we have

$$\lim_{\omega \to \infty} [G^A(z)]_{nn} = \left(1 - |F_0^{A(0)}\rangle\langle F_0^{A(0)}|\right) \frac{1}{z - H_0^A} \delta_{n0}. \tag{7.17}$$

As a consequence,

$$\lim_{\omega \to \infty} |F^A\rangle = \beta|F^{A(0)}\rangle. \tag{7.18}$$

The central result of the HFFT can therefore be stated as follows. In the intense-field, high-frequency limit, the Floquet vector describing the atom interacting with the laser field in the K–H frame is simply an eigenstate of the Hamiltonian H_0^A. We will explore the consequences of this important result in the following sections.

The Hamiltonian H_0^A plays a central role in determining the behavior of the atom at high laser intensities and high laser frequencies. It is therefore convenient to introduce the set of eigenvalues of H_0^A, which we denote by $w_j^{(0)}$, and their corresponding eigenvectors, $|u_j\rangle$, so that we have the eigenvalue equation

$$H_0^A|u_j\rangle = w_j^{(0)}|u_j\rangle. \tag{7.19}$$

We note that the eigenvalues $w_j^{(0)}$ are real since the operator H_0^A is Hermitian. We also remark that H_0^A depends on the laser field only through the excursion amplitude α_0.

We will now consider the case in which ω is large but finite. Making use of Equation (7.16), we can view Equation (7.14) as a power series expansion in ω^{-1}. Defining the Green's operator

$$G^{A(+)}(E^A) = \frac{1}{E^A - H^A + i\epsilon}, \quad \epsilon \to 0^+, \tag{7.20}$$

and setting $\beta = 1$, to first order in ω^{-1} the Floquet vector is given by

$$|F^{A(1)}\rangle = |F^{A(0)}\rangle + Q\,G^{A(+)}(E^{(0)})\,Q\,V^A(\alpha_0)|F^{A(0)}\rangle, \tag{7.21}$$

with the corresponding Floquet quasi-energy given by

$$E^{(1)} = E^{(0)} + \langle F^{A(0)}|V^A(\alpha_0)Q\,G^{A(+)}(E^{(0)})\,Q\,V^A(\alpha_0)|F^{A(0)}\rangle. \tag{7.22}$$

Let us denote by $|F_i^{A(0)}\rangle$ a particular zeroth-order Floquet vector with components

$$|F_{ni}^{A(0)}\rangle = |u_i\rangle\delta_{n0} \tag{7.23}$$

and energy $E_i^{(0)} = w_i^{(0)}$. We will now separate the first-order quasi-energy (7.22) into its real and imaginary parts. This can be accomplished by using the fact that

$$\lim_{\epsilon \to 0^+} \frac{1}{x - x_0 + i\epsilon} = P\left(\frac{1}{x - x_0}\right) - i\pi \delta(x - x_0), \tag{7.24}$$

where P denotes the principal value. Making use of this result and the closure relation

$$\sum_j |u_j\rangle\langle u_j| = I \tag{7.25}$$

yields the following expression for the diagonal components of the Green's operator (7.20):

$$[G^{A(+)}(E_i^{(0)})]_{nn} = \sum_j \left[P\left(\frac{1}{E_i^{(0)} + n\omega - w_j^{(0)}}\right) - i\pi \delta\left(E_i^{(0)} + n\omega - w_j^{(0)}\right) \right] |u_j\rangle\langle u_j|$$

$$= \sum_j P\left(\frac{|u_j\rangle\langle u_j|}{E_i^{(0)} + n\omega - w_j^{(0)}}\right) - i\pi |u_{k_n}^{(-)}\rangle\langle u_{k_n}^{(-)}|. \tag{7.26}$$

Here we have introduced the scattering solutions of Equation (7.19),

$$H_0^A u_{k_n}^{(\pm)}(\mathbf{r}) = (E_i^{(0)} + n\omega) u_{k_n}^{(\pm)}(\mathbf{r}), \tag{7.27}$$

which behave for $r \to \infty$ as a (modified) plane wave corresponding to the wave vector $\mathbf{k}_n$, with

$$\frac{k_n^2}{2} = E_i^{(0)} + n\omega, \tag{7.28}$$

plus an outgoing (+) or incoming (−) spherical wave, respectively, normalized to a delta function in wave vector space.

Using Equation (7.26), the Floquet quasi-energy (7.22) to order ω^{-1} can be expressed as

$$E_i^{(1)} = E_i^{(0)} + \Delta_i^{(1)} - i\frac{\Gamma_i^{(1)}}{2}, \tag{7.29}$$

where the quantity $\Delta_i^{(1)}$ is the AC Stark shift correction to the unperturbed energy $E_i^{(0)} = w_i^{(0)}$, namely

$$\Delta_i^{(1)} = \sum_{\substack{s=-\infty \\ s \neq 0}}^{\infty} \sum_{j \neq i} P \frac{|\langle u_j | V_s(\alpha_0) | u_i \rangle|^2}{E_i^{(0)} + s\omega - w_j^{(0)}} \tag{7.30}$$

and $\Gamma_i^{(1)}$ is the total energy width of the level, given by

$$\Gamma_i^{(1)} = \sum_{n=n_0}^{\infty} \Gamma_{ni}^{(1)}$$

$$= 2\pi \sum_{n=n_0}^{\infty} k_n \int d\Omega |\langle u_{k_n}^{(-)} |V_n(\alpha_0)|u_i\rangle|^2, \qquad (7.31)$$

where the quantity $\Gamma_{ni}^{(1)}$ is, in a.u., the n-photon ionization rate of the atom from the state $|u_i\rangle$. In this equation n_0 is the smallest integer such that the left-hand side of Equation (7.28) is non-negative, so that the corresponding channel is open. We remark that the width $\Gamma_i^{(1)}$ depends on α_0 and ω. However, the only dependence on ω is due to the fact that the quantity k_n satisfies the energy conserving relation (7.28). If, for a fixed value of α_0, we take the limit $\omega \to \infty$ (so that $k_n \to \infty$), the rate (7.31) tends to zero because of the rapid oscillations of $u_{k_n}^{(-)*}(\mathbf{r})$ in the integral. In addition, in the limit that $\omega \to \infty$, one has $\Delta_i^{(1)} \to 0$, so that the complex Floquet quasi-energy $\mathsf{E}_i^{(1)}$ becomes real, and reduces to $w_i^{(0)}$.

The iteration scheme carried out above to first order in ω^{-1} can be pursued to higher orders. It can be shown [1] that the terms introduced by each iteration in the expressions of the energy shifts become increasingly smaller provided that

$$\omega \gg |w_i^{(0)}(\alpha_0)|. \qquad (7.32)$$

Here $w_i^{(0)}(\alpha_0)$ is the lowest eigenvalue of Equation (7.19) within the manifold of states to which the initial state $|u_i\rangle$ belongs. The reason for the convergence of the iteration scheme is that, as seen above, only the Green's operators $G_0^{A(+)}(E_i^{(0)}+n\omega)$ with $n \neq 0$ depend on the angular frequency ω, and that their dominant behavior for $\omega \to \infty$, at fixed α_0, is given by Equation (7.16). The iteration scheme therefore proceeds essentially in powers of ω^{-1} at fixed α_0.

We note that the condition (7.32) is a *high-frequency requirement*, which implicitly allows one-photon ionization. On the other hand, the value of α_0 may be chosen arbitrarily but must remain finite. The condition (7.32) is sufficient for the validity of the iteration scheme, but may not be necessary. We emphasize that the energy $w_i^{(0)}(\alpha_0)$ corresponds to the laser-field-modified Hamiltonian H_0^A. As we shall see below, the quantity $|w_i^{(0)}(\alpha_0)|$ is a decreasing function of α_0, a fact which eases the fulfilment of the condition (7.32) at high values of α_0.

Since there is no restriction on the value of α_0 (except for being finite), the HFFT discussed above applies for $\alpha_0 \ll 1$, which is the domain of perturbation theory, as well as for $\alpha_0 \gg 1$, where the theory is non-perturbative. Because $\alpha_0 = \mathcal{E}_0/\omega^2$, it is clear that the value of α_0 depends more sensitively on the angular frequency ω than on the intensity I. By choosing very high values of ω, one would more readily satisfy the validity criterion (7.32), but on the other hand this choice would lead to small values of α_0 corresponding to the perturbative domain. In order to reach the interesting, non-perturbative regime where α_0 is large, it is desirable to choose ω as low as possible but compatible with the condition (7.32), so that the intensity should not be prohibitively high.

We point out that the condition (7.32) has a simple semi-classical interpretation. Indeed, it requires that the variation of the external perturbation be fast with respect to the relevant orbital period of the electron. In that case, one should expect that, to a good approximation, the electron moving in the time-dependent, oscillating potential $V[\mathbf{r} + \boldsymbol{\alpha}(t)]$ will effectively experience its time-averaged potential $V_0(\alpha_0, \mathbf{r})$ given by Equation (7.2).

In the foregoing discussion, we have assumed that the dipole approximation is valid. Clearly its applicability can be called into question for sufficiently high laser frequencies. However, as we will see in Section 7.3.4, for moderately high frequencies it is the value of the laser intensity that will determine when the dipole approximation breaks down.

7.2 Structure of atomic hydrogen in intense, high-frequency laser fields

We shall now apply the HFFT to study how intense, high-frequency laser fields modify the structure of atomic hydrogen [11–14]. In this section, we will follow the account given by Gavrila [1], where additional details can be found.

To lowest order in the iteration, that is in the high-frequency limit, the TDSE for a single-active-electron atom interacting with a spatially homogeneous, monochromatic laser field of arbitrary polarization reduces, in the K–H frame, to the time-independent equation

$$\left[\left(\frac{\mathbf{p}^2}{2} + V_0(\alpha_0, \mathbf{r}) \right) - \mathsf{E}^A \right] F_0^{A(0)}(\mathsf{E}^A; \mathbf{r}) = 0, \tag{7.33}$$

for the zeroth Floquet component $F_0^{A(0)}(\mathbf{r})$. In this approximation, the wave function of the system is the stationary state

$$\Psi^A(\mathbf{r}, t) = F_0^{A(0)}(\mathsf{E}^A; \mathbf{r}) \exp(-i\mathsf{E}^A t). \tag{7.34}$$

It follows from Equation (7.33) that in the high-frequency limit the atom is *stable*, namely it cannot decay by multiphoton ionization. Its structure depends on the angular frequency ω and the intensity I only through the excursion amplitude α_0. Since the dressed potential $V_0(\alpha_0, \mathbf{r})$ may differ appreciably from $V(r)$, Equation (7.33) describes the *radiative distortion* of the electron probability density. As we have seen in Section 7.1, ionization becomes possible in the first order of the iteration, which, in addition to a real shift of the quasi-energy E^A, also yields an imaginary part to this quasi-energy and expressions for the n-photon ionization rates.

It is also worth noting that the approximate solution of the TDSE in the high-frequency limit given by Equation (7.34) is expressed in the K–H (accelerated) frame. The solution of the TDSE in the laboratory frame and in the velocity gauge is given by

$$\Psi^V(\mathbf{r}, t) = \exp[-i\boldsymbol{\alpha}(t) \cdot \mathbf{p}]\Psi^A(\mathbf{r}, t)$$

$$= \Psi^A[\mathbf{r} - \boldsymbol{\alpha}(t), t]. \tag{7.35}$$

In the high-frequency limit we have, by using Equation (7.34),

$$\Psi^V(\mathbf{r}, t) = F_0^A[E^A; \mathbf{r} - \boldsymbol{\alpha}(t)]\exp(-iE^A t), \tag{7.36}$$

and we see that the electronic probability density is given in the laboratory frame by

$$|\Psi^V(\mathbf{r}, t)|^2 = |F_0^A[E^A; \mathbf{r} - \boldsymbol{\alpha}(t)]|^2, \tag{7.37}$$

so that it oscillates like a free classical electron.

7.2.1 The dressed potential classification of states

In the case of atomic hydrogen, for which the field-free potential $V(r)$ is the Coulomb potential $V_c(r) = -1/r$, and for a linearly polarized (L) monochromatic laser field whose electric field is given by Equation (2.118) with $\varphi = 0$, the dressed potential (7.2) becomes

$$V_0^L(\alpha_0, \mathbf{r}) = -(T)^{-1} \int_0^T \frac{dt}{|\mathbf{r} + \hat{\boldsymbol{\epsilon}}\alpha_0\cos(\omega t)|}. \tag{7.38}$$

This is the time average of the electrostatic potential exerted on the electron by the proton, which in the K–H frame oscillates harmonically along the polarization axis with the angular frequency of the laser field ω and with the excursion amplitude α_0.

Since in the present case of linear polarization the problem has azimuthal symmetry about the polarization axis (which we choose to be the z-axis), the laser field does not couple states with different magnetic quantum numbers. As a result, we may restrict $|w_i^{(0)}(\alpha_0)|$ in Equation (7.32) to the manifold of states having the same magnetic quantum number as the initial state.

Returning to Equation (7.38) and recalling that the z-axis is taken along the polarization vector $\hat{\boldsymbol{\epsilon}}$, the dressed Coulomb potential $V_0^L(\alpha_0, \mathbf{r})$ can also be expressed as

$$V_0^L(\alpha_0, \mathbf{r}) = -\pi^{-1} \int_{-\alpha_0}^{\alpha_0} \frac{d\zeta}{|\mathbf{r} - \hat{\boldsymbol{\epsilon}}\zeta|(\alpha_0^2 - \zeta^2)^{1/2}}. \tag{7.39}$$

This potential can be considered as the electrostatic potential due to a distribution of charges

$$\sigma(z) = \frac{1}{\pi(\alpha_0^2 - z^2)^{1/2}} \tag{7.40}$$

extending from $-\alpha_0$ to α_0 along the polarization axis, the total charge being

$$\int_{-\alpha_0}^{\alpha_0} \sigma(\zeta)d\zeta = 1. \tag{7.41}$$

The charge density $\sigma(z)$ can be interpreted as a time average of the oscillating proton charge, and we see from Equation (7.40) that there is an accumulation of charge near the turning points of the proton oscillation, where its motion is slowest.

The dressed Coulomb potential (7.39) can be recast in the form

$$V_0^{\mathrm{L}}(\alpha_0, \mathbf{r}) = -\frac{2}{\pi(r_+ r_-)^{1/2}} K\left[\left(\frac{1 - \hat{\mathbf{r}}_+ \cdot \hat{\mathbf{r}}_-}{2}\right)^{1/2}\right], \tag{7.42}$$

where $\mathbf{r}_\pm = \mathbf{r} \pm \hat{\boldsymbol{\epsilon}}\alpha_0$ and K denotes the complete elliptic integral of the first kind [15]. We note that $V_0^{\mathrm{L}}(\alpha_0, \mathbf{r})$ has $r^{-1/2}$ singularities at the end points $\pm\hat{\boldsymbol{\epsilon}}\alpha_0$ of the charge distribution and a logarithmic singularity along the segment connecting them, due to the behavior of the function $K(x)$ when $x \to 1$ [16]. We also remark that, since $K(0) = \pi/2$, the dressed Coulomb potential (7.42) reduces at large r to the field-free Coulomb potential $V_{\mathrm{c}}(r) = -1/r$. A graphical representation of the quantity $\alpha_0 |V_0^{\mathrm{L}}(\alpha_0, \mathbf{r})|$ is given in Fig. 7.1 as a function of the position in a plane passing through the polarization vector $\hat{\boldsymbol{\epsilon}}$. The saddle in the figure is due to the rise of $|V_0^{\mathrm{L}}(\alpha_0, \mathbf{r})|$ near the line of singularities extending along the polarization axis from $\hat{\boldsymbol{\epsilon}}\alpha_0$ to $-\hat{\boldsymbol{\epsilon}}\alpha_0$ and their increasing strength towards the end points.

Turning to circular polarization (C), and choosing now the z-axis along the laser propagation direction, the dressed potential corresponding to the Coulomb potential $V_{\mathrm{c}}(r)$ can be written as

$$V_0^{\mathrm{C}}(\alpha_0, \mathbf{r}) = -\frac{2}{\pi \tilde{r}_+} K\left(\left[1 - \left(\frac{\tilde{r}_-}{\tilde{r}_+}\right)^2\right]^{1/2}\right), \tag{7.43}$$

where now $\tilde{r}_+$ and $\tilde{r}_-$ denote the longest and shortest distances, respectively, from the point $\mathbf{r}$ to the circle of charges (in the plane containing the z-axis and the point $\mathbf{r}$). Again, because the function $K(x)$ has a logarithmic singularity for $x \to 1$, $V_0^{\mathrm{C}}(\alpha_0, \mathbf{r})$ has a logarithmic singularity on the circle of charges.

We note that $V_0^{\mathrm{L}}(\alpha_0, \mathbf{r})$ and $V_0^{\mathrm{C}}(\alpha_0, \mathbf{r})$ have the same symmetry properties. Recalling that the z-axis has been taken along the laser polarization direction for linear polarization and along the laser propagation direction for circular polarization,

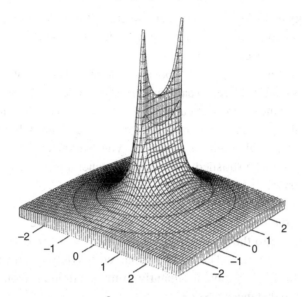

Figure 7.1. The quantity $\alpha_0 |V_0^L(\alpha_0, \mathbf{r})|$, in a.u., as a function of the position in a plane passing through the polarization axis. Distances in the base plane are measured in units of α_0. The maximum value of $\alpha_0 |V_0^L(\alpha_0, \mathbf{r})|$ represented in the figure is 10 a.u. (From M. Gavrila and J. Z. Kaminski, *Phys. Rev. Lett.* **52**, 613 (1984).)

they are axially symmetric around the z-axis. There are also even functions with respect to the parity transformation, namely an inversion (reflection) through the origin of the coordinate system, which is the centre of $V_c(r)$. The absence of the spherical symmetry implies that only the projection of the orbital angular momentum on the z-axis (with the associated magnetic quantum number m) and the parity are conserved quantities. In fact, the eigenvalues depend on $|m|$ rather than on m, so that there is a two-fold degeneracy for $m \neq 0$. The symmetry of the problem suggests a classification scheme for the states which is similar to that for homonuclear diatomic molecules [17]. Following Pont *et al.* [11], we shall denote the states with $|m| = 0, 1, 2, \ldots$ by the symbols $\sigma, \pi, \delta, \ldots$ The even (gerade) states will be labeled by g and the odd (ungerade) states by u. To distinguish a particular state within a manifold with given $|m|$ and parity, we shall label it by the quantum numbers (n, l) of the unperturbed state to which it is connected continuously in the field-free limit $(\alpha_0 = 0)$.

The energy eigenfunctions $u_j(\mathbf{r})$ of the dressed Hamiltonian in the K–H frame, H_0^A, will be denoted more explicitly by $u_{nlm}(\mathbf{r})$. For both linear and circular polarization, they have the form

$$u_{nlm}(\mathbf{r}) = (2\pi)^{-1/2} f_{nlm}(r, \theta) \exp(im\phi) \qquad (7.44)$$

and their parity is that of l. We note that reflections through the origin multiply these energy eigenfunctions by $(-1)^l$, reflections with respect to the z-axis by $(-1)^m$ and reflections with respect to the x-y plane by $(-1)^{l+m}$.

It will be shown below that, for both small and large values of α_0, the eigenvalue equation (7.19) can be solved in part analytically. As a result, "correlation diagrams" can be drawn for the eigenvalue curves $w_j^{(0)}(\alpha_0)$, which connect the eigenvalues for large and small α_0 belonging to the same symmetry manifold. If all the symmetries of the problem are taken into account, the von Neumann–Wigner "non-crossing rule" [17, 18] must be satisfied. This leads in general to avoided crossings of the eigenvalue curves.

We also note that the high-lying states of each manifold exhibit a Rydberg character. The reason for this behavior is that for $r \geq \alpha_0$ the dressed potentials (7.42) and (7.43) reduce to the Coulomb potential $V_c(r) = -1/r$ corrected by terms in $r^{-(2s+1)}$, with $s = 1, 2, \ldots$ The Rydberg character is only manifest for those states whose wave functions lie predominantly in the Coulomb region of the potential, namely at radial distances $r \gg \alpha_0$.

7.2.2 Small-α_0 limit

For small α_0, the dressed potential $V_0(\alpha_0, \mathbf{r})$ differs little from the Coulomb potential $V_c(r) = -1/r$. As a consequence, one can apply time-independent perturbation theory with respect to α_0 in order to obtain the corrections to the field-free atomic hydrogen energy levels. To perform this perturbative calculation, we shall use as a basis the hydrogen atom eigenfunctions

$$\langle \mathbf{r} | nlm \rangle \equiv \psi_{nlm}(\mathbf{r}) = R_{nl}(r) Y_{lm}(\hat{\mathbf{r}}), \tag{7.45}$$

where $R_{nl}(r)$ are radial hydrogenic eigenfunctions with $Z = 1$ and $Y_{lm}(\hat{\mathbf{r}})$ are spherical harmonics.

Writing $H_0^A = (\mathbf{p}^2/2 + V_c) + (V_0 - V_c)$, and remembering that the field-free atomic hydrogen energy levels are degenerate for $n \geq 2$, we must (except for the case $n = 1$) diagonalize to lowest order in α_0 the matrix of the perturbation $H' = V_0 - V_c$ in each subspace characterized by a given $n \geq 2$. Because the matrix elements of V_0 are diagonal with respect to the magnetic quantum number m, which is a conserved quantity, we only have to calculate the energy shift,

$$\Delta w_{nlm,nl'm'} = \langle nlm | V_0 - V_c | nl'm' \rangle, \tag{7.46}$$

for all possible values of l and l'.

In order to perform this calculation, we introduce the Fourier transform of the Coulomb potential $V_c(r) = -1/r$, namely

$$\hat{V}_c(\mathbf{k}) = (2\pi)^{-3/2} \int \exp(-i\mathbf{k} \cdot \mathbf{r}) \left(-\frac{1}{r}\right) d\mathbf{r}. \tag{7.47}$$

The integral on the right-hand side of this equation can be evaluated by considering the Coulomb potential $V_c(r)$ as a limiting case of the screened Coulomb potential

$$V(r) = -\frac{1}{r}\exp(-\beta r), \quad \beta > 0, \tag{7.48}$$

and by taking the limit $\beta \to 0^+$ at the end of the calculation. Thus we write

$$\hat{V}_c(\mathbf{k}) = (2\pi)^{-3/2} \lim_{\beta \to 0^+} \int \exp(-i\mathbf{k}\cdot\mathbf{r} - \beta r)\left(-\frac{1}{r}\right)d\mathbf{r}. \tag{7.49}$$

Performing the integral and taking the limit $\beta \to 0^+$, we find that

$$\hat{V}_c(k) = -\left(\frac{2}{\pi}\right)^{1/2}\frac{1}{k^2}. \tag{7.50}$$

Let us now return to the expression (7.2) of the dressed potential $V_0(\alpha_0, \mathbf{r})$. By inserting in it the inverse Fourier transform of the Coulomb potential $V_c(r)$, namely

$$V_c(r) = (2\pi)^{-3/2}\int \exp(i\mathbf{k}\cdot\mathbf{r})\hat{V}_c(k)d\mathbf{k}, \tag{7.51}$$

and using Equations (2.116) and (7.50), we have, for arbitrary polarization,

$$V_0(\alpha_0, \mathbf{r}) = (2\pi)^{-5/2}\int d\mathbf{k}\,\hat{V}_c(k)\int_0^{2\pi} d\chi$$
$$\times \exp\left\{i\mathbf{k}\cdot\left[\mathbf{r} + \frac{\alpha_0}{(1+\xi^2)^{1/2}}[\hat{\boldsymbol{\epsilon}}_1\cos(\chi+\varphi) - \xi\hat{\boldsymbol{\epsilon}}_2\sin(\chi+\varphi)]\right]\right\}. \tag{7.52}$$

For small values of α_0, one may expand the exponential in the above equation in powers of α_0. To order α_0^2, one obtains in this way

$$V_0(\alpha_0, \mathbf{r}) - V_c(r) = \frac{\alpha_0^2}{4(1+\xi^2)}\left[U_1(\mathbf{r}) + \xi^2 U_2(\mathbf{r})\right], \tag{7.53}$$

where

$$U_\mu(\mathbf{r}) = (2\pi^2)^{-1}\int \exp(i\mathbf{k}\cdot\mathbf{r})\frac{(\hat{\boldsymbol{\epsilon}}_\mu\cdot\mathbf{k})^2}{k^2}d\mathbf{k}, \quad \mu = 1, 2. \tag{7.54}$$

The objects $U_\mu(\mathbf{r})$ are singular at $\mathbf{r} = 0$ and have a well defined meaning only under an integral, as for example in Equation (7.46). It has been shown [19,20] that Equation (7.54) defines the distributions

$$U_\mu(\mathbf{r}) = \frac{3\pi}{4}\delta(\mathbf{r}) - \left[\frac{2P_2(\hat{\boldsymbol{\epsilon}}_\mu\cdot\mathbf{r})}{r^3}\right]', \tag{7.55}$$

where $P_2(x) = (3x^2 - 1)/2$ is a Legendre polynomial and the prime indicates that an integral containing the second term on the right-hand side of Equation (7.55) must be evaluated first over the region outside a sphere of radius ρ around the origin, and then the limit $\rho \to 0$ must be taken.

Using Equations (7.46), (7.53) and (7.55), one obtains, for linear and circular polarizations,

$$\Delta w_{nlm,nl'm'} = \alpha_0^2 \left[A \frac{\pi}{3} |\psi_{n00}|^2 \delta_{l,0} \delta_{l',0} - \frac{B}{2} \langle nl |r^{-3}|nl'\rangle \langle lm |P_2|l'm\rangle \right], \quad (7.56)$$

where the constants A and B are given by

$$A = 1, \quad B = 1, \qquad \text{for linear polarization,}$$
$$A = 2, \quad B = -1, \qquad \text{for circular polarization.} \quad (7.57)$$

We note that the first term on the right-hand side of Equation (7.56) is non-vanishing only when $l = l' = 0$, while the second term on the right-hand side of Equation (7.56) should be omitted in this case. Indeed, the second term on the right-hand side of Equation (7.56) contains the radial matrix element $\langle nl |r^{-3}|nl'\rangle$ and the angular one $\langle lm |P_2|l'm\rangle$. The latter vanishes unless l and l' have the same parity and $|l - 2| \leqslant l' \leqslant l + 2$. Therefore, only the radial matrix elements with $l' = l \neq 0$ and $l' = l \pm 2$ ($l \geq 2$) must be considered. However, due to the selection rule [21]

$$\langle nl |r^{-3}|nl \pm 2\rangle = 0, \quad (7.58)$$

the matrix Δw of Equation (7.56) is diagonal with respect to l and l' to order α_0^2. Moreover, since

$$|\psi_{nlm}(0)|^2 = \frac{1}{\pi n^3} \delta_{l,0}, \quad (7.59)$$

$$\langle nl |r^{-3}|nl\rangle = \frac{1}{n^3 l(l + 1/2)(l + 1)} \quad (7.60)$$

and

$$\langle lm |P_2|lm\rangle = \frac{l(l + 1) - 3m^2}{(2l - 1)(2l + 3)}, \quad (7.61)$$

we have

$$\Delta w_{nlm,nl'm'} = (\Delta w)_{nl|m|} \delta_{ll'}, \quad (7.62)$$

where the energy shift (AC Stark shift) is given by

$$(\Delta w)_{nl|m|} = \alpha_0^2 \left[A \frac{1}{3n^2} \delta_{l,0} - B(1 - \delta_{l,0}) \frac{l(l + 1) - 3m^2)}{n^3 l(l + 1)(2l - 1)(2l + 1)(2l + 3)} \right], \quad (7.63)$$

and the constants A and B are given by Equation (7.57). The result, (7.63), derived by Pont, Walet and Gavrila [13] in the K–H frame, agrees with the AC Stark shift calculated by Ritus [22] in the laboratory frame.

It follows from Equations (7.46), (7.62) and (7.63) that, for small values of α_0, the field-free hydrogen atom eigenfunctions are still approximate eigenfunctions of the problem. This allows, by continuity with respect to α_0, the use of the quantum numbers n and l to label the exact eigenstates for all values of α_0.

7.2.3 Large-α_0 limit

Let us now analyze the eigenvalue equation (7.19) for large values of α_0. We shall first examine the case of linear polarization, and show that the bound-state eigenfunctions $u_j(\mathbf{r})$ corresponding to the lowest eigenenergies $w_j^{(0)}$ of the dressed Hamiltonian H_0^A split into two practically non-overlapping parts, an effect called *atomic dichotomy* [11].

We assume that, for a given, large value of α_0, the wave function $u_j(\mathbf{r})$ is concentrated near the two end points, $\pm\alpha_0\hat{\epsilon}$, of the linear charge distribution generating the dressed potential $V_0^L(\alpha_0, \mathbf{r})$ of Equation (7.42). Then $u_j(\mathbf{r})$ will have significant values only in the regions of space such that $r_-/\alpha_0 \ll 1$ and $r_+/\alpha_0 \ll 1$, where the dressed potential $V_0^L(\alpha_0, \mathbf{r})$ reduces to

$$V_0^L(\alpha_0, \mathbf{r}) \simeq \begin{cases} \tilde{V}_0(\alpha_0, \mathbf{r}_-), & \text{for } r_-/\alpha_0 \ll 1, \\ \tilde{V}_0(\alpha_0, -\mathbf{r}_+), & \text{for } r_+/\alpha_0 \ll 1, \end{cases} \tag{7.64}$$

where

$$\tilde{V}_0(\alpha_0, \mathbf{r}) = -\frac{1}{\pi}\left(\frac{2}{\alpha_0 r}\right)^{1/2} K\left[\left(\frac{1-\hat{\mathbf{r}}\cdot\hat{\epsilon}}{2}\right)^{1/2}\right] \tag{7.65}$$

will be called the "end-point potential." Denoting by Θ the angle between the vectors $\hat{\mathbf{r}}$ and $\hat{\epsilon}$, we have

$$\tilde{V}_0(\alpha_0, \mathbf{r}) = -\frac{1}{\pi}\left(\frac{2}{\alpha_0 r}\right)^{1/2} K[\sin(\Theta/2)], \tag{7.66}$$

and we remark that $\tilde{V}_0(\alpha_0, \mathbf{r})$ has an overall $r^{-1/2}$ radial dependence and is not spherically symmetric.

According to the assumption we have made, in the neighborhood of the end point $\alpha_0\hat{\epsilon}$ the dressed potential $V_0(\alpha_0, \mathbf{r})$ can be replaced in the eigenvalue equation (7.19) by the simple expression $\tilde{V}_0(\alpha_0, \mathbf{r}_-)$. Introducing the scaled variable $\mathbf{r}' = \alpha_0^{-1/3}\mathbf{r}_-$, Equation (7.19) becomes

$$\left[-\frac{1}{2}\nabla_{r'}^2 - \frac{1}{\pi}\left(\frac{2}{r'}\right)^{1/2} K[\sin(\Theta/2)]\right] v_j(\mathbf{r}') = W_j v_j(\mathbf{r}'), \tag{7.67}$$

where

$$W_j = \alpha_0^{2/3} w_j^{(0)} \tag{7.68}$$

and

$$v_j(\mathbf{r}') = C_j u_j(\alpha_0^{1/3} \mathbf{r}' + \alpha_0 \hat{\boldsymbol{\epsilon}}), \tag{7.69}$$

with the normalization constant C_j chosen such that

$$\int |v_j(\mathbf{r}')|^2 \, d\mathbf{r}' = 1. \tag{7.70}$$

It is worth noting that Equation (7.67) does not depend on α_0. A bound-state (square integrable) eigenfunction $v_j(\mathbf{r}')$ extends essentially over a finite region in the space of the variable $\mathbf{r}'$. Its asymptotic behavior for $r' \to \infty$ is governed by the factor $\exp[-(2|W_j|)^{1/2} r']$. As a result, the corresponding eigenfunction $u_j(\mathbf{r}')$ has significant values only in regions such that $r_-/\alpha_0 = \alpha_0^{-2/3} r' \ll 1$, and is exponentially small elsewhere. A similar reasoning can be made for the region in the vicinity of the end point $-\alpha_0 \hat{\boldsymbol{\epsilon}}$, such that $r_+/\alpha_0 = \alpha_0^{-2/3} r' \ll 1$. Thus, in agreement with the assumption made initially, and to lowest order in α_0^{-1}, the dressed potential $V_0(\alpha_0, \mathbf{r})$ can support bound states whose eigenfunctions are localized in the neighborhood of its end points $\pm \alpha_0 \hat{\boldsymbol{\epsilon}}$, and are vanishingly small elsewhere. This property, valid for any bound state provided α_0 is sufficiently large, is called *atomic dichotomy*, since the electron cloud (which oscillates in the laboratory frame) splits into two disjoint parts.

The end-point potential $\tilde{V}_0(\alpha_0, \mathbf{r})$ given by Equation (7.66) is no longer even under the parity transformation, but it still has axial symmetry about the polarization axis, so that the eigenvalues W_j depend only on $|m|$. To label the energy levels in each manifold $|m|$, an additional index is needed, which will be designated by s. The eigenvalues W_j will therefore be denoted by $W_{j|m|s}$. The eigenfunctions $v_j(\mathbf{r}')$ depend on m and s and will be designated by $v_{jms}(\mathbf{r}')$. From Equations (7.68) and (7.69) we also have for large α_0 the scaling laws

$$w_{j|m|s}^{(0)} \simeq \alpha_0^{-2/3} W_{j|m|s}^{(0)} \tag{7.71}$$

and

$$u_{jms}(\mathbf{r}) \simeq N_{jms}(\alpha_0) v_{jms}(\mp \alpha_0^{-1/3} \mathbf{r}_\pm), \tag{7.72}$$

where $N_{jms}(\alpha_0)$ is a normalization constant for the wave function $u_{jms}(\mathbf{r})$, and the vector $\mathbf{r}$ is either in the neighborhood of $\alpha_0 \hat{\boldsymbol{\epsilon}}$ (upper sign) or $-\alpha_0 \hat{\boldsymbol{\epsilon}}$ (lower sign). We note that it follows from Equation (7.71) that the binding energies of the bound states decrease for large α_0 as $\alpha_0^{-2/3}$.

It is also worth noticing that two eigenfunctions $u_{jns}^{g,u}$ of well defined parity can be written in the neighborhood of both end points $\pm\alpha_0\hat{\epsilon}$ as

$$u_{jms}^{g,u}(\mathbf{r}) \simeq (2\alpha_0)^{-1/2}\left[v_{jms}(\alpha_0^{-1/3}\mathbf{r}_-) \pm v_{jms}(-\alpha_0^{-1/3}\mathbf{r}_+)\right], \qquad (7.73)$$

where the plus sign corresponds to the gerade (g) state and the minus sign to the ungerade (u) state. These eigenfunctions are essentially normalized to unity, since the two terms on the right-hand side of Equation (7.73) have a vanishingly small overlap integral, and the normalization condition (7.70) has been imposed on the wave functions v_{jms}. The fact that there are two linearly independent eigenfunctions corresponding to the same eigenvalue $w_{j|m|s}^{(0)}$ shows that in the limit $\alpha_0 \to \infty$ there is a *gerade–ungerade degeneracy*. At large but finite values of α_0, the energies $w_{j|m|s}^{(0)g,u}(\alpha_0)$ of the two eigenfunctions $u_{jms}^{g,u}(\mathbf{r})$ are different, corresponding to a *gerade–ungerade energy level splitting* which decreases exponentially at large α_0. We also note that the eigenfunctions $u_{jms}(\mathbf{r})$ *dilate* as $\alpha_0^{1/3}$ when α_0 increases. However, this dilation does not compensate for the fact that the end points $\pm\alpha_0\hat{\epsilon}$ move apart as α_0 becomes larger.

As an illustration of the above discussion, we show in Fig. 7.2 the high-frequency limit energies of the hydrogen atom first few lower-lying eigenvalues belonging to the two symmetry manifolds σ_g and σ_u as functions of α_0, calculated by Pont, Walet and Gavrila [13] for linear polarization. The energy curves belonging to the same manifold form a correlation diagram connecting the field-free ($\alpha_0 = 0$)

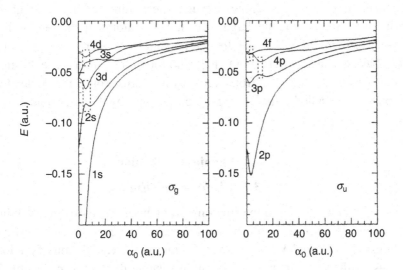

Figure 7.2. High-frequency limit energies of the first few eigenstates of atomic hydrogen belonging to the σ_g and σ_u symmetries, as a function of α_0 (in a.u.) for linear polarization. (From M. Pont, N. Walet and M. Gavrila, *Phys. Rev. A* **41**, 477 (1990).)

atom eigenvalues to their dichotomized (large-α_0) counterparts. For all values of α_0, the correlation curves must satisfy the von Neumann–Wigner non-crossing rule [17, 18]. Assuming the angular frequency ω to be fixed at some given value large enough to satisfy the condition (7.32), and remembering that $\alpha_0 = \mathcal{E}_0/\omega^2$ (in a.u.), it follows that the correlation curves yield the dependence of the energy eigenvalues on $I^{1/2}$, as this quantity varies from zero towards high values. As seen from Fig. 7.2, a general feature of the correlation curves is their tendency to increase towards zero as α_0 becomes large. This increase can be either monotonic (e.g. (1s) σ_g) or preceded by a minimum (e.g. (2s)σ_g, (3d)σ_g, (2p)σ_u). In the range $\alpha_0 < 40$ a.u. the increase is quite sharp for the lower-lying states. For example, the binding energy of the ground state (1s)σ_g drops from its value of 0.5 a.u. at $\alpha_0 = 0$ to about 0.05 a.u. at $\alpha_0 = 40$ a.u. This behavior is in contrast to the case of low frequencies, where the ionization potential of the ground state increases as α_0 increases.

The high-frequency limit normalized energy eigenfunction of the ground state (1s)σ_g of atomic hydrogen in the K–H frame is displayed in Fig. 7.3 for various values of α_0 in the case of linear polarization. As seen from this figure, with increasing α_0 the wave function undergoes radiative stretching along the z-axis (which coincides with the polarization axis), following the elongation of the line of charges (of length $2\alpha_0$) generating the dressed potential $V_0^L(\alpha_0, \mathbf{r})$. As α_0 reaches 30 a.u., two pronounced maxima appear around the end points $\pm\alpha_0\hat{\boldsymbol{\epsilon}}$ of the line of charges, and when $\alpha_0 = 70$ a.u. dichotomy is complete. As it splits for increasing values of α_0, the wave function also spreads out about the end points, as predicted by Equation (7.72). We note in this context the change of scale in the pictures of Fig. 7.3 for growing α_0.

Finally, we remark that the large-α_0 behavior of atomic hydrogen for the case of *circular polarization*, where the dressed potential $V_0^C(\alpha_0, \mathbf{r})$ is given by Equation (7.43), has been analyzed by Pont [12], who found that in this case the eigenfunctions are practically different from zero only in a torus around the circle of charges. This property is called *toroidal shaping* of the atom by the laser field.

7.3 Atomic stabilization

7.3.1 Adiabatic stabilization

In Section 7.2, we analyzed the structure problem of the hydrogen atom in strong, high-frequency laser fields by solving the zeroth-order Equation (7.33) involving the "dressed" potential $V_0(\alpha_0, \mathbf{r})$. We now turn to the calculation of the ionization rates, using the first order of the iteration scheme discussed in Section 7.1, and follow the treatment of Pont and Gavrila [23]. The starting point is Equation (7.31), which contains the continuum eigenfunction $u_{\mathbf{k}_n}^{(-)}(\mathbf{r})$. Using the fact that in the high-frequency limit all ionized electrons have large kinetic energies with respect to the

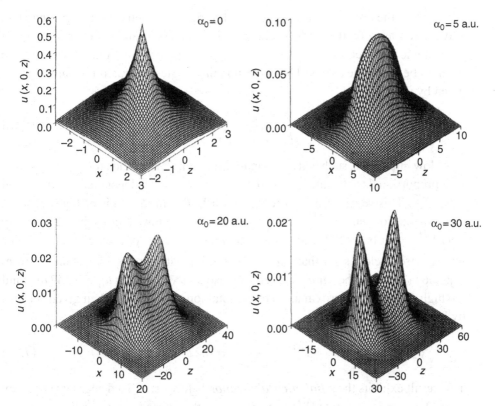

Figure 7.3. High-frequency limit normalized energy eigenfunction of the ground state $(1s)\sigma_g$ of atomic hydrogen in the K–H frame for increasing values of α_0 in the case of linear polarization. The wave function $u(x, 0, z)$ (in a.u.) is plotted in the x-z plane, where the z-axis is chosen along the unit polarization vector, which is the axis of symmetry of the dressed potential $V_0^L(\alpha_0, \mathbf{r})$. The x-axis is arbitrary due to cylindrical symmetry. There is a change of scale in the pictures for the various values of α_0. (From M. Pont, N. Walet and M. Gavrila, *Phys. Rev. A* **41**, 477 (1990).)

ground-state binding energy, Pont and Gavrila used the first Born approximation, in which $u_{\mathbf{k}_n}^{(-)}(\mathbf{r})$ is replaced by a plane wave, namely

$$u_{\mathbf{k}_n}^{(-)}(\mathbf{r}) \simeq (2\pi)^{-3/2} \exp(i\mathbf{k}_n \cdot \mathbf{r}). \tag{7.74}$$

This greatly simplifies the computation of the ionization rates. Thus, for the ground state $(1s)\sigma_g$ and in the case of circular polarization, Pont and Gavrila found that the differential n-photon ionization rate is given (in a.u.) by

$$\frac{\mathrm{d}\Gamma_n^{(1)}}{\mathrm{d}\Omega} = \frac{4}{k_n^3}(u_0^c)^2 J_n^2(k_n\alpha_0\sin\theta), \tag{7.75}$$

where u_0^c is the constant value (due to cylindrical symmetry) of the ground-state wave function u_0 on the circle of charges, J_n is a Bessel function of order n and θ is the angle between the vector $\mathbf{k}_n$ and the propagation direction, which we have taken to be along the z-axis. The corresponding angle-integrated ionization rate is given by

$$\Gamma_n^{(1)} = \frac{8\pi}{k_n^4 \alpha_0} (u_0^c)^2 \int_0^{2k_n\alpha_0} J_{2n}(x)\mathrm{d}x. \tag{7.76}$$

At low intensities, such that α_0 is small, Equation (7.76) yields for $\Gamma_1^{(1)}$ an expression proportional to the intensity I, the angular frequency dependence being given by $\omega^{-9/2}$. This result is in accordance with LOPT in the limit of high frequencies. In addition, since $\Gamma_n^{(1)}$ is proportional to I^n, to lowest order in the intensity only $\Gamma_1^{(1)}$ contributes to the total ionization rate $\Gamma^{(1)}$, and one has $\Gamma^{(1)} \simeq \Gamma_1^{(1)}$. For intense laser fields, such that $k_n\alpha_0 \gg 1$, the ionization rates $\Gamma_n^{(1)}$ obtained from Equation (7.76) can be summed over n. If also $\alpha_0 \gg 1$, one has $u_0^c \simeq 0.147/\alpha_0$, and the high-frequency approximation to the total ionization rate is then given (in a.u.) by [12]

$$\Gamma^{(1)} \simeq \frac{0.223}{\alpha_0^3 \omega^2} = 0.223 \frac{\omega^4}{\mathcal{E}_0^3}. \tag{7.77}$$

This result exhibits the *adiabatic stabilization* behavior: at fixed angular frequency ω, the total ionization rate $\Gamma^{(1)}$ *decreases* with the intensity as $I^{-3/2}$. We also note that, at fixed intensity I, the total ionization rate $\Gamma^{(1)}$ increases with the angular frequency as ω^4.

The lifetime of the atom is given in a.u. by $\tau = 1/\Gamma^{(1)}$. The intensity dependence of the lifetimes of the hydrogen atom in the ground state, for a circularly polarized laser field, as computed by Pont and Gavrila [23] is shown in Fig. 7.4 for various values of the angular frequency ω. For the larger values of ω considered ($1 \le \omega \le 8$), the high-frequency condition (7.32) is approximately satisfied, so that the lifetime curves can be followed from small to large intensities. In the log-log plot of Fig. 7.4, these curves have a linearly descending LOPT regime, a linearly ascending stabilization regime and an intermediate regime with a minimum marking the breakdown of LOPT and the onset of the stabilization regime at some critical intensity which, for the values of ω considered, lies between 1 a.u. and 10^3 a.u. Also shown in Fig. 7.4 are three lifetime curves corresponding to lower values of ω ($\omega = 0.125$, 0.25 and 0.5 a.u.) for which, at small intensities (corresponding to small values of α_0), the high-frequency condition (7.32) is not satisfied. Nevertheless, even for these values of ω, if I (or α_0) is increased sufficiently, condition (7.32) will be satisfied, and the HFFT predicts the existence of states whose lifetimes are given by the three curves shown.

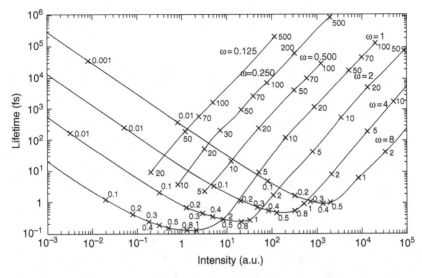

Figure 7.4. Lifetime of the ground state of atomic hydrogen (in femtoseconds) interacting with a circularly polarized laser field as a function of the intensity (in a.u.) for the indicated values of the angular frequency ω (in a.u.). The results were obtained using the HFFT. The numbers next to the points on the curves are the corresponding values of α_0. The descending parts of the curves correspond to LOPT; the ascending parts correspond to adiabatic stabilization. (From M. Pont and M. Gavrila, *Phys. Rev. Lett.* **65**, 2362 (1990).)

The HFFT also predicts the existence of adiabatic stabilization for elliptic polarization. In particular, for linear polarization, the high-frequency behavior of the total ionization rate is given for $\alpha_0 \gg 1$ by [1,3]

$$\Gamma^{(1)} \simeq \frac{1}{\alpha_0^2 \omega^2}[0.00746\log(\alpha_0^2\omega^3) + 0.285]. \qquad (7.78)$$

The existence of adiabatic stabilization in the non-relativistic, dipole approximation has been confirmed by *ab initio* Floquet calculations. These include Sturmian-Floquet [24] and *R*-matrix–Floquet (RMF) [25] calculations, and also the close-coupling–Floquet computations of Marte and Zoller [26], Dimou and Faisal [27] and Lefebvre and Stern [28]. The results obtained from these various Floquet calculations, which agree well with each other, are more accurate than the HFFT results discussed above, because they are obtained by solving the Floquet equations exactly. In addition, no approximation is made in the description of the ionized electron.

As an example, we show the RMF total ionization rate for H(1s) in a linearly polarized laser field of angular frequency $\omega = 0.65$ a.u. in Fig. 7.5(a). The RMF results [25], which are in excellent agreement with those obtained by using the

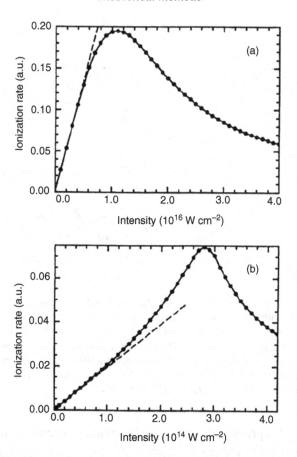

Figure 7.5. Total ionization rate (in a.u.) as a function of intensity for (a) H(1s) in a linearly polarized laser field of angular frequency $\omega = 0.65$ a.u. and (b) for H(2s) in a linearly polarized laser field of angular frequency $\omega = 0.184$ a.u. The solid lines refer to the R-matrix–Floquet calculations, with the circles corresponding to the calculated results. The lowest (in both cases, first-order) perturbative results are given by the broken lines. (From M. Dörr *et al.*, *J. Phys. B* **26**, L275 (1993).)

Sturmian-Floquet method [24], are seen to increase linearly at low intensities, as predicted by LOPT since the absorption of one photon is sufficient to ionize the atom. However, the RMF (and the Sturmian-Floquet) results have a maximum at an intensity near $10^{16}\,\mathrm{W\,cm^{-2}}$, and then decrease with increasing intensity, thus exhibiting the adiabatic stabilization behavior. In Fig. 7.5(b), the total ionization rate is shown for H(2s) in a linearly polarized laser field of angular frequency $\omega = 0.184$ a.u. (KrF laser). Again, the RMF results agree at low intensities with the LOPT values, then have a maximum at an intensity of about $3 \times 10^{14}\,\mathrm{W\,cm^{-2}}$

(much lower than for the ground state), and display the high-frequency adiabatic stabilization behavior at high intensities.

In the Floquet approach that we have used so far, two main assumptions have been made. Firstly, we started from the non-relativistic, time-dependent Schrödinger equation (TDSE) in the dipole approximation. We shall discuss at the end of this section non-dipole and relativistic effects, which become important at very high intensities. Secondly, a monochromatic laser field of constant intensity was taken to be present at all times. In fact, atoms interact with laser pulses of finite duration. These pulses must be sufficiently short, otherwise the neutral atoms would not be able to survive the rising edge of the pulse and enter the stabilization regime [29]. Indeed, they would ionize while crossing the valley-shaped part of their lifetime curve in Fig. 7.4, which acts as a "death valley" where their lifetime is extremely short. On the other hand, the laser pulses must be long enough so that the atom adiabatically remains in the ground state in the K–H frame. By comparing Floquet calculations with time-dependent ones, it has been shown that these criteria can both be fulfilled [30–32].

Due to the adiabaticity condition and the high-frequency, high-intensity requirements, the experimental verification of adiabatic stabilization for the ground state of atomic hydrogen is a very difficult task. However, Pont and Shakeshaft [30] and Vos and Gavrila [14] suggested that adiabatic stabilization could be observed with the existing experimental capabilities for Rydberg states of atoms. The advantage of considering Rydberg states is that, even for a relatively small angular frequency ω, the binding energy of a sufficiently high-lying state will be smaller than ω, so that the high-frequency condition (7.32) should be satisfied. The laser field must be chosen to have linear polarization, and the Rydberg state must have a high value of $|m|$, where m is the magnetic quantum number with respect to the polarization axis, for the following two reasons. Firstly, atomic states with high values of $|m|$ are essentially hydrogenic, so that the calculation of the lifetime of the states is simplified. Secondly, for linear polarization, the selection rule $\Delta m = 0$ holds, which implies that the initial state can couple only to states having the same value of m. In this case, the quantity $|w_i^{(0)}(\alpha_0)|$ appearing in Equations (7.32) will refer to the m-manifold to which the initial state belongs, and is of the order of magnitude of the binding energy of the lowest state of the manifold. In the absence of the laser field, the lowest state of quantum number m is a "circular" state with $|m| = l = n - 1$. Its binding energy is $|w_i^{(0)}(\alpha_0 = 0)| = 2^{-1}(|m| + 1)^{-2}$. By choosing $|m|$ to be high enough, $|w_i^{(0)}(\alpha_0)| < |w_i^{(0)}(0)|$ can be made small with respect to the angular frequencies of available intense lasers.

The calculations of Vos and Gavrila [14] were performed for the case $|m| = 5$. The lowest-lying states associated with this value of $|m|$ in the field-free case are the "circular" states of quantum numbers $n = 6$, $l = 5$ (ungerade case) and the $n = 7$, $l = 6$ (gerade case). Their decay rates were evaluated according to Equation (7.31), in which the first Born approximation (7.74) was used for the continuum eigenfunction $u_{k_n}^{(-)}(\mathbf{r})$. The computations were performed for the two angular frequencies $\omega = 1.17\,\mathrm{eV} = 0.043$ a.u. (corresponding to the Nd:YAG laser) and $\omega = 2.0\,\mathrm{eV} = 0.073$ a.u. (which is that of the experiments described in Section 7.3.3). Both frequencies are high enough so that the condition (7.32) is satisfied. The dependence of the lifetimes of the states on the laser intensity is similar to those displayed in Fig. 7.4 for the ground state. However, there is an important difference. For the circular Rydberg states, the minimum lifetimes are much larger than the duration of the high-intensity short laser pulses which were available at that time (about 50–100 fs), so that the neutral atom has a high probability of passing through the death valley without being ionized. For example, in the case of a Rydberg state evolving from $n = 7$, $l = 6$ at an angular frequency $\omega = 0.073$ a.u., the minimum lifetime was calculated to be 2000 fs [14]. Moreover, if the turn-on time of the laser pulse is not shorter than 100 fs, the laser-field variation can be considered to be adiabatic, which means that the atom will remain in its adiabatically evolving initial state.

Another difference between adiabatic stabilization for Rydberg states and the ground state is that the onset of stabilization occurs for the Rydberg states at lower intensities (below 0.1 a.u.) and, in the stabilization regime, their lifetimes increase faster as the laser intensity is increased.

Several more accurate Floquet calculations of Rydberg-state lifetimes have been performed [33–37]. These calculations, which are in good agreement with each other, have yielded lifetime curves similar in shape to those of Vos and Gavrila [14], but with much larger values for the minimum lifetimes, so that the situation is more favorable for the experimental verification of adiabatic stabilization.

Until now, we have considered only the hydrogen atom and Rydberg atoms, which are quasi-hydrogenic. The HFFT theory has been generalized and applied to two-electron atoms and ions [38–40]. An interesting prediction of the HFFT is that H^-, which in the field-free case has only one bound state, acquires light-induced states (LIS) [39]. Such states were introduced in Section 4.4.4. Another prediction of the HFFT is that, for large values of α_0, two-electron atoms and ions undergo dichotomy, as in the case of one-electron atoms [38, 40], and that in the dichotomy regime they are also stabilized [40]. Exotic atomic structures have also been predicted using the HFFT [41–43]. In particular, it was shown that in an intense, high-frequency laser field, a proton can bind more than two electrons, in contrast to the field-free case.

Light-induced states have also been found in one-dimensional model atomic systems described by short-range attractive potentials. Examples include the Gaussian potential [44–46] and the zero-range potential [47–49]. Generally, one-dimensional short-range potentials give rise to LIS, the number of which increases as α_0 increases. This feature can be understood by considering a square-well potential

$$V(x) = \begin{cases} -V_0, & |x| < a, \\ 0, & |x| > a, \end{cases} \tag{7.79}$$

where the positive constant V_0 is the depth of the well and a is its range. In the K–H frame, the corresponding static dressed potential is denoted by $V_0(\alpha_0, x)$. This potential becomes increasingly elongated and shallow as α_0 increases. Its range extends from $x = -(\alpha_0 + a)$ to $x = \alpha_0 + a$, while its depth at $x = 0$ is approximately proportional to α_0^{-1} for α_0 sufficiently large. For any α_0 and x, $V_0(\alpha_0, x) < W(\alpha_0, x)$, where $W(\alpha_0, x)$ is the field-free square-well potential

$$W(\alpha_0, x) = \begin{cases} -W_0, & |x| < R, \\ 0, & |x| > R, \end{cases} \tag{7.80}$$

with $R = |\alpha_0 - a|$ and $W_0 = V_0(\alpha_0, x = 0)$. The number of bound states supported by $W(\alpha_0, x)$ is proportional to $(W_0 R^2)^{1/2}$ [50] and increases without bound as α_0 increases. It follows that the number of bound states supported by the dressed square-well potential $V_0(\alpha_0, x)$ also increases without limit. This argument can be extended to any one-dimensional dressed potential corresponding to a field-free potential that has a short range and is attractive everywhere. This is not true for potentials having an attractive Coulomb tail because the infinite accumulation of bound states at the continuum threshold prevents additional ones from appearing at high intensities [51,52].

As pointed out by Potvliege [51,52], there is an important difference between the one-dimensional and three-dimensional models with regard to LIS for short-range potentials of finite depth. While in the one-dimensional case the number of LIS increases without limit when $\alpha_0 \to \infty$, this number tends to zero in the three-dimensional case. We remark that three-dimensional short-range potentials are often used to model negative ions such as H^-. However, in the high-frequency, high-intensity regime a two-electron description is needed [39–41].

7.3.2 Time-dependent studies of stabilization

We now turn to studies of atomic stabilization performed by integrating numerically the TDSE in the dipole approximation, as discussed in Chapter 5. In such studies, the initial state is usually taken to be an eigenstate $|\psi_i\rangle$ of the field-free atomic Hamiltonian. The laser pulse is then applied at time $t = t_0$, and the state vector $|\Psi_i(t)\rangle$ evolving from the initial state is propagated in time from t_0 until the end of the pulse at time $t_1 = t_0 + \tau_p$. The ionization probability P_{ion} is then given by Equation (5.58). We recall that P_{ion} depends on the initial atomic state and on the parameters of the laser pulse, namely its peak electric-field strength $\mathcal{E}_0$, carrier angular frequency ω, shape and duration. Within this context, *dynamic stabilization* refers to the phenomenon whereby the ionization probability P_{ion} at the end of a laser pulse of fixed envelope and carrier frequency does not approach unity as the peak laser intensity is increased. Instead, beyond some critical intensity, P_{ion} starts decreasing with increasing peak laser intensity or remains stable at some value less than unity.

Using the one-dimensional model atom described in Section 5.3 based on the soft-core Coulomb potential (5.72), Eberly and co-workers [53–58] have studied the dependence of the ionization probability P_{ion} on the peak electric-field strength $\mathcal{E}_0$ and carrier angular frequency ω. They found that indeed, for high values of ω, the ionization probability P_{ion} is an increasing function of $\mathcal{E}_0$, up to a certain critical value, and then begins to decrease. This is illustrated in Fig. 7.6, which shows the

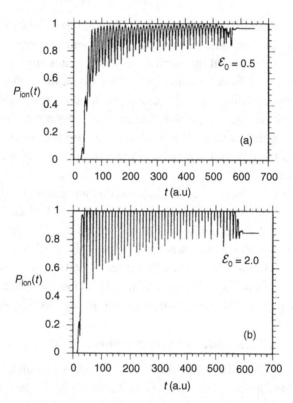

Figure 7.6. Ionization probability as a function of time (in a.u) for the ground state of a one-dimensional model atom, described by the soft-core Coulomb potential (5.72), interacting with laser pulses of carrier angular frequency $\omega = 2$ a.u. and peak electric-field strengths (a) $\mathcal{E}_0 = 0.5$ a.u and (b) $\mathcal{E}_0 = 2$ a.u. The laser pulses are turned on and off over 4.25 optical cycles = 13.4 a.u. The ionization probability is indicated at the end of the pulse. (From C. K. Law, Q. Su and J. H. Eberly, *Phys. Rev. A* **44**, 7844 (1991).)

results obtained by Law, Su and Eberly [56] for laser pulses of carrier angular frequency $\omega = 2$ a.u., turned on and off over 4.25 optical cycles, and having peak electric-field strengths $\mathcal{E}_0 = 0.5$ a.u. and 2 a.u., respectively. It is apparent from Fig. 7.6 that P_{ion} is greater for $\mathcal{E}_0 = 0.5$ a.u. than for $\mathcal{E}_0 = 2$ a.u., thereby demonstrating the dynamic stabilization of the model atom. In a subsequent calculation, Su, Irving and Eberly [58] evaluated P_{ion} for a laser pulse of carrier angular frequency $\omega = 0.8$ a.u. having a $\sin^2$ turn-on and turn-off of 5 optical cycles, and a flat top of 40 optical cycles in between. Their calculation displayed four ionization regimes. The first one, at low intensities, was the LOPT regime (in this case first-order perturbation theory since only one photon is required to ionize the system), where P_{ion} was found to be roughly proportional to the intensity I, as expected. This was followed by the death-valley regime at intensities of about 1 a.u., where P_{ion} was close to unity. Then came the dynamic stabilization regime, where P_{ion} decreased slowly with I in an oscillatory way. Finally, at still higher intensities ($I \geq 10^2$ a.u.), the calculations indicated a breakdown of stabilization, such that P_{ion} increased again, and to which we shall return at the end of this chapter. The influence on the stabilization dynamics of excited states that are populated during rapid laser pulse turn-ons has been analyzed by Reed, Knight and Burnett [59] and Vivirito and Knight [60].

Dynamic stabilization has also been obtained in other one-dimensional model calculations for potentials of finite range [61–63]. In particular, Grobe and Fedorov [61, 62] have demonstrated that, for very large excursion amplitudes, the initial evolution of the wave packet can be modeled by the spreading of a free wave packet in the K–H frame, as long as the value of α_0 during the turn-on of the laser pulse is larger than the width of the spreading wave packet. Patel, Kylstra and Knight [63] have studied the ionization probability P_{ion} for the potential (5.75). Their calculations, carried out for laser pulses with $\omega = 1$ a.u. and a $\sin^2$ pulse envelope, confirmed the existence of a dynamic stabilization regime, except for very short pulses, where P_{ion} was found to increase sharply with $\mathcal{E}_0$ and then to stay nearly constant. This behavior will be analyzed below.

As pointed out in Section 5.3, one-dimensional models are very convenient for performing "numerical experiments," but they also have drawbacks. Regarding stabilization, Ménis *et al.* [64] have shown that, in comparison with the three-dimensional Coulomb potential $V_c(r) = -1/r$, the soft-core Coulomb potential (5.72) can lead to an overestimation of stabilization due to the softening of the Coulomb singularity at the origin.

Using a two-dimensional model of atomic hydrogen, in which the soft-core Coulomb potential (5.78) was employed (with $Z = 1$ and $a = 0.8$ a.u.), Protopapas, Lappas and Knight [65] and Patel *et al.* [66] have studied stabilization for arbitrary laser polarizations. They used laser pulses with a two-cycle turn-on, which is short enough to ensure that the atom passes through death valley quickly enough. Their calculations were performed at the angular frequency $\omega = 1$ a.u. They found that stabilization increased as a function of the ellipticity, being highest for circular polarization. A simple explanation of this observation is that, when the electron wave packet misses the atomic core, increasingly as the absolute value of the ellipticity parameter increases, the cycle-averaged interaction strength with the atomic core decreases. As a result, the binding energies of the time-averaged dressed potential $V_0(\alpha_0, \mathbf{r})$ in the K–H frame decrease with increasing ellipticity, so that the high-frequency condition (7.32) is more readily satisfied when the ellipticity is larger, implying that circular polarization leads to the highest degree of stabilization.

A theoretical demonstration of dynamic stabilization for the hydrogen atom (in three dimensions), obtained from the numerical solution of the TDSE in the dipole approximation, was first given by Kulander, Schafer and Krause [67,68]. They used the hybrid representation discussed in Section 5.4, in which the time-dependent wave function of the system is first expanded in spherical harmonics, then discretized in the radial variable r and propagated in time. The signature of ionization was taken to be the decrease of the probability in a finite volume surrounding the atomic nucleus. Following the turn-on of the laser field, an approximately exponential decay of this probability was observed, from which an atomic ionization rate was extracted. By analyzing its peak electric-field dependence, it was found that this ionization rate starts to decrease above a critical intensity, which characterizes the stabilization of the atom. This behavior is illustrated in Fig. 7.7, where the ionization rate obtained by Kulander, Schafer and Krause [68] for a hydrogen atom in a laser field of carrier angular frequency $\omega = 1$ a.u. is plotted as a function of α_0. As seen from Fig. 7.7, the ionization rate first rises linearly, passes through a maximum near $\alpha_0 = 1.2$ a.u., and then decreases. This general behavior is similar to the adiabatic stabilization behavior predicted by the Sturmian-Floquet method [24] and the R-matrix–Floquet method [25] and illustrated in Fig. 7.5. Also shown in Fig. 7.7 are the ionization rates calculated by Pont and Gavrila [23] using the high-frequency Floquet theory. The peak ionization rate obtained by Kulander, Schafer and Krause [68] was 1.3×10^{15} s^{-1}, which is a significant fraction of the angular frequency, and corresponds to a lifetime of less than 1 fs or approximately 5 optical cycles. These results therefore confirm the conclusions based on the Floquet method and predict that an atom can survive the turn-on of a short, intense, high-frequency pulse, thereby allowing the stabilization regime to be reached. The stabilization of the hydrogen atom in the ground state has also been studied in the dipole approximation by Horbatsch [69], who solved the TDSE in the K–H frame.

As another illustration of the stabilization of H(1s), we show in Fig. 7.8 the results obtained by Latinne, Joachain and Dörr [70], who solved the TDSE in the laboratory frame for linearly polarized laser pulses having a carrier angular

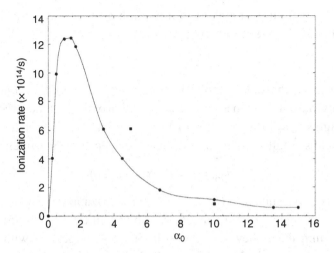

Figure 7.7. Ionization rate as a function of $\alpha_0 = \mathcal{E}_0/\omega^2$ (in a.u.) for a hydrogen atom in a laser field of angular frequency $\omega = 1$ a.u. (filled circles). Also shown are the ionization rates calculated by Pont and Gavrila by using the HFFT for two values of α_0 (filled squares). (From K. C. Kulander, K. J. Schafer and J. L. Krause, *Adv. At., Mol. Opt. Phys. Suppl.* **1**, 247 (1992).)

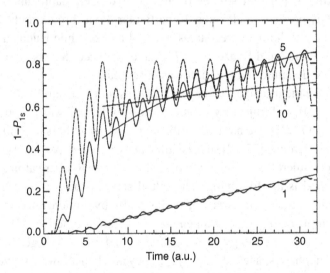

Figure 7.8. Ionized fraction versus time (in a.u.) for H(1s) in linearly polarized laser pulses of carrier angular frequency $\omega = 2$ a.u. and peak electric-field strengths $\mathcal{E}_0 = 1$ a.u., 5 a.u. and 10 a.u., as indicated next to the curves. The thin lines give the corresponding Floquet results. (From O. Latinne, C. J. Joachain and M. Dörr, *Europhys. Lett.* **26**, 333 (1994).)

frequency $\omega = 2$ a.u. and a short turn-on with a two-cycle ramp. In Fig. 7.8 the ionized fraction is given approximately by the quantity

$$P_{\text{ion}} \simeq 1 - P_{1s}(t) = 1 - \langle \psi_{1s} | \Psi(t) \rangle, \tag{7.81}$$

where $|\Psi(t)\rangle$ is the state vector which has evolved from the initial ground state $|\psi_{1s}\rangle$. This ionized fraction is plotted as a function of time for the three peak electric-field strengths $\mathcal{E}_0 = 1$ a.u., 5 a.u. and 10 a.u. The Floquet results at the corresponding electric-field strengths are fitted as decreasing exponentials, namely

$$P_{\text{ion}}^{\text{F}}(t) = P_0 \exp[-\Gamma(\mathcal{E}_0)t], \tag{7.82}$$

where P_0 is a constant and $\Gamma(\mathcal{E}_0)$ is the Floquet ionization rate (in a.u.) corresponding to the peak electric-field strength $\mathcal{E}_0$. The quantities $P_{\text{ion}}^{\text{F}}(t)$ are shown in Fig. 7.8 as the thin lines; they are seen to be in good agreement with the results of the time-dependent calculation. Both calculations exhibit stabilization, the ionized fraction corresponding to $\mathcal{E}_0 = 10$ a.u. being inferior to that for $\mathcal{E}_0 = 5$ a.u. The extra beat structure in the curve for $\mathcal{E}_0 = 10$ a.u. is due to interference between the 1s and higher Floquet states (mainly the 2s state), which are populated during the short turn-on of the pulse.

Other time-dependent studies of atomic hydrogen stabilization have focused on the behavior of excited states. In particular, Pont, Proulx and Shakeshaft [71] have considered the field-free states $n = 3$, 4 and 5, while Huens and Piraux [72] and Gajda, Piraux and Rzążewski [73] have analyzed the case $n = 2$. In all these calculations, dynamic stabilization was observed.

Finally, we mention classical Monte Carlo studies of stabilization. The method, known as "classical trajectory Monte Carlo simulation" was proposed by Abrines and Percival [74,75] and has been utilized by several authors [76–80]. In particular, Leopold and Percival [77,78] successfully analyzed the experiments of Bayfield and Koch [81] on microwave ionization of highly excited Rydberg atoms. The principle of the method is the following. The initial state of the system (for instance, the ground state of the atom) is modeled classically by a microcanonical ensemble of electrons in phase space, with an energy distribution function $\rho(E) \sim \delta(E - E_i)$, where $E_i < 0$ is the energy of the initial bound state. A large number of initial "samples" in phase space are employed. They are then allowed to evolve according to the classical equations of motion, and the analogs of quantum averages are obtained by averaging over the ensemble at later times. The ionization probability at the end of a laser pulse is taken to be the fraction of the classical trajectories having positive energies, corresponding to electrons escaping from the atom. By analyzing the dependence of the ionization probability on the peak electric-field strength of the laser pulse, classical aspects of the stabilization dynamics can be investigated.

The classical Monte Carlo method has been applied by several authors to investigate stabilization for one-dimensional or three-dimensional model atoms, in the *non-relativistic, dipole* approximation [64, 82–90]. These studies demonstrated that "classical atoms" also exhibit stabilization, with the stability being greater in one-dimensional models using a soft-core potential than in the three-dimensional Coulomb case. Classical *relativistic* Monte Carlo calculations will be discussed in Section 7.3.4

7.3.3 Experimental results

Two experiments, guided by the theoretical work on the adiabatic stabilization of Rydberg states described in Section 7.3.1, have been performed. The most favorable candidate, given the experimental conditions at that time, was selected to be the neon atom. The circular 5g ($n = 5, l = m = 4$) Rydberg state was chosen as the test case. The first experiment, carried out by de Boer *et al.* [91,92] involved two stages: the preparation of the initial Rydberg state and the detection of its ionization by an intense laser field. In the second experiment, performed by van Druten *et al.* [93], a third stage was added to monitor the atomic population left in the initial Rydberg state. This second experiment, which confirmed the first one, will now be described. The experiment involved the application of a sequence of three laser pulses, with the laser beams being collinearly overlapped in space and focused in a magnetic bottle electron spectrometer [94] filled with neon gas. The first pulse, or "preparation pulse," consisted of tightly focused, intense circularly polarized ultra-violet laser light (photon energy $\hbar\omega = 4.34$ eV, angular frequency $\omega = 0.159$ a.u., intensity $I \simeq 2 \times 10^{14}$ W cm^{-2}, pulse duration $\simeq 1$ ps). Its role was to excite the neon atoms from the ground state to the 5g Rydberg state. The second pulse, or "main pulse," driving the ionization process under study, was a short ($\simeq$90 fs), linearly polarized, intense ($I \simeq 10^{14}$ W cm^{-2}) pulse of red laser light (photon energy $\hbar\omega = 2$ eV, angular frequency $\omega = 0.073$ a.u.). The high-frequency condition (7.32) is satisfied for this main pulse and for the $m = 4$ manifold, whose largest binding energy is 0.02 a.u. The main pulse could be delayed by a variable time τ_d with respect to the preparation pulse. Finally, the third pulse, or "probe pulse," was a weakly focused, low-intensity long pulse of green laser light (photon energy $\hbar\omega = 2.33$ eV, angular frequency $\omega = 0.086$ a.u., intensity $I \leq 10^{10}$ W cm^{-2}), whose duration (5 ns) was sufficiently long to ionize all the atomic population in the 5g and neighboring states surviving the main laser pulse. It was delayed by 14 ns with respect to the preparation pulse.

The neon 5g Rydberg state was produced from the ground state by absorption of five photons from the preparation pulse. In a single-active-electron (SAE) approximation, and because of the $\Delta m = +1$ selection rule for absorption of left circularly polarized light, this process corresponds to a transition from the (2p, $m = -1$) orbital to the (5g, $m = 4$) orbital. The quantization axis is the propagation direction of the preparation pulse. In the weak-field limit, five photons are just sufficient to reach the continuum. However, during the turn-on of the preparation pulse, there is an increase of the ionization potential of the ground state. As a consequence, the ng Rydberg series is shifted into five-photon resonance with the ground state. Atomic population is transferred to these states, and a large fraction of it survives the preparation pulse. Thus, along with the 5g state, the 6g, 7g, ... states are also populated, although to a lesser extent.

When performing high-intensity laser experiments, there is always a spatial range of intensities in the focal region, which significantly complicates the interpretation of the results. This difficulty was circumvented in the present case by choosing the focal region of the main pulse to be much larger than that of the preparation pulse. The excited neon atoms were therefore interacting only with the central part of the main pulse, which had an approximately constant intensity profile. In this way, the problem of averaging over focal intensities was avoided. On the other hand, the fact that the laser beams corresponding to the preparation pulse and the main pulse were parallel created the following problem: the quantization axis of the magnetic states prepared by the preparation pulse was parallel to the propagation direction of the laser beams, and therefore perpendicular to the linear polarization direction of the main pulse, contrary to the requirements of the theory. This problem was solved by using the magnetic field of the electron spectrometer, which was perpendicular to the laser beams. In this magnetic field of 0.9 T, the $m = 4$ excited atomic states underwent Larmor precession, which rotated their quantization axis, allowing it to become perpendicular to the laser beam propagation direction. Thus, after one-quarter of the 80 ps precession period, the rotation angle was $\pi/2$ radians, and the 5g state had predominantly $m = 4$ character along a quantization axis parallel to the linear polarization direction of the main pulse, defining an optimum delay time $\tau_d = 20$ ps for observing the stabilization phenomenon.

A typical electron energy spectrum measured by van Druten *et al.* [93] is displayed in Fig. 7.9. The solid curve corresponds to the case when all three laser pulses were applied in succession, with $\tau_d = 20$ ps. Two series of photoelectron peaks are seen. One is due to the main pulse and the other to the probe pulse. The peaks correspond to one-photon absorption originating from the various (ng) states populated by the preparation pulse. The energy difference of the corresponding peaks is equal to that between the photons of the probe and the main laser pulses, namely 2.33 eV $-$ 2 eV $=$ 0.33 eV. The dashed curve in Fig. 7.9 corresponds to the spectrum obtained for $\tau_d = -5$ ps without the probe pulse, the negative value of τ_d meaning that the main pulse preceded the preparation pulse. Thus, the dashed curve represents the electron yield from the ground state due to the main pulse, and hence

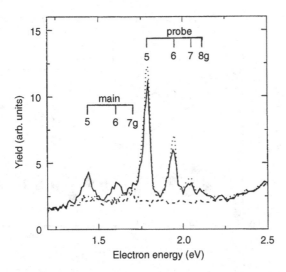

Figure 7.9. Photoelectron spectra of neon after interacting with the preparation, main and probe pulses, for a time delay $\tau_d = 20$ ps between the preparation and main pulses. The electron yield (in arbitrary units) is plotted as a function of the electron energy (in eV). For comparison, the spectra for $\tau_d = -5$ ps without (dashed curve) and with (dotted curve) the probe pulse are also shown. The expected electron energies for one-photon ionization from the ng states by the main and probe pulses are indicated. (From N. J. van Druten *et al.*, *Phys. Rev. A* **55**, 622 (1997).)

gives the background signal under the main-pulse absorption peaks in the three-pulse experiment. The dotted curve in Fig. 7.9 corresponds to the spectrum obtained for $\tau_d = -5$ ps with the probe pulse; it gives the electron yield in the absence of the main pulse, and therefore the total atomic population created in the 5g state. By subtracting the dashed curve from the full curve under the main-pulse peaks, and the full curve from the dotted curve under the probe-pulse peaks, van Druten *et al.* [93] could determine, respectively, the fraction of the 5g atomic population which had been ionized, and that which had survived.

These two fractions are shown in Fig. 7.10. We first remark that, within the error bars, the two fractions add up to unity, which indicates that there are no significant decay channels other than one-photon ionization, in agreement with the theoretical calculations. We also note that, above a critical peak intensity of the main pulse of about 6×10^{13} W cm^{-2} (corresponding to a fluence of 5 J cm^{-2}), hardly any additional ionization is observed when the peak intensity is increased to the maximum of about 2×10^{14} W cm^{-2}. Instead, the (one-photon) ionization yield from the main pulse saturates at about 25% of the initial atomic population, and conversely some 70% of it is left in the initial 5g state. These results are seen to be in marked contrast with the behavior expected from LOPT (in this case Fermi's

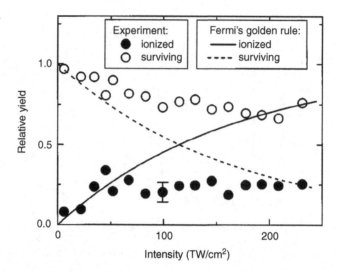

Figure 7.10. Measured ionized fraction (solid circles) and surviving fraction (open circles) of the 5g atomic population of neon, following the interaction with the main laser pulse, as a function of the intensity of that pulse. The theoretical results based on LOPT for the ionized (full curve) and surviving (dashed curve) fraction are also shown. (From N. J. van Druten *et al.*, *Phys. Rev. A* **55**, 622 (1997).)

golden rule for one-photon ionization), and provide a clear indication of adiabatic stabilization.

A theoretical analysis of the experiment of van Druten *et al.* [93] has been made by Piraux and Potvliege [32], using a one-electron model of neon. They calculated the ionization probability at the end of the pulse under the experimental conditions, using two methods. The first was an adiabatic approximation employing the Floquet theory to calculate the decay rate. The second method was the numerical integration of the TDSE. Several pulse shapes were investigated. Virtually identical results for the ionization probability were found for the same pulse shape using both methods, showing the adiabatic character of the atomic evolution. The fraction of the atomic population left in the initial state at the end of a 90 fs pulse is shown in Fig. 7.11, where it is compared with the surviving fraction measured directly by van Druten *et al.* [93] and estimated indirectly as the complement to unity of the measured ionized fraction. The solid line in Fig. 7.11 represents the results of Floquet calculations for a laser pulse with a $sech^2$ shape, assuming that the neon atom was initially in a pure $5g(m = 4)$ state, while in the experiment the atoms were initially in a superposition of several states with a small (at most 13%) admixture of states having $5g(m = 3)$ character. The importance of this admixture was tested by repeating the calculation for an incoherent superposition of the $5g(m = 4)$ and the $5g(m = 3)$ states in a 87/13 ratio. The resulting survival probability, corresponding

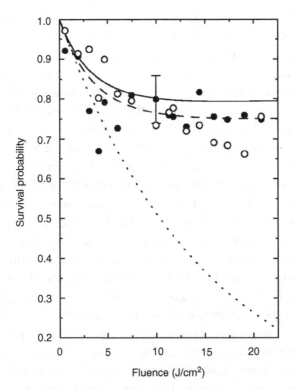

Figure 7.11. Survival probability of the 5g population of neon, as a function of the fluence (in J cm^{-2}). Open circles: surviving fraction measured by van Druten *et al.* [93]. Solid circles: ionized fraction [93]. Solid curve: non-perturbative survival probability calculated for a pure 5g($m = 4$) initial state. Dashed curve: same as the solid curve, but for an incoherent superposition of the 5g($m = 4$) state (87%) and the 5g($m = 3$) state (13%). Dotted curve: prediction of LOPT for a pure 5g($m = 4$) initial state. (From B. Piraux and R. M. Potvliege, *Phys. Rev. A* **57**, 5009 (1998).)

to the dashed line, is smaller than for the pure 5g($m = 4$) state, since ionization proceeds faster for the $m = 3$ state, but the difference is small. As seen from Fig. 7.11, the calculated probabilities and the experimental data are in fair quantitative agreement, thus confirming adiabatic stabilization. We also note that the LOPT results disagree with the experimental values for fluences larger than 5 J cm^{-2}.

7.3.4 Breakdown of stabilization

In order to observe stabilization, the laser pulses must be short enough so that the atom can survive the rising edge of the pulse, but also sufficiently long so that the atom adiabatically remains in the ground state in the K–H frame. As we have seen, it is possible to fulfil this condition, even for relatively short laser pulses. However, for intense, ultra-short laser pulses, this adiabaticity condition is not satisfied [60, 61, 63, 67, 68]. The system no longer stays in a single Floquet state, as assumed in the HFFT previously discussed, and its evolution must now be analyzed in terms of a multi-state Floquet theory (see Section 4.2.4). As the electric-field strength is increased, the atom will start feeling the turn-on of the laser pulse as a shock, and will make transitions to excited Floquet states, a process called "shake-up" [3]. In particular, excited states of the static dressed potential $V_0(\alpha_0, \mathbf{r})$ in the K–H frame, given by Equation (7.2), become populated. As a result, dichotomy becomes less apparent [59, 67], but stabilization still persists because ionization from higher Floquet states occurs with reduced rates. Indeed, all binding energies become smaller at high intensities, and it follows that more shake-up does not necessarily imply a larger ionization probability. However, at the end of the laser pulse, there will be atomic population in the field-free states associated with these discrete excited Floquet states. If the peak pulse intensity is further increased, atomic population will be transferred directly into the continuum during the turn-on of the laser pulse, which means that continuum Floquet states will be populated. When the laser pulse is turned off, the atoms will experience a second shock; some of this population may be transferred back to bound states of the atom, however most will remain in the continuum. This process is called "shake-off" [3].

As a consequence of shake-off, the ionization probability can once again begin to increase with increasing peak laser intensity. However, keeping the pulse envelope function constant, this probability will approximately converge to a fixed value that is independent of the laser angular frequency for asymptotically large peak laser intensities. Let us focus on the particular case of a linearly polarized laser pulse. We recall that using a one-dimensional model, Grobe and Fedorov [61, 62] have demonstrated that, for very large excursion amplitudes α_0, the initial evolution of the wave packet can be modeled by the spreading of a free wave packet in the K–H frame. As confirmed by Patel, Kylstra and Knight [63], this approximation works well as long as the width of the spreading wave packet along the laser polarization axis is much less than $2\alpha_0$. This regime can always be reached with a laser pulse of sufficiently high intensity. Let us now calculate the ionization probability within the sudden approximation. In the laboratory frame and working in the velocity gauge, the survival probability of an atom initially in its ground state is given by

$$P_{\text{survival}} = 1 - P_{\text{ion}}$$

$$= \sum_{j \in \text{bound states}} \left| \int d\mathbf{k} \exp[-iE_k(t_1 - t_0) - i\mathbf{k} \cdot \boldsymbol{\alpha}(t_1, t_0)] \langle \psi_j | \mathbf{k} \rangle \langle \mathbf{k} | \psi_0 \rangle \right|^2 .$$

$$(7.83)$$

We note that the experimental constraints imposed by the optical media generating the laser pulse lead, as discussed in Section 2.2, to the requirement that the displacement $\boldsymbol{\alpha}(t_0, t_1)$, given by Equation (2.77), of a classical electron initially at rest at the origin, must vanish. As a consequence, the survival probability depends only on the pulse duration. Patel, Kylstra and Knight [63] found that the approximation (7.83) gave a good estimate of the survival probability of a one-dimensional model atom interacting with a sufficiently intense, high-frequency laser field. They also pointed out that, in this regime, shake-off ionization implies that electrons are ejected at *low* energies. This has been illustrated by Førre *et al.* [95], who have calculated the ejected electron energy distribution for atomic hydrogen interacting with a five-cycle pulse of angular frequency $\omega = 2$ a.u. as a function of the peak electric-field strength. They showed that at moderately high laser intensities ATI peaks corresponding to multiphoton ionization are present. However, as the intensity is increased and the stabilization regime is entered, these ATI peaks are suppressed and electrons are preferentially ejected with low energies.

We shall now analyze the very high-intensity regime, where *non-dipole effects* (e.g. magnetic-field effects) and *relativistic effects* (at still higher intensities) become important. Let us begin by discussing *classical* calculations. We first consider the motion of a classical, free electron interacting with a linearly polarized monochromatic laser field propagating in the positive z-direction, with the electric-field vector $\boldsymbol{\mathcal{E}}(\omega t)$ and magnetic-field vector $\boldsymbol{\mathcal{B}}(\omega t)$ oscillating along the x-axis and the y-axis, respectively (see Section 2.8.1). As we have seen in Section 2.8, if this electron has no drift velocity, it will move along a "figure-of-eight" trajectory in the x-z plane, with the axis of the eight in the direction of the electric-field vector. However, the case of interest for our purposes is that of a classical, free electron, initially at rest, which interacts with a laser pulse of finite duration propagating in the z-direction. In this case, the electron acquires a drift velocity in that direction, which increases from zero to a maximum and then decreases to zero, [96–98]. As a result, the electron is left at rest at the end of the laser pulse. It is not displaced along the laser polarization direction. However, the electron is displaced along the laser propagation direction. Katsouleas and Mori [99] pointed out that in the case of an atomic electron this displacement reduces the probability of recapture of the electron by the parent core at the end of the laser pulse, and hence has a *destabilizing effect*.

Classical relativistic Monte Carlo studies of the interaction of atoms with intense laser fields were initiated by Kyrala [97] and pursued by Keitel and Knight [98], who investigated the dynamics of a hydrogen atom interacting with a high-frequency, high-intensity laser pulse. The results obtained by Keitel and Knight for a ten-cycle laser pulse of carrier angular frequency $\omega = 5$ a.u. are illustrated in Fig. 7.12. They show that stabilization breaks down as soon as non-dipole effects become important, namely when the magnetic-field component of the laser pulse is effective

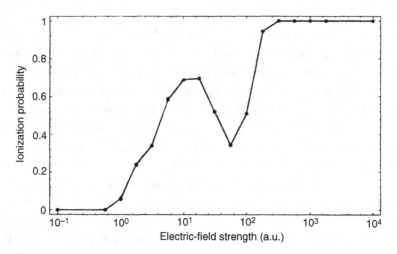

Figure 7.12. Ionization probability of atomic hydrogen as a function of the peak electric-field strength in a.u. (on a logarithmic scale) for a ten-cycle laser pulse of angular frequency $\omega = 5$ a.u. The results were obtained from three-dimensional, relativistic classical Monte Carlo simulations. (From C. H. Keitel and P. L. Knight, *Phys. Rev. A* **51**, 1420 (1995).)

in inducing a motion of the electron in the propagation direction. We see that in this very high-frequency case the ionization probability exhibits the stabilization behavior between the peak electric-field strengths $\mathcal{E}_0 = 10$ a.u. and 100 a.u., but approaches unity rapidly when the magnetic-field force becomes substantial.

We now turn to quantum calculations investigating the influence of non-dipole effects on atomic stabilization. Vázquez de Aldana and Roso [100], Kylstra *et al.* [101, 102], and Vázquez de Aldana *et al.* [103] have solved numerically, on a uniform grid, the TDSE for a two-dimensional model of atomic hydrogen, based on the soft-core Coulomb potential of the form (5.78) with $Z = 1$ and $a = 0.80$ a.u., interacting with a short laser pulse and taking into account non-dipole effects. The probability that the atom does not ionize during the laser pulse (its survival probability) was determined by projecting the wave function at the end of the pulse onto the field-free bound states of the atom.

In Fig. 7.13, the survival probability is displayed as a function of the peak electric-field strength $\mathcal{E}_0$, for an angular frequency $\omega = 1$ a.u. Results obtained in the dipole approximation (labeled D) and non-dipole results (labeled ND) are shown. It is seen that the dipole approximation breaks down when $\mathcal{E}_0 > 10$ a.u., corresponding to peak laser intensities $I > 3.5 \times 10^{18}$ W cm^{-2}. For $\mathcal{E}_0 = 10$ a.u., the displacement per optical cycle in the laser propagation direction is about 1 a.u. This displacement is comparable to the width of the electron wave packet along the z-direction and is sufficient to reduce the influence of the atomic potential, thereby disrupting the

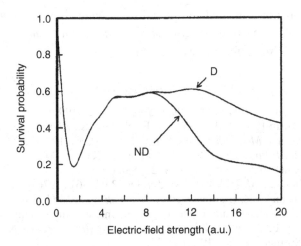

Figure 7.13. Survival probability, as a function of the peak electric-field strength $\mathcal{E}_0$ of the laser pulse, of a two-dimensional model of atomic hydrogen. A 12-cycle trapezoidal pulse (3-cycle turn-on and turn-off, 6-cycle constant intensity) with a carrier angular frequency $\omega = 1$ a.u. was used. Dipole (D) and non-dipole (ND) results are shown. (From J. R. Vázquez de Aldana *et al.*, *Phys. Rev. A* **64**, 013411 (2001).)

electron dynamics required for stabilization to occur. Indeed, as we have seen above, atomic stabilization relies on the formation of a wave packet of bound K-H states that is localized along the laser polarization axis between the turning points of the motion of a classical electron in the laser field. Beyond the dipole approximation, when non-dipole effects are taken into account, the magnetic-induced drift in the propagation direction results in the electron wave packet moving away from the nucleus. This, in turn, leads to a marked decrease of the survival probability of the atom, corresponding to a breakdown of stabilization.

It should be stressed that this breakdown of stabilization is a dynamical effect. Indeed, it is possible to formulate a relativistic HFFT and show that the resulting static K–H potential supports bound states [104]. By making use of a smoother $\sin^2$ pulse envelope of angular frequency $\omega = 1$, Førre *et al.* [95] have also calculated the ionization probability of a two-dimensional model hydrogen atom. They also found that the dipole approximation begins to break down at peak field strengths of about 10 a.u. However, the differences between the dipole and non-dipole ionization probabilities were smaller than those shown in Fig. 7.13. Førre *et al.* also solved the full three-dimensional non-dipole TDSE for atomic hydrogen. Their results indicate that non-dipole effects are smaller in three dimensions. It is interesting to point out that non-dipole effects result in electrons that are ejected opposite to the laser propagation direction [103, 105]. This phenomenon has been analyzed by Førre *et al.* [105].

We therefore conclude that atomic stabilization in high-frequency, high-intensity laser fields is limited to a *window of intensities*, where *non-dipole effects* (and a-fortiori *relativistic effects*) are negligible.

References

[1] M. Gavrila, *Adv. At. Mol. Opt. Phys. Suppl.* **1**, 435 (1992).
[2] M. Gavrila. In P. G. Burke and C. J. Joachain, eds., *Photon and Electron Collisions with Atoms and Molecules* (New York: Plenum Press, 1997), p. 47.
[3] M. Gavrila, *J. Phys. B: At. Mol. Opt. Phys.* **35**, R147 (2002).
[4] W. C. Henneberger, *Phys. Rev. Lett.* **21**, 838 (1968).
[5] J. I. Gersten and M. H. Mittleman, *J. Phys. B* **9**, 2561 (1976).
[6] M. Gavrila and J. Z. Kaminski, *Phys. Rev. Lett.* **52**, 613 (1984).
[7] M. Gavrila, Electron-atom interactions in intense, high-frequency laser fields. In F. Ehlotzky, ed., *Fundamentals of Laser Interactions*, Lecture Notes in Physics 229 (Berlin: Springer-Verlag, 1985), p. 3.
[8] M. Gavrila, M. Offerhaus and J. Kaminski, *Phys. Lett. A* **118**, 331 (1986).
[9] M. J. Offerhaus, J. Z. Kaminski and M. Gavrila, *Phys. Lett. A* **112**, 151 (1985).
[10] J. van de Ree, J. Kaminski and M. Gavrila, *Phys. Rev. A* **37**, 4536 (1988).
[11] M. Pont, N. Walet, M. Gavrila and C. W. McCurdy, *Phys. Rev. Lett.* **61**, 939 (1988).
[12] M. Pont, *Phys. Rev. A* **40**, 5659 (1989).
[13] M. Pont, N. Walet and M. Gavrila, *Phys. Rev. A* **41**, 477 (1990).
[14] R. J. Vos and M. Gavrila, *Phys. Rev. Lett.* **68**, 170 (1992).
[15] M. Abramowitz and I. A. Stegun, *Handbook of Mathematical Functions* (New York: Dover, 1970).
[16] I. S. Gradshteyn and I. M. Ryzhik, *Table of Integrals, Series and Products* (New York: Academic Press, 1980).
[17] B. H. Bransden and C. J. Joachain, *Physics of Atoms and Molecules*, 2nd edn (Harlow, UK: Prentice Hall-Pearson, 2003).
[18] J. von Neumann and E. P. Wigner, *Zeit. Phys.* **30**, 467 (1929).
[19] H. A. Bethe and E. E. Salpeter, *Quantum Mechanics of One- and Two-Electron Atoms* (Berlin: Springer-Verlag, 1957).
[20] C. Frahm, *Am. J. Phys.* **51**, 826 (1983).
[21] S. Pasternack and R. M. Sternheimer, *J. Math. Phys.* **3**, 1280 (1962).
[22] V. I. Ritus, *Sov. Phys. JETP* **24**, 1041 (1967).
[23] M. Pont and M. Gavrila, *Phys. Rev. Lett.* **65**, 2362 (1990).
[24] M. Dörr, R. M. Potvliege, D. Proulx and R. Shakeshaft, *Phys. Rev. A* **43**, 3729 (1991).
[25] M. Dörr, P. G. Burke, C. J. Joachain, C. J. Noble, J. Purvis and M. Terao-Dunseath, *J. Phys. B* **26**, L275 (1993).
[26] P. Marte and P. Zoller, *Phys. Rev. A* **43**, 1512 (1991).
[27] L. Dimou and F. H. M. Faisal, *Phys. Rev. A* **46**, 4442 (1992).
[28] R. Lefebvre and B. Stern, *Int. J. Quant. Chem.* **84**, 552 (2001).
[29] P. Lambropoulos, *Phys. Rev. Lett.* **55**, 2141 (1985).
[30] M. Pont and R. Shakeshaft, *Phys. Rev. A* **44**, R4110 (1991).
[31] J. Zakrzewski and D. Delande, *J. Phys. B* **28**, L667 (1995).
[32] B. Piraux and R. M. Potvliege, *Phys. Rev. A* **57**, 5009 (1998).
[33] R. M. Potvliege and P. H. G. Smith, *Phys. Rev. A* **48**, R46 (1993).
[34] A. Scrinzi, N. Elander and B. Piraux, *Phys. Rev. A* **48**, R2527 (1993).
[35] A. Buchleitner and D. Delande, *Phys. Rev. Lett.* **71**, 3633 (1993).
[36] L. Dimou and F. H. M. Faisal, *Phys. Rev. A* **49**, 4564 (1994).

[37] M. Baik, M. Pont and R. Shakeshaft, *Phys. Rev. A* **51**, 3117 (1995).
[38] M. H. Mittleman, *Phys. Rev. A* **42**, 5645 (1990).
[39] H. G. Muller and M. Gavrila, *Phys. Rev. Lett.* **71**, 1693 (1993).
[40] M. Gavrila and J. Shertzer, *Phys. Rev. A* **53**, 3431 (1996).
[41] E. van Duijn, M. Gavrila and H. G. Muller, *Phys. Rev. Lett.* **77**, 3759 (1996).
[42] E. van Duijn and H. G. Muller, *Phys. Rev. A* **56**, 2182 (1996).
[43] E. van Duijn and H. G. Muller, *Phys. Rev. A* **56**, 2192 (1996).
[44] J. N. Bardsley and M. J. Comella, *Phys. Rev. A* **39**, 2252 (1989).
[45] G. Yao and S. I. Chu, *Phys. Rev. A* **45**, 6735 (1992).
[46] M. Marinescu and M. Gavrila, *Phys. Rev. A* **53**, 2513 (1996).
[47] T. P. Grozdanov, P. S. Krstic and M. H. Mittleman, *Phys. Lett. A* **149**, 144 (1990).
[48] A. Sanpera, Q. Su and L. Roso-Franco, *Phys. Rev. A* **47**, 2312 (1993).
[49] M. Boca, C. Chirilă, M. Stroe and V. Florescu, *Phys. Lett. A* **286**, 410 (2001).
[50] B. H. Bransden and C. J. Joachain, *Quantum Mechanics*, 2nd edn (Harlow, UK: Prentice Hall-Pearson, 2000).
[51] R. M. Potvliege, *Phys. Rev. A* **62**, 013403 (2000).
[52] R. M. Potvliege, *Physica. Scripta* **68**, C18 (2003).
[53] J. Javanainen, J. Eberly and Q. Su, *Phys. Rev. A* **38**, 3430 (1988).
[54] Q. Su, J. H. Eberly and J. Javanainen, *Phys. Rev. Lett.* **64**, 862 (1990).
[55] Q. Su and J. Eberly, *J. Opt. Soc. Am. B* **7**, 564 (1990).
[56] C. K. Law, Q. Su and J. H. Eberly, *Phys. Rev. A* **44**, 7844 (1991).
[57] J. H. Eberly, R. Grobe, C. K. Law and Q. Su, *Adv. At. Mol. Opt. Phys. Suppl.* **1**, 301 (1992).
[58] Q. Su, B. P. Irving and J. Eberly, *Laser Phys.* **7**, 1 (1997).
[59] V. C. Reed, P. L. Knight and K. Burnett, *Phys. Rev. Lett.* **67**, 1415 (1991).
[60] S. Vivirito and P. L. Knight, *J. Phys. B* **28**, 4357 (1995).
[61] R. Grobe and M. Fedorov, *Phys. Rev. Lett.* **68**, 2592 (1992).
[62] R. Grobe and M. V. Fedorov, *J. Phys. B* **26**, 1181 (1993).
[63] A. Patel, N. J. Kylstra and P. L. Knight, *J. Phys. B* **32**, 5759 (1999).
[64] T. Ménis, R. Taïeb, V. Véniard and A. Maquet, *J. Phys. B* **25**, L263 (1992).
[65] M. Protopapas, D. G. Lappas and P. L. Knight, *Phys. Rev. Lett.* **79**, 4550 (1997).
[66] A. Patel, M. Protopapas, D. G. Lappas and P. L. Knight, *Phys. Rev. A* **58**, R2652 (1998).
[67] K. C. Kulander, K. J. Schafer and J. L. Krause, *Phys. Rev. Lett.* **66**, 2601 (1991).
[68] K. C. Kulander, K. J. Schafer and J. C. Krause, *Adv. At. Mol. Opt. Phys. Suppl.* **1**, 247 (1992).
[69] M. Horbatsch, *Phys. Rev. A* **44**, R5346 (1991).
[70] O. Latinne, C. J. Joachain and M. Dörr, *Europhys. Lett.* **26**, 333 (1994).
[71] M. Pont, D. Proulx and R. Shakeshaft, *Phys. Rev. A* **44**, 4486 (1991).
[72] E. Huens and B. Piraux, *Phys. Rev. A* **47**, 1568 (1993).
[73] M. Gajda, B. Piraux and K. Rzążewski, *Phys. Rev. A* **50**, 2528 (1994).
[74] R. Abrines and I. C. Percival, *Proc. Phys. Soc.* **88**, 861 (1966).
[75] R. Abrines and I. C. Percival, *Proc. Phys. Soc.* **88**, 873 (1966).
[76] I. C. Percival and D. Richards, *Adv. At. Mol. Phys.* **11**, 1 (1975).
[77] J. C. Leopold and I. C. Percival, *Phys. Rev. Lett.* **41**, 944 (1978).
[78] J. C. Leopold and I. C. Percival, *J. Phys. B* **12**, 709 (1979).
[79] G. Bandarage and R. Parson, *Phys. Rev. A* **41**, 5878 (1990).
[80] G. Bandarage, A. Maquet, T. Ménis, R. Taïeb, V. Véniard and J. Cooper, *Phys. Rev. A* **46**, 380 (1992).
[81] J. E. Bayfield and P. M. Koch, *Phys. Rev. Lett.* **33**, 258 (1974).
[82] J. Grochmalicki, M. Lewenstein and K. Rzążewski, *Phys. Rev. Lett.* **66**, 1038 (1991).

[83] R. Grobe and C. K. Law, *Phys. Rev. A* **44**, R4114 (1991).

[84] M. Gajda, J. Grochmalicki, M. Lewenstein and K. Rzążewski, *Phys. Rev. A* **46**, 1638 (1992).

[85] F. Benvenuto, G. Casati and D. L. Shepelyansky, *Phys. Rev. A* **45**, R7670 (1992).

[86] F. Benvenuto, G. Casati and D. L. Shepelyansky, *Phys. Rev. A* **47**, R786 (1993).

[87] F. Benvenuto, G. Casati and D. L. Shepelyansky, *Z. Phys. B* **94**, 481 (1994).

[88] J. Bestle, V. M. Akulin and W. P. Schleich, *Phys. Rev. A* **48**, 746 (1993).

[89] D. L. Shepelyansky, *Phys. Rev. A* **50**, 575 (1994).

[90] K. Rzążewski, M. Lewenstein and P. Salières, *Phys. Rev. A* **49**, 1196 (1994).

[91] M. P. de Boer, J. H. Hoogenraad, R. B. Vrijen, L. D. Noordam and H. G. Muller, *Phys. Rev. Lett.* **71**, 3263 (1993).

[92] M. P. de Boer, J. H. Hoogenraad, R. B. Vrijen, R. C. Constantinescu, L. D. Noordam and H. G. Muller, *Phys. Rev. A* **50**, 4085 (1994).

[93] N. J. van Druten, R. C. Constantinescu, J. M. Schins, H. Nieuwenhuize and H. G. Muller, *Phys. Rev. A* **55**, 622 (1997).

[94] P. Kruit and F. H. Read, *J. Phys. E* **16**, 313 (1983).

[95] M. Førre, S. Selstø, J. P. Hansen and L. B. Madsen, *Phys. Rev. Lett.* **95**, 043601 (2005).

[96] E. S. Sarachik and G. T. Schappert, *Phys. Rev. D* **1**, 2738 (1970).

[97] G. A. Kyrala, *J. Opt. Soc. Am. B* **4**, 731 (1987).

[98] C. H. Keitel and P. L. Knight, *Phys. Rev. A* **51**, 1420 (1995).

[99] T. Katsouleas and W. B. Mori, *Phys. Rev. Lett.* **70**, 1561(C) (1993).

[100] J. R. Vázquez de Aldana and L. Roso, *Opt. Exp.* **5**, 144 (1999).

[101] N. J. Kylstra, R. A. Worthington, A. Patel, P. L. Knight, J. R. Vázquez de Aldana and L. Roso, *Phys. Rev. Lett.* **85**, 1835 (2000).

[102] N. J. Kylstra, R. M. Potvliege, R. A. Worthington *et al.* In B. Piraux and K. Rzążewski, eds., *Super-Intense Laser-Atom Physics* (Dordrecht: Kluwer Academic, 2001), p. 345.

[103] J. R. Vázquez de Aldana, N. J. Kylstra, L. Roso, P. L. Knight, A. S. Patel, and R. A. Worthington, *Phys. Rev. A* **64**, 013411 (2001).

[104] P. S. Krstić and M. H. Mittleman, *Phys. Rev. A* **42**, 4037 (1990).

[105] M. Førre, J. P. Hansen, L. Kocbach and L. B. Madsen, *Phys. Rev. Lett.* **97**, 043601 (2006).

Part III

Multiphoton atomic physics

8

Multiphoton ionization

In this chapter, we focus our attention on the multiphoton ionization (MPI) of atoms interacting with intense laser fields with wavelengths in the infra-red to visible part of the spectrum. Section 8.1 is devoted to multiphoton single ionization. We begin by giving an overview of key early experiments, in particular those exploring the phenomenon of "above-threshold ionization" (ATI). We then discuss general features of ATI spectra and consider how these features can be understood within the framework of the semi-classical model. We conclude the section by examining how two-color processes can be used to study MPI. In Section 8.2, we analyze multiphoton double ionization, a process which has attracted considerable attention due to the prominent role played by electron correlation effects. Detailed reviews of atomic multiphoton ionization and ATI have been given by Joachain [1], DiMauro and Agostini [2], Protopapas, Keitel and Knight [3], Joachain, Dörr and Kylstra [4], Kylstra, Joachain and Dörr [5], Dörner *et al.* [6], Becker *et al.* [7] and Lewenstein and L'Huillier [8].

8.1 Multiphoton single ionization

As noted in Section 1.3, multiphoton ionization (MPI) was first observed in 1963 by Damon and Tomlinson [9] and also investigated in 1965 by Voronov and Delone [10] and Hall, Robinson and Branscomb [11]. In the following two decades, a number of experiments were performed to study various aspects of MPI, and results were obtained concerning the dependence of the ionization yields on the laser intensity, absolute MPI cross sections and the resonantly enhanced multiphoton ionization (REMPI) phenomenon. In all these experiments, the ions produced were taken as the signature of the MPI process.

At sufficiently low laser intensities $I \ll I_a$, where $I_a \simeq 3.5 \times 10^{16} \, \text{W cm}^{-2}$, perturbation theory can be used to analyze non-resonant multiphoton single ionization

of atomic systems [12–14]. As we have seen in Section 3.1, lowest-order perturbation theory (LOPT) predicts that the transition rate for the absorption of n photons is
proportional of I^n. Hence, if n_0 denotes the minimum number of photons required
to ionize an atomic system, the LOPT ionization rate is proportional to I^{n_0}. Thus,
if one defines an *index of non-linearity*

$$K = \frac{\partial (\log W_{\text{ion}})}{\partial (\log I)} \tag{8.1}$$

such that the total ionization rate W_{ion} is proportional to I^K, then LOPT predicts
that $K = n_0$.

As an example, let us consider the experiments of L'Huillier *et al.* [15] in
which the production of singly and multiply charged ions by multiphoton absorption in noble-gas atoms was investigated using 50 ps laser pulses at a wavelength
of 1064 nm, generated by a Nd:YAG laser, in the intensity range from 10^{13} to
10^{14} W cm^{-2}. The laser pulses were focused into a vacuum chamber pumped to
4×10^{-9} Torr and then filled with the noble gas to a pressure of 5×10^{-5} Torr.
The ions resulting from the interaction of laser pulses with the atoms in the focal
volume were extracted by using a transverse electric field of 1 kV cm^{-1}, separated
by a time-of-flight spectrometer and then detected in an electron multiplier. The
number of ions produced was measured as a function of the laser intensity. The
data obtained by L'Huillier *et al.* [15] for single and double ionization of helium
are shown in Fig. 8.1(a), where the number of helium ions produced is plotted
in a log-log plot as a function of the laser intensity. In Fig. 8.1(b), the 22-photon
and 68-photon absorption processes leading, respectively, to He$^+$ and He^{2+} ions
are schematically represented. Denoting by N^+ the number of singly charged ions
detected before the onset of saturation, L'Huillier *et al.* [15] deduced the index
of non-linearity K from the experimental data, displayed in log-log plots of the
type shown in Fig. 8.1(a), by determining the slope $\partial (\log N^+)/\partial (\log I)$ using least
squares fitting. As seen from Table 8.1, the values of K they deduced from their
data agree with the LOPT prediction $K = n_0$ within the experimental errors.

The vertical broken line in Fig. 8.1(a) indicates the saturation intensity I_s that,
for a given pulse duration, is defined as the intensity for which the corresponding
ionization probability is close to unity. In other words, when the saturation intensity
is reached, the ionization rate times the pulse duration is close to unity and the
atomic system will be ionized with high probability. The saturation effect therefore
leads to the depletion of atoms in the ionization volume [16], and, as seen from
Fig. 8.1(a), a marked change in the intensity dependence occurs above I_s. It is
important to remark that in order to study the behavior of atoms interacting with
laser pulses at a given intensity, it is necessary to ensure that this intensity is lower
than the saturation intensity I_s. This can be achieved by reducing the pulse duration.

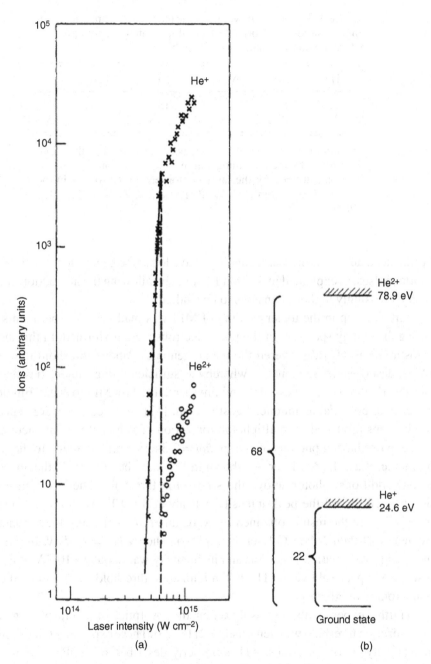

Figure 8.1. (a) Log-log plot of the variation of the number of helium ions measured as a function of the laser intensity I. The vertical broken line indicates the saturation intensity I_s. (b) Schematic representation of 22-photon and 68-photon processes leading to He$^+$ and He^{2+} ions, respectively. (From A. L'Huillier *et al.*, *J. Phys. B* **16**, 1363 (1983).)

Table 8.1. *Experimental values of the index of non-linearity K for singly charged ions formed in a multiphoton ionization process, before the onset of saturation, for noble gases.*

	He	Ne	Ar	Kr	Xe
n_0	22	19	14	13	11
K	21 ± 2	19 ± 2	13.6 ± 0.7	12.9 ± 0.5	10.8 ± 0.5

The minimum number of photons necessary to ionize the atom, n_0, is given by the next integer above the ionization potential of the atom, divided by the laser photon energy (1.165 eV in the present case). (From A. L'Huillier *et al.*, *J. Phys. B* **16**, 1363 (1983).)

Fortunately, advances in laser technology have led to the generation of increasingly shorter pulses, as explained in Section 1.2, thereby allowing the interaction of atoms with increasingly higher intensities to be studied.

A crucial step in the understanding of MPI was made when experiments measuring the energy spectrum of the photoelectrons were performed. In this manner, Agostini *et al.* [17] discovered the phenomenon of "above-threshold ionization" (ATI), described in Section 1.3, whereby at sufficiently high intensities the atom can absorb photons in excess of the minimum required for MPI to occur. Employing a retarding potential technique, Agostini *et al.* [17] analyzed the energy spectrum of electrons produced by multiphoton ionization of xenon atoms interacting with a laser pulse having photons of energy $\hbar\omega = 2.34$ eV, corresponding to the second harmonic of a Nd:YAG laser. As shown in Fig. 8.2, they observed the absorption of one additional photon above the six-photon threshold. The laser pulse duration was 10 ns and the peak intensity was about 8×10^{12} W cm^{-2}. Also shown in Fig. 8.2 are the results obtained by Agostini *et al.* [17] using the fundamental frequency of their Nd:YAG laser, with photon energy $\hbar\omega = 1.17$ eV. In this case, the laser pulse duration was 12 ns and its intensity was about 4×10^{13} W cm^{-2}. No distinct ATI peaks above the 11-photon ionization threshold were observed due to ponderomotive effects.

In further experiments, also using xenon, the absorption of several photons above the ionization threshold was demonstrated [18–23]. These experiments showed that, at relatively weak intensities, ATI is properly described by LOPT. The intensity dependence of the ATI peaks follows the power law I^{n_0+s} predicted by LOPT, where n_0 is the minimum number of photons required to ionize the atom and s is the number of excess photons (see Section 1.3). However, in studying the multiphoton ionization of xenon using a Nd:YAG laser generating pulses of 10 ns

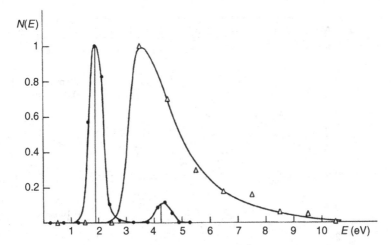

Figure 8.2. Energy spectra of electrons produced by multiphoton ionization of xenon atoms, for two photon energies. Triangles: $\hbar\omega = 1.17\,\text{eV}$, $I = 4 \times 10^{13}\,\text{W cm}^{-2}$. Circles: $\hbar\omega = 2.34\,\text{eV}$, $I = 8 \times 10^{12}\,\text{W cm}^{-2}$. (From P. Agostini *et al.*, *Phys. Rev. Lett.* **42**, 1127 (1979).)

duration and photon energy $\hbar\omega = 1.17$ eV (so that $n_0 = 11$), Kruit *et al.* [22] observed that, for laser intensities of about 10^{13} W cm^{-2}, important departures from the perturbative behavior occurred. Indeed, as the intensity increases, the intensity dependence of the ATI peaks did not follow the I^{n_0+s} power law and in addition the low-energy peaks were reduced in magnitude, as explained in Section 1.3. This behavior is illustrated in Fig. 8.3, which shows the results of Kruit *et al.* [22], and in Fig. 1.4(b), which displays those obtained by Petite, Agostini and Muller [23].

An important trend in high-intensity laser–atom physics is the use of increasingly shorter laser pulses. In the case of ATI, the first significant step in this direction was taken by Freeman *et al.* [24, 25]. Using sub-picosecond pulses, they showed that the ponderomotive effects that are present in ATI spectra obtained with longer pulses [26] could be suppressed. In particular, the ATI peaks were found to exhibit a substructure (see Fig. 1.6) because the intensity-dependent AC Stark shifts bring different states of the atom into multiphoton resonance during the laser pulse, as explained in Sections 1.3 and 4.4.3.

Sub-picosecond pulses have also been used by Rottke *et al.* [27–30] to measure the energies and angular distributions of the photoelectrons produced during the MPI of atomic hydrogen. The laser used in these experiments was a mode-locked dye laser pumped by the second harmonic of a Nd:YAG laser, giving pulses with wavelengths in the range 596–616 nm, intensities of the order of 10^{14} W cm^{-2}

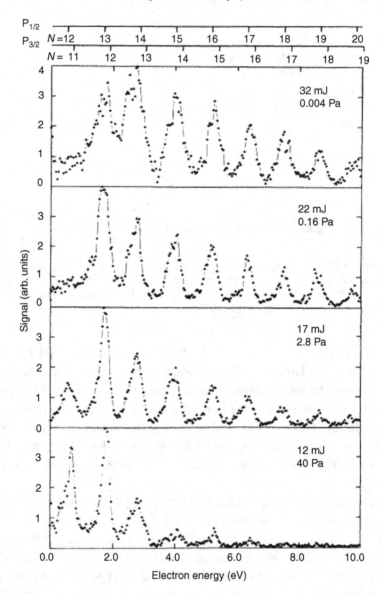

Figure 8.3. Energy spectra of electrons produced by multiphoton ionization of xenon atoms by laser pulses of photon energy $\hbar\omega = 1.17\,\text{eV}$. The pulse energy E_{pulse} (in mJ) and pressure (in Pa) at which each spectrum is taken is given. The estimated intensity is given by E_{pulse} (in mJ) $\times 2 \times 10^{12}\,\text{W cm}^{-2}$. (From P. Kruit *et al.*, *Phys. Rev. A* **28**, 248 (1983).)

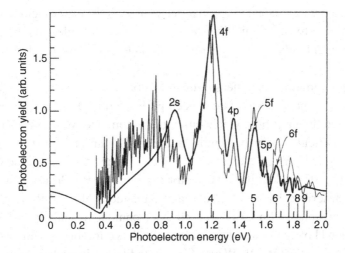

Figure 8.4. Yield of photoelectrons, into the lowest open channel, as a function of the photoelectron energy, for ionization of H(1s) by linearly polarized laser pulses of carrier wavelength $\lambda = 608$ nm, peak intensity $I = 6.6 \times 10^{13}$ W cm^{-2} and whose duration is 0.5 ps. Thick line: theoretical calculations using the Sturmian-Floquet method [31]. Thin line: experimental data [32]. (From R. M. Potvliege and R. Shakeshaft, *Adv. At. Mol. Opt. Phys. Suppl.* **1**, 373 (1992).)

and a duration of 0.5 ps. In Fig. 8.4, the results of these experiments are compared with non-perturbative Sturmian-Floquet calculations [31] (see Section 4.6). The low-energy part of the photoelectron spectrum for the multiphoton ionization of H(1s) by linearly polarized laser pulses of carrier wavelength $\lambda = 608$ nm, peak intensity 6.6×10^{13} W cm^{-2} and a duration of 0.5 ps is shown. In order to compare the Sturmian-Floquet results with the experimental ATI spectrum, the photoelectron yield was obtained as a function of the electron energy by dividing the range of intensities spanned by the pulse into small intervals within which the intensity was considered to be constant. The yield for each intensity interval, as a function of position with respect to the laser focus, was obtained by multiplying the Floquet ionization rate by the time during which the intensity was within that interval. The yield for the whole pulse was then calculated by integrating the spatially resolved yields over space. The intensity distribution of photoelectrons obtained in this way was converted into an energy distribution under the assumption that, in each intensity interval, the photoelectrons were all emitted with a kinetic energy equal to the difference between the real part of the quasi-energy (counted from the ionization threshold) and the energy the electron must have gained from the laser field, by multiphoton absorption, in order to be released in the lowest open ATI channel. In addition, it was assumed that most photoelectrons are emitted from the dressed initial state and in the lowest ATI channel. As shown by Fig. 8.4, this procedure gives results that are in reasonably good agreement with the experimental data [32]. The peaks are due to Stark-shift-induced resonances. A detailed comparison between the experimental data and the Sturmian-Floquet theory results has been given by Rottke *et al.* [30].

A direct, simultaneous measurement of both the energy and the angular distributions of the photoelectrons has been performed by Helm and Dyer [33]. They investigated the multiphoton ionization of helium in a strong sub-picosecond laser field. In particular, at wavelengths between 310 nm and 330 nm and intensities between 8×10^{13} W cm^{-2} and 5×10^{14} W cm^{-2}, they found that resonant enhancement via the AC Stark-shifted six-photon resonant states 1s3d and 1s3s is a dominant path, as described by Perry, Szöke and Kulander [34] and Rudolph *et al.* [35]. Schyja, Lang and Helm [36] also used this technique to study the effect of channel closing in ATI of xenon with laser pulses of 100 fs duration, 800 nm wavelength and intensities around 2×10^{13} W cm^{-2} generated by a Ti:sapphire laser with chirped pulse amplifier at a 1 kHz repetition rate.

Several experimental studies of MPI of noble gases in the tunneling regime, $\gamma_K \lesssim 1$, where the Keldysh parameter γ_K is defined by Equation (1.8), have been performed [37–40]. In particular, Augst *et al.* [39] studied tunneling ionization of noble gases using a CPA Nd:glass laser system delivering pulses of wavelength $\lambda = 1053$ nm having durations of about 1 ps and intensities up to the mid 10^{16} W cm^{-2}. They determined the critical intensities I_c at which over-the-barrier ionization (OBI) occurs. The evolution from the multiphoton ionization regime ($\gamma_K \gtrsim 1$) to the tunneling regime ($\gamma_K \lesssim 1$) has been studied by Mevel *et al.* [40], who measured electron energy spectra for the ionization of noble gases by 617 nm, 100 fs laser pulses having intensities ranging from about 10^{13} W cm^{-2} to more than 5×10^{16} W cm^{-2}. They found that, at the lowest intensities, the ionization process is clearly of multiphoton character and the AC Stark shift-induced resonances are dominant. At intensities about ten times larger, the resonances are no longer apparent, but the ATI structure in the photoelectron spectrum is still present. Eventually, this structure is also lost (see Fig. 1.8), as expected, when OBI takes place, so that the ionization process is concentrated in times of the order of one optical period.

The availability of laser systems delivering high-intensity, very short pulses at kilohertz repetition rates has allowed measurements of photoelectron energy spectra to be made with high sensitivity [41–49]. In this way, new features of the ATI phenomenon have been uncovered. In particular, Paulus *et al.* [43] found a "plateau" in the ATI photoelectron energy spectra of noble-gas atoms. Their laser system consisted of a colliding-pulse mode-locked dye oscillator, whose pulses were amplified in a two-stage dye amplifier pumped by a copper vapor laser at a repetition rate of 6.2 kHz. After compression in a prism sequence, laser pulses having an energy up to $20\,\mu$J and a duration of about 40 fs were obtained at a wavelength $\lambda = 630$ nm. The laser beam was focused into a vacuum chamber, giving rise to intensities up to about 5×10^{14} W cm^{-2}. The background pressure in the vacuum chamber was kept below 10^{-8} Torr and the target gas pressure in the chamber was 10^{-7} to 10^{-8} Torr. Photoelectrons were analyzed with a high-resolution time-of-flight spectrometer. The results of their experiment are shown in Fig. 1.9.

A plateau is also observed in theoretical ATI spectra. As an example, we show in Fig. 8.5 two ATI spectra, obtained by Paulus *et al.* [45], who solved numerically the time-dependent Schrödinger equation (TDSE) for atomic hydrogen in the dipole approximation, for linearly polarized laser pulses having peak intensities of 0.43×10^{14} W cm^{-2} and 1.05×10^{14} W cm^{-2}, respectively, and a carrier wavelength of 630 nm. The duration of the sin^2 pulse used in the calculation was 30 optical cycles, which is equal to 63 fs. As seen from Fig. 8.5, a plateau feature is seen for the higher laser intensity. It is more pronounced than the plateaus observed experimentally, which may be attributed to the fact that the calculations did not take into account the intensity variation in the laser focus.

As we have seen in Chapter 6, the existence of the plateau and other basic features of ATI spectra can be explained [50] by using the semi-classical recollision model [51, 52], which is applicable when the laser field intensity is high and its frequency is low, so that $\gamma_K \lesssim 1$. We recall that, according to this model, in the first step an

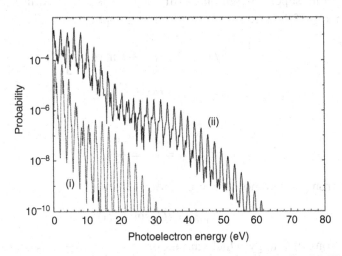

Figure 8.5. Two ATI spectra, calculated by solving numerically the TDSE, for multiphoton ionization of H(1s) by linearly polarized laser pulses of carrier wavelength $\lambda = 630$ nm and duration 63 fs. (i) Peak intensity of 0.43×10^{14} W cm^{-2}. (ii) Peak intensity of 1.05×10^{14} W cm^{-2}. (From G. G. Paulus *et al.*, *J. Phys. B* **29**, L249 (1996).)

electron is assumed to be detached at some time t_d. Due to the large oscillation amplitude of the electron in the intense laser field, the electron is described classically and the influence of the atomic potential on the electron's motion is neglected in the first approximation. Assuming that the laser field is described as a spatially homogeneous electric field $\mathcal{E}(t)$, and taking the electron drift velocity at the time t_d to be zero, from Equation (2.72) its velocity at the later time t is (in a.u.)

$$\mathbf{v}(t) = \mathbf{A}(t) - \mathbf{A}(t_d), \tag{8.2}$$

where $\mathbf{A}(t)$ is the corresponding vector potential of the laser field. The electron kinetic energy is then

$$E_{\text{kin}}(t, t_d) = \frac{1}{2}[\mathbf{A}(t) - \mathbf{A}(t_d)]^2. \tag{8.3}$$

Let us focus on the particular case in which the laser field is modeled by a spatially homogeneous, monochromatic, linearly polarized electric field of the form

$$\mathcal{E}(t) = \hat{\boldsymbol{\epsilon}}\mathcal{E}_0\cos(\omega t), \tag{8.4}$$

so that the electron velocity is given by

$$\mathbf{v}(t) = -\hat{\boldsymbol{\epsilon}}\frac{\mathcal{E}_0}{\omega}[\sin(\omega t) - \sin(\omega t_d)]. \tag{8.5}$$

The acceleration of the ejected electron in the laser field results in an oscillatory quiver velocity superimposed on a drift velocity, as discussed in Section 2.2. The kinetic energy of the electron, averaged over an optical cycle, is given by

$$\frac{1}{2}\langle v^2(t)\rangle = 2U_p\left[\frac{1}{2} + \sin^2(\omega t_d)\right]$$

$$= E_{\text{quiver}} + E_{\text{drift}}, \tag{8.6}$$

where

$$E_{\text{quiver}} = U_p = \frac{\mathcal{E}_0^2}{4\omega^2} \tag{8.7}$$

is the electron ponderomotive energy and

$$E_{\text{drift}} = 2U_p\sin^2(\omega t_d) \tag{8.8}$$

is the electron drift energy, which can take on values between zero and $2U_p$. Equation (8.6) predicts a maximum cut-off energy of the ATI spectrum of $3U_p$, including one U_p of quiver energy which is normally lost for pulses short enough so that the laser electric field is turned off before the electron leaves the laser focus. On the contrary, for long pulses, all or part of the quiver energy U_p of the electron may be converted into kinetic energy due to the ponderomotive force as the electron leaves the laser focus. If we restrict our attention to ATI spectra obtained with short laser pulses, the measured energy of the photoelectrons is given by the drift energy (8.8), whose maximum value is $2U_p$. We note that electrons ejected from the atom at a phase ωt_d, such that the magnitude of the electric field (8.4) has a maximum, will not gain any drift energy from the laser field. In contrast, electrons released into the laser field when the electric field amplitude vanishes gain the maximum drift energy of $2U_p$.

The simple classical argument given above, which predicts the existence of a cut-off energy of $2U_p$ for the photoelectrons, does not hold exactly due to the quantum-mechanical uncertainty associated with the "ionization time" of the electron. Indeed, it was shown in Section 6.3 that the strong-field approximation (SFA) transition amplitude corresponding to direct ionization of an electron with a final drift energy greater than $2U_p$ is non-zero, but falls off exponentially with increasing electron energy. This cut-off law is in qualitative agreement with experiments and numerical solutions of the TDSE, both of which show a drop in the yield of photoelectrons in the *low-order* part of the ATI spectrum at about $2U_p$ [50]. The cut-off becomes more apparent as the laser intensity increases and its frequency decreases (within the limits of validity of the dipole approximation), that is when the Keldysh parameter γ_K becomes smaller. Under these conditions the dynamics of the ionized electron becomes more "classical" as the quiver amplitude α_0 of the electron increases.

The foregoing classical arguments indicate that the global shape of the low-order part of the ATI spectrum will depend on the time during the laser cycle at which the electron is ejected. This was also seen to be the case in our analysis of the SFA direct ionization amplitude in Section 6.3. Most of the electrons are ejected near the maxima of the magnitude of the electric field, and will therefore acquire little drift energy. As a result, in the case of linear polarization, the majority of the photoelectrons will have low energies. We will return to this point below, where we will consider ATI in an elliptically polarized laser field. It is also worth pointing out that the features of the low-order ATI spectra are, to a large extent, independent of the atomic species, since the influence of the atom is limited to the production, via tunneling, of a "free" electron in the laser field.

We now return to the ATI plateau measured by Paulus *et al.* [43]. Classically, photoelectron energies in excess of $2U_p$, corresponding to "super-ponderomotive" or "hot" electrons, can only be acquired through an additional interaction of the photoelectron with its parent core. In particular, the backward-scattering of an ionized electron that returns to its parent core can give rise to a large change of the drift velocity if the scattering takes place when the magnitude of the electric field is small, as discussed in Section 6.5. This leads to a high-order ATI cut-off energy of $10U_p$ (see Section 6.5). The existence of this cut-off has been confirmed by experiments and theory [43,50,53–56]. In contrast to low-order ATI spectra, the details of the interaction between the recolliding electron and its parent core depend on the atomic species. Therefore, different atoms can exhibit different structures in the high-order part of the ATI spectrum [46–49,57].

Experiments on ATI have also been performed using ultra-short laser pulses. Grasbon *et al.* [56] recorded ATI photoelectron spectra of noble-gas atoms ionized with intense few-cycle pulses. Their laser system consisted of a 10 fs Ti:sapphire oscillator, a multipass amplifier and a prism compressor. The laser pulses were then spectrally broadened in a gas-filled hollow fiber and compressed to a duration of 7 to 8 fs by means of chirped mirrors. The pulse energy was about $400 \, \mu J$ at a repetition rate of 1 kHz. Using focusing mirrors, intensities up to $5 \times 10^{14} \, W \, cm^{-2}$ could be achieved.

We show in Fig. 8.6 ATI photoelectron energy spectra of argon obtained by Grasbon *et al.* [56] at different intensities (Fig. 8.6(a)) and for various pulse durations (Fig. 8.6(b)). Figure 8.6(a) demonstrates the main features that are observed in ATI spectra of noble gases interacting with short, intense, low-frequency laser pulses. Firstly, electrons are ejected preferentially with energies less than $2U_p$. Secondly, beyond $2U_p$, the electron yield decreases over many orders of magnitude, and it then levels out into a plateau that persists to approximately $10U_p$. In Fig. 8.6(b), it can be seen that the plateau-like structure at high electron energies persists as the laser pulse duration is decreased. This is consistent with the semi-classical recollision model.

A third important feature is the resonance-like enhancements seen, particularly for the ATI spectrum obtained using a 40 fs pulse, in the ATI plateau around 20 eV.

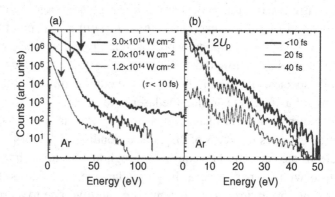

Figure 8.6. (a) ATI spectra of argon at different intensities. Arrows indicate the energy position $2U_p$. (b) ATI spectra of argon for different pulse durations. The spectrum corresponding to 40 fs was taken from Paulus *et al.* [54]. (From F. Grasbon *et al.*, *Phys. Rev. Lett.* **91**, 173003 (2003).)

These prominent groups of ATI peaks were originally studied experimentally in atomic hydrogen by Paulus *et al.* [45] and in argon by Hertlein, Bucksbaum and Muller [46] and theoretically by Muller and Kooiman [57]. Highly resolved ATI photoelectron energy spectra have also been measured in xenon by Hansch, Walker and van Woerkom [47,48] and in argon by Nandor *et al.* [49]. Marked enhancements of up to an order of magnitude of regions of the ATI spectrum were observed for small changes (a few percent) of the peak intensity of the laser pulse.

A detailed theoretical study of resonant enhancement of high-order ATI peaks in the photoelectron spectrum of argon at a carrier wavelength of 789.3 nm has been made by Muller [58], who integrated the TDSE using the single-active-electron (SAE) approximation discussed in Section 5.5. He concluded that these ATI peaks in the back-scattering region of the spectrum are almost exclusively due to resonances. From his analysis of the wave function, he deduced two types of resonant states: (i) high-angular-momentum states that remain far from the nucleus and decay mainly by emission of low-energy electrons and (ii) states that are located near the polarization axis, which decay through collisions with the ionic core and lead to the enhancement of high-energy photoelectrons. A detailed comparison of experimental ATI spectra in argon (obtained by using 120 fs laser pulses having a carrier wavelength of 800 nm and a repetition rate of 1 kHz) and the corresponding numerical simulations of Muller [58] has been made by Nandor *et al.* [49]. As an example, we show in Fig. 8.7 a comparison between experiment and theory for sample high energy ATI spectra at various laser intensities. The excellent agreement between the experimental and theoretical results found by Nandor *et al.* [49] shows that the SAE approximation describes the physics of high-intensity multiphoton single ionization to a high-degree of accuracy. The observed resonant enhancements are therefore not due to electron correlation effects [49, 57–59].

The ATI resonance phenomenon has also been studied within the SFA by Paulus *et al.* [54] and Kopold *et al.* [55] using a model atom described by a zero-range potential. We recall that the SFA does not incorporate excited bound states of the atom. They showed that channel closings could be responsible for the observed resonant enhancements. As discussed in Chapter 1, the effective ionization potential of the atom in a low-frequency laser field increases by approximately the ponderomotive energy. At laser intensities just below a channel closing, electrons are emitted with nearly zero drift momentum. These electrons can rescatter from their respective parent cores a number of times, leading to enhancements in the plateau region of ATI spectra. As this mechanism can only lead to enhancements if the laser pulse duration is sufficiently long, no enhancements should be observed in ATI spectra from ultra-short laser pulses. As seen in Fig. 8.6(b), Grasbon *et al.* [56] found that indeed these resonance-like features are suppressed when argon atoms interact with ultra-short laser pulses.

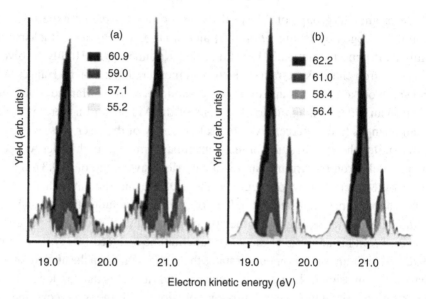

Figure 8.7. Sample high-kinetic-energy spectra for various laser intensities, given in units of $10^{12}\,\mathrm{W\,cm^{-2}}$, obtained for argon atoms interacting with laser pulses having a carrier wavelength of 800 nm. The energy range corresponds to 12th and 13th ATI order. (a) Experimental data; (b) calculations. (From M. J. Nandor *et al.*, *Phys. Rev. A* **60**, R1771 (1999).)

Wassaf *et al.* [60, 61] have solved the one-dimensional TDSE for the soft-core Coulomb potential (5.73) as well as for the short-range soft-core Yukawa and Pöschl–Teller potentials. Using Floquet quasi-energy spectra and classical simulations to analyze the enhancement phenomenon observed in their calculated ATI spectra, they concluded that (i) a channel-closing mechanism can explain ATI enhancements for an atom modeled by a short-range potential and (ii) multiphoton resonances are at the origin of the enhancements found in ATI spectra of an atom modeled by a long-range potential. Using the Sturmian-Floquet method in the SAE approximation, Potvliege and Vučić [62, 63] calculated the quasi-energy spectrum of argon interacting with a linearly polarized 800 nm laser field with peak intensities up to $7 \times 10^{13}\,\mathrm{W\,cm^{-2}}$. From this quasi-energy spectrum they were able to relate the plateau ATI enhancements to AC Stark shift-induced resonances, thereby supporting the interpretation that these resonances are responsible for the observed enhancements in the ATI spectra of argon.

Instead of analyzing ATI yields as a function of the photoelectron energy, one can investigate ATI angular distributions where the electron yield is measured as a function of the photoelectron emission angle θ_f, with the photoelectron energy as a parameter. Experiments have revealed the presence of sharp peaks in the angular distributions of photoelectrons emitted at certain energies [42, 64]. These angular peaks are sometimes called "rescattering rings" because the angular distributions are symmetric about the polarization axis of a linearly polarized laser field.

The semi-classical recollision model is able to shed light onto the origin of these peaks. When the Keldysh parameter γ_K is small, the angular distribution of electrons corresponding to *low-order* ATI are largely determined by the momentum distribution of the tunneling wave packets, and are strongly peaked along and opposite to the laser polarization axis. Approximate tunneling electron momentum distributions have been given by Delone and Krainov [65]. The *high-order* ATI electrons, which have rescattered from their parent core, display angular distributions that depend on the energy of the detected electron [42, 64]. The form of the angular distributions is determined by the species of the parent core from which the returning electron scatters, as well as by the dynamics of the scattered electron in the laser field. As noted above, a classical electron that is back-scattered by its parent core can acquire a final drift energy of up to $10U_p$. These electrons gain their additional drift velocity by being accelerated approximately an additional half-cycle by the electric-field component of the laser field. Let us assume that the laser field is linearly polarized and that an electron scatters elastically at time t_r by some angle θ with respect to the laser polarization vector $\hat{\epsilon}$. For $\theta > 0$, the electron has a transverse velocity component, thereby limiting the extent to which it can be accelerated again by the laser field. This implies that, for a fixed final electron kinetic energy E_{drift}, scattering is not allowed classically beyond some angle $\theta_{\max}$. As the energy E_{drift} decreases, the maximum allowed scattering angle $\theta_{\max}$ increases. Writing the vector potential describing a spatially homogeneous and linearly polarized laser field in the form $\mathbf{A}(t) = \hat{\epsilon} A(t)$, the relationship between the energy of an electron that is ionized at time t_d and the angle θ into which it scatters at the return time t_r is given by (in a.u.) [66]

$$E_{\text{drift}}(t_r, t_d, \theta) = \frac{1}{2}\left(A^2(t_d) + 2A(t_r)[A(t_r) - A(t_d)](1 \pm \cos\theta)\right). \quad (8.9)$$

In this equation, the upper sign corresponds to the case in which $A(t_d) > A(t_r)$, while the lower sign holds for $A(t_d) < A(t_r)$. From Equation (2.72), the observed emission angle of the photoelectron θ_f is found to be

$$\theta_f(t_r, t_d, \theta) = \tan^{-1}\frac{\sin\theta|A(t_d) - A(t_r)|}{\cos\theta|A(t_d) - A(t_r)| - A(t_r)}. \quad (8.10)$$

Thus, for a given electron drift energy E_{drift}, the electric-field component of the laser pulse maps all allowed scattering angles θ into the two intervals ($\pm\theta_{f,\max}$) and ($\pi \pm \theta_{f,\max}$), where $\theta_{f,\max}$ is the maximum value of θ_f obtained from Equation (8.10). Further analysis [66, 67] shows that this behavior can be viewed as an example of rainbow scattering [68], and it has been suggested that when electron emission into the angle about $\theta_f = 0$ is suppressed in some energy region due to quantum interference effects, rainbow scattering in ATI could be observed. In this case, the cut-offs in the angular distributions are not dominated by direct ionization around $\theta_f \simeq 0$ and become visible as side-lobes in polar plots of the angular distribution of the photoelectrons [42,64].

Let us now turn our attention to ATI by atoms interacting with elliptically polarized laser fields. We first note that, within the semi-classical model, the drift energy of an electron interacting with a spatially homogeneous electric-field with ellipticity parameter ξ is such that

$$\frac{2\xi^2}{1+\xi^2}U_p \leq E_{\text{drift}} \leq \frac{2}{1+\xi^2}U_p. \tag{8.11}$$

In particular, while for linear polarization the semi-classical model shows that the electron drift energy can vary between zero and $2U_p$, as shown above, for circular polarization its drift energy is restricted to the value U_p. As a result, the low-order ATI photoelectron spectra are very different for linearly and circularly polarized laser fields, respectively [69–71].

Detailed measurements of ATI electron spectra in an elliptically polarized laser field have been carried out by Paulus *et al.* [72,73] using their femtosecond laser system described above. As an example, we show in Fig. 8.8 the ellipticity distributions (EDs) they obtained for xenon atoms interacting with laser pulses of peak intensity of approximately $0.8 \times 10^{14}\,\text{W}\,\text{cm}^{-2}$. The ATI electron yields for certain energies were measured as a function of the ellipticity parameter ξ. The form of the ED is very similar for all energies in the plateau region. The electron yield drops very quickly for increasing values of $|\xi|$. The EDs of the plateau region are readily explained within the framework of the semi-classical recollision model. The electron yield in the plateau region, for a particular laser field ellipticity and electron drift energy, can be estimated by calculating the overlap of the initial ground-state wave function and the wave packet describing the returning electron. The result of this analysis is given by the bold curve in Fig. 8.8, which is in excellent agreement with the experimental data.

For energies below the plateau region, corresponding to ionized electrons that leave the laser field without rescattering, the EDs change dramatically. The most conspicuous effect is a local maximum or "wing" in the electron yield for circular polarization. A simple analysis of the semi-classical model [50,72] shows that electrons released in the elliptically polarized field near its maximum tend to dodge the large component of the electric-field. As the ellipticity grows, they move increasingly away from their parent core in the direction of the small component. This result also follows from a quantum-mechanical analysis [74]. When circular polarization is approached, this effect decreases since all directions tend to become

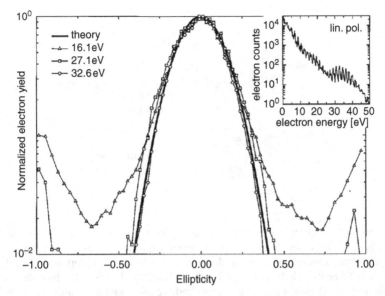

Figure 8.8. Ellipticity distributions obtained from xenon atoms interacting with 50 fs laser pulses having a carrier wavelength of 630 nm and an intensity of about 0.8×10^{14} W cm^{-2}. Results are shown for several photoelectron energies. The inset displays the ATI spectrum for linear polarization. (From G. G. Paulus *et al.*, *Phys. Rev. Lett.* **80**, 484 (1998).)

equivalent, so that the electron yield rises again. There is no reason for such a rise to occur for rescattered electrons. Indeed, as seen from Fig. 8.8, it is not observed for the EDs of the plateau electrons.

ATI experiments with ultra-short few-cycle laser pulses demonstrating the influence of the carrier-envelope phase (CEP) of the pulse on the emission of photoelectrons have also been performed. In particular, Paulus *et al.* [75] measured an anticorrelation in the number of photoelectrons emitted in opposite directions, as discussed in Section 1.3 (see Fig. 1.10). As mentioned in Section 1.3, Paulus *et al.* [76] were able to use few-cycle laser pulses with stabilized CEP to control the direction of emission of photoelectrons by changing the CEP of the laser pulse. This provides a tool for the measurement of the CEP of few-cycle laser pulses by means of the "stereo ATI" spectrometer shown in Fig. 8.9.

More recently, Kling *et al.* [77] have achieved sub-femtosecond control of the ATI photoelectron emission in the noble gases Ar, Xe and Kr in intense few-cycle laser pulses, with full angular resolution. Their experimental data were obtained by using a velocity-map imaging (VMI) technique [78, 79], in which electrons are projected onto a two-dimensional position-sensitive detector. They demonstrated that the VMI technique is not only a powerful tool for measuring the energy and

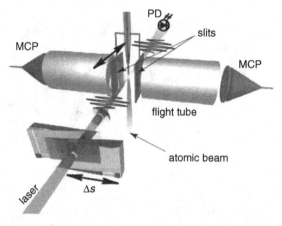

Figure 8.9. "Stereo ATI" spectrometer. Two opposing electrically and magneti-
cally shielded time-of-flight spectrometers are mounted in an ultra-high vacuum
apparatus. Xenon atoms fed in through a nozzle from the top are ionized in the
focus of a few-cycle laser beam. The focal length is 250 mm (the lens shown in the
diagram is actually a concave mirror) and the pulse energy is attenuated to 20 μJ.
The laser polarization is linear and parallel to the flight tubes. A photodiode (PD)
and microchannel plates (MCP) detect the laser pulses and photoelectrons, respectively. The laser repetition rate is 1 kHz, and 50 electrons per pulse are recorded
at each MCP. The displacement Δs of a pair of glass wedges is used to optimize
dispersion and adjust the carrier-envelope phase. (From G. G. Paulus *et al.*, *Phys.
Rev. Lett.* **91**, 253004 (2003).)

angular distributions of the emitted electrons, but also provides a way to determine
the CEP of few-cycle laser pulses from these distributions.

Kling *et al.* [77] also performed TDSE simulations using the SAE approximation, which they compared with their experimental data for argon atoms. The
angle-resolved photoelectron spectrum was computed up to an energy of 4 a.u. and
used to determine volume-integrated spectra for comparison with the experimental
results. In particular, they studied the dependence of asymmetries in the energy
and angular distributions of the photoelectrons on the CEP of the laser pulse. As
an illustration, we show in Figs. 8.10(a) and (b) two momentum maps calculated
for CEPs given, respectively, by $\varphi = 0$ and $\varphi = \pi/2$, for conditions corresponding to the experiment, namely 5.8 fs laser pulses having a carrier wavelength of
760 nm and a peak intensity of 10^{14} W cm^{-2}. The photoelectron energy distributions in the upward and downward directions along the laser polarization axis were
obtained by integrating the volume-integrated angle-resolved photoelectron spectrum within 40° along the laser polarization axis, and are shown in Figs. 8.10(c)
and (d). Figure 8.10 shows a dependence of the angle-resolved photoelectron

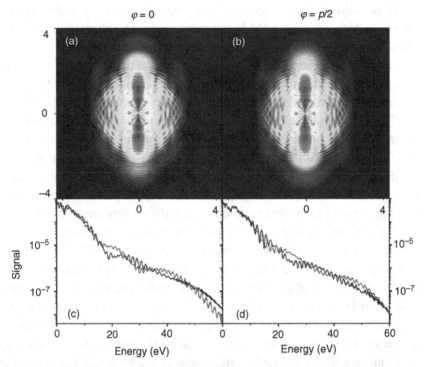

Figure 8.10. (a), (b) Two-dimensional momentum representation obtained from the numerical solution of the TDSE for argon interacting with 5.8 fs laser pulses having a peak intensity of 10^{14} W cm^{-2}. The laser field is linearly polarized, with the polarization axis along the vertical axis. The results have been obtained by volume integrating single-intensity calculations. (c), (d) Electron kinetic energy spectra obtained by integrating over an angular range of 40° along the laser polarization direction. Black curves: upward emission; gray curves: downward emission. In (a) and (c) the carrier-envelope phase is $\varphi = 0$, while in (b) and (d) $\varphi = \pi/2$. (From M. F. Kling *et al.*, *New J. Phys.* **10**, 025024 (2008).)

spectra on the CEP. It is worth noting that the CEP-dependent changes are most important in the high-energy region of the spectrum, beyond $2U_\mathrm{p}$.

As mentioned in Section 1.3, *two-color* MPI experiments can allow the "bound–free" and the "free–free" steps of ATI to be distinguished. Early two-color experiments include those of Muller, van Linden van den Heuvel and van der Wiel [80], who used photons of 1.17 eV and 4 eV to ionize xenon atoms, and of Tate, Papaioannou and Gallagher [81], who used microwaves to drive the electron quiver motion and laser pulses in the optical region for ionization. Two-color experiments with harmonics have been performed by Chen, Yin and Elliot [82]. Using the fundamental frequency of a dye laser, they ionized mercury atoms with five photons, the third one being resonant with a bound state. They then added the third harmonic, which was generated in the same vapor. The phase difference between the fundamental and the third harmonic was controlled by changing the vapor pressure. They observed ionization suppression, due to the fact that three photons of the fundamental lead to the same intermediate resonance as one photon of the third harmonic, and these two ionization pathways can interfere destructively. Muller *et al.* [83] used the fundamental and the second harmonic of a Nd:YAG laser to ionize krypton atoms. In this case, a specific energy can be reached through pathways containing even as well as odd numbers of photons, so that the final state need not be of definite parity. As pointed out by Muller [84], such experiments can allow one to measure not only the amplitude of the various peaks in the photoelectron spectrum, but also their phases by observing their interferences with a reference wave, for example created by ionization with another laser pulse.

Two other interesting types of experiments on two-color ATI, discussed by Agostini [85], are laser-assisted Auger decay and the laser-assisted photoelectric effect. Although the physics of these two processes is different, they have in common the fact that they both "promote" a bound electron into the continuum that is "dressed" by a strong laser field. In the case of laser-assisted Auger decay, a high-energy photon removes an inner-shell electron from an atom, thus creating a vacancy (hole) in this inner shell. A radiationless transition, called an Auger transition, can then occur, in which one electron from a level with higher energy fills the vacancy, with the simultaneous ejection of a second electron, called an Auger electron, so that a doubly charged ion is produced. We remark that radiationless (Auger) transitions compete with radiative transitions, in which the initial vacancy is filled by an electron having higher energy accompanied by the emission of a photon. Laser-assisted Auger decay has been studied experimentally in argon atoms by Schins *et al.* [86, 87]. Part of an intense, sub-picosecond, infra-red Ti:sapphire laser beam of wavelength $\lambda = 800$ nm and intensity of about 10^{12} W cm^{-2} was focused on a metal target to produce sub-picosecond, broadband X-ray radiation, which induced LMM Auger transitions in argon as follows. One X-ray photon removes a p electron from the L-shell. An electron from the M-shell then fills the L-shell vacancy, while a second M-shell electron is emitted in the continuum as an Auger electron having a well defined energy, independent of the absorbed photon. The Auger electrons are detected by using a time-of-flight spectrometer. The other part of the intense infra-red laser beam was directly focused onto the argon atoms, causing the absorption and emission of several photons from the infra-red laser field. One therefore expects the electron energy spectrum to consist of a central line at the Auger electron energy and several "sidebands" on both sides, separated by the infra-red photon energy. In reality, the Auger "line" contains several components due to the fine structure of both the singly charged ion excited state and the doubly charged ion ground state.

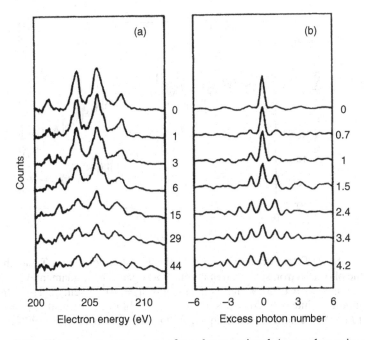

Figure 8.11. Electron energy spectra from laser-assisted Auger decay in argon. (a) "Raw spectra" for increasing infra-red intensities from top to bottom. The labels indicate the experimental intensities of the dressing laser field, in units of $10^{10}\,\mathrm{W\,cm^{-2}}$. (b) The same spectra after deconvolution from the zero-intensity spectra. The number on the right of the spectrum gives the calculated number of sidebands. (From J. Schins *et al.*, *Phys. Rev. Lett.* **73**, 2180 (1994).)

The corresponding raw spectra displayed in Fig. 8.11(a) must therefore be deconvoluted to extract the electron energy spectra reduced to a central line accompanied by sidebands separated by the infra-red photon energy, as shown in Fig. 8.11(b).

A similar situation arises in the case of the laser-assisted photoelectric effect, where atoms interact simultaneously with the infra-red (fundamental-frequency) pulse and ultra-violet high harmonics of a Ti:sapphire laser. Here, only outer-shell electrons are involved and, for long infra-red driving pulses, sidebands can be detected in the energy spectrum of the primary electrons [88, 89]. This process was investigated theoretically by Taïeb, Véniard and Maquet [90] who performed a two-color calculation in which a hydrogen atom interacts with an intense infra-red Ti:sapphire laser pulse having photon energy, $\hbar\omega_L = 1.55\,\mathrm{eV}$ and a weaker ultra-violet 13th harmonic whose photon energy, $\hbar\omega_H = 20.15\,\mathrm{eV}$, is large enough to ionize the atom with a single photon. As seen from Fig. 8.12, the laser-assisted single-photon ionization process strongly modifies the ATI spectrum.

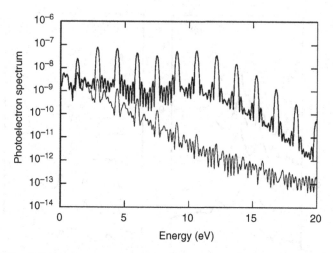

Figure 8.12. Effect of the presence of the 13th harmonic of a Ti:sapphire laser on the photoelectron spectrum of atomic hydrogen. The fundamental laser photon energy is $\hbar\omega_L = 1.55\,\text{eV}$ and its intensity is $I_L = 10^{13}\,\text{W cm}^{-2}$. Thick line: 13th harmonic of intensity $I_H = 3 \times 10^8\,\text{W cm}^{-2}$. Thin line: intensity $I_H = 0$. (From R. Taïeb, V. Véniard and A. Maquet, *J. Opt. Soc. Am. B* **13**, 363 (1996).)

In further work, Véniard, Taïeb and Maquet [91] considered a similar scheme in which both the infra-red driving laser pulse and its harmonic pulse are relatively weak. Their analysis showed that the magnitudes of the photoelectron peaks are strongly dependent on the difference of phases between successive harmonics of the driving Ti:sapphire laser field. This observation forms the basis of important applications of two-color ATI processes to cross-correlation measurements of ultra-short XUV [89] or X-ray pulses [85] that can be synchronized with a driving infra-red pulse. Such measurements will be discussed in Section 9.4.

8.2 Multiphoton double ionization

8.2.1 One-photon, double ionization of helium

Before discussing atomic multiphoton double ionization, it is of interest to recall some important features of the process of double ionization of helium by *one photon*, namely

$$\hbar\omega + \text{He}(1^1\text{S}) \rightarrow \text{He}^{2+} + 2e^-. \tag{8.12}$$

We shall assume that the two outgoing electrons move at non-relativistic velocities and use the dipole approximation. Atomic units will be used, unless otherwise stated.

Let us denote by $\mathbf{k}_A$ and $\mathbf{k}_B$ the momenta of the two outgoing electrons and by $E_{k_A} = k_A^2/2$ and $E_{k_B} = k_B^2/2$ their corresponding energies. If $\mathbf{Q}$ is the recoil momentum of the He^{2+} nucleus, we have, from momentum conservation,

$$\mathbf{Q} = -(\mathbf{k}_A + \mathbf{k}_B) = -\mathbf{K}, \tag{8.13}$$

where $\mathbf{K} = \mathbf{k}_A + \mathbf{k}_B$ is the center-of-mass momentum of the two-electron subsystem and we have set the photon momentum equal to zero in accordance with the dipole approximation. From energy conservation, we have

$$\hbar\omega - I_P = E_{k_A} + E_{k_B}, \tag{8.14}$$

where $I_P = 2.90372$ a.u. is the ionization potential of the helium atom, and the recoil energy of the He^{2+} nucleus has been neglected.

In 1967, Byron and Joachain [92] proved that this process is a very sensitive probe of *electron correlation effects*. Following their treatment, we consider the triple differential cross section (TDCS) for the two outgoing electrons to emerge into the solid angles $d\Omega_A$ and $d\Omega_B$, centered, respectively, about the directions (θ_A, ϕ_A) and (θ_B, ϕ_B). The TDCS is given in the dipole approximation and in the velocity gauge, for linearly polarized electromagnetic radiation, by

$$\frac{d^3\sigma_V^{2+}}{d\Omega_A d\Omega_B dE} = \frac{4\pi^2 \alpha k_A k_B}{\omega} \left| \hat{\boldsymbol{\epsilon}} \cdot \langle \psi_{\mathbf{k}_A,\mathbf{k}_B}^{(-)}(\mathbf{r}_1, \mathbf{r}_2) | (\nabla_{\mathbf{r}_1} + \nabla_{\mathbf{r}_2}) | \psi_i(\mathbf{r}_1, \mathbf{r}_2) \rangle \right|^2, \tag{8.15}$$

where $\alpha = e^2/(4\pi\epsilon_0\hbar c) \simeq 1/137$ is the fine-structure constant. In the above expression, $\psi_i(\mathbf{r}_1, \mathbf{r}_2)$ denotes the initial wave function corresponding to the ground state of helium and $\psi_{\mathbf{k}_A,\mathbf{k}_B}^{(-)}(\mathbf{r}_1, \mathbf{r}_2)$ is the final-state wave function describing the two outgoing electrons, and having an incoming spherical wave behavior. In Equation (8.15), the prefactor depends on the normalization of the final-state wave function, which is chosen to be normalized on the wave vector (or momentum in a.u.) scale.

Let us now examine more closely the TDCS (8.15). First of all, we note that, if the interaction between the two electrons is completely neglected, so that $\psi_i(\mathbf{r}_1, \mathbf{r}_2)$ is a product and $\psi_{\mathbf{k}_A,\mathbf{k}_B}^{(-)}$ is a symmetrized product of hydrogenic wave functions with $Z = 2$, then these matrix elements vanish. As a result, the one-photon, double ionization process (8.12) cannot occur in the absence of the electron–electron interaction. This is clearly not the case for single ionization, which would therefore be the only possible ionization process if the electron–electron interaction were "switched off." It follows from this reasoning that the one-photon, double ionization process (8.12) is expected to be delicately dependent on electron correlation effects. To demonstrate this property, Byron and Joachain [92] calculated the ratio of the double to single ionization total cross sections,

$$R(\omega) = \frac{\sigma_{\text{tot}}^{2+}(\omega)}{\sigma_{\text{tot}}^{+}(\omega)} \tag{8.16}$$

in the asymptotic limit $\omega \to \infty$. In this limit, the evaluation of the integrals required to calculate the total cross sections σ_{tot}^+ and σ_{tot}^{2+} is simplified, since final-state electron correlations do not play any role. Byron and Joachain [92] calculated the asymptotic ratio $R(\omega \to \infty)$ using various approximations for the helium ground-state wave function ψ_i. Their best result,

$$R(\omega \to \infty) = 1.66\%, \tag{8.17}$$

was obtained with the very accurate 39-parameter, Hylleraas-type wave function of Kinoshita [93], which gives for the ionization potential of helium the value $I_P = 2.90372$ a.u. On the other hand, the Hartree–Fock ground state wave function, which gives for the ionization potential of helium the result $I_P = 2.86167$ a.u., yields for $R(\omega \to \infty)$ the value 0.51%, which is smaller than the best value 1.66% by more than a factor of three. Thus, the calculation of $R(\omega \to \infty)$ gives a striking illustration of the importance of initial-state electron correlation effects in one-photon, double ionization phenomena. The theoretical value of 1.66% obtained for $R(\omega \to \infty)$ by Byron and Joachain has been confirmed by subsequent calculations [94, 95] and by several experiments using synchrotron radiation. In particular, Levin *et al.* [96] measured in 1991 the value $R(\omega) = (1.6 \pm 0.3)\%$ at a photon energy $\hbar\omega = 2.8$ keV, and Spielberger *et al.* [97] found in 1995 that $R(\omega) = (1.72 \pm 0.12)\%$ at $\hbar\omega = 7$ keV.

The calculation of double photoionization cross sections (triple, double, single, total) for helium at *finite* photon energies is a difficult problem involving, in particular, the accurate determination of the final-state wave function $\psi_{\mathbf{k}_A, \mathbf{k}_B}^{(-)}$. In 1993, Maulbetsch and Briggs [98] and Proulx and Shakeshaft [99] succeeded in calculating angular distributions which agreed well with the experimental data of Schwarzkopf *et al.* [100]. Later, Pont and Shakeshaft [101] were able to calculate the total one-photon, double ionization cross section of helium in the near-threshold region. Their theoretical results were found to be in excellent agreement with the absolute data from experiments using synchrotron radiation, performed by Kossmann, Schmidt and Andersen [102] and by Bizau and Wuilleumier [103].

8.2.2 Multiphoton double ionization of atoms

We now turn to the multiphoton double ionization process

$$n\hbar\omega + A(i) \to A^{2+}(f) + 2e^-. \tag{8.18}$$

Neglecting again the recoil energy of the $A^{2+}(f)$ ion, we have, from energy conservation,

$$n\hbar\omega + E_i = E_f + E_{k_A} + E_{k_B}, \tag{8.19}$$

where E_i and E_f denote, respectively, the bound-state energies of the atom A in the initial state i and of the ion A^{2+} in the final state f.

The first experimental evidence suggesting a departure from sequential double ionization (SDI) was obtained by L'Huillier *et al.* [104, 105] in noble gases. In particular, they investigated the interaction of Xe atoms with 50 ps laser pulses having a carrier wavelength of 1064 nm and peak intensity in the range 10^{13}–10^{14} W cm^{-2} and a carrier wavelength of 532 nm and peak intensity in the 10^{12} W cm^{-2} range. The deviation from SDI showed up as the premature appearance of the signal corresponding to Xe^{2+} ions, manifested as a "knee" in the log-log plot of the ionic species as a function of the laser intensity. In order to explain this feature, L'Huillier *et al.* [104, 105] assumed the existence of non-sequential double ionization (NSDI) from the ground state of the atom, in addition to SDI. This problem was further examined by Charalambidis *et al.* [106], who studied double multiphoton ionization of Kr and Xe in the wavelength range from 527 to 531 nm. They concluded that the knee did not result from NSDI. Instead, they suggested that it involved the excitation of two electrons into a manifold of doubly excited states, followed by the ejection of one electron, leaving the ion in an excited state which is subsequently ionized during the interaction with the laser pulse through a lower-order multiphoton process, thus resulting in double ionization. This double ionization process therefore occurs in two steps, and each time this is possible a knee can appear in the plot of the doubly ionized species as a function of the laser intensity. L'Huillier *et al.* [15] observed similar knees in several of the noble gases at the wavelength 1064 nm. In view of the pulse durations and intensities used in these measurements, Lambropoulos, Maragakis and Zhang [107] pointed out that it could not be concluded that NSDI was the cause of the observed knees.

A different situation was found to occur in the experiments performed by Fittinghoff *et al.* [108]. They irradiated helium atoms with linearly polarized 120 fs laser pulses having a carrier wavelength of 614 nm and intensities up to 10^{16} W cm^{-2}. At these laser wavelengths and intensities the Keldysh parameter $\gamma_K < 1$, so that one is in the tunneling ionization regime. The experiments of Fittinghoff *et al.* were refined by Walker *et al.* [44], who measured ion yields for helium interacting with linearly polarized 160 fs laser pulses having a carrier wavelength of 780 nm. Their experimental data are shown in Fig. 1.11. Looking at this figure, we see that they were able to measure ion production over a range of 12 orders of magnitude. A striking feature of the data is the very strong dependence of the single and double ionization yields of helium on the laser intensity over the intensity range from 10^{14} W cm^{-2} to 10^{16} W cm^{-2}. Another very important feature of the experimental results of Walker *et al.* [44], already noted in Section 1.3, is the existence of two distinct intensity regimes. The first one occurs at high intensities, above the saturation intensity of He$^+$, where SDI takes place predominantly, in agreement with the single-active-electron (SAE) approximation. This means that the He^{2+} ions are produced solely from the interaction of the laser pulse with He$^+$

ions, previously created by the laser pulse interacting with the He atom. The second regime appears at lower intensities, below the saturation intensity of He^+, where a distinct "knee" in the production of He^{2+} ions can be seen. In this regime, the very significant departure of the measured He^{2+} ion yield from the SAE prediction clearly demonstrates that the double ionization process occurs in a non-sequential way, in other words as NSDI. The wavelength dependence of NSDI of helium was studied by Kondo *et al.* [109], who measured He^{2+} yields using linearly polarized 200 fs laser pulses with a carrier wavelength of 745 nm (in the tunneling regime) and 440 fs laser pulses with a carrier wavelength of 248 nm (in the multiphoton regime). Their experimental data at the longer wavelength of 745 nm exhibited a significant knee, as in the measurements of Fittinghoff *et al.* [108] and Walker *et al.* [44], whereas at the wavelength of 248 nm Kondo *et al.* [109] found no sign of NSDI. The direct multiple ionization of neon, argon and xenon has also been studied by Larochelle, Talebpour and Chin [110] using 200 fs Ti:sapphire laser pulses of carrier wavelength $\lambda = 800$ nm and intensities up to about 10^{16} W cm^{-2}.

The large enhancement of NSDI yields in helium at low intensities, giving rise to the knee seen in Fig. 1.11, has been the subject of many theoretical investigations. This feature has been relatively well reproduced by theoretical models, including at least some amount of electron correlation. Several of these models will now be discussed.

We begin by considering an approach proposed by Watson *et al.* [111], which allowed the three steps of the semi-classical recollision model [51, 52] to play a role. This approach relied on the assumption that, at least in this context, a distinction could be made between an "inner" electron and an "outer" electron labeled 1 and 2, respectively. Such a distinction implies significant separation between the two electrons, as a result of which exchange effects were neglected. The two-electron spatio-temporal part of the wave function, $\Psi(\mathbf{r}_1, \mathbf{r}_2, t)$, was therefore written approximately as the product $\Psi_1(\mathbf{r}_1, t)\Psi_2(\mathbf{r}_2, t)$. The outer electron of the helium ground state was the only electron allowed to interact directly with the laser field. The release of the outer electron wave packet proceeded as in the first step of the semi-classical recollision model. In the second step, this wave packet was propagated under the combined influences of the laser field and the parent ion He^+. Although the inner electron was not allowed to interact directly with the laser field, it was allowed to interact with the outer electron during the latter's excursion in the laser field. This allowed the third step of the recollision model to take place, in which the outer electron could collide with the inner one and eject it from the atom, thus inducing an (e, 2e) reaction resulting in double ionization. As in the SAE approximation, the outer electron (2) experienced the time-independent Hartree–Fock potential $V_2(r_2)$ due to the inner electron and the He^{2+} nucleus. However, the inner electron (1) moved in a time-dependent potential determined by the He^{2+}

nucleus and the outer electron, given by

$$V_1(\mathbf{r}_1, t) = -\frac{2}{r_1} + \int d\mathbf{r}_2 \frac{1}{r_{12}} |\Psi_2(\mathbf{r}_2, t)|^2, \tag{8.20}$$

where $r_{12} = |\mathbf{r}_1 - \mathbf{r}_2|$. In the calculations, the electron–electron interaction $1/r_{12}$ was expanded in Legendre polynomials according to Equation (5.120), and only the monopole ($l = 0$) and the dipole ($l = 1$) terms were retained. The electron correlations included in the model of Watson *et al.* [111] allow the role of recollisions in the ejection of the inner electron to be investigated. The calculations carried out by Watson *et al.* produced a knee in plots of double ionization yields versus laser peak intensity, thereby demonstrating the importance of recollisions in NSDI.

van der Hart and Burnett [112] have also proposed a model for multiphoton double ionization in helium which is based on the three-step semi-classical recollision model. In the first step, in which an electron tunnels out of the atom, the Ammosov–Delone–Krainov (ADK) tunneling rate [113] given by Equation (6.39) was used. In the second step, the motion of the ejected electron in the presence of the laser field and its parent He$^+$ ion was analyzed. In the direction parallel to the laser polarization axis, this motion was treated classically, as in the original work of Corkum [51], giving the energy with which the electron returns to the parent ion. However, this second step was refined by allowing for the quantum-mechanical spreading of the electronic wave packet in the direction perpendicular to the laser polarization axis, which gave the width of the returning wave packet. The knowledge of the return energy of the electron and of the width of the returning wave packet allowed the collision behavior between the electron and its parent He$^+$ ion to be analyzed. This constitutes the third step of the process. During the collision, NSDI may occur if the returning electron imparts a sufficient amount of energy to ionize an electron of its parent ion. Alternatively, the returning electron can transfer only enough energy to promote an electron of its parent ion to an excited state, from which it is subsequently detached by the laser field. The latter process has been termed resonantly enhanced sequential ionization (RESI). In both cases, a non-elastic scattering process leads to the ejection of a second electron. The probability that the returning electron caused the emission of a second electron was assumed to be given by the ratio between the total cross section for all non-elastic processes and the total cross section for the recollision process. In this non-relativistic regime, the two He electrons remain in a singlet spin state. Thus, within the recollision model for multiphoton double photoionization of helium, one only needs to take into account the singlet scattering cross sections. At the carrier wavelength of 780 nm, and for intensities ranging from 5×10^{14} W cm^{-2} to 10^{15} W cm^{-2}, van der Hart and Burnett [112] found a ratio for double to single ionization four times smaller than the experimental value [114]. In subsequent calculations, van der Hart [115] analyzed in more detail the third step of the process by separating the electron recollision cross section into excitation and ionization contributions. He carried out calculations at a carrier wavelength of 780 nm over an intensity range extending from 1.5×10^{14} W cm^{-2} to 7×10^{14} W cm^{-2}, and found that at the lower intensities the RESI process was dominant, while the NSDI grew as the intensity increased, becoming comparable with excitation at the upper end of the intensity range considered. van der Hart [115] also performed calculations at a carrier wavelength of 390 nm, over a range of intensities from 7.5×10^{14} W cm^{-2} to 2.25×10^{15} W cm^{-2}. In this case, he found that the NSDI process never exceeded one-quarter of the total over the entire intensity range considered.

Yudin and Ivanov [116] also used the three-step semi-classical recollision model to perform calculations similar to those of van der Hart and Burnett [112] and van der Hart [115]. At a carrier wavelength of 780 nm, and in the intensity range from 2×10^{14} W cm^{-2} to 10^{15} W cm^{-2}, they also found that, at the lowest intensities considered, RESI was the dominant process. Yudin and Ivanov [116] also took into account repeated returns of the initially ionized electron to the parent ion. They found that, in this way, the ratio of He^{2+} to He$^+$ ions increased, so that good agreement for this ratio with the experimental results of Walker *et al.* [44] was obtained.

Another theoretical approach to the multiphoton double ionization of helium, termed "intense-field many-body S-matrix theory" (IMST), has been developed by Becker and Faisal [117–121]. This approach can be viewed within the framework of the SFA discussed in Chapter 6. They obtained results in agreement (within an order of magnitude) with the double ionization yields in helium measured by Walker *et al.* [44] over a range of intensities extending to the saturation intensity. Becker and Faisal [120] also used the IMST approach to calculate the ion yield for single and double ionization as well as the ratio of double to single ionization for noble-gas atoms in Ti:sapphire laser pulses. Overall, they obtained good agreement with the experimental data [44, 110, 114, 122].

The multiphoton double ionization of helium has also been analyzed by direct integration of the TDSE. Parker *et al.* [123] reported calculations of single and double ionization rates of helium at a carrier wavelength of 390 nm. They compared the ratio of He^{2+} to He$^+$ ion yields with the measurements of Sheehy *et al.* [114] for laser intensities ranging from 0.35×10^{15} W cm^{-2} to 2.2×10^{15} W cm^{-2}. The calculated and measured ratios are plotted in Fig. 8.13 as a function of the laser intensity. All experimental data points have been shifted along the intensity axis by 50% ($I \rightarrow 1.5I$), resulting in excellent agreement with theory. It should be noted that a 50% uncertainty in the experimental determination of the intensity is not atypical in strong laser field experiments. Calculated single and double ionization

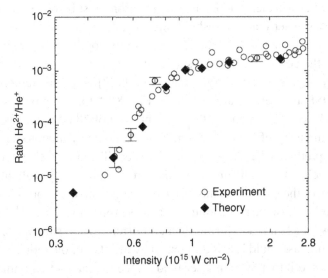

Figure 8.13. Ratio of He^{2+} to He^+ ion yields, as a function of the laser field intensity (in units of 10^{15} W cm^{-2}). The experimental data points of Sheehy *et al.* [114] are represented by the open circles and the theoretical values of Parker *et al.* [123] by solid diamonds. A few characteristic error bars are shown. All experimental data points have been shifted along the intensity axis in the direction of higher intensity by 50%. (From J. S. Parker *et al.*, *J. Phys. B* **33**, L691 (2000).)

rates for helium atoms interacting with a laser pulse having a carrier wavelength of 390 nm, in the intensity range from 0.4×10^{15} W cm^{-2} to 2.2×10^{15} W cm^{-2}, were also reported by Parker, Moore and Taylor [124]. The TDSE for helium has also been solved by Muller [125] for double ionization at carrier wavelengths of 390 nm and 800 nm.

An important feature of the semi-classical recollision model [51, 52] discussed in Section 6.5 is the time delay of approximately three-quarters of an optical period between the release of an initially bound electron into the laser field and its subsequent recollision with its parent core (or, in other words, the time taken by the second step of the recollision process). In particular, let us consider an electron that ionizes in the laser field by quasi-static tunneling at time t_d and returns to its parent core at time t_r where it recollides non-elastically. We have seen in Section 6.4 that the maximum energy of the returning electron at recollision is $3.17U_p$. If the kinetic energy of the returning electron is greater than $I_P(t_r)$, where from Section 6.1.2,

$$I_P(t_r) = I_P - 2(Z|\mathcal{E}(t_r)|)^{1/2} \tag{8.21}$$

is the time-dependent effective ionization potential of the parent core in the presence of the laser field, it can impart a sufficient fraction of its kinetic energy to the parent core for NSDI to occur [126]. This implies that for helium interacting with a laser field of carrier wavelength 780 nm, the electron must be released into a laser field of intensity greater than a threshold intensity, I_{TI}, of about 3×10^{14} W cm^{-2} such that collisional ionization of the He$^+$(1s) ion is possible. At a carrier wavelength of 390 nm the corresponding threshold intensity is about $I_{TI} = 10^{15}$ W cm^{-2}. The time-delay results obtained by Parker *et al.* [127] from the numerical integration of the TDSE for helium are shown in Fig. 8.14 for laser pulses having carrier wavelengths of 780 nm and 390 nm. In the case of 780 nm laser light, beyond the threshold intensity of 3×10^{14} W cm^{-2}, the time delay monotonically approaches the value predicted by the semi-classical recollision model. By contrast, at the carrier wavelength of 390 nm, where the threshold intensity is about 10^{15} W cm^{-2}, the validity of the recollision model can be called into question. We remark that the quiver amplitude of the recolliding electron is proportional to $\lambda U_p^{1/2}$ so that, at fixed U_p, it is twice as large at 780 nm than at 390 nm. As a result, an electron detached into a 780 nm laser field is less influenced by its interaction with its parent core than if it is released into a 390 nm laser field. Since the interaction with its parent core is neglected in the semi-classical recollision model, at least in its original form, it is reasonable to expect this model to be more accurate in the long-wavelength and high-intensity limits, as confirmed by the examination of Fig. 8.14.

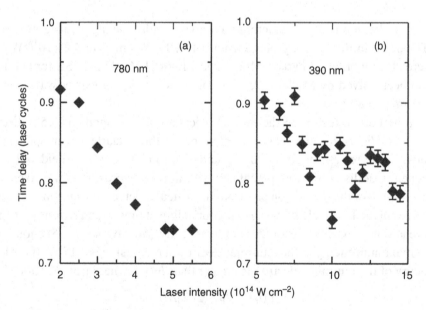

Figure 8.14. Calculated time delay between single and double ionization of helium, as a function of the laser field intensity (in units of 10^{14} W cm^{-2}), for carrier wavelengths of (a) 780 nm and (b) 390 nm. (From J. S. Parker *et al.*, *J. Phys. B* **36**, L393 (2003).)

It is interesting to note that Fittinghoff *et al.* [128] have measured the polarization dependence of the double ionization of helium by 120 fs laser pulses having a carrier wavelength of 614 nm. While for linearly polarized pulses their data exhibited a knee, no such enhancement was found for circularly polarized pulses. These results are consistent with the semi-classical recollision model.

Thus far we have discussed experiments in which ion yields were measured as a function of the laser field intensity at different wavelengths. Key advances in the experimental study of multiphoton double ionization have occurred with the development of energy- and momentum-resolved measurement techniques. For example, using 100 fs linearly polarized laser pulses having a carrier wavelength of 780 nm, Lafon *et al.* [129] were able, by means of electron time-of-flight detection in coincidence with residual-ion detection, to measure energy-resolved single- and double-ionization electron yields of helium in the strong-field limit. As seen from Fig. 8.15, the electron distributions from double ionization were found to extend to significantly higher energies than those arising from single ionization. They were able to explain their results within the framework of the semi-classical recollision model of NSDI by assuming that, during the (e, 2e) recollision process, one of the two ejected electrons is emitted in the backward direction.

Another important experimental advance in the study of multiphoton single and multiple ionization of atoms has been made by Weber *et al.* [130] and Moshammer *et al.* [131] using the cold target recoil ion momentum spectroscopy (COLTRIMS) technique. In these experiments, information on the *momentum* carried by the

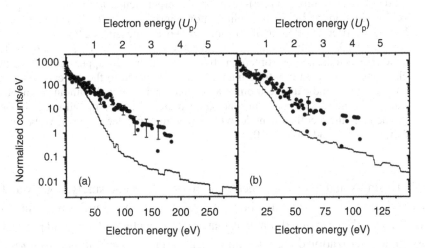

Figure 8.15. Energy spectra of electrons detected in coincidence with He^+ (line) and He^{2+} ions (circles) from helium interacting with 100 fs laser pulses having a carrier wavelength of 780 nm and a peak intensity of (a) 8×10^{14} W cm^{-2} and (b) 4×10^{14} W cm^{-2}. (From R. Lafon *et al.*, *Phys. Rev. Lett.* **86**, 2762 (2001).)

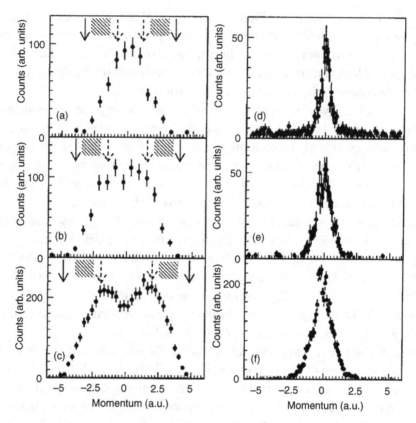

Figure 8.16. Distribution of recoil momenta of He^{2+} ions (i) parallel to the laser field polarization direction ((a)–(c)) and (ii) perpendicular to the laser field polarization direction ((d)–(f)) for helium atoms interacting with 220 fs laser pulses having a carrier wavelength of 800 nm and of peak intensities 2.9×10^{14} W cm^{-2}, (a) and (d); 3.8×10^{14} W cm^{-2}, (b) and (e); 6.6×10^{14} W cm^{-2}, (c) and (f). The solid arrows indicate the momentum $4(U_p)^{1/2}$, which is the upper bound if the two electrons are ionized in a time interval short compared to the optical cycle. The dashed arrows indicate the momenta in a recollision model without momentum transfer in the (e, 2e) collision. The hatched area indicates the allowed classical momenta in the recollision model with momentum transfer. (From T. Weber *et al.*, *Phys. Rev. Lett.* **84**, 443 (2000).)

residual singly and doubly charged helium ions [130] and singly, doubly and triply charged neon ions [131] was obtained. The most important finding, illustrated in Fig. 8.16 for the case of helium, was that the doubly charged ion yield parallel to the laser field polarization direction exhibits a prominent *double-hump structure*. These results were confirmed by further measurements of this kind, performed in neon [132, 133]. Measurements of the momentum distribution of He^{2+}, Ne^{2+} and Ar^{2+} ions carried out by de Jesus *et al.* [134] showed that the contrast of the double-hump

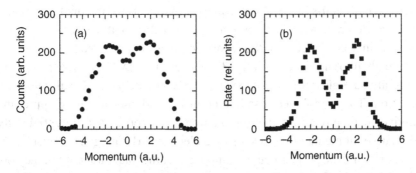

Figure 8.17. Distribution of recoil momenta of He^{2+} ions parallel to the laser field polarization direction: (a) experimental data as in Fig. 8.16(c); (b) calculated values of the parallel component of the vector $-(\mathbf{k}_A + \mathbf{k}_B)$. (From A. Becker and F. H. M. Faisal, *Phys. Rev. Lett.* **84**, 3546 (2000).)

structure depends strongly on the atomic species, a point to which we will return below.

Let us first consider why the existence of a double-hump structure in the momentum yield distribution parallel to the laser polarization axis is consistent with a recollision mechanism being responsible for the NSDI observed in experiments. In order that a recolliding electron returns with a high enough kinetic energy, it must have been ionized at a time t_d close to a maximum of the magnitude of the electric-field of the laser pulse and return at a time t_r when this electric-field magnitude is small. Just after impact ionization, both electrons will have very low kinetic energies. However, they can be subsequently accelerated by the electric-field component of the laser field and thereby acquire a large drift momentum leading to a double-peaked final ion momentum distribution with a pronounced minimum at zero momentum. This feature is observed most prominently in neon [134, 135]. In contrast, argon displays only a weak minimum in the intensity range from 3×10^{14} W cm^{-2} to 6×10^{14} W cm^{-2} of the driving Ti:sapphire laser pulse.

Using the IMST approach, Becker and Faisal [136] were able to reproduce the double-hump structure found experimentally [130] in the distribution of the component of the recoil momentum of the He^{2+} ions parallel to the laser field polarization direction. This is illustrated in Fig. 8.17, where the data of Weber *et al.* [130] corresponding to Fig. 8.16(c) are compared to the values of the parallel component of the vector $\mathbf{Q} = -(\mathbf{k}_A + \mathbf{k}_B)$, calculated by Becker and Faisal [136]. We recall that, in the dipole approximation, the momentum of the (long-wavelength) laser photon is neglected so that the conservation of momentum requires that the recoil momentum $\mathbf{Q}$ of the He^{2+} ion satisfies Equation (8.13). Becker and Faisal also found that the double-hump feature of the parallel component of the distribution of the recoil momentum of the He^{2+} ions collapsed into a single-hump structure in the absence of final-state Gordon–Volkov dressing. Thus, the final-state interaction of the two electrons with the laser field is responsible for the observed double-hump structure, in accordance with the recollision picture discussed above.

Popruzhenko and Goreslavskii [137] also used the IMST approach, together with a generalized Keldysh model [138], to study non-sequential multiphoton double ionization. They employed a saddle-point method and tunneling approximations to derive an analytical distribution function for the process, differential in the total and relative momenta of the two ejected electrons. By integrating over the relative momentum and the perpendicular component of the total momentum, they obtained a distribution for the parallel component of the total momentum, which exhibits a double-hump structure and a width in reasonable agreement with the experimental data of Weber *et al.* [130] and the IMST calculations of Becker and Faisal [136] for helium, and with the measurements of Moshammer *et al.* [131] for neon. Kopold *et al.* [139] also performed IMST calculations for NSDI, which differed from those of Becker and Faisal [136] and of Popruzhenko and Goreslavskii [137] by treating the electron collision with its parent core (the third step of the recollision model) as a zero-range (contact) rather than a Coulomb interaction. Further calculations of this kind were carried out by Goreslavskii *et al.* [140] for neon and argon.

In contrast to neon, the measured ion recoil momentum distribution parallel to the laser polarization direction in argon, and to a lesser extent in helium, contain significant contributions at small momenta. In order to understand the double ionization mechanism leading to small recoil momenta, let us recall that an electron that is ionized at a time close to the maximum of the electric-field component of the laser field acquires little drift momentum, leading to a narrow momentum distribution along the laser polarization axis that is centered about zero. This is clearly observed in the momentum distributions of the yields of *singly* charged helium and neon ions [130, 131]. The observation of a relatively large proportion of ions having very low momenta along the laser field polarization direction suggests significant contributions from the RESI process. We recall that when a returning electron scatters inelastically and only imparts enough of its kinetic energy to promote an electron of its parent core to an excited state, this excited electron can be readily ionized approximately a half-cycle later during the subsequent extremum of the electric-field component of the laser. Because this electron is ionized at an electric-field extremum, its dominant contribution to the momentum distribution along the polarization direction will be around zero momentum, and will thus tend to "fill in the valley" between the peaks that are due to the direct non-sequential ionization process described above. It is interesting to note that, using a completely classical model, Ho *et al.* [141] were able to distinguish between electron trajectories that can be attributed to an NSDI mechanism and those that correspond to sequential ionization. In further work, Wang and Eberly [142] found, using the same classical approach, that the contributions from the two different double ionization mechanisms could be clearly distinguished using an elliptically polarized driving laser pulse.

Further insight into the NSDI and RESI processes can be gained by measuring the momentum of both ejected electrons in coincidence. The first investigations of this type were carried out by Weber *et al.* [130] and Feuerstein *et al.* [143] on argon. Using a Ti:sapphire laser of wavelength 795 nm producing 25 fs pulses, Feuerstein *et al.* measured the momentum of each of the two ejected electrons *parallel* to the laser field polarization direction over a range of peak laser field intensities from 2×10^{14} W cm^{-2} to 2×10^{15} W cm^{-2}. Using the kinematical constraints imposed by the recollision model, they were able to separate the contributions from NSDI and RESI. In particular, an electron pair ionized simultaneously at time t_r with negligible drift momentum will result in each of the two electrons acquiring a momentum parallel to the laser polarization direction that is equal to $k_{A,B}^{\parallel} = 2(U_p)^{1/2} \sin(\omega t_r)$. As a consequence, in the $k_A^{\parallel}$-$k_B^{\parallel}$ plane, strongly positively correlated momentum pairs are signatures of recollision ionization, whereas the momentum difference, $k_A^{\parallel} - k_B^{\parallel}$, yields information about the delay in time between the emission of the two electrons.

The NSDI process can also be probed by measuring the momentum of the ejected electron pair *perpendicular* to the laser polarization direction. Such studies have been carried out by Weckenbrock *et al.* [144] in argon and by Weckenbrock *et al.* [145] in neon using 40 fs Ti:sapphire laser pulses of peak intensity 1.9×10^{14} W cm^{-2}. When two electrons are ejected simultaneously in an NSDI process, the Coulomb interaction between the two electrons will influence their final-state dynamics. Along the polarization direction, this interaction is difficult to discern due to the dominant role of the electric-field component of the laser pulse on the dynamics of the electrons. In contrast, measurements of the momentum of the two electrons perpendicular to the laser polarization direction provide a sensitive probe of electron–electron dynamics. Weckenbrock *et al.* found that electrons ejected simultaneously have opposite momenta components perpendicular to the laser field polarization direction. Further electron coincidence experiments by Staudte *et al.* [146] and Rudenko *et al.* [147] in helium have revealed features that can be directly associated with field-free (e, 2e) collision events.

Let us turn our attention again to the RESI process. If the contribution of RESI to the total double ionization yield is important, then the momentum distribution along the laser polarization direction should depend on the pulse duration. Indeed, for sufficiently short pulses, the RESI process cannot occur because only a single recollision takes place. Double ionization can then only occur via the NSDI process, resulting in a distinct minimum at zero momentum of the recoil momentum of the ion parallel to the laser polarization direction. The contribution of RESI to the double ionization of neon and argon was studied experimentally by Rudenko *et al.* [148] using intense 25 fs and 7 fs Ti:sapphire laser pulses of carrier wavelength 795 nm. For neon, no dependence on the laser pulse duration was observed. This is consistent with the good agreement that was obtained between the experiments of Moshammer *et al.* [131] and the IMST calculations of Becker and Faisal [121] that did not include RESI processes. However, for argon, Rudenko *et al.* found, in particular for the production of Ar^{3+} and Ar^{4+}, that RESI processes contribute significantly to the multiple ionization yields.

At sufficiently high laser intensities the sequential ionization process becomes the dominant double ionization mechanism [44]. As the peak laser intensity is decreased, double ionization occurs predominantly via recollisional NSDI, while at even lower peak intensities the relative importance of RESI increases [134, 135]. Let us now focus our attention on double ionization in laser fields whose peak intensity is *not* sufficiently high to ensure that the kinetic energy of an ejected electron that returns to its parent core at time t_r is greater than the effective ionization potential $I_P(t_r)$ of the ion, given by Equation (8.21). In other words, we assume that $I < I_{TI}$. In this intensity regime, recollisional NSDI is not possible within the semi-classical model, but RESI can still occur. It is interesting to remark that no threshold effect has been observed in the total double ionization yield at $I = I_{TI}$ [44]. As the laser field peak intensity decreases below I_{TI}, the validity of the semi-classical recollision model must be called into question as the Keldysh parameter γ_K becomes comparable to or greater than unity, implying a transition to the multiphoton ionization regime. Experiments performed in this regime [149–152] have shown, in fact, that the semi-classical recollision description of double ionization can no longer account for many of the observed features of the energy and momentum spectra of the ejected electrons. For example, Chaloupka *et al.* [149] and Rudati *et al.* [150] found pronounced resonance structures in the electron momentum spectra while Liu *et al.* [152] observed strongly correlated back-to-back electron emission along the laser polarization direction. Parker *et al.* [151] have calculated electron energy distributions at a carrier wavelength of 390 nm for singly and doubly ionized helium at laser intensities of 8×10^{14} W cm^{-2} and 1.1×10^{15} W cm^{-2}. Their results were found to be in good agreement with measured distributions. Parker *et al.* also found a $5U_p$ cut-off in the total electron momentum distribution.

Finally, we remark that, with the advent of free electron lasers which will deliver intense ultra-violet and X-ray laser radiation, the study of multiphoton double ionization processes in atoms will be considerably enlarged. In particular, double ionization processes involving only a few photons, which have been the subject of a number of theoretical investigations [153–160], will be amenable to experimental study. This is the case, for example, of the *double-above-threshold ionization*

(DATI) process, which has been studied theoretically at XUV wavelengths by Parker *et al.* [161]. We also note that, in the wavelength range from 1.0 nm to 0.1 nm, the photon energies span the K-shell ionization potentials of neon through argon, and that at high laser intensities *double* K-shell vacancies in such atoms can be produced.

References

[1] C. J. Joachain, Theory of laser-atom interactions. In R. M. More, ed., *Laser Interactions with Atoms, Solids and Plasmas* (New York: Plenum Press, 1994), p. 39.

[2] L. F. DiMauro and P. Agostini, *Adv. At. Mol. Phys.* **35**, 79 (1995).

[3] M. Protopapas, C. H. Keitel and P. L. Knight, *Rep. Progr. Phys.* **60**, 389 (1997).

[4] C. J. Joachain, M. Dörr and N. J. Kylstra, *Adv. At. Mol. Opt. Phys.* **42**, 225 (2000).

[5] N. J. Kylstra, C. J. Joachain and M. Dörr. In D. Batani, C. J. Joachain, S. Martellucci and A. N. Chester, eds., *Atoms, Solids and Plasmas in Super-intense Laser Fields* (New York: Kluwer Academic–Plenum Publishers, 2001), p. 15.

[6] R. Dörner, T. Weber, M. Weckenbrock *et al.*, *Adv. At. Mol. Opt. Phys.* **48**, 1 (2002).

[7] W. Becker, F. Grasbon, R. Kopold, D. B. Milosevic, G. G. Paulus and H. Walther, *Adv. At. Mol. Opt. Phys.* **48**, 35 (2002).

[8] M. Lewenstein and A. L'Huillier, Principles of single atom physics: high-order harmonic generation, above-threshold ionization and non-sequential ionization. In T. Brabec, ed., *Strong Field Laser Physics*, Springer Series in Optical Sciences 134 (New York: Springer, 2009), p. 147.

[9] E. K. Damon and R. G. Tomlinson, *Appl. Opt.* **2**, 546 (1963).

[10] G. S. Voronov and N. B. Delone, *Sov. Phys. JETP Lett.* **1**, 66 (1965).

[11] J. L. Hall, E. J. Robinson and L. M. Branscomb, *Phys. Rev. Lett.* **14**, 1013 (1965).

[12] Y. Gontier and M. Trahin, *Phys. Rev. A* **4**, 1896 (1971).

[13] M. Aymar and M. Crance, *J. Phys. B* **14**, 3585 (1981).

[14] M. Crance, *Phys. Rep.* **114**, 117 (1987).

[15] A. L'Huillier, L. A. Lompré, G. Mainfray and C. Manus, *J. Phys. B* **16**, 1363 (1983).

[16] M. Cervenan and N. Isenor, *Opt. Commun.* **13**, 175 (1975).

[17] P. Agostini, F. Fabre, G. Mainfray, G. Petite and N. K. Rahman, *Phys. Rev. Lett.* **42**, 1127 (1979).

[18] P. Agostini, M. Clement, F. Fabre and G. Petite, *J. Phys. B* **14**, L491 (1981).

[19] F. Fabre, P. Agostini, G. Petite and M. Clement, *J. Phys. B* **14**, L677 (1981).

[20] P. Kruit, J. Kimman and M. J. van der Wiel, *J. Phys. B* **14**, L597 (1981).

[21] F. Fabre, G. Petite, P. Agostini and M. Clement, *J. Phys. B* **15**, 1353 (1982).

[22] P. Kruit, J. Kimman, H. G. Muller and M. J. van der Wiel, *Phys. Rev. A* **28**, 248 (1983).

[23] G. Petite, P. Agostini and H. G. Muller, *J. Phys. B* **21**, 4097 (1988).

[24] R. R. Freeman, P. H. Bucksbaum, H. Milchberg, S. Darack, D. Schumacher and M. E. Geusic, *Phys. Rev. Lett.* **59**, 1092 (1987).

[25] R. R. Freeman, P. H. Bucksbaum, W. E. Cooke, G. Gibson, T. J. McIlrath and L. D. van Woerkom, *Adv. At. Mol. Opt. Phys. Suppl.* **1**, 43 (1992).

[26] R. R. Freeman, T. J. McIlrath, P. H. Bucksbaum and M. Bashkansky, *Phys. Rev. Lett.* **57**, 3156 (1986).

[27] H. Rottke, B. Wolff, M. Tapernon, D. Feldmann and K. H. Welge. In F. Ehlotzky, ed., *Fundamentals of Laser Interactions II*, Lecture Notes in Physics 339 (Berlin: Springer-Verlag, 1989).

[28] H. Rottke, B. Wolff, M. Brickwedde, D. Feldmann and K. H. Welge, *Phys. Rev. Lett.* **64**, 404 (1990).

[29] H. Rottke, B. Wolff, M. Brickwedde, D. Feldmann and K. H. Welge. In G. Mainfray and P. Agostini, eds., *Multiphoton Processes* (Paris: Commissariat à l'Energie Atomique, 1991).

[30] H. Rottke, B. Wolff-Rottke, D. Feldmann *et al.*, *Phys. Rev. A* **49**, 4837 (1994).

[31] M. Dörr, R. M. Potvliege and R. Shakeshaft, *Phys. Rev. A* **41**, 558 (1990).

[32] M. Dörr, D. Feldman, R. M. Potvliege *et al.*, *J. Phys. B* **25**, L275 (1992).

[33] H. Helm and M. J. Dyer, *Phys. Rev. A* **49**, 2726 (1994).

[34] M. D. Perry, A. Szöke and K. C. Kulander, *Phys. Rev. Lett.* **63**, 1058 (1989).

[35] H. Rudolph, X. Tang, H. Bachau, P. Lambropoulos and E. Cormier, *Phys. Rev. Lett.* **66**, 3241 (1991).

[36] V. Schyja, T. Lang and H. Helm, *Phys. Rev. A* **57**, 3692 (1998).

[37] F. Yergeau, S. L. Chin and P. Lavigne, *J. Phys. B* **20**, 723 (1987).

[38] S. L. Chin, W. Xiong and P. Lavigne, *J. Opt. Soc. Am. B* **4**, 853 (1987).

[39] S. Augst, D. Strickland, D. Meyerhofer, S. L. Chin and J. H. Eberly, *Phys. Rev. Lett.* **63**, 2212 (1989).

[40] E. Mevel, P. Breger, R. Trainham *et al.*, *Phys. Rev. Lett.* **70**, 406 (1993).

[41] K. J. Schafer, B. Yang, L. F. DiMauro and K. C. Kulander, *Phys. Rev. Lett.* **70**, 1599 (1993).

[42] G. G. Paulus, W. Nicklich and H. Walther, *Europhys. Lett.* **27**, 267 (1994).

[43] G. G. Paulus, W. Nicklich, H. Xu, P. Lambropoulos and H. Walther, *Phys. Rev. Lett.* **72**, 2851 (1994).

[44] B. Walker, B. Sheehy, L. F. DiMauro, P. Agostini, K. J. Schafer and K. Kulander, *Phys. Rev. Lett.* **73**, 1227 (1994).

[45] G. G. Paulus, W. Nicklich, F. Zacher, P. Lambropoulos and H. Walther, *J. Phys. B* **29**, L249 (1996).

[46] M. P. Hertlein, P. H. Bucksbaum and H. G. Muller, *J. Phys. B* **30**, L197 (1997).

[47] P. Hansch, M. A. Walker and L. D. van Woerkom, *Phys. Rev. A* **55**, R2535 (1997).

[48] P. Hansch, M. A. Walker and L. D. van Woerkom, *Phys. Rev. A* **57**, R709 (1998).

[49] M. J. Nandor, M. A. Walker, L. D. van Woerkom and H. G. Muller, *Phys. Rev. A* **60**, R1771 (1999).

[50] G. G. Paulus and H. Walther. In D. Batani, C. J. Joachain, S. Martellucci and A. N. Chester, eds., *Atoms, Solids and Plasmas in Super-Intense Laser Fields* (New York: Kluwer Academic–Plenum Publishers, 2001), p. 285.

[51] P. B. Corkum, *Phys. Rev. Lett.* **71**, 1994 (1993).

[52] K. C. Kulander, K. J. Schafer and J. L. Krause. In B. Piraux, A. L'Huillier and K. Rzążewski, eds., *Super-Intense Laser-Atom Physics* (New York: Plenum Press, 1993).

[53] E. Cormier and P. Lambropoulos, *J. Phys. B* **30**, 77 (1997).

[54] G. G. Paulus, F. Grasbon, H. Walther, R. Kopold and W. Becker, *Phys. Rev. A* **64**, 021401(R) (2001).

[55] R. Kopold, W. Becker, M. Kleber and G. G. Paulus, *J. Phys. B* **35**, 217 (2002).

[56] F. Grasbon, G. G. Paulus, H. Walther *et al.*, *Phys. Rev. Lett.* **91**, 173003 (2003).

[57] H. G. Muller and F. C. Kooiman, *Phys. Rev. Lett.* **81**, 1207 (1998).

[58] H. G. Muller, *Phys. Rev. A* **60**, 1341 (1999).

[59] H. G. Muller, *Phys. Rev. Lett.* **83**, 3158 (1999).

[60] J. Wassaf, V. Véniard, R. Taïeb and A. Maquet, *Phys. Rev. Lett.* **90**, 013003 (2003).

[61] J. Wassaf, V. Véniard, R. Taïeb and A. Maquet, *Phys. Rev. A* **67**, 053405 (2003).

[62] R. M. Potvliege and S. Vučić, *Phys. Rev. A* **74**, 023412 (2006).

[63] R. M. Potvliege and S. Vučić, *Physica Scripta* **74**, C55 (2006).

[64] B. Yang, K. J. Schafer, B. Walker, K. C. Kulander, P. Agostini and L. F. DiMauro, *Phys. Rev. Lett.* **71**, 3770 (1993).

[65] N. B. Delone and V. P. Krainov, *J. Opt. Soc. Am. B* **8**, 1207 (1991).

[66] G. Paulus, W. Becker, W. Nicklich and H. Walther, *J. Phys. B* **27**, L703 (1994).

[67] M. Lewenstein, K. C. Kulander, K. J. Schafer and P. H. Bucksbaum, *Phys. Rev. A* **51**, 1495 (1995).

[68] B. H. Bransden and C. J. Joachain, *Physics of Atoms and Molecules*, 2nd edn (Harlow, UK: Prentice Hall-Pearson, 2003).

[69] P. H. Bucksbaum, M. Bashkansky, R. R. Freeman, T. J. McIlrath and L. F. DiMauro, *Phys. Rev. Lett.* **56**, 2590 (1986).

[70] U. Mohideen, M. H. Sher, H. W. K. Tom *et al.*, *Phys. Rev. Lett.* **71**, 509 (1993).

[71] H. R. Reiss, *Phys. Rev. A* **54**, R1765 (1996).

[72] G. G. Paulus, F. Zacher, H. Walther, A. Lohr, W. Becker and M. Kleber, *Phys. Rev. Lett.* **80**, 484 (1998).

[73] G. G. Paulus, F. Grasbon, A. Dreischuh, H. Walther, R. Kopold and W. Becker, *Phys. Rev. Lett.* **84**, 3791 (2000).

[74] S. P. Goreslavskii and S. V. Popruzhenko, *Sov. Phys. JETP Lett.* **38**, 661 (1996).

[75] G. G. Paulus, F. Grasbon, H. Walther *et al.*, *Nature* **414**, 182 (2001).

[76] G. G. Paulus, F. Lindner, H. Walther *et al.*, *Phys. Rev. Lett.* **91**, 253004 (2003).

[77] M. F. Kling, J. Rauschenberger, A. Verhoef *et al.*, *New J. Phys* **10**, 025024 (2008).

[78] D. W. Chandler and P. L. Houston, *J. Chem. Phys.* **87**, 1445 (1987).

[79] A. T. Eppink and D. H. Parker, *Rev. Sci. Instrum.* **68**, 3477 (1997).

[80] H. G. Muller, H. B. van Linden van den Heuvel and M. J. van der Wiel, *J. Phys. B* **19**, L733 (1986).

[81] D. A. Tate, D. G. Papaioannou and T. F. Gallagher, *Phys. Rev. A* **42**, 5703 (1990).

[82] C. Chen, Y. Yin and D. Elliott, *Phys. Rev. Lett.* **64**, 507 (1989).

[83] H. G. Muller, P. H. Bucksbaum, D. W. Schumacher and A. Zavriyev, *J. Phys. B* **23**, 2761 (1990).

[84] H. G. Muller, *Commun. At. Mol. Phys.* **24**, 355 (1990).

[85] P. Agostini, In D. Batani, C. J. Joachain, S. Martellucci and A. N. Chester, eds., *Atoms, Solids and Plasmas in Super-Intense Laser Fields* (New York: Kluwer Academic–Plenum Publishers, 2001), p. 59.

[86] J. M. Schins, P. Breger, P. Agostini *et al.*, *Phys. Rev. Lett.* **73**, 2180 (1994).

[87] J. M. Schins, P. Breger, P. Agostini *et al.*, *Phys. Rev. A* **52**, 1272 (1995).

[88] A. Bouhal, R. Evans, G. Grillon *et al.*, *J. Opt. Soc. Am. B* **14**, 950 (1997).

[89] E. S. Toma, H. G. Muller, P. M. Paul *et al.*, *Phys. Rev. A* **62**, 061801 (2000).

[90] R. Taïeb, V. Véniard and A. Maquet, *J. Opt. Soc. Am. B* **13**, 363 (1996).

[91] V. Véniard, R. Taïeb and A. Maquet, *Phys. Rev. A* **54**, 721 (1996).

[92] F. W. Byron and C. J. Joachain, *Phys. Rev.* **164**, 1 (1967).

[93] T. Kinoshita, *Phys. Rev.* **105**, 1490 (1957).

[94] A. Dalgarno and H. Sadeghpour, *Phys. Rev. A* **46**, R3591 (1992).

[95] L. Andersson and J. Burgdörfer, *Phys. Rev. Lett.* **71**, 201 (1993).

[96] J. Levin, D. Lindle, N. Keller *et al.*, *Phys. Rev. Lett.* **67**, 968 (1991).

[97] L. Spielberger, O. Jagutzki, R. Dörner *et al.*, *Phys. Rev. Lett.* **74**, 4615 (1995).

[98] F. Maulbetsch and J. Briggs, *J. Phys. B* **26**, 1679 (1993).

[99] D. Proulx and R. Shakeshaft, *Phys. Rev. A* **48**, R875 (1993).

[100] O. Schwartzkopf, B. Krässig, V. Schmidt, F. Maulbetsch and J. Briggs, *J. Phys. B* **27**, L347 (1994).

[101] M. Pont and R. Shakeshaft, *J. Phys. B* **28**, L571 (1995).

[102] H. Kossmann, V. Schmidt and T. Andersen, *Phys. Rev. Lett.* **60**, 1266 (1988).

[103] J. Bizau and F. J. Wuilleumier, *J. Electron Spectrosc. Relat. Phenom.* **71**, 205 (1995).
[104] A. L'Huillier, L. A. Lompré, G. Mainfray and C. Manus, *Phys. Rev. Lett.* **48**, 1814 (1982).
[105] A. L'Huillier, L. A. Lompré, G. Mainfray and C. Manus, *Phys. Rev. A* **27**, 2503 (1983).
[106] D. Charalambidis, P. Lambropoulos, H. Schröder *et al.*, *Phys. Rev. A* **50**, R2822 (1994).
[107] P. Lambropoulos, P. Maragakis and J. Zhang, *Phys. Rep.* **305**, 203 (1998).
[108] N. Fittinghoff, P. R. Bolton, B. Chang and K. C. Kulander, *Phys. Rev. Lett.* **69**, 2642 (1992).
[109] K. Kondo, A. Sagisaka, T. Tamida, Y. Nabekawa and S. Watanabe, *Phys. Rev. A* **48**, R2531 (1993).
[110] S. F. J. Larochelle, A. Talebpour and S. L. Chin, *J. Phys. B* **31**, 1201 (1998).
[111] J. B. Watson, A. Sanpera, K. Burnett, D. G. Lappas and P. L. Knight, *Phys. Rev. Lett.* **78**, 1884 (1997).
[112] H. W. van der Hart and K. Burnett, *Phys. Rev. A* **62**, 013407 (2000).
[113] M. V. Ammosov, N. B. Delone and V. P. Krainov, *Sov. Phys. JETP* **64**, 1191 (1986).
[114] B. Sheehy, R. Lafon, M. Widmer *et al.*, *Phys. Rev. A* **58**, 3942 (1998).
[115] H. W. van der Hart, *J. Phys. B* **33**, L699 (2000).
[116] G. Yudin and M. Ivanov, *Phys. Rev. A* **63**, 033404 (2001).
[117] A. Becker and F. H. M. Faisal, *J. Phys. B* **29**, L197 (1996).
[118] F. H. M. Faisal and A. Becker, *Laser Phys.* **7**, 684 (1997).
[119] A. Becker and F. H. M. Faisal, *Phys. Rev. A* **59**, R1742 (1999).
[120] A. Becker and F. H. M. Faisal, *J. Phys. B* **32**, L335 (1999).
[121] A. Becker and F. H. M. Faisal, *J. Phys. B* **38**, R1 (2005).
[122] A. Talebpour, C.-Y. Chien, Y. Liang, S. Larochelle and S. L. Chin, *J. Phys. B* **30**, 1721 (1997).
[123] J. S. Parker, L. R. Moore, D. Dundas and K. T. Taylor, *J. Phys. B* **33**, L691 (2000).
[124] J. S. Parker, L. R. Moore and K. T. Taylor, *Opt. Express* **8**, 436 (2001).
[125] H. G. Muller, *Opt. Express* **8**, 417 (2001).
[126] E. Eremina, X. Liu, H. Rottke *et al.*, *J. Phys. B* **36**, 3269 (2003).
[127] J. S. Parker, B. J. S. Doherty, K. J. Meharg and K. T. Taylor, *J. Phys. B* **36**, L393 (2003).
[128] N. Fittinghoff, P. R. Bolton, B. Chang and K. C. Kulander, *Phys. Rev. A* **49**, 2174 (1994).
[129] R. Lafon, J. L. Chaloupka, B. Sheehy *et al.*, *Phys. Rev. Lett.* **86**, 2762 (2001).
[130] T. Weber, M. Weckenbrock, A. Staudte *et al.*, *Phys. Rev. Lett.* **84**, 443 (2000).
[131] R. Moshammer, B. Feuerstein, W. Schmitt *et al.*, *Phys. Rev. Lett.* **84**, 447 (2000).
[132] R. Moshammer, B. Feuerstein, J. Crespo López-Urrutia *et al.*, *Phys. Rev. A* **65**, 035401 (2002).
[133] R. Moshammer, J. Ullrich, B. Feuerstein *et al.*, *J. Phys. B* **36**, L113 (2003).
[134] V. L. B. de Jesus, B. Feuerstein, K. Zrost *et al.*, *J. Phys. B* **37**, L161 (2004).
[135] A. Rudenko, T. Ergler, K. Zrost *et al.*, *Phys. Rev. A* **78**, 015403 (2008).
[136] A. Becker and F. H. M. Faisal, *Phys. Rev. Lett.* **84**, 3546 (2000).
[137] S. Popruzhenko and S. Goreslavskii, *J. Phys. B* **34**, L329 (2001).
[138] L. V. Keldysh, *Sov. Phys. JETP* **20**, 1307 (1965).
[139] R. Kopold, W. Becker, H. Rottke and W. Sandner, *Phys. Rev. Lett.* **85**, 3781 (2000).
[140] S. P. Goreslavskii, S. V. Popruzhenko, R. Kopold and W. Becker, *Phys. Rev. A* **64**, 053402 (2001).
[141] P. J. Ho, R. Panfili, S. L. Haan and J. H. Eberly, *Phys. Rev. Lett.* **94**, 093002 (2005).
[142] X. Wang and J. H. Eberly, *Phys. Rev. Lett.* **103**, 103007 (2009).
[143] B. Feuerstein, R. Moshammer, D. Fischer *et al.*, *Phys. Rev. Lett.* **87**, 043003 (2001).
[144] M. Weckenbrock, A. Becker, S. Kammer *et al.*, *Phys. Rev. Lett.* **91**, 123004 (2003).

[145] M. Weckenbrock, D. Zeidler, A. Staudte *et al.*, *Phys. Rev. Lett.* **92**, 213002 (2004).
[146] A. Staudte, C. Ruiz, M. Schöffler *et al.*, *Phys. Rev. Lett.* **99**, 263002 (2007).
[147] A. Rudenko, V. L. B. de Jesus, T. Ergler *et al.*, *Phys. Rev. Lett.* **99**, 263003 (2007).
[148] A. Rudenko, K. Zrost, B. Feuerstein *et al.*, *Phys. Rev. Lett.* **93**, 253001 (2004).
[149] J. L. Chaloupka, J. Rudati, R. Lafon, P. Agostini, K. C. Kulander and L. F. DiMauro, *Phys. Rev. Lett.* **90**, 033002 (2003).
[150] J. Rudati, J. L. Chaloupka, P. Agostini, K. C. Kulander and L. F. DiMauro, *Phys. Rev. Lett.* **92**, 203001 (2004).
[151] J. S. Parker, B. J. S. Doherty, K. T. Taylor, K. D. Schultz, C. I. Blaga and L. F. DiMauro, *Phys. Rev. Lett.* **96**, 133001 (2006).
[152] Y. Liu, S. Tschuch, A. Rudenko *et al.*, *Phys. Rev. Lett.* **101**, 053001 (2008).
[153] L. A. A. Nikolopoulos and P. Lambropoulos, *J. Phys. B* **34**, 545 (2001).
[154] L. Feng and H. W. van der Hart, *J. Phys. B* **36**, L1 (2003).
[155] S. Laulan and H. Bachau, *Phys. Rev. A* **68**, 013409 (2003).
[156] S. X. Hu, J. Colgan and L. A. Collins, *J. Phys. B* **38**, L35 (2005).
[157] A. S. Kheifets and I. A. Ivanov, *J. Phys. B* **39**, 1731 (2006).
[158] E. Foumouo, K. Lagmago, G. Edah and B. Piraux, *Phys. Rev. A* **74**, 063409 (2006).
[159] D. A. Horner, F. Morales, T. N. Rescigno, F. Martin and C. W. McCurdy, *Phys. Rev. A* **76**, 030701 (2007).
[160] R. Shakeshaft, *Phys. Rev. A* **76**, 063405 (2007).
[161] J. S. Parker, L. R. Moore, K. J. Meharg, D. Dundas and K. T. Taylor, *J. Phys. B* **34**, L69 (2001).

9

Harmonic generation and attosecond pulses

This chapter is devoted to the study of harmonic generation in atoms and the physics of attosecond pulses, also called attophysics, which are two major topics in the study of high-intensity laser–atom interactions. We start in Section 9.1 by reviewing important experiments, with particular emphasis on high-order harmonic generation, which is a very interesting probe of the behavior of atoms interacting with intense laser fields. In Section 9.2, we discuss harmonic generation calculations, first at the microscopic (single-atom response) level and then at the macroscopic level. The main properties of harmonics and some of their applications are discussed in Section 9.3. Finally, in Section 9.4, we examine how attosecond pulses can be produced and used to investigate the dynamics of atoms at unprecedented time and space scales. Reviews of harmonic generation have been given by L'Huillier, Schafer and Kulander [1], L'Huillier *et al.* [2], Joachain [3], Salières *et al.* [4], Protopapas, Keitel and Knight [5], Joachain, Dörr and Kylstra [6], Brabec and Krausz [7], Salières [8, 9] and Salières and Christov [10]. Attosecond physics has been reviewed by Agostini and DiMauro [11], Scrinzi *et al.* [12], Kienberger *et al.* [13], Niikura and Corkum [14], Krausz and Ivanov [15], Lewenstein and L'Huillier [16] and Scrinzi and Muller [17].

9.1 Experiments

In this section, we give an overview of key experiments which have been performed in the field of harmonic generation and the production of attosecond pulses. Optical harmonic generation was discovered in 1961 by Franken *et al.* [18], who produced the second harmonic of a ruby laser in a quartz crystal. They demonstrated in this way that atoms could absorb simultaneously two optical photons having each the energy $\hbar\omega$ and emit a photon having the energy $2\hbar\omega$. This discovery opened new perspectives for the extension of coherent light sources to higher frequencies. In 1967, New and Ward [19] observed the third harmonic of a ruby laser in noble gases.

We recall that only odd harmonics are observed in an isotropic gaseous medium having inversion symmetry. Although higher harmonics could be generated in the following years, the observation of very high harmonic orders was prevented by the rapidly decreasing frequency conversion efficiency as the order increased. This is due to the fact that in the weak-field, perturbative regime, the probability of absorbing q photons simultaneously falls off quickly when q increases. Detailed reviews of early work on harmonic generation in gases have been given by Hanna, Yuratich and Cotter [20], Reintjes [21], Shen [22], Arkhipkin and Popov [23] and Delone and Krainov [24].

The situation changed considerably with the progress in short-pulse, intense-laser technology, which allowed the investigation of the behavior of atoms interacting with strong laser pulses having focused intensities from 10^{13} to 10^{15} W cm^{-2} and durations in the picosecond range. When such laser pulses were focused into a noble-gas jet, the spectrum of the radiation emitted in the propagation direction exhibited the phenomenon of *high-order harmonic generation* (HHG). In particular, experiments performed by McPherson *et al.* [25] in 1987 and by Rosman *et al.* [26] in 1988 demonstrated the generation of up to the 17th harmonic of a 248 nm KrF laser in a neon vapor. Soon afterwards, using a Nd:YAG laser at a wavelength of 1064 nm and intensities of about 10^{13} W cm^{-2}, the 33rd harmonic in argon was observed by Ferray *et al.* [27] and by Li *et al.* [28], who also generated up to the 29th harmonic in krypton and the 21st harmonic in xenon. A schematic picture of the experimental arrangement used at Saclay is shown in Fig. 9.1. The number of photons produced by each laser shot by Lompré *et al.* [29] in a 15 Torr xenon vapor is shown in Fig. 9.2 as a function of the harmonic order for several laser intensities between 5×10^{12} W cm^{-2} and 3×10^{13} W cm^{-2}. At the lowest intensity, the harmonic signal is seen to decrease with the order. As the intensity increases, a plateau followed by an abrupt cut-off appears. The length of the plateau increases with the laser intensity, up to an intensity (in this case about 1.3×10^{13} W cm^{-2}) above which the medium becomes ionized with a probability close to unity. In further experiments performed in neon, the 25th harmonic of a KrF laser was observed by Sarakura *et al.* [30] and the 53rd harmonic of a 1053 nm Nd:glass laser was observed by L'Huillier *et al.* [31].

Since the discovery of the plateau, a number of experiments have been carried out in order to investigate its extension into very high harmonic orders. This was made possible by continuing advances in laser technology, which allowed the production of very intense, short laser pulses. In particular, Macklin, Kmetec and Gordon [32] reported the generation of up to the 109th harmonic in neon, using 125 fs pulses from a 806 nm Ti:sapphire laser at an intensity near 10^{15} W cm^{-2}. Also in 1993, L'Huillier and Balcou [33] detected high-order harmonics in noble gases, using 1 ps pulses from a 1053 nm Nd:glass laser at intensities between 10^{14} and

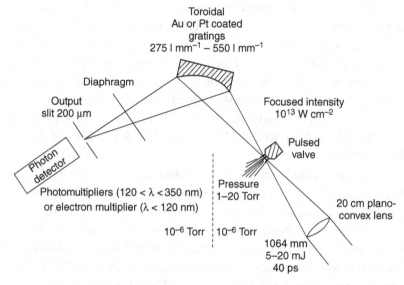

Figure 9.1. Experimental arrangement for the detection of VUV light employed in Saclay. (From A. L'Huillier *et al.*, *Adv. At. Mol. Opt. Phys. Suppl.* **1**, 139 (1992).)

$10^{15}\,\mathrm{W\,cm^{-2}}$. Their results for xenon, argon, neon and helium at an intensity of about $1.5 \times 10^{15}\,\mathrm{W\,cm^{-2}}$ are shown in Fig. 1.12. As noted in Section 1.4, the plateau observed by L'Huillier and Balcou is particularly long for helium and neon. They were able to detect up to the 135th harmonic in neon, without apparently reaching the plateau limit.

In later experiments using ultra-short, few-cycle laser pulses, with peak intensities above $10^{15}\,\mathrm{W\,cm^{-2}}$ [34–40], very high harmonic frequencies and harmonic orders have been observed, as mentioned in Section 1.4. The harmonic angular frequencies have exceeded 300ω and extended into the water window [36, 37], namely the region of the electromagnetic spectrum for which water is transparent. For example, Chang *et al.* [37] observed discrete harmonic peaks up to the 221st order, corresponding to a wavelength of 3.6 nm, and coherent soft X-ray radiation down to 2.7 nm in helium (see Fig. 9.3), using a driving Ti:sapphire laser producing 800 nm, 26 fs pulses with peak intensities around $6 \times 10^{15}\,\mathrm{W\,cm^{-2}}$. The generation of coherent X-rays in the water window, below the carbon K-edge at a wavelength of 4.4 nm, was reported at about the same time in helium by Spielmann *et al.* [36] using a Ti:sapphire laser delivering 5 fs pulses. The same group [38] observed 500 eV photons (corresponding to the 323rd harmonic order) at a wavelength of 2.5 nm, also in helium. More recently, Chen *et al.* [40] have demonstrated fully phase-matched high harmonic emission spanning the water window spectral region.

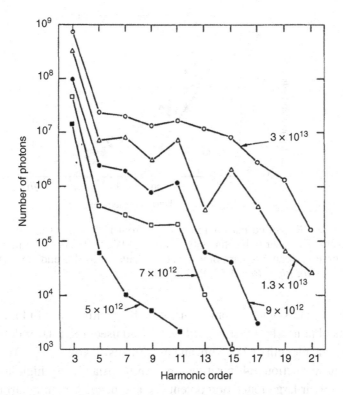

Figure 9.2. Number of photons per laser shot produced in xenon at a wavelength of 1064 nm, as a function of the harmonic order. The values of the laser intensity (in W cm^{-2}) are indicated next to the curves. (From L.A. Lompré *et al.*, *J. Opt. Soc. Am. B* **7**, 754 (1990).)

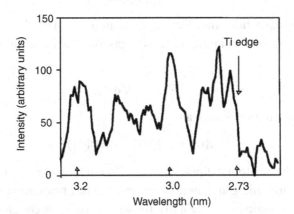

Figure 9.3. Harmonic emission spectrum of helium interacting with a 26 fs laser pulse of wavelength $\lambda = 800$ nm and intensity $I = 6 \times 10^{15}$ W cm^{-2}, filtered through a 0.2 μm Ti filter. (From Z. Chang *et al.*, *Phys. Rev. Lett.* **79**, 2967 (1997).)

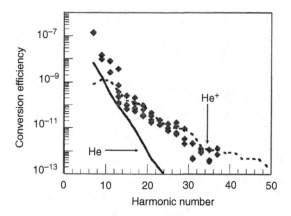

Figure 9.4. Measured harmonic emission from He$^+$ ions at a wavelength $\lambda = 248.6$ nm. The laser pulse intensity is 4×10^{17} W cm^{-2} and its duration is 380 fs. The curves display the harmonic spectra calculated for He and He$^+$. (From S.G. Preston *et al.*, *Phys. Rev. A* **53**, R31 (1996).)

Finally, we consider harmonic generation by positive ions. In Fig. 9.4, we show the results obtained by Preston *et al.* [41], who used a KrF laser delivering 380 fs pulses at a wavelength of 248.6 nm and peak intensities of 4×10^{17} W cm^{-2}. As was pointed out in Section 1.4, positive ions can withstand very high laser intensities because of their large ionization potentials, and hence can emit harmonic photons of high energy. Photon emission by positive ions will be analyzed in Section 9.2, particularly in connection with the investigation of non-dipole effects.

9.2 Calculations

In response to a laser field, the electrons in an atom oscillate, the source of harmonic generation being the polarization of the gaseous medium induced by the laser field [2],

$$\mathcal{P}(t) = \mathcal{N}\mathbf{d}(t), \tag{9.1}$$

where $\mathcal{N}$ is the atomic density and

$$\mathbf{d}(t) = \langle \Psi(t)| -e\mathbf{R}|\Psi(t)\rangle \tag{9.2}$$

is the laser-induced atomic dipole moment. As noted in Section 1.4, the theoretical treatment of harmonic generation by an intense laser beam focused into a gaseous medium has two main aspects. Firstly, the *microscopic, single-atom* response to the laser field must be analyzed. This means that the harmonic emission spectra emitted by individual atoms must be calculated for a range of driving laser intensities. The theoretical methods required to perform these calculations have been discussed in

Part II of this book. Secondly, these single-atom spectra must be combined to obtain the *macroscopic* harmonic fields generated from the *coherent* emission of all the atoms of the generating medium. The properties of the harmonic emission result from the interplay between the single-atom response and the propagation of the harmonic fields in the non-linear medium.

9.2.1 Single-atom response

9.2.1.1 Basic equations

In weak laser fields, the electrons oscillate essentially at the fundamental angular frequency ω of the driving laser. In stronger laser fields the electrons also oscillate at additional higher angular frequencies, thereby causing the emission of radiation at these angular frequencies. To analyze this phenomenon, we shall consider first the case of a spatially homogeneous and monochromatic driving laser field. As discussed in Chapter 4, we can then expand the state vector $|\Psi(t)\rangle$ in a Floquet–Fourier series as follows:

$$|\Psi(t)\rangle = \exp(-\mathrm{i}\mathsf{E}t/\hbar) \sum_{n=-\infty}^{\infty} \exp(-\mathrm{i}n\omega t)|F_n\rangle, \qquad (9.3)$$

where $|F_n\rangle$ are the harmonic components of $|\Psi(t)\rangle$. We recall that the complex quasi-energy E of the Siegert state corresponding to the state vector $|\Psi(t)\rangle$ is given by

$$\mathsf{E} = E_i + \Delta - \mathrm{i}\Gamma/2, \qquad (9.4)$$

where E_i is the energy of the initial unperturbed state, Δ is the AC Stark shift of that state and Γ is its induced ionization width. Inserting the Floquet–Fourier expansion (9.3) into the expression (9.2) of the laser-induced atomic dipole moment, we have

$$\mathbf{d}(t) = \exp(-\Gamma t/\hbar) \sum_{m=-\infty}^{\infty} \sum_{n=-\infty}^{\infty} \langle F_m|-e\mathbf{R}|F_n\rangle \exp[-\mathrm{i}(n-m)\omega t]. \quad (9.5)$$

Writing $q = n - m$ and defining the quantities

$$\mathbf{d}_{qn} = \langle F_{n-q}|-e\mathbf{R}|F_n\rangle \qquad (9.6)$$

and

$$\mathbf{d}_q = \sum_{n=-\infty}^{\infty} \mathbf{d}_{qn}, \qquad (9.7)$$

we see that Equation (9.5) becomes

$$\mathbf{d}(t) = \exp(-\Gamma t/\hbar) \left\{ \mathbf{d}_0 + \sum_{q=1}^{\infty} [\mathbf{d}_q \exp(-iq\omega t) + \text{c.c.}] \right\}, \qquad (9.8)$$

where c.c. denotes the complex conjugate. Since $\mathbf{d}(t)$ is real, it follows that $\mathbf{d}_{-q} = \mathbf{d}_q^*$, and we can rewrite Equation (9.8) in the form

$$\mathbf{d}(t) = \exp(-\Gamma t/\hbar) \left\{ \mathbf{d}_0 + 2 \sum_{q=1}^{\infty} \text{Re} \left[\mathbf{d}_q \exp(-iq\omega t) \right] \right\}. \qquad (9.9)$$

Let $\psi_i(X) \equiv \langle X | \psi_i \rangle$, where X denotes the ensemble of the coordinates of the atomic electrons, be the initial field-free atomic state. We choose the energy scale so that, in the field-free limit, $F_0(X) \equiv \langle X | F_0 \rangle$ reduces to $\psi_i(X)$. If $\psi_i(X)$ has a *definite parity*, $\Pi = \pm 1$, the function $F_m(X) \equiv \langle X | F_m \rangle$ has parity $\Pi(-1)^m$, since $| F_m \rangle$ corresponds to the harmonic component of an atom which has absorbed m (real or virtual) photons, each of parity (-1). Hence $\langle F_{n-q} | X \rangle \langle X | F_n \rangle$ has parity $\Pi(-1)^{n-q}(-1)^n \Pi = (-1)^q$ and therefore $\mathbf{d}_{qn}$ and $\mathbf{d}_q$ vanish unless q is *odd*. The harmonics are then emitted at angular frequencies

$$\Omega = q\omega, \quad q = 3, 5, 7, \ldots, \qquad (9.10)$$

which are odd multiples of the angular frequency ω of the driving laser field.

We remark that if the ionization width Γ is such that $\Gamma \ll q\hbar\omega$, the exponential decay factor $\exp(-\Gamma t/\hbar)$ due to multiphoton ionization can be neglected in Equation (9.9). The laser-induced atomic dipole moment then takes the simpler form

$$\mathbf{d}(t) = 2 \sum_{q=1}^{\infty} \text{Re} \left[\mathbf{d}_q \exp(-iq\omega t) \right]. \qquad (9.11)$$

In this case, we can readily calculate the Fourier transform of the dipole moment, with the result

$$\mathbf{d}(\Omega) = (2\pi)^{-1/2} \int_{-\infty}^{\infty} \exp(-i\Omega t) \mathbf{d}(t) dt$$

$$= (2\pi)^{1/2} \mathbf{d}_q^* \delta(\Omega - q\omega), \qquad (9.12)$$

and we note that $\mathbf{d}(-\Omega) = \mathbf{d}^*(\Omega)$.

While only the laser-induced atomic dipole moment is required to obtain the polarization $\mathcal{P}(t)$, it is interesting to calculate the rate of spontaneous emission of harmonic photons about the angular frequency $\Omega = q\omega$ by a single atom. Let us assume that the harmonic photons are linearly polarized with a polarization vector

$\hat{\epsilon}_H$ and are emitted into a solid angle $d\hat{n}$, where $\hat{n}$ is a unit vector pointing along the direction of emission, which is orthogonal to $\hat{\epsilon}_H$. The corresponding rate of photon emission by the atom in the laser field is then given by [42]

$$W_H(\Omega, \hat{\epsilon}_H)d\hat{n}\,d\Omega = \frac{\Omega^3}{2\pi c^3 \hbar}\left(\frac{1}{4\pi\epsilon_0}\right)|\hat{\epsilon}_H \cdot \mathbf{d}_q|^2\,d\hat{n}. \tag{9.13}$$

The angular distribution of emitted photons depends only on the angle between the driving laser field polarization direction and the polarization of the emitted photons. In particular, the emission rate along the laser field propagation direction and the emission rate opposite to the laser field propagation direction are the same.

The total emission rate of harmonic photons in a small interval $d\Omega$ about the angular frequency $\Omega = q\omega$ is obtained by integrating over all observation directions and summing over the two directions of polarization, with the result

$$W_H(\Omega)\,d\Omega = \frac{4\Omega^3}{3c^3\hbar}\left(\frac{1}{4\pi\epsilon_0}\right)|\mathbf{d}_q|^2. \tag{9.14}$$

The probability of emitting a photon per laser optical period is therefore given by

$$P_H = \frac{8\pi}{3c^3\hbar\omega}\left(\frac{1}{4\pi\epsilon_0}\right)\sum_{q=1}^{\infty}\Omega^3|\mathbf{d}_q|^2. \tag{9.15}$$

We also remark that for a monochromatic laser field the Fourier transform $\mathbf{a}(\Omega)$ of the acceleration of the laser-induced atomic dipole moment, as defined by Equation (1.16), is related to the Fourier transform of $\mathbf{d}(t)$ by

$$\mathbf{a}(\Omega) = -(2\pi)^{1/2}\Omega^2\mathbf{d}_q^*\,\delta(\Omega - q\omega)$$
$$\equiv (2\pi)^{1/2}\mathbf{a}_q^*\,\delta(\Omega - q\omega). \tag{9.16}$$

This result allows us to express the quantum-mechanical analog of the Larmor formula as

$$P_H(\Omega = q\omega)d\Omega = \hbar\Omega W_H(\Omega)d\Omega$$
$$= \frac{4}{3c^3}\left(\frac{1}{4\pi\epsilon_0}\right)|\mathbf{a}_q|^2. \tag{9.17}$$

Taking the finite lifetime of the atom into account, and assuming that this lifetime is sufficiently long so that $\Gamma \ll \hbar\omega$, the harmonic emission spectrum is approximately given by a series of Lorentzian profiles, each having width Γ:

$$W_H(\Omega)d\Omega = \left(\frac{1}{4\pi\epsilon_0}\right)\sum_{q=1}^{\infty}\frac{4\Omega^3}{3\pi c^3\hbar}\frac{\Gamma}{\Gamma^2 + (\Omega - q\omega)^2}|\mathbf{d}_q|^2\,d\Omega. \tag{9.18}$$

We now consider a driving laser *pulse*. The corresponding laser field will be assumed to be spatially homogeneous (dipole approximation), but not monochromatic. In order to obtain the polarization of the medium, the laser-induced atomic dipole moment $\mathbf{d}(t)$ can be calculated from Equation (9.2), where the state vector $|\Psi(t)\rangle$ is obtained by solving the TDSE in the dipole approximation.

The probability of emitting a photon of angular frequency Ω is given by

$$P(\Omega)d\Omega = \frac{4}{3c^3\hbar\Omega}\left(\frac{1}{4\pi\epsilon_0}\right)|\mathbf{a}(\Omega)|^2\,d\Omega, \tag{9.19}$$

where $\mathbf{a}(\Omega)$ is given for a laser pulse of duration τ_p by

$$\mathbf{a}(\Omega) = (2\pi)^{-1/2}\int_0^{\tau_p}\exp(-i\Omega t)\mathbf{a}(t)dt. \tag{9.20}$$

The laser-induced atomic dipole moment acceleration can be obtained either from $\mathbf{d}(t)$, namely

$$\mathbf{a}(t) = \frac{d^2}{dt^2}\mathbf{d}(t) = -e\frac{d^2}{dt^2}\langle\Psi(t)|\mathbf{R}|\Psi(t)\rangle, \tag{9.21}$$

or from the laser-induced atomic dipole moment velocity $\mathbf{v}(t)$,

$$\mathbf{a}(t) = \frac{d}{dt}\mathbf{v}(t) = -\frac{e}{m}\frac{d}{dt}[\langle\Psi(t)|\mathbf{P}|\Psi(t)\rangle + e\mathbf{A}(t)], \tag{9.22}$$

where $\mathbf{P}$ is the total momentum operator defined by Equation (2.120). Alternatively, the dipole acceleration can be obtained directly as

$$\mathbf{a}(t) = \frac{e}{m}[\langle\Psi(t)|\nabla_{\mathbf{R}}V|\Psi(t)\rangle + e\boldsymbol{\mathcal{E}}(t)]. \tag{9.23}$$

Here, V, given by Equation (2.107), denotes the sum of all Coulomb interactions within the atom. The latter form is particularly interesting from the numerical point of view because it emphasizes the part of the wave function close to the nucleus. In the low-frequency regime, this is precisely, according to the semi-classical recollision model, where photon emission through recombination is expected to occur.

Integrating Equation (9.20) by parts twice, we see that

$$\mathbf{a}(\Omega) = (2\pi)^{-1/2}[\mathbf{v}(\tau_p) + i\Omega\mathbf{d}(\tau_p)]\exp(-i\Omega\tau_p) - (2\pi)^{-1/2}\Omega^2$$

$$\times\int_0^{\tau_p}\exp(-i\Omega t)\mathbf{d}(t)dt$$

$$= (2\pi)^{-1/2}[\mathbf{v}(\tau_p) + i\Omega\mathbf{d}(\tau_p)]\exp(-i\Omega\tau_p) - \Omega^2\mathbf{d}(\Omega). \tag{9.24}$$

At the end of the pulse, only excited bound states and low-energy continuum states can contribute to the laser-induced dipole moment of the atom. Hence the boundary terms at $t = \tau_p$ in Equation (9.24) will only affect the low-energy part of the emission spectrum. When these terms are omitted, the Fourier transform of the laser-induced atomic dipole moment acceleration can be calculated directly from the Fourier transform $\mathbf{d}(\Omega)$ of $\mathbf{d}(t)$.

Equation (9.14) for the photon emission rate by an atom interacting with a monochromatic laser field can also be obtained from Equation (9.19) by assuming that this laser field is adiabatically switched on in the far past and adiabatically switched off in the far future.

As we have seen in Section 6.4, and will discuss further in Section 9.4, there is a direct correlation between the frequency of a given emitted harmonic photon and the moment in time during the laser pulse that it is emitted. The Fourier analysis described above cannot be used to study this relationship because a frequency spectrum contains no information concerning the localization in time of a given frequency component. This aspect of the HHG process can be investigated by performing a time-frequency, or wavelet, analysis of the laser-induced atomic dipole moment acceleration $\mathbf{a}(t)$.

Let us consider a sufficiently localized function g whose Fourier transform we denote by $\hat{g}$. We introduce two parameters, t_0 and Δt, where t_0 corresponds to some point in time during the laser pulse and Δt is a time interval centered about t_0. We can now define a set of functions, or wavelets,

$$g_{t_0, \Delta t}(t) = (\Delta t)^{-1/2} g\left(\frac{t - t_0}{\Delta t}\right), \tag{9.25}$$

corresponding to a translation and a dilation ($\Delta t > 1$) or contraction ($\Delta t < 1$) of the original function g. The continuous wavelet transform of the laser-induced atomic dipole moment acceleration is then defined to be

$$\mathbf{a}_{t_0, \Delta t} = \int_0^{\tau_p} g_{t_0, \Delta t}^*(t)\mathbf{a}(t)dt. \tag{9.26}$$

Using the fact that

$$(2\pi)^{-1/2} \int_{-\infty}^{\infty} \exp(i\Omega t) g^*\left(\frac{t - t_0}{\Delta t}\right) dt = \Delta t \exp(i\Omega t_0)\hat{g}^*(\Omega \Delta t), \tag{9.27}$$

we may also write

$$\mathbf{a}_{t_0, \Delta t} = (\Delta t)^{1/2} \int_{-\infty}^{\infty} \exp(i\Omega t_0)\hat{g}^*(\Omega \Delta t)\mathbf{a}(\Omega)d\Omega. \tag{9.28}$$

As we have assumed that g is a localized function, the "width" of the function $\hat{g}$ is determined by $\Delta\Omega = (\Delta t)^{-1}$. In order that the inverse wavelet transform exists, it is necessary that $\hat{g}(0) = 0$, or, equivalently,

$$\int_{-\infty}^{\infty} g(s)ds = 0. \tag{9.29}$$

This implies that g must be an oscillating function.

A convenient choice for g is the modulated Gaussian, or Morlet, wavelet [43,44] given by the function

$$g(s) = (2\pi)^{-1/2} \exp(i\beta s) \exp\left(-s^2/2\right), \qquad (9.30)$$

whose Fourier transform is given by

$$\hat{g}(u) = (2\pi)^{-1/2} \exp\left[-\frac{1}{2}(u - \beta)^2\right], \qquad (9.31)$$

where $\beta = \Omega_0/\Delta\Omega$ must be chosen to be large enough for Equation (9.29) to be satisfied to a good approximation in numerical calculations. In particular, when analyzing the emission of photons with angular frequencies centered about $\Omega_0 = q\omega$, the time window must be chosen such that $q\omega\Delta t \gg 1$. For values of q corresponding to high-order harmonics, harmonic emission on time scales that are much smaller than an optical cycle can be readily analyzed. The wavelet transform is a powerful tool for uncovering the time-profiles of harmonic emission and has been used, for example, to establish the link between the semi-classical recollision mechanism of HHG and harmonic emission obtained from "exact" or approximate calculations of $\mathbf{a}(t)$ for atoms interacting with intense, low-frequency laser fields.

9.2.1.2 Perturbative calculations

If the laser field is sufficiently weak, the single-atom harmonic emission rates can be calculated using perturbation theory. As discussed in Chapter 3, the state vector $|\Psi(t)\rangle$ can be expressed in terms of a power series expansion in the electric-field strength $\mathcal{E}_0$. The lowest-order perturbation theory (LOPT) expression for the quantity $\mathbf{d}_q$ is then given by

$$\mathbf{d}_q = \left(-\frac{e\mathcal{E}_0}{2}\right)^q \langle \psi_0 | - e\mathbf{R} | \bar{\psi}_0^{(q)} \rangle, \qquad (9.32)$$

where $|\psi_0\rangle$ is the ground state of the atom and $|\bar{\psi}_0^{(q)}\rangle$ is the LOPT state corresponding to the absorption of q photons, given by Equation (3.151). The LOPT rate for the emission of the harmonic q is proportional to I^q, where I is the laser field intensity.

Perturbative calculations of harmonic generation can also be performed by expanding the polarization of the medium $\mathcal{P}(t)$ in powers of the driving electric-field. The coefficients of this expansion are proportional to the *susceptibilities* $\chi^{(q)}$, with $q = 1, 3, 5, \ldots$ Here, $\chi^{(1)}$ is the linear susceptibility, while $\chi^{(3)}, \chi^{(5)}, \ldots$ are non-linear susceptibilities that are responsible for harmonic emission at angular frequencies $\Omega = q\omega$, with $q = 3, 5, \ldots$.

Following Manakov *et al.* [45, 46] and Gontier and Trahin [47], a number of perturbative calculations of single-atom harmonic emission have been performed for hydrogenic systems [48–51]. The validity of LOPT calculations can be tested by computing the next-order correction to the qth-order susceptibility $\chi^{(q)}$. In particular, for the case of harmonic emission in atomic hydrogen by a Nd:YAG laser of wavelength $\lambda = 1064$ nm, Pan, Taylor and Clark [51] found that LOPT fails to describe harmonic generation in the plateau region and beyond at intensities as low as 2×10^{11} W cm^{-2}.

9.2.1.3 Non-perturbative calculations

We now describe calculations of harmonic generation by a single atom in the non-perturbative regime. We begin by considering the Sturmian-Floquet calculations of Potvliege and Shakeshaft [52] performed for atomic hydrogen in a spatially homogeneous, linearly polarized and monochromatic driving laser field at a wavelength of 1064 nm. The corresponding harmonic generation rates for emission into the direction of incident propagation, summed over the polarizations of the emitted radiation, are displayed in Fig. 9.5 as a function of the harmonic order, for several intensities ranging from 10^{12} to 2×10^{13} W cm^{-2}. Both LOPT (dashed lines) and non-perturbative (solid lines) results are shown. The rates are normalized to have the same value for the third harmonic. The non-perturbative rates exhibit a behavior similar to the experimental results, with a decrease for the first few orders, a plateau (instead of a large bump obtained by using LOPT) and a cut-off. The width of the plateau and the emission rates increase with the laser intensity.

The R-matrix–Floquet (RMF) theory, discussed in Section 4.7, has been applied to calculate harmonic generation rates for atoms interacting with spatially homogeneous, linearly polarized, monochromatic laser fields. Gebarowski *et al.* [53] have discussed the RMF computational techniques required to calculate the laser-induced atomic dipole moment $\mathbf{d}(t)$. These techniques have been applied by Gebarowski, Taylor and Burke [54] to study harmonic generation in magnesium, for the case when three photons are needed to ionize this atom in the perturbative limit. As illustrated in Fig. 9.6, there is a strong enhancement by several orders of magnitude in the third harmonic generation rate in the vicinity of the resonance with the 3p3d ^{1}P^o auto-ionizing state, where the detuning $\Delta = \omega - \omega_r$ is close to zero. Here ω_r denotes the angular frequency for which there is a three-photon resonance with the auto-ionizing state. The non-symmetrical shape of the third harmonic generation curves suggests that interference can occur between two excitation paths leading to a continuum level which is then coupled to the ground state through the emission of a photon of energy $\hbar\Omega = 3\hbar\omega$. It is worth noting that the peak position of the third harmonic generation spectrum moves away from $\Delta = 0$ (no detuning from resonance) with increasing laser field intensity. Another interesting feature of Fig. 9.6 is the sensitivity of the overall shape of the third harmonic generation rate to the laser field intensity.

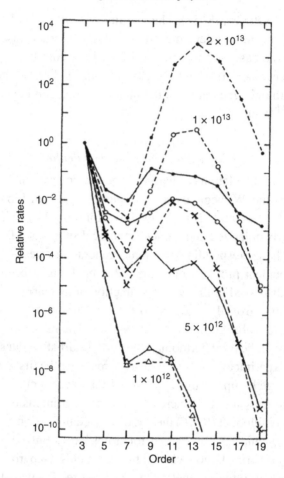

Figure 9.5. Harmonic generation rates as a function of the harmonic order for
H(1s) in a spatially homogeneous, linearly polarized and monochromatic laser
field of wavelength 1064 nm. These rates correspond to emission into the direction of propagation and are summed over polarizations of the emitted radiation.
They are normalized to have the same value for the third harmonic. Dashed curves
refer to results obtained in lowest-order perturbation theory, while solid curves
correspond to non-perturbative Sturmian-Floquet calculations, at the following
intensities. Open triangles: 10^{12} W cm^{-2}; crosses: 5×10^{12} W cm^{-2}; open circles: 10^{13} W cm^{-2}; filled circles: 2×10^{13} W cm^{-2}. (From R. M. Potvliege and R.
Shakeshaft, *Phys. Rev.* **40**, 3061 (1989).)

More recently, Plummer and Noble [55] have used the RMF theory to study
the resonant enhancement of harmonic generation in argon at the KrF laser wavelength $\lambda = 248$ nm. A strong resonant enhancement of the third harmonic generation
rate was found at a laser intensity of 7.5×10^{12} W cm^{-2}, due to the relative ponderomotive AC Stark shifts of energies of the argon ^{1}S^e ground state and the ^{1}P^o

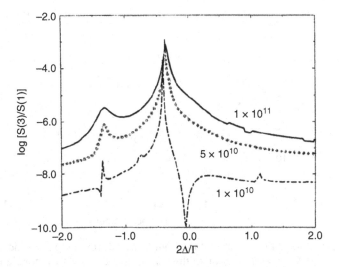

Figure 9.6. The third harmonic generation spectrum (scaled by the fundamental power spectrum) of magnesium near the 3p3d $^1P^o$ auto-ionizing resonance at the laser field intensities $I = 10^{10}$ W cm^{-2} (broken curves), $I = 5 \times 10^{10}$ W cm^{-2} (dotted curves) and $I = 10^{11}$ W cm^{-2} (solid curves) as a function of the scaled detuning $2\Delta / \Gamma = (\omega - \omega_r)/(\Gamma/2)$, where ω_r is the angular frequency for which there is a three-photon resonance with the auto-ionizing state in the weak-field limit and ω is the angular frequency of the driving laser field. (From R. Gebarowski, K. T. Taylor and P. G. Burke, *J. Phys. B* **30**, 2505 (1997).)

excited state. Resonant enhancement of the fifth and seventh harmonic emission rates was also found. These calculations demonstrate the importance of resonances on harmonic generation rates in multielectron atoms.

Extensive studies of harmonic generation by a single atom or ion in the presence of an intense laser field have been carried out by solving numerically the time-dependent Schrödinger equation (TDSE) in the dipole approximation [56–63]. As an example, we show in Fig. 9.7 the harmonic spectra calculated by Krause, Schafer and Kulander [61] for helium interacting with laser pulses of carrier wavelength λ = 527 nm and peak intensities between 10^{14} W cm^{-2} and 6×10^{14} W cm^{-2}. These spectra were obtained by solving the TDSE in the single-active-electron (SAE) approximation [62, 63], as explained in Section 5.5. The output of the calculation was the laser-induced atomic dipole moment $\mathbf{d}(t)$. As we have seen above, the modulus squared of the Fourier transform of $\mathbf{d}(t)$, namely the quantity $|\mathbf{d}(\Omega)|^2$, shown in Fig. 9.7, is proportional to the single-atom harmonic spectrum. The theoretical results displayed in Fig. 9.7 exhibit the typical features of high-order harmonic spectra generated by atoms in intense laser fields: a rapid decline over the first few harmonics followed by a plateau of relatively constant harmonic intensities and an

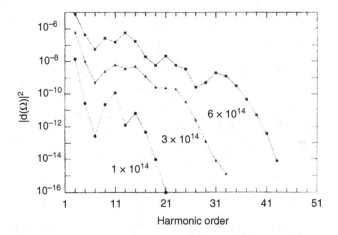

Figure 9.7. Modulus squared of the Fourier transform of the laser-induced dipole moment, $|\mathbf{d}(\Omega)|^2$, as a function of the harmonic order $q = \Omega/\omega$ for He interacting with spatially homogeneous, linearly polarized laser pulses of carrier wavelength $\lambda = 527$ nm and peak intensities $I = 10^{14}$ W cm^{-2} (circles), $I = 3 \times 10^{14}$ W cm^{-2} (triangles) and $I = 6 \times 10^{14}$ W cm^{-2} (squares). (From J. L. Krause, K. J. Schafer and K. C. Kulander, *Phys. Rev. Lett.* **68**, 3535 (1992).)

abrupt cut-off. As the intensity of the driving laser field increases, the harmonic intensities and the extent of the plateau increase, with the cut-off in the spectrum starting at about the 33rd harmonic at the intensity of 6×10^{14} W cm^{-2}.

The analysis of their results led Krause, Schafer and Kulander [61] to make an important breakthrough in the theoretical understanding of high-order harmonic generation. They discovered that the cut-off angular frequency ω_c is given quite accurately by the relation (1.18), namely $\omega_c \simeq (I_P + 3U_p)/\hbar$. The explanation of this general cut-off law was found soon afterwards within the framework of the semi-classical recollision model by Corkum [64] and Kulander, Schafer and Krause [65], as mentioned in Section 1.4 and discussed in Section 6.4. We recall that, according to this model, in an intense, low-frequency laser field, the active electron is detached from its parent core by tunneling. This electron then oscillates in the laser field. If it returns to the vicinity of its parent core, it can radiatively recombine to the ground state of the atom. A photon emitted in this way has an energy equal to the ionization potential I_P plus the kinetic energy acquired from the laser field by the returning electron. Since, as shown in Section 6.4, the maximum kinetic energy that a classical electron can gain in the laser field is $3.17U_p$, the cut-off angular frequency ω_c of the harmonic spectrum predicted by the semi-classical recollision model is given by

$$\omega_c = (I_P + 3.17U_p)/\hbar. \tag{9.33}$$

This prediction is in good agreement with experiment.

Krause, Schafer and Kulander [61] also calculated harmonic spectra for the He^+ ion. They found that this positive ion can produce harmonics comparable in strength to those obtained from neutral He, and that the emission from He^+ extends to much higher order. This can be understood by using Equation (9.33) and recalling that positive ions have high ionization potentials I_P.

As discussed in Section 6.4, a quantum-mechanical formulation of the semi-classical recollision model of HHG, based on the "strong-field approximation" (SFA) has been developed by Lewenstein *et al.* [66]. This approach has been widely applied to analyze HHG generation and has been shown to describe well the main features of photon emission spectra [66–70]. We recall that, in its original formulation [66], the SFA includes only a single bound (initial) state, and the influence of the atomic potential on the motion of the ionized electron in the laser field is neglected.

The SFA theory of Lewenstein *et al.* [66] is applicable at low frequencies and high intensities, where the Keldysh parameter γ_K is less than unity. As shown in Section 6.4, in this regime the SFA laser-induced atomic dipole moment can be evaluated using the saddle-point method. The result is a semi-analytical formula for the harmonic emission amplitude, which allows harmonic spectra to be readily calculated and analyzed over a wide range of experimental conditions.

An important prediction of the SFA, discussed in Section 6.4, is that the harmonic dipole moment $\mathbf{d}_q$ has a phase ϕ_q related to the modified action acquired by a detached electron along its trajectory in the laser field. We recall that this phase contains a contribution that is given by the product $U_p\tau$, where τ is the time spent by the electron in the continuum. Because the ponderomotive energy U_p of the electron is proportional to the intensity I of the driving laser field, the dipole phase ϕ_q varies linearly with I, with a slope depending on τ. We have also seen in Section 1.4 and Section 6.4 that since the tunneling probability increases exponentially with the electric-field strength, the electrons are ejected close to the maxima of $|\mathcal{E}(t)|$. For each harmonic below the cut-off angular frequency ω_c, two trajectories lead to the same photon energy. These trajectories give the main contributions to each harmonic in the plateau region. The electrons ionized very early after the peak of the electric-field strength spend almost one full optical cycle in the continuum before recombining with their parent core, corresponding to "long" trajectories. Electrons ionized slightly later spend only about half an optical cycle in the continuum, corresponding to "short" trajectories. For a long laser pulse, this process is repeated periodically at each half optical cycle, giving rise to the harmonic spectrum of discrete lines separated by $2\hbar\omega$.

When calculating absolute photon emission yields, the accuracy of the SFA can be improved by taking into account, in an approximate way, the influence of

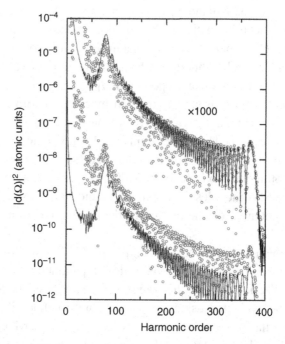

Figure 9.8. Modulus squared of the Fourier transform of the laser-induced dipole moment, $|\mathbf{d}(\Omega)|^2$, as a function of the harmonic order $q = \Omega/\omega$ for a He^+ ion interacting with a two-cycle laser pulse of carrier wavelength $\lambda = 800\,\text{nm}$ and peak intensity $I = 10^{16}\,\text{W}\,\text{cm}^{-2}$. Solid lines depict the predictions of the SFA with (upper curves) and without (lower curves) the Coulomb corrections in the ionization and recombination amplitudes. For clarity, the results with the Coulomb corrections are shifted upwards by a factor 1000. Circles show the spectrum resulting from the numerical, "exact" solution of the TDSE, and a copy of this spectrum shifted upwards by a factor of 1000. (From C. C. Chirilă, C. J. Joachain, N. J. Kylstra and R. M. Potvliege, *Phys. Rev. Lett.* **93**, 243603 (2004).)

the Coulomb interaction on the electron during the ionization and recombination processes. Such Coulomb corrections to the SFA have been proposed and tested by several authors [71–76]. As an illustration of the SFA and the corrections to it, we consider the case of a He^+ ion interacting with a two-cycle Ti:sapphire laser pulse of carrier wavelength $\lambda = 800\,\text{nm}$ and peak intensity $I = 10^{16}\,\text{W}\,\text{cm}^{-2}$ [75]. In Fig. 9.8, the quantity $|\mathbf{d}(\Omega)|^2$ is shown as a function of the harmonic order $q = \Omega/\omega$ for q up to 400. The predictions of the SFA, with and without the Coulomb corrections in the ionization and recombination amplitudes, are compared with the result of a numerical integration of the TDSE. The low-energy spectrum, for $q < 40$, is dominated by bound–bound transitions. The rest of the spectrum arises from the three-step recollision mechanism, and can be understood by analyzing the classical trajectories of the ionized electrons [77]. Except near the minima of the oscillations, the Coulomb corrections significantly improve the agreement of the SFA results with the "exact" numerical results.

We now turn to the influence of non-dipole effects on high-order harmonic generation [78–84]. As we have seen before, these effects, due to the magnetic-field component of the laser field, manifest themselves at very high laser intensities. In particular, using the non-dipole Volkov wave functions obtained in Section 2.8, we showed in Section 6.6 that it is possible to generalize the SFA to include the non-dipole effects in the calculation of harmonic generation spectra.

As an example, we shall consider non-dipole effects in photon emission by a positive ion interacting with very intense, few-cycle Ti:sapphire laser pulses of carrier wavelength $\lambda = 800$ nm. Such effects have already been discussed in Sections 1.4 and 6.6. They become important for peak intensities in excess of 10^{17} W cm^{-2} [79, 82–84]. At these intensities, the displacement of the ionized electron wave packet in the propagation direction due to the magnetic drift is large enough to reduce considerably the overlap of the returning wave packet with the parent core. This is illustrated in Fig. 1.15 for the case of a Be^{3+} ion interacting with a four-cycle pulse of carrier wavelength $\lambda = 800$ nm and peak intensity $I = 3.6 \times 10^{17}$ W cm^{-2}. The reduction in the emission of photons polarized along the polarization direction of the driving laser field is particularly striking, as is the modification of the plateau structure of the spectrum. Such effects are readily understood within the framework of the semi-classical recollision model. Another interesting non-dipole effect is the (weak) emission of photons polarized along the propagation direction of the driving laser field. In addtion, positive ions interacting with very intense ultra-short laser pulses have been predicted to generate attosecond X-ray pulses [82–84].

As a last illustration of harmonic generation by a single atomic system, we shall discuss the interaction of super-intense laser fields with relativistic ions [75]. The development of a new accelerator complex at the GSI Helmholtzzentrum für Schwerionenforschung in Darmstadt will allow relatively dense bunches of positive ions, in arbitrary charge states, to be accelerated to Lorentz factors $\gamma = (1 - v^2/c^2)^{-1/2}$ up to about 25, where v is the speed of the positive ion. In addition, it will be possible to irradiate these ions with ultra-intense laser pulses produced by the PHELIX laser facility developed on the same site, which has a wavelength of 1053.7 nm. In what follows, we shall assume that the laser beam is oriented in such a way that it counter-propagates with respect to the direction of motion of the beam of ions. Because of the relativistic Doppler effect [85], the ions are then exposed in their frame of reference (moving with respect to the laboratory frame with a speed $v \simeq c$) to a laser field of angular frequency $\omega_I = (1 + v/c)\gamma\omega_L$ and electric-field strength $\mathcal{E}_I = (1 + v/c)\gamma\mathcal{E}_L$, where ω_L and $\mathcal{E}_L$ are, respectively, the carrier angular frequency and peak electric-field strength of the laser pulse in the laboratory frame. An important question is the efficacy, in these conditions, of multiphoton processes such as above-threshold ionization, multiple ionization and harmonic generation which arise from the semi-classical recollision mechanism. As we have seen in Section 6.6, these three-step processes are inhibited at high laser intensities by the non-dipole effects, the magnetic component of the laser pulse tending to push the ionized electron in the direction of propagation of the pulse, away from the atomic core. This inhibition is particularly severe for electrons returning to the core with a high relative velocity. However, as Chirilă *et al.* [75] have shown, in the conditions that can be achieved at the GSI this inhibition is much attenuated, for the same relative velocity, by the Doppler boost in frequency due to the relativistic motion of the ions.

The calculations performed by Chirilă *et al.* [75] were carried out by using the non-dipole generalization of the SFA [79,82–84] discussed in Section 6.6, corrected for the effects of the Coulomb potential of the core on the ionization and recombination amplitudes. In order to examine how the relativistic Doppler effect modifies three-step recollision processes such as harmonic generation, they compared photon emission by two ions exposed to the same laboratory laser field. The first ion was assumed to be at rest in the laboratory frame, while the second ion was taken to be moving in the direction opposite to the propagation direction of the laser field, modeled as a linearly polarized, monochromatic electromagnetic field of wavelength $\lambda_L = 1053.7$ nm and constant intensity $I_L = 10^{17}$ W cm^{-2} in the laboratory frame. This intensity is high enough for the magnetic-field-induced suppression of recollision to be significant, so that non-dipole effects must be taken into account, but not so high as to invalidate the non-relativistic character of the non-dipole SFA [79,82–84]. The ponderomotive energy $U_p = e^2 \mathcal{E}_L^2/(4m\omega_L^2) = e^2 \mathcal{E}_I^2/(4m\omega_I^2) = 10.4$ keV. The maximum energy of the photons emitted by the ion, in the ion's rest frame, is $I_P + 3.17U_p$. It is worth noting that neither the ponderomotive energy, nor the Keldysh parameter $\gamma_K = [I_P/(2U_p)]^{1/2}$ depend on the speed of the ions. The first ion is a sodium-like Ar^{7+} ion ($Z = 3.247$, $I_P = 143$ eV), which is assumed to be at rest in the laboratory frame ($\gamma = 1$). For the laser parameters chosen here, this ion has an ionization probability of about 2% per optical cycle. The laser intensity is therefore close to the saturation intensity. The corresponding Keldysh parameter is $\gamma_K = 0.08$, well into the tunneling regime. The second ion is a hydrogenic Ne^{9+} ion ($Z = 10$, $I_P = 1.36$ keV), assumed to be moving in the laboratory frame with a Lorentz factor $\gamma = 15$. In this case, the ionization probability per optical cycle is about 1%. The Keldysh parameter is $\gamma_K = 0.26$, small enough for the SFA to be adequate. We also note that the Ne^{9+} ion was chosen because at the Lorentz factor $\gamma = 15$ it ionizes in the laser field at about the same rate as the Ar^{7+} ion at $\gamma = 1$, which facilitates the comparison. Because of the Doppler boost in intensity, the Ar^{7+} ion would be promptly ionized if moving with $\gamma = 15$ in the direction opposite to the laser pulse. Indeed, in this ion's rest frame, the laser intensity is almost 9×10^{19} W cm^{-2} when $\gamma = 15$.

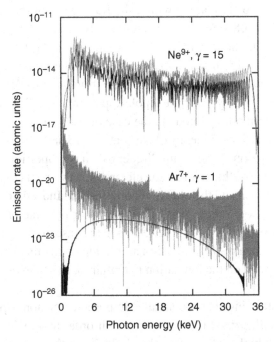

Figure 9.9. Rate of emission of harmonic photons in the direction of propagation of the driving laser pulse (in a.u.) as a function of the photon energy (in keV) for (a) a Ne^{9+} ion moving at a velocity corresponding to a Lorentz factor $\gamma = 15$ in the laboratory frame and (b) an Ar^{7+} ion at rest, as obtained using the Coulomb-corrected SFA. The rate and the photon energy are given in the ion's rest frame. For each ion, the upper curve shows the results obtained within the dipole approximation and the lower curve shows the results obtained without making this approximation. The laser field is monochromatic and linearly polarized. In the laboratory frame, its intensity is $10^{17}\,\mathrm{W\,cm^{-2}}$ and its wavelength is 1053.7 nm. (From C. C. Chirilă, C. J. Joachain, N. J. Kylstra and R. M. Potvliege, *Phys. Rev. Lett.* **93**, 243603 (2004).)

The spectra of the harmonic photons emitted by these two ions, calculated using the Coulomb-corrected SFA, are shown in Fig. 9.9. Two sets of results are displayed for each ion: (i) those obtained by using the dipole approximation (where the effect of the magnetic-field component of the laser field is neglected) and (ii) those obtained by taking the magnetic field effects into account within the framework of the non-dipole SFA [79, 82–84]. In the case of the Ar^{7+} ion, at rest in the laboratory frame ($\gamma = 1$), the magnetic field reduces the intensity of harmonic emission by about three orders of magnitude in the high-energy part of the spectrum. We recall that this dramatic reduction of harmonic emission, as well as the modification of the plateau structure of the spectrum, are characteristic non-dipole effects, which can be explained within the framework of the semi-classical recollision model [79, 80, 82–84]. The difference between the dipole and the non-dipole spectra is much smaller for the Ne^{9+} ion moving in the laboratory frame with a Lorentz factor $\gamma = 15$. In particular, and despite the higher intensity, there is no significant decrease in the efficiency of the harmonic emission as compared to the calculation performed by using the dipole approximation. The much larger intensity of emission of harmonic photons in the case of Ne^{9+} at $\gamma = 15$ is also striking.

The higher emission efficiency of Ne^{9+} at $\gamma = 15$ has a double origin. Firstly, it is due to the larger recombination amplitude, which is proportional to $Z^{5/2}$. Secondly, it is a consequence of the fact that the electronic wave packet spreads less between the time of ionization and the time of recombination. Indeed, in the ion's rest frame, the difference between these two times is smaller by a factor 30 for $\gamma = 15$. The relative weakness of the non-dipole effects in the case of Ne^{9+} at $\gamma = 15$ originates from the dependence of the ionization amplitude on the magnetic-field component of the laser field and on the ionization potential I_P, as discussed in detail by Chirilă *et al.* [75].

It is worth stressing that the smaller importance of non-dipole effects at large Lorentz factors illustrated in Fig. 9.9 for high-order harmonic generation can also be expected for high-order above-threshold ionization or non-sequential double ionization, as it arises from the ionization stage of the process, which is identical for all these three-step recollision processes. We emphasize that this smaller importance of non-dipole effects is not a relativistic effect per se. The speed of the ions intervenes only through the Doppler effect, which makes it possible to ionize electrons from multicharged positive ions at high laser intensities without imposing a large magnetic drift on these electrons. Clearly, identical results would be obtained by irradiating positive ions at rest by a laser field of sufficiently high intensity and high frequency.

9.2.2 Propagation in the medium

In a macroscopic medium, different atoms experience different peak laser field intensities and phases. The observed harmonic signal arises by adding coherently the laser field induced dipole emission contributions from the individual atoms. This must be done by taking propagation effects into account, as these effects can destroy the coherence unless constructive phase matching occurs [1, 7, 68]. We note that coherent emission by the atoms can only occur along the propagation direction of the driving laser, where the harmonic signal depends quadratically on the atomic density. This is in contrast with other directions where the signal varies linearly with the atomic density [86–88].

Three important propagation effects act to limit the harmonic conversion efficiency that can be obtained. The first is absorption of the harmonics by the atoms, leading to core excited states. The second is defocusing due to the free-electron density profile in the interaction region. The density is highest at the center of the focal region where the peak laser pulse intensity is highest. The third and most important propagation effect is dephasing. Dephasing occurs when a phase mismatch occurs between the harmonics emitted at different points along the propagation direction of the driving pulse. The distance at which this mismatch is equal to π (destructive interference) is called the coherence length. A number of effects can lead to dephasing. These include the presence of free electrons ionized by the pulse, the fact that the focused driving laser pulse has a curved wave front that gives rise to the Gouy phase shift, and that the laser-induced atomic dipole moments depend on the laser pulse intensity and hence on the position of the atom in the focal region.

The macroscopic response of the medium is governed by Maxwell's equations. These equations can be combined to form a single equation, which is of second order in both the time and spatial derivatives. It describes the electric-field vector of the driving laser pulse and the emitted harmonic pulses and is given by

$$\nabla \times (\nabla \times \mathcal{E}) + \frac{\sigma}{\epsilon_0 c^2} \frac{\partial \mathcal{E}}{\partial t} + \frac{1}{c^2} \frac{\partial^2 \mathcal{E}}{\partial t^2} = -\frac{1}{\epsilon_0 c^2} \frac{\partial^2 \mathcal{P}}{\partial t^2}, \tag{9.34}$$

where σ is the conductivity of the medium. This equation is solved after making the slowly varying envelope approximation and the paraxial approximation, as discussed in detail by Salières *et al.* [68] and Brabec and Krausz [7]. We will give an overview of the coherence properties of harmonics in Section 9.3, and in Section 9.4 we will discuss a key theoretical result, obtained by Antoine, L'Huillier and Lewenstein [89], concerning the role of propagation effects inside the medium.

9.3 Properties of harmonics and applications

In this section, we outline the main properties of the harmonic emission: conversion efficiency, spatial and temporal coherence, chirp, tunability and polarization. Detailed discussions of this subject may be found in the review articles of Salières [8, 9] and Salières and Christov [10].

9.3.1 Conversion efficiency

The optimization of the conversion efficiency of harmonics is important for applications that require a high photon flux. It follows from the cut-off law (9.33) that the most energetic harmonic photons are produced in light noble gases, such as helium and neon. However, the conversion efficiency in heavier noble gases is higher in their allowed spectral range, so that the choice of the best generating gas depends on the required photon energy. In order to optimize phase matching, and to increase the transverse size of the generating medium, the driving laser should be focused as weakly as possible. As an example, we show in Fig. 9.10 the optimal number of harmonic photons per pulse and the corresponding conversion efficiencies, as a function of the photon wavelength, measured by Hergott *et al.* [90] using 60 fs, 30 mJ, 800 nm pulses from a Ti:sapphire laser loosely focused with a 2 m focal length lens in neon, argon and xenon jets. Also shown in Fig. 9.10 are the results obtained by Hergott *et al.* [90] using a 5 m focal length lens. This allowed the full energy of the laser to be used without significantly ionizing the xenon gas medium, thereby achieving an energy greater than 1 μJ at the 15th harmonic frequency. The conversion efficiency reached 10^{-5} for harmonics in the plateau of xenon, 7×10^{-6} in argon and close to 10^{-7} in neon. Higher photon energies have been obtained using

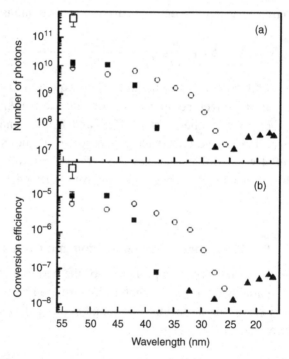

Figure 9.10. (a) Optimal number of harmonic photons and (b) corresponding conversion efficiencies as a function of the photon wavelength, generated by a 60 fs, 30 mJ, Ti:sapphire laser pulse of wavelength $\lambda = 800$ nm in neon (triangles), argon (circles) and xenon (squares), obtained with a focal length lens of 2 m. The open squares correspond to the data obtained with a 5 m focal length lens. (From J. F. Hergott *et al.*, *Phys. Rev. A* **66**, 021801(R) (2002).)

ultra-short laser pulses having peak intensities larger than $10^{15}\,\mathrm{W\,cm^{-2}}$ [34–40]. In helium, the spectral range has been extended to an energy of 500 eV, corresponding to a wavelength of 2.5 nm reaching into the water window, as mentioned above. In fact, for these short pulses and high photon energies, the harmonic spectra are no longer resolved and resemble a "white" spectrum, as the harmonics are so broad that they overlap.

In addition to the gas jet, several geometries for the generating medium have been investigated: hollow core fibers [39, 91] or cells filled with gas [92–94]. The dependence of the conversion efficiency on the atomic density in the interaction region has also been studied [95]. Since harmonic generation is a coherent process, it is expected to be proportional to the square of the atomic density. However, this simple scaling is modified by the dispersion of the medium and by the increase of absorption with increasing atomic density.

In general, the conversion efficiency of harmonics is very sensitive to the experimental conditions, and, as seen from Fig. 9.10, is strongly dependent on the spectral range considered. We remark that Serrat and Biegert [96] have shown theoretically that HHG conversion efficiency can be improved using a DC electric-field periodic in space in the interaction region.

9.3.2 Coherence

The spatial and temporal coherence properties of the emitted harmonics are determined to some extent by the driving laser field. However, as mentioned above, the intensity-dependent dipole phase strongly influences phase matching in the generating medium, and may significantly distort the harmonic spatial profiles [4,97,98]. As an illustration, we show in Fig. 9.11 the results obtained by Salières, L'Huillier and Lewenstein [98], who studied the evolution of the 39th harmonic generated in neon for different positions of the laser focus relative to the gas jet. The good agreement between the experimental data and the numerical simulations allow us to interpret this evolution as follows. If the laser is focused before the gas jet (dashed curve), phase matching is achieved on-axis, resulting in a narrow profile. In contrast, when the laser is focused after the jet (dotted curve), phase matching can only be realized off-axis, giving an annular profile. This behavior arises from the interplay between the phase due to focusing and the phase of the laser-induced atomic dipole moment. It is worth pointing out that this phenomenon can also be used to generate very regular, nearly Gaussian spatial profiles with a divergence smaller than that of the driving laser pulse [99].

The ionization of the generating medium also distorts the harmonic coherence, as it gives rise to free electron dispersion that degrades the harmonic beam quality [100, 101]. This degradation grows with increasing pressure, since free electron

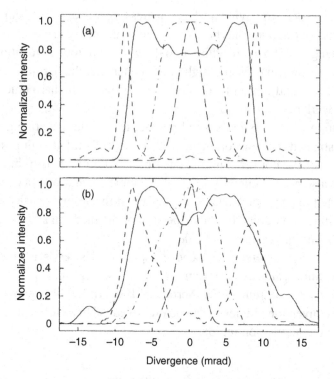

Figure 9.11. Evolution of the spatial profile of the 39th harmonic of neon as a function of the position of the laser focus relative to the gas jet. (a) Experiment; (b) theory. Results for the laser focus well before the gas jet (dashed curve), closer to the gas jet (dot-dashed curve and solid curves, successively) and after the gas jet (short-dashed curve) are shown. (From P. Salières, A. L'Huillier and M. Lewenstein, *Phys. Rev. Lett.* **74**, 3776 (1995).)

dispersion leads to laser defocusing and a modification of phase matching in the medium.

Both phenomena contribute to the loss of spatial coherence of the harmonic beam, particularly in the focal region where the laser intensity is high. However, it is possible to find conditions such that a high degree of coherence is maintained over the entire harmonic beam, as shown by the good fringe contrast observed with a Young two-slits setup [102, 103] or with a Fresnel mirror interferometer [104].

The temporal coherence of harmonics produced by focusing 100 fs laser pulses of carrier wavelength around 790 nm and peak intensity of the order of 10^{14} W cm^{-2} has been investigated by Bellini *et al.* [105] and Lynga *et al.* [106], who measured the visibility of the interference fringes produced when two spatially separated harmonic sources interfere in the far field, as a function of the time delay between the two sources. They found, in general, long coherence times, comparable to the expected pulse durations of the harmonics.

The temporal coherence of the harmonic beam can also be degraded by the intensity-dependent phase of the laser-induced atomic dipole moment and ionization-induced free electron dispersion. To characterize the temporal coherence, the pulse duration must be compared with the coherence time, which is inversely proportional to the spectral width. This was done by Salières *et al.* [69], who measured the spectral width $\Delta\omega$ and the pulse duration Δt of the 21st harmonic emission in argon and found good temporal coherence. However, this coherence is degraded in some generating conditions. In particular, if the laser beam is focused after the gas jet, the harmonic spectrum broadens considerably and symmetrically, while the harmonic pulse duration increases slightly, indicating a chirp of the emission, to which we now turn our attention.

9.3.3 Harmonic chirp

The existence of a frequency variation in time (chirp) of the emitted harmonic pulses was first discussed by Salières, L'Huillier and Lewenstein [98], and investigated in detail by Varjú *et al.* [107]. The causes of this chirp can be understood from the viewpoint of the semi-classical recollision model, as will be discussed in more detail in Section 9.5. At this point, we simply remark that, as we have seen in Section 6.4, the phases of the harmonic dipole moments contain a contribution given by $U_p\tau$, where τ is the time the electron spends in the continuum. The importance of this intensity-dependent phase increases as the electron trajectory leading to the harmonic emission becomes longer. Since longer electron trajectories are favored when the laser is focused after the gas jet, this position leads to a larger chirp, and therefore a larger spectral width. This is consistent with measurements of the coherence times inside the harmonic beam performed by Bellini *et al.* [105] and Lynga *et al.* [106].

9.3.4 Other properties

The tunability of the harmonic radiation is an important requirement for many applications. In addition to the coarse tunability which can be obtained by changing the harmonic order, a fine adjustment of the wavelength can be achieved by generating the harmonics with a laser mixed with an optical parametric amplifier (OPA) [108, 109] or with a chirped laser [34]. We also note that by using ultra-short, very intense laser pulses, one obtains a white spectrum [36–38], so that complete tunability is obtained.

Another feature of the harmonic radiation is its polarization. A linearly polarized driving laser field induces atomic dipole moments along the polarization axis of the laser field. As a result, the emitted harmonic radiation is also linearly polarized in the same direction. There is no harmonic emission when the driving laser field is circularly polarized. In fact, the harmonic emission decreases as soon as a small ellipticity is introduced in the polarization of the driving laser field [110–113]. This can be easily understood within the semi-classical model of HHG: a classical electron ejected with an initial momentum along the instantaneous electric-field vector of an elliptically polarized field will never return to the core.

Finally, we remark that propagation in the medium leads to harmonic pulses that are more collimated than the driving laser pulse, thereby allowing the harmonic pulses to be separated geometrically from the driving laser pulse [93, 114].

9.3.5 Applications of harmonics

A number of applications of HHG have been reported that exploit the properties of this source of XUV laser light. In particular, the ultra-short duration of the harmonic pulses and their synchronization with the driving laser pulse are well suited for performing table-top pump-probe experiments with the aim of studying ultra-fast processes in atomic and molecular physics [115–118], solid state physics [119–123] and plasma physics [124, 125].

9.4 Attosecond pulses

The possibility of generating sub-femtosecond pulses, also called attosecond pulses ($1 \text{ as} = 10^{-18}$ s), by harmonic generation has attracted considerable interest in recent years. In this section, we shall first discuss attosecond pulse trains, then single attosecond pulses and finally some applications. Detailed reviews of the physics of attosecond pulses have been written by Agostini and DiMauro [11], Scrinzi *et al.* [12], Kienberger *et al.* [13], Niikura and Corkum [14] and Krausz and Ivanov [15].

9.4.1 Attosecond pulse trains

We begin this section by explaining why, from the point of view of the semi-classical recollision model of HHG discussed in Section 6.4, harmonics photons can be expected to be emitted in the form of short pulses having widths much less than the laser field period $T = 2\pi/\omega$. We consider an atom interacting with a linearly polarized infra-red laser pulse and, for the moment, assume that the pulse duration is sufficiently long so that the laser field can be taken to be monochromatic. As was shown in Section 6.4, the SFA predicts that the dominant contributions to the laser-induced atomic dipole moment arise from two groups of electron trajectories. Their classical counterparts are referred to, respectively, as the short and the long trajectories. Let us focus on the classical trajectories that return to the parent core with kinetic energies that fall within a small interval about E_{kin}. These trajectories correspond to electrons that are ionized at times about t_{d_1} and t_{d_2}. They return to their parent core at times $t_{r_1} = t_{d_1} + \tau_1$ and $t_{r_2} = t_{d_2} + \tau_2$, respectively, where τ_1 and τ_2 are the times the electrons spend in the continuum. If the returning electron recombines with its parent core, a photon of energy $\hbar\Omega = I_P + E_{kin}$ is emitted. The recombination times are therefore correlated with the energy of the emitted photon so that, in the semi-classical picture, the emission of photons of energy within an interval about $\hbar\Omega$ would be expected to occur as short bursts around the times t_{r_1} and t_{r_2}. As a consequence of the periodicity of the laser field, each trajectory gives rise to photon emission once per half laser cycle and therefore to two bursts per cycle. In other words, photons are emitted as trains of ultra-short pulses.

Of course, quantum mechanically neither the ionization time nor the recombination time can be precisely determined, and hence uncertainty exists about the energy of the emitted photon. For long interaction times, interference ensures that only "energy-conserving" photons with angular frequencies $\Omega = q\omega$, $q = 1, 3, 5, \ldots$, are emitted. Extending the basic idea of the semi-classical recollision model then implies that groups of harmonics are emitted together as short bursts. Let us now establish the conditions that must be fulfilled for a group of neighboring harmonics to produce a temporally localized pulse of radiation whose width is much smaller than the driving laser field period $T = 2\pi/\omega$. We recall that the power spectrum of the emitted harmonics, as given, for example by Equation (9.19), gives no information about the temporal properties of the harmonic emission. As we shall see, temporal information requires knowledge of the harmonic phases.

We will denote by $\mathcal{E}_H(t)$ the electric-field component of a spatially homogeneous laser field that is made up of a superposition of odd harmonics of the driving laser field. Taking this field to be linearly polarized with polarization vector $\hat{\epsilon}$, we express the electric-field component of the harmonic field as $\mathcal{E}_H(t) = \hat{\epsilon}\mathcal{E}_H$, where

$$\mathcal{E}_H(t) = \sum_{n=n_1}^{n_2} a_{2n-1} \cos[(2n-1)\omega t + \varphi_{2n-1}], \tag{9.35}$$

with $1 \leq n_1 < n_2$. The function $\mathcal{E}_H(t)$ is periodic, with period T. Let us assume that the harmonics are phase-locked, namely

$$\varphi_{2n+1} - \varphi_{2n-1} = \varphi, \tag{9.36}$$

where φ is a constant phase, and

$$a_{2n+1} \simeq a_{2n-1}, \qquad n = n_1, n_2, \ldots \tag{9.37}$$

As the number of harmonics is increased in Equation (9.35), the function $\mathcal{E}_H(t)$ will become localized at the times $t_{\max} = (2\pi m - \varphi)/\omega$, $m = 0, \pm 1, \pm 2, \ldots$, with pronounced peaks occurring at these times. The width of these peaks is approximately $\pi/[2\omega(n_2 - n_1)]$, while their amplitude is approximately given by the sum of the coefficients a_{2n-1}, with $n = n_1, \ldots, n_2$. The phase φ determines the position of the maxima of $\mathcal{E}_H(t)$ with respect to the driving laser field.

It follows that a group of harmonics satisfying the criteria that they all have (i) comparable amplitudes and (ii) neighboring phases that differ by at most a constant, will give rise to a train of attosecond pulses. This is the same principle upon which the operation of a mode-locked laser is based. By locking the phase of N spectral modes within the gain bandwidth, a sequence of pulses, separated in time by the cavity round-trip time and having a width proportional to N^{-1}, are produced.

As discussed in Section 9.1 and illustrated in Fig. 1.12, experimental HHG spectra exhibit a broad plateau, thereby satisfying the first condition. In order to establish experimentally that the harmonics are emitted temporally as attosecond trains, it must be verified that the second condition is also satisfied. Using the SFA, Antoine, L'Huillier and Lewenstein [89] showed that this was *not* the case at the single-atom level. Nevertheless, by carrying out a time-frequency analysis, they found that the temporal profile obtained from the superposition of ten plateau harmonics yielded two harmonic pulses per half optical cycle. One of the pulses is due to the short trajectories, while the other arises from the long trajectories. This feature implies that each of the two groups of trajectories, taken separately, gives rise to harmonics that are nearly phase-locked. However, the harmonic phases have a contribution that is proportional to the laser field intensity, which varies over time as well as over space in the interaction region. As a consequence, it should not be expected a priori that a constant phase difference between harmonics would be maintained in the atomic medium. In fact, calculations of the temporal profile of the emitted harmonics have shown that it is possible, by adjusting experimental parameters, to maintain the phase difference of harmonics generated by either the short or long electron trajectories [89, 126]. For example, Antoine, L'Huillier and Lewenstein [89] found that by placing the interaction region before the laser focus, the phases of the harmonics generated by the short trajectories could be locked. A detailed analysis of the influence of the spatio-temporal laser intensity distribution corresponding to realistic experimental conditions has been made by Gaarde and Schafer [126]. They concluded that there exist conditions such that the production of phase-locked harmonics is robust.

We will now consider how the theoretical prediction of the generation of attosecond pulse trains by atoms interacting with intense many-cycle infra-red laser pulses can be verified experimentally. In particular, the "reconstruction of attosecond beating by interference of two-photon transitions" (RABITT) [11, 127] method for measuring the relative phases of harmonics will be discussed.

RABITT is a cross-correlation technique that measures the modulation of photoelectrons as a function of the delay between the driving infra-red laser pulse and the generated harmonics. In order to understand the idea underlying this approach [127], let us focus on the case in which the infra-red and XUV pulses are sufficiently long, so that the probability that an atom interacting with the two pulses will eject a photoelectron of energy E_f can be characterized by an ionization rate. In addition, let us assume that the intensity of both pulses is sufficiently weak so that the interaction with the atom can be described using perturbation theory. The dominant contributions to the total ionization rate will then come from the absorption of a single photon from the XUV pulse. Recalling that in the long-pulse limit, the XUV pulse is composed of a series of odd harmonics of the angular frequency ω of the driving laser field, peaks in the photoelectron energy spectrum will occur at the energies

$$E_f = \hbar\Omega - I_P = \hbar\omega_q - I_P, \tag{9.38}$$

where $\omega_q = q\omega$, $q_0 \leq q = q_0, q_0 + 2, \ldots$ and q_0 is the smallest odd integer such that $E_f > 0$. In the presence of the infra-red laser pulse, there is some probability that the atom will, in addition, absorb or emit an infra-red photon. This will lead to sidebands in the photoelectron energy spectrum, with peaks at

$$\tilde{E}_f = \hbar\omega_q \pm \hbar\omega - I_P. \tag{9.39}$$

Focusing on a particular sideband energy, which we denote by $\tilde{E}_f^q$, we have that

$$\tilde{E}_f^q = \hbar\omega_{q+2} - \hbar\omega - I_P$$
$$= \hbar\omega_q + \hbar\omega - I_P. \tag{9.40}$$

It is seen that each sideband has contributions from two underlying processes: the absorption of a photon of energy $\hbar\omega_q$ from the XUV pulse and the absorption of an infra-red photon of energy $\hbar\omega$, and the absorption of an XUV photon of energy $\hbar\omega_{q+2}$ and the emission of an infra-red photon of energy $\hbar\omega$. These two ionization paths lead to the same final state, and therefore cannot be distinguished. By introducing a delay Δt between the infra-red and the XUV pulses, the interference between the two ionization paths can be controlled.

The ionization rate for electrons ejected with energy $\tilde{E}_f^q$ is given by LOPT as

$$W(\tilde{E}_f^q, \Delta t) = \sum_f \left[|A_{f,i}^{\text{direct}}|^2 + |A_{f,i}^{\text{inter}}|^2 \cos(2\omega\Delta t + \phi_q - \phi_{q+2} + \phi_f) \right], \tag{9.41}$$

where $|A_{f,i}^{\text{direct}}|$ and $|A_{f,i}^{\text{inter}}|$ are, respectively, the magnitudes of the direct and inter-fering ionization amplitudes. The sum is over the final states of the atom, and ϕ_f is an atomic phase. If this atomic phase is known, the phase difference $\phi_q - \phi_{q+2}$ can be determined by measuring the modulations of $W(\tilde{E}_f^q, \Delta t)$ as a function of Δt.

Paul *et al.* [128] have applied the RABITT technique to measure the relative phases of the 11th to 19th harmonics generated by the interaction of argon atoms with a 40 fs, 800 nm Ti:sapphire laser pulse having a peak intensity of about $10^{14}\,\mathrm{W\,cm^{-2}}$. They found that the harmonics were locked in phase, and formed a train of 250 as pulses. The RABITT method does not provide information concerning the absolute phase of the harmonics, which is necessary to determine at what point in time during the optical cycle the harmonics are emitted. In Section 9.4.3, we shall discuss a later experiment performed by Dinu *et al.* [129] who, by using the RABITT method and evaluating all the phases in the experiment, were able to determine the absolute harmonic phases.

At the level of the single-atom response, a fundamental limit exists concerning the width of the attosecond pulses that can be generated. This can be readily seen from the viewpoint of the semi-classical recollision model. We recall that the main contributions to the emission in the plateau region of a harmonic of a given angular frequency $\Omega = q\omega$ come from two electron trajectories, the short and the long trajectories discussed above. We denote by t_{r_1} and t_{r_2}, respectively, the corresponding electron return times at which, in the semi-classical recollision model, a photon can be emitted. Let us now consider the emission of the harmonic $(q+2)\omega$. The contribution to this higher harmonic from a short trajectory occurs at a time *later* than t_{r_1} (see Fig. 6.5). Conversely, the contribution to this harmonic from a long trajectory occurs *earlier* in time than t_{r_2}. It follows that the short trajectories give rise to a *positive* chirp of harmonic emission, while the long trajectories lead to a *negative* chirp. Experimentally, either the short trajectory or the long trajectory can be favored by placing the laser focus before or after, respectively, the gas jet. In this way, either positively or negatively chirped attosecond pulse trains can be generated [69].

As a consequence of this chirp, the phase differences between neighboring harmonics will be frequency-dependent [130, 131]. The inclusion of additional harmonics beyond a critical number has the effect of increasing the width of the pulses in the attosecond pulse train. This phenomenon has been studied experimentally by Mairesse *et al.* [130] in neon. They measured the harmonic chirp and found good overall agreement with results obtained from calculations of the single-atom response.

Finally, we point out that an additional chirp of the harmonic emission can occur when the laser pulse is sufficiently short, as discussed by Varjú *et al.* [107]. This chirp arises because of the temporal variation of the driving laser intensity. Indeed, this temporal variation modifies the ionization and return times, from one optical cycle to the next, of the trajectories that correspond to electrons that return to their parent core with a given kinetic energy.

9.4.2 Single attosecond pulses

Of considerable interest, in particular from the point of view of applications, is the possibility of generating a single attosecond pulse. From our discussion in Section 9.4.1, it is clear that such an attosecond pulse could, in principle, be produced by engineering a driving laser pulse that ensures that only a single recollision of an ionized electron with its parent core is possible. Let us now consider two approaches for generating such a driving pulse.

The first one makes use of a "polarization gate." We recall that the probability that an ejected electron returns to the core is maximum when the driving pulse is linearly polarized, and harmonic generation efficiency decreases rapidly as the magnitude of the ellipticity parameter of the infra-red driving laser field is increased from linear to circular polarization. By using a polarization gate so that the polarization of the driving pulse varies rapidly from circular to linear, and then back again to linear, the window in time during which harmonics will be emitted can be reduced significantly. Only the electrons that are ejected and recombine when the driving pulse is linearly polarized will produce harmonics.

This technique can be used in principle to confine temporally harmonic generation to a single attosecond pulse [132, 133]. Initial experimental investigations of this proposal have been carried out by Kovacev *et al.* [134] and Tcherbakoff *et al.* [135] by creating from the original linearly polarized driving pulse two time-delayed driving pulses with left- and right-circular polarization. These two pulses were recombined, resulting in a pulse whose polarization is circular at the beginning and end of the pulse, and is linearly polarized when the two pulses overlap. Using a 35 fs Ti:sapphire pulse, Tcherbakoff *et al.* [135] were able to reduce the temporal emission by argon to 7 fs. In a later experiment, Shan, Ghimire and Chang [136] were able to infer an emission time of 200 as using an 8 fs driving pulse with a carrier wavelength of 750 nm.

A more direct route to producing attosecond pulses involves using an extremely short, few-cycle driving pulse. Theoretical work [7, 137] has demonstrated that the generation of a single XUV pulse is then possible. If the carrier-envelope phase (CEP) is chosen in such a way that a few-cycle "cosine-like" pulse (see Fig. 1.3(a)) is produced, the highest harmonics are generated by electrons that are ejected and recombine during the laser optical cycle having the highest instantaneous intensity. Therefore, the highest harmonics are emitted within a narrow window of time. The application of a spectral filter that removes the lower harmonics will then isolate a single attosecond pulse. A necessary requirement is a laser system that is capable of producing CEP-stabilized "cosine-like" pulses. We note that if a few-cycle "sine-like" pulse is used, two attosecond pulses separated temporally by half the laser optical cycle would be generated. Ultra-short pulses can also be used to produce a polarization gate [138].

Let us now turn our attention to the characterization of single attosecond pulses. Due to their extremely short duration and the fact that they have spectral components in the XUV, it is difficult to determine their properties [11]. A technique that has been applied successfully is based on the cross-correlation of the attosecond pulse with an intense femtosecond infra-red pulse. The latter can be taken to be the same driving pulse that was used to generate the attosecond pulse. Thus, in a typical attosecond experiment, an intense, few-cycle, infra-red pulse is focused onto a gas jet. Both the generated XUV attosecond pulse and the driving infra-red pulse are filtered and focused onto a second gas jet and the photoelectron spectrum recorded as a function of the relative time delay between the XUV attosecond pulse and the driving infra-red pulse [11].

In order to analyze the resulting photoelectron spectrum, let us describe the photoionization process using a modified semi-classical description of the interaction of the atom with the weak attosecond XUV pulse and the intense femtosecond infra-red pulse. Instead of the intense infra-red pulse ionizing the atom, as was the case in the semi-classical description of MPI and HHG, it is now the attosecond XUV pulse that ionizes the atom. The dynamics of the ejected electrons are then governed by their interaction with the intense femtosecond infra-red pulse. Because the duration of the attosecond pulse is much shorter than the infra-red pulse, ionization is well localized in time with respect to the infra-red pulse. This allows properties of the attosecond pulse to be deduced from the recorded photoionization spectra.

We describe the electric-field component of the linearly polarized, infra-red (IR) pulse as a spatially homogeneous electric-field of the form

$$\mathcal{E}_{IR}(t) = \hat{\epsilon} \mathcal{E}_{IR} F_{IR}(t) \cos(\omega t + \varphi_{IR}), \qquad (9.42)$$

where the pulse envelope function $F_{IR}(t)$ is taken to be zero for $t \leq 0$ and $t \geq \tau_{IR}$. In addition, we approximate the vector potential of the infra-red laser field as

$$\mathbf{A}_{IR}(t) = -\hat{\epsilon} (\mathcal{E}_{IR}/\omega) F_{IR}(t) \sin(\omega t + \varphi_{IR}). \qquad (9.43)$$

We express the electric-field component of the XUV pulse as

$$\mathcal{E}_{XUV}(t) = \hat{\epsilon} \mathcal{E}_{XUV}(t), \qquad (9.44)$$

and denote by τ_{XUV} the duration of the XUV pulse, so that $\mathcal{E}_{XUV}(t)$ is zero for $t < 0$ and $t > \tau_{XUV}$. We will assume that this pulse is delayed by the amount of time $\Delta t > 0$ with respect to the infra-red pulse.

Let us consider the case in which the duration of the XUV pulse is such that $\tau_{\text{XUV}} \ll T = 2\pi/\omega$. For the moment, we will take the electron to be detached with velocity v_0 at precisely the time $t_d = \Delta t$ when the XUV pulse interacts with the atom. From Equation (2.72), the velocity of the electron at times $t \geq \Delta t$ is, in a.u.,

$$\mathbf{v}(t) = \mathbf{A}_{\text{IR}}(t) - \mathbf{A}_{\text{IR}}(\Delta t) + \mathbf{v}_0. \tag{9.45}$$

Provided that the electron does not recollide with its parent core, the kinetic energy of the photoelectron at the end of the infra-red laser pulse is then given by

$$
\begin{aligned}
E_f(\Delta t) &= \frac{1}{2}[\mathbf{v}_0 - \mathbf{A}_{\text{IR}}(\Delta t)]^2 \\
&= E_0 + 2U_p(\Delta t)\sin^2(\omega\Delta t + \varphi_{\text{IR}}) + [8E_0 U_p(\Delta t)]^{1/2} \\
&\quad \times \sin(\omega\Delta t + \varphi_{\text{IR}})\cos\theta,
\end{aligned}
\tag{9.46}
$$

where $E_0 = v_0^2/2$,

$$U_p(t) = \frac{[\mathcal{E}_{\text{IR}} F_{\text{IR}}(t)]^2}{4\omega^2} \tag{9.47}$$

is the "instantaneous ponderomotive energy" of a free electron in the laser field and θ is the angle between $\mathbf{v}_0$ and the infra-red laser field polarization vector $\hat{\boldsymbol{\epsilon}}$. From Equation (9.45), the angle θ_f between the momentum of the electron at the end of the infra-red laser pulse and $\hat{\boldsymbol{\epsilon}}$ is given by

$$\tan\theta_f = \frac{v_0\sin\theta}{v_0\cos\theta - \hat{\boldsymbol{\epsilon}}\cdot\mathbf{A}_{\text{IR}}(\Delta t)}. \tag{9.48}$$

This allows Equation (9.46) to be expressed in terms of the angle θ_f. We emphasize that it is assumed that the infra-red pulse does not influence the ionization process and the XUV pulse does not influence the dynamics of the ejected photoelectron.

Referring to Equation (9.46), it is seen that information concerning the infra-red pulse can be directly obtained from the energy distribution of photoelectrons that are ejected perpendicular to $\hat{\boldsymbol{\epsilon}}$ as a function of the XUV pulse delay Δt. This allows the electric-field of the infra-red laser pulse to be reconstructed up to an overall sign.

Information about the duration of the XUV pulse can be obtained if we relax the assumption that the electron is detached at precisely the time Δt. Instead we will assume that detachment occurs with equal probability within the window of time between $\Delta t - \tau_{\text{XUV}}/2$ and $\Delta t + \tau_{\text{XUV}}/2$. Taking again the velocity of the electron ejected by the XUV pulse to be v_0, the average kinetic energy of the photoelectron at the end of the infra-red laser pulse is now given by

$$E_f(\Delta t) = \frac{1}{2\tau_{\text{XUV}}} \int_{\Delta t - \tau_{\text{XUV}}/2}^{\Delta t + \tau_{\text{XUV}}/2} [\mathbf{v}_0 - \mathbf{A}_{\text{IR}}(t')]^2 dt'$$

$$\simeq E_0 + U_{\text{p}}(\Delta t) \left[1 - \frac{1}{\omega\tau_{\text{XUV}}} \cos[2(\omega\Delta t + \varphi_{\text{IR}})] \sin(\omega\tau_{\text{XUV}}) \right]$$

$$+ \frac{2}{\omega\tau_{\text{XUV}}} [8E_0 U_{\text{p}}(\Delta t)]^{1/2} \sin(\omega\Delta t + \varphi_{\text{IR}}) \sin(\omega\tau_{\text{XUV}}/2) \cos\theta. \quad (9.49)$$

As in Equation (9.46), the energy distribution of photoelectrons ejected perpendicular to the infra-red laser field polarization vector $\hat{\boldsymbol{\epsilon}}$ exhibits a modulation of angular frequency 2ω as a function of the delay Δt of the XUV pulse. However, the modulation contrast decreases rapidly (approximately as τ_{XUV}^{-2}) as the duration of the XUV pulse increases. A similar analysis shows that the width of the ejected photoelectron energy distribution also modulates as a function of the delay of the XUV pulse Δt, and the contrast decreases as τ_{XUV} increases. This fact was used by Drescher *et al.* [114] and Hentschel *et al.* [139] to estimate the duration of the XUV pulse generated by the interaction of a 7 fs pulse having a 770 nm carrier wavelength with neon atoms. By focusing the outgoing infra-red pulse and the generated XUV pulse on krypton atoms and measuring the resulting width of the photoelectron energy spectrum as a function of the delay Δt, Hentschel *et al.* [139] concluded that XUV pulses of full-width at half-maximum duration (650 ± 150) as were being produced. This simple classical analysis was extended by Kitzler *et al.* [140], who carried out quantum-mechanical calculations of the dependence of the photoelectron spectrum on the duration of the XUV pulse.

We will now turn our attention to the "frequency-resolved optical gating" (FROG) technique. Let us return to Equation (9.46), which gives, within the semiclassical model, the final kinetic energy of a photoelectron ejected with velocity $\mathbf{v}_0$ at time Δt. The key observation is that the moment in time that the photoelectron is ejected is mapped, or "streaked," onto its final kinetic energy [141]. By measuring the energy distribution of photoelectrons ejected parallel to the electric-field polarization direction as a function of the delay Δt of the XUV pulse with respect to the infra-red laser pulse [15, 142], the FROG technique allows both the infra-red and XUV pulses to be reconstructed simultaneously [143, 144].

We will focus on the particular case of a one-electron atomic system. In the absence of the infra-red laser pulse, the probability of the electron being ejected by the XUV pulse with wave vector $\mathbf{k}_f$ can be calculated to good approximation using first-order time-dependent perturbation theory in the first Born approximation. The latter entails ignoring the Coulomb interaction in the final state. Describing the XUV laser–atom interaction in the length gauge, and referring to Equation (6.47), the probability that a photoelectron having an energy within the interval $(E_f, E_f + \mathrm{d}E_f)$ will be ejected into the solid angle $\mathrm{d}\hat{\mathbf{k}}_f$ centered about $\hat{\mathbf{k}}_f$ is

$$P(E_f, \hat{\mathbf{k}}_f) = (2E_f)^{1/2} |T_{\mathbf{k}_f,0}|^2 \, \mathrm{d}E_f \, \mathrm{d}\hat{\mathbf{k}}_f, \tag{9.50}$$

where

$$T_{\mathbf{k}_f,0} = -\int_0^{\tau_{\mathrm{XUV}}} \hat{\boldsymbol{\epsilon}} \cdot \mathbf{d}(\mathbf{k}_f) \mathcal{E}_{\mathrm{XUV}}(t) \exp[\mathrm{i}(E_f + I_{\mathrm{P}})t] \mathrm{d}t \tag{9.51}$$

and

$$\mathbf{d}(\mathbf{k}_f) = -\langle \mathbf{k}_f | \mathbf{r} | \psi_0 \rangle \tag{9.52}$$

is the dipole transition matrix element between the initial atomic ground state $| \psi_0 \rangle$ having ionization potential I_{P} and the final plane wave state $| \mathbf{k}_f \rangle$.

The transition amplitude (9.51) is readily generalized to the case in which the atom also interacts with the infra-red laser pulse if, as in our classical analysis, the assumption is made that only the absorption of XUV photons leads to photoionization. We then invoke the SFA and describe the ejected electron as a "free" particle that interacts only with the infra-red laser pulse. With these approximations, the plane-wave final state in Equation (9.51) is replaced with the Gordon–Volkov state in the length gauge (2.172). The resulting transition amplitude corresponding to an electron being ejected with momentum $\mathbf{k}_f$ is

$$T_{\mathbf{k}_f,0} = -\int_{\Delta t}^{t_1} \hat{\boldsymbol{\epsilon}} \cdot \mathbf{d}[\boldsymbol{\pi}(t)] \mathcal{E}_{\mathrm{XUV}}(t - \Delta t) \exp[\mathrm{i}\tilde{S}(\mathbf{k}_f, t_1, t)] \mathrm{d}t, \tag{9.53}$$

with $\boldsymbol{\pi}(t) = \mathbf{k}_f + \mathbf{A}_{\mathrm{IR}}(t)$ and $t_1 = \mathrm{Max}(\tau_{\mathrm{XUV}} + \Delta t, \tau_{\mathrm{IR}})$. The modified action, $\tilde{S}(\mathbf{k}_f, t_1, t)$, is given by Equation (6.76). Introducing the phase,

$$\varphi(t) = -\frac{1}{2} \int_t^{t_1} \mathrm{d}t' \left[2\mathbf{k}_f \cdot \mathbf{A}_{\mathrm{IR}}(t') + \mathbf{A}_{\mathrm{IR}}^2(t') \right], \tag{9.54}$$

and neglecting a constant overall phase factor, we may now express Equation (9.53) as

$$T_{\mathbf{k}_f,0} = -\int_{\Delta t}^{t_1} \hat{\boldsymbol{\epsilon}} \cdot \mathbf{d}[\boldsymbol{\pi}(t)] \exp[\mathrm{i}\varphi(t)] \mathcal{E}_{\mathrm{XUV}}(t - \Delta t) \exp[\mathrm{i}(E_f + I_{\mathrm{P}})t] \mathrm{d}t, \tag{9.55}$$

so that the photoelectron ionization probability (9.50) is given by

$$P(E_f, \hat{\mathbf{k}}_f) = (2E_f)^{1/2} \left| \int_{\Delta t}^{t_1} \hat{\boldsymbol{\epsilon}} \cdot \mathbf{d}[\boldsymbol{\pi}(t)] \exp[\mathrm{i}\varphi(t)] \mathcal{E}_{\mathrm{XUV}}(t - \Delta t) \right.$$
$$\left. \times \exp[\mathrm{i}(E_f + I_{\mathrm{P}})t] \mathrm{d}t \right|^2 \mathrm{d}E_f \, \mathrm{d}\hat{\mathbf{k}}_f. \tag{9.56}$$

Focusing on the particular case in which electrons are ejected parallel to $\hat{\epsilon}$ Equation (9.56) depends on two quantities, namely the kinetic energy of the ejected electron E_f and the time-delay Δt. This allows us to interpret $P(E_f, \hat{\mathbf{k}}_f)$ as a spectrogram that gives the dependence of the ejected electron energy density on Δt [143, 144]. This spectrogram can be analyzed using the FROG technique.

Let us assume that some unknown laser pulse, whose electric-field component we denote by $\mathcal{E}(t) = \hat{\epsilon}\mathcal{E}(t)$, is analyzed temporally using a gating function, $G(t)$. The form of this function can be quite general. However, its amplitude or phase must vary on a time scale that is comparable or smaller than the characteristic time scales of the unknown pulse. The power spectrum of the output of the gating function is measured as a function of the time delay Δt of the unknown laser pulse. The result is the following spectrogram:

$$S(\bar{\omega}, \Delta t) = \left| \int_{-\infty}^{\infty} G(t)\mathcal{E}(t - \Delta t)\exp(i\bar{\omega}t)dt \right|^2. \tag{9.57}$$

For a given two-dimensional data set $S(\bar{\omega}, \Delta t)$, the FROG technique can be applied to obtain an estimate of the pulse function $\mathcal{E}(t)$ and (or) the gating function $G(t)$ using an iterative algorithm [143, 144].

Comparing the photoelectron ionization probability (9.56) with Equation (9.57) allows us to establish the following correspondences:

$$\bar{\omega} = E_f + I_p,$$
$$G(t) = \hat{\epsilon} \cdot \mathbf{d}[\boldsymbol{\pi}(t)]\exp[i\varphi(t)],$$
$$\mathcal{E}(t) = \mathcal{E}_{\text{XUV}}(t). \tag{9.58}$$

Hence, by measuring the photoelectron distribution as a function of Δt, the FROG technique can be used to recover the pulse function $\mathcal{E}_{\text{XUV}}(t)$ and the gating function $G(t)$ [143,144]. In our case, $G(t)$ is primarily a phase gate: the dipole matrix element is a slowly varying function of $\mathbf{A}_{\text{IR}}(t)$, and hence of time, while $\varphi(t)$ varies rapidly.

The FROG technique has proven to be a powerful tool for both characterizing attosecond XUV pulses and analyzing the resulting dynamics when atomic systems interact with these pulses. For example, a FROG analysis was used by Sansone *et al.* [138], who concluded that a single 130 as XUV pulse was generated when a 5 fs driving infra-red laser pulse of carrier wavelength 750 nm interacted with argon atoms (see Fig. 1.14). Additional examples are discussed in the review of Krausz and Ivanov [15].

9.4.3 Applications of attosecond pulses: attophysics

The investigation of the motion of electrons in atoms and molecules directly in the time domain using attosecond pulses constitutes the subject of *attophysics*. The goal is to explore the electronic structure of matter with atomic-scale resolution in space (at the angström level) and time (at the sub-femtosecond level) and to control its dynamics on the attosecond time scale. In what follows, we shall describe three experiments which illustrate how attosecond pulses can monitor or control the electronic motion in atoms.

The first experiment, on *time-resolved Auger decay*, was performed by Drescher *et al.* [145]. They used an intense infra-red 7 fs few-cycle laser pulse having a carrier wavelength of 750 nm, originating from a Ti:sapphire laser, to produce by harmonic generation in neon a XUV laser pulse of photon energy 97 eV and duration $\tau_X \simeq 900$ as. This XUV pulse was then employed to photoionize a 3d core electron in krypton in a separate target jet. This photoionization process creates a hole (or vacancy) in the M-shell, which then undergoes Auger decay: a bound electron from a level with higher energy spontaneously fills the hole and a second electron (the Auger electron) is ejected. Since the Auger decay of the hole is spontaneous, it has a decay lifetime τ_h, while the photoionization process is prompt and reflects approximately in time the temporal intensity profile of the exciting XUV pulse profile having the duration τ_X, as shown in Fig. 9.12. The aim of the experiment was to measure in real time the Auger lifetime of the hole by monitoring the time of ejection of the Auger electron in the continuum after the 900 as photoionizing XUV pulse was applied. To this end, the electron energy spectra following the excitation of the core were recorded as a function of the time delay Δt between the 900 as

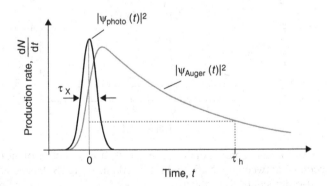

Figure 9.12. Production rates of the photoelectrons and of the Auger electrons, as a function of time. The former is determined by the duration τ_X of the XUV pulse and the latter by the hole lifetime τ_h. (From M. Drescher *et al.*, *Nature* **419**, 803 (2002).)

XUV pulse and the intense infra-red 7 fs Ti:sapphire laser pulse. The improved transmittivity and spectral resolution (0.5 eV) of their time-of-flight spectrometer at lower kinetic energies favored the $M_{4,5}N_1N_{2,3}$ Auger group for a detailed analysis of the Auger electron emission. The sideband area A_{sb} of the first-order sideband of the $M_{4,5}N_1N_{2,3}$ Auger line was evaluated from the normalized electron spectra by combined Gaussian peak-fitting to the whole $M_{4,5}N_1N_{2,3}$ group. The evolution of A_{sb} as a function of the time delay Δt is shown in Fig. 9.13(a), while Fig. 9.13(b) displays the spectral broadening δw of the width of the prompt 4p photoelectron emission line recorded at the same instants. Since the duration $\tau_X = 900$ as of the XUV pulse is much shorter than the duration $\tau_{IR} = 7$ fs of the infra-red Ti:sapphire laser pulse, this spectral broadening calibrates the response function. The resulting deconvolution of the data gave for the Auger lifetime of the hole the value

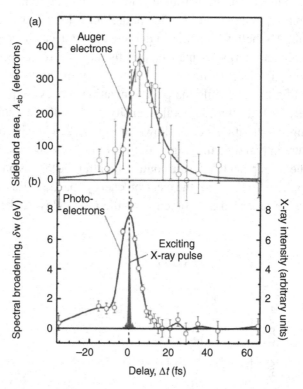

Figure 9.13. Emission profile of the Auger electron as a function of the delay Δt (in femtoseconds) between the XUV and infra-red pulses. (a) Sideband area (circles) of the first-order sideband of the lowest energy $M_{4,5}N_1N_{2,3}$ Auger line extracted from the photoelectron spectra. The solid line is a fit of an exponential decay curve convoluted with the pulse profile. Because $\tau_{XUV} \ll \tau_{IR}$, the spectral broadening of the prompt photoemission shown in (b) calibrates the response function. (From M. Drescher *et al.*, *Nature* **419**, 803 (2002).)

$\tau_h = 7.9^{+1.0}_{-0.9}$ fs. The corresponding Auger line width $\hbar/\tau_h$ is (84 ± 10) meV, in good agreement with the value (88 ± 4) meV obtained from measurements in the energy domain [146].

The second experiment that we shall consider was performed by Dinu *et al.* [129] about *attosecond timing of electron dynamics in high-order harmonic generation.* We first recall that according to the semi-classical recollision model [64,65], high-energy photons are generated when an electron is first detached from its parent core by tunneling ionization, is accelerated by the laser field and then returns to its parent core where it recombines radiatively.

In order to test experimentally the details of this process, Dinu *et al.* [129] used the RABITT method, which determines the timing of the attosecond XUV bursts with respect to the generating infra-red pump cycle. They used a Ti:sapphire laser delivering at 1 kHz an infra-red (800 nm) beam that a mask divided into an outer, annular (pump) part and an inner (probe) part. The pump part was focused into an argon jet. As explained above in the discussion of the RABITT technique, the infra-red probe part generates sidebands in the photoelectron spectrum when combined with the XUV pulses. By evaluating all the phase shifts between the ionization and harmonic generation steps (including the Gouy phase at the focus, the dispersion of the gas medium and the metallic reflection at the mirror), Dinu *et al.* found that the attosecond pulses occurred (190 ± 20) as after each maximum of the magnitude of the driving infra-red electric field, twice per optical cycle. For the 15th harmonic (chosen as the middle order of the harmonic comb being considered), the calculated short and long recombination times, respectively, are 1380 as and 2320 as. The first result is in good agreement with the experimental result (1523 ± 150) as, which excludes the long trajectory. This is a confirmation that the attosecond pulse train arises from the short trajectory in this experiment. The qualitative agreement with the semi-classical recollision model is appealing since argon ionization does not take place in a pure tunneling regime, the value of the Keldysh parameter being $\gamma_K = 1.14$. More refined models [137] predict similar timings for the short trajectory. In addition, Gaarde and Schafer [126,147] have shown in a numerical study that the phase locking is highly favored by the short trajectory, in agreement with the experimental results. Thus, the experiment of Dinu *et al.* [129] demonstrates that it is possible, by determining precisely all the phase delays between the harmonic generation point and the detection apparatus, to measure the timing of the attosecond pulses in high-harmonic generation with respect to the pump field cycle. The experiment of Dinu *et al.* [129] has been extended by Aseyev *et al.* [148] in krypton and xenon in a set-up using an in-line geometry, where the XUV and infra-red beams are not refocused, thus creating the optimal conditions for the determination of the relative phases of the high-order harmonics. In addition, a velocity-mapping detector was used, which yielded the

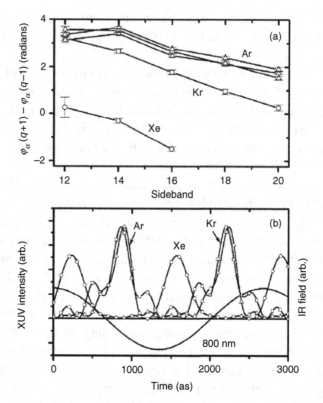

Figure 9.14. (a) Extracted phase differences, $\varphi_{2n+1} - \varphi_{2n-1}$, between adjacent harmonics for experiments performed in argon (triangles), in xenon (circles) and in krypton (squares). (b) XUV temporal shape for argon, krypton and xenon targets obtained from the harmonic phase differences shown in (a). The harmonic amplitudes required to reconstruct the XUV emission profiles were obtained from photoelectron spectra recorded in the absence of the infra-red laser field. (From S. A. Aseyev, Y. Ni, L. J. Frasinski, H. G. Muller and M. J. J. Vrakking, *Phys. Rev. Lett.* **91**, 223902 (2003).)

angular distribution of the various electron peaks. Their results are shown in Fig. 9.14. In particular, they uncovered a difference between xenon and the other two gases in the timing: while argon and krypton are ionized in the tunneling regime, xenon is ionized in the multiphoton regime, where the recollision time may be strongly affected by intermediate resonances [149].

As a third illustration of attophysics experiments, we shall now consider a *double-slit scheme in the time domain* realized by Lindner *et al.* [150]. The double-slit scheme has played a crucial role in the development of optics and quantum mechanics. In optics, its history goes back to the double-slit experiment performed by Young in 1803. In quantum mechanics, double-slit experiments with electrons have been carried out in 1956 by Möllenstedt and Dücker [151] and in 1961 by Jönsson [152].

Of particular importance for the interpretation of quantum mechanics have been experiments with a single electron at any time in the apparatus [153–155], in which the accumulation of the interference pattern has been observed.

In the experiment of Lindner *et al.* [150], argon atoms were ionized by intense few-cycle laser pulses having a carrier wavelength of 760 nm. Photoelectrons emitted in opposite directions ("left" and "right") were detected by two opposing electron detectors placed symmetrically with respect to the laser focus. The laser field was linearly polarized, parallel to an axis defined by the electron detectors. The interference of temporally separated wave packets leads to a fringe pattern in the energy domain, since time and energy are conjugate variables. Therefore the electron kinetic energy must be measured, and this was done in the present case by a time-of-flight (TOF) method. The carrier-envelope phase of the laser field, and thus its temporal evolution, were controlled by delaying the envelope of the laser pulse with respect to the carrier. For intense laser fields, the first step of the ionization process can be described by field ionization. As a result, an attosecond window (or slit) is generated in time per half-cycle close to the maximum of the magnitude of the electric-field, as shown in Fig. 9.15. By using phase-controlled few-cycle laser pulses [156], it is possible to manipulate the temporal evolution of the laser field, and therefore gradually to open or close the slits, and control "which-way" information. Depending on the laser field, one or two half-cycles (or anything in

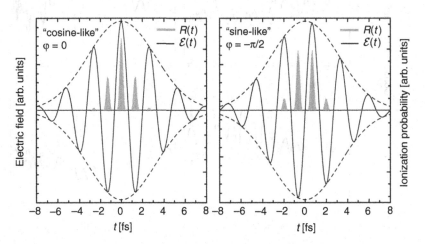

Figure 9.15. Temporal variation of the electric-field function $\mathcal{E}(t) = F(t)\cos(\omega t + \varphi)$ of few-cycle laser pulses with phase $\varphi = 0$ ("cosine-like") and $\varphi = -\pi/2$ ("sine-like"). The field ionization probability $R(t)$, calculated at the experimental parameters, is also indicated. Note that an electron ionized at $t = t_0$ will not necessarily be detected in the opposite direction of the electric-field $\mathcal{E}(t_0)$ due to deflection in the oscillating field. (From F. Lindner *et al.*, *Phys. Rev. Lett.* **95**, 040401 (2005).)

between) contribute to the electron amplitude for a given direction and electron energy. This corresponds to a varying degree of which-way information and thus to a varying contrast of the interference fringes. The temporal slits leading to electrons of given final momentum are spaced by approximately the laser period, resulting in a fringe spacing close to the photon energy. Photoelectron spectra measured with 6 fs laser pulses having a peak intensity of 10^{14} W cm^{-2} are displayed in Fig. 9.16 as a function of the phase φ. The spectra recorded at the left and the right detectors are shown for $\pm$"cosine-like" and $\pm$"sine-like" pulses, as defined in Fig. 9.16. As expected from the foregoing discussion, interference fringes with varying visibility were observed. The highest visibility was observed for $-$"sine-like" pulses in the positive (right) direction. For the same pulses, the visibility was found to be very low in the opposite direction. Changing the phase by π interchanges the role of left and right, as expected. The results can be explained by assuming that, for ☐"sine-like" pulses, there are two slits and no which-way information for the positive direction and just one slit and (almost) complete which-way information in the

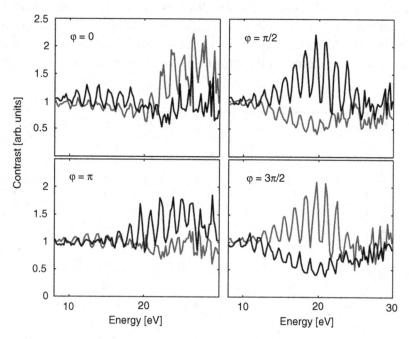

Figure 9.16. Photoelectron spectra of argon measured with 6 fs, 760 nm laser pulses of intensity 10^{14} W cm^{-2} as a function of the phase φ (in radians). The gray curves are spectra recorded with the left detector (negative direction), while the black curves correspond to the positive direction. For $\varphi = \pi/2$, the fringes exhibit maximum visibility for electron emission to the right, while in the opposite direction minimum fringe visibility is observed. In addition, the fringe positions are shifted. (From F. Lindner *et al.*, *Phys. Rev. Lett.* **95**, 040401 (2005).)

negative direction. The fact that the interference pattern does not entirely disappear is due to the pulse duration, which is still a little too long to create a perfect single slit in time.

Under the conditions of the experiment, each argon atom emits at most one electron, the probability for double ionization being several orders of magnitude smaller than for single ionization. The various possibilities for reaching a given final state lead to interference. For "sine-like" pulses, these possibilities correspond to a double slit in time in one direction and to a single slit in time in the other, and are created for each argon atom separately by the few-cycle laser pulse. Thus, despite the fact that there is more than one argon atom in the laser focus, the experiment operates under single-electron conditions. From the number of visible fringes, Lindner *et al.* [150] deduced that the temporal slits extended over about 500 as, thereby demonstrating interferometry on the sub-femtosecond time scale.

References

[1] A. L'Huillier, K. J. Schafer and K. C. Kulander, *J. Phys. B* **24**, 3315 (1991).

[2] A. L'Huillier, L. A. Lompré, G. Mainfray and C. Manus, *Adv. At. Mol. Opt. Phys. Suppl.* **1**, 139 (1992).

[3] C. J. Joachain, Theory of laser-atom interactions. In R. M. More, ed., *Laser Interactions with Atoms, Solids and Plasmas* (New York: Plenum Press, 1994), p. 39.

[4] P. Salières, A. L'Huillier, P. Antoine and M. Lewenstein, *Adv. At. Mol. Opt. Phys.* **41**, 83 (1999).

[5] M. Protopapas, C. H. Keitel and P. L. Knight, *Rep. Progr. Phys.* **60**, 389 (1997).

[6] C. J. Joachain, M. Dörr and N. J. Kylstra, *Adv. At. Mol. Opt. Phys.* **42**, 225 (2000).

[7] T. Brabec and F. Krausz, *Rev. Mod. Phys.* **72**, 545 (2000).

[8] P. Salières. In D. Batani, C. J. Joachain, S. Martellucci and A. N. Chester, eds, *Atoms, Solids and Plasmas in Super-Intense Laser Fields* (New York: Kluwer Academic – Plenum Publishers, 2001), p. 83.

[9] P. Salières. In D. Batani, C. J. Joachain and S. Martellucci, eds, *Atoms and Plasmas in Super-Intense Laser Fields* 88 (Bologna: Italian Physical Society, 2004), p. 303.

[10] P. Salières and I. Christov, Macroscopic effects in high-order harmonic generation. In T. Brabec, ed., *Strong Field Laser Physics*, Springer Series in Optical Sciences 134 (New York: Springer, 2009), p. 241.

[11] P. Agostini and L. F. DiMauro, *Rep. Prog. Phys.* **67**, 813 (2004).

[12] A. Scrinzi, M. Ivanov, R. Kienberger and D. Villeneuve, *J. Phys. B* **39**, R1 (2006).

[13] R. Kienberger, M. Uiberacker, M. F. Kling and F. Krausz, *J. Mod. Opt. Suppl. 1* **54**, 141 (2007).

[14] H. Niikura and P. B. Corkum, *Adv. At. Mol. Opt. Phys.* **54**, 511 (2007).

[15] F. Krausz and M. Ivanov, *Rev. Mod. Phys.* **81**, 163 (2009).

[16] M. Lewenstein and A. L'Huillier, Principles of single atom physics: high-order harmonic generation, above-threshold ionization and non-sequential ionization. In T. Brabec, ed., *Strong Field Laser Physics*, Springer Series in Optical Sciences, 134 (New York: Springer, 2009), p. 147.

[17] A. Scrinzi and H. G. Muller, Attosecond pulses: generation, detection and applications. In T. Brabec, ed., *Strong Field Laser Physics*, Springer Series in Optical Sciences 134 (New York: Springer, 2009), p. 281.

[18] P. A. Franken, A. E. Hill, C. W. Peters and G. Weinreich, *Phys. Rev. Lett.* **7**, 118 (1961).

[19] G. H. C. New and J. F. Ward, *Phys. Rev. Lett.* **19**, 556 (1967).
[20] D. C. Hanna, M. A. Yuratich and D. Cotter, *Non-linear Optics of Free Atoms and Molecules* (Berlin: Springer-Verlag, 1979).
[21] J. Reintjes, *Non-linear Optical Parametric Processes in Liquids and Gases* (New York: Academic Press, 1984).
[22] Y. R. Shen, *The Principles of Non-linear Optics* (New York: Wiley, 1984).
[23] V. G. Arkhipkin and A. K. Popov, *Sov. Phys. Usp.* **30**, 952 (1987).
[24] N. B. Delone and V. P. Krainov, *Fundamentals of Non-linear Optics of Atomic Gases* (New York: Wiley, 1988).
[25] A. McPherson, G. Gibson, H. Jara *et al*, *J. Opt. Soc. Am. B* **4**, 595 (1987).
[26] R. Rosman, G. Gibson, K. Boyer *et al*, *J. Opt. Soc. Am. B* **5**, 1237 (1988).
[27] M. Ferray, A. L'Huillier, X. F. Li, L. A. Lompré, G. Mainfray, and C. Manus, *J. Phys. B* **21**, L31 (1988).
[28] X. F. Li, A. L'Huillier, M. Ferray, L. A. Lompré and G. Mainfray, *Phys. Rev. A* **39**, 5751 (1989).
[29] L. A. Lompré, A. L'Huillier, P. Monot, M. Ferray, G. Mainfray and C. Manus, *J. Opt. Soc. Am. B* **7**, 754 (1990).
[30] N. Sarakura, K. Hata, T. Adachi, R. Nodomi, M. Watanabe and S. Watanabe, *Phys. Rev. A* **43**, 1669 (1991).
[31] A. L'Huillier, L. A. Lompré, G. Mainfray and C. Manus. In G. Mainfray and P. Agostini, eds., *Proceedings of the Fifth International Conference on Multiphoton Processes*. (Paris: Commisariat à l'Energie Atomique, 1990).
[32] J. J. Macklin, J. D. Kmetec and C. L. Gordon, *Phys. Rev. Lett.* **70**, 766 (1993).
[33] A. L'Huillier and P. Balcou, *Phys. Rev. Lett.* **70**, 774 (1993).
[34] J. Zhou, J. Peatross, M. M. Murnane, H. C. Kapteyn and I. P. Christov, *Phys. Rev. Lett.* **76**, 752 (1996).
[35] I. P. Christov, J. P. Zhou, J. Peatross, A. Rundquist, M. M. Murnane and H. C. Kapteyn, *Phys. Rev. Lett.* **77**, 1743 (1996).
[36] C. Spielmann, N. H. Burnett, S. Sartania *et al.*, *Science* **278**, 661 (1997).
[37] Z. Chang, A. Rundquist, H. Wang, M. M. Murnane and H. C. Kapteyn, *Phys. Rev. Lett.* **79**, 2967 (1997).
[38] M. Schnürer, C. Spielmann, P. Wobrauschek *et al.*, *Phys. Rev. Lett.* **80**, 3236 (1998).
[39] A. Rundquist, C. G. Durfee, Z. H. Chang *et al.*, *Science* **280**, 1412 (1998).
[40] M. C. Chen, P. Arpin, T. Popmintchev *et al.*, *Phys. Rev. Lett.* **105**, 173901 (2010).
[41] S. G. Preston, A. Sanpera, M. Zepf *et al.*, *Phys. Rev. A* **53**, R31 (1996).
[42] B. H. Bransden and C. J. Joachain, *Physics of Atoms and Molecules*, 2nd edn. (Harlow, UK: Prentice Hall–Pearson, 2003).
[43] P. Antoine, B. Piraux and A. Maquet, *Phys. Rev. A* **51**, R1750 (1995).
[44] K. J. Schafer and K. C. Kulander, *Phys. Rev. Lett.* **78**, 638 (1997).
[45] N. L. Manakov, V. D. Ovsyannikov and L. P. Rapoport, *Sov. J. Quantum. Electron.* **5**, 22 (1975).
[46] N. L. Manakov and V. D. Ovsyannikov, *Sov. Phys. JETP* **52**, 895 (1980).
[47] Y. Gontier and M. Trahin, *J. Quantum Electron.* **QE-18**, 1137 (1982).
[48] B. Gao and A. F. Starace, *Phys. Rev. A* **39**, 4550 (1989).
[49] L. Pan, K. T. Taylor and C. W. Clark, *Phys. Rev. A* **39**, 4894 (1989).
[50] R. M. Potvliege and R. Shakeshaft, *Z. Phys. D* **11**, 93 (1989).
[51] L. Pan, K. T. Taylor and C. W. Clark, *J. Opt. Soc. Am. B* **7**, 509 (1990).
[52] R. M. Potvliege and R. Shakeshaft, *Phys. Rev. A* **40**, 3061 (1989).
[53] R. Gebarowski, P. G. Burke, K. T. Taylor, M. Dörr, M. Bensaid and C. J. Joachain, *J. Phys. B* **30**, 1837 (1997).

[54] R. Gebarowski, K. T. Taylor and P. G. Burke, *J. Phys. B* **30**, 2505 (1997).
[55] M. Plummer and C. J. Noble, *J. Phys. B* **35**, L51 (2002).
[56] K. C. Kulander and B. W. Shore, *Phys. Rev. Lett.* **62**, 534 (1989).
[57] K. C. Kulander and B. W. Shore, *J. Opt. Soc. Am. B* **7**, 502 (1990).
[58] P. L. DeVries, *J. Opt. Soc. Am. B* **7**, 517 (1990).
[59] K. J. LaGattuta, *Phys. Rev. A* **36**, 3827 (1990).
[60] A. L'Huillier, K. J. Schafer and K. C. Kulander, *J. Phys. B* **24**, 3215 (1991).
[61] J. L. Krause, K. J. Schafer and K. C. Kulander, *Phys. Rev. Lett.* **68**, 3535 (1992).
[62] J. L. Krause, K. J. Schafer and K. C. Kulander, *Phys. Rev. A* **45**, 4998 (1992).
[63] K. C. Kulander, K. J. Schafer and J. C. Krause, *Adv. At. Mol. Opt. Phys. Suppl.* **1**, 247 (1992).
[64] P. B. Corkum, *Phys. Rev. Lett.* **71**, 1994 (1993).
[65] K. C. Kulander, K. J. Schafer and J. L. Krause. In B. Piraux, A. L'Huillier and K. Rzążewski, eds., *Super-Intense Laser-Atom Physics* (New York: Plenum Press, 1993), p. 95.
[66] M. Lewenstein, P. Balcou, M. Y. Ivanov, A. L'Huillier and P. B. Corkum, *Phys. Rev. A* **49**, 2117 (1994).
[67] W. Becker, S. Long and J. K. McIver, *Phys. Rev. A* **50**, 1540 (1994).
[68] P. Salières, A. L'Huillier, P. Antoine and M. Lewenstein, *Adv. At. Mol. Opt. Phys.* **41**, 83 (1999).
[69] P. Salières, B. Carré, L. Le Déroff *et al.*, *Science* **292**, 902 (2001).
[70] N. Milosevic, A. Scrinzi and T. Brabec, *Phys. Rev. Lett.* **88**, 093905 (2002).
[71] M. Y. Ivanov, T. Brabec and N. Burnett, *Phys. Rev. A* **54**, 742 (1996).
[72] V. P. Krainov, *J. Opt. Soc. Am. B* **14**, 425 (1997).
[73] A. Lohr, M. Kleber, R. Kopold and W. Becker, *Phys. Rev. A* **55**, R4003 (1997).
[74] D. B. Milosevic and F. Ehlotzky, *Phys. Rev. A* **57**, 5002 (1998).
[75] C. C. Chirilă, C. J. Joachain, N. J. Kylstra and R. M. Potvliege, *Phys. Rev. Lett* **93**, 243603 (2004).
[76] C. C. Chirilă and R. M. Potvliege, *Phys. Rev. A* **71**, 021402(R) (2005).
[77] R. M. Potvliege, N. J. Kylstra and C. J. Joachain, *J. Phys. B* **33**, L743 (2000).
[78] M. W. Walser, C. H. Keitel, A. Scrinzi and T. Brabec, *Phys. Rev. Lett.* **85**, 5082 (2000).
[79] N. J. Kylstra, R. M. Potvliege and C. J. Joachain, *J. Phys. B* **34**, L55 (2001).
[80] D. B. Milosevic, S. X. Hu and W. Becker, *Phys. Rev. A* **63**, R011403 (2001).
[81] M. Dammasch, M. Dörr, U. Eichmann, E. Lenz and W. Sandner, *Phys. Rev. A* **64**, 061402(R) (2001).
[82] N. J. Kylstra, R. M. Potvliege and C. J. Joachain, *Laser Phys.* **12**, 409 (2002).
[83] C. C. Chirilă, N. J. Kylstra, R. M. Potvliege and C. J. Joachain, *Phys. Rev. A* **66**, 063411 (2002).
[84] C. J. Joachain, N. J. Kylstra and R. M. Potvliege, *J. Mod. Opt.* **50**, 313 (2003).
[85] J. D. Jackson, *Classical Electrodynamics*, 3rd edn (New York: Wiley, 1998).
[86] B. Sundaram and P. W. Milonni, *Phys. Rev. A* **41**, R6571 (1990).
[87] J. H. Eberly and M. V. Fedorov, *Phys. Rev. A* **45**, 4706 (1992).
[88] F. I. Gauthey, C. H. Keitel, P. L. Knight and A. Maquet, *Phys. Rev. A* **52**, 525 (1995).
[89] P. Antoine, A. L'Huillier and M. Lewenstein, *Phys. Rev. Lett.* **77**, 1234 (1996).
[90] J. F. Hergott, M. Kovacev, H. Merdji *et al.*, *Phys. Rev. A* **66**, 021801(R) (2002).
[91] E. Constant, D. Garzella, P. Bréger *et al.*, *Phys. Rev. Lett.* **82**, 1668 (1999).
[92] Y. Tamaki, J. Itatani, Y. Nagata, M. Obara and K. Midorikawa, *Phys. Rev. Lett.* **82**, 1422 (1999).
[93] M. Schnürer, Z. Cheng, M. Hentschel *et al.*, *Phys. Rev. Lett.* **83**, 722 (1999).
[94] C. Delfin, C. Altucci, F. De Filippo *et al.*, *J. Phys. B* **32**, 5397 (1999).

[95]   C. Altucci, T. Starczewski, E. Mével, C. G. Wahlström, B. Carré and A. L'Huillier *J. Opt. Soc. Am. B* **13**, 148 (1996).
[96]   C. Serrat and J. Biegert, *Phys. Rev. Lett.* **104**, 073901 (2010).
[97]   J. Peatross and D. D. Meyerhofer, *Phys. Rev. A* **51**, R906 (1995).
[98]   P. Salières, A. L'Huillier and M. Lewenstein, *Phys. Rev. Lett.* **74**, 3776 (1995).
[99]   P. Salières, T. Ditmire, K. S. Budil, M. D. Perry and A. L'Huillier, *J. Phys. B* **27**, L217 (1994).
[100] J. W. G. Tisch, R. A. Smith, J. E. Muffett, M. Ciarrocca, J. P. Marangos and M. H. R. Hutchinson, *Phys. Rev. A* **49**, R28 (1994).
[101] L. Le Déroff, P. Salières and B. Carré, *Opt. Lett.* **23**, 1544 (1998).
[102] T. Ditmire, E. T. Gumbrell, R. A. Smith, J. W. G. Tisch, D. D. Meyerhofer and M. H. R. Hutchinson, *Phys. Rev. Lett.* **77**, 4756 (1996).
[103] R. A. Bartels, A. Paul, H. Green *et al.*, *Science* **297**, 376 (2002).
[104] L. Le Déroff, P. Salières and B. Carré, *Phys. Rev. A* **61**, 043802 (2000).
[105] C. Bellini, C. Lyngå, A. Tozzi *et al.*, *Phys. Rev. Lett.* **81**, 297 (1998).
[106] C. Lynga, M. B. Gaarde, C. Delfin *et al.*, *Phys. Rev. A* **60**, 4823 (1999).
[107] K. Varjú, Y. Mairesse, B. Carré *et al.*, *J. Mod. Opt.* **52**, 379 (2005).
[108] H. Eichmann, A. Egbert, S. Nolte *et al.*, *Phys. Rev. A* **51**, R3414 (1996).
[109] M. B. Gaarde, P. Antoine, A. Persson, B. Carré, A. L'Huillier and C. G. Wahlström, *J. Phys. B* **29**, L163 (1996).
[110] K. S. Budil, P. Salières, A. L'Huillier, T. Ditmire and M. D. Perry, *Phys. Rev. A* **48**, R3437 (1993).
[111] P. B. Corkum, N. H. Burnett and M. Y. Ivanov, *Opt. Lett.* **19**, 1870 (1994).
[112] P. Dietrich, N. H. Burnett, M. Ivanov and P. B. Corkum, *Phys. Rev. A* **50**, R3585 (1994).
[113] Y. Liang, M. V. Ammosov and S. L. Chin, *J. Phys. B* **27**, 1269 (1994).
[114] M. Drescher, M. Hentschel, R. Kienberger *et al.*, *Science* **291**, 1923 (2001).
[115] J. Larsson, E. Mevel, R. Zerne, A. L'Huillier, C. G. Wahlström and S. Svanberg, *J. Phys. B* **28**, L53 (1995).
[116] D. Xenakis, O. Faucher, D. Charalambidis and C. Fotakis, *J. Phys. B* **29**, L457 (1996).
[117] M. Gisselbrecht, D. Descamps, C. Lynga, A. L'Huillier, C. G. Wahlström and M. Meyer, *Phys. Rev. Lett.* **82**, 4607 (1999).
[118] A. S. Sandhu, E. Gagnon, R. Santra *et al.*, *Science* **322**, 1081 (2008).
[119] R. Haight and D. R. Peale, *Phys. Rev. Lett.* **70**, 3982 (1993).
[120] R. Haight and P. F. Seidler, *Appl. Phys. Lett.* **65**, 517 (1994).
[121] F. Quéré, S. Guizard, P. Martin *et al.*, *Phys. Rev. B* **61**, 9883 (2000).
[122] C. La-O-Vorakiat, M. Siemens, M. M. Murnane *et al.*, *Phys. Rev. Lett.* **103**, 257402 (2009).
[123] M. E. Siemens, Q. Li, R. Yang *et al.*, *Nature Mater.* **9**, 26 (2010).
[124] W. Theobald, R. Hässner, C. Wülker and R. Sauerbrey, *Phys. Rev. Lett.* **77**, 298 (1996).
[125] D. Descamps, J. F. Hergott, H. Merdji *et al.*, *Opt. Lett.* **25**, 135 (2000).
[126] M. B. Gaarde and K. J. Schafer, *Phys. Rev. Lett.* **89**, 213901 (2002).
[127] V. Véniard, R. Taïeb and A. Maquet, *Comm. At. Mol. Phys.* **33**, 53 (1996).
[128] P. M. Paul, E. S. Toma, P. Breger *et al.*, *Science* **292**, 1689 (2001).
[129] L. C. Dinu, H. G. Muller, S. Kazamias *et al.*, *Phys. Rev. Lett.* **91**, 063901 (2003).
[130] Y. Mairesse, A. de Bohan, L. J. Frasinski *et al.*, *Science* **302**, 1540 (2003).
[131] S. Kazamias and P. Balcou, *Phys. Rev. A* **69**, 063416 (2004).
[132] M. Ivanov, P. B. Corkum, T. Zuo and A. Bandrauk, *Phys. Rev. Lett.* **74**, 2933 (1995).
[133] V. T. Platonenko and V. V. Strelkov, *J. Opt. Soc. Am. B* **16**, 435 (1999).
[134] M. Kovacev, Y. Mairesse, E. Priori *et al.*, *Eur. Phys. J. D* **26**, 79 (2003).

[135] O. Tcherbakoff, E. Mével, D. Descamps, J. Plumridge and E. Constant, *Phys. Rev. A* **68**, 043804 (2003).

[136] B. Shan, S. Ghimire and Z. Chang, *J. Mod. Opt.* **52**, 277 (2005).

[137] I. P. Christov, M. M. Murnane and H. C. Kapteyn, *Phys. Rev. Lett.* **78**, 1251 (1997).

[138] G. Sansone, E. Benedetti, F. Calegari *et al.*, *Science* **314**, 443 (2006).

[139] M. Hentschel, R. Kienberger, C. Spielmann *et al.*, *Nature* **414**, 509 (2001).

[140] M. Kitzler, N. Milosevic, A. Scrinzi, F. Krausz and T. Brabec, *Phys. Rev. Lett.* **88**, 173904 (2002).

[141] J. Itatani, F. Quéré, G. L. Yudin, M. Y. Ivanov, F. Krausz and P. B. Corkum, *Phys. Rev. Lett.* **88**, 173903 (2002).

[142] R. Kienberger, E. Gouliemakis, M. Uiberacker *et al.*, *Nature* **427**, 817 (2004).

[143] Y. Mairesse and F. Quéré, *Phys. Rev. A* **71** (2005).

[144] F. Quéré, Y. Mairesse and J. Itatani, *J. Mod. Opt.* **52**, 339 (2005).

[145] M. Drescher, F. Krausz, M. Hentschel *et al.*, *Nature* **419**, 803 (2002).

[146] M. Jurvansuu, A. Kivimki and S. Aksela, *Phys. Rev. A* **64**, 012502 (2001).

[147] M. B. Gaarde and K. J. Schafer, *Phys. Rev. A* **65**, 031406(R) (2002).

[148] S. A. Aseyev, Y. Ni, L. J. Frasinski, H. G. Muller and M. J. J. Vrakking, *Phys. Rev. Lett.* **91**, 223902 (2003).

[149] H. G. Muller, *Opt. Express* **8**, 417 (2001).

[150] F. Lindner, M. Schätzel, H. Walther *et al.*, *Phys. Rev. Lett.* **95**, 040401 (2005).

[151] G. Möllenstedt and H. Dücker, *Z. Phys.* **145**, 377 (1956).

[152] C. Jönsson, *Z. Phys.* **161**, 454 (1961).

[153] P. G. Merli, G. F. Misiroli and G. Pozzi, *Am. J. Phys.* **44**, 306 (1976).

[154] A. Tonomura, J. Endo, T. Matsuda, T. Kawasaki and H. Ezawa, *Am. J. Phys.* **57**, 117 (1989).

[155] S. Frabboni, G. C. Gazzadi and G. Pozzi, *Am. J. Phys.* **75**, 1053 (2007).

[156] A. Baltuška, Th. Udem, M. Uiberacker *et al.*, *Nature* **421**, 611 (2003).

10

Laser-assisted electron–atom collisions

In contrast to laser-induced processes such as multiphoton ionization and high-harmonic generation, which can occur only in the presence of a laser field, laser-assisted processes are phenomena that are modified by the presence a laser field. Examples of laser-assisted processes include electron–atom collisions and atom–atom collisions in the presence of a laser field. The former will constitute the subject of this chapter.

Laser-assisted electron–atom collisions are interesting for several reasons. Firstly, from the experimental point of view, they allow the observation of multi-photon processes at relatively moderate laser intensities, as we shall see below. Secondly, from the theoretical point of view, they require original methods for describing the influence of the laser field on the collisions. Indeed, the presence of the laser field introduces in the problem several new parameters (such as the laser frequency, intensity and polarization), the variation of which allows deeper insight into the fundamental processes to be obtained. Depending on the values of these parameters, laser-assisted electron–atom collision cross sections exhibit different dominant features and display new effects such as laser-induced resonances. Since we shall consider multiphoton processes at laser intensities beyond the limit of validity of perturbation theory, we will need theoretical methods which possess the required non-perturbative features. Thirdly, laser-assisted electron–atom (ion) collisions are also important in several applied areas such as plasma heating by electromagnetic waves [1–4].

The basic electron–atom collision processes in the presence of a laser field which we shall consider in this chapter are the following.

(1) Laser-assisted "elastic" collisions:

$$ e^- + A(i) + n\hbar\omega \longrightarrow e^- + A(i). \tag{10.1} $$

(2) Laser-assisted inelastic collisions, also known as "simultaneous electron–photon excitation (SEPE) or de-excitation (SEPD)":

$$e^- + A(i) + n\hbar\omega \longrightarrow e^- + A(f). \qquad (10.2)$$

(3) Laser-assisted single ionization (e, 2e) collisions, also known as "simultaneous electron–photon ionization (SEPI)":

$$e^- + A(i) + n\hbar\omega \longrightarrow A^+(f) + 2e^-. \qquad (10.3)$$

Here $A(i)$ and $A(f)$ denote an atom A in the initial state i and the final state f, respectively, and $A^+(f)$ means that the ion A^+ is in the final state f. In the above processes, the number n can take on the values $0, \pm1, \pm2, \ldots$ If n is greater than zero, n laser photons are absorbed during the collision process, whereas if n is less than zero, $|n|$ photons are emitted. Hence, these processes are sometimes called *inverse bremsstrahlung* (when $n > 0$) and *stimulated bremsstrahlung* (when $n < 0$). If $n = 0$ the collision process occurs without net absorption or emission of photons in the presence of the laser field. We remark that, whereas n can be arbitrarily large, it cannot decrease below a certain negative or zero integer $n_{\min}$ ($n_{\min} \leq 0$), which is the minimal value of n for which the final channel is open, so that the process is energetically allowed.

The processes (10.1)–(10.3) are particular cases of transitions for which the electron–atom system under investigation is initially and finally in a continuum state. Such transitions are therefore called "free–free transitions" (FFT). We note that the laser-assisted (e, 2e) process (10.3) is the simplest electron–atom reaction in the presence of a laser field. More complicated reactions, such as laser-assisted electron–atom multiple ionization, still await further investigations, and will not be considered here. In what follows we shall also restrict our attention to processes in which the electrons move at non-relativistic velocities.

The layout of this chapter is as follows. Section 10.1 is devoted to a description of the key features of the experiments on electron–atom collisions in the presence of a laser field. In Section 10.2, we begin our theoretical analysis by considering first the simpler problem of the elastic scattering of an electron by a potential in the presence of a laser field. Using the Floquet method, we derive the Floquet–Lippmann–Schwinger equations for the T-matrix elements [5, 6], from which the first Born result of Bunkin and Fedorov[7] and the low-frequency (or soft-photon) approximation of Kroll and Watson [8] are obtained. We also discuss the high-frequency theory of Gavrila and Kaminski [9] and the laser-assisted Coulomb scattering of electrons. In Section 10.3, we generalize our study to laser-assisted collisions of electrons with real atoms having an internal structure. We analyze successively *elastic, inelastic* and *single ionization* (e, 2e) laser-assisted electron–atom collisions. Two particularly successful theoretical methods will be considered

in detail. The first one is the *semi-perturbative theory* of Byron and Joachain [10]
which is suitable for analyzing fast electron–atom collisions in the presence of
relatively strong laser fields, and allows us to calculate how the "dressing" of the
atomic states by the laser field can influence the collision cross sections. The second
one is an extension of the semi-perturbative theory which treats the interaction
between the laser field and the atom in a non-perturbative way by using the Floquet
method [11, 12]. This approach is particularly useful for the study of resonant
processes, such that the laser frequency is close to a transition frequency in the
atom. We pursue our analysis of laser-assisted electron collisions with real atoms
by discussing the application of the fully non-perturbative *R*-matrix–Floquet theory
[13, 14] to these processes [15–18]. Finally, we consider doubly resonant processes
in laser-assisted electron–atom collisions, in which double poles of the *S*-matrix
occur[19, 20].

The subject of laser-assisted electron–atom collisions has been reviewed by
Joachain [21, 22], Gavrila [23], Francken and Joachain [24], Mason [25] and
Ehlotzky, Jaron and Kaminski [26].

10.1 Experiments

The experimental study of laser-assisted electron–atom collisions requires three
coincident beams: an electron beam, an atomic beam and a laser beam. A typical
experimental arrangement is shown in Fig. 10.1. The scattering geometry is chosen
here so that the atomic beam is perpendicular to the plane defined by the incident
electron beam and the laser beam, which coincides with the scattering plane.

The incident electron beam is produced by an electron source in which the
electrons are accelerated in an electric field that determines their energy. This
electron beam must be well collimated and nearly monoenergetic, with a small
spread (in direction and magnitude) of the incident electron momenta about the
momentum $\mathbf{p}_i = \hbar \mathbf{k}_i$. The corresponding incident energy of the electrons is then
$E_{k_i} = \hbar^2 k_i^2 /(2m)$, with a small energy spread (or resolution) $\Delta E \ll E_{k_i}$. This energy
spread is due partly to stray electrostatic fields penetrating the focusing lenses, and
partly to space charges generated by the beam.

The atomic target is a gaseous beam which must be of small size, with uniform
high density. For laser-assisted electron–atom collisions, experimental work has
been confined to noble gas targets which are easier to handle, and for which the
field-free cross sections are well known.

The laser beam is produced by a laser whose pulse parameters (photon energy,
intensity, polarization and duration) must be chosen carefully. As we shall see in
Section 10.2.2, the FFT cross sections in first Born approximation are strongly

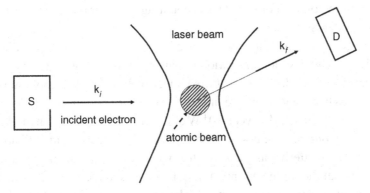

Figure 10.1. Schematic diagram of a typical experimental set-up to study laser-assisted electron–atom collisions. An incident beam of electrons of wave vector $\mathbf{k}_i$ is produced by an electron source S. The atomic beam is perpendicular to the plane defined by the incident beam and the laser beam, represented here in the plane of the figure. The scattered electrons of wave vector $\mathbf{k}_f$, having absorbed or emitted $|n|$ photons, are detected by an electron detector D, in the same plane, which therefore coincides with the scattering plane defined by the wave vectors $\mathbf{k}_i$ and $\mathbf{k}_f$.

enhanced at low frequencies, being proportional to $\omega^{-4|n|}$ as $\omega \to 0$. Most of the experimental results have been obtained using CO_2 lasers ($\omega = 0.0043$ a.u.).

The intensity of the laser field must be sufficiently high to observe multiphoton FFT. For example, Weingartshofer *et al.* [27–29] demonstrated for the first time the existence of multiphoton FFT by using a pulsed CO_2 laser yielding about 10^8 W cm^{-2} in the focal spot. In fact, intensities of the order of magnitude 10^8–10^{12} W cm^{-2} are required to observe multiphoton processes in laser-assisted electron–atom collisions, depending on the laser frequency. It should be noted that very intense laser fields cannot be used, because in such laser fields the atomic target is ionized by multiphoton ionization, so that the laser-assisted electron–atom process cannot be observed.

The polarization of the laser field is also an important parameter. As we shall see below, FFT differential cross sections depend sensitively on the orientation of the polarization vector $\hat{\epsilon}$ with respect to the momentum transfer vector

$$\mathbf{\Delta} = \hbar(\mathbf{k}_i - \mathbf{k}_f) \tag{10.4}$$

of the electron–atom collision. Indeed, we will show that, for moderately intense laser fields the differential cross section has a maximum when $\hat{\epsilon}$ is parallel to $\mathbf{\Delta}$, while a minimum occurs when the two vectors are perpendicular.

While laser-assisted electron–atom collision have attracted considerable interest on the theoretical side, only a few experiments have been performed in this area. Since in these experiments three beams are used that intersect simultaneously in the interaction region (see Fig. 10.1), the counting rates for the laser-assisted collision events of interest are small.

We begin by considering laser-assisted "elastic" electron–atom collisions of the type (10.1). The first demonstration of this process was provided by Andrick and Langhans [30,31]. They used a 10.6 eV electron beam with a resolution of 40 meV an argon gas beam and a low-frequency CO_2 laser of photon energy $\hbar\omega = 0.117$ eV yielding low intensities between 10^4 W cm^{-2} and 10^5 W cm^{-2}. Under these circumstances, only *one-photon* laser-assisted "elastic" electron–atom collisions could be detected. The scattering angle θ was fixed at 160°, so that the polarization vector $\hat{\epsilon}$ was nearly parallel to the momentum transfer $\mathbf{\Delta}$, and the laser-assisted differential cross section was therefore enhanced. The results of Andrick and Langhans [30,31] could be interpreted satisfactorily using perturbation theory to treat the interaction of the projectile electron with the laser field, while neglecting target-dressing effects.

In a subsequent experiment Langhans [32] observed *resonance effects* in laser-assisted "elastic" electron–atom collisions with the transfer of one photon ($n = \pm 1$). Let us suppose, for example, that with the laser off a resonance occurs in the field-free elastic scattering cross section at an incident electron energy E_r. With the laser on, the scattered electrons, having *lost* the energy $\hbar\omega$ of one photon, are detected, their energy being given by $E_{k_f} = E_{k_i} - \hbar\omega$. If the incident electron energy E_{k_i} is increased, it will pass at one point the resonance energy ($E_{k_i} = E_r$) followed by E_{k_f} when $E_{k_f} = E_r$. The low-frequency theory of resonant FFT [33,34], which will be discussed in Section 10.2, predicts that the laser-assisted cross section will exhibit a resonance when either the initial electron energy E_{k_i} or the final electron energy E_{k_f} coincides with E_r. As a result, *two* resonances are expected to be detected in the scattered electron energy spectrum, situated $\hbar\omega$ apart, for every field-free elastic scattering resonance at the energy E_r. This was confirmed by the experimental data of Langhans [32], who measured the laser-assisted differential cross section corresponding to $n = -1$ at a scattering angle $\theta = 160°$ in the energy region of the $^2P_{1/2}$ and $^2P_{3/2}$ resonances of argon at 11.270 eV and 11.098 eV, respectively.

In subsequent experiments, Andrick and Bader [35] and Bader [36] have extended the Langhans experiment to the region of smaller scattering angles (in the range 20° to 70°) where the counting rates are much smaller. They used as target atoms the noble gases He, Ne and Ar. In general, satisfactory agreement was found with the low-frequency theory of resonant FFT.

Wallbank *et al.* [37], using incident electrons of energy $E_{k_i} = 10.55$ eV and Ar atoms as the target, have studied one-photon FFT in the intensity regime where the first-order perturbation theory cross section, proportional to the laser intensity I, breaks down. They showed that, for laser intensities $I > 2 \times 10^6$ W cm^{-2}, the cross section increases more slowly than linearly with I, in agreement with the prediction of the non-perturbative low-frequency theory of FFT, which we shall consider in Section 10.2.

Weingartshofer *et al.* [27–29] have demonstrated the existence of multiphoton FFT by using a pulsed CO_2 laser with a peak intensity of about 10^8 W cm^{-2}, much higher than in previous laser-assisted electron–atom collision experiments. The target consisted of argon atoms, and the incident electron energy E_{k_i} was first 11.72 eV and then 15.80 eV, with respective scattering angles θ of 153° and 155°. By improving the characteristics of their experiments, they could observe as many as 11-photon free–free absorption and emission transitions. This is illustrated in Fig. 1.16. It is worth noting that, because of the long recording times, the measurements were only performed at final electron energies E_{k_f}, differing from E_{k_i} by integer or half-odd-integer values of the photon energy. As pointed out in Section 1.5, the relative intensities of two successive FFT peaks in Fig. 1.16 are of the same order of magnitude, which is typical of a non-perturbative regime.

Let us now consider laser-assisted inelastic electron–atom collisions of the type (10.2). Several experiments have been performed to study the simultaneous electron–photon excitation (SEPE) process in which an incident electron of energy E_{k_i} together with one or several photons cause an excitation of the atom. In particular, the excitation of helium from its 1^1S ground state to its metastable 2^3S state has been investigated by Mason and Newell [38, 39] with a single-mode CO_2 laser yielding an intensity of 10^4 W cm^{-2}, by Wallbank *et al.* [40, 41] with a pulsed multi-mode CO_2 laser delivering an intensity of 10^8 W cm^{-2}, by Luan, Hippler and Lutz [42] with a multi-mode Nd:YAG laser giving an intensity of 10^{10} W cm^{-2} and by Mason and Newell [43] with both linearly and circularly polarized CO_2 laser pulses having intensities in the range 10^4 to 10^5 W cm^{-2}. In all these SEPE experiments the laser-assisted excitation cross section was measured by the time-of-flight detection of the excited metastable helium atoms. The experimental cross sections clearly demonstrate the existence of the SEPE process. Indeed, the production of the 2^3S state of helium was observed *below* the field-free threshold excitation energy of $E_{k_i} = 19.8$ eV.

We now turn to laser-assisted (e, 2e) experiments of the type (10.3). Höhr *et al.* [44] have studied the electron impact single ionization of helium in the presence of a Nd:YAG laser pulse of wavelength 1064 nm and intensity 4×10^{12} W cm^{-2}. A 1 keV pulse electron beam overlapped the laser beam at the position of a helium target beam. Höhr *et al.* observed distinct differences in the doubly and triply differential cross sections compared to the corresponding field-free cross sections.

Finally, we remark that a detailed comparison between the experimental data and the theoretical calculations must take into account the actual experimental conditions, such as beam focusing and pulsed and (possibly) multi-mode operation of the laser. The spatial and temporal inhomogeneities of laser beams and their influence on laser-assisted electron–atom measurements have been investigated theoretically by Bivona *et al.* [45] and Francken and Joachain [24].

10.2 Laser-assisted potential scattering

10.2.1 Basic theory

We begin our theoretical analysis by considering the simple case in which the target atom is modeled by a center of force, and hence does not interact with the laser field. The problem is then reduced to the non-relativistic scattering of an electron by a (real) potential $V(\mathbf{r})$ in the presence of a laser field. The potential will be taken to be of finite range. Coulomb scattering in the presence of a laser field is discussed in Section 10.2.4.

We shall treat the laser field classically as a spatially homogeneous (dipole approximation), linearly polarized electric field $\boldsymbol{\mathcal{E}}(t)$. We also suppose that the pulse duration is sufficiently long so that, on average, the laser intensity does not vary much from its peak intensity on time scales that are of the order of typical collision times. In this limit, the laser field can be taken to be monochromatic. The electric field will be written in the form of Equation (2.118) with $\varphi = -\pi/2$, namely

$$\boldsymbol{\mathcal{E}}(t) = \hat{\boldsymbol{\epsilon}}\mathcal{E}_0 \sin(\omega t), \tag{10.5}$$

the corresponding vector potential being

$$\mathbf{A}(t) = \hat{\boldsymbol{\epsilon}}A_0 \cos(\omega t), \tag{10.6}$$

with $A_0 = \mathcal{E}_0/\omega$.

The laser-assisted electron–potential scattering dynamics that we study is governed by the time-dependent Schrödinger equation (TDSE), which we write in the dipole approximation and in the position representation as follows:

$$i\hbar\frac{\partial}{\partial t}\Psi(\mathbf{r}, t) = [H_0 + H_{\text{int}}(t)]\Psi(\mathbf{r}, t), \tag{10.7}$$

where

$$H_0 = \frac{\mathbf{p}^2}{2m} + V(\mathbf{r}) \tag{10.8}$$

is the field-free Hamiltonian and

$$H_{\text{int}}(t) = \frac{e}{m}\mathbf{A}(t)\cdot\mathbf{p} + \frac{e^2}{2m^2}\mathbf{A}^2(t) \tag{10.9}$$

is the Hamiltonian describing the interaction between the electron and the laser field. In the above equation, $\mathbf{p} = -i\hbar\nabla$.

As we have seen in Section 2.4, the $\mathbf{A}^2(t)$ term can be removed from the TDSE (10.7) by performing a gauge transformation, giving for the wave function $\Psi^{\mathrm{V}}(\mathbf{r}, t)$ in the velocity gauge the TDSE

$$i\hbar \frac{\partial}{\partial t} \Psi^{\mathrm{V}}(\mathbf{r}, t) = \left[\frac{\mathbf{p}^2}{2m} + V(\mathbf{r}) + H_{\mathrm{int}}^{\mathrm{V}}(t) \right] \Psi^{\mathrm{V}}(\mathbf{r}, t), \qquad (10.10)$$

where

$$H_{\mathrm{int}}^{\mathrm{V}}(t) = \frac{e}{m} \mathbf{A}(t) \cdot \mathbf{p} \qquad (10.11)$$

is the electron–laser field interaction in the velocity gauge.

The TDSE (10.10) must be solved subject to the appropriate scattering boundary conditions. Initially, when the electron is far enough from the scattering center, it can be treated as a "free" electron in the laser field with $V(r) = 0$. We recall that in Section 2.7 the Gordon–Volkov wave functions in the velocity gauge were obtained as solutions of the TDSE (2.163) describing free electrons in a laser field. We have, from Equation (2.167),

$$\chi_{\mathbf{k}}^{\mathrm{V}}(\mathbf{r}, t) = (2\pi)^{-3/2} \exp\{i\mathbf{k} \cdot [\mathbf{r} - \boldsymbol{\alpha}(t)] - iE_k t/\hbar\}, \qquad (10.12)$$

where $E_k = \hbar^2 k^2/(2m)$ and, according to Equation (2.113),

$$\boldsymbol{\alpha}(t) = \frac{e}{m} \int_{-\infty}^{t} \mathbf{A}(t')\mathrm{d}t'. \qquad (10.13)$$

For the particular case of a linearly polarized, monochromatic, homogeneous laser field described by the electric field (10.5), the vector $\boldsymbol{\alpha}(t)$ is given by

$$\boldsymbol{\alpha}(t) = \boldsymbol{\alpha}_0 \sin(\omega t), \qquad (10.14)$$

where $\boldsymbol{\alpha}_0 = \hat{\boldsymbol{\epsilon}} \alpha_0$ and $\alpha_0 = e\mathcal{E}_0/(m\omega^2)$ is the quiver amplitude of the electron in the laser field. Thus, the non-relativistic Gordon–Volkov wave function (10.12) in the velocity gauge becomes

$$\chi_{\mathbf{k}}^{\mathrm{V}}(\mathbf{r}, t) = (2\pi)^{-3/2} \exp\{i(\mathbf{k} \cdot \mathbf{r} - \mathbf{k} \cdot \boldsymbol{\alpha}_0 \sin(\omega t) - E_k t/\hbar)\}. \qquad (10.15)$$

We see that the effect of the laser field on the free electron is characterized by the dimensionless quantity

$$\mathbf{k} \cdot \boldsymbol{\alpha}_0 = \zeta \, \hat{\boldsymbol{\epsilon}} \cdot \hat{\mathbf{k}}, \qquad (10.16)$$

where

$$\zeta = k\alpha_0 = \frac{e\mathcal{E}_0 k}{m\omega^2}. \qquad (10.17)$$

Clearly, a perturbative treatment of the laser–electron interaction, in which the quantity $\exp[-i\mathbf{k} \cdot \boldsymbol{\alpha}_0 \sin(\omega t)]$ in the Gordon–Volkov wave function (10.15) is expanded in a power series, will only be valid if $\zeta \ll 1$. From Equations (10.16) and (10.17) we see that such a perturbative treatment will fail for strong laser fields, low frequencies and (or) fast electrons. In all these cases, the full Gordon–Volkov wave function (which in the non-relativistic limit considered here treats the electron interaction with the laser field exactly) must be used.

As an example, let us consider the laser-assisted electron–atom experiments of Weingartshofer *et al.* [29] discussed in Chapter 1. In this case, the laser photon energy is $\hbar\omega = 0.117$ eV, the average laser intensity $\bar{I} = 10^8$ W cm^{-2} (corresponding to a value $\mathcal{E}_0 = 2.7 \times 10^5$ V cm^{-1} of the electric-field strength) and for an incident electron energy of 15.8 eV we have $k \simeq 1$. Thus $\zeta \simeq 4$, and the electron interaction with the laser field must be treated by using Gordon–Volkov wave functions.

We now return to the Gordon–Volkov wave function (10.15). Using the generating function of the ordinary Bessel functions [46],

$$\exp(ix\sin\gamma) = \sum_{m=-\infty}^{\infty} \exp(im\gamma)J_m(x), \tag{10.18}$$

with $x = \mathbf{k}\cdot\boldsymbol{\alpha}_0$ and $\gamma = -\omega t$, one can also write this Gordon–Volkov wave function in the Floquet–Fourier form

$$\chi_{\mathbf{k}}^{V}(\mathbf{r}, t) = \exp(-iE_k t/\hbar) \sum_{m=-\infty}^{\infty} \exp(-im\omega t)u_{k,m}(\mathbf{r}) \tag{10.19}$$

with

$$u_{k,m}(\mathbf{r}) = (2\pi)^{-3/2} J_m(\mathbf{k}\cdot\boldsymbol{\alpha}_0)\exp(i\mathbf{k}\cdot\mathbf{r}). \tag{10.20}$$

The full TDSE (10.10) must now be solved, subject to the boundary conditions discussed above. In order to do this, it is convenient to rewrite the TDSE in the form of a time-dependent Lippmann–Schwinger equation [47] for the state vector $|\Psi_{\mathbf{k}_i}^{V(+)}\rangle$. That is

$$|\Psi_{\mathbf{k}_i}^{V(+)}(t)\rangle = |\chi_{\mathbf{k}_i}^{V}(t)\rangle + \int_{-\infty}^{t} dt'\, K_F^{V(+)}(t, t')V|\Psi_{\mathbf{k}_i}^{V(+)}(t')\rangle, \tag{10.21}$$

where $|\chi_{\mathbf{k}_i}^{V}(t)\rangle$ is a particular Gordon–Volkov state corresponding to the initial electron momentum $\mathbf{k}_i$. In Equation (10.21), $K_F^{V(+)}(t, t')$ is the causal (time-retarded) propagator (see Section 2.6) which, according to Equations (2.156) and (2.179), is given in the velocity gauge by

$$K_F^{V(+)}(t, t') = -\frac{i}{\hbar}\Theta(t - t')\int d\mathbf{k}\, |\chi_{\mathbf{k}}^{V}(t)\rangle\langle\chi_{\mathbf{k}}^{V}(t')|, \tag{10.22}$$

where $\Theta(x)$ is the Heaviside step function defined by Equation (2.157).

Scattering information is obtained by calculating the S-matrix elements, which, for any choice of gauge, are given by [47]

$$S_{\mathbf{k}_f,\mathbf{k}_i} = \lim_{t\to+\infty} \langle \Psi_{\mathbf{k}_f}^{(-)}(t) | \Psi_{\mathbf{k}_i}^{(+)}(t) \rangle$$

$$= \delta(\mathbf{k}_f - \mathbf{k}_i) - \frac{i}{\hbar} \int_{-\infty}^{\infty} dt \, \langle \chi_{\mathbf{k}_f}(t) | V | \Psi_{\mathbf{k}_i}^{(+)}(t) \rangle. \qquad (10.23)$$

Introducing the transition matrix, or T-matrix, elements

$$\mathcal{T}_{\mathbf{k}_f,\mathbf{k}_i} = \int_{-\infty}^{\infty} dt \, \langle \chi_{\mathbf{k}_f}(t) | V | \Psi_{\mathbf{k}_i}^{(+)}(t) \rangle, \qquad (10.24)$$

we can also write Equation (10.23) in the form

$$S_{\mathbf{k}_f,\mathbf{k}_i} = \delta(\mathbf{k}_f - \mathbf{k}_i) - \frac{i}{\hbar}\mathcal{T}_{\mathbf{k}_f,\mathbf{k}_i}. \qquad (10.25)$$

It is worth noting that the first term on the right-hand side of Equation (10.25) corresponds to elastic scattering in the forward direction, without exchange of photons.

We have assumed above that the laser pulse durations are long enough so that, on average, the laser intensity does not vary much from its peak intensity on time scales of the order of typical collision times. As a result, we can describe the scattering process in terms of *time-independent* transition rates. For the monochromatic laser field given by Equation (10.5), the Hamiltonian of the system is periodic in time with period $T = 2\pi/\omega$, and we can write the state vector $|\Psi_{\mathbf{k}_i}^{V(+)}(t)\rangle$ in the velocity gauge in Floquet–Fourier form as

$$|\Psi_{\mathbf{k}_i}^{V(+)}(t)\rangle = \exp(-iE_{k_i}t/\hbar) \sum_{m=-\infty}^{\infty} \exp(-im\omega t) |F_{\mathbf{k}_i,m}^{(+)}\rangle. \qquad (10.26)$$

Similarly, following the notation introduced in Equations (10.20), we can express the Gordon–Volkov state $|\chi_{\mathbf{k}_f}^V(t)\rangle$ in Floquet–Fourier form as

$$|\chi_{\mathbf{k}_f}^V(t)\rangle = \exp(-iE_{k_f}t/\hbar) \sum_{m=-\infty}^{\infty} \exp(-im\omega t) |u_{\mathbf{k}_f,m}\rangle. \qquad (10.27)$$

Inserting the Floquet–Fourier expansions (10.26) and (10.27) into Equation (10.24), we find that the S-matrix elements can be written as

$$S_{\mathbf{k}_f,\mathbf{k}_i} = \delta(\mathbf{k}_f - \mathbf{k}_i) - 2\pi i \sum_{n=-\infty}^{\infty} \delta(E_{k_f} - E_{k_i} - n\hbar\omega) T_{\mathbf{k}_f,\mathbf{k}_i}^n, \qquad (10.28)$$

where

$$T_{\mathbf{k}_f,\mathbf{k}_i}^n = \sum_{m=-\infty}^{\infty} \langle u_{\mathbf{k}_f,m-n} | V | F_{\mathbf{k}_i,m}^{(+)} \rangle \qquad (10.29)$$

is the transition matrix element corresponding to the transition $\mathbf{k}_i \to \mathbf{k}_f$ with the exchange of n photons. The delta function $\delta(E_{k_f} - E_{k_i} - n\hbar\omega)$ in Equation (10.28) ensures energy conservation for the transition, so that

$$E_{k_f} = E_{k_i} + n\hbar\omega, \qquad (10.30)$$

with positive values of n corresponding to photon absorption, negative values to photon emission and $n = 0$ to scattering without net absorption or emission of photons. We note that the energy of the scattered electron must satisfy the following condition:

$$E_{k_f} \geq 0. \qquad (10.31)$$

In the case of photon absorption ($n > 0$), this condition is always fulfilled, so that n can be arbitrarily large. On the contrary, in the case of photon emission ($n < 0$), the condition (10.31), together with Equation (10.30) imply that

$$-n\hbar\omega \leq E_{k_i}. \qquad (10.32)$$

As a result, the value of n cannot be lower than the negative, or zero, $n_{\min}$, which is the minimal value of n for which the condition (10.32) is satisfied. It follows that

$$T^n_{\mathbf{k}_f, \mathbf{k}_i} = 0 \qquad (10.33)$$

for $n < n_{\min}$.

The differential cross section for the non-relativistic scattering of an electron by a potential, accompanied by the transfer of $|n|$ photons, is given by

$$\frac{d\sigma^n}{d\Omega} = (2\pi)^4 \frac{m^2}{\hbar^4} \frac{k_f(n)}{k_i} |T^n_{\mathbf{k}_f, \mathbf{k}_i}|^2, \qquad (10.34)$$

where

$$k_i = \left(\frac{2m}{\hbar^2} E_{k_i}\right)^{1/2} \qquad (10.35)$$

and

$$k_f(n) = \left[\frac{2m}{\hbar^2}(E_{k_i} + n\hbar\omega)\right]^{1/2}. \qquad (10.36)$$

In Equation (10.34), the solid angle $d\Omega$ is centered about the direction (θ, ϕ), where the scattering angle θ is the angle between the wave vectors $\mathbf{k}_i$ and $\mathbf{k}_f$ and ϕ is the azimuthal angle of the scattered electron (see Fig. 10.2).

Introducing the scattering amplitude

$$f^n_{\mathbf{k}_f, \mathbf{k}_i} = -\frac{(2\pi)^2 m}{\hbar^2} T^n_{\mathbf{k}_f, \mathbf{k}_i}, \qquad (10.37)$$

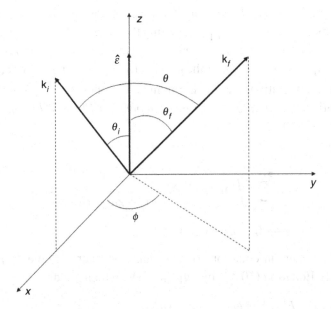

Figure 10.2. Scattering geometry for electron scattering by a potential in the presence of a linearly polarized laser field. The polarization vector $\hat{\epsilon}$ defines the z-axis. The incident electron wave vector $\mathbf{k}_i$ is in the x-z plane, and has polar angles θ_i and $\phi_i = 0$. The scattered electron wave vector $\mathbf{k}_f$ has polar angles θ_f and $\phi_f = \phi$. The angle between the vectors $\mathbf{k}_i$ and $\mathbf{k}_f$ is the scattering angle θ. We note that $\theta_f = \theta$ when $\theta_i = 0$, namely when the polarization vector $\hat{\epsilon}$ is parallel to $\mathbf{k}_i$.

where the choice of phase is a conventional one, we can write Equation (10.34) in the form

$$\frac{d\sigma^n}{d\Omega} = \frac{k_f(n)}{k_i} |f^n_{\mathbf{k}_f,\mathbf{k}_i}|^2. \tag{10.38}$$

Henceforth we shall use atomic units (a.u.), unless stated otherwise. We then have

$$f^n_{\mathbf{k}_f,\mathbf{k}_i} = -(2\pi)^2 T^n_{\mathbf{k}_f,\mathbf{k}_i}, \tag{10.39}$$

and Equation (10.34) becomes

$$\frac{d\sigma^n}{d\Omega} = (2\pi)^4 \frac{k_f(n)}{k_i} |T^n_{\mathbf{k}_f,\mathbf{k}_i}|^2, \tag{10.40}$$

while the form of Equation (10.38) is unchanged. In addition, we now have

$$k_i = \left(2E_{k_i}\right)^{1/2} \tag{10.41}$$

and

$$k_f(n) = \left[2(E_{k_i} + n\omega)\right]^{1/2}. \tag{10.42}$$

Following Kylstra and Joachain [5,6], we shall now obtain an integral equation
for the T-matrix elements $T^n_{\mathbf{k}_f,\mathbf{k}_f}$. Inserting the Floquet–Fourier expansions (10.26)
and (10.27) into the time dependent Lippmann–Schwinger equation (10.21) and
using the expression (10.22) for the propagator $K_{\mathrm{F}}^{V(+)}(t,t')$, we find, after performing the time-integration, that the harmonic components $|F^{(+)}_{\mathbf{k}_i,m}\rangle$ of the state vector
(10.26) satisfy the *infinite set of time-independent Floquet–Lippmann–Schwinger
equations* (FLSE):

$$
|F^{(+)}_{\mathbf{k}_i,m}\rangle = |u_{\mathbf{k}_i,m}\rangle
$$
$$
+ \sum_{r,s=-\infty}^{\infty} \int d\mathbf{k}\,|u_{\mathbf{k},m-r}\rangle \frac{1}{E_{k_i}+r\omega-E_k+i\epsilon}\langle u_{\mathbf{k},s-r}|V|F^{(+)}_{\mathbf{k}_i,s}\rangle,
$$
$$
\epsilon \to 0^+, \quad m=0\pm1,\pm2,\dots \tag{10.43}
$$

In order to obtain an equation for the transition matrix elements $T^n_{\mathbf{k}_f,\mathbf{k}_i}$, we first
premultiply Equation (10.43) by $\langle u_{\mathbf{k}_f,m-n}|V$, which yields

$$
\langle u_{\mathbf{k}_f,m-n}|V|F^{(+)}_{\mathbf{k}_i,m}\rangle = \langle u_{\mathbf{k}_f,m-n}|V|u_{\mathbf{k}_i,m}\rangle
$$
$$
+ \sum_{r,s=-\infty}^{\infty} \int d\mathbf{k}\,\langle u_{\mathbf{k}_f,m-n}|V|u_{\mathbf{k},m-r}\rangle \frac{1}{E_{k_i}+r\omega-E_k+i\epsilon}\langle u_{\mathbf{k},s-\mathbf{r}}|V|F^{(+)}_{\mathbf{k}_i,s}\rangle.
$$
$$
\tag{10.44}
$$

We now define the quantities

$$
V^{n-r}_{\mathbf{k}',\mathbf{k}} = \sum_{m=-\infty}^{+\infty} \langle u_{\mathbf{k}',m-n}|V|u_{\mathbf{k},m-r}\rangle
$$
$$
= \sum_{m=-\infty}^{+\infty} J_{m+r-n}(\mathbf{k}'\cdot\boldsymbol{\alpha}_0)J_m(\mathbf{k}\cdot\boldsymbol{\alpha}_0)\langle\mathbf{k}'|V|\mathbf{k}\rangle, \tag{10.45}
$$

where we have used Equation (10.20). The summation over the index m can be
carried out by making use of the Bessel function sum rule [46]

$$
J_n(x-y) = \sum_{m=-\infty}^{\infty} J_{n+m}(x)J_m(y). \tag{10.46}
$$

Using the fact that $J_n(x)=J_{-n}(-x)$, we obtain

$$
V^{n-r}_{\mathbf{k}',\mathbf{k}} = J_{n-r}(\boldsymbol{\Delta}_{\mathbf{k}',\mathbf{k}}\cdot\boldsymbol{\alpha}_0)\langle\mathbf{k}'|V|\mathbf{k}\rangle, \tag{10.47}
$$

with $\boldsymbol{\Delta}_{\mathbf{k}',\mathbf{k}}=\mathbf{k}-\mathbf{k}'$. In particular, the wave vector transfer (or momentum transfer
in atomic units) of the collision is given by

$$
\boldsymbol{\Delta} = \boldsymbol{\Delta}_{\mathbf{k}_f,\mathbf{k}_i} = \mathbf{k}_i - \mathbf{k}_f. \tag{10.48}
$$

This quantity depends on the number n of photons exchanged with the laser field, since $\mathbf{k}_f = \hat{\mathbf{k}}_f k_f(n)$.

From Equation (10.29), the on-the-energy shell T-matrix elements $T^n_{\mathbf{k}_f, \mathbf{k}_i}$ can be obtained by solving the Floquet–Lippmann–Schwinger equations (FLSE)

$$T^n_{\mathbf{k}_f, \mathbf{k}_i} = V^n_{\mathbf{k}_f, \mathbf{k}_i} + \sum_{r=-\infty}^{\infty} \int d\mathbf{k} \, V^{n-r}_{\mathbf{k}_f, \mathbf{k}} \frac{1}{E_{k_i} + r\omega - E_k + i\epsilon} T^r_{\mathbf{k}, \mathbf{k}_i}. \tag{10.49}$$

For $n < n_{\min}$, the quantities $T^n_{\mathbf{k}_f, \mathbf{k}_i}$ vanish. From time-reversal invariance, the T-matrix elements satisfy

$$T^n_{\mathbf{k}_f, \mathbf{k}_i} = T^{-n}_{\mathbf{k}_i, \mathbf{k}_f}. \tag{10.50}$$

Indeed, the process whereby an incident electron having wave vector $\mathbf{k}_i$ that absorbs (emits) $|n|$ photons and has the wave vector $\mathbf{k}_f$ in the final state is the same as the reverse process in which an electron with initial wave vector $\mathbf{k}_f$ emits (absorbs) $|n|$ photons and has the final wave vector $\mathbf{k}_i$ in the final state.

When solving numerically the FLSE (10.49), the size of the problem is effectively given by the number of Fourier components retained in the Floquet–Fourier expansion (10.26) times the number of grid points in three dimensions retained in the evaluation of the integral. At low electron energies, it is convenient to perform a partial wave expansion of the FLSE, since only a limited number of partial waves contribute to the differential cross section, and the angular integrations can then efficiently be carried out [6]. Using this approach, Kylstra and Joachain [5, 6] have obtained the differential cross section for laser-assisted low-energy electron scattering by several central potentials in a low-frequency CO_2 laser field ($\omega = 0.0043$ a.u.). Their results will be discussed below, where they will be compared to those obtained by using a low-frequency approximation.

10.2.2 Born series

In general, the evaluation of the scattering amplitude $f^n_{\mathbf{k}_f, \mathbf{k}_i}$ (or the T-matrix element $T^n_{\mathbf{k}_f, \mathbf{k}_i}$) is a difficult task. We now examine various approximation methods for laser-assisted potential scattering, and in particular some limiting cases in which important simplifications occur. Atomic units will be used.

We begin by considering the *Born series*, which is a perturbative expansion of the wave function or the scattering amplitude (and therefore of related quantities such as the S-matrix or T-matrix elements) in powers of the interaction potential $V(\mathbf{r})$. The Born series for the state vector in the velocity gauge $|\Psi^{V(+)}_{\mathbf{k}_i}(t)\rangle$ is obtained by solving the time-dependent Lippmann–Schwinger equation (10.21) by iteration, starting with the Gordon–Volkov state $|\chi^V_{\mathbf{k}_i}(t)\rangle$ as the initial approximation. Upon substitution of this Born series for $|\Psi_{\mathbf{k}_i}^{V(+)}(t)\rangle$ on the right-hand side of Equation (10.23), we obtain for the S-matrix element $S_{\mathbf{k}_f,\mathbf{k}_i}$ the Born series

$$S_{\mathbf{k}_f,\mathbf{k}_i} = \delta(\mathbf{k}_f - \mathbf{k}_i) + \sum_{j=1}^{\infty} \bar{S}_{\mathbf{k}_f,\mathbf{k}_i}^{Bj}, \qquad (10.51)$$

where the time-independent quantity

$$\bar{S}_{\mathbf{k}_f,\mathbf{k}_i}^{Bj} = -i \int_{-\infty}^{\infty} dt_{j-1} \cdots \int_{-\infty}^{\infty} dt_1 \int_{-\infty}^{\infty} dt_0$$
$$\times \langle \chi_{\mathbf{k}_f}^{V}(t_{j-1}) | V K_F^{V(+)}(t_{j-1}, t_{j-2}) V \cdots V K_F^{V(+)}(t_1, t_0) V | \chi_{\mathbf{k}_i}^{V}(t_0) \rangle$$
$$(10.52)$$

for $j > 1$ contains j times the interaction potential V and $(j-1)$ times the propagator $K_F^{V(+)}$, and

$$S_{\mathbf{k}_f,\mathbf{k}_i}^{B1} = -i \int_{-\infty}^{\infty} dt \, \langle \chi_{\mathbf{k}_f}^{V}(t) | V | \chi_{\mathbf{k}_i}^{V}(t) \rangle. \qquad (10.53)$$

We shall call $\bar{S}_{\mathbf{k}_f,\mathbf{k}_i}^{Bj}$ the term of order j of the Born series for the S-matrix element. We also define the Nth-order Born approximation to the S-matrix element $S_{\mathbf{k}_f,\mathbf{k}_i}$ as

$$S_{\mathbf{k}_f,\mathbf{k}_i}^{BN} = \delta(\mathbf{k}_f - \mathbf{k}_i) + \sum_{j=1}^{N} \bar{S}_{\mathbf{k}_f,\mathbf{k}_i}^{Bj}. \qquad (10.54)$$

In a similar way, we can write for the T-matrix element $T_{\mathbf{k}_f,\mathbf{k}_i}^{n}$ defined by Equation (10.29) the Born series

$$T_{\mathbf{k}_f,\mathbf{k}_i}^{n} = \sum_{j=1}^{\infty} \bar{T}_{\mathbf{k}_f,\mathbf{k}_i}^{n,Bj}, \qquad (10.55)$$

where the quantity $\bar{T}_{\mathbf{k}_f,\mathbf{k}_i}^{n,Bj}$ contains j times the interaction potential V. The Nth-order Born approximation to the T-matrix element $T_{\mathbf{k}_f,\mathbf{k}_i}^{n}$ is given by

$$T_{\mathbf{k}_f,\mathbf{k}_i}^{n,BN} = \sum_{j=1}^{N} \bar{T}_{\mathbf{k}_f,\mathbf{k}_i}^{n,Bj}. \qquad (10.56)$$

For example, the first Born approximation to $T_{\mathbf{k}_f,\mathbf{k}_i}^{n}$ is $T_{\mathbf{k}_f,\mathbf{k}_i}^{n,B1} = \bar{T}_{\mathbf{k}_f,\mathbf{k}_i}^{n,B1}$, and the second Born approximation is given by

$$T_{\mathbf{k}_f,\mathbf{k}_i}^{n,B2} = \bar{T}_{\mathbf{k}_f,\mathbf{k}_i}^{n,B1} + \bar{T}_{\mathbf{k}_f,\mathbf{k}_i}^{n,B2}. \qquad (10.57)$$

Using Equations (10.28), (10.51) and (10.55), we also find that

$$\bar{S}^{Bj}_{\mathbf{k}_f,\mathbf{k}_i} = -2\pi i \sum_{n=-\infty}^{\infty} \delta(E_{k_f} - E_{k_i} - n\omega)\bar{T}^{n,Bj}_{\mathbf{k}_f,\mathbf{k}_i}. \tag{10.58}$$

Explicit expressions for the terms $\bar{T}^{n,Bj}_{\mathbf{k}_f,\mathbf{k}_i}$ can be obtained in two equivalent ways. The first one is to use Equation (10.52) and employ the relation (10.18) to perform the time integrations. The second one is to solve the Floquet–Lippmann–Schwinger equations (10.49) by iteration, starting with the matrix element $V^n_{\mathbf{k}_f,\mathbf{k}_i}$ as the initial approximation. We find in this way that in first Born approximation ($j = 1$) we have

$$\bar{T}^{n,B1}_{\mathbf{k}_f,\mathbf{k}_i}(\mathbf{\Delta}) = J_n(\mathbf{\Delta}\cdot\boldsymbol{\alpha}_0)\langle\mathbf{k}_f|V|\mathbf{k}_i\rangle, \tag{10.59}$$

where we recall that $\mathbf{\Delta} = \mathbf{k}_i - \mathbf{k}_f$ is the wave vector transfer of the collision and

$$\langle\mathbf{k}_f|V|\mathbf{k}_i\rangle = (2\pi)^{-3}\int \exp(i\mathbf{\Delta}\cdot\mathbf{r})V(\mathbf{r})d\mathbf{r}. \tag{10.60}$$

Defining the Fourier transform of the potential $V(\mathbf{r})$ to be

$$\hat{V}(\mathbf{q}) = (2\pi)^{-3/2}\int \exp(-i\mathbf{q}\cdot\mathbf{r})V(\mathbf{r})d\mathbf{r}, \tag{10.61}$$

we have

$$\langle\mathbf{k}_f|V|\mathbf{k}_i\rangle = (2\pi)^{-3/2}\hat{V}(-\mathbf{\Delta}) \tag{10.62}$$

and hence

$$\bar{T}^{n,B1}_{\mathbf{k}_f,\mathbf{k}_i}(\mathbf{\Delta}) = (2\pi)^{-3/2}J_n(\mathbf{\Delta}\cdot\boldsymbol{\alpha}_0)\hat{V}(-\mathbf{\Delta}). \tag{10.63}$$

For $j \geq 2$, we have

$$\bar{T}^{n,Bj}_{\mathbf{k}_f,\mathbf{k}_i}(\mathbf{\Delta}) = (2\pi)^{-3j/2}\sum_{L_1=-\infty}^{\infty}\cdots\sum_{L_{j-1}=-\infty}^{\infty}\int d\mathbf{k}_1\cdots\int d\mathbf{k}_{j-1}$$

$$\times\frac{J_{L_{j-1}}(\mathbf{\Delta}_{\mathbf{k}_f,\mathbf{k}_{j-1}}\cdot\boldsymbol{\alpha}_0)\hat{V}(-\mathbf{\Delta}_{\mathbf{k}_f,\mathbf{k}_{j-1}})\cdots J_{L_1}(\mathbf{\Delta}_{\mathbf{k}_1,\mathbf{k}_i}\cdot\boldsymbol{\alpha}_0)\hat{V}(-\mathbf{\Delta}_{\mathbf{k}_1,\mathbf{k}_i})}{(E_{k_i} + L_{j-1}\omega - E_{k_{j-1}} + i\epsilon)\cdots(E_{k_i} + L_1\omega - E_{k_1} + i\epsilon)},$$

$$\epsilon \to 0^+, \tag{10.64}$$

with $E_{k_s} = k_s^2/2$ and $s = 1, 2, \ldots, j - 1$.

A Born series similar to that given by Equation (10.55) for the transition matrix element $T^n_{\mathbf{k}_f,\mathbf{k}_i}$ may also be defined for the scattering amplitude $f^n_{\mathbf{k}_f,\mathbf{k}_i}$. Using Equations (10.39) and (10.55), we have

$$f^n_{\mathbf{k}_f,\mathbf{k}_i} = \sum_{j=1}^{\infty} \bar{f}^{n,Bj}_{\mathbf{k}_f,\mathbf{k}_i} \tag{10.65}$$

with

$$\bar{f}^{n,Bj}_{\mathbf{k}_f,\mathbf{k}_i} = -(2\pi)^2 \bar{T}^{n,Bj}_{\mathbf{k}_f,\mathbf{k}_i}. \tag{10.66}$$

We also define the Nth-order Born approximation to the scattering amplitude as

$$f^{n,BN}_{\mathbf{k}_f,\mathbf{k}_i} = \sum_{j=1}^{N} \bar{f}^{n,Bj}_{\mathbf{k}_f,\mathbf{k}_i}. \tag{10.67}$$

Thus, for example, the first Born approximation to $f^n_{\mathbf{k}_f,\mathbf{k}_i}$ is $f^{n,B1}_{\mathbf{k}_f,\mathbf{k}_i} = \bar{f}^{n,B1}_{\mathbf{k}_f,\mathbf{k}_i}$ and the second Born approximation is $f^{n,B2}_{\mathbf{k}_f,\mathbf{k}_i} = \bar{f}^{n,B1}_{\mathbf{k}_f,\mathbf{k}_i} + \bar{f}^{n,B2}_{\mathbf{k}_f,\mathbf{k}_i}$.

Although no detailed studies of this problem have been made, it is expected, on the basis of field-free potential scattering studies [47], that the Born series for laser-assisted potential scattering converges at all impact energies when the potential $V(\mathbf{r})$ is weak enough, namely when the associated potential $-|V(\mathbf{r})|$ (which is always attractive) does not support any bound states. The Born series is also expected to be convergent at incident electron energies that are sufficiently high with respect to the potential energy.

10.2.3 First Born approximation

Let us now examine in more detail the *first Born approximation* for laser-assisted potential scattering. We begin by observing that, for field-free potential scattering, the first Born T-matrix element $T^{B1}_{\mathbf{k}_f,\mathbf{k}_i}(\mathbf{\Delta})$ and scattering amplitude $f^{B1}_{\mathbf{k}_f,\mathbf{k}_i}(\mathbf{\Delta})$ are given, respectively, by

$$T^{B1}_{\mathbf{k}_f,\mathbf{k}_i}(\mathbf{\Delta}) = \langle \mathbf{k}_f | V | \mathbf{k}_i \rangle$$
$$= (2\pi)^{-3/2} \hat{V}(-\mathbf{\Delta}) \tag{10.68}$$

and

$$f^{B1}_{\mathbf{k}_f,\mathbf{k}_i}(\mathbf{\Delta}) = -(2\pi)^2 T^{B1}_{\mathbf{k}_f,\mathbf{k}_i}(\mathbf{\Delta}). \tag{10.69}$$

Using Equations (10.39), (10.63), (10.68) and (10.69), we can write the corresponding laser-assisted quantities as

$$T^{n,B1}_{\mathbf{k}_f,\mathbf{k}_i}(\mathbf{\Delta}) = J_n(\mathbf{\Delta} \cdot \boldsymbol{\alpha}_0) T^{B1}_{\mathbf{k}_f,\mathbf{k}_i}(\mathbf{\Delta}) \tag{10.70}$$

and

$$f^{n,B1}_{\mathbf{k}_f,\mathbf{k}_i}(\mathbf{\Delta}) = J_n(\mathbf{\Delta} \cdot \boldsymbol{\alpha}_0) f^{B1}_{\mathbf{k}_f,\mathbf{k}_i}(\mathbf{\Delta}), \tag{10.71}$$

and we see that in these formulae the laser-field-dependent part factorizes out in the form of the Bessel function $J_n(\mathbf{\Delta} \cdot \boldsymbol{\alpha}_0)$.

It follows from Equations (10.38) and (10.71) that the differential cross section for laser-assisted electron scattering by a potential is given in first Born approximation by

$$\frac{d\sigma^{n,\text{B1}}}{d\Omega} = \frac{k_f(n)}{k_i} J_n^2(\boldsymbol{\Delta} \cdot \boldsymbol{\alpha}_0) \frac{d\sigma^{\text{B1}}}{d\Omega}(\boldsymbol{\Delta}), \tag{10.72}$$

where

$$\frac{d\sigma^{\text{B1}}}{d\Omega}(\boldsymbol{\Delta}) = \left| f_{\mathbf{k}_f,\mathbf{k}_i}^{\text{B1}}(\boldsymbol{\Delta}) \right|^2 \tag{10.73}$$

is the first Born differential cross section for field-free potential scattering. The remarkably simple result (10.72), valid if the first Born approximation provides an accurate description of the electron scattering by the potential $V(\mathbf{r})$, was first obtained by Bunkin and Fedorov [7]. We see that it contains three factors: the phase-space factor $k_f(n)/k_i$, the square of a Bessel function, which depends on the laser parameters through $\boldsymbol{\alpha}_0$, and the field-free first Born differential cross section, which is independent of the laser parameters. It is worth noting that the Bessel function $J_n(\boldsymbol{\Delta} \cdot \boldsymbol{\alpha}_0)$ contains the effect of the laser field in a non-perturbative way.

It is instructive to examine two particular cases of the Bunkin–Fedorov result (10.72). The first one is when

$$E_{k_i} = k_i^2/2 \gg n\omega \tag{10.74}$$

for all relevant values of n. Then, it follows from Equation (10.42) that $k_f(n)/k_i \simeq 1$, and the Bunkin–Fedorov formula (10.72) reduces to

$$\frac{d\sigma^{n,\text{B1}}}{d\Omega} \simeq J_n^2(\boldsymbol{\Delta} \cdot \boldsymbol{\alpha}_0) \frac{d\sigma^{\text{B1}}}{d\Omega}(\boldsymbol{\Delta}). \tag{10.75}$$

Using this result and setting $x = \boldsymbol{\Delta} \cdot \boldsymbol{\alpha}_0$, we deduce, with the help of the sum rule [46]

$$\sum_{n=-\infty}^{\infty} J_n^2(x) = 1, \tag{10.76}$$

that the first Born differential cross sections for laser-assisted potential scattering satisfy the approximate sum rule

$$\sum_n \frac{d\sigma^{n,\text{B1}}}{d\Omega} \simeq \frac{d\sigma^{\text{B1}}}{d\Omega}, \tag{10.77}$$

where the sum on n runs over all relevant values such that $k_f(n) \geq 0$.

A second interesting special case of the Bunkin–Fedorov formula (10.72) is the weak-field limit $\mathcal{E}_0 \ll 1$ so that, for fixed ω, one has $\alpha_0 \ll 1$. To lowest order in α_0, it is then found that the LOPT result for the quantity $f_{\mathbf{k}_f,\mathbf{k}_i}^{n,\mathrm{B1}}(\boldsymbol{\Delta})$ is given by

$$f_{\mathbf{k}_f,\mathbf{k}_i}^{n,\mathrm{B1}}(\boldsymbol{\Delta}) = \frac{(\boldsymbol{\Delta}\cdot\boldsymbol{\alpha}_0)^{|n|}}{2^{|n|}|n|!} f_{\mathbf{k}_f,\mathbf{k}_i}^{\mathrm{B1}}(\boldsymbol{\Delta}) \begin{cases} 1, & n > 0 \\ (-1)^n, & n < 0 \end{cases} + \mathcal{O}(\alpha_0^{|n|+2}), \quad (10.78)$$

where the factor 1 must be taken for absorption $(n > 0)$ and $(-1)^n$ for emission $(n < 0)$. The corresponding LOPT differential cross sections are given by

$$\frac{d\sigma^{n,\mathrm{B1}}}{d\Omega} = \frac{k_f(n)}{k_i} \frac{(\boldsymbol{\Delta}\cdot\boldsymbol{\alpha}_0)^{2|n|}}{2^{2|n|}(|n|!)^2} \frac{d\sigma^{\mathrm{B1}}}{d\Omega}(\boldsymbol{\Delta}). \quad (10.79)$$

We note that, since $\alpha_0 = \mathcal{E}_0/\omega^2$, these differential cross sections, as well as the corresponding total cross sections (integrated over the scattering angles), are proportional to $I^{|n|}$. Using Equations (10.42) and (10.79), we remark that, in the limit $\omega \to 0$ (so that $k_f(n) \to k_i$), the LOPT differential cross sections and the corresponding total cross sections are proportional to $\omega^{-4|n|}$, a result mentioned in Section 10.1. We also note that the LOPT differential cross sections (10.79) have a maximum when the polarization vector $\hat{\boldsymbol{\epsilon}}$ (and hence $\boldsymbol{\alpha}_0$) is parallel to the wave vector transfer $\boldsymbol{\Delta}$ and vanish when $\hat{\boldsymbol{\epsilon}}$ is perpendicular to $\boldsymbol{\Delta}$.

According to the LOPT result (10.79), the first Born laser-assisted differential and total cross sections should grow with the intensity like $I^{|n|}$. We note, however, from the Bunkin–Fedorov result (10.72) and the behavior of $J_n^2(x)$, where $x = \boldsymbol{\Delta}\cdot\boldsymbol{\alpha}$, that this is only true if the weak-field condition $\alpha_0 \ll 0$ is satisfied. In contrast, for large $|x|$, the function $J_n^2(x)$ oscillates with an amplitude falling off slowly to zero as $|x| \to \infty$. Indeed, for $|n|$ fixed and $|x| \to \infty$, one has

$$J_{|n|}(x) \to \left(\frac{2}{\pi x}\right)^{1/2} \cos\left(x - \frac{|n|\pi}{2} - \frac{\pi}{4}\right). \quad (10.80)$$

Let us now return to the Bunkin–Fedorov result (10.72) which, as we have emphasized above, contains the effect of the laser field in a non-perturbative way. On the other hand, the Bunkin–Fedorov formula is obtained by treating the electron–target interaction potential $V(\mathbf{r})$ to first order of perturbation theory by using the first Born approximation. It yields accurate results only if the first Born approximation gives a good description of field-free electron scattering by the potential $V(\mathbf{r})$. This means that the potential $-|V(\mathbf{r})|$ must be weak enough (so that it has no bound states) or that the electron kinetic energy is sufficiently large compared to its potential energy [47]. It should be noted, however, that for field-free potential scattering the Born series for the scattering amplitude converges towards its first Born term $f_{\mathbf{k}_f,\mathbf{k}_i}^{\mathrm{B1}}$ $(= \bar{f}_{\mathbf{k}_f,\mathbf{k}_i}^{\mathrm{B1}})$ for large k_i (i.e. large incident energies E_{k_i}) and for all momentum transfers (i.e. for all scattering angles) only for a limited class of potentials [48]. This class includes Yukawa (screened Coulomb) potentials of the form

$$V(r) = V_0 \frac{\exp(-\beta r)}{r} \tag{10.81}$$

as well as arbitrary superpositions of Yukawa potentials and also related potentials of the form

$$V(r) = \sum_{i=1}^{M} V_{0,i} r^{n_i} \exp(-\beta_i r), \tag{10.82}$$

where $n_1 = -1$, $n_i \geq -1$ for $i \geq 2$ and $\beta_i > 0$.

10.2.4 Low-frequency approximation

A second limiting case for which simplifications occur is the *low-frequency* approximation (LFA), or *soft-photon* limit, such that $\omega \ll E_{k_i}$. In that limit, Kroll and Watson [8] have obtained the simple result

$$f_{\mathbf{k}_f,\mathbf{k}_i}^{n,\text{LFA}} = J_n(\mathbf{\Delta} \cdot \boldsymbol{\alpha}_0) f_{\mathbf{k}_f^*,\mathbf{k}_i^*}(E^*, \mathbf{\Delta}) \tag{10.83}$$

for the scattering amplitude corresponding to non-relativistic laser-assisted electron scattering by a finite range potential. Here $f_{\mathbf{k}_f^*,\mathbf{k}_i^*}(E^*, \mathbf{\Delta})$ is the *exact* field-free scattering amplitude for elastic scattering from the initial wave vector $\mathbf{k}_i^*$ to the final wave vector $\mathbf{k}_f^*$. The amplitude $f_{\mathbf{k}_f,\mathbf{k}_i}^{n,\text{LFA}}$ is assumed to vary smoothly in the energy range in the vicinity of E^*. The modifications required when this is not the case, for example when this energy range contains a resonance, will be considered later. The new wave vectors $\mathbf{k}_i^*$ and $\mathbf{k}_f^*$ are *shifted wave vectors* associated to the corresponding physical wave vectors $\mathbf{k}_i$ and $\mathbf{k}_f$ by the relations

$$\mathbf{k}_{i,f}^* = \mathbf{k}_{i,f} + \mathbf{s}, \tag{10.84}$$

where

$$\mathbf{s} = \frac{n\omega}{\mathbf{\Delta} \cdot \boldsymbol{\alpha}_0} \boldsymbol{\alpha}_0. \tag{10.85}$$

The wave vector transfer between the shifted wave vectors is equal to the wave vector transfer between the physical wave vectors defined by Equation (10.48). That is,

$$\mathbf{\Delta} = \mathbf{k}_i - \mathbf{k}_f = \mathbf{k}_i^* - \mathbf{k}_f^*. \tag{10.86}$$

It is readily verified that both shifted wave vectors $\mathbf{k}_i^*$ and $\mathbf{k}_f^*$ correspond to th same energy, which we have denoted by E^*. Thus, using Equations (10.30) an (10.48), we have

$$
\begin{aligned}
E^* = \frac{1}{2}\left(\mathbf{k}_i^*\right)^2 &= \frac{1}{2}\left(\mathbf{k}_f^*\right)^2 \\
&= E_{k_i} + \frac{\mathbf{k}_i \cdot \boldsymbol{\alpha}_0}{\boldsymbol{\Delta} \cdot \boldsymbol{\alpha}_0} n\omega + \frac{\alpha_0^2}{2\left(\boldsymbol{\Delta} \cdot \boldsymbol{\alpha}_0\right)^2}(n\omega)^2 \\
&= E_{k_i} + \mathbf{k}_i \cdot \mathbf{s} + \frac{1}{2}s^2.
\end{aligned} \tag{10.87}
$$

The amplitude $f_{\mathbf{k}_f^*,\mathbf{k}_i^*}(E^*, \boldsymbol{\Delta})$ appearing in Equation (10.83) is therefore completely defined. We note that it is *on the energy shell*, so that scattering in the presence of the laser field can be described in terms of scattering in the absence of the laser field. This is a generalization of a result derived by Low [49] in the *weak field-limit* for *single-photon processes*, which is know as the *Low theorem*.

Using Equations (10.38) and (10.83), one obtains the Kroll–Watson LFA differential cross section for non-relativistic laser-assisted electron scattering by a short-range potential:

$$
\frac{\mathrm{d}\sigma^{n,\mathrm{LFA}}}{\mathrm{d}\Omega} = \frac{k_f(n)}{k_i} J_n^2(\boldsymbol{\Delta} \cdot \boldsymbol{\alpha}_0) \frac{\mathrm{d}\sigma}{\mathrm{d}\Omega}(E^*, \boldsymbol{\Delta}), \tag{10.88}
$$

where

$$
\begin{aligned}
\frac{\mathrm{d}\sigma}{\mathrm{d}\Omega}(E^*, \boldsymbol{\Delta}) &\equiv \frac{\mathrm{d}\sigma}{\mathrm{d}\Omega}(\mathbf{k}_f^*, \mathbf{k}_i^*) \\
&= \left| f_{\mathbf{k}_f^*,\mathbf{k}_i^*}(E^*, \boldsymbol{\Delta}) \right|^2
\end{aligned} \tag{10.89}
$$

is the exact elastic differential cross section corresponding to the on-the-energy-shell field-free electron scattering transition $\mathbf{k}_i^* \to \mathbf{k}_f^*$.

The Kroll–Watson formula (10.88) is remarkable since it is non-perturbative in both the electron interaction with the laser field and in the electron–target interaction potential $V(\mathbf{r})$. This is in contrast with the Bunkin–Fedorov result (10.72), which includes the electron interaction with the laser field in a non-perturbative way, but treats the electron–target interaction potential $V(\mathbf{r})$ in first Born approximation. We also note that the Kroll–Watson formula reduces to the Bunkin–Fedorov result when the first Born approximation can be used to calculate the field-free differential cross section.

It is worth noting that the vector $\mathbf{s}$ defined by Equation (10.85) becomes infinite when $n \neq 0$ and $\boldsymbol{\Delta} \cdot \boldsymbol{\alpha}_0 = 0$. The latter condition implies that the wave vector transfer $\boldsymbol{\Delta}$ is perpendicular to the laser polarization vector $\hat{\boldsymbol{\epsilon}}$. We see that when $n \neq 0$ and $\boldsymbol{\Delta} \cdot \boldsymbol{\alpha}_0 = 0$, the shifted wave vectors $\mathbf{k}_i^*$ and $\mathbf{k}_f^*$ become infinite. Moreover, under these conditions the Bessel function $J_n(\Delta \cdot \alpha_0)$ vanishes, and so does the Kroll–Watson LFA differential cross section (10.88). Kroll and Watson [8] have discussed this effect in classical terms.

In what follows, we shall call *critical scattering geometries* those for which $\Delta \cdot \alpha_0 = 0$. Consider, for example, the geometrical arrangement in which the laser polarization vector $\hat{\epsilon}$ is parallel to the incident electron wave vector $\mathbf{k}_i$. As a result, for photon absorption ($n = 1, 2, 3, \ldots$) the wave vector transfer Δ is perpendicular to $\hat{\epsilon}$ (and therefore to α_0) at the particular scattering angle θ_0 given by

$$\theta_0 = \cos^{-1}[k_i / k_f(n)], \quad n > 0. \tag{10.90}$$

We also note that in another geometrical arrangement such that the laser polarization vector $\hat{\epsilon}$ bisects the scattering angle θ, the quantity $\Delta \cdot \alpha_0$ nearly vanishes for all scattering angles. We will see below that, while the Kroll–Watson approximation is quite accurate near the critical geometries, the exact scattering cross sections do not vanish there.

The original derivation of Equation (10.88) given by Kroll and Watson [8] is based on a solution of the time-dependent Lippmann–Schwinger equation (10.21), obtained by summing formally the Born series in the low-frequency limit. Since then, many proofs of the Kroll–Watson formula have been given, either within the semi-classical theory used here [6, 33, 50, 51] or in quantum electrodynamics [52, 53], where the Kroll–Watson formula is related to a result of Bloch and Nordsieck [54, 55] for soft-photon spontaneous bremsstrahlung. In what follows, we derive the Kroll–Watson formula (10.88) by following the method of Kylstra and Joachain [6], which is based on the Floquet–Lippmann–Schwinger equations (FLSE) obtained above.

We start with the FLSE (10.49), which we extend off the energy shell, to write

$$T_{\mathbf{k}_f, \mathbf{k}_i}^{(+)n}(E) = V_{\mathbf{k}_f, \mathbf{k}_i}^n + \sum_{r=-\infty}^{\infty} \int d\mathbf{k} \, V_{\mathbf{k}_f, \mathbf{k}}^{n-r} \frac{1}{E + r\omega - E_k + i\epsilon} T_{\mathbf{k}, \mathbf{k}_i}^{(+)r}(E). \tag{10.91}$$

Using Equations (10.20) and (10.29), the off-the-energy-shell T-matrix elements can be expressed as

$$T_{\mathbf{k}_f, \mathbf{k}_i}^{(+)n}(E) = \sum_{m=-\infty}^{\infty} \langle u_{\mathbf{k}_f, m-n} | V | F_{\mathbf{k}_i, m}^{(+)}(E) \rangle$$

$$= \sum_{m=-\infty}^{\infty} \langle u_{\mathbf{k}_f, m-n} | T^{(+)}(E) | u_{\mathbf{k}_i, m} \rangle$$

$$= \sum_{m=-\infty}^{\infty} J_{m-n}(\mathbf{k}_f \cdot \boldsymbol{\alpha}_0) J_m(\mathbf{k}_i \cdot \boldsymbol{\alpha}_0) \langle \mathbf{k}_f | T^{(+)}(E) | \mathbf{k}_i \rangle$$

$$= J_n(\boldsymbol{\Delta} \cdot \boldsymbol{\alpha}_0) \langle \mathbf{k}_f | T^{(+)}(E) | \mathbf{k}_i \rangle, \tag{10.92}$$

so that, instead of working with the T-matrix elements evaluated with the "dressed" plane waves, we use those evaluated with plane wave states. With this result, we obtain, after dividing by $J_n(\boldsymbol{\Delta} \cdot \boldsymbol{\alpha}_0)$, the following form of the FLSE:

$$\langle \mathbf{k}_f | T^{(+)}(E) | \mathbf{k}_i \rangle = \langle \mathbf{k}_f | V | \mathbf{k}_i \rangle$$

$$+ \frac{1}{J_n(\boldsymbol{\Delta} \cdot \boldsymbol{\alpha}_0)} \sum_{r=-\infty}^{\infty} \int d\mathbf{k} \, \langle \mathbf{k}_f | V | \mathbf{k} \rangle \frac{J_{n-r}(\boldsymbol{\Delta}_{\mathbf{k}_f,\mathbf{k}} \cdot \boldsymbol{\alpha}_0) J_r(\boldsymbol{\Delta}_{\mathbf{k},\mathbf{k}_i} \cdot \boldsymbol{\alpha}_0)}{E + r\omega - E_k + i\epsilon}$$

$$\times \langle \mathbf{k} | T^{(+)}(E) | \mathbf{k}_i \rangle. \tag{10.93}$$

For low frequencies, we expand the energy denominator in the integration kernel in Equation (10.93) as

$$\frac{1}{E + r\omega - E_k + i\epsilon} = \frac{1}{E - E_k + i\epsilon} \left(1 - r\omega \frac{1}{E - E_k + i\epsilon} \right) + \mathcal{O}(\omega^2). \tag{10.94}$$

Calling $\langle \mathbf{k}_f | T^{(+)(N)}(E) | \mathbf{k}_i \rangle$ the T-matrix element which contains contributions to order ω^N, the zeroth-order approximation can be obtained by retaining only the first term on the right-hand side of Equation (10.94). Then the summation over the index r can be carried out with the result

$$\langle \mathbf{k}_f | T^{(+)(0)}(E) | \mathbf{k}_i \rangle = \langle \mathbf{k}_f | V | \mathbf{k}_i \rangle + \int d\mathbf{k} \, \langle \mathbf{k}_f | V | \mathbf{k} \rangle \frac{1}{E - E_k + i\epsilon}$$

$$\times \langle \mathbf{k} | T^{(+)(0)}(E) | \mathbf{k}_i \rangle, \tag{10.95}$$

so that the T-matrix element $\langle \mathbf{k}_f | T^{(+)(0)}(E_{k_i}) | \mathbf{k}_i \rangle$ simply satisfies the field-free Lippmann–Schwinger equation. To first order in ω, the low-frequency FLSE is

$$\langle \mathbf{k}_f | T^{(+)(1)}(E) | \mathbf{k}_i \rangle = \langle \mathbf{k}_f | V | \mathbf{k}_i \rangle + \int d\mathbf{k} \, \langle \mathbf{k}_f | V | \mathbf{k} \rangle \frac{1}{E - E_k + i\epsilon}$$

$$\times \langle \mathbf{k} | T^{(+)(1)}(E) | \mathbf{k}_i \rangle$$

$$- \frac{\omega}{J_n(\boldsymbol{\Delta} \cdot \boldsymbol{\alpha}_0)} \sum_{r=-\infty}^{\infty} \int d\mathbf{k} \, \langle \mathbf{k}_f | V | \mathbf{k} \rangle \tag{10.96}$$

$$\times \frac{J_{n-r}(\boldsymbol{\Delta}_{\mathbf{k}_f,\mathbf{k}} \cdot \boldsymbol{\alpha}_0) r J_r(\boldsymbol{\Delta}_{\mathbf{k},\mathbf{k}_i} \cdot \boldsymbol{\alpha}_0)}{(E - E_k + i\epsilon)^2} \langle \mathbf{k} | T^{(+)(1)}(E) | \mathbf{k}_i \rangle.$$

The Bessel function recurrence relation [46]

$$J_{m-1}(x) + J_{m+1}(x) = \frac{2m}{x} J_m(x) \tag{10.97}$$

and the Bessel function sum rule (10.46) can now be used to obtain the sum rule

$$\sum_{m=-\infty}^{\infty} m J_{n-m}(x) J_m(y) = \frac{ny}{x-y} J_n(x-y). \tag{10.98}$$

Using this result to carry out the sum over the index r yields

$$\langle \mathbf{k}_f | T^{(+)(1)}(E) | \mathbf{k}_i \rangle = \langle \mathbf{k}_f | V | \mathbf{k}_i \rangle$$
$$+ \int d\mathbf{k} \langle \mathbf{k}_f | V | \mathbf{k} \rangle \left(\frac{1}{E - E_k + i\epsilon} + \frac{\mathbf{\Delta}_{\mathbf{k},\mathbf{k}_i} \cdot \mathbf{s}}{(E - E_k + i\epsilon)^2} \right) \langle \mathbf{k}_i | T^{(+)(1)}(E) | \mathbf{k}_a \rangle, \tag{10.99}$$

where, from Equation (10.85),

$$\mathbf{\Delta}_{\mathbf{k},\mathbf{k}_i} \cdot \mathbf{s} = n\omega \frac{\mathbf{\Delta}_{\mathbf{k},\mathbf{k}_i} \cdot \boldsymbol{\alpha}_0}{\mathbf{\Delta} \cdot \boldsymbol{\alpha}_0}. \tag{10.100}$$

Provided that $|\mathbf{\Delta} \cdot \boldsymbol{\alpha}_0| \gg n$, we can rewrite Equation (10.99) as

$$\langle \mathbf{k}_f | T^{(+)(1)}(E) | \mathbf{k}_i \rangle = \langle \mathbf{k}_f | V | \mathbf{k}_i \rangle$$
$$+ \int d\mathbf{k} \langle \mathbf{k}_f | V | \mathbf{k} \rangle \frac{1}{E - E_k - \mathbf{\Delta}_{\mathbf{k},\mathbf{k}_i} \cdot \mathbf{s} + i\epsilon} \langle \mathbf{k} | T^{(+)(1)}(E) | \mathbf{k}_i \rangle, \tag{10.101}$$

so that

$$\langle \mathbf{k}_f | T^{(+)(1)}(E_{k_i}) | \mathbf{k}_i \rangle = \langle \mathbf{k}_f | V | \mathbf{k}_i \rangle$$
$$+ 2 \int d\mathbf{k} \langle \mathbf{k}_f | V | \mathbf{k} \rangle \frac{1}{|\mathbf{k}_i + \mathbf{s}|^2 - |\mathbf{k} + \mathbf{s}|^2 + i\epsilon} \langle \mathbf{k} | T^{(+)(1)}(E_{k_i}) | \mathbf{k}_i \rangle. \tag{10.102}$$

In order to interpret this equation, we use Equation (10.60) to note that

$$\langle \mathbf{k}_f | V | \mathbf{k}_i \rangle = \langle \mathbf{k}_f + \mathbf{s} | V | \mathbf{k}_i + \mathbf{s} \rangle \tag{10.103}$$

and we shift the intermediate wave vector according to $\mathbf{k}' = \mathbf{k} - \mathbf{s}$. Equation (10.102) now becomes

$$\langle \mathbf{k}_f | T^{(+)(1)}(E_{k_i}) | \mathbf{k}_i \rangle = \langle \mathbf{k}_f + \mathbf{s} | V | \mathbf{k}_i + \mathbf{s} \rangle$$
$$+ 2 \int d\mathbf{k}' \langle \mathbf{k}_f + \mathbf{s} | V | \mathbf{k}' \rangle \frac{1}{|\mathbf{k}_i + \mathbf{s}|^2 - (k')^2 + i\epsilon} \langle \mathbf{k}' - \mathbf{s} | T^{(+)(1)}(E_{k_i}) | \mathbf{k}_i \rangle. \tag{10.104}$$

It can now be concluded that the T-matrix element $\langle \mathbf{k}_f | T^{(+)(1)}(E_{k_i}) | \mathbf{k}_i \rangle$ satisfies the same Lippmann–Schwinger equation as the on-shell field-free T-matrix element with the wave vectors shifted by $\mathbf{s}$:

$$\langle \mathbf{k}_f | T^{(+)(1)}(E_{k_i}) | \mathbf{k}_i \rangle = \langle \mathbf{k}_f + \mathbf{s} | T^{(+)(0)}(E^*) | \mathbf{k}_i + \mathbf{s} \rangle$$
$$= \langle \mathbf{k}_f^* | T^{(+)(0)}(E^*) | \mathbf{k}_i^* \rangle. \tag{10.105}$$

The shifted wave vectors $\mathbf{k}^*_{i,f}$ and the shifted energy E^* are defined, respectively, by Equations (10.84) and (10.87). To verify the validity of the result (10.105), we insert it into Equation (10.104) to obtain

$$\langle \mathbf{k}^*_f | T^{(+)(0)}(E^*) | \mathbf{k}^*_i \rangle = \langle \mathbf{k}^*_f | V | \mathbf{k}^*_i \rangle + \int d\mathbf{k}' \, \langle \mathbf{k}^*_f | V | \mathbf{k}' \rangle \frac{1}{E^* - E_{k'} + i\epsilon}$$
$$\times \langle \mathbf{k}' | T^{(+)(0)}(E^*) | \mathbf{k}^*_i \rangle. \tag{10.106}$$

Using the result (10.105), together with Equations (10.39) and (10.92), we obtain for the laser-assisted scattering amplitude (to first order in ω) the Kroll–Watson LFA formula (10.83), which leads to the LFA differential cross section (10.88). It is worth stressing that these results are obtained by keeping the leading *two* terms (of order zero and one) in the expansion of the quantity $\langle \mathbf{k}_f | T^{(+)(N)} | \mathbf{k}_i \rangle$ in powers of ω. By retaining only the leading term (of order zero), we obtain a simplified, zeroth-order Kroll–Watson LFA formula for the laser-assisted scattering amplitude, namely

$$f^{n,\text{LFA0}}_{\mathbf{k}_f, \mathbf{k}_i} = J_n(\mathbf{\Delta} \cdot \boldsymbol{\alpha}_0) f_{\mathbf{k}_f, \mathbf{k}_i}(E_{k_i}, \mathbf{\Delta}), \tag{10.107}$$

which yields for the corresponding differential cross section the zeroth-order Kroll–Watson LFA result

$$\frac{d\sigma^{n,\text{LFA0}}}{d\Omega} = \frac{k_f(n)}{k_i} J^2_n(\mathbf{\Delta} \cdot \boldsymbol{\alpha}_0) \frac{d\sigma}{d\Omega}(E_{k_i}, \mathbf{\Delta}), \tag{10.108}$$

where

$$\frac{d\sigma}{d\Omega}(E_{k_i}, \mathbf{\Delta}) \equiv \frac{d\sigma}{d\Omega}(\mathbf{k}_f, \mathbf{k}_i)$$
$$= \left| f_{\mathbf{k}_f, \mathbf{k}_i}(E_{k_i}, \mathbf{\Delta}) \right|^2. \tag{10.109}$$

is the exact field-free differential cross section for the elastic electron scattering transition $\mathbf{k}_i \rightarrow \mathbf{k}_f$.

Let us assume that the condition (10.74) is satisfied, so that $k_f(n)/k_i \simeq 1$ for all relevant values of n. Then, using the Bessel function sum rule (10.76), it follows that the zeroth-order Kroll–Watson LFA differential cross sections (10.108) satisfy the sum rule [33]

$$\sum_n \frac{d\sigma^{n,\text{LFA0}}}{d\Omega} = \frac{d\sigma}{d\Omega}(\mathbf{k}_f, \mathbf{k}_i), \tag{10.110}$$

where again the sum on the index n runs over all values for which $k_f(n) \geq 0$.

In deriving the Kroll–Watson result (10.83), it was assumed that the finite-range potential $V(\mathbf{r})$ is such that the scattering amplitude $f_{\mathbf{k}^*_f \mathbf{k}^*_i}(E^*, \mathbf{\Delta})$ varies smoothly over the energy range in the vicinity of E^*. If this is not the case, as for example when the scattering amplitude contains a resonance in this energy range, the Kroll–Watson result must be modified. Krüger and Jung [33] have derived in this case the following formula, valid to lowest order in ω:

$$f_{\mathbf{k}_f,\mathbf{k}_i}^{n,\text{LFA0}} = \sum_j J_{n-j}(-\mathbf{k}_f \cdot \boldsymbol{\alpha}_0) J_j(\mathbf{k}_i \cdot \boldsymbol{\alpha}_0) f_{\mathbf{k}_f,\mathbf{k}_i}(E_{k_j}, \boldsymbol{\Delta}), \qquad (10.111)$$

where the off-the-energy-shell field-free scattering amplitude $f_{\mathbf{k}_f,\mathbf{k}_i}(E_{k_j}, \boldsymbol{\Delta})$ is evaluated at the energy $E_{k_j} = E_{k_i} + j\omega$. If one can assume that the field-free scattering amplitude is a slowly varying function about E_{k_i}, we can replace $f_{\mathbf{k}_f,\mathbf{k}_i}(E_{k_j}, \boldsymbol{\Delta})$ by $f_{\mathbf{k}_f,\mathbf{k}_i}(E_{k_i}, \boldsymbol{\Delta})$. Then, using the Bessel function sum rule (10.46), it is seen that the Krüger and Jung formula (10.111) reduces to the zeroth-order Kroll–Watson LFA result (10.107). Higher-order corrections in ω have been calculated by Mittleman [56] and Milosevic and Krstic [57]. The accuracy of Equation (10.111) has been tested by Jung and Krüger [34] by performing multichannel calculations for a square well potential. They found that the soft-photon approximation for the free–free scattering amplitude accounts for the occurrence of isolated resonances by making use only of knowledge of the scattering amplitude in the absence of the laser field. They also showed that if the laser wavelength is long compared to the range of the interaction potential $V(\mathbf{r})$, the agreement of the soft-photon approximation (10.111) with accurate model calculations is excellent. Using this result, Jung and Taylor [58] have shown that, for laser-assisted electron scattering by a one-dimensional potential, one can suppress the non-resonant background in the scattering cross sections and isolate the contributions that vary rapidly with the incoming electron energy. In this way, it should become possible to measure pure Breit–Wigner peaks and threshold effects, even if the interferences between the resonances and the non-resonant background in field-free collisions is strong.

Further generalizations of the Kroll–Watson formula (10.83) have been made to allow for arbitrary polarization of the monochromatic laser field and for non-dipole effects [52,59,60]. Off-the-energy-shell effects have been investigated either within the framework of the impulse approximation [61–63] or by using the second Born approximation [64–67]. A relativistic generalization of the Kroll–Watson result (10.83) using Dirac–Volkov waves has been given by Kaminski [68]. Relativistic potential scattering in a laser field has also been considered by Rosenberg [69] and by Rosenberg and Zhou [70,71].

A detailed comparison between the Kroll–Watson LFA and an "exact" solution of the FLSE, obtained by performing a partial wave analysis, has been made by Kylstra and Joachain [5, 6] for laser-assisted low-energy electron scattering by various finite-range potentials in a CO_2 laser field ($\omega = 0.0043$ a.u.). As an example, we compare in Fig. 10.3 the FLSE results (solid lines) and the LFA results (broken lines) for the laser-assisted differential cross section corresponding to the elastic scattering

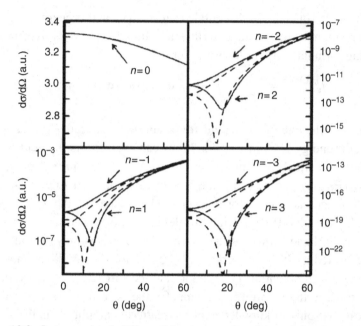

Figure 10.3. Laser-assisted differential cross section (in a.u.) for the scattering of 3.4 eV electrons by the static potential (10.112) felt by an incident electron in the field of a hydrogen atom, accompanied by the absorption ($n > 0$) or emission ($n < 0$) of one, two and three photons, as a function of the scattering angle θ (in degrees). Also shown in the upper-left corner is the differential cross section corresponding to laser-assisted scattering without net absorption or emission of photons ($n = 0$). The laser field is spatially homogeneous, monochromatic and linearly polarized. The polarization vector $\hat{\boldsymbol{\epsilon}}$, which defines the z-axis, is parallel to the incident electron wave vector $\mathbf{k}_i$. The laser field angular frequency is $\omega = 0.0043$ a.u., and its intensity is $I = 1.2 \times 10^5\,\mathrm{W\,cm^{-2}}$, so that $\alpha_0 = 0.1$ a.u. Solid lines: FLSE results. Broken lines: LFA results. For $n = 0$ the FLSE and LFA results cannot be distinguished. (From N. J. Kylstra and C. J. Joachain, *Phys. Rev. A* **58**, R26 (1998).)

of 3.4 eV electrons by atomic hydrogen in the static, no-exchange approximation. The corresponding static potential is given in atomic units by

$$V(r) = -\left(1 + \frac{1}{r}\right)\exp(-2r). \qquad (10.112)$$

It is apparent from Fig. 10.3 that, for the laser parameters and incident electron energy considered, the Kroll–Watson LFA remains valid to a good approximation. No differences can be discerned between the LFA and FLSE results for the case $n = 0$, corresponding to laser-assisted elastic scattering without net absorption or emission of photons. For laser-assisted "elastic" scattering accompanied by the absorption or emission of $|n|$ photons (with $|n| = 1, 2, 3$), and for scattering angles larger than about $40°$, the LFA is a very good approximation. Even for small-angle scattering, the LFA reproduces in a reasonable way the FLSE results, with the FLSE differential cross sections being somewhat larger. We recall that, since the laser polarization vector $\hat{\epsilon}$ is chosen to be parallel to the incident electron momentum $\mathbf{k}_i$, "critical scattering geometries" (such that $\mathbf{\Delta} \cdot \boldsymbol{\alpha}_0 = 0$) occur for photon absorption (with $n = 1, 2, 3, \ldots$ and $k_f(n) > k_i$) at scattering angles θ_0 such that $\cos \theta_0 = k_i / k_f(n)$. At these scattering angles, the LFA differential cross sections vanish, while the FLSE do not. We also note that the minima predicted by the FLSE are shifted to larger angles than those given by the LFA. Finally, we remark that the "exact" FLSE values [5] shown in Fig. 10.3 have been confirmed by R-matrix–Floquet calculations for laser-assisted electron–atomic hydrogen scattering [15], which will be discussed in Section 10.3.1.3.

Let us now examine how the Kroll–Watson LFA differential cross sections obtained from Equation (10.108) compare with experiment. A direct application of this formula does not lead to agreement with the experimental data of Weingartshofer *et al.* [29]. The most likely reason for this discrepancy is that, when comparing the theory with the data, one must take into account the actual experimental conditions. In particular, as we discussed in Section 2.1, one must include the spatial and temporal inhomogeneities of the laser field due to the laser pulse shape and the beam focusing. These characteristics can be included in the theory [24,45] by introducing an appropriate distribution function $P(\mathbf{r}, t, I)$ associated with the laser field intensity I, and by performing an adiabatic average of the results given by the Kroll–Watson LFA formula (10.108). Reasonable agreement is then obtained with the experimental values.

10.2.5 High-frequency approximation

A third case for which simplifications occur in laser-assisted electron scattering by a potential is when the laser field is strong and has a high frequency (HF). This case has been investigated by Gavrila and colleagues [9, 72–74] using the high-frequency approach formulated in the Kramers–Henneberger (K–H) frame discussed in Section 7.1.

Let us start from Equation (7.1), which is the set of coupled Floquet equations written in the accelerated (K–H) frame. Using atomic units,

$$\left(E_n - H_0^A \right) F_n^A(\mathbf{E}^A; \mathbf{r}) = \sum_{\substack{s=-\infty \\ s \neq n}}^{\infty} V_{n-s}(\boldsymbol{\alpha}_0, \mathbf{r}) F_s^A(\mathbf{E}^A; \mathbf{r}), \qquad (10.113)$$

where $E_n = \mathsf{E}^A + n\omega$, E^A is the quasi-energy of the state and the Hamiltonian

$$H_0^A = \frac{\mathbf{p}^2}{2} + V_0(\alpha_0, \mathbf{r}) \tag{10.114}$$

contains the static "dressed" potential $V_0(\alpha_0, \mathbf{r})$, given in the present case by

$$V_0(\alpha_0, \mathbf{r}) = (T)^{-1} \int_0^T V[\mathbf{r} + \boldsymbol{\alpha}_0 \sin(\omega t)] dt, \tag{10.115}$$

where $T = 2\pi/\omega$ is the laser field optical period. As discussed in Chapter 7, the quantities $V_n(\alpha_0, \mathbf{r})$ are the coefficients of the Fourier expansion of the oscillating potential $V[\mathbf{r} + \boldsymbol{\alpha}_0 \sin(\omega t)]$. We have also seen in Chapter 7 that the potential $V_0(\alpha_0, \mathbf{r})$ governs the atomic structure in a high-frequency laser field. In the present context, this potential determines the properties of the scattering states.

Let us consider the scattering solutions $u_{\mathbf{k}}^{(\pm)}(\mathbf{r})$ associated with the Hamiltonian H_0^A, which satisfy the Schrödinger equation

$$H_0^A u_{\mathbf{k}}^{(\pm)}(\mathbf{r}) = w_k^{(0)} u_{\mathbf{k}}^{(\pm)}(\mathbf{r}). \tag{10.116}$$

The solutions $u_{\mathbf{k}}^{(+)}(\mathbf{r})$ and $u_{\mathbf{k}}^{(-)}(\mathbf{r})$ exhibit *outgoing* (+) or *ingoing* (−) spherical wave behavior for $r \to \infty$.

Gavrila and Kaminski [9] have shown that in the high-frequency regime, such that

$$\omega \gg |w_0^{(0)}(\alpha_0)| \tag{10.117}$$

and

$$\omega \gg w_{k_i}^{(0)}, \tag{10.118}$$

where $|w_0^{(0)}(\alpha_0)|$ is the ionization potential of the ground state of the potential $V_0(\alpha_0, \mathbf{r})$, the leading term in ω^{-1} for the elastic scattering amplitude can be obtained from $u_{\mathbf{k}_i}^{(+)}(\mathbf{r})$. Similarly, to lowest non-vanishing order in ω^{-1}, the HF free–free transition amplitudes are given for $n \neq 0$ by

$$f_{\mathbf{k}_f, \mathbf{k}_i}^{n, \mathrm{HF}} = -(2\pi)^2 \langle u_{\mathbf{k}_f}^{(-)} | V_n | u_{\mathbf{k}_i}^{(+)} \rangle. \tag{10.119}$$

The conditions (10.117) and (10.118) imply that only absorption ($n > 0$) can occur, while emission is excluded.

This approach has been applied by Gavrila, Offerhaus and Kaminski [73] to analyze laser-assisted electron scattering by a Yukawa (screened Coulomb) potential at high frequencies. Offerhaus, Kaminski and Gavrila [72] and van de Ree, Kaminski and Gavrila [74] have studied the Coulomb case. We shall discuss their results in Section 10.2.6, which is devoted to the Coulomb scattering of electrons in the presence of a laser field.

10.2.6 Laser-assisted Coulomb scattering of electrons

Let us consider the laser-assisted scattering of electrons by the Coulomb potential

$$V_c(r) = -\frac{Z}{r}. \tag{10.120}$$

This problem is more complicated than for a finite-range potential. Indeed, since the Coulomb potential has an infinite range, the electron interaction with this potential can never be ignored, even for $r \to \infty$. As a result, the Gordon–Volkov wave functions discussed in Section 2.7 cannot be used to describe laser-dressed asymptotic electron states. A first approximation consists in using Coulomb-distorted Gordon–Volkov wave functions of the form (in the velocity gauge)

$$\chi_{\mathbf{k}}^{V,c}(\mathbf{r}, t) = (2\pi)^{-3/2} \exp\{i[\mathbf{k} \cdot (\mathbf{r} - \boldsymbol{\alpha}(t)) - E_k t - \eta_k \log(kr - \mathbf{k} \cdot \mathbf{r})]\}, \tag{10.121}$$

where $E_k = k^2/2$ and $\eta_k = -Z/k$ is the Sommerfeld parameter corresponding to the wave number k.

The full scattering wave function $\Psi_{\mathbf{k}_i}^{(+)}(\mathbf{r}, t)$ must also be modified to account for the presence of the Coulomb potential. To this end, let us denote by $\psi_{\mathbf{k}_i}^{c(+)}(Z, \mathbf{r})$ and $\psi_{\mathbf{k}_i}^{c(-)}(Z, \mathbf{r})$ the Coulomb wave function describing an electron of wave vector $\mathbf{k}_i$ and energy E_{k_i} moving in a Coulomb field $-Z/r$ and exhibiting an outgoing (+) or incoming (−) spherical wave asymptotic behavior. These Coulomb wave functions are given by Equation (3.140) and are normalized to a delta function in wave vector space according to Equation (3.139). A first approximation to the laser-modified Coulomb wave function, which we call $\Psi_{\mathbf{k}_i}^{V,c(+)}$, consists in writing

$$\Psi_{\mathbf{k}_i}^{V,c(+)}(\mathbf{r}, t) = \psi_{\mathbf{k}_i}^{c(+)}(Z, \mathbf{r}) \exp\left\{-i\left[\mathbf{k}_i \cdot \boldsymbol{\alpha}(t) + E_{k_i} t\right]\right\}. \tag{10.122}$$

Using the state (10.122) for the evaluation of the S-matrix element (10.23), and a Coulomb-distorted Gordon–Volkov state $\chi_{\mathbf{k}_f}^{V,c}(\mathbf{r}, t)$ of the form (10.121) with $\mathbf{k}$ replaced by $\mathbf{k}_f$ as the asymptotic final state, we obtain for laser-assisted Coulomb scattering of electrons the differential cross section

$$\frac{d\sigma^n}{d\Omega} = J_n^2(\boldsymbol{\Delta} \cdot \boldsymbol{\alpha}_0) \frac{d\sigma_c}{d\Omega}, \tag{10.123}$$

where $d\sigma_c/d\Omega$ is the exact field-free Coulomb differential cross section, given by [47] the Rutherford formula

$$\frac{d\sigma_c}{d\Omega} = |f_c(\theta)|^2 = \frac{Z^2}{16 E_{k_i}^2 \sin^4(\theta/2)}. \tag{10.124}$$

In this formula, f_c is the field-free Coulomb scattering amplitude, given by

$$f_c(\theta) = -\frac{\eta_{k_i}}{2k_i \sin^2(\theta/2)} \exp\left\{-i\eta_{k_i} \log[\sin^2(\theta/2)]\right\} \exp(2i\sigma_0), \tag{10.125}$$

with

$$\exp(2i\sigma_0) = \frac{\Gamma(1+i\eta_{k_i})}{\Gamma(1-i\eta_{k_i})},$$
(10.126)

and Γ is Euler's gamma function.

A more detailed analysis of this problem has been given by Rosenberg [75] and Rosenberg and Zhou [76], who have proved that Equation (10.123) is valid to lowest order in ω/E_{k_i} provided the scattering angles close to the forward direction are excluded. Higher-order corrections to the low-frequency result (10.123), in particular at small scattering angles, have been investigated [69, 76–78]. We also mention earlier investigations of Coulomb scattering in a laser field by Banerji and Mittleman [79] and Saez and Mittleman [80], where the effect of truncating the Coulomb potential (10.120) at a finite distance was investigated.

The result (10.123) has been confirmed in the low-frequency and low-intensity limits by Gavrila, Maquet and Véniard [81, 82], who performed a calculation of stimulated two-photon FFT in a Coulomb potential. Two-photon FFT in a Coulomb potential has also been studied by Florescu and Djamo [83] and by Florescu and Florescu [84].

Rapoport [85] has derived approximate wave functions for a charged particle interacting with a Coulomb potential and a laser field. In the case of laser-assisted Coulomb scattering of electrons, Rapoport found a more accurate wave function than that given by Equation (10.122). This leads to corrections to the result (10.123) that become important when α_0 is larger than the impact parameter b, which corresponds to large scattering angles.

Electron scattering by a Coulomb potential in the presence of a strong laser field of high frequency has been studied by Offerhaus, Kaminski and Gavrila [72] and van de Ree, Kaminski and Gavrila [74] using the high-frequency Floquet theory discussed in Chapter 7 and Section 10.2.5. Since the dressed potential V_0 is a Coulomb potential modified at finite distances, it is useful to apply the two-potential formalism [47]. The differential cross section for laser-modified elastic scattering without net absorption or emission of photons ($n = 0$) can then be written in the form

$$\frac{d\sigma^0}{d\Omega} = \left| f_c(\theta) + \tilde{f}_{\mathbf{k}_f, \mathbf{k}_i}(\alpha_0, \theta) \right|^2.$$
(10.127)

The amplitude $\tilde{f}_{\mathbf{k}_f, \mathbf{k}_i}$ is due to the departure of V_0 from the Coulomb form due to the laser field.

As an example, we show in Fig. 10.4 the results obtained by van de Ree, Kaminski and Gavrila [74] for electron–proton scattering in a strong spatially homogeneous, monochromatic and linearly polarized high-frequency laser field. The corresponding Coulomb potential is therefore $V_c = -1/r$. The electrons were assumed to have

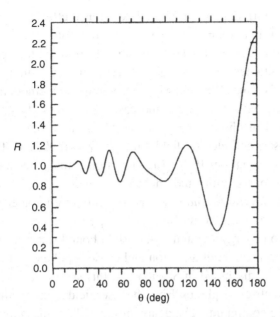

Figure 10.4. Ratio R defined by Equation (10.128), corresponding to laser-assisted electron–proton scattering, as a function of the scattering angle θ (in degrees). The laser field is spatially homogeneous, monochromatic and linearly polarized. The polarization vector $\hat{\epsilon}$, which defines the z-axis, is parallel to the incident electron wave vector $\mathbf{k}_i$. The incident electron energy is $E_{k_i} = 0.005$ a.u. and $\alpha_0 = 5$ a.u. (From J. van de Ree, J. Z. Kaminski and M. Gavrila, *Phys. Rev. A* **37**, 4536 (1988).)

the low energy $E_{k_i} = 0.005$ a.u., and the value $\alpha_0 = 5$ a.u. was chosen. The incident electron wave vector is parallel to the laser polarization vector $\hat{\epsilon}$, which defines the z-axis. The quantity displayed in Fig. 10.4 is

$$R = \frac{d\sigma^0/d\Omega}{d\sigma_c/d\Omega}, \tag{10.128}$$

which is the ratio of the differential cross section (10.127) for laser-modified Coulomb scattering (with $n = 0$) to the field-free Coulomb differential cross section given by the Rutherford formula (10.124), both differential cross sections being calculated for $Z = 1$.

We note from Fig. 10.4 that in the case of forward scattering, such that $\theta \to 0$, we have $R \to 1$. This is due to the fact that $|f_c| \to \infty$ as $\theta \to 0$, while $\tilde{f}$ remains finite. At small θ, R exhibits oscillations due to the interference between the two amplitudes f_c and $\tilde{f}$ in Equation (10.127), and the fact that the Coulomb amplitude f_c contains an oscillating θ-dependent phase factor, as seen from Equation (10.125). For large scattering angles, $\theta \geq 90°$, the Coulomb interference pattern is distorted and the ratio R may differ substantially from unity. Thus, we see that Coulomb scattering in the presence of a strong high-frequency laser field may deviate considerably from the Rutherford formula, and exhibits interference effects. In this regime of laser parameters, multiphoton FFT cross sections are small.

The Floquet equations for laser-assisted electron–proton scattering have been solved in the Kramers–Henneberger (K–H) frame by Dimou and Faisal [86, 87] and Collins and Csanak [88] using the close-coupling method. In particular, Dimou and Faisal [86] have considered the case of a circularly polarized monochromatic laser field of photon energy $\hbar\omega = 0.236$ a.u. $= 6.42$ eV (corresponding to an excimer laser) and electric-field strength $\mathcal{E}_0 = 0.005$ a.u. They calculated the ratio R given by Equation (10.128) as a function of the incident electron energy, for fixed directions of the initial and final electron wave vector. A remarkable result is the presence of several *resonances*. At a given laser photon energy (in the present case $\hbar\omega = 6.42$ eV) which matches the energy difference between the (positive) incident electron energy E_{k_i} and a (negative) bound state energy of the hydrogen atom, the electron can emit a photon and cause it to be captured temporarily into a bound state. As discussed in Section 4.4, this state is a decaying dressed state. As such, it will absorb a photon from the laser field, thereby allowing the electron to escape to the continuum. The delay introduced by this *capture–escape* process shows up as a resonance in the scattering signal, and the position and width of the resonance are determined by the position and width of the decaying dressed state. Dimou and Faisal also found that inverse bremsstrahlung cross sections for the absorption of one photon ($n = 1$) are dominated by capture–escape resonances.

Non-perturbative calculations of laser-assisted electron–proton scattering using the R-matrix–Floquet (RMF) theory of multiphoton processes have been performed by Dörr *et al.* [89]. The scattering amplitudes for "elastic" scattering accompanied by the transfer of $|n|$ photons between the atomic system (here the hydrogen atom) and the laser field is given according to the two-potential formula [47] as

$$f^n_{\mathbf{k}_f,\mathbf{k}_i}(\theta) = f_\text{c}(\theta)\delta_{n,0} + \hat{f}^n_{\mathbf{k}_f,\mathbf{k}_i}(\theta), \qquad (10.129)$$

where f_c is again the field-free Coulomb scattering amplitude, and $\hat{f}^n_{\mathbf{k}_f,\mathbf{k}_i}$ is the additional scattering amplitude due to the presence of the laser field. The corresponding differential cross section for "elastic" scattering accompanied by the transfer of n photons between the H atom and the laser field is

$$\frac{\text{d}\sigma^n}{\text{d}\Omega} = \left| f_\text{c}(\theta)\delta_{n,0} + \hat{f}^n_{\mathbf{k}_f,\mathbf{k}_i}(\theta) \right|^2. \qquad (10.130)$$

Dörr *et al.* have calculated such differential cross sections, as well as the corresponding total cross sections, for a spatially homogeneous, monochromatic and linearly polarized laser field. In agreement with Dimou and Faisal [86], they found that the laser field induces resonances due to the temporary capture of the projectile electron into atomic hydrogen bound states, accompanied by the exchange of

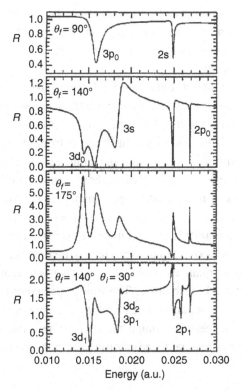

Figure 10.5. The ratio R defined by Equation (10.128), corresponding to laser-assisted electron–proton scattering, as a function of the incident electron energy (in a.u.). The laser field is spatially homogeneous, monochromatic and linearly polarized. The polarization vector $\hat{\boldsymbol{\epsilon}}$ lies along the z-axis. The laser field angular frequency is $\omega = 0.074$ a.u. and its intensity is $I = 10^{12}$ W cm^{-2}. The incident electron wave vector $\mathbf{k}_i$ is in the direction $\Omega_i \equiv (\theta_i, \phi_i = 0)$ and the scattered electron wave vector $\mathbf{k}_f$ is in the direction $\Omega_f \equiv (\theta_f, \phi_f = \phi)$, as shown in Fig. 10.2. For the first three panels, $\theta_i = 0$ while $\theta_f = 90°$, $140°$ and $175°$, respectively. For the lowest panel $\theta_i = 30°$, $\theta_f = 140°$ and $\phi = 0$. (From M. Dörr *et al.*, Journal of Physics B **28**, 3545 (1995).)

photons with the laser field. Structures corresponding to different sub-levels can also appear. This is illustrated in Fig. 10.5, where the ratio R defined by Equation (10.128) is shown for a spatially homogeneous, monochromatic and linearly polarized laser field of angular frequency $\omega = 0.074$ a.u. (corresponding to a dye laser of wavelength $\lambda = 616$ nm) and intensity $I = 10^{12}$ W cm^{-2}, the scattering geometry being defined by Fig. 10.2.

In the first three panels of Fig. 10.5, the incoming angle is taken to be $\theta_i = 0$, so that the polarization vector $\hat{\boldsymbol{\epsilon}}$ (which defines the z-axis) is parallel to $\mathbf{k}_i$, while the outgoing angle θ_f (which is equal to the scattering angle θ) has the values $\theta_f = 90°$, $140°$ and $175°$, respectively. The results are independent of the azimuthal angle $\phi_f (= \phi)$ due to cylindrical symmetry about the z-axis. We note the one-photon resonances with the $n = 3$ manifold (the s and d_0 angular momentum components are mixed, the dominant component being indicated) and the two-photon resonance with the $n = 2$ manifold.

If the incoming angle θ_i is chosen to be different from zero, the cylindrical symmetry around the z-axis is lost and states with the magnetic quantum number $m \neq 0$ contribute to the cross section. However, since space inversion is still preserved with a linearly polarized field, the contributions from m and $-m$ are equal. Resonances corresponding to states having the same values of $|m|$ appear at the same position with the same width. However, their combined effect on the differential cross section depends on the scattering geometry. In the lowest panel of Fig. 10.5 one has taken $\theta_i = 30°, \theta_f = 140°$ and $\phi_f (= \phi) = 0$. The $2p_1$ resonance is now visible, while the $n = 3$ manifold can exhibit up to six substructures. However, the 3s, $3p_1$ and $3d_2$ resonances overlap and can hardly be distinguished. The same is true for the $3d_0$ and $3d_1$ resonances. The little spike at an electron energy of 0.0188 a.u. is due to the narrow $3d_2$ resonance. The $n = 2$ resonances can be resolved much better than the $n = 3$ resonances because they are connected to the continuum by two photons and are therefore much narrower. The relatively large shifts of the states are due to near-resonant one-photon couplings between the $n = 2$ and the $n = 3$ manifolds.

10.3 Laser-assisted collisions of electrons with real atoms

Let us now consider collisions of electrons with real atoms in the presence of a laser field. Instead of being modeled by a center of force, real atoms have an internal structure which we shall now take into account. In fact, three types of interactions must be considered. Firstly, the interaction between the unbound electron(s) and the target atom takes place, as in the field-free case. Secondly, the laser field interacts with the unbound electron(s). Thirdly, the laser field interacts with the target atom and hence "dresses" the atomic target states.

In this section we shall first analyze laser-assisted *elastic* electron–atom scattering. Next, we shall discuss laser-assisted *inelastic* electron–atom collisions, and then *single ionization* (e, 2e) reactions in the presence of a laser field. We shall conclude this section by considering doubly resonant processes in laser-assisted electron–atom collisions, in which double poles of the S-matrix occur.

10.3.1 Laser-assisted "elastic" electron–atom collisions

In this section we study laser-assisted "elastic" collisions between non-relativistic electrons and target atoms. During these collisions, $|n|$ photons are transferred between the electron–atom system and the laser field. Thus, if $\mathbf{k}_i$ and $\mathbf{k}_f$ denote, respectively, the wave vectors of the unbound (free) electron before and after the collision and $E_{k_i} = \hbar^2 k_i^2/(2m)$ and $E_{k_f} = \hbar^2 k_f^2/(2m)$ are the corresponding energies, we have, from energy conservation,

$$E_{k_f} = E_{k_i} + n\hbar\omega \qquad (10.131)$$

with $n = 0, \pm 1, \pm 2, \ldots$ The case $n = 0$ corresponds to pure elastic scattering in the presence of the laser field. The cases $n < 0$ and $n > 0$ correspond to stimulated and inverse bremsstrahlung, respectively, with the atom remaining after the collision in its initial state, which we shall take to be the ground state $|0\rangle$.

The first studies of electron collisions with real atoms in the presence of a laser field were performed by Gersten and Mittleman [90] and Mittleman [91–94]. They chose atomic hydrogen as the target atom and treated the laser–atom interaction by using perturbation theory. They used the close-coupling Floquet method and the optical potential approach [47] to investigate the long-range behavior of the polarization potential due to the laser-induced dipole moment of the atom, giving rise to the "dressing" of atomic states by the laser field. Other early investigations concerning the dressing of atomic target states by the laser field in laser-assisted electron–atom scattering include the work of Zon [95], Deguchi, Taylor and Yaris [96], Beilin and Zon [97] and Golovinskii [98].

The importance of the laser–target interaction in laser-assisted electron–atom scattering was also discussed by Lami and Rahman [99, 100], who considered the scattering of fast electrons by atomic hydrogen and treated the interaction of the laser field with both the unbound electron and the target atom to first-order of perturbation theory. More detailed first-order perturbative calculations on single-photon FFT in atomic hydrogen were performed by Dubois, Maquet and Jetzke [101] and extended to two-photon FFT by Kracke *et al.* [102]. Also, within the framework of perturbation theory, circular dichroism in laser-assisted electron–atom scattering was studied by Manakov, Marmo and Volovich [103].

We emphasize at this point that, since the laser field cannot be too strong in order to avoid concomitant multiphoton ionization, the electric-field strength $\mathcal{E}_0$ must be such that $\mathcal{E}_0 \ll \mathcal{E}_a$, where $\mathcal{E}_a \simeq 5.1 \times 10^9$ V cm^{-1} is the atomic unit of electric-field strength. It is therefore in general a good approximation to treat the interaction of the laser field with the target atom to first order in perturbation theory, except for resonant cases which require a more elaborate treatment and will be considered at a later stage. On the other hand, we have seen in Section 10.2 that, even for moderately strong laser fields, it is inadequate to treat the interaction between the unbound electron(s) and the laser field in a perturbative way when the laser frequency is low and (or) the projectile electron is fast.

10.3.1.1 Semi-perturbative theory

We shall now discuss a semi-perturbative theory proposed by Byron and Joachai[10], which takes into account the above considerations and has been applied to variety of laser-assisted electron–atom processes. The theory deals with fast inci dent electrons ($E_{k_i} \geq 100\,\text{eV}$), so that the electron–atom interaction can be treate perturbatively by using the Born series [47]. The laser field is described as a spatiall homogeneous, monochromatic electric-field. For simplicity, we shall assume tha it is linearly polarized, so that, as in Equation (10.5), we have $\mathcal{E}(t) = \hat{\epsilon}\mathcal{E}_0 \sin(\omega t)$ The corresponding vector potential $\mathbf{A}(t)$ is given by Equation (10.6). According to the foregoing discussion, the interaction between this laser field and the unbound electron is treated exactly, as in the case of potential scattering, by using a Gordon-Volkov wave function. On the other hand, the laser–atom interaction is treated by first-order time-dependent perturbation theory. As we have seen above, this approximation is accurate provided that (i) $\mathcal{E}_0 \ll \mathcal{E}_a$, where $\mathcal{E}_a$ is the atomic unit of electric-field strength and (ii) the laser frequency is not close to resonance with an atomic transition frequency between the initial or final state of the atom (which are identical for elastic scattering) and an atomic intermediate state.

We shall now examine how the semi-perturbative approach described above can be used to analyze laser-assisted fast "elastic" electron–atom collisions, accompanied by the transfer of $|n|$ photons (with $n = 0, \pm 1, \pm 2, \ldots$) between the electron–atom system and the laser field [10, 104]. We begin by considering direct (no exchange) "elastic" collisions. We denote by $\mathbf{r}_0$ the position vector of the projectile electron and by $\mathbf{r}_1, \mathbf{r}_2, \ldots, \mathbf{r}_N$ the position vectors of the N atomic electrons. The target is assumed to be a neutral atom, so that $N = Z$. The Hamiltonian of the electron–atom system in the presence of the laser field can be written in the direct arrangement channel as

$$H(t) = H_{\mathrm{F}}(t) + H_{\mathrm{T}}(t) + V_{\mathrm{d}}, \tag{10.132}$$

where $H_{\mathrm{F}}(t)$ is the Hamiltonian of the "free" (unbound) electron in the presence of the laser field, $H_{\mathrm{T}}(t)$ is the Hamiltonian of the target atom in the presence of the laser field and V_{d} is the electron–atom interaction potential in the direct arrangement channel. We have explicitly

$$H_{\mathrm{F}}(t) = \frac{1}{2m}[\mathbf{p}_0 + e\mathbf{A}(t)]^2 \tag{10.133}$$

with $\mathbf{p}_0 = -i\hbar\nabla_{\mathbf{r}_0}$,

$$H_{\mathrm{T}}(t) = \sum_{i=1}^{N}\left[\frac{1}{2m}[\mathbf{p}_i + e\mathbf{A}(t)]^2 - \frac{Ze^2}{(4\pi\epsilon_0)r_i}\right] + \sum_{i<j=1}^{N}\frac{e^2}{(4\pi\epsilon_0)r_{ij}} \tag{10.134}$$

with $r_{ij} = |\mathbf{r}_i - \mathbf{r}_j|$ and

$$V_{\mathrm{d}} = -\frac{Ze^2}{(4\pi\epsilon_0)r_0} + \sum_{j=1}^{N} \frac{e^2}{(4\pi\epsilon_0)r_{0j}} \tag{10.135}$$

with $r_{0j} = |\mathbf{r}_0 - \mathbf{r}_j|$. In writing Equations (10.134) and (10.135) we have neglected spin effects (spin-orbit, spin-spin interactions) which are indeed very small for the light atoms (atomic hydrogen and helium) considered here.

The wave function corresponding to the non-relativistic motion of a "free" electron in the presence of a laser field described by the electric-field (10.5) is obtained by solving the TDSE

$$i\hbar\frac{\partial}{\partial t}\chi(\mathbf{r}_0, t) = H_{\mathrm{F}}(t)\chi(\mathbf{r}_0, t)$$

$$= \frac{1}{2m}[\mathbf{p}_0 + e\mathbf{A}(t)]^2\chi(\mathbf{r}_0, t), \tag{10.136}$$

where, according to Equation (10.6), $\mathbf{A}(t) = \hat{\boldsymbol{\epsilon}}A_0\cos(\omega t)$, with $A_0 = \mathcal{E}_0/\omega$.

To solve Equation (10.136), we perform the gauge transformation

$$\chi^{\mathrm{V}}(\mathbf{r}_0, t) = \exp\left[\frac{\mathrm{i}}{\hbar}\frac{e^2}{2m}\int_{-\infty}^{t}\mathbf{A}^2(t')\mathrm{d}t'\right]\chi(\mathbf{r}_0, t), \tag{10.137}$$

which gives for $\chi^{\mathrm{V}}(\mathbf{r}_0, t)$ the TDSE in the velocity gauge

$$i\hbar\frac{\partial}{\partial t}\chi^{\mathrm{V}}(\mathbf{r}_0, t) = \left[\frac{\mathbf{p}_0^2}{2m} + \frac{e}{m}\mathbf{A}\cdot\mathbf{p}_0\right]\chi^{\mathrm{V}}(\mathbf{r}_0, t). \tag{10.138}$$

This equation can be solved as in Section 2.7 to yield the Gordon–Volkov wave function in the velocity gauge (see Equations (2.167) and (10.14))

$$\chi_{\mathbf{k}}^{\mathrm{V}}(\mathbf{r}_0, t) = (2\pi)^{-3/2}\exp[\mathrm{i}(\mathbf{k}\cdot\mathbf{r}_0 - \mathbf{k}\cdot\boldsymbol{\alpha}_0\sin(\omega t) - E_kt/\hbar)], \tag{10.139}$$

where $\mathbf{k}$ is the electron wave vector and $E_k = \hbar^2k^2/(2m)$ is its kinetic energy. We recall that $\boldsymbol{\alpha}_0 = \hat{\boldsymbol{\epsilon}}\alpha_0$, where $\alpha_0 = e\mathcal{E}_0/(m\omega^2)$ is the excursion amplitude of the electron in the laser field, and we note that the Gordon–Volkov wave function (10.139) has been normalized to a delta function in wave vector space.

The next step consists in obtaining the "dressed" states of the target atom in the laser field. To this end, we must solve the TDSE

$$i\hbar\frac{\partial}{\partial t}\Phi(X, t) = H_{\mathrm{T}}(t)\Phi(X, t), \tag{10.140}$$

where we recollect that $X \equiv (q_1, q_2, \ldots, q_N)$ denotes the ensemble of the atomic electron coordinates and $q_i \equiv (\mathbf{r}_i, \sigma_i)$ represents the space and spin coordinates of electron i.

Since we are considering laser fields which, although strong by laboratory stan-
dards, are nevertheless such that $\mathcal{E}_0 \ll \mathcal{E}_a$, the TDSE (10.140) can be solved by using
first-order, time-dependent perturbation theory. To first order in $\mathcal{E}_0$, the dressed target wave functions are given in the length gauge by Equation (3.193). Now, as we
shall perform the calculations of the semi-perturbative theory in the velocity gauge,
we must obtain the dressed target wave functions also in that gauge. As explained
in Section 3.2.3, this can be done by starting from the expression (3.193) of the
dressed target wave functions in the length gauge, and performing two successive
gauge transformations to end up in the velocity gauge. The resulting dressed target
wave functions are given in the velocity gauge by Equation (3.199). We recall that
in this equation, the gauge factor $\exp[(i/\hbar)(e^2 N/(2m)) \int_{-\infty}^t \mathbf{A}^2(t')dt']$, which is
independent of the electron coordinates, plays no role in calculating any scattering
amplitude. The *Göppert-Mayer gauge factor* $\exp[-i\mathbf{a}(t) \cdot \mathbf{R}]$ plays no role in the
calculation of *direct* scattering amplitudes, but must be taken into account in the
treatment of laser-assisted electron–atom *exchange* collisions.

We recall that the dressed target wave functions that we have obtained are not
valid for angular frequencies ω in the vicinity of a Bohr angular frequency $\omega_{j'j}$. We
shall return to this problem below and obtain the required dressed wave functions
by using the Floquet method. We also note that if $\omega \ll \omega_{j'j}$ for j fixed and all j',
then one can deduce from Equation (3.199) the low-frequency approximation

$$\Phi_j^V(X, t) = \exp(-iE_j t/\hbar) \exp\left[\frac{i}{\hbar}\frac{e^2 N}{2m} \int_{-\infty}^t \mathbf{A}^2(t')dt'\right] \exp[-i\mathbf{a}(t) \cdot \mathbf{R}]$$

$$\times \left\{ \psi_j(X) + \frac{\mathcal{E}_0}{\hbar} \sin(\omega t) \sum_{j'} \omega_{j'j}^{-1} M_{j'j} \psi_{j'}(X) \right\}. \quad (10.141)$$

Finally, we remark that if one specializes Equations (3.199) or (10.141) to the
ground state, and if this ground state is spherically symmetric (which is the case
considered below), then because of dipole selection rules the summation over the
index j' in these two equations runs only over discrete and continuum field-free
P states.

We shall now use atomic units. To first order in the electron–atom interaction
potential V_d, that is in the first Born approximation, the S-matrix element for the
direct (no-exchange) elastic scattering transition $(\mathbf{k}_i, 0 \to \mathbf{k}_f, 0)$ from the ground
state of the atom, in the presence of a laser field, is given by

$$S_{el}^{B1} = -i \int_{-\infty}^{+\infty} dt \langle \chi_{\mathbf{k}_f}^V(\mathbf{r}_0, t)\Phi_0^V(X, t)|V_d(\mathbf{r}_0, X)|\chi_{\mathbf{k}_i}^V(\mathbf{r}_0, t)\Phi_0^V(X, t)\rangle, \quad (10.142)$$

where $\Phi_0^V(X, t)$ is the wave function of the dressed ground state in the velocity
gauge, obtained by using Equation (3.199) with $j = 0$. We assume that the field-free

target atom ground state is an S state. In the following we shall denote its state vector by $|\psi_0\rangle$ and the state vectors of the intermediate field-free P-states by $|\psi_{jP}\rangle$. Using Equations (10.139), (3.199) and (10.142), together with the generating function of the ordinary Bessel functions given by Equation (10.18), we find after performing the time integration that

$$S_{el}^{B1} = (2\pi)^{-1} i \sum_{n=-\infty}^{+\infty} \delta(E_{k_f} - E_{k_i} - n\omega) f_{el}^{n,B1}(\mathbf{\Delta}), \qquad (10.143)$$

where $\mathbf{\Delta} = \mathbf{k}_i - \mathbf{k}_f$ is the wave vector transfer (or momentum transfer in atomic units) and $f_{el}^{n,B1}(\mathbf{\Delta})$, the first Born approximation to the elastic scattering amplitude with the transfer of $|n|$ photons, is given by

$$f_{el}^{n,B1}(\mathbf{\Delta}) = J_n(\mathbf{\Delta} \cdot \boldsymbol{\alpha}_0) f_{el}^{B1}(\mathbf{\Delta}) + i\mathcal{E}_0 J_n'(\mathbf{\Delta} \cdot \boldsymbol{\alpha}_0)$$
$$\times \sum_j \frac{\omega_{j0}}{\omega_{j0}^2 - \omega^2} \left\{ f_{0,jP}^{B1}(\mathbf{\Delta}) M_{jP,0} + M_{0,jP} f_{jP,0}^{B1}(\mathbf{\Delta}) \right\}, \qquad (10.144)$$

with the summation on the index j running over P states belonging to the discrete and the continuum part of the spectrum. The first Born differential cross section for laser-assisted elastic scattering is given by

$$\frac{d\sigma_{el}^{n,B1}}{d\Omega} = \frac{k_f(n)}{k_i} |f_{el}^{n,B1}|^2, \qquad (10.145)$$

where from the energy-conservation relation (10.131) we have

$$k_f(n) = (k_i^2 + 2n\omega)^{1/2}. \qquad (10.146)$$

In obtaining Equation (10.144), we have used the relation [46]

$$J_n'(x) = \frac{1}{2}[J_{n-1}(x) - J_{n+1}(x)] \qquad (10.147)$$

and we have defined the amplitudes

$$f_{el}^{B1}(\mathbf{\Delta}) = -\frac{2}{\Delta^2} \langle \psi_0 | \tilde{V}_d(\mathbf{\Delta}, X) | \psi_0 \rangle, \qquad (10.148)$$

$$f_{jP,0}^{B1}(\mathbf{\Delta}) = -\frac{2}{\Delta^2} \langle \psi_{jP} | \tilde{V}_d(\mathbf{\Delta}, X) | \psi_0 \rangle \qquad (10.149)$$

and

$$f_{0,jP}^{B1}(\mathbf{\Delta}) = -\frac{2}{\Delta^2} \langle \psi_0 | \tilde{V}_d(\mathbf{\Delta}, X) | \psi_{jP} \rangle, \qquad (10.150)$$

which are, respectively, the field-free first Born amplitudes for elastic scattering and for the $0 \to jP$ and $jP \to 0$ transitions. In writing down Equations (10.148) (10.150), we have used the Bethe result [105]

$$\int \exp(i\boldsymbol{\Delta} \cdot \mathbf{r}_0) V_d(\mathbf{r}_0, X) d\mathbf{r}_0 = \frac{4\pi}{\Delta^2} \tilde{V}_d(\boldsymbol{\Delta}, X) \qquad (10.151)$$

with

$$\tilde{V}_d(\boldsymbol{\Delta}, X) = \sum_{k=1}^{N} \exp(i\boldsymbol{\Delta} \cdot \mathbf{r}_k) - N. \qquad (10.152)$$

We note that the Göppert-Mayer gauge factor $\exp[-i\mathbf{a}(t) \cdot \mathbf{R}]$ in Equation (3.199) plays no role in calculating $f_{el}^{n,B1}$, or indeed any *direct* scattering amplitude, as stated above.

Francken and Joachain [106] have shown that by expanding the Bessel functions in Equation (10.144) to first order in α_0 (and therefore in $\mathcal{E}_0$), one finds results which are identical to the first Born values of Dubois, Maquet and Jetzke [101], obtained by dressing both the projectile electron and the target atom to first order in the electric-field strength $\mathcal{E}_0$. In this case, of course, only the processes corresponding to the transfer of one photon ($n = \pm 1$) must be considered. Francken and Joachain also pointed out that the importance of target-dressing effects increases with the laser frequency.

Let us now analyze the result (10.144) in more detail. The first term on the right-hand side, which we shall call *"electronic,"* is analogous to the result (10.71) obtained by studying laser-assisted potential scattering in the first Born approximation [7]. If only this term were retained, the first Born differential cross section for laser-assisted elastic scattering would be given by the purely "electronic" expression (which ignores the "dressing" of the target atom by the laser field)

$$\left(\frac{d\sigma_{el}^{n,B1}}{d\Omega} \right)_{\text{no dressing}} = \frac{k_f(n)}{k_i} J_n^2(\boldsymbol{\Delta} \cdot \boldsymbol{\alpha}_0) \frac{d\sigma_{el}^{B1}}{d\Omega}(\boldsymbol{\Delta}), \qquad (10.153)$$

where

$$\frac{d\sigma_{el}^{B1}}{d\Omega}(\boldsymbol{\Delta}) = |f_{el}^{B1}(\boldsymbol{\Delta})|^2 \qquad (10.154)$$

is the field-free first Born elastic differential cross section. In the above expression, the field-free first Born elastic scattering amplitude is given by

$$f_{el}^{B1}(\boldsymbol{\Delta}) = -(2\pi)^{-1} \int \exp(i\boldsymbol{\Delta} \cdot \mathbf{r}) V_s(\mathbf{r}) d\mathbf{r}, \qquad (10.155)$$

where

$$V_s = \langle \psi_0 | V_d | \psi_0 \rangle \qquad (10.156)$$

is the static potential corresponding to the ground state $|\psi_0\rangle$ and V_d is the direct electron–atom interaction potential (10.135). We remark that if the scattering potential $V(\mathbf{r})$ of Section 10.2 is chosen to be the static potential $V_s(\mathbf{r})$, then the field-free first Born scattering amplitude $f^{\text{B1}}_{\mathbf{k}_f,\mathbf{k}_i}(\boldsymbol{\Delta})$ given by Equation (10.69) coincides with $f^{\text{B1}}_{\text{el}}(\boldsymbol{\Delta})$, and therefore the "no dressing" expression (10.153) for the laser-assisted first Born differential cross section is then identical with the Bunkin–Fedorov result (10.72).

The second term on the right-hand side of Equation (10.144), which will be called "atomic," accounts for the *dressing effect* due to the dipole distortion of the target atom by the laser field. Since the field-free first Born scattering amplitudes $f^{\text{B1}}_{0,j\text{P}}$ and $f^{\text{B1}}_{j\text{P},0}$, which correspond to allowed S $\leftrightarrow$ P transitions, are very large at small momentum transfers, we expect this second term to be important for small $\boldsymbol{\Delta}$. In contrast, the rapid decrease of S $\leftrightarrow$ P field-free first Born scattering amplitudes with increased $\boldsymbol{\Delta}$ compared with the field-free first Born elastic scattering amplitude $f^{\text{B1}}_{\text{el}}$ [47] implies that dressing effects in Equation (10.144) will be small in the large momentum transfer region.

In the case of scattering by an atomic hydrogen target, the summation in the second term on the right-hand side of Equation (10.144) can be performed exactly, using known expressions for the first Born scattering amplitudes and the dipole coupling matrix elements. For more complex atoms, however, this procedure is prohibitively difficult. Instead, one can use the *closure approximation*, in which the target energy differences $E_j - E_0$ are replaced by an average excitation energy $\bar{E}$. As a result, the summation on the intermediate field-free P states can be performed analytically, and Equation (10.144) reduces to

$$f^{n,\text{B1}}_{\text{el}}(\boldsymbol{\Delta}) = J_n(\boldsymbol{\Delta} \cdot \boldsymbol{\alpha}_0) f^{\text{B1}}_{\text{el}}(\boldsymbol{\Delta})$$

$$+ \mathcal{E}_0 \frac{4\bar{E}}{\Delta^2(\bar{E}^2 - \omega^2)} J'_n(\boldsymbol{\Delta} \cdot \boldsymbol{\alpha}_0) \hat{\boldsymbol{\epsilon}} \cdot \boldsymbol{\nabla}_{\boldsymbol{\Delta}} \langle \psi_0 | \tilde{V}_d(\boldsymbol{\Delta}, X) | \psi_0 \rangle. \quad (10.157)$$

It is interesting to examine the *low-frequency (soft-photon)* limit of the results obtained above. In this limit the dressed ground-state wave function given in the velocity gauge by Equation (10.141) with $j = 0$ becomes (in a.u.)

$$\Phi^{\text{V}}_0(X, t) = \exp(-iE_0 t) \exp\left[i\frac{N}{2} \int_{-\infty}^{t} \mathbf{A}^2(t')dt' \right] \exp[-i\mathbf{a}(t) \cdot \mathbf{R}]$$

$$\times [\psi_0(X) + \mathcal{E}_0 \sin(\omega t) \sum_j \omega_{j0}^{-1} M_{j\text{P},0} \psi_{j\text{P}}(X)]. \quad (10.158)$$

Introducing the wave function

$$\tilde{\psi}_0(X) = \sum_j \omega_{j0}^{-1} M_{j\text{P},0} \psi_{j\text{P}}(X), \quad (10.159)$$

which has the symmetry of a P state, the first Born amplitude for laser-assisted elastic scattering becomes, in the low-frequency limit,

$$f_{el}^{n,B1}(\mathbf{\Delta}) = J_n(\mathbf{\Delta}\cdot\boldsymbol{\alpha}_0)f_{el}^{B1}(\mathbf{\Delta})$$
$$+ i\mathcal{E}_0 J_n'(\mathbf{\Delta}\cdot\boldsymbol{\alpha}_0)[f^{B1}(\tilde{\psi}_0 \to \psi_0) + f^{B1}(\psi_0 \to \tilde{\psi}_0)], \quad (10.160)$$

where the quantity

$$f^{B1}(\psi_0 \to \tilde{\psi}_0) = -\frac{2}{\Delta^2}\langle\tilde{\psi}_0|\tilde{V}_d(\mathbf{\Delta}, X)|\psi_0\rangle \quad (10.161)$$

is the first Born amplitude corresponding to an S → P transition from the state $|\psi_0\rangle$ to the "state" $|\tilde{\psi}_0\rangle$. We also see that in the low-frequency limit the closure result (10.157) becomes

$$f_{el}^{n,B1}(\mathbf{\Delta}) = J_n(\mathbf{\Delta}\cdot\boldsymbol{\alpha}_0)f_{el}^{B1}(\mathbf{\Delta})$$
$$+ \mathcal{E}_0\frac{4}{\Delta^2\bar{E}}J_n'(\mathbf{\Delta}\cdot\boldsymbol{\alpha}_0)\hat{\boldsymbol{\epsilon}}\cdot\nabla_{\mathbf{\Delta}}\langle\psi_0|\tilde{V}_d(\mathbf{\Delta}, X)|\psi_0\rangle. \quad (10.162)$$

For the case of an atomic hydrogen target one can obtain a closed-form expression for the "state" $|\tilde{\psi}_0\rangle$. Denoting by

$$H_h = \frac{\mathbf{p}_1^2}{2} - \frac{1}{r_1} \quad (10.163)$$

the hydrogen atom Hamiltonian, we have

$$(H_h - E_0)\tilde{\psi}_0(\mathbf{r}_1) = -\hat{\boldsymbol{\epsilon}}\cdot\mathbf{r}_1\,\psi_0(r_1), \quad (10.164)$$

where $\psi_0(r_1) = \pi^{-1/2}\exp(-r_1)$ is the field-free ground-state wave function of atomic hydrogen. The solution of Equation (10.164) is readily shown to be

$$\tilde{\psi}_0(\mathbf{r}_1) = -\pi^{-1/2}\hat{\boldsymbol{\epsilon}}\cdot\mathbf{r}_1\left(1 + \frac{r_1}{2}\right)\exp(-r_1). \quad (10.165)$$

Using this expression in Equations (10.160) and (10.161), we find that

$$f_{el}^{n,B1}(\mathbf{\Delta}) = J_n(\mathbf{\Delta}\cdot\boldsymbol{\alpha}_0)\frac{2(\Delta^2+8)}{(\Delta^2+4)^2} - \mathcal{E}_0 J_n'(\mathbf{\Delta}\cdot\boldsymbol{\alpha}_0)\frac{192\,\hat{\boldsymbol{\epsilon}}\cdot\mathbf{\Delta}}{\Delta^2(\Delta^2+4)^3}\left(1 + \frac{8}{\Delta^2+4}\right). \quad (10.166)$$

In obtaining this result we have used the fact that the first Born scattering amplitude for field-free elastic electron–atomic hydrogen scattering is given by [48]

$$f_{el}^{B1} = \frac{2(\Delta^2+8)}{(\Delta^2+4)^2}, \quad (10.167)$$

and we have also used the relation

$$f^{B1}(\psi_0 \to \tilde{\psi}_0) = -i\frac{96\,\hat{\boldsymbol{\epsilon}}\cdot\mathbf{\Delta}}{\Delta^2(\Delta^2+4)^3}\left(1 + \frac{8}{\Delta^2+4}\right). \quad (10.168)$$

It is interesting to compare the result (10.166) with the low-frequency closure result (10.162), which, for an atomic hydrogen target, yields

$$f_{el}^{n,B1}(\mathbf{\Delta}) = J_n(\mathbf{\Delta} \cdot \boldsymbol{\alpha}_0)\frac{2(\Delta^2 + 8)}{(\Delta^2 + 4)^2} - \mathcal{E}_0 J_n'(\mathbf{\Delta} \cdot \boldsymbol{\alpha}_0)\frac{256}{\bar{E}}\frac{\hat{\boldsymbol{\epsilon}} \cdot \mathbf{\Delta}}{\Delta^2(\Delta^2 + 4)^3}. \tag{10.169}$$

We note that for the choice $\bar{E} = (4/9)$ a.u. (which gives the correct value $\bar{\alpha} = 4.5$ a.u. for the static dipole polarizability of atomic hydrogen in the closure approximation), the agreement between the soft-photon "exact" result (10.166) and the soft-photon closure result (10.169) is excellent for values of the momentum transfer $\Delta \lesssim 1$, where the dressing term is most important. This agreement is independent of both the orientation of the polarization vector $\hat{\boldsymbol{\epsilon}}$ and the number of emitted or absorbed photons. This gives support for using the closure approximation to study dressing effects in more complicated target atoms at small momentum transfers.

We also remark that for moderate values of $|n|$ one has $k_f(n) \simeq k_i$ in the soft-photon limit. Provided that the n-dependence of $\mathbf{\Delta}$ and $\mathbf{\Delta} \cdot \boldsymbol{\alpha}_0$ can be neglected in Equation (10.144), we then have

$$f_{el}^{-n,B1}(\mathbf{\Delta}) = (-1)^n f_{el}^{n,B1}(\mathbf{\Delta}), \tag{10.170}$$

where we have used the fact that $J_{-n}(x) = (-1)^n J_n(x)$. As a result, the laser-assisted first Born differential cross sections will be nearly the same for the absorption $(n > 0)$ or the stimulated emission $(n < 0)$ of a given number $|n|$ of photons. This conclusion, however, is not valid for all geometrical configurations because it depends on the fact that the argument $\mathbf{\Delta} \cdot \boldsymbol{\alpha}_0$ of the Bessel functions in Equation (10.144) is such that

$$\mathbf{\Delta} \cdot \boldsymbol{\alpha}_0 \gg n\omega\alpha_0. \tag{10.171}$$

In a geometry such that $\mathbf{\Delta}$ is perpendicular (or nearly perpendicular) to $\boldsymbol{\alpha}_0$, we can have $\mathbf{\Delta} \cdot \boldsymbol{\alpha}_0 \ll n\omega\alpha_0$, so that the n-dependence of $\mathbf{\Delta} \cdot \boldsymbol{\alpha}_0$ can no longer be neglected.

From the above discussion, the laser-assisted cross sections are expected to depend weakly on the sign of n when $\hat{\boldsymbol{\epsilon}}$ is parallel to $\mathbf{\Delta}$. The same is true when $\hat{\boldsymbol{\epsilon}}$ is perpendicular to $\mathbf{k}_i$, at least for the small-angle region. In contrast, when $\hat{\boldsymbol{\epsilon}}$ is parallel to $\mathbf{k}_i$, the quantity $\mathbf{\Delta} \cdot \boldsymbol{\alpha}_0$ will pass through zero at a small scattering angle in the $n > 0$ case (absorption) when k_f is slightly larger than k_i. The criterion for this to happen is clearly that $\mathbf{\Delta} \cdot \mathbf{k}_i = 0$, or, in other words, $k_i - k_f \cos\theta = 0$, where θ is the scattering angle. In the low-frequency limit, this equation is readily solved to give

$$\theta \simeq \left(\frac{n\omega}{E_{k_i}}\right)^{1/2} \tag{10.172}$$

since θ is small. Consequently, important asymmetries between stimulated and inverse bremsstrahlung are expected in this geometry for which $\hat{\boldsymbol{\epsilon}}$ is parallel to

$\mathbf{k}_i$, as noticed by Dubois, Maquet and Jetzke [101]. The polarization dependence of laser-assisted "elastic" electron–atomic hydrogen first Born differential cross section has also been studied by Fainstein and Maquet [107], who showed that, as in the case of linear polarization considered by Byron and Joachain [10] and Byron Francken and Joachain [104], the dressing of the target by a circularly polarized laser field introduces important modifications of laser-assisted differential cross sections at small scattering angles.

We also point out that in the soft-photon limit and in the geometrical arrangement for which Equation (10.170) is valid (for example when $\hat{\boldsymbol{\epsilon}}$ is parallel to $\boldsymbol{\Delta}$ or is perpendicular to $\mathbf{k}_i$ and θ is small), one finds the sum rule

$$\sum_n \frac{\mathrm{d}\sigma_{\mathrm{el}}^{n,\mathrm{B1}}}{\mathrm{d}\Omega} = \frac{\mathrm{d}\sigma_{\mathrm{el}}^{\mathrm{B1}}}{\mathrm{d}\Omega} + \mathcal{O}(\mathcal{E}_0^2), \tag{10.173}$$

where the sum on n runs over all values of n such that $k_f(n) \geq 0$, and we have made use of Equation (10.76) and of the sum rule

$$\sum_{n=-\infty}^{+\infty} J'_n(x) J_n(x) = 0. \tag{10.174}$$

It should be noted that the correction of order $\mathcal{E}_0^2$ in Equation (10.173) contains two kinds of contributions. The first one is in fact proportional to $\mathcal{E}_0^2/\Delta^2$ and arises from the dressing of the atomic P states considered here; this term is always significant when dressing effects are large. The second contribution to the correction in Equation (10.173) is due to the dressing of atomic S and D states obtained in second order of perturbation theory, and is directly proportional to $\mathcal{E}_0^2$; this second contribution does not produce Δ^{-2} singularities and can therefore be neglected in the present context.

Let us now consider the second Born contribution to the S-matrix element for the direct (no exchange) elastic scattering transition ($\mathbf{k}_i, 0 \to \mathbf{k}_f, 0$). That is,

$$\bar{S}_{\mathrm{el}}^{\mathrm{B2}} = -\mathrm{i} \int_{-\infty}^{+\infty} \mathrm{d}t \int_{-\infty}^{+\infty} \mathrm{d}t' \langle \chi_{\mathbf{k}_f}^{\mathrm{V}}(\mathbf{r}_0, t) \Phi_0^{\mathrm{V}}(X, t) | V_{\mathrm{d}}(\mathbf{r}_0, X)$$
$$\times \tilde{K}_{\mathrm{F}}^{\mathrm{V}(+)}(\mathbf{r}_0, X, t; \mathbf{r}'_0, X', t') V_{\mathrm{d}}(\mathbf{r}'_0, X') | \chi_{\mathbf{k}_i}^{\mathrm{V}}(\mathbf{r}'_0, t') \Phi_0^{\mathrm{V}}(X', t') \rangle, \tag{10.175}$$

where $\tilde{K}_{\mathrm{F}}^{\mathrm{V}(+)}$ is the causal propagator given in the position representation by

$$\tilde{K}_{\mathrm{F}}^{\mathrm{V}(+)}(\mathbf{r}_0, X, t; \mathbf{r}'_0, X', t')$$
$$= -\mathrm{i}\Theta(t-t') \sum_j \int \mathrm{d}\mathbf{q} \, \chi_{\mathbf{q}}^{\mathrm{V}}(\mathbf{r}_0, t) \chi_{\mathbf{q}}^{\mathrm{V}*}(\mathbf{r}'_0, t') \Phi_j^{\mathrm{V}}(X, t) \Phi_j^{\mathrm{V}*}(X', t'). \tag{10.176}$$

Here Θ is the Heaviside step function (2.157), $\chi_{\mathbf{q}}^{\text{V}}$ is a Gordon–Volkov wave function in the velocity gauge of the type given by Equation (10.139), and Φ_j^{V} is a dressed target wave function in the velocity gauge given by Equation (3.199).

The integrations on the time variables in Equation (10.175) can be performed by using Equations (10.18), (10.139) and (3.199). Working through first order in $\mathcal{E}_0$ for the target dressed states, one finds that $\bar{S}_{\text{el}}^{\text{B2}}$ is the sum of two terms which are, respectively, of zeroth and first order in $\mathcal{E}_0$. The zeroth-order term $\bar{S}_{\text{el},0}^{\text{B2}}$ is given by

$$
\bar{S}_{\text{el},0}^{\text{B2}} = \frac{\mathrm{i}}{2\pi^3} \sum_{n=-\infty}^{+\infty} \delta(E_{k_f} - E_{k_i} - n\omega) \sum_{n'=-\infty}^{+\infty} \sum_j
$$

$$
\int d\mathbf{q} \frac{J_{n-n'}(\boldsymbol{\Delta}_f \cdot \boldsymbol{\alpha}_0) J_{n'}(-\boldsymbol{\Delta}_i \cdot \boldsymbol{\alpha}_0)}{\Delta_i^2 \Delta_f^2} \frac{\langle \psi_0 | \tilde{V}_d(\boldsymbol{\Delta}_f, X) | \psi_j \rangle \langle \psi_j | \tilde{V}_d(-\boldsymbol{\Delta}_i, X) | \psi_0 \rangle}{E_q - E_{k_i} + \omega_{j0} - n'\omega - \mathrm{i}\epsilon},
$$
$$
\epsilon \to 0^+, \quad (10.177)
$$

where $E_q = q^2/2$, $\boldsymbol{\Delta}_i = \mathbf{q} - \mathbf{k}_i$ and $\boldsymbol{\Delta}_f = \mathbf{q} - \mathbf{k}_f$. Through order ω^2, the summation over n' can be carried out by shifting the momenta $\mathbf{k}_i, \mathbf{k}_f$ and $\mathbf{q}$ according to [50]

$$
\mathbf{k}_i \to \mathbf{k}_i^* = \mathbf{k}_i + \mathbf{s}, \quad (10.178)
$$

$$
\mathbf{k}_f \to \mathbf{k}_f^* = \mathbf{k}_f + \mathbf{s} \quad (10.179)
$$

and

$$
\mathbf{q} \to \mathbf{q}^* = \mathbf{q} + \mathbf{s}, \quad (10.180)
$$

where $\mathbf{s} = n\omega\boldsymbol{\alpha}_0/(\boldsymbol{\Delta} \cdot \boldsymbol{\alpha}_0)$, as in Equation (10.85). Using the fact that

$$
\boldsymbol{\Delta} = \mathbf{k}_i - \mathbf{k}_f = \mathbf{k}_i^* - \mathbf{k}_f^*, \quad (10.181)
$$

$$
\boldsymbol{\Delta}_i = \mathbf{q} - \mathbf{k}_i = \mathbf{q}^* - \mathbf{k}_i^* \quad (10.182)
$$

and

$$
\boldsymbol{\Delta}_f = \mathbf{q} - \mathbf{k}_f = \mathbf{q}^* - \mathbf{k}_f^*, \quad (10.183)
$$

we find that

$$
\bar{S}_{\text{el},0}^{\text{B2}} = (2\pi)^{-1} \mathrm{i} \sum_{n=-\infty}^{+\infty} \delta(E_{k_f} - E_{k_i} - n\omega) \bar{f}_{\text{el},0}^{n,\text{B2}}(\boldsymbol{\Delta}), \quad (10.184)
$$

where

$$
\bar{f}_{\text{el},0}^{n,\text{B2}}(\boldsymbol{\Delta}) = J_n(\boldsymbol{\Delta}_f \cdot \boldsymbol{\alpha}_0) \bar{f}_{\text{el}}^{\text{B2}}(\boldsymbol{\Delta}). \quad (10.185)
$$

Here,

$$
\bar{f}_{\text{el}}^{\text{B2}}(\boldsymbol{\Delta}) = \frac{2}{\pi^2} \sum_j \int d\mathbf{q}^* \frac{\langle \psi_0 | \tilde{V}_d(\boldsymbol{\Delta}_f, X) | \psi_j \rangle \langle \psi_j | \tilde{V}_d(-\boldsymbol{\Delta}_i, X) | \psi_0 \rangle}{\Delta_i^2 \Delta_f^2 (q^{*2} - k_i^{*2} + 2\omega_{j0} - \mathrm{i}\epsilon)} \quad (10.186)
$$

is the field-free second Born elastic scattering amplitude evaluated at the shifted
momenta $\mathbf{k}_i^*$ and $\mathbf{k}_f^*$.

Following a similar procedure, the contribution to $\bar{S}_{\mathrm{el}}^{\mathrm{B2}}$, which is of first order in
$\mathcal{E}_0$, is given by

$$\bar{S}_{\mathrm{el},1}^{\mathrm{B2}} = (2\pi)^{-1}\mathrm{i} \sum_{n=-\infty}^{+\infty} \delta(E_{k_f} - E_{k_i} - n\omega) \bar{f}_{\mathrm{el},1}^{n,\mathrm{B2}}(\mathbf{\Delta}), \tag{10.187}$$

where

$$\bar{f}_{\mathrm{el},1}^{n,\mathrm{B2}}(\mathbf{\Delta}) = \mathrm{i}\mathcal{E}_0 J_n'(\mathbf{\Delta}\cdot\boldsymbol{\alpha}_0)[T_1(\mathbf{\Delta}) + T_2(\mathbf{\Delta}) + T_3(\mathbf{\Delta})], \tag{10.188}$$

and the quantities $T_i(\mathbf{\Delta})$ with $i = 1, 2, 3$ are given by

$$T_1(\mathbf{\Delta}) = \frac{2}{\pi^2}\sum_j\sum_{j'}$$

$$\int d\mathbf{q}^* \frac{\langle\psi_0|\tilde{V}_{\mathrm{d}}(\mathbf{\Delta}_f, X)|\psi_{j'}\rangle\langle\psi_{j'}|\tilde{V}_{\mathrm{d}}(-\mathbf{\Delta}_i, X)|\psi_j\rangle M_{j0}}{\Delta_i^2\Delta_f^2(q^{*2} - k_i^{*2} + 2\omega_{j'0} - \mathrm{i}\epsilon)\omega_{j0}}, \tag{10.189}$$

$$T_2(\mathbf{\Delta}) = \frac{2}{\pi^2}\sum_j\sum_{j'}$$

$$\int d\mathbf{q}^* \frac{M_{0j'}\langle\psi_{j'}|\tilde{V}_{\mathrm{d}}(\mathbf{\Delta}_f, X)|\psi_j\rangle\langle\psi_j|\tilde{V}_{\mathrm{d}}(-\mathbf{\Delta}_i, X)|\psi_0\rangle}{\omega_{j'0}\Delta_i^2\Delta_f^2(q^{*2} - k_i^{*2} + 2\omega_{j0} - \mathrm{i}\epsilon)} \tag{10.190}$$

and

$$T_3(\mathbf{\Delta}) = \frac{4}{\pi^2}\sum_j\sum_{j'}$$

$$\int d\mathbf{q}^* \frac{\langle\psi_0|\tilde{V}_{\mathrm{d}}(\mathbf{\Delta}_f, X)|\psi_{j'}\rangle M_{j'j}\langle\psi_j|\tilde{V}_{\mathrm{d}}(-\mathbf{\Delta}_i, X)|\psi_0\rangle}{\Delta_i^2\Delta_f^2(q^{*2} - k_i^{*2} + 2\omega_{j'0} - \mathrm{i}\epsilon)(q^{*2} - k_i^{*2} + 2\omega_{j0} - \mathrm{i}\epsilon)}. \tag{10.191}$$

In obtaining these equations, we have neglected ω with respect to ω_{j0}, $\omega_{j'0}$ and
unity.

Equations (10.189)–(10.191) are written in a form which emphasizes the fact
that these expressions are really *third-order* terms when one considers the interaction to be the sum of the electron–atom interaction $V_{\mathrm{d}}(\mathbf{r}_0, X)$ and the interaction
of the laser field with the atom. In each of Equations (10.189)–(10.191) there are
three interaction matrix elements and two "propagators." In the first two of these
equations, one of the "propagators" degenerates simply to ω_{j0}^{-1} and $\omega_{j'0}^{-1}$ because
of the momentum-conserving delta functions associated with the laser–atom inter-
action matrix element when the laser–atom interaction occurs in the initial or final

scattering matrix element. When the laser–atom interaction occurs in the intermediate scattering matrix element, one obtains Equation (10.191), where both "propagators" have their original form, but with the same intermediate momentum $\mathbf{q}^*$ because of the momentum-conserving delta function.

In terms of our foregoing discussion of $S_{\text{el}}^{\text{B1}}$, Equations (10.189) and (10.190) can be rewritten more suggestively by using the definition (10.159) of $\tilde{\psi}_0(X)$ to obtain

$$T_1(\boldsymbol{\Delta}) = \frac{2}{\pi^2} \sum_j \int d\mathbf{q}^* \frac{\langle \psi_0 | \tilde{V}_d(\boldsymbol{\Delta}_f, X) | \psi_j \rangle \langle \psi_j | \tilde{V}_d(-\boldsymbol{\Delta}_i, X) | \tilde{\psi}_0 \rangle}{\Delta_i^2 \Delta_f^2 (q^{*2} - k_i^{*2} + 2\omega_{j0} - i\epsilon)} \qquad (10.192)$$

and

$$T_2(\boldsymbol{\Delta}) = \frac{2}{\pi^2} \sum_j \int d\mathbf{q}^* \frac{\langle \tilde{\psi}_0 | \tilde{V}_d(\boldsymbol{\Delta}_f, X) | \psi_j \rangle \langle \psi_j | \tilde{V}_d(-\boldsymbol{\Delta}_i, X) | \psi_0 \rangle}{\Delta_i^2 \Delta_f^2 (q^{*2} - k_i^{*2} + 2\omega_{j0} - i\epsilon)}. \qquad (10.193)$$

Using Equation (10.188), we see that T_1 and T_2 yield second Born-type corrections to the two terms in square brackets on the right-hand side of Equation (10.160) which describe the effects of the ground-state dressing in lowest order. The terms T_1 and T_2 are standard second Born expressions which can be evaluated by using well known methods [108].

The quantity T_3 given by Equation (10.191) is more complicated than T_1 and T_2. It appears that using the closure approximation is the only practical means of evaluation. If, in Equation (10.191), we replace ω_{j0} and $\omega_{j'0}$ by an average excitation energy $\bar{E}$, then the summation over intermediate states can be performed, and one finds that

$$T_3(\boldsymbol{\Delta}) = \frac{4}{\pi^2} \int d\mathbf{q}^* \frac{\langle \psi_0 | (\hat{\boldsymbol{\epsilon}} \cdot \mathbf{D}) \tilde{V}_d(\boldsymbol{\Delta}_f, X) \tilde{V}_d(-\boldsymbol{\Delta}_i, X) | \psi_0 \rangle}{\Delta_i^2 \Delta_f^2 (q^{*2} - k_i^{*2} + 2\bar{E} - i\epsilon)^2}$$

$$= -\frac{2}{\pi^2} \frac{d}{d\bar{E}} \int d\mathbf{q}^* \frac{\langle \psi_0 | (\hat{\boldsymbol{\epsilon}} \cdot \mathbf{D}) \tilde{V}_d(\boldsymbol{\Delta}_f, X) \tilde{V}_d(-\boldsymbol{\Delta}_i, X) | \psi_0 \rangle}{\Delta_i^2 \Delta_f^2 (q^{*2} - k_i^{*2} + 2\bar{E} - i\epsilon)}, \qquad (10.194)$$

where $\mathbf{D} = -\mathbf{R}$ is the electric dipole moment of the atom (in a.u.). Since we have assumed that $|\psi_0\rangle$ is an S state, it follows that $\hat{\boldsymbol{\epsilon}} \cdot \mathbf{D} |\psi_0\rangle$ has the symmetry of a P state, so that T_3 is simply the derivative with respect to $\bar{E}$ of a second-order S–P amplitude. In fact, in the closure approximation, $|\tilde{\psi}_0\rangle$ is given by

$$|\tilde{\psi}_0\rangle = \hat{\boldsymbol{\epsilon}} \cdot \mathbf{D}\bar{E}^{-1} |\psi_0\rangle, \qquad (10.195)$$

and therefore in this approximation the terms T_1 and T_2 differ from T_3 only by a constant of order unity.

We recall that the two terms enclosed in square brackets in Equation (10.160) are S–P amplitudes and therefore vary like Δ^{-1} at small scattering angles [48]. It is also known from the study of second-order corrections to S–P amplitudes [108] that these corrections tend to a value of order k_i^{-1} as Δ becomes small and thus are rather unimportant at small scattering angles. However, this is precisely the angular range of interest to display target-dressing effects since when $\mathcal{E}_0 \ll \mathcal{E}_a$ the factor Δ^{-1} coming from the first-order amplitude is necessary to provide a significant dressing effect. Therefore, in obtaining small-angle results, the term $\bar{f}_{\text{el},1}^{n,\text{B2}}$ can be neglected. We remark, however, that for values of $\mathcal{E}_0$ of order unity (in a.u.) these second-order S–P terms would not be negligible. In fact, since second-order S–P amplitudes vary like $ik_i^{-2}\Delta^{-1}$ in the large-angle region [108], these terms will provide the main correction to $f_{\text{el}}^{n,\text{B1}}$ in this angular range.

Higher Born contributions to f_{el}^n, the elastic scattering amplitude with the transfer of $|n|$ photons, can be analyzed in a similar way. One finds that these contributions can be split into a part proportional to $J_n(\Delta \cdot \alpha_0)$, which corrects the undressed scattering amplitude, and a part correcting the dressing term, which is proportional to $J_n'(\Delta \cdot \alpha_0)$. We recall that a consistent calculation of the field-free elastic direct scattering amplitude through order k_i^{-2} can be performed by using the eikonal-Born series (EBS) direct scattering amplitude [108, 109]

$$f_{\text{el}}^{\text{EBS}} = f_{\text{el}}^{\text{B1}} + \bar{f}_{\text{el}}^{\text{B2}} + \bar{f}_{\text{el}}^{\text{G3}}, \tag{10.196}$$

where $\bar{f}_{\text{el}}^{\text{G3}}$ is the third-order term in the expansion of the Glauber elastic scattering amplitude in powers of the electron–atom interaction potential V_d [47]. We shall therefore write the direct scattering amplitude in the presence of a laser field as

$$f_{\text{el}}^n(k_i^*, \Delta) = J_n(\Delta \cdot \alpha_0) f_{\text{el}}^{\text{EBS}}(k_i^*, \Delta)$$
$$+ i\mathcal{E}_0 J_n'(\Delta \cdot \alpha_0) \sum_j \frac{\omega_{j0}}{\omega_{j0}^2 - \omega^2} [f_{0,j\text{P}}^{\text{B1}}(\Delta) M_{j\text{P},0} + M_{0,j\text{P}} f_{j\text{P},0}^{\text{B1}}(\Delta)]. \tag{10.197}$$

We note that if the closure approximation is used to evaluate the sum on the right-hand side of this equation, then one has

$$f_{\text{el}}^n(k_i^*, \Delta) = J_n(\Delta \cdot \alpha_0) f_{\text{el}}^{\text{EBS}}(k_i^*, \Delta)$$
$$+ \mathcal{E}_0 \frac{4\bar{E}}{\Delta^2(\bar{E}^2 - \omega^2)} J_n'(\Delta \cdot \alpha_0) \hat{\epsilon} \cdot \nabla_\Delta \langle \psi_0 | \tilde{V}_\text{d}(\Delta, X) | \psi_0 \rangle. \tag{10.198}$$

We also remark that, in the soft-photon approximation, the result (10.197) becomes

$$f_{\text{el}}^n(k_i^*, \Delta) = J_n(\Delta \cdot \alpha_0) f_{\text{el}}^{\text{EBS}}(k_i, \Delta)$$
$$+ i\mathcal{E}_0 J_n'(\Delta \cdot \alpha_0) [f^{\text{B1}}(\tilde{\psi}_0 \to \psi_0) + f^{\text{B1}}(\psi_0 \to \tilde{\psi}_0)], \tag{10.199}$$

while the closure result (10.198) takes the form

$$f_{el}^n(k_i, \Delta) = J_n(\Delta \cdot \alpha_0) f_{el}^{EBS}(k_i, \Delta)$$

$$+ \mathcal{E}_0 \frac{4}{\Delta^2 \bar{E}} J_n'(\Delta \cdot \alpha_0) \hat{\epsilon} \cdot \nabla_\Delta \langle \psi_0 | \tilde{V}_d(\Delta, X) | \psi_0 \rangle, \qquad (10.200)$$

where the momentum shift of Equations (10.178)–(10.180) has been neglected since it is important only at small momentum transfers where the laser-assisted scattering amplitude f_{el}^n is dominated by the dressing term.

Let us now consider exchange effects in laser-assisted "elastic" electron–atom collisions, within the framework of the semi-perturbative theory. Since we are dealing here with fast electrons, and we are interested in small scattering angles where target-dressing can be important, we shall, in the spirit of the eikonal-Born series (EBS) method [108, 109], consider only the leading term of g_{el}^n, the exchange amplitude for elastic scattering with the transfer of $|n|$ photons.

We begin by considering the case of elastic exchange collisions of electrons by the ground state of atomic hydrogen, in the presence of a laser field. The corresponding first Born S-matrix element is given (in a.u.) by [110]

$$S_{el,ex}^{B1} = -i \int_{-\infty}^{+\infty} dt \, \langle \chi_{\mathbf{k}_f}^V(\mathbf{r}_1, t) \Phi_0^V(\mathbf{r}_0, t) | V_{ex}(\mathbf{r}_0, \mathbf{r}_1) | \chi_{\mathbf{k}_i}^V(\mathbf{r}_0, t) \Phi_0^V(\mathbf{r}_1, t) \rangle, \qquad (10.201)$$

where

$$V_{ex}(\mathbf{r}_0, \mathbf{r}_1) = \frac{1}{r_{01}} - \frac{1}{r_0} \qquad (10.202)$$

is the "prior" form of the interaction potential, namely the potential acting in the *initial* arrangement channel of the collision. An equivalent expression of $S_{el,ex}^{B1}$ can be written down in which one uses the "post" form of V_{ex}, that is the potential acting in the *final* arrangement channel of the collision, obtained by replacing the term $(-1/r_0)$ by $(-1/r_1)$ on the right-hand side of Equation (10.202). In what follows, we only retain in V_{ex} the electron–electron interaction term $1/r_{01}$, which plays the dominant role in electron exchange collisions [47]. After performing the time integration in Equation (10.201), we obtain, to first order in $\mathcal{E}_0$,

$$S_{el,ex}^{B1} = (2\pi)^{-1} i \sum_{n=-\infty}^{+\infty} \delta(E_{k_f} - E_{k_i} - n\omega) g_{el}^{n,B1}, \qquad (10.203)$$

where $g_{el}^{n,B1}$ is the dominant part of the first Born elastic exchange amplitude with the transfer of $|n|$ photons, arising from the electron–electron interaction term $1/r_{01}$. The quantity $g_{el}^{n,B1}$ is the sum of two terms, which are, respectively, of zeroth and first order in $\mathcal{E}_0$. The zeroth-order term $g_{\text{el},0}^{n,\text{B1}}$ is given by

$$
g_{\text{el},0}^{n,\text{B1}} = -(2\pi)^{-1} \sum_{n'=-\infty}^{+\infty} \mathrm{i}^{n'} J_{n-n'}(\boldsymbol{\Delta} \cdot \boldsymbol{\alpha}_0)
$$

$$
\times \int \mathrm{d}\mathbf{r}_0 \, \mathrm{d}\mathbf{r}_1 \exp(-\mathrm{i}\mathbf{k}_f \cdot \mathbf{r}_1)\psi_0(r_0)\frac{J_{n'}(\boldsymbol{\beta}_0 \cdot \mathbf{r}_{01})}{r_{01}}\exp(\mathrm{i}\mathbf{k}_i \cdot \mathbf{r}_0)\psi_0(r_1)
$$

$$(10.204)$$

with $\boldsymbol{\beta}_0 = \omega\boldsymbol{\alpha}_0$ and $\mathbf{r}_{01} = \mathbf{r}_0 - \mathbf{r}_1$. Using the integral representation of the Bessel functions [46]

$$
J_n(x) = (2\pi)^{-1} \int_0^{2\pi} \exp(-\mathrm{i}n\phi + \mathrm{i}x \sin \phi)\mathrm{d}\phi \qquad (10.205)
$$

one obtains [110]

$$
g_{\text{el},0}^{n,\text{B1}} = -(2\pi)^{-2} \sum_{n'=-\infty}^{+\infty} \mathrm{i}^{n'} J_{n-n'}(\boldsymbol{\Delta} \cdot \boldsymbol{\alpha}_0) \int_0^{2\pi} \mathrm{d}\phi \exp(-\mathrm{i}n'\phi)
$$

$$
\times \int \mathrm{d}\mathbf{r}_0 \, \mathrm{d}\mathbf{r}_1 \exp(-\mathrm{i}\mathbf{q}_f \cdot \mathbf{r}_1)\psi_0(r_0)\frac{1}{r_{01}}\exp(\mathrm{i}\mathbf{q}_i \cdot \mathbf{r}_0)\psi_0(r_1), \qquad (10.206)
$$

where $\mathbf{q}_i = \mathbf{k}_i + \boldsymbol{\beta}_0 \sin \phi$ and $\mathbf{q}_f = \mathbf{k}_f + \boldsymbol{\beta}_0 \sin \phi$. The integral on the variables $\mathbf{r}_0$ and $\mathbf{r}_1$ is a standard one appearing in the calculation of the field-free first-order elastic exchange amplitude for electron–atomic hydrogen collisions [47]. It can be evaluated exactly by using a parameterization technique due to Feynman [111], but the result is rather cumbersome. Since we are considering the case of fast electrons, we can readily obtain the leading term of this integral for large k_i by using the Ochkur approximation [112]. One finds in this way that in the large k_i limit

$$
g_{\text{el},0}^{n,\text{B1}} = (2\pi)^{-1} \sum_{n'=-\infty}^{+\infty} \mathrm{i}^{n'} J_{n-n'}(\boldsymbol{\Delta} \cdot \boldsymbol{\alpha}_0) \int_0^{2\pi} \mathrm{d}\phi \exp(-\mathrm{i}n'\phi)g_{\text{el}}^{\text{Och}}(q_i, \Delta),
$$

$$(10.207)$$

where [47,112]

$$
g_{\text{el}}^{\text{Och}}(q_i, \Delta) = -\frac{32}{q_i^2(\Delta^2 + 4)^2} \qquad (10.208)
$$

is the field-free Ochkur amplitude for elastic exchange electron–atomic hydrogen scattering corresponding to an incident wave vector $\mathbf{q}_i$ and a momentum transfer $\boldsymbol{\Delta} = \mathbf{k}_i - \mathbf{k}_f = \mathbf{q}_i - \mathbf{q}_f$. Because we are dealing here with fast electrons, we can write

$$
\frac{1}{q_i^2} = \frac{1}{|\mathbf{k}_i + \boldsymbol{\beta}_0 \sin \phi|^2} \simeq \frac{1}{k_i^2}\left[1 - \frac{2\mathbf{k}_i \cdot \boldsymbol{\beta}_0 \sin \phi}{k_i^2}\right]. \qquad (10.209)
$$

Upon substitution of this result into Equations (10.208) and then (10.207), the integral on the ϕ variable can be performed. Moreover, the summation on n' reduces to the terms $n' = 0, \pm 1$. Using the Bessel function recurrence relation (10.97) we find that

$$g_{\text{el},0}^{n,\text{B1}} \simeq J_n(\boldsymbol{\Delta} \cdot \boldsymbol{\alpha}_0) \left\{ -\frac{32}{k_i^2(\Delta^2+4)^2} \left[1 - \frac{2n\omega}{k_i^2} \frac{\mathbf{k}_i \cdot \boldsymbol{\alpha}_0}{\boldsymbol{\Delta} \cdot \boldsymbol{\alpha}_0} + \mathcal{O}(\omega^2) \right] \right\} \quad (10.210)$$

and therefore

$$g_{\text{el},0}^{n,\text{B1}} \simeq J_n(\boldsymbol{\Delta} \cdot \boldsymbol{\alpha}_0) g_{\text{el}}^{\text{Och}}(k_i^*, \Delta), \quad (10.211)$$

where $g_{\text{el}}^{\text{Och}}(k_i^*, \Delta)$ is given for an atomic hydrogen target by Equation (10.208) with $\mathbf{q}_i$ replaced by $\mathbf{k}_i^*$. A result identical in form to that of Equation (10.211) can be derived for complex atoms. As in the case of the direct scattering amplitude f_{el}^n, we shall neglect the momentum shift $\mathbf{k}_i \to \mathbf{k}_i^*$ in the exchange scattering amplitude $g_{\text{el},0}^{n,\text{B1}}$ since this shift is only important at small momentum transfers, where the quantity $g_{\text{el},0}^{n,\text{B1}}$ is much smaller than the dressing term in Equation (10.197) for f_{el}^n.

The contribution to $g_{\text{el}}^{n,\text{B1}}$, which is of first order in $\mathcal{E}_0$, is given by

$$g_{\text{el},1}^{n,\text{B1}} = i\mathcal{E}_0(2\pi)^{-1} \sum_{n'=-\infty}^{+\infty} i^{n'} J_{n-n'}(\boldsymbol{\Delta} \cdot \boldsymbol{\alpha}_0) \int d\mathbf{r}_0 \, d\mathbf{r}_1 \exp(-i\mathbf{k}_f \cdot \mathbf{r}_1)$$

$$\times \left[\frac{J_{n'}(\boldsymbol{\beta}_0 \cdot \mathbf{r}_{01})}{r_{01}} \right] \exp(i\mathbf{k}_i \cdot \mathbf{r}_0)[\psi_0(\mathbf{r}_0)\tilde{\psi}_0(\mathbf{r}_1) + \tilde{\psi}_0(\mathbf{r}_0)\psi_0(\mathbf{r}_1)] \quad (10.212)$$

and can be studied in a way similar to that used for the zeroth-order term $g_{\text{el},0}^{n,\text{B1}}$. Using for $\tilde{\psi}_0$ the closure approximation (10.195), one finds that

$$g_{\text{el},1}^{n,\text{B1}} \simeq -\mathcal{E}_0(2\pi)^{-1} \bar{E}^{-1} \sum_{n'=-\infty}^{+\infty} i^{n'} J_{n-n'}(\boldsymbol{\Delta} \cdot \boldsymbol{\alpha}_0)$$

$$\times \int_0^{2\pi} d\phi \exp(-in'\phi) \hat{\boldsymbol{\epsilon}} \cdot [\nabla_{\mathbf{q}_f} - \nabla_{\mathbf{q}_i}] g_{\text{el}}^{\text{Och}}(q_i, \Delta)$$

$$\simeq -\mathcal{E}_0(2\pi)^{-1} \bar{E}^{-1} \sum_{n'=-\infty}^{+\infty} i^{n'} J_{n-n'}(\boldsymbol{\Delta} \cdot \boldsymbol{\alpha}_0)$$

$$\times \int_0^{2\pi} d\phi \exp(-in'\phi) \hat{\boldsymbol{\epsilon}} \cdot (2\mathbf{q}_i q_i^{-2}) g_{\text{el}}^{\text{Och}}(q_i, \Delta). \quad (10.213)$$

Proceeding as in the case of $g_{\text{el},0}^{n,\text{B1}}$, we obtain, in the closure approximation,

$$g_{\text{el},1}^{n,\text{B1}} \simeq 2\mathcal{E}_0 \bar{E}^{-1} \hat{\boldsymbol{\epsilon}} \cdot (\mathbf{k}_i k_i^{-2}) g_{\text{el},0}^{n,\text{B1}}, \quad (10.214)$$

and we see that for laser fields such that $\mathcal{E}_0 \ll 1$ (in a.u.) and for fast electrons $(k_i \gg 1)$ the term $g_{el,1}^{n,B1}$ can be neglected with respect to $g_{el,0}^{n,B1}$.

We shall now illustrate the semi-perturbative theory discussed above by considering first the case of an atomic hydrogen target. Using Equation (10.199) for the direct soft-photon laser-assisted elastic scattering amplitude f_{el}^n, with $f^{B1}(\psi_0 \to \tilde{\psi}_0)$ given by Equation (10.168), together with Equation (10.211) for the corresponding elastic laser-assisted exchange amplitude $g_{el}^{n,B1}$ (with no momentum shift, so that $\mathbf{k}_i^*$ is replaced by $\mathbf{k}_i$), Byron, Francken and Joachain [104] have evaluated the differential cross section for fast electron–atomic hydrogen "elastic" collisions with the transfer of $|n|$ photons:

$$\frac{d\sigma_{el}^n}{d\Omega} = \frac{k_f(n)}{k_i} \left[\frac{1}{4} \left| f_{el}^n + g_{el}^{n,B1} \right|^2 + \frac{3}{4} \left| f_{el}^n - g_{el}^{n,B1} \right|^2 \right], \tag{10.215}$$

where it is assumed that both the incident electron beam and the hydrogen atom beam are unpolarized, and that no measurement is made to distinguish the final spin states. We note that at low frequencies and small scattering angles, $d\sigma_{el}^n/d\Omega$ is nearly identical for the cases of absorption ($n > 0$) and emission ($n < 0$) of a given number of photons in those geometries for which $\hat{\boldsymbol{\epsilon}}$ is taken to be parallel to $\boldsymbol{\Delta}$ or perpendicular to $\mathbf{k}_i$. However, as mentioned above, the sign of n can strongly influence small-angle laser-assisted differential cross sections when $\hat{\boldsymbol{\epsilon}}$ is parallel to $\mathbf{k}_i$.

In Fig. 10.6, the differential cross section given by Equation (10.215) for the case $n = 1$ (absorption of one laser photon) is displayed for an incident electron energy of 100 eV, a laser photon energy of 2 eV (corresponding to a wavelength of 620 nm), an electric-field strength $\mathcal{E}_0 = 0.02$ a.u. (10^8 V cm^{-1}) and a geometry in which $\hat{\boldsymbol{\epsilon}}$ is parallel to $\boldsymbol{\Delta}$. Also shown in Fig. 10.6 are the results obtained by neglecting the target-dressing effect, that is by omitting the second term on the right-hand side of Equation (10.199). We see that the angular distribution displayed in Fig. 10.6 is radically modified at small angles by the dressing term. The dressed differential cross section exhibits a dramatic enhancement near the forward direction, due to the fact that the dressing term takes the form of an S–P amplitude, which behaves like Δ^{-1} when Δ is small [47]. In fact, it can be shown [104], using Equation (10.197), that in the geometry for which $\hat{\boldsymbol{\epsilon}}$ is parallel to $\boldsymbol{\Delta}$ one has for any atom and for $n \neq 0$

$$f_{el}^n(\theta = 0) = J_n(\Delta \alpha_0) \left[f_{el}^{EBS}(\theta = 0) - \frac{4\bar{\alpha} E_{k_i}}{|n|} \right], \tag{10.216}$$

where $\bar{\alpha}$ is the static dipole polarizability of the target atom. Thus, at small angles, the modification of the angular distribution due to the dressing of the target by the laser field is always important provided $|n|$ is not too large, and increases with $\bar{\alpha}$ and E_{k_i}. Also shown in Fig. 10.6 are the results obtained by using Equation

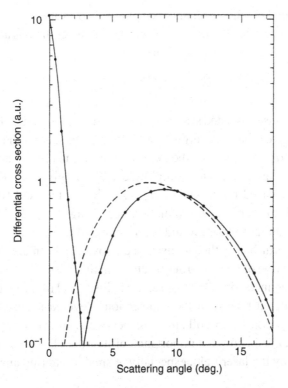

Figure 10.6. Differential cross section (in a.u.), as given by Equation (10.215), for "elastic" electron–atomic hydrogen scattering with the absorption of one photon at an incident electron energy of 100 eV, as a function of the scattering angle θ (in degrees). The laser field is spatially homogeneous, monochromatic and linearly polarized, with the polarization vector $\hat{\epsilon}$ taken to be parallel to the momentum transfer $\mathbf{\Delta}$. The laser photon energy is $\hbar\omega = 2$ eV and the electric-field strength is $\mathcal{E}_0 = 0.02$ a.u. (10^8 V cm^{-1}). Full curve: full calculation using Equation (10.199) for f_{el}^n and Equation (10.211) for $g_{\text{el}}^{n,\text{B1}}$; the dots on the full curve refer to the same calculation performed using the closure approximation of Equation (10.200) to evaluate f_{el}^n, with an average excitation energy $\bar{E} = (4/9)$ a.u. Broken curve: results obtained by neglecting the dressing of the target, i.e. by omitting the second term on the right-hand side of Equation (10.199). (From F. W. Byron, P. Francken and C. J. Joachain, *J. Phys. B* **20**, 5487 (1987).)

(10.200), in which the dressing term is evaluated in the closure approximation by choosing the average excitation energy $\bar{E} = (4/9)$ a.u., which gives the exact value $\bar{\alpha} = 4.5$ a.u., for the static dipole polarizability of the hydrogen atom in the closure approximation. As expected from the discussion following Equation (10.169), the agreement between the "exact" and the closure results is seen to be excellent throughout the small-angle region.

Byron, Francken and Joachain [104] have also evaluated the differential cros[
section for the case of fast "elastic" electron–helium collisions accompanied by th[
transfer of $|n|$ photons using the expression

$$\frac{d\sigma_{el}^n}{d\Omega} = \frac{k_f(n)}{k_i}\left|f_{el}^n - g_{el}^{n,B1}\right|^2.$$ (10.217

Here the direct laser-assisted scattering amplitude f_{el}^n is given by Equation (10.200)
with an average excitation energy $\bar{E} = 1.15$ a.u., which gives the exact static dipol[
polarizability $\bar{\alpha} = 1.38$ a.u. of the helium atom in the closure approximation. Th[
laser-assisted elastic exchange amplitude $g_{el}^{n,B1}$ is given by Equation (10.211)
where k_i^* is replaced by k_i and where now g_{el}^{Och} is the field-free Ochkur ampli
tude for elastic exchange electron–helium scattering. They obtained results that ar[
qualitatively similar to those found for atomic hydrogen.

The semi-perturbative theory discussed above has been applied not only to th[
"elastic" scattering of fast electrons by atoms in the presence of a laser field, als[
to laser-assisted inelastic [113] and (e, 2e) collisions [114, 115]. These application[
will be considered below, in the sub-sections devoted to laser-assisted inelasti[
and (e, 2e) electron–atom collisions, respectively. In every case, the results tha[
have been obtained show the importance of the role played by the dressing of the
atomic states by the laser field, especially at small scattering angles and high lase[
frequencies.

10.3.1.2 Extensions of the semi-perturbative theory

We shall now examine how the semi-perturbative theory can be improved. We still
assume that the projectile electron is fast, so that a perturbative treatment of the
projectile–target interaction (using the Born series) is adequate. The only cause of
concern is therefore the first-order treatment of the interaction between the laser field
and the target atom (i.e. the treatment of target-dressing effects to first order in the
electric-field strength $\mathcal{E}_0$). The reason is two-fold. Firstly, if $\mathcal{E}_0$ is increased (within
reasonable limits, otherwise the target atom would be ionized too rapidly for the
laser-assisted electron–atom experiment to be performed), it is expected that higher-
order terms in the laser field–target atom interaction could become significant.
Secondly, and more importantly, a first-order treatment of the laser field–target atom
interaction is inadequate in the immediate vicinity of a resonance, namely when the
laser photon energy matches the excitation energy of an intermediate state, as seen
from Equation (3.199) with $j = 0$. Perturbation theory will then exhibit a spurious
divergence at the resonance. There is no divergence when the states which are
resonantly coupled by the laser field are treated non-perturbatively, for example by
representing the atom by a two-state model in the rotating wave approximation [116,
117]. Unnikrishnan [118] has proposed to improve this simple two-state model,

which is clearly not valid off resonance, by including the first-order corrections arising from the counter-rotating terms and by treating the coupling of the two resonant states to the other, non-resonant states by first-order perturbation theory. A more general and accurate treatment of the laser field–target atom interaction is that of Francken and Joachain [11]. Their method consists in treating a few target states (namely, those which are resonantly coupled by the laser field) "exactly," using the Floquet approach, while the coupling of the laser field with the remaining target states is treated perturbatively. The approaches of Unnikrishnan [118] and Francken and Joachain [11] belong to the category of essential states methods, discussed in Section 3.3.

The Floquet theory is clearly a natural framework for developing a non-perturbative treatment of the dressing of the target by the laser field. It is appropriate to describe laser-assisted electron–atom collisions in which the duration of one optical cycle of the laser field is much shorter than (i) the temporal scale of the variation of the laser intensity experienced by the target atoms as the laser pulse passes by; (ii) the lifetime of the target atoms in the laser field; and (iii) the duration of the pulses of incident electrons that are directed onto the target atoms. Condition (i) has been fulfilled in experiments carried out so far in laser-assisted electron-atom collisions. However, together with condition (ii), it sets an upper limit to the laser pulse peak intensity that can be considered for a given angular frequency and pulse duration. In what follows, we shall assume that the decaying atom can be characterized by a single Floquet state before the collision occurs; this would not be possible if the intensity were varying too rapidly. Moreover, representing the atom by a single (pure) state vector implies that the duration of the laser pulses is limited not only by the rate of photoionization, but also by the rate of fluorescence, since spontaneous decay destroys the coherence of the initial Floquet state [119]. We shall consider the case where the laser pulse is shorter than the fluorescence lifetime of the atom, and assume that the rate of photoionization is larger than the rate of fluorescence. The Floquet theory used below for studying target-dressing effects is therefore not suited to the cases where the atom can decay by fluorescence before the collision occurs or where the laser bandwidth is significant. Condition (iii) amounts to imposing that the energy width of the projectile electron wave packet is small compared to the energy of the photons and to the differences in energy of the relevant target states, so that the incident electron wave packet can be approximated, as in field-free time-independent collision theory, by a monoenergetic beam of infinite duration. This makes it possible to distinguish processes in which different numbers of photons have been absorbed or emitted. Since it is not possible to define when the collision takes place if there is no uncertainty in the energy of the projectile, one should not expect the collision cross sections to depend on the phase of the laser field when condition (iii) is fulfilled.

We now discuss Born–Floquet calculations performed by Dörr *et al.* [12, 120] These calculations generalize the semi-perturbative theory of Byron and Joachain [10] and go also beyond the work of Francken and Joachain [11] since all the states of the hydrogen atom target are coupled "exactly" to the laser field and multiphoton ionization is taken into account by using the non-Hermitian Floquet method. The interactions between the laser field and both the unbound electron and the target atom are treated in a non-perturbative way: the former by using a Gordon–Volkov wave function and the latter by using the non-Hermitian Floquet method. The interaction between the unbound electron and the target atom is treated by using the first Born approximation. The Born–Floquet theory is therefore physically meaningful for fast incident electrons, provided that the laser field does not act for a long time before the collision takes place, since no allowance is made for fluorescence. It is also applicable to cases where the semi-perturbative theory [10] is questionable, namely at higher laser intensities or for resonant laser fields.

Dörr *et al.* [12, 120] have applied their Born–Floquet theory to the case of an atomic hydrogen target, for which the contribution of the entire spectrum (bound and continuum) to the dressing of the atomic states can be taken into account by expanding the wave function of the target atom, dressed by the laser field and decaying by multiphoton ionization, on a discrete basis of complex Sturmian functions, as explained in Section 4.6. They studied the case of incident electrons having high impact energies, $E_{k_i} \geq 500\,\text{eV}$, where electron exchange effects can be neglected [47].

As an illustration of the Born–Floquet theory, we show in Fig. 10.7 the differential cross sections for laser-assisted "elastic" scattering of electrons by atomic hydrogen with the absorption of n photons ($n = 0, 1, 2$), as a function of the electric-field strength, and for a fixed scattering angle $\theta = 0.5°$. The laser field is spatially homogeneous, monochromatic and linearly polarized, with the polarization vector $\hat{\epsilon}$ parallel to the momentum transfer $\boldsymbol{\Delta}$. It has a wavelength of 620 nm, corresponding to a photon energy of 2 eV. The incident electron energy is $E_{k_i} = 500\,\text{eV}$, which is large enough for a first Born treatment of the projectile electron–target atom potential V_d to be adequate [47]. We remark that at an electric-field strength $\mathcal{E}_0 = 0.0377$ a.u. (corresponding to an intensity of $5.0 \times 10^{13}\,\text{W}\,\text{cm}^{-2}$), multiphoton ionization of H(1s) is fully non-perturbative [121]. The lifetime of the atom in the laser field is 3.2 ps. Despite these extreme conditions, the differential cross sections calculated using the Born–Floquet theory with non-perturbative target-dressing and those calculated using the semi-perturbative theory with target-dressing taken into account to first order in $\mathcal{E}_0$ are in good agreement. This illustrates how different the importance of certain harmonic components is in the scattering case compared to the multiphoton ionization case. A detailed analysis [12] reveals that the open-channel

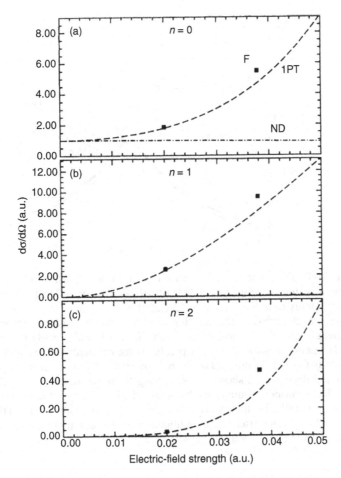

Figure 10.7. Differential cross section (in a.u.) for "elastic" electron–atomic hydrogen scattering with the absorption of (a) no photons, (b) one photon and (c) two photons, at an incident electron energy of 500 eV, as a function of the electric-field strength (in a.u.). The laser field is spatially homogeneous, monochromatic and linearly polarized, with the polarization vector $\hat{\epsilon}$ taken to be parallel to the momentum transfer $\mathbf{\Delta}$. The laser photon energy is $\hbar\omega = 2$. The Born–Floquet results obtained when the target atom is dressed non-perturbatively by using the Floquet method are represented by solid squares (F). The broken curves (1PT) correspond to results obtained when the target atom is dressed in first-order perturbation theory by using the first-Born dressed scattering amplitude (10.144). The dot-dashed curve (ND) refers to the results obtained by neglecting the dressing of the target. (From M. Dörr *et al.*, *Phys. Rev. A* **49**, 4852 (1994).)

part of the Floquet wave function does not contribute much to the differential cross section, at least in the absence of resonances, even in cases where the multiphoton ionization rate is large. In other words, the projectile electron is scattered essentially as if the atom were not decaying. By contrast, it is important to retain a sufficiently

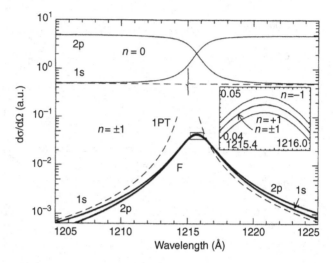

Figure 10.8. Differential cross section (in a.u.) for "elastic" electron–atomic hydrogen scattering with the transfer of $|n|$ photons ($n = 0, \pm 1$), at an incident electron energy of 500 eV and a fixed scattering angle ($\theta = 10°$), as a function of the laser wavelength (in Å), in the vicinity of the one-photon 1s–2p resonance. The laser field is spatially homogeneous, monochromatic and linearly polarized, with the polarization vector $\hat{\boldsymbol{\epsilon}}$ taken to be parallel to the momentum transfer $\boldsymbol{\Delta}$. The laser intensity is $I = 10^{10}\,\mathrm{W\,cm^{-2}}$. The Born–Floquet results with non-perturbative Floquet target-dressing are shown for scattering from the adiabatic 1s or 2p dressed states. The semi-perturbative results obtained when the target atom is dressed in first-order perturbative theory (1PT) are shown only for scattering from the 1s state and $n = 0$ or 1. The inset is a magnification of the region delimited by the box. (From M. Dörr *et al.*, *Phys. Rev. A* **49**, 4852 (1994).)

large number of angular momentum components l in the wave function; in the present case values of l up to 7 must be included in order to obtain differential cross sections accurate to within 1%. It is also worth noting that the contribution of the *continuum* states of the field-free atom to the *closed*-channel part of the Floquet wave function is considerable.

The Born–Floquet approach is particularly useful when the laser field brings the initial state into resonance with another state. This is illustrated in Fig. 10.8. In the weak-field limit, the 1s and the 2p states are in resonance through a one-photon dipole coupling at a wavelength of 121.5 nm. While the laser-assisted "elastic" electron–atomic hydrogen scattering cross section diverges at this wavelength when the target atom is dressed by using first-order perturbation theory, it remains finite when the coupling between these two states is taken into account non-perturbatively. Off resonance, at 122.5 nm, the ionization widths of the 1s and 2p states are 4.36×10^{-10} a.u. and 7.66×10^{-8} a.u., respectively. The ionization width of the 1s state at the resonance wavelength of 121.5 nm is 2.8×10^{-8} a.u. This last number should be compared to the natural width of the unperturbed 2p level, 1.5×10^{-8} a.u. We see that loss of coherence is not a cause of concern in this particular case, since photoionization is faster than spontaneous decay. If the intensity is kept constant and the wavelength is increased adiabatically, the Floquet state corresponding to the dressed 1s state below 121.5 nm loses its character as the resonance is passed, and takes on a 2p character; conversely, the dressed 2p state acquires the character of a dressed 1s state above 121.5 nm. The character interchange of the two states, which is associated with an *avoided crossing* in the real part of their quasi-energies [122], manifests itself clearly in the cross section for elastic scattering without a net exchange of photons, as seen in Fig. 10.8. We note that the crossing of the curves actually occurs at a slightly longer wavelength than in the zero-field limit, because of the shift and width of the dressed states. On the other hand, the cross sections for scattering from the dressed 1s state or from the dressed 2p state with a net exchange of one photon are also very close near resonance. The $|n| = 1$ Born–Floquet results consist of four curves, namely two curves for $n = 1$ (one for each dressed state) and two curves for $n = -1$. The results for $n = \pm 1$ are very close. At the crossing, two of the four curves osculate: this is why there are only three curves in the inset of Fig. 10.8. As expected, and for both $n = 0$ and $|n| = 1$, it is only in the immediate vicinity of the resonance that there is a substantial difference between the results obtained with target-dressing treated by using first-order perturbation theory and those obtained with non-perturbative Floquet target-dressing.

The Born–Floquet theory has also been used to analyze the differential cross sections for laser-assisted "elastic" collisions of 500 eV electrons by atomic hydrogen in the neighborhood of a two-photon resonance between the 1s and 2s states [12] and of a three-photon resonance between the 1s state and the 2p state [120]. In both cases, the dressed target states cannot be obtained by using first-order perturbation theory. In contrast with the one-photon and two-photon cases, the crossing between the two quasi-energy curves is a *true crossing* for the 1s–2p three photon resonance – that is, the real parts of the quasi-energies intersect. In all three resonant cases (one-, two- and three-photon resonances), the Born–Floquet differential cross sections are strongly enhanced at the resonance.

10.3.1.3 *R-matrix–Floquet calculations*

Let us now consider "elastic" collisions of slow electrons with atoms in the presence of a moderately strong laser field. In this case, a non-perturbative treatment of the three types of interactions (unbound electron–target atom, unbound electron–laser field and target atom–laser field) is required. The *R*-matrix–Floquet (RMF) theory [13, 14] discussed in Section 4.7 provides such a treatment. It is interesting to note that the semi-perturbative theory described in Section 10.3.1.1 and the RMF theory are complementary. As pointed out in Section 1.5, the former breaks down for slow incident electrons, where the Born series cannot be used to treat the electron–atom interaction [47]. The latter is difficult to apply for fast incident electrons, where many partial waves are required to calculate the cross section accurately.

The implementation of the RMF theory for the study of laser-assisted collisions of slow electrons with real atoms has been discussed by Terao-Dunseath and Dunseath [17]. They showed that the calculations can be simplified by using channel functions built from laser field-dressed target states. Charlo *et al.* [15] studied electron–atomic hydrogen "elastic" collisions in a CO_2 laser field of photon energy $\hbar\omega = 0.117$ eV, corresponding to an angular frequency of $\omega = 0.0043$ a.u. Their results are in excellent agreement with those obtained by Kylstra and Joachain [5] who solved the Floquet–Lippmann–Schwinger equations (FLSE) for laser-assisted low-energy electron–H(1s) "elastic" scattering in the static approximation, as we have seen in Section 10.2. Charlo *et al.* [15] also found good agreement between their RMF results and those obtained by applying the low-frequency (soft-photon) approximation (LFA) of Kroll and Watson [8] to the laser-assisted "elastic" scattering of electrons by H(1s) in the static approximation. Similar conclusions also hold for the "elastic" scattering of low-energy electrons by helium atoms in the presence of a CO_2 laser field, where the RMF results of Dunseath and Terao-Dunseath [18] are in excellent agreement with the FLSE calculations of Kylstra and Joachain [6] and in good agreement with the values obtained by using the LFA approximation [8].

10.3.2 *Laser-assisted inelastic electron–atom collisions*

We now turn to the study of inelastic electron–atom collisions in the presence of a laser field. Let $\mathbf{k}_i$ and $\mathbf{k}_f$ be the wave vectors of the unbound electron before and after the collision, $E_{k_i} = \hbar^2 k_i^2/(2m)$ and $E_{k_f} = \hbar^2 k_f^2/(2m)$ being the corresponding energies. Let i and f denote the initial and final atomic bound states, respectively, having energies E_i and E_f. For an inelastic transition $(\mathbf{k}_i, i \rightarrow \mathbf{k}_f, f)$ accompanied by the transfer of $|n|$ photons between the electron–atom system and the laser field, we have, from energy conservation,

$$E_{k_f} = E_{k_i} - \hbar\omega_{fi} + n\hbar\omega, \qquad (10.218)$$

where the Bohr angular frequency ω_{fi} is given by $\omega_{fi} = (E_f - E_i)/\hbar$. The first theoretical investigation of the SEPE process (10.2) was made by Göppert-Mayer [123] within the framework of perturbation theory, well before the invention of the laser.

We begin our discussion by considering the case of fast electrons. Early theoretical investigations of laser-assisted excitation of atoms by high-energy electrons were performed by neglecting the laser dressing of the target atom and using the first Born approximation to treat the projectile electron–target atom direct interaction. In particular, Bhakar and Choudhury [124] and Mohan and Chand [125, 126] have carried out such calculations for atomic hydrogen as the target. Similar calculations have been performed by Cavaliere and Leone [127] for alkali atoms. Laser-assisted inelastic collisions of fast electrons with atoms have also been considered by Pert [128], Fedorov and Yudin [129] and Pundir and Mathur [130, 131].

10.3.2.1 Semi-perturbative theory

The semi-perturbative theory of Byron and Joachain [10] has been applied to laser-assisted inelastic collisions of fast electrons by atomic hydrogen and helium by Francken, Attaouti and Joachain [113]. Using the first Born approximation to treat the electron–atom interaction potential V_d, the first Born S-matrix element for the direct excitation transition $(\mathbf{k}_i, 0 \to \mathbf{k}_f, f)$ from the ground state of the atom to a final atomic bound state f of energy E_f, in the presence of a laser field, is given in atomic units by

$$S_{f,0}^{B1} = -i \int_{-\infty}^{+\infty} dt \langle \chi_{\mathbf{k}_f}^V(\mathbf{r}_0, t) \Phi_f^V(X, t) | V_d(\mathbf{r}_0, X) | \chi_{\mathbf{k}_i}^V(\mathbf{r}_0, t) \Phi_0^V(X, t) \rangle, \quad (10.219)$$

where $\Phi_f^V(X, t)$ is the dressed atomic wave function (3.199) with $j = f$. After performing the time integration, one obtains

$$S_{f,0}^{B1} = (2\pi)^{-1} i \sum_{n=-\infty}^{+\infty} \delta(E_{k_f} - E_{k_i} + \omega_{f0} - n\omega) f_{f,0}^{n,B1}(\boldsymbol{\Delta}), \quad (10.220)$$

where $f_{f,0}^{n,B1}(\boldsymbol{\Delta})$ is the first Born approximation to the scattering amplitude corresponding to the excitation transition $(\mathbf{k}_i, 0 \to \mathbf{k}_f, f)$ with the transfer of $|n|$ photons.

Francken and Joachain [24] have investigated the modifications which are required to extend the above analysis to intermediate electron energies, where higher-order terms in the *direct* electron–atom interaction potential V_d as well as *exchange* effects must be taken into account, following the EBS theory.

The excitation of atomic hydrogen from its ground state to the $n = 2$ level under the simultaneous action of fast electron impact and a spatially homogeneous, monochromatic, linearly polarized laser field has also been considered by

Bhattacharya, Sinha and Sil [132], who constructed the dressed wave functions of the hydrogen atom by using first-order perturbation theory in the parabolic coordinate representation of the unperturbed eigenfunctions of the H atom. The advantage of using parabolic coordinates in this case, already recognized earlier by Cavaliere, Leone and Ferrante [133], is that the dressed wave functions can be obtained in terms of a finite number of Laguerre polynomials instead of an infinite sum as in the usual perturbative treatment. The laser-assisted excitation of atoms by fast electrons has also been considered by Rahman and Faisal [134–136] and by Jetzke *et al.* [137] and Jetzke, Broad and Maquet [138], who treated the interaction with the laser field to first order of perturbation theory.

10.3.2.2 *Extensions of the semi-perturbative theory*

Let us now turn to the excitation of atoms by fast electrons in the presence of a nearly resonant laser field. As in the case of laser-assisted "elastic" collisions considered in Section 10.3.1, the semi-perturbative theory, which treats the interaction of the laser field with the target atom to first order of perturbation theory, must be modified in the immediate vicinity of a resonance. This problem has been studied by Francken and Joachain [11], who assumed that the laser field is spatially homogeneous, monochromatic and linearly polarized, and that its photon energy is chosen to be close to the energy difference between two excited final states. Both the laser field–unbound electron and laser field–target atom interactions were treated in a non-perturbative way, while the electron–atom interaction was treated by using the first Born approximation. This treatment of the resonant excitation problem requires a solution of the TDSE (10.140) for the target atom in the laser field which goes beyond the first-order perturbative treatment used in the semi-perturbative theory of Byron and Joachain [10]. Since it is necessary only to treat exactly the resonant part of the interaction, Francken and Joachain [11] made use of the two orthogonal projection operators P and Q introduced in Section 3.3. We recall that P projects on the subspace $\mathcal{H}_P$ of the states which are included exactly in the calculation. This subspace contains in the present case the initial (ground) state and the final states that are resonantly coupled by the interaction. The TDSE (10.140) describing the target atom in the laser field is then approximated, in a first step, by the following simplified equation:

$$i\hbar\frac{\partial}{\partial t}P\Phi = (PH_TP)P\Phi \tag{10.221}$$

in which only the few dominant target states are coupled. Since the Hamiltonian $H_T(t)$ is periodic in time, one can use the Floquet method to seek solutions of

Equation (10.221) having the form

$$P\Phi_j^{\mathrm{V}}(X,t) = \exp(-i\mathsf{E}_j^{\mathrm{V}} t/\hbar)\exp\left[\frac{i}{\hbar}\frac{e^2 N}{2m}\int_{-\infty}^{t} \mathbf{A}^2(t')dt'\right]$$

$$\times \exp[-i\mathbf{a}(t)\cdot\mathbf{R}]\sum_{j'\in\mathcal{H}_P}\sum_{N=-\infty}^{+\infty} C_{j'j}^N \exp(-iN\omega t)\psi_{j'}(X) \quad (10.222)$$

in the velocity gauge, with $\psi_{j'}(X)$ denoting a target eigenfunction of energy $E_{j'}$ in the absence of the laser field. The quasi-energies $\mathsf{E}_j^{\mathrm{V}}$ and the coefficients $C_{j'j}^N$ can be found by solving the eigenvalue problem

$$(E_{j'} - N\hbar\omega)C_{j'j}^N - i\frac{\mathcal{E}_0}{2}\sum_{k\in\mathcal{H}_P} M_{j'k}(C_{kj}^{N-1} - C_{kj}^{N+1}) = \mathsf{E}_j^{\mathrm{V}} C_{j'j}^N. \quad (10.223)$$

Finally, a correction to the approximate wave function $P\Phi_j^{\mathrm{V}}(X,t)$ can be found by using first-order perturbation theory to treat the coupling to all the states that are not included in the subspace $\mathcal{H}_P$. Thus, we obtain

$$\Phi_j^{\mathrm{V}}(X,t) = P\Phi_j^{\mathrm{V}}(X,t) + Q\Phi_j^{\mathrm{V}}(X,t), \quad (10.224)$$

where $P\Phi_j^{\mathrm{V}}(X,t)$ is given by Equation (10.222) and the first-order approximation to $Q\Phi_j^{\mathrm{V}}(X,t)$ is a solution of the equation

$$i\hbar\frac{\partial}{\partial t}Q\Phi = (QH_{\mathrm{T}}P)P\Phi \quad (10.225)$$

and is given by

$$Q\Phi_j^{\mathrm{V}}(X,t) \simeq \exp(-i\mathsf{E}_j^{\mathrm{V}} t/\hbar)\exp\left[\frac{i}{\hbar}\frac{e^2 N}{2m}\int_{-\infty}^{t}\mathbf{A}^2(t')dt'\right]\exp[-i\mathbf{a}(t)\cdot\mathbf{R}]\frac{i\mathcal{E}_0}{2\hbar}$$

$$\times \sum_{j'\in\mathcal{H}_P}\sum_{k\in\mathcal{H}_P} M_{kj'}\sum_{N=-\infty}^{+\infty}\frac{C_{j'j}^{N-1} - C_{j'j}^{N+1}}{E_k - \mathsf{E}_j^{\mathrm{V}} - N\hbar\omega}\exp(-iN\omega t)\psi_k(X).$$

$$(10.226)$$

The laser-dressed atomic wave functions (10.224) can then be introduced into the first Born S-matrix element (10.219) and the corresponding first Born scattering amplitude $f_{f,0}^{n,\mathrm{B1}}$ can be evaluated as in the case of the semi-perturbative theory.

Francken and Joachain [11] have applied this approach to analyze the laser-assisted excitation of the 2^1S and 2^1P states of helium by fast electrons of incident energy $E_{k_i} = 500\,\mathrm{eV}$, in the first Born approximation. Subsequently, Smith and Flannery [139, 140] have used the Floquet method (including a few target states in the calculation) in a study of laser-assisted 1s–2s and 1s–2p excitation of atomic

hydrogen by intermediate energy electrons, the electron–atom interaction bein
taken into account either within the framework of the multichannel eikonal fo
malism [141, 142] or in the first Born approximation. The Born–Floquet theo
developed by Dörr *et al.* [12, 120] and discussed in Section 10.3.1 for the case c
laser-assisted "elastic" electron–atom collisions, has been applied by Vučić [14?
to study the excitation of the 2s and 2p states of atomic hydrogen by fast electron
in the presence of a laser field.

Purohit and Mathur [144] have analyzed the electron impact excitation of th
metastable 2s state of atomic hydrogen in the presence of a resonant laser field
They also studied the optically forbidden transitions 2s → 3s, 4s, 3d and 4d. Th
laser field was assumed to be circularly polarized, and its interaction with the hydro
gen atom was treated by using the rotating wave approximation. The laser photor
energy was chosen to be equal to the energy difference between the 3p and 2
states of the H atom. The role of the laser field in the excitation of individual mag
netic substates ($m_f = 0, \pm1, \pm2$) was studied. Purohit and Mathur [144] founc
that for these optically forbidden transitions a significant increase of the excita
tion cross section is obtained through joint collisions with electrons and photons
If the laser angular frequency matches that of an atomic resonance, a two-leve
model of the target atom can be used. This was done in previous work by Hahn
and Hertel [145], Gazazian [116], Mittleman [93, 94, 146] and Pundir and Mathur
[147]. The two-level atomic model was also used in the work of Purohit and Mathur
[144] to describe the resonantly laser-dressed hydrogen atom. Electron exchange
effects were neglected and the electron–atom interaction V_d in the direct arrangement channel was treated in the first Born approximation. The cross sections for the
transitions 2s → 3s, 4s, 3d and 4d were calculated at laser intensities ranging from
10^6 W cm^{-2} to 10^{11} W cm^{-2}, and for incident electron energies between 50 and
200 eV. The conclusion of Purohit and Mathur [144] about the important increase
in the cross sections corresponding to laser-assisted optically forbidden transitions
from the initial metastable 2s state of atomic hydrogen is in agreement with earlier
results of Rahman and Faisal [135, 136] for laser-assisted electron impact excitations from the ground state (1s) of atomic hydrogen and of Pundir and Mathur [147]
for such excitations from the ground state (2s) of lithium and the ground state (3s)
of sodium.

10.3.2.3 Low-energy simultaneous electron–photon excitation

Let us now consider laser-assisted electron–atom collisions at low electron energies.
Of particular interest is the simultaneous electron–photon excitation (SEPE) process
in helium

$$e^- + He(1^1S) + n\hbar\omega \rightarrow e^- + He(2^3S), \tag{10.227}$$

which has been the subject of several experimental studies [38–43].

A number of theoretical investigations of the SEPE process at low electron impact energies have also been made [16, 148–156]. In the soft-photon case, such as with a CO_2 laser, Geltman and Maquet [151] have proposed a simple extension of the low-frequency approximation of Kroll and Watson [8] based on the following arguments. They first noted that for laser intensities of $10^8 \, W \, cm^{-2}$ (used in the SEPE experiments with CO_2 lasers) the target-dressing effects are only significant at small scattering angles [10, 101, 104, 106, 113, 138], and would lead to negligibly small corrections of the *total* laser-assisted cross section. If then one neglects the dressing of the target atom by the laser field, the first Born laser-assisted differential cross section for the SEPE excitation process $(\mathbf{k}_i, 0 \rightarrow \mathbf{k}_f, f)$, summed over all photon transfers, is given in atomic units by

$$\frac{\mathrm{d}\sigma_{\mathrm{SEPE}}^{\mathrm{B1}}}{\mathrm{d}\Omega}(E_{k_i}) = \sum_n J_n^2(\boldsymbol{\Delta} \cdot \boldsymbol{\alpha}_0) \frac{\mathrm{d}\sigma_{f,0}^{\mathrm{B1}}}{\mathrm{d}\Omega}(E_{k_i} + n\omega), \qquad (10.228)$$

where

$$\frac{\mathrm{d}\sigma_{f,0}^{\mathrm{B1}}}{\mathrm{d}\Omega}(E_{k_i} + n\omega) = \frac{k_f(n)}{k_i} |f_{f,0}^{\mathrm{B1}}|^2 \qquad (10.229)$$

is the corresponding first Born approximation for the field-free excitation process. In Equation (10.228), the summation on the index n runs over all values of n for which $k_f(n) \geq 0$, with

$$\frac{k_f^2}{2} = E_{k_f} = E_{k_i} - \omega_{f0} + n\omega \qquad (10.230)$$

and $\omega_{f0} = E_t$ is the excitation threshold energy.

Equation (10.228) is a generalization of the Bunkin–Fedorov formula (10.72) derived in Section 10.2 for laser-assisted potential scattering. We recall, however, that the first Born approximation for an excitation cross section is very inaccurate near threshold, and becomes a reasonable approximation only at much higher energies. For the case of laser-assisted potential scattering in the soft-photon limit, we have seen in Section 10.2 that Kroll and Watson [8] derived the low-frequency approximation formula (10.88), which relates the *exact* differential cross section with and without the laser field. Although an exact relation of this kind has not been obtained for laser-assisted electron collisions with real atoms, and in particular for inelastic processes, Geltman and Maquet [151] conjectured that Equation (10.228) holds using the *exact* differential cross section and not only its first Born approximation. That is,

$$\frac{\mathrm{d}\sigma_{\mathrm{SEPE}}}{\mathrm{d}\Omega}(E_{k_i}) = \sum_n J_n^2(\boldsymbol{\Delta} \cdot \boldsymbol{\alpha}_0) \frac{\mathrm{d}\sigma_{f,0}}{\mathrm{d}\Omega}(E_{k_i} + n\omega), \qquad (10.231)$$

where $d\sigma_{f,0}/d\Omega$ is the exact differential cross section for the field-free electron impact excitation process $0 \to f$. The corresponding relation for total cross section valid to lowest order in ω, is given by

$$\sigma_{\text{SEPE}}^{\text{tot}}(E_{k_i}) = \int d\Omega \sum_n J_n^2(\mathbf{\Delta} \cdot \boldsymbol{\alpha}_0) \frac{d\sigma_{f,0}}{d\Omega}(E_{k_i} + n\omega) \tag{10.232}$$

and was adopted by Geltman and Maquet [151] as a model to interpret the experimental data of Mason and Newell [38] and Wallbank *et al.* [40, 41] concerning the SEPE process (10.227) in helium.

The actual measured quantities in the Mason and Newell [38] and Wallbank *et al.* [40, 41] experiments are the differences in the respective metastable 2^3S yield M_{on} and M_{off} between laser-on and laser-off conditions. In terms of the electron beam energy distribution $f(E)$ and monochromatic beam total cross sections, these yields are proportional to the corresponding average total cross sections $\bar{\sigma}_{\text{SEPE}}^{\text{tot}}$ and $\bar{\sigma}_{f,0}^{\text{tot}}$ for the 1^1S $\to 2^3$S excitation process as

$$M_{\text{on}}(E_{\text{m}}) \propto \bar{\sigma}_{\text{SEPE}}^{\text{tot}}(E_{\text{m}}) = \int_0^\infty \sigma_{\text{SEPE}}^{\text{tot}}(E) f(E) dE \tag{10.233}$$

and

$$M_{\text{off}}(E_{\text{m}}) \propto \bar{\sigma}_{f,0}^{\text{tot}}(E_{\text{m}}) = \int_0^\infty \sigma_{f,0}^{\text{tot}}(E) f(E) dE, \tag{10.234}$$

where E_{m} is the energy at which the function $f(E)$ has its maximum value.

As an example, we show in Fig. 10.9 the laser-assisted total cross sections for the SEPE process (10.227), calculated by Geltman and Maquet [151] for a CO_2 laser of angular frequency $\omega = 0.0043$ a.u., and for the two laser intensities $I_1 = 4 \times 10^4$ W cm^{-2} and $I_2 = 10^5$ W cm^{-2}, as a function of the incident electron energy E_{k_i}. The laser field is spatially homogeneous, monochromatic and linearly polarized, with its polarization vector $\hat{\boldsymbol{\epsilon}}$ parallel to the incident electron wave vector $\mathbf{k}_i$. Geltman and Maquet adopted for the field-free total electron-impact excitation cross section of He(2^3S) the values of Fon *et al.* [157] obtained by using the close-coupling and R-matrix methods. Also shown in Fig. 10.9 is the difference cross section,

$$\Delta\bar{\sigma}(E_{\text{m}}) = \bar{\sigma}_{\text{SEPE}}^{\text{tot}}(E_{\text{m}}) - \bar{\sigma}_{f,0}^{\text{tot}}(E_{\text{m}}), \tag{10.235}$$

obtained by taking the function $f(E)$ in Equations (10.233) and (10.234) to be a Maxwell distribution having its maximum value at E_{m} and a full-width at half-maximum (FWHM) equal to 35 meV, as quoted by Mason and Newell [38]. The resulting $\Delta\bar{\sigma}(E_{\text{m}})$ is in qualitative agreement with the experimental data of Mason and Newell [38] for the quantity

$$\Delta M(E_{\text{m}}) = M_{\text{on}}(E_{\text{m}}) - M_{\text{off}}(E_{\text{m}}). \tag{10.236}$$

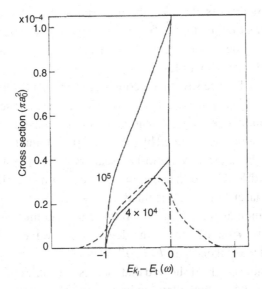

Figure 10.9. Laser-assisted total cross section $\sigma_{\text{SEPE}}^{\text{tot}}$ (in units of πa_0^2), calculated by using Equation (10.232), for the SEPE process (10.227) in helium, as a function of energy (in units of ω from threshold). The laser field is spatially homogeneous, monochromatic and linearly polarized, with the polarization vector $\hat{\boldsymbol{\epsilon}}$ taken to be parallel to the incident electron wave vector $\mathbf{k}_i$, for two laser intensities (4×10^4 W cm^{-2} and 10^5 W cm^{-2}) indicated next to the corresponding full curves. Dot-dashed curve: field-free total cross section for the electron impact excitation of the 2^3S state. Dashed curve: Maxwell-convoluted difference cross section $\Delta\bar{\sigma}$ as a function of $E_{k_i} = E_{\text{m}}$ for a laser intensity $I = 4 \times 10^4$ W cm^{-2} and FWHM = 35 meV. (From S. Geltman and A. Maquet, *J. Phys. B* **22**, L419 (1989).)

Maquet and Cooper [152] have shown that the approach proposed by Geltman and Maquet [151], based on Equation (10.231), is valid in the soft-photon limit if the intensity of the laser field is not too high (typically, less than 10^8 W cm^{-2} for a CO$_2$ laser). The relation (10.231) has also been analyzed by Chichkov [153] and Mittleman [155]. It has been used by Fainstein, Maquet and Fon [156] to calculate laser-assisted differential cross sections for simultaneous electron–photon excitation of the 2^3S and 2^1S states of helium for electron impact energies near threshold, in the case of a CO$_2$ laser with an intensity $I = 10^8$ W cm^{-2}. The field-free differential cross sections were obtained from a 29-state R-matrix calculation which is in good agreement with experiment [158, 159]. Fainstein, Maquet and Fon [156] found that important modifications in the differential cross sections are produced when the laser is turned on.

The main limitation of the Geltman–Maquet approximation based on Equation (10.231) is that it relies on the assumption that the target states are unaffected by the laser field. As discussed above, the important role of target-dressing effects at higher laser intensities and higher frequencies, particularly on the differential cross sections, has been demonstrated in several calculations at high electron impact energies [10, 101, 104, 106, 113, 138].

In Section 10.3.1.3, we saw that a consistent treatment of low-energy electron atom collisions in the presence of a laser field, which treats all three types of interactions (unbound electron(s)–target atom, laser field–unbound electron(s), laser field–target atom) is provided by the R-matrix–Floquet (RMF) theory [13, 14]. This method has been used by Terao-Dunseath *et al.* [16] to study SEPE processes in helium in a Nd:YAG laser field, at incident electron energies near the He(1s2*l*) thresholds. The laser field was assumed to be spatially homogeneous, monochromatic (with an angular frequency $\omega = 0.043$ a.u.) and linearly polarized, with the polarization vector $\hat{\boldsymbol{\epsilon}}$ parallel to the incident electron wave vector $\mathbf{k}_i$. The intensity of the laser field was taken to be $I = 10^{10}\,\mathrm{W\,cm^{-2}}$.

Terao-Dunseath *et al.* [16] showed that there is a strong AC Stark mixing between the 2^3S and 2^3P states, which plays an important role in determining detailed cross sections for excitation into the triplet states. As an example, we display in Fig. 10.10 the total cross sections obtained by Terao-Dunseath *et al.* [16] for SEPE of He($2^3\tilde{\mathrm{S}}$), where here the symbol "$\sim$" is used to distinguish the field-dressed atomic states from the field-free states when strong AC Stark mixing occurs, the convention being to label a field-dressed atomic state by the quantum numbers of its dominant component. Also shown in Fig. 10.10 (bold solid curve) is the field-free total cross section for electron impact excitation of the 2^3S state of helium. These total cross sections are shown for electron energies ranging from 0.65 a.u. to 0.78 a.u. The most striking feature of Fig. 10.10 is the tall, isolated resonance in the total cross section with absorption of one photon. This resonance occurs below the 2^3S threshold, at an energy of 0.7062 a.u., very close to the He$^-$(1s2s^2 ^{2}S) resonance in the field-free elastic electron–helium total cross section at 0.7063 a.u. [160]. It may be understood as follows: the incident electron is captured in the He$^-$ resonant state, and then a photon is absorbed so that the helium atom is left in an excited state after the collision. Other paths leading to excitation of the target atom with net absorption of one photon are possible and are included in the calculation.

Also shown in Fig. 10.10 is the total cross section for excitation of He($2^3\tilde{\mathrm{S}}$) with no net exchange of photons. It is similar in shape to the field-free total cross section but is about 40% lower in magnitude. A resonance is visible at an energy of 0.7492 a.u., which corresponds to the position of the He$^-$ resonance shifted upwards by one photon energy. It may be explained in the following way: during the collision, the e$^-$–He system emits one photon and is temporarily captured in the He$^-$ resonance. It then absorbs one photon, leading to excitation of the helium atom with no net exchange of photons. Even though this is a two-step process, it

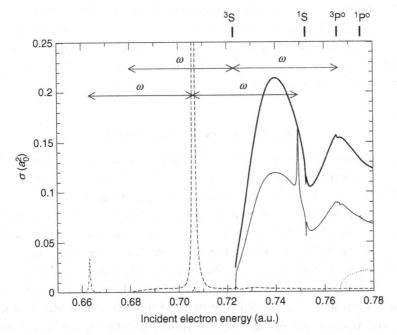

Figure 10.10. Total cross sections (in units of πa_0^2) for excitation of the He($2^3\tilde{S}$) state as a function of the incident electron energy (in a.u.). The laser field is spatially homogeneous, monochromatic and linearly polarized, with the polarization vector $\hat{\epsilon}$ taken to be parallel to the incident electron wave vector $\mathbf{k}_i$. The laser photon energy is $\hbar\omega = 1.17$ eV (Nd:YAG laser) and its intensity is $I = 10^{10}$ W cm^{-2}. Solid curve: excitation with no net exchange of photons ($n = 0$). Dashed curve: excitation with absorption of one photon ($n = 1$). Dash-dotted curve: excitation with absorption of two photons ($n = 2$). Dotted curve: excitation with emission of one photon ($n = -1$). Bold solid curve: field-free total cross section for electron impact excitation of He(2^3S). The field-free excitation thresholds are indicated by vertical bars. (From M. Terao-Dunseath *et al.*, *J. Phys. B* **34**, L263 (2001).)

accounts for about 40% of the total cross section at resonance. The height of this peak is seven times smaller than that at 0.7062 a.u., corresponding to the direct formation of the He$^-$ resonance state followed by the absorption of one photon.

The total cross section for excitation of He(2^3S) with absorption of two photons, also displayed in Fig. 10.10, is of the order of $10^{-5}a_0^2$ and is therefore almost negligible, except at resonance. At an energy of 0.706 a.u., the total cross section corresponding to capture into the He$^-$ resonance, followed by absorption of two photons is small, being about $5 \times 10^{-3}a_0^2$ at its maximum. The two-step process, where one photon is absorbed before the formation of the He$^-$ intermediate state and one photon after, is more important, the corresponding signature peak at 0.663 a.u. being seven times taller than that at 0.706 a.u.

The threshold for excitation with net emission of one photon corresponds to th: of the 2^3S state shifted upwards by ω. The He$^-$ resonance gives rise to a very sma structure as at least three photons must be exchanged (two photons emitted and on absorbed). Finally, we remark that excitation with net emission of one photon : more important than with net absorption.

10.3.3 Laser-assisted (e, 2e) collisions

As in the previous two subsections, the early theoretical investigations on laser assisted (e, 2e) collisions neglected the dressing of the target atom by the laser field Mohan and Chand [161] studied electron impact ionization of atomic hydrogen i. the presence of a laser field by treating the problem in the Kramers–Henneberge frame. They derived a low-frequency cross section consisting of the field-free cros section times the square of a Bessel function. A more general treatment of thi problem was given by Banerji and Mittleman [162], who derived a low-frequency expression of the Kroll–Watson type for the triple differential cross section (TDCS) In their derivation, the ejected electron was described by a modified Coulomk wave function, as proposed by Jain and Tzoar [163], having the form, in the lengtl gauge

$$\Psi_{\mathbf{k}}^c(Z, \mathbf{r}, t) = \exp(-iE_k t)\exp[-i\mathbf{k} \cdot \boldsymbol{\alpha}(t)]\psi_{\mathbf{k}}^{c(-)}(Z, \mathbf{r}), \qquad (10.237)$$

where $\mathbf{k}$ is the wave vector of the electron, $E_k = k^2/2$ its energy, $\boldsymbol{\alpha}(t)$ is given by Equation (2.113), $\psi_{\mathbf{k}}^{c(-)}(Z, \mathbf{r})$ is a Coulomb wave function given by Equation (3.140) with incoming $(-)$ spherical wave asymptotic behavior and atomic units are used. Coulomb effects in laser-assisted (e, 2e) collisions were considered by Mittleman [164] using a truncated Coulomb potential of the form

$$V(r) = -\frac{Z}{r}\Theta(r_0 - r), \qquad (10.238)$$

where Θ is the Heaviside step function (2.157) and r_0 is the cut-off radius. The case of a laser field which is resonant with one of the atomic transitions was analyzed by Fiordilino and Mittleman [165].

The influence of a laser field on the dynamics of fast (e, 2e) collisions has been the subject of several theoretical investigations. The (e, 2e) kinematical arrangement which was selected in these studies is the "Ehrhardt asymmetric coplanar geometry" [166], such that a fast electron of wave vector $\mathbf{k}_i$ and energy $E_{k_i} = k_i^2/2$ is incident on the target, and a fast ("scattered") electron of wave vector $\mathbf{k}_A$ and energy $E_{k_A} = k_A^2/2$ is detected in coincidence with a slow ("ejected") electron of wave vector $\mathbf{k}_B$ and energy $E_{k_B} = k_B^2/2$, the three wave vectors $\mathbf{k}_i$, $\mathbf{k}_A$ and $\mathbf{k}_B$ being in the same plane. Moreover, the scattering angle θ_A of the fast ("scattered") electron is fixed and small, while the angle θ_B of the slow ("ejected") electron is varied. The reasons for the choice of the Ehrhardt asymmetric coplanar geometry are that (i) at high incident electron energies most of the (e, 2e) collisions occur in this kinematical regime and (ii) accurate experimental and theoretical results are available in this geometry for field-free (e, 2e) reactions involving simple target atoms such as atomic hydrogen and helium [167].

In particular, Cavaliere, Ferrante and Leone [168] have analyzed laser-assisted (e, 2e) collisions in atomic hydrogen at incident electron energies of a few keV. They used the first Born approximation to treat the projectile electron–hydrogen atom target interaction. The incident and scattered electrons were described by Gordon–Volkov wave functions, while the ejected electron was represented by a modified Coulomb wave function of the form of Equation (10.237), as proposed by Jain and Tzoar [163]. In subsequent work, Cavaliere *et al.* [169] showed that important differences arise between the field-free TDCS and the corresponding laser-assisted TDCS for single-photon absorption. The (e, 2e) reaction in atomic hydrogen in the presence of a resonant laser field was studied in another investigation by Cavaliere, Ferrante and Leone [170]. They assumed that the 1s–2p transition in atomic hydrogen was strongly coupled by a laser field, in such a way that the other atomic states may be neglected and the hydrogen atom can be described as a two-level system. In this case, the TDCS exhibits a resonant mixture of the transitions of the bound electron from the s-state and p-states to the continuum during the laser-assisted collision. Further calculations on laser-assisted (e, 2e) collisions in atomic hydrogen, using an approach similar to that of Cavaliere *et al.* [168, 169] were performed by Burlon, Cavaliere and Ferrante [171] and by Mandal and Ghosh [172]. Zarcone, Moores and McDowell [173], using the same approximation, extended the calculations to laser-assisted (e, 2e) collisions in helium. In order to allow the comparison of theoretical laser-assisted (e, 2e) TDCS with future experimental data, Zangara *et al.* [174] took into account the spatial and temporal inhomogeneities of the laser pulse.

We shall now discuss the work of Joachain *et al.* [114] in which the semi-perturbative theory of Byron and Joachain [10] was generalized to laser-assisted fast (e, 2e) collisions in atomic hydrogen. As before, the laser field was treated as a spatially homogeneous, monochromatic and linearly polarized electric-field $\mathcal{E}(t) = \hat{\epsilon}\mathcal{E}_0 \sin(\omega t)$. For the (e, 2e) kinematical arrangement, the Ehrhardt asymmetric coplanar geometry was chosen, since accurate experimental [175] and theoretical [176–179] results were available in this geometry for the corresponding field-free (e, 2e) reaction $e^- + H(1s) \rightarrow H^+ + 2e^-$.

Remembering that, in the Ehrhardt geometry, exchange effects between the projectile and target electrons are small, and that a perturbative treatment of the direct interaction $V_d = 1/r_{01} - 1/r_0$ between the fast projectile electron and the target hydrogen atom is justified [167], one can start from the first Born ionization S-matrix element, which for the present laser-assisted (e, 2e) reaction is given by

$$S_{\text{ion}}^{\text{B1}} = -i \int_{-\infty}^{+\infty} dt \langle \chi_{\mathbf{k}_A}^V(\mathbf{r}_0, t) \Phi_{\mathbf{k}_B}^V(\mathbf{r}_1, t) | 1/r_{01} - 1/r_0 | \chi_{\mathbf{k}_i}^V(\mathbf{r}_0, t) \Phi_0^V(\mathbf{r}_1, t) \rangle,$$

(10.239)

where $\chi_{\mathbf{k}_i}^V(\mathbf{r}_0, t)$ and $\chi_{\mathbf{k}_A}^V(\mathbf{r}_0, t)$ are Gordon–Volkov wave functions describing in the velocity gauge, the projectile and scattered electron in the laser field respectively.

The wave functions $\Phi_0^V(\mathbf{r}_1, t)$ and $\Phi_{\mathbf{k}_B}^V(\mathbf{r}_1, t)$ appearing in Equation (10.239) describe, in the velocity gauge, the dressed states of the hydrogen atom in the laser field, the first one corresponding to the initial ground state and the second one to final continuum state in which a slow electron of wave vector $\mathbf{k}_B$ has been ejected. In what follows it will be assumed that the laser field is such that $\mathcal{E}_0 \ll \mathcal{E}_a$, where $\mathcal{E}_a$ is the atomic unit of electric-field strength. The dressed atomic ground state $\Phi_0^V(\mathbf{r}_1, t)$ can then be obtained by using first-order, time-dependent perturbation theory, as explained previously, and is given by Equation (3.199) with $j = 0$. The dressed continuum wave function in Equation (10.239) is given by

$$\Phi_{\mathbf{k}_B}^V(\mathbf{r}_1, t) = \exp(-iE_{\mathbf{k}_B}t) \exp\left[\frac{i}{2} \int_{-\infty}^t \mathbf{A}^2(t') dt'\right]$$

$$\times \exp[-i\mathbf{a}(t) \cdot \mathbf{r}_1] \exp[-i\mathbf{k}_B \cdot \boldsymbol{\alpha}_0 \sin(\omega t)]$$

$$\times \left\{ \psi_{\mathbf{k}_B}^{c(-)}(Z=1, \mathbf{r}_1) - i\frac{\mathcal{E}_0}{2} \sum_j \left[\frac{\exp(i\omega t)}{E_j - E_{\mathbf{k}_B} + \omega} - \frac{\exp(-i\omega t)}{E_j - E_{\mathbf{k}_B} - \omega} \right] \right.$$

$$\left. \times M_{j\mathbf{k}_B} \psi_j(\mathbf{r}_1) + i\mathbf{k}_B \cdot \boldsymbol{\alpha}_0 \sin(\omega t) \psi_{\mathbf{k}_B}^{c(-)}(Z=1, \mathbf{r}_1) \right\},$$

(10.240)

where the matrix element $M_{j\mathbf{k}_B}$ is given by

$$M_{j\mathbf{k}_B} = -\hat{\boldsymbol{\epsilon}} \cdot \langle \psi_j | \mathbf{r}_1 | \psi_{\mathbf{k}_B}^{c(-)}(Z=1) \rangle.$$

(10.241)

The dressed continuum wave function (10.240) of Joachain *et al.* [114] is a generalization of the wave function (10.237) of Jain and Tzoar [163]. This generalization takes into account to first order in $\mathcal{E}_0$ the role of all the target states in "dressing" the ejected electron wave function.

Using the expressions of the Gordon–Volkov wave functions $\chi_{\mathbf{k}_i}^V(\mathbf{r}_0, t)$ and $\chi_{\mathbf{k}_A}^V(\mathbf{r}_0, t)$, and those of the dressed atomic hydrogen states $\Phi_0^V(\mathbf{r}_1, t)$ and $\Phi_{\mathbf{k}_B}^V(\mathbf{r}_1, t)$ one finds, after performing the time integration, that the S-matrix element (10.239) can be rewritten as

$$S_{\text{ion}}^{\text{B1}} = (2\pi)^{-1} i \sum_{n=-\infty}^{+\infty} \delta(E_{k_A} + E_{k_B} - E_{k_i} - E_0 - n\omega) f_{\text{ion}}^{n,\text{B1}},$$

(10.242)

where $f_{\text{ion}}^{n,\text{B1}}$, the first Born approximation to the (e, 2e) scattering amplitude with the transfer of $|n|$ photons, is given by

$$f_{\text{ion}}^{n,\text{B1}} = f_I + f_{II} + f_{III} \tag{10.243}$$

with

$$f_I = -\frac{2}{\Delta^2} \langle \psi_{\mathbf{k}_B}^{c(-)} | \exp(i\boldsymbol{\Delta} \cdot \mathbf{r}_1) | \psi_0 \rangle J_n(\rho), \tag{10.244}$$

$$f_{II} = i\frac{\mathcal{E}_0}{\Delta^2} \sum_j \langle \psi_{\mathbf{k}_B}^{c(-)} | \exp(i\boldsymbol{\Delta} \cdot \mathbf{r}_1) | \psi_j \rangle M_{j0}$$

$$\times \left[\frac{J_{n+1}(\rho)}{E_j - E_0 + \omega} - \frac{J_{n-1}(\rho)}{E_j - E_0 - \omega} \right] \tag{10.245}$$

and

$$f_{III} = i\frac{\mathcal{E}_0}{\Delta^2} \sum_j \langle \psi_j | \exp(i\boldsymbol{\Delta} \cdot \mathbf{r}_1) | \psi_0 \rangle M_{j,\mathbf{k}_B}^* \left[\frac{J_{n+1}(\rho)}{E_j - E_{k_B} - \omega} - \frac{J_{n-1}(\rho)}{E_j - E_{k_B} + \omega} \right]$$

$$+ \frac{1}{\Delta^2} \mathbf{k}_B \cdot \boldsymbol{\alpha}_0 \langle \psi_{\mathbf{k}_B}^{c(-)} | \exp(i\boldsymbol{\Delta} \cdot \mathbf{r}_1) | \psi_0 \rangle [J_{n+1}(\rho) - J_{n-1}(\rho)]. \tag{10.246}$$

In the above equations, $\boldsymbol{\Delta} = \mathbf{k}_i - \mathbf{k}_A$ is the wave vector transfer (i.e. the momentum transfer in a.u.), and we have defined $\rho = (\boldsymbol{\Delta} - \mathbf{k}_B) \cdot \boldsymbol{\alpha}_0$. The first Born TDCS corresponding to the laser-assisted (e, 2e) collision with the transfer of $|n|$ photons is given by

$$\frac{d\sigma_{\text{ion}}^{n,\text{B1}}}{d\Omega_A d\Omega_B dE} = \frac{k_A k_B}{k_i} |f_{\text{ion}}^{n,\text{B1}}|^2. \tag{10.247}$$

We note that the results of Cavaliere, Ferrante and Leone [168] can be recovered from the foregoing treatment if only the first term f_I is kept on the right-hand side of Equation (10.243). In the case of one-photon absorption ($n = 1$) or stimulated emission ($n = -1$) processes, lowest-order perturbation theory (LOPT) results can also be obtained from Equations (10.243)–(10.247) by retaining only the first term in the expansion of the Bessel functions in powers of the electric-field strength $\mathcal{E}_0$.

The amplitudes f_{II} and f_{III} contain infinite sums running over the entire atomic spectrum. These sums can be accurately computed by using the method of Dalgarno and Lewis [180].

As an illustration of the calculations performed by Joachain *et al.* [114], we show in Fig. 10.11 their results, obtained from Equation (10.247) for the case $n = 1$, concerning the effect of the electric-field strength $\mathcal{E}_0$ at a fixed angular frequency $\omega = 0.043$ a.u. (photon energy $\hbar\omega = 1.17$ eV), on the first Born TDCS. This figure displays the angular distributions of the ejected electrons at two different electric-field strengths, $\mathcal{E}_0 = 5 \times 10^7$ V cm^{-1} (see Fig. 10.11(a)) and $\mathcal{E}_0 = 10^6$ V cm^{-1} (see

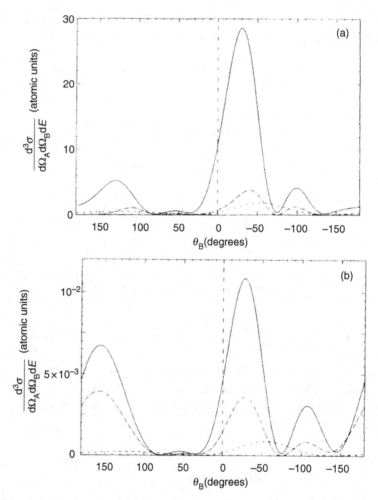

Figure 10.11. Triple differential cross section (in a.u.) for the ionization of atomic hydrogen from the ground state by electron impact in the presence of a laser field, for the case $n = 1$, as a function of the ejected electron angle θ_B (in degrees). The values of θ_B are displayed following the convention used by Ehrhardt *et al.* [166]. The incident electron energy is $E_{k_i} = 250$ eV, the ejected electron energy is $E_{k_B} = 5$ eV and the scattering angle is $\theta_A = 3°$. The laser field is spatially homogeneous, monochromatic and linearly polarized, with the polarization vector $\hat{\epsilon}$ taken to be parallel to the incident electron wave vector $\mathbf{k}_i$. The laser photon energy is $\hbar\omega = 1.17$ eV. Panel (a) corresponds to a non-perturbative regime, the electric-field strength being $\mathcal{E}_0 = 5 \times 10^7$ V cm^{-1}, while panel (b) corresponds to a perturbative regime with $\mathcal{E}_0 = 10^6$ V cm^{-1}. Solid curve: complete calculation using the scattering amplitude (10.243). Dashed line: results obtained by using the simplified approach of Cavaliere, Ferrante and Leone [168]. Dotted line: results obtained when the dressing of the target is completely neglected. (From C. J. Joachain *et al.*, *Phys. Rev. Lett.* **61**, 165 (1988).)

Fig. 10.11(b)), which correspond, respectively, to a non-perturbative regime and a perturbative regime. Also included in Fig. 10.11 are the TDCS obtained from the calculations of Cavaliere, Ferrante and Leone [168], which take into account the dressing of the ejected electron in a simplified way by using the wave function (10.237) of Jain and Tzoar [163]. For comparison, the angular distributions obtained when the dressing of the target is completely neglected are also shown; the corresponding curves are then homothetic to the field-free first Born TDCS.

It is apparent from the examination of Fig. 10.11 that the laser parameters can have a strong influence on the dynamics of laser-assisted (e, 2e) reactions, and that the "dressing" of the target can be very important. Joachain *et al.* [114] also showed that such "dressing" effects can be dramatically enhanced by suitable adjustment of the laser angular frequency and (or) the energy of the ejected electron. A detailed account of the theory has been given subsequently by Martin *et al.* [115], where several illustrative examples of laser-assisted (e, 2e) reactions in atomic hydrogen were presented. In particular, Martin *et al.* [115], analyzed the influence of the choice of the laser polarization direction on the angular distribution of the ejected electrons. This subject has also been discussed by Taïeb *et al.* [181] who expanded the "dressed" atomic hydrogen wave functions on a basis of Sturmian functions, which allowed them to take into account accurately the contribution of the continuum spectrum to the dressing of the atomic states. The results obtained from these investigations indicate that laser polarization effects can lead to important modifications of the TDCS for laser-assisted (e, 2e) collisions.

The semi-perturbative treatment of laser-assisted (e, 2e) reactions, described above for the case of an atomic hydrogen target, has been extended by Joachain *et al.* [182], Khalil *et al.* [183], Makhoute *et al.* [184] and Bouzidi *et al.* [185] to laser-assisted (e, 2e) reactions in helium, for which the possibilities of performing experiments are more favorable. These studies also showed that the choice of the laser parameters can strongly influence the dynamics of such processes, and that target-dressing effects can be important. The first experimental kinematically complete measurements of laser-assisted (e, 2e) reactions in helium, performed by Höhr *et al.* [44], confirmed the existence of distinct differences in TDCS between laser-on and laser-off conditions.

10.3.4 *Double poles of the S-matrix in laser-assisted electron–atom collisions*

It has been pointed out by Goldberger and Watson [186] that no general principle requires that poles of the S-matrix that are associated with resonances be simple. A multiple pole of the S-matrix implies that the time evolution of the resonant, or quasi-stationary, state does not follow the usual exponential decay law, but is modified by a polynomial in time [186, 187]. In a scattering context, a multiple pole

results in a resonant profile that does not have the Breit–Wigner form [188], bu
for example, in the case of a double pole, is a double-peaked function [189, 190]
No experimental evidence has yet been found that points to a resonant proces
associated with a multiple pole of the S-matrix. Thus, while being a property that th
S-matrix can possess, multiple poles would appear to be precluded from occurring

However, one can take a different point of view and ask whether degeneracie
can occur among the resonant states of a system. If that is indeed the case, th
multiple poles of the S-matrix can arise due to the degeneracies of the resonan
states. Such degeneracies, which are not due to symmetries, are termed *accidenta*
In particular, if the resonant states can be manipulated by using external parameter
for example by applying external fields, degeneracies can be made to occur b
adjusting these parameters. For example, Stark mixing in an atom [190] and th
decay of Rabi oscillations in a two-level system [191] can induce degeneracie
that lead to double-pole decay. Similarly, we saw in Section 4.7 that laser-induce
degenerate states (LIDS) occur when the ground state (or an excited state) of a
atom is strongly coupled to an auto-ionizing state [192]. They can also occur in th
two-color ionization of atoms using commensurable frequencies [193], or when two
auto-ionizing states are strongly coupled by a laser field [194]. In all these cases
the realization of the double-pole decay requires the preparation of the resonan
state on a time scale which is small compared to the lifetime of that state. This
time scale is typically in the femtosecond range, so that the external field must be
applied very rapidly.

Kylstra and Joachain [19, 20] have suggested that this liability can be avoided
in a scattering context, where these degeneracies correspond to double poles of the
laser-assisted electron–atom S-matrix. They considered two scenarios. The first
one is the case in which the energy of the projectile electron is such that there is a
resonance due to an auto-ionizing state of the electron–target atom system and that
the angular frequency of the laser field is tuned to a resonance between this auto-
ionizing state and a quasi-bound state (a state that would be bound in the absence of
the laser field) of the electron–target atom system. Here the effect of the laser field
is to introduce a resonance in the same channel as the auto-ionizing resonance. This
is an example of overlapping, or interacting, resonances [195, 196], with the degree
of the interaction between the resonances being determined by the laser intensity
and angular frequency. The second case is one in which the laser angular frequency
is chosen in such a way that two auto-ionizing states of the electron–target atom
system are resonantly coupled.

In what follows we shall only discuss the first case. We assume that the background scattering is negligible, so that the total cross section is due only to the
resonant process. Using the Feshbach projection operator formalism [47, 197, 198]
and the rotating wave approximation for solving the Floquet–Lippmann–Schwinger

equations, the total cross section, which depends on the angular frequency ω and intensity I of the laser field and the energy E_{k_i} of the projectile electron, is given (in a.u.) by

$$\sigma_{\text{tot}}(E_{k_i}, I, \omega) = \frac{2\pi g}{E_{k_i}} \frac{[\Gamma_a + \Gamma_g]^2 [E_{k_i} - E_W(I, \omega)]^2}{|E_{k_i} - z_1|^2 |E_{k_i} - z_2|^2}, \qquad (10.248)$$

where g contains the channel-dependent statistical weight factors. The width acquired by the ground state in a bare continuum is $\Gamma_g = I\gamma_g$, which is proportional to the laser intensity I and Γ_a is the natural width of the auto-ionizing resonance. All the atomic parameters are taken to be constants in the neighborhood of the resonance. The detuning is given by $\delta = E_a - E_g - \omega$, with the energies E_g and E_a being, respectively, the field-free position of the ground state (or an excited bound state) and the auto-ionizing state. At the window energy

$$E_W = \frac{\Gamma_a(E_g + \omega) + \Gamma_g(E_a - q\Gamma_a)}{\Gamma_a + \Gamma_g}, \qquad (10.249)$$

where q is the Fano resonance shape parameter [199], the two resonant processes interfere fully destructively and the resonant contribution to the total cross section vanishes. We note that within the context of this model, the pole z_2 (for $q > 0$) becomes purely real when the detuning is equal to the trapping detuning $\delta_T = q(\Gamma_g - \Gamma_a)/2$. At this trapping detuning, the expression for the total cross section reduces to the Breit–Wigner formula for a single, isolated resonance.

As was discussed in Section 4.7, degeneracies in the complex energy spectrum of the dressed states, or LIDS, occur when the quasi-energies become degenerate. These degeneracies occur at the following critical detunings:

$$\delta_c^{\pm} = \Gamma_a(q \pm \sqrt{q^2 + 1}) \qquad (10.250)$$

and intensities

$$I_c^{\pm} = \frac{(\delta_c^{\pm})^2}{\gamma_g \Gamma_a}. \qquad (10.251)$$

At the degeneracy, or double pole, the total cross section is a symmetric function about the window energy in the neighborhood of the resonance. For this particular case, where background scattering has been neglected, the symmetry of the total cross section about the window energy is a unique signature of the double pole. The more general situation of a resonant process accompanied by background scattering has been discussed by Kylstra and Joachain [20].

In order to illustrate the effect of the resonant coupling on the total cross section due to the laser field, we show in Fig. 10.12 the total cross section (10.248) as a function of the incident electron energy E_{k_i} for four different laser detunings near

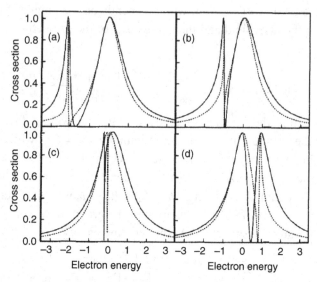

Figure 10.12. Resonant contribution of the total cross-section (10.248), in arbitrary units, as a function of the electron energy, in units of Γ_c^- and measured with respect to the position of the auto-ionizing level, is plotted for four different detunings and two intensities. The dotted curves are for $I = I_c^-/8$ and the solid curves for $I = I_c$. The four plots correspond to the following detunings: (a) $\delta = 2\Gamma_c^-$; (b) $\delta = \Gamma_c^-$; (c) $\delta = 0$; (d) $\delta = \delta_c^-$. The solid curve in plot (d) shows the resonant cross section at the double pole. (From N. J. Kylstra and C. J. Joachain, *Europhys. Lett.* **36**, 657 (1996).)

resonance and two laser intensities. The solid curves correspond to $I = I_c^-$ while the broken curves are for the lower intensity $I = I_c^-/8$. At the lower intensity, the effect of the resonant coupling due to the laser field appears as a sharp, asymmetric feature that modifies the Breit–Wigner profile of the auto-ionizing resonance. This feature is particularly sharp in plot (c) where the detuning of the laser field is close to δ_T, the trapping detuning. The solid curve in plot (d) exhibits the signature of the double pole: the total cross section is symmetric about the window energy.

References

[1] C. G. Morgan, *Rep. Progr. Phys.* **38**, 621 (1975).
[2] L. Schlessinger and J. Wright, *Phys. Rev. A* **20**, 1934 (1979).
[3] L. Schlessinger and J. Wright, *Phys. Rev. A* **22**, 909 (1980).
[4] L. Schlessinger and J. Wright, *Phys. Rev. A* **24**, 2541 (1981).
[5] N. J. Kylstra and C. J. Joachain, *Phys. Rev. A* **58**, R26 (1998).
[6] N. J. Kylstra and C. J. Joachain, *Phys. Rev. A* **60**, 2255 (1999).
[7] F. V. Bunkin and M. V. Fedorov, *Sov. Phys. JETP* **22**, 844 (1966).
[8] N. M. Kroll and K. M. Watson, *Phys. Rev. A* **8**, 804 (1973).
[9] M. Gavrila and J. Z. Kaminski, *Phys. Rev. Lett.* **52**, 613 (1984).
[10] F. W. Byron and C. J. Joachain, *J. Phys. B* **17**, L295 (1984).

[11] P. Francken and C. J. Joachain, *Phys. Rev. A* **41**, 3770 (1990).

[12] M. Dörr, C. J. Joachain, R. M. Potvliege and S. Vučić, *Phys. Rev. A* **49**, 4852 (1994).

[13] P. G. Burke, P. Francken and C. J. Joachain, *Europhys. Lett.* **13**, 617 (1990).

[14] P. G. Burke, P. Francken and C. J. Joachain, *J. Phys. B* **24**, 761 (1991).

[15] D. Charlo, M. Terao-Dunseath, K. M. Dunseath and J. M. Launay, *J. Phys. B* **31**, L539 (1998).

[16] M. Terao-Dunseath, K. M. Dunseath, D. Charlo, A. Hibbert and R. Allen, *J. Phys. B* **34**, L263 (2001).

[17] M. Terao-Dunseath and K. M. Dunseath, *J. Phys. B* **35**, 125 (2002).

[18] K. M. Dunseath and M. Terao-Dunseath, *J. Phys. B* **37**, 1305 (2004).

[19] N. J. Kylstra and C. J. Joachain, *Europhys. Lett.* **36**, 657 (1996).

[20] N. J. Kylstra and C. J. Joachain, *Phys. Rev. A* **57**, 412 (1998).

[21] C. J. Joachain, Electron-atom collisions in a strong laser field. In F. Ehlotzky, ed., *Fundamentals of Laser Interactions*, Lecture Notes in Physics 229 (Berlin: Springer-Verlag, 1985), p. 37.

[22] C. J. Joachain, Theory of laser-atom interactions. In R. M. More, ed., *Laser Interactions with Atoms, Solids and Plasmas* (New York: Plenum Press, 1994), p. 39.

[23] M. Gavrila, Free-free transitions of electron-atom systems in intense radiation fields. In F. A. Gianturco, ed., *Collision Theory for Atoms and Molecules* (New York: Plenum Press, 1989), p. 139.

[24] P. Francken and C. J. Joachain, *J. Opt. Soc. Am. B* **7**, 554 (1990).

[25] N. J. Mason, *Rep. Progr. Phys.* **56**, 1275 (1993).

[26] F. Ehlotzky, A. Jaron and J. Z. Kaminski, *Phys. Rep.* **297**, 63 (1998).

[27] A. Weingartshofer, J. K. Holmes, G. Caudle, E. M. Clarke and H. Kruger, *Phys. Rev. Lett.* **39**, 269 (1977).

[28] A. Weingartshofer, E. Clarke, J. Holmes and C. Jung, *Phys. Rev. A* **19**, 2371 (1979).

[29] A. Weingartshofer, J. K. Holmes, J. Sabbagh and S. L. Chin, *J. Phys. B* **16**, 1805 (1983).

[30] D. Andrick and L. Langhans, *J. Phys. B* **9**, L459 (1976).

[31] D. Andrick and L. Langhans, *J. Phys. B* **11**, 2355 (1978).

[32] L. Langhans, *J. Phys. J* **11**, 2361 (1978).

[33] H. Krüger and C. Jung, *Phys. Rev. A* **17**, 1706 (1978).

[34] C. Jung and H. Krüger, *Z. Phys. A* **287**, 7 (1978).

[35] D. Andrick and H. Bader, *J. Phys. B* **17**, 4549 (1984).

[36] H. Bader, *J. Phys. B* **19**, 2177 (1986).

[37] B. Wallbank, V. Conners, J. K. Holmes and A. Weingartshofer, *J. Phys. B* **20**, L833 (1987).

[38] N. J. Mason and W. R. Newell, *J. Phys. B* **20**, L323 (1987).

[39] N. J. Mason and W. R. Newell, *J. Phys. B* **22**, 777 (1989).

[40] B. Wallbank, J. K. Holmes, L. L. Blanc and A. Weingartshofer, *Z. Phys. D* **10**, 467 (1988).

[41] B. Wallbank, J. K. Holmes and A. Weingartshofer, *J. Phys. B* **23**, 2997 (1990).

[42] S. Luan, R. Hippler and H. O. Lutz, *J. Phys. B* **24**, 3241 (1991).

[43] N. J. Mason and W. R. Newell, *J. Phys. B* **23**, L179 (1990).

[44] C. Höhr, A. Dorn, B. Najjari, D. Fischer, C. Schröter and J. Ullrich, *Phys. Rev. Lett.* **94**, 153201 (2005).

[45] S. Bivona, R. Burlon, R. Zangara and G. Ferrante, *J. Phys. B* **18**, 3149 (1985).

[46] M. Abramowitz and I. A. Stegun, *Handbook of Mathematical Functions* (New York: Dover, 1970).

[47] C. J. Joachain, *Quantum Collision Theory*, 3rd edn. (Amsterdam: North Holland, 1983).

[48] B. H. Bransden and C. J. Joachain, *Physics of Atoms and Molecules*, 2nd edn (Harlow, UK: Prentice Hall–Pearson, 2003).

[49] F. E. Low, *Phys. Rev.* **110**, 974 (1958).

[50] M. H. Mittleman, *Phys. Rev. A* **19**, 134 (1979).

[51] P. S. Krstić and D. B. Milosević, *J. Phys. B* **20**, 3487 (1987).

[52] E. J. Kelsey and L. Rosenberg, *Phys. Rev. A* **19**, 756 (1979).

[53] J. Bergou and F. Ehlotzky, *Phys. Rev. A* **33**, 3054 (1986).

[54] F. Bloch and A. Nordsieck, *Phys. Rev.* **52**, 54 (1937).

[55] A. Nordsieck, *Phys. Rev.* **52**, 59 (1937).

[56] M. H. Mittleman, *Phys. Rev. A* **20**, 1965 (1979).

[57] D. B. Milosevic and P. S. Krstic, *J. Phys. B* **20**, 2843 (1987).

[58] C. Jung and H. S. Taylor, *Phys. Rev. A* **23**, 1115 (1981).

[59] L. Rosenberg, *Phys. Rev. A* **20**, 275 (1979).

[60] C. Leone, P. Cavaliere and G. Ferrante, *J. Phys. B* **17**, 1027 (1984).

[61] D. B. Milosevic and F. Ehlotzky, *J. Phys. B* **30**, 2999 (1997).

[62] L. B. Madsen and K. Taulbjerg, *J. Phys. B* **31**, 4701 (1998).

[63] L. W. Garland, A. Jaron, J. Z. Kaminski and R. M. Potvliege, *J. Phys. B* **35**, 286 (2002).

[64] F. Trombetta and G. Ferrante, *J. Phys. B* **22**, 3381 (1989).

[65] J. Sun, S. Zhang, Y. Jiang and G. Yu, *Phys. Rev. A* **58**, 2225 (1998).

[66] S. Zhang, X. Qian, Y. Jiang and J. Sun, *Phys. Lett.* **17**, 496 (2000).

[67] S. Hokland and L. B. Madsen, *Eur. Phys. J. D* **29**, 209 (2004).

[68] J. Z. Kaminski, *J. Phys. A* **18**, 3365 (1985).

[69] L. Rosenberg, *Phys. Rev. A* **26**, 132 (1982).

[70] L. Rosenberg and F. Zhou, *J. Phys. A* **24**, 631 (1991).

[71] F. Zhou and L. Rosenberg, *Phys. Rev. A* **45**, 7818 (1992).

[72] M. J. Offerhaus, J. Z. Kaminski and M. Gavrila, *Phys. Lett. A* **112**, 151 (1985).

[73] M. Gavrila, M. Offerhaus and J. Kaminski, *Phys. Lett. A* **118**, 331 (1986).

[74] J. van de Ree, J. Z. Kaminski and M. Gavrila, *Phys. Rev. A* **37**, 4536 (1988).

[75] L. Rosenberg, *Phys. Rev. A* **20**, 457 (1979).

[76] L. Rosenberg and F. Zhou, *Phys. Rev. A* **46**, 7093 (1992).

[77] I. Y. Berson, *Sov. Phys. JETP* **53**, 891 (1982).

[78] M. H. Mittleman, *Phys. Rev. A* **38**, 82 (1988).

[79] J. Banerji and M. H. Mittleman, *Phys. Rev. A* **26**, 3706 (1982).

[80] L. F. Saez and M. H. Mittleman, *Phys. Rev. A* **29**, 2228 (1984).

[81] M. Gavrila, A. Maquet and V. Véniard, *Phys. Rev. A* **32**, 2537 (1985).

[82] M. Gavrila, A. Maquet and V. Véniard, *Phys. Rev. A* **42**, 236 (1990).

[83] V. Florescu and V. Djamo, *Phys. Lett. A* **119**, 73 (1986).

[84] A. Florescu and V. Florescu, *Phys. Lett. A* **226**, 280 (1997).

[85] L. P. Rapoport, *Sov. Phys. JETP* **78**, 284 (1994).

[86] L. Dimou and F. H. M. Faisal, *Phys. Rev. Lett.* **59**, 872 (1987).

[87] L. Dimou and F. H. M. Faisal, *Phys. Rev. A* **46**, 4442 (1992).

[88] L. Collins and G. Csanak, *Phys. Rev. A* **44**, 5343 (1991).

[89] M. Dörr, M. Terao-Dunseath, P. G. Burke, C. J. Joachain, C. J. Noble and J. Purvis, *J. Phys. B* **28**, 3545 (1995).

[90] J. I. Gersten and M. H. Mittleman, *J. Phys. B* **9**, 2561 (1976).

[91] M. H. Mittleman, *Phys. Rev. A* **14**, 1338 (1976).

[92] M. H. Mittleman, *Phys. Rev. A* **16**, 1549 (1977).

[93] M. H. Mittleman, *Phys. Rev. A* **18**, 685 (1978).

[94] M. H. Mittleman, *Phys. Rev. A* **21**, 79 (1980).

[95] B. Zon, *Sov. Phys. JETP* **46**, 65 (1977).
[96] K. Deguchi, H. S. Taylor and R. Yaris, *J. Phys. B* **12**, 613 (1979).
[97] E. Beilin and B. Zon, *J. Phys. B* **16**, L159 (1983).
[98] P. Golovinskii, *Sov. Phys. JETP* **68**, 1346 (1988).
[99] A. Lami and N. K. Rahman, *J. Phys. B* **14**, L523 (1981).
[100] A. Lami and N. K. Rahman, *J. Phys. B* **16**, L201 (1983).
[101] A. Dubois, A. Maquet and S. Jetzke, *Phys. Rev. A* **34**, 1888 (1986).
[102] G. Kracke, J. Briggs, A. Dubois, A. Maquet and V. Véniard, *J. Phys. B* **27**, 3241 (1994).
[103] N. Manakov, S. Marmo and V. Volovich, *Phys. Lett. A* **204**, 42 (1995).
[104] F. W. Byron, P. Francken and C. J. Joachain, *J. Phys. B* **20**, 5487 (1987).
[105] H. A. Bethe, *Ann. Physik* **5**, 325 (1930).
[106] P. Francken and C. J. Joachain, *Phys. Rev. A* **35**, 1590 (1987).
[107] P. Fainstein and A. Maquet, *J. Phys. B* **27**, 5563 (1994).
[108] F. W. Byron and C. J. Joachain, *Phys. Rep.* **34**, 233 (1977).
[109] F. W. Byron and C. J. Joachain, *Phys. Rev. A* **8**, 1267 (1973).
[110] G. Ferrante, C. Leone and F. Trombetta, *J. Phys. B* **15**, L475 (1982).
[111] R. Feynman, *Phys. Rev.* **76**, 769 (1949).
[112] V. Ochkur, *Sov. Phys. JETP* **18**, 503 (1964).
[113] P. Francken, Y. Attaourti and C. J. Joachain, *Phys. Rev. A* **38**, 1785 (1988).
[114] C. J. Joachain, P. Francken, A. Maquet, P. Martin and V. Véniard, *Phys. Rev. Lett.* **61**, 165 (1988).
[115] P. Martin, V. Véniard, A. Maquet, P. Francken and C. J. Joachain, *Phys. Rev. A* **39**, 6178 (1989).
[116] A. Gazazian, *J. Phys. B* **9**, 3197 (1976).
[117] J. I. Gersten and M. H. Mittleman, *Phys. Rev. A* **13**, 123 (1976).
[118] K. Unnikrishnan, *Phys. Rev. A* **38**, 2317 (1988).
[119] M. H. Mittleman, *Introduction to the Theory of Laser-Atom Interactions*, 2nd edn (New York: Plenum Press, 1993).
[120] M. Dörr, C. J. Joachain, R. M. Potvliege and S. Vučić, *Z. Phys. D* **29**, 245 (1994).
[121] M. Dörr, R. M. Potvliege and R. Shakeshaft, *Phys. Rev. A* **41**, 558 (1990).
[122] Y. Gontier and M. Trahin, *Phys. Rev. A* **19**, 264 (1979).
[123] M. Göppert-Mayer, *Ann. Phys.* **9**, 273 (1931).
[124] B. S. Bhakar and B. J. Choudhury, *J. Phys. B* **7**, 1866 (1974).
[125] M. Mohan and P. Chand, *Phys. Lett. A* **63**, 245 (1977).
[126] M. Mohan and P. Chand, *Phys. Lett. A* **63**, 257 (1977).
[127] P. Cavaliere and C. Leone, *Nuovo Cimento Lett.* **25**, 301 (1979).
[128] G. Pert, *J. Phys. B* **11**, 1105 (1978).
[129] M. V. Fedorov and G. L. Yudin, *Sov. Phys. JETP* **54**, 1061 (1981).
[130] R. Pundir and K. Mathur, *Phys. Lett. A* **95**, 148 (1983).
[131] R. Pundir and K. Mathur, *Z. Phys. D* **1**, 385 (1986).
[132] M. Bhattacharya, C. Sinha and N. Sil, *Phys. Rev. A* **44**, 1884 (1995).
[133] P. Cavaliere, C. Leone and G. Ferrante, *Nuovo Cimento D* **4**, 79 (1984).
[134] N. K. Rahman and F. H. M. Faisal, *Phys. Lett. A* **57**, 426 (1976).
[135] N. K. Rahman and F. H. M. Faisal, *J. Phys. B* **9**, L275 (1976).
[136] N. K. Rahman and F. H. M. Faisal, *J. Phys. B* **11**, 2003 (1978).
[137] S. Jetzke, F. H. M. Faisal, R. Hippler and H. O. Lutz, *Z. Phys. A* **315**, 271 (1984).
[138] S. Jetzke, J. Broad and A. Maquet, *J. Phys. B* **20**, 2887 (1987).
[139] P. G. H. Smith and M. R. Flannery, *J. Phys. B* **24**, L489 (1991).
[140] P. G. H. Smith and M. R. Flannery, *J. Phys. B* **24**, 1021 (1992).
[141] M. R. Flannery and K. McCann, *J. Phys. B* **7**, L233 (1974).

[142] M. R. Flannery and K. McCann, *J. Phys. B* **7**, 2518 (1974).
[143] S. Vučić, *Phys. Rev. A* **51**, 4754 (1995).
[144] S. Purohit and K. Mathur, *Phys. Rev. A* **45**, 6502 (1992).
[145] L. Hahn and I. Hertel, *J. Phys. B* **5**, 1995 (1972).
[146] M. H. Mittleman, *Phys. Rev. A* **19**, 99 (1979).
[147] R. Pundir and K. Mathur, *J. Phys. B* **18**, 523 (1985).
[148] V. Buimistrov, *Phys. Lett. A* **30**, 136 (1969).
[149] M. Conneely and S. Geltman, *J. Phys. B* **14**, 4847 (1981).
[150] I. Beigman and B. Chichkov, *JETP Lett.* **46**, 395 (1987).
[151] S. Geltman and A. Maquet, *J. Phys. B* **22**, L419 (1989).
[152] A. Maquet and J. Cooper, *Phys. Rev. A* **41**, 1724 (1990).
[153] B. Chichkov, *J. Phys. B* **23**, L333 (1990).
[154] A. Cionga and V. Florescu, *Phys. Rev. A* **45**, 5282 (1992).
[155] M. H. Mittleman, *J. Phys. B* **26**, 2709 (1993).
[156] P. Fainstein, A. Maquet and W. Fon, *J. Phys. B* **28**, 2723 (1995).
[157] W. C. Fon, K. A. Berrington, P. G. Burke and A. E. Kingston, *J. Phys. B* **14**, 292 (1981).
[158] W. C. Fon, K. P. Lim and P. M. J. Sawey, *J. Phys. B* **26**, 305 (1993).
[159] W. C. Fon, K. P. Lim, K. Ratnavelu and P. M. J. Sawey, *J. Phys. B* **27**, 1561 (1994).
[160] G. Schulz, *Phys. Rev. Lett.* **10**, 104 (1963).
[161] M. Mohan and P. Chand, *Phys. Lett. A* **65**, 399 (1978).
[162] J. Banerji and M. H. Mittleman, *J. Phys. B* **14**, 3717 (1981).
[163] M. Jain and N. Tzoar, *Phys. Rev. A* **18**, 538 (1978).
[164] M. H. Mittleman, *J. Phys. B* **16**, 1089 (1983).
[165] E. Fiordilino and M. H. Mittleman, *J. Phys. B* **16**, 2205 (1983).
[166] H. Ehrhardt, K. Jung, G. Knoth and P. Schlemmer, *Z. Phys. D* **1**, 3 (1986).
[167] F. W. Byron and C. J. Joachain, *Phys. Rep.* **179**, 211 (1989).
[168] P. Cavaliere, G. Ferrante and C. Leone, *J. Phys. B* **13**, 4495 (1980).
[169] P. Cavaliere, C. Leone, R. Zangara and G. Ferrante, *Phys. Rev. A* **24**, 910 (1981).
[170] P. Cavaliere, G. Ferrante and C. Leone, *J. Phys. B* **15**, 475 (1982).
[171] R. Burlon, P. Cavaliere and G. Ferrante, *Nuovo Cimento D* **4**, 19 (1984).
[172] S. Mandal and A. Ghosh, *Phys. Rev. A* **30**, 2759 (1984).
[173] M. Zarcone, D. Moores and M. McDowell, *J. Phys. B* **16**, L11 (1983).
[174] R. Zangara, P. Cavaliere, C. Leone and G. Ferrante, *J. Phys. B* **15**, 3881 (1982).
[175] H. Ehrhardt, G. Knoth, P. Schlemmer and K. Jung, *Phys. Lett. A* **110**, 92 (1985).
[176] F. W. Byron, C. J. Joachain and B. Piraux, *Phys. Lett. A* **99**, 427 (1983).
[177] F. W. Byron, C. J. Joachain and B. Piraux, *Phys. Lett. A* **109**, 299 (1984).
[178] F. W. Byron, C. J. Joachain and B. Piraux, *J. Phys. B* **18**, 3203 (1985).
[179] E. Curran and H. Walters, *J. Phys. B* **20**, 337 (1987).
[180] A. Dalgarno and J. T. Lewis, *Proc. Roy. Soc. Lond. A* **233**, 70 (1955).
[181] R. Taïeb, V. Véniard, A. Maquet, S. Vučić and R. M. Potvliege, *J. Phys. B* **24**, 3229 (1991).
[182] C. J. Joachain, A. Makhoute, A. Maquet and R. Taïeb, *Z. Phys. D* **23**, 397 (1992).
[183] D. Khalil, A. Maquet, R. Taïeb, C. J. Joachain and A. Makhoute, *Phys. Rev. A* **56**, 4918 (1997).
[184] A. Makhoute, D. Khalil, A. Maquet, C. J. Joachain and R. Taïeb, *J. Phys. B* **32**, 3255 (1999).
[185] M. Bouzidi, A. Makhoute, D. Khalil, A. Maquet and C. J. Joachain, *J. Phys. B* **34**, 737 (2001).
[186] M. L. Goldberger and K. M. Watson, *Phys. Rev.* **136**, B1472 (1964).

187] J. S. Bell and C. J. Goebel, *Phys. Rev.* **138**, B1198 (1965).

188] G. Breit and E. P. Wigner, *Phys. Rev.* **49**, 519 (1936).

189] M. L. Goldberger and K. M. Watson, *Collision Theory* (New York: Wiley, 1964).

190] K. E. Lassila and V. Ruuskanen, *Phys. Rev. Lett.* **17**, 490 (1966).

191] P. L. Knight, *Phys. Lett. A* **72**, 309 (1979).

192] O. Latinne, N. J. Kylstra, M. Dörr *et al.*, *Phys. Rev. Lett.* **74**, 46 (1995).

193] M. Pont, R. M. Potvliege, R. Shakeshaft and P. H. G. Smith, *Phys. Rev. A* **46**, 555 (1992).

194] N. J. Kylstra, H. W. van der Hart, P. G. Burke and C. J. Joachain, *J. Phys. B* **31**, 3089 (1998).

195] F. H. Mies, *Phys. Rev.* **175**, 164 (1968).

196] J.-P. Connerade and A. M. Lane, *Rep. Prog. Phys.* **51**, 1439 (1988).

197] H. Feshbach, *Ann. Phys. NY* **5**, 357 (1958).

198] H. Feshbach, *Ann. Phys. NY* **19**, 287 (1962).

199] U. Fano, *Phys. Rev.* **124**, 1866 (1961).

Appendix
Atomic units and conversion factors

In this book, we use rationalized MKSA (SI) units to display the explicit dependence on fundamental constants, and atomic units (a.u.) to simplify the equations.

We recall that the permittivity of free space ϵ_0 and the permeability of free space μ_0 are related by the formula

$$\epsilon_0 \mu_0 c^2 = \kappa^2,\tag{A.1}$$

where c is the velocity of light in vacuo and κ is a coefficient depending on the system of units. In the SI system, one has

$$\kappa = 1,$$
$$c = 2.99792 \times 10^8\, \mathrm{m\,s^{-1}},$$
$$\mu_0 = 4\pi \times 10^{-7}\, \mathrm{H\,m^{-1}} = 1.25664 \times 10^{-6}\, \mathrm{H\,m^{-1}},$$
$$\epsilon_0 = \frac{1}{\mu_0 c^2} = 8.85419 \times 10^{-12}\, \mathrm{F\,m^{-1}}.\tag{A.2}$$

The atomic units are defined in Table A.1. We note that $m = e = \hbar = a_0 = 1$ in a.u., while the fine structure constant $\alpha = 1/137.036$ is dimensionless. As a consequence, one has (with $\kappa = 1$)

$$c = \frac{1}{\alpha}\ \mathrm{a.u.} = 137.036\ \mathrm{a.u.},$$
$$\mu_0 = \frac{4\pi}{c^2} = 4\pi\alpha^2\ \mathrm{a.u.},$$
$$\epsilon_0 = \frac{1}{4\pi}\ \mathrm{a.u.}\tag{A.3}$$

A few useful conversion factors are given in Table A.2.

Table A.1. Atomic units

Quantity	Unit	Physical significance	Value in SI units
Mass	m	electron mass	$9.10938 \times 10^{-31}\,\mathrm{kg}$
Charge	e	absolute value of electron charge	$1.60218 \times 10^{-19}\,\mathrm{C}$
Angular momentum	$\hbar$	Planck's constant divided by 2π	$1.05457 \times 10^{-34}\,\mathrm{J\,s}$
Length	a_0	Bohr radius of first Bohr orbit of atomic hydrogen (with infinite nuclear mass)	$5.29177 \times 10^{-11}\,\mathrm{m}$
Velocity	$v_0 = \alpha c$	magnitude of electron velocity in first Bohr orbit	$2.18769 \times 10^{6}\,\mathrm{m\,s^{-1}}$
Momentum	mv_0	magnitude of electron momentum in first Bohr orbit	$1.99285 \times 10^{-24}\,\mathrm{kg\,m\,s^{-1}}$
Time	a_0/v_0	time required for electron in first Bohr orbit to travel one Bohr radius	$2.41888 \times 10^{-17}\,\mathrm{s}$
Angular frequency	v_0/a_0	angular frequency of electron in first Bohr orbit	$4.13414 \times 10^{16}\,\mathrm{s^{-1}}$
Energy	$e^2/(4\pi\epsilon_0 a_0)$	twice the ionization potential of atomic hydrogen (with infinite nuclear mass)	$4.35974 \times 10^{-18}\,\mathrm{J}$ $(= 27.2114\,\mathrm{eV})$
Electric-field strength	$e/(4\pi\epsilon_0 a_0^2)$	electric-field strength experienced by an electron in first Bohr orbit	$5.14221 \times 10^{11}\,\mathrm{V\,m^{-1}}$

Table A.2. Conversion factors

Energy (in J) $= 6.24151 \times 10^{18}$ energy (in eV)
Energy (in eV) $= 1.60218 \times 10^{-19}$ energy (in J)
Energy (in eV) $= 1.23984 \times 10^{-6}/[\lambda$ (in m)]
Energy (in a.u.) $= 4.55633 \times 10^{-8}/[\lambda$ (in m)]
I (in $\mathrm{W\,cm^{-2}}) = 3.50945 \times 10^{16}\,[\mathcal{E}_0$ (in a.u.)]^2
U_p (in eV) $= 9.33729 \times 10^{-2}\,I$ (in $\mathrm{W\,cm^{-2}}$) [λ (in m)]^2
U_p (in a.u.) $= 3.43139 \times 10^{-3}\,I$ (in $\mathrm{W\,cm^{-2}}$) [λ (in m)]^2
α_0 (in a.u.) $= 2.57128 \times 10^{6}\,[I$ (in $\mathrm{W\,cm^{-2}}$)]$^{1/2}$ [λ (in m)]^2

I is the laser field intensity, $\mathcal{E}_0$ is the electric-field strength, U_p is the electron ponderomotive energy and α_0 is the electron excursion amplitude in the laser field

Author index

Author index

Subject index

Printed in the United States
By Bookmasters